Graduate Texts in Physics

Graduate Texts in Physics publishes core learning/teaching material for graduate and advanced-level undergraduate courses on topics of current and emerging fields within physics, both pure and applied. These textbooks serve students at the MS- or PhD-level and their instructors as comprehensive sources of principles, definitions, derivations, experiments and applications (as relevant) for their mastery and teaching, respectively. International in scope and relevance, the textbooks correspond to course syllabi sufficiently to serve as required reading. Their didactic style, comprehensiveness and coverage of fundamental material also make them suitable as introductions or references for scientists entering, or requiring timely knowledge of, a research field.

More information about this series at http://www.springer.com/series/8431

Oded Regev • Orkan M. Umurhan • Philip A. Yecko

Modern Fluid Dynamics
for Physics and Astrophysics

 Springer

Oded Regev
Technion, Haifa, Israel

Orkan M. Umurhan
NASA Ames Research Center
Moffett Field, CA, USA

Philip A. Yecko
The Cooper Union
New York, USA

ISSN 1868-4513   ISSN 1868-4521   (electronic)
Graduate Texts in Physics
ISBN 978-1-4939-7992-9   ISBN 978-1-4939-3164-4   (eBook)
DOI 10.1007/978-1-4939-3164-4

Springer New York Heidelberg Dordrecht London

Printed on acid-free paper

Springer Science+Business Media LLC New York is part of Springer Science+Business Media (www.springer.com)

We dedicate this book to our teacher, colleague, and friend

Edward A. Spiegel

Foreword

Panta rhei
Heraclitus of Ephesus (c. 535–c. 475 BC)

In the last hundred years or so, there has been substantial development in fluid dynamics, though it seems to have been "left behind" by physicists, because of a shift in their focus to fundamental quantum mechanics and particle physics. The study of this classical subject was dropped from most curricula of physics departments the world over. This split physicists and astrophysicists to those concentrating on cosmology and high-energy physics, in their quest for the "Holy Grail" of quantum gravity, and those who took on more classical unsolved problems. Alongside other topics like plasma physics, kinetic theory, and dynamical systems theory, fluid dynamics (FD) and magnetohydrodynamics (MHD) are today disciplines formulated as branches of physics. Therefore, their study is essential not only for the physics graduate student, who chooses to do his or her research in a field related to fluid dynamics, but also for graduate students in a number of additional physical disciplines, as well as for their supervisors and scientists, many of whom have had to learn the subject on their own because FD was not part of their formal studies when they themselves were graduate students. There is little doubt, at least in our opinion, that FD and MHD are indispensable for, e.g., a mathematical or condensed matter physicist, astrophysicist, geophysicist, and biophysicist and should be taught as a compulsory graduate course.

In this book, we concentrate on FD with one chapter on MHD. In our view, it would be impossible to cover all the topics of FD and MHD in sufficient depth. Thus we had to make a subjective choice as to what can be omitted without missing our main goal: endowing the relevant PhD student with solid knowledge of FD and a basic one of MHD. Moreover, both FD and MHD, as disciplines by themselves, are advancing rapidly in their evermore sophisticated experimental and observational aspects, as well as in both the utilization of powerful digital computers for the solution of problems and the expansion of analytical tools applied toward developing a better understanding of the underlying nonlinear phenomena and mechanisms. For example, there are new ideas on transition to turbulence via transiently growing stable linear modes and perhaps linear, or nonlinear, secondary instabilities, new approaches to turbulence itself, which still remain an enigma of fluid dynamics despite intensive research efforts and advances and expanded use of asymptotic approximation methods, which give analytical or semi-analytical results to complement numerical treatments. Advancing the understanding of the *nonlinear* aspects of FD is important to relevant physical and astrophysical study, and this is generally achieved with the aid of numerical solutions of governing nonlinear equations. Consequently, it has become an ever-increasing modern practice to simulate flows on computers, and as such, we briefly discuss (in Appendix B) some important considerations that should be taken into account when developing numerical codes. It is in this sense that the word "modern" in the title of this work should be understood: the subject obviously remains classical with those forays into nonlinearity and digital computing considered as a "modern" flavor.

Because of the diversity of various university programs around the world offering graduate study in physics and astrophysics (e.g., departments of physics, astrophysics, and sometimes even mathematics and geophysics), we do not presume to suggest at what point in the training of the graduate student a thorough course in FD, like the one based on this book, should be included, leaving such deliberations to curriculum committees. Before courses in FD were routinely adopted by astronomy and astrophysics departments and even physics departments, books on the topic were typically self-contained and drew examples from a variety of fields. For example, Landau and Lifshitz included *Fluid Mechanics* in their course of theoretical physics, and that volume, the second edition of which was actually written by Lifshitz and Pitaevskii, quickly acquired the status of a classic and has been used by physics and astrophysics students for (often enigmatic) self-study.

Close to the turn of the millennium, the Department of Physics of the Technion-Israel Institute of Technology yielded (rather reluctantly) to the pressure of a growing number of faculty members, whose research (and of their students) was in astrophysics, soft condensed matter, and biophysics, to offer a course in FD. One of us (O.R.) was tasked in preparing the course entitled *The physics of Fluids* (a nineteenth-century subject, according to some who consider anything not "quantum" as not belonging to modern physics) to be taught to senior undergraduate and graduate students in the department. The course turned out to be an exceptional success among senior undergraduate and graduate students. Students of astrophysics and also of several fields of condensed matter, biophysics, and mathematical physics, plus a substantial group of engineering students, as well as some established researchers in several fields, attended. This book is largely based on that course. Owing to its success, this course began to be offered regularly, approximately every 2 years, by the Technion Department of Physics.

To the best of our knowledge, most astronomy, astrophysics, and geophysics departments (which prefer books using their own approach and nomenclature) include nowadays in their curricula a serious course on fluid dynamics, and this is also the case in most physics departments. As mentioned above, this book is on a senior undergraduate or graduate level, depending, of course, on the different institutions' programs and syllabi. It contains too much material for a one-semester course, and the various departments will have to decide what will be the scope of specific senior undergraduate and graduate courses. It is assumed that the students are mathematically equipped with a working knowledge of analysis, including intermediate-level ordinary and partial differential equations, as well as a good command of vector calculus and linear algebra. In physics, the student will benefit by having already taken some courses more advanced than introductory undergraduate courses in mechanics and electricity and magnetism. A working knowledge in undergraduate thermal physics is also essential.

Fluid dynamics has grown into a vast subject, and any one of our chapters could be expanded into a full-length book. The constraint of finite size means, of course, that there is no escape from being brief in our expositions and from omitting many (often important) subjects. This has also forced us, as mentioned above, into making difficult choices as to which subjects to include and the most difficult choices: what to leave out. Naturally, our choices were biased by our own personal opinions, but we have tried to include as many subjects as possible that are essential for a research physicist or astrophysicist using fluid dynamics.

Preface

This book was originally intended to be a textbook for a one-semester graduate course (open also for senior undergraduates) in departments of physics, astrophysics, and astronomy (the latter for students specializing in theory). We do not exclude, obviously, any other departments, in interdisciplinary programs (so popular nowadays), from adopting the book, or parts thereof, as a course textbook. As mentioned in the *Foreword*, this book grew out of a course taught by one of us in a department of physics. It seems that in our efforts to be as complete as possible in the choice of the topics, the book includes now, almost certainly, significantly more material than can be covered in a one-semester course.

There are therefore two possibilities: Departments for whom many aspects of fluid dynamics are important may decide to base a two-semester course on the book. Alternatively, the instructor may choose the material that he or she considers essential and omit other topics. We have tried to make the chapters as self-contained as possible, but it is, obviously, inappropriate to see them as ten separate treatises. In any case, it is our hope that this book will serve, in addition to being a specific text for this or that course, as a general modern reference book on the subject of fluid dynamics as it is seen today at the beginning of the twenty-first century. Even though the book is voluminous, each of its chapters covers mainly the basics of the topic it deals with. We have decided, rather than giving the usual, called-for in review papers, conference proceedings, articles, and even textbooks, an abundance of citations in the body of the text to conclude each chapter with *Bibliographical Notes*, wherein we cite and briefly describe essential literature used in our discussions. We also often give recommendations for further reading and deeper exploration. We apologize for omitting many (often important) books and articles, as we realized quickly the impossibility of including all the citations on any subject.

The book opens in Chap. 1 with fundamentals, setting the stage for how to describe a fluid in a *continuum approximation*, not forgetting, though, that all matter is made of *discrete* microscopic particles. The issue of deriving a FD continuous description from kinetic theory is completely omitted due to its considerable complexity and inherent difficulties. We chose to use the more simplistic but concise approaches, dividing the fluid into very small elements, approximating fluid "particles," on which Newtonian dynamics is applied. Alternatively, the fluid properties are described as *fields* defined on spatial coordinates and time. These two descriptions, *Lagrangian* and *Eulerian*, are shown to be equivalent. The other approach, starting from the kinetic equations for the *microscopic* particles and manipulating them mathematically and physically toward the "fluid limit," may perhaps be more rigorous, but it is lengthy and plagued by problems of principle that are solved "ad hoc." Chapter 2 deals with restricted cases of flows and develops various important properties of some of them. Most of these properties are given, as is usual in our science, in the form of conservation laws. We then move on to a particularly important kind of flow, namely, that in which the viscosity, a property analogous to friction in mechanics, is important or even dominant. Thus we abandon *ideal* (nondissipative) fluids. The microscopic viscosity coefficient is dimensional, so to decide when it is important, we assess it by the size of the celebrated nondimensional *Reynolds number* Re, whose inverse indicates the relative importance of viscous effects relative to inertial ones. During our discussion, several other nondimensional numbers, which have import in fluid dynamics, appear as well. We move on to deal with waves in incompressible fluids, e.g., liquids, like water in Chap. 4. The chapter starts with a mainly mathematical primer on waves, in general, and continues with various linear and nonlinear wave phenomena, including solitons. Chapter 5 deals with rotating fluids, whose importance in geophysics cannot be overstated as both the atmosphere and the oceans reside on a rotating planet—the Earth. But this is not a book on geophysical fluid dynamics, so only the basics are described, together with examples of laboratory rotating flows and of one such astrophysical flow, an accretion disk. The next topic concentrates on the effects of compressibility, and its subject is sometimes referred to as *gas dynamics*. Supersonic flows are dealt with and the formation of shocks and the nature of explosions, detonations, and deflagrations are discussed. As such, these are subjects which will interest astrophysicists, strong explosions experts, and perhaps aircraft and missile engineers, who must know more than just our text. We move then to the topic of hydrodynamic stability (Chap. 7) including some relatively recent findings, like non-normal growth, followed, in Chap. 8, by the interesting and complicated issue of nonlinear development of instabilities. Nontrivial approximation methods are used to unravel the often fascinating behavior of patterns arising in the nonlinear development of instabilities.

This naturally leads to the issue of turbulence, which is sometimes the end of the route of linear instability arising above some critical value of a control parameter, whose progressive increase leads to nonlinear development and sometimes to a transition to turbulence. Turbulent flow is not a well-understood phenomenon, to say the least, despite many efforts to advance in its understanding by trying to devise a well-founded theory. Serious efforts started in earnest around 1900 and occupied the attention of the leading hydrodynamicists of that time. Richard Feynman, who needs no introduction, famously declared that turbulence is "the most important unsolved problem of classical physics." In Chap. 9, we aim to describe what the turbulence problem is and discuss some of the different efforts to tackle it, including the recent ones which arose after the early 1980s, with the hope fed by the flourishing of dynamical systems theory. Chapter 10 is relatively self-contained, as it deals with the rudiments of magnetohydrodynamics. The hydrodynamics of ionized, but locally neutral, fluids flowing in nonrelativistic velocities can be formulated in a consistent framework. It contains, in addition to the usual fluid dynamical fields, the magnetic field (or the electrical current density) as an additional unknown function. Of course, there are additional terms, e.g., the Lorentz force as a body force, and equations, e.g., the induction equation for the magnetic field, with new microscopic parameters, e.g., conductivity, and therefore the equations and their solutions are even more complicated than that in pure FD. We have included this chapter mainly for the astrophysicists as the magnetic field often plays a role in the objects they study, but it is our opinion that basic concepts of magnetohydrodynamics should be known also to a physics PhD student. After all, many see controlled fusion as our only energetic hope for the future. Of course, MHD is only the tip of the iceberg which is plasma physics.

Finally, even though the vision of this book is to think physics and try to find a simple solvable problem, on the way to the full problem, before one starts a big computer simulation, we do not think that *correct* simulations are useless. On the contrary simulations must be sound, based on codes that have been verified and validated, and one should use the simulations like lab experiments—finding what the effect of various input changes is. If the entire simulation of a fluid dynamical problem is so big that a calculation of only one case is feasible, we have a problem. In Appendix B, a primer of computational fluid dynamics (CFD) methods is given, stressing the caveats of convergence, accuracy, and stability. We also list a number of useful references for various methods.

In an effort to make this book useful as a course textbook, we have included, in each chapter, a number of problems, some of them relevant to real systems. The level of the problems varies, and solving them leads to a deeper understanding of the text and its continuity. We encourage students to solve the problems assigned and hope that lecturers will choose those problems that are most relevant to their lecture scope and level. In addition, we would like to mention an Internet site, known as NCFMF, web.mit.edu/hml/ncfmf.html, which contains a large number of films on the topic of fluid dynamics. Even though the films are outdated, some of them contain very valuable parts. The scenes depicting laboratory demonstrations, experiments, or natural phenomena, which are usually short segments of the often lengthy films, are the most useful for our purposes.

Haifa, Israel   Oded Regev
Moffett Field, CA, USA   Orkan M. Umurhan
New York, NY, USA   Philip A. Yecko

Acknowledgments

I have learned from my good teachers,
but most of all, from my students.

based on the Mishna, free translation

We acknowledge the help of our colleagues and friends who read chapters of the book and commented on them. As usual, we are responsible for all the material in the book, but the comments of Dov Levine, Giora Shaviv, and Attay Kovetz were important, and we thank them for the time they devoted to reading parts of the book and for their useful comments and suggestions. We single out Michael Mond, who read two entire chapters (Chaps. 7 and 10) (his area of expertise), and we have taken into account his important comments.

We thank Lloyd Miller for his masterful drawings and patience. Additional technical editorial help was provided to us by R. Korec and C. Rice in the NY Springer office and by K. Natarayan and finally E. Ahmad, who took care of the final copy editing. We thank them all.

Our students, undergraduate and graduate, while learning fluid dynamics on different levels, asked many questions; some of them lead us to deeper understanding.

Without the support, encouragement, and care of Tom Spicer, a senior Springer editor, it would have been impossible to ever complete this book. For this we all owe him a debt of profound gratitude. His professionalism combined with flexibility and patient attention to our questions is particularly acknowledged. One of us (O.R.) would like to thank Gabriel Seiden, his teaching assistant in the graduate course, for the notes which started it all and from whom he learned quite a lot. Dr. Seiden was also kind enough to read carefully eight chapters of the book, and his comments were useful.

Last, but not least, we are grateful to our families: Judy, Tamar, and Daniel; Susanna, Gabriel, and Elodie; and Burcu and Erkin. The children for just being there and the wives for moral support and understanding during the often "turbulent" times in the years of our writing this book.

Contents

Chapter 1
Fundamentals

The fundament upon which all our knowledge and learning rests is the inexplicable.

Arthur Schopenhauer (1788–1860)

1.1 Continuum Description of Fluids

Fluid dynamics, which is considered in this book to be a branch of physics, consists of the description and study of gas and liquid dynamics in a *macroscopic* way, and the fluid is considered to be a continuous medium. But any matter is made of microscopic particles and therefore a continuous description must only be an approximation. Even an infinitessimally small element of a continuous fluid, what is commonly referred to as a *fluid particle* and is considered to be a point mass, contains, no doubt, an extremely large number of true microscopic particles (see below). These really microscopic particles (atoms, molecules, and so forth), comprising fluids, obey statistical kinetic equations, which are the simplest for dilute gases. In principle, fluid dynamics can be derived using those equations. Indeed, this approach is sometimes used, e.g., in reference [3] in the *Bibliographical Notes* of this chapter, to lay the foundations of fluid dynamics. The procedure is, however, lengthy and far from being trivial. Instead of taking this approach, we choose to describe, in this book, fluids as continua, *ab initio*. We should remark, here at the outset, that a continuum description is formally a mathematical one. As said above, it is an approximation and has to be treated with care in particular, e.g., at fluid interfaces or shocks (sharp for continua) or in boundary conditions (i.e., no-slip), which may be ideal for mathematics, but not so for physics.

A continuous description of the kind we want to adopt can be a good approximation only as long as the mean free path between microscopic particle collisions Λ_{mfp} is extremely small, as compared to a typical size of the system L. The ratio Λ_{mfp}/L,

© Springer Science+Business Media, LLC 2016

O. Regev et al., *Modern Fluid Dynamics for Physics and Astrophysics*,
Graduate Texts in Physics, DOI 10.1007/978-1-4939-3164-4_1

which incidentally is of the same order of magnitude as the microscopic collision time to the hydrodynamical one, that is, the bulk motion time, goes under the name of the *Knudsen* number (Kn). A necessary (but unclear if also sufficient) condition for the description of a fluid as a continuum is thus Kn ≪ 1.

We find it useful to mention at this point that in the continuous approximation of matter, a *fluid* is defined by its being unable to stay in equilibrium when *shearing stresses* (a concept that will be defined below) are applied to it. That is, shearing stresses cause sustained motion of fluids, and this property defines the latter and separates them from solids.

1.2　Kinematics of Fluid Motion

There are two different approaches to mathematically formulating continuum fluid kinematics and later on, also its dynamics. In what follows, we shall explain, in detail, the mathematical and physical meaning of these approaches. Historically they are associated (by name) with two of the most distinguished mathematicians and physicists of the eighteenth century: J.-L. Lagrange and L. Euler. Of course, their two approaches ultimately describe physical objects that are equivalent to one another, as well-defined mathematical manipulations transform one perspective into the other. In certain practical situations, however, one approach may be superior to the other, mainly for practical reasons, both in terms of the mathematical formulation of a given problem and its interpretation.

There exists a well-defined relationship between the Lagrangian and Eulerian approaches, where the term *descriptions* is commonly used instead of "approaches." The connection between the two when the continuum assumption is invoked must, in our opinion, be established with considerable pedantry in order to justify the fundamental soundness of the subject, both mathematically and physically. Students feel especially at ease, in our experience, when the basics are completely clarified and not left out from the discussion. Most books do not discuss these "first principles" of continuum fluid dynamics, but there are some that do, e.g., references [5] and [6] in this chapter's *Bibliographical Notes*. We realize that there may be readers who are less interested in such fundamentals and are, instead, more concerned with applications. From our perspective as scientists and educators, we do not recommend working with equations whose origins and the approximations leading to them are not completely clear. Nonetheless, one may choose to skip over these matters and move directly to Sects. 1.3 and 1.4.

Continuing our exposition of the fundamentals, we note that the aim is to express the laws of fluid dynamics (we shall hereafter often use the acronym FD) using differential and integral calculus. This calls for an even more stringent approximation than just Kn ≪ 1. Taking a very small fluid element $\mathscr{V}_\varepsilon$ (here and throughout this book finite volumes will usually be denoted by $\mathscr{V}$), we assume that the fluid properties in this element change negligibly, compared to the changes occurring on the larger, sizable scales, comprising the fluid system itself.

We actually assume that these tiny, with respect to scales of change of the fluid properties, elements satisfy

$$\Lambda_{\text{mfp}} \ll (\mathscr{V}_\varepsilon)^{1/3} \ll L,$$

that is, the number of particles in them is necessarily very large. The fluid *parcel*, often referred to as a *fluid particle* (we shall use these two terms interchangeably, according to convenience), is precisely the fluid volume element $\mathscr{V}_\varepsilon$, taken conceptually to its limit of a point. Thus at any point in space, there is a fluid parcel, or particle, which is assumed to maintain its identity while the fluid is in motion. In this way, the fluid particles (parcels) are dense in space available to them, that is, it is assumed that they occupy all the space that is taken by the fluid.

1.2.1   Lagrangian Description

It is perhaps worth mentioning, for the sake of historical justice, that both descriptions are due to Euler, and we use the name of Lagrange, just as a convenience. The notion of fluid particles and the action of following them, with the help of appropriate equations, is called the *Lagrangian* description. We would like to describe the motion of the dense set of fluid particles as obeying particle dynamics. This description has to yield a trajectory for every fluid particle, but to accomplish that we have to be able to distinguish between different particles. Even more fundamentally, we have to assume that a fluid particle preserves its identity. This is not a trivial assumption, as such a fluid parcel is an element of matter, containing microscopic particles that, in reality, are exchanged with its surrounding. To overcome this difficulty, we may proceed by assuming an *ideal fluid* (defined later in this chapter as one whose elements do not experience any dissipative processes). Later on in our discussion, these neglected dissipative/exchange processes can be parametrized as coefficients of viscosity, thermal conduction, and so on. Still, the idea that the fluid particle keeps its identity cannot, in reality, continue indefinitely long. So, we must limit ourselves in this Lagrangian description to a finite amount of time that is (hopefully) long enough for our purposes. A useful mental picture to hold, which serves to distinguish different fluid particles, is one where each fluid particle has a label, i.e., a unique "name" defining it. The label can be of any kind, but perhaps the most straightforward labeling is one marking a particle by its *position* at some initial time, say, $t = 0$. In this way, the fluid particles are labelled by just their initial position vector, $\mathbf{a}$, say, as expressed in some fixed inertial coordinate frame. Thus, the fluid *flow* or motion can be fully described, as required, by expressing the motion of the fluid parcels with time—their trajectories. The trajectories are expressed by the function $\mathbf{x}(\mathbf{a}, t)$, which is supposed to follow in time the position of the fluid particle that was at $\mathbf{a}$ at time $t = 0$, i.e.,

$$\mathbf{x}(\mathbf{a}, t = 0) = \mathbf{a}.$$

The function $\mathbf{x}(\mathbf{a}, t)$, which describes the flow, can be perceived as a continuous (continuously changing with time) differentiable mapping of the three-dimensional Euclidian space onto itself

$$\mathbf{x}(\mathbf{a}, t) = \Phi_t(\mathbf{a}). \tag{1.1}$$

When we consider the flow as a differentiable map, a basic mathematical construction is the Jacobian matrix of the mapping

$$\mathbb{J}_{ik}(t) \equiv \left(\frac{\partial x_i}{\partial a_k} \right)(t), \tag{1.2}$$

where we have explicitly noted the dependence on time, which serves as a continuous parameter of the mapping.

A remark regarding our notation of Cartesian coordinates is now in order. The position vector of a point is denoted $\mathbf{x}$ and, as it is in the previous equation, we use indices $i = 1, 2, 3$ for the Cartesian x_1, x_2, x_3 components of vectors. This will be usually used throughout this book. Alongside it however, depending upon convenience and making sure that there is no room for confusion, the other convention for coordinate components (x, y, z) will appear. The corresponding Eulerian velocity field (see below) is $\mathbf{u} = (u_1, u_2, u_3)$, but the velocity components (u, v, w) will also be used frequently. Regarding unit vectors in the three Cartesian directions, we shall use interchangeably, according to convenience, other common notational conventions as well, and again hopefully avoiding confusion $\hat{\mathbf{x}}_1 = \hat{\mathbf{x}}; \ \hat{\mathbf{x}}_2 = \hat{\mathbf{y}}; \ \hat{\mathbf{x}}_3 = \hat{\mathbf{z}}$, where the "hat" symbol is used to denote a *unit* vector. $\hat{\mathbf{n}}$ will be a unit normal vector to a surface at a point but for consistency and convenience we shall not distinguish between $\hat{\mathbf{n}}$ and $\mathbf{n}$.

The above Jacobian *matrix* $\mathbb{J}$ is usually called in FD the *displacement gradient matrix*, and its determinant simply the *Jacobian*,

$$J_t \equiv \det \mathbb{J}(t), \tag{1.3}$$

which has been labelled here by the subscript t to indicate time dependence. It will be seen later on to play an important role in various useful relations. It suffices to remark that $J \neq 0$ for all t guarantees the invertibility of the continuous (in the parameter t) mapping. This guarantees that one may find (in principle) the original position of a fluid particle that happens to be at the position $\mathbf{x}$ at time t:

$$\mathbf{a} = \Phi_t^{-1}(\mathbf{x}).$$

This is consistent with a kind of time reversibility (a property characteristic of point masses in Newtonian mechanics), which here also includes the statement that trajectories of fluid particles do not intersect. Note that since we are not dealing with relativistic physics in this book, we may carelessly not distinguish a matrix element index notation with a tensor element, i.e., J_{ki}.

Thus, in the Lagrangian description the free variables of any function, defining some property of the fluid, are $\mathbf{a}$, the initial position, serving here as the label of the fluid particle, and time. For example, the property f, say, of the fluid particle, whose label is $\mathbf{a}$, is given at time t by the value of the function $f(\mathbf{a}, t)$. A key concept in fluid kinematics is the *Lagrangian time derivative*. In the Lagrangian description, the derivative is taken in t for $\mathbf{a}$ fixed. For example, if we are talking about the property f, we have

$$\left(\frac{Df}{Dt}\right) \equiv \left(\frac{\partial f}{\partial t}\right)_{\mathbf{a}}. \tag{1.4}$$

Here it is stressed that the time derivative is for *the same* fluid particle labelled by $\mathbf{a}$, that is, *following* its motion. In this way, D/Dt is introduced as the customary notation of the *Lagrangian* time derivative, known also as the *advective* or *substantial* or *material* derivatives. The *velocity* of a fluid element, which is at position $\mathbf{x}$, at time t, is denoted here by an overdot, that is,

$$\dot{\mathbf{x}}(\mathbf{a}, t) = \frac{D\mathbf{x}}{Dt}. \tag{1.5}$$

It is also convenient to designate the same quantity for the fluid particle by the function,

$$\mathbf{v}(\mathbf{a}, t) \equiv \dot{\mathbf{x}}, \tag{1.6}$$

say, distinguishing it from $\mathbf{u}$, which will soon be needed for the Eulerian description, and reminding one that it is the Lagrangian velocity at time t and of the particle labelled $\mathbf{a}$. This definition will be useful later, when we transform to the Eulerian description. The acceleration $\dot{\mathbf{v}}$ of a fluid particle is similarly

$$\dot{\mathbf{v}}(\mathbf{a}, t) = \ddot{\mathbf{x}} = \frac{D}{Dt}\left(\frac{D\mathbf{x}}{Dt}\right),$$

and this quantity naturally lends itself to the application of the laws of classical dynamics based on Newton's laws. Indeed, the Lagrangian description is physically more natural as it considers the evolution of a dense collection of fluid particles, each obeying the laws of classical mechanics. If the motion is one-dimensional, as is sometimes the case, the Lagrangian description is generally preferred. The Lagrangian description is most satisfying, as we shall see, from a Newtonian dynamics framework, but it maps the nonlinearity of the continuum onto parcel/particle trajectories, which may be unsatisfying, because there is an infinite number of them, and each may be chaotic, see, e.g., reference [8], in many flows. The need to follow fluid parcels calls for cumbersome techniques including the translation of forces because of the above-mentioned mapping of the nonlinearity of the continuum on trajectories. It is thus generally preferable, except for special cases, like one-dimensional flows, see Sect. 1.6.2, to translate fluid particle dynamics into the dynamics of fluid fields, as we introduce next.

1.2.2   Eulerian Description

This approach is based on the description of fluid motion by functions of a fixed inertial coordinate system and time. For example, we may look at the fields $\rho(\mathbf{x},t)$, $P(\mathbf{x},t)$, $T(\mathbf{x},t)$, i.e., density, pressure, temperature, and possibly other physical characteristic fields, $f(\mathbf{x},t)$, say, at point $\mathbf{x}$ and time t. This is conceptually and practically simpler than the Lagrangian approach in multidimensional flows. The quantities $\mathbf{x}$ and t take the role of coordinates and the fluid properties are fields. The key is to understand that the Eulerian fluid field at time t and position $\mathbf{x}$ is actually the relevant property of the fluid particle that at time t *happens* to be at position $\mathbf{x}$, that is, the particle whose label is

$$\mathbf{a} = \Phi_t^{-1}(\mathbf{x}). \tag{1.7}$$

Similarly, any flow property, $f(\mathbf{x},t)$, at a particular position changes with time, but it cannot be attributed to a definite fluid particle. Any difficulty in "translation" from the more physically natural Lagrangian "language" to the Eulerian one can be done by using (1.1) and its inverse. Specifically, the Eulerian *velocity* may be identified with the particle one, only by formally identifying the particle

$$\mathbf{u}(\mathbf{x},t) = \mathbf{v}[\Phi_t^{-1}(\mathbf{x}),t], \tag{1.8}$$

where we have written $\mathbf{v}(\mathbf{a},t)$, defined in Eq. (1.6), to indicate the Lagrangian velocity. The Eulerian time derivative is naturally $(\partial/\partial t)_{\mathbf{x}}$, that is, taken holding $\mathbf{x}$ fixed. It is of practical interest to relate the Lagrangian time derivative to Eulerian derivatives, i.e., the time derivative *following* the motion, identified with the time derivative for a fluid particle, as acting on an Eulerian field, $f(\mathbf{x},t)$. We have

$$\frac{D}{Dt}f(\mathbf{x},t) = \left\{ \frac{\partial}{\partial t} f[\mathbf{x}(\mathbf{a},t),t] \right\}_{\mathbf{a}}, \tag{1.9}$$

and may use, on the right-hand side of this expression, the chain rule of differentiation. This gives

$$\frac{D}{Dt}f(\mathbf{x},t) = \left(\frac{\partial \mathbf{x}}{\partial t}\right)_{\mathbf{a}} \cdot \nabla_{\mathbf{x}} f + \left(\frac{\partial f}{\partial t}\right)_{\mathbf{x}}. \tag{1.10}$$

Symbolically, we may thus write the operator equation

$$\frac{D}{Dt} = \frac{\partial}{\partial t} + (\mathbf{u} \cdot \nabla), \tag{1.11}$$

where we have dropped the subscripts and will henceforth understand partial derivatives as those acting on Eulerian fields. The term containing $\mathbf{u}$ appearing here is the so-called *advective* term (see below). This velocity is identical to the Eulerian velocity field $\mathbf{u}(\mathbf{x},t)$. The operator equation is written in such a way as to allow operation also on vector fields.

A little reflection furnishes the physical meaning of the various terms in Eq. (1.11). On the left-hand side, we have the Lagrangian (following the flow) rate of change. On the right-hand side, the first term is the Eulerian time derivative, that is, the local rate of change. Regarding the second term, consider a fluid property $f(\mathbf{x},t)$, say, an arbitrary Eulerian field. Then this second term on the right-hand side can be written as

$$\mathbf{u} \cdot \nabla f = u\,\hat{\mathbf{u}} \cdot \nabla f = u\frac{\partial f}{\partial s},$$

where $u = |\mathbf{u}|$ and $\hat{\mathbf{u}} = \mathbf{u}/u.\ \partial/\partial s \equiv \hat{\mathbf{u}} \cdot \nabla$ is the directional spatial derivative. So, the second term is the rate of change just because of the motion and following it, i.e., the advective rate of change.

An important concept regarding a flow may be defined here. The unit vector field $\hat{\mathbf{u}}(\mathbf{x},t)$ at each point and time defines a directional family of curves, called *streamlines*, whose property is that they are tangent to the Eulerian velocity vector field. Obtaining these curves is not always a simple matter, even if the velocity is known explicitly. It involves the solution of the differential equations

$$\frac{dx_1}{u_1} = \frac{dx_2}{u_2} = \frac{dx_3}{u_3}, \tag{1.12}$$

which follow from elementary curve theory in multivariable calculus. The collection of streamlines illustrates the entire flow at a particular moment, but they may change in time. Only when the flow is *steady*, that is, the velocity field is time independent, the streamlines are also constant in time. *Streaklines* are another tool that may help visualize a flow. A streakline is the collection of the current positions of fluid particles that have passed through a given fixed point in the fluid, $\mathbf{p}$, say, at a continuous succession of times, t_p in the time interval $0 \leq t_p \leq t$. For a flow starting at $t = 0$, this can be mathematically expressed as the curve composed of the points

$$\mathbf{x} = \mathbf{x}\left[\mathbf{a}(\mathbf{p},t_p),t\right], \qquad 0 \leq t_p \leq t. \tag{1.13}$$

Thus, streamlines and streaklines are conceptually different lines. In addition, Lagrangian trajectories of different fluid particles are sometimes called *pathlines*. Can you think of a flow in which they all coincide? The goal of Problem 1.2 is to clarify these concepts by examples of simple flows (Fig. 1.1 illustrates some pathlines of a real flow).

Fig. 1.1 Pathlines of different air fluid particles marked by glowing residue from a campfire, as they are advected out by the heated air around the campfire. This particular photograph of the "Kabarneeme" campfire site in Estonia was taken using long exposure time so as to show the pathlines. (*Author: Abrget47j, licensed under Creative Commons Attribution-Share Alike 3.0 Unported—http://creativecommons.org/licenses/by-sa/3.0/deed.en*)

1.2.3 Rate of Deformation and Rotation

Useful kinematical information can be obtained by defining the tensor quantity, called the *velocity gradient tensor*, $\mathcal{U}_{ik}$, in the following way:

$$\mathcal{U}_{ik} \equiv \frac{\partial u_i}{\partial x_k} = \frac{1}{2}\left(\frac{\partial u_i}{\partial x_k} + \frac{\partial u_k}{\partial x_i}\right) + \frac{1}{2}\left(\frac{\partial u_i}{\partial x_k} - \frac{\partial u_k}{\partial x_i}\right),\tag{1.14}$$

where a customary splitting of a tensor into its symmetrical and antisymmetric parts has been done. Defining now the symmetric tensor on the right-hand side of Eq. (1.14) as $\mathcal{D}_{ik}$, that is,

$$\mathcal{D}_{ik} = \frac{1}{2}\left(\frac{\partial u_i}{\partial x_k} + \frac{\partial u_k}{\partial x_i}\right),\tag{1.15}$$

and the antisymmetric one, similarly, as $\mathcal{O}_{ik}$, we get upon multiplication by dx_k and summation according to the Einstein convention, i.e., a sum from 1 to 3 is assumed over any repeated index in an expression, that

$$du_i = \mathcal{D}_{ik}\,dx_k + \mathcal{O}_{ik}\,dx_k.\tag{1.16}$$

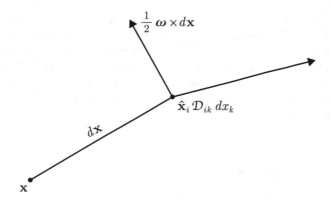

Fig. 1.2 A schematic depiction of the deformation and rotation rates of dx. The differential element dx, shown grossly enlarged, for the sake of clarity, connecting the *two black circles* representing dots, in which the *left one* is considered fixed, is subject to rotation along the perpendicular vector direction and distortion—stretching along the direction of the other vector

Henceforth, the application of Einstein's summation convention will be the default and we shall indicate if it should *not* be effected. Now multiplying through by the unit vector in the i direction $\hat{\mathbf{x}}_i$, we get, in mixed notation,

$$d\mathbf{u} = \hat{\mathbf{x}}_i \, \mathscr{D}_{ik} \, dx_k + \frac{1}{2}\omega \times d\mathbf{x}, \qquad (1.17)$$

where $\omega \equiv \nabla \times \mathbf{u}$ is the *vorticity*, a quantity which will be used frequently in this book. The second term can readily be verified by writing out all the components explicitly and remembering that the vector product can be written $(\mathbf{A} \times \mathbf{B})_i = \varepsilon_{ijk}A_jB_k$, where ε_{ilk} is the Levi-Civita symbol[1] (and see Problem 1.3). Figure 1.2 illustrates that $\mathscr{D}_{ik}$ is responsible for the *deformation rate*, while the second tensor can be interpreted as inducing a *rotation rate* of a differential length element.

1.3   Dynamics of Fluid Motion

The functions $\mathbf{u}(\mathbf{x},t)$ in Eulerian description and $\mathbf{x}(\mathbf{a},t)$ in Lagrangian one are the most basic equivalent dynamical characterizations of a fluid flow. The density is also needed for the flow dynamics to be fully described. The basic dynamics consists of two conservation equations: that of mass, where we assume that matter is neither destroyed nor created, and of momentum, in which it is assumed that momentum changes only due to the application of force according to Newton's second law.

---

[1] $\varepsilon_{ijk} = 1$ if (i,j,k) is an even permutation of $(1,2,3)$, $= -1$ if the permutation is odd and $= 0$ if any two indices are repeated.

We know that energy has also to be conserved and this makes the full set of fluid equations quite involved, because the addition of an energy equation calls for the inclusion of internal (thermal) energy. Thermodynamic variables have to be introduced and various complications, like energy production and dissipation and, in addition, issues related to the equation of state, have to be addressed. In this section, we shall concentrate solely on the dynamics and will devote a separate section to the energy equation.

1.3.1 Forces and Stresses

Fluids of different sorts can have very complicated relations between forces acting *on* them and stresses developing *in* them, depending on the electromagnetic properties of the matter they are composed of, for example. In this book, we cannot cover all the possibilities and limit ourselves to fluids that are generally considered in most classical books (e.g., [1, 2], and so on). The properties of these fluids will be uncovered during our derivations of the equations of motion for them. Consider a *finite* volume element of a fluid, $\mathscr{V}$. We distinguish between two types of forces that can act on this element. The first one, which we call body force, includes all long-range forces. Classical examples of body forces are the gravitational and Lorentz force. We include the value of these forces, per unit mass, acting at a given position and time and denote them symbolically by $\mathbf{b}(\mathbf{x}, t)$. Thus, a vector component of the total body force on the entire element is the integral over the volume of the element:

$$(F^b)_j = \int_{\mathscr{V}} \rho \, b_j \, d^3 x, \tag{1.18}$$

where the superscript distinguishes between body and surface (see below) forces.

The second type includes the short-range interatomic or molecular forces that, in the case of a solid and liquid, hold it together. Those second kinds of forces may act on a fluid element $\mathscr{V}$, which is not considered here as a small fluid particle or infinitesimal element, by its surrounding medium, and they can do it only by their action on the *surface* of $\mathscr{V}$. Consider a single component of surface force per unit volume $(f^s)_j$, acting in a point in the fluid volume element $\mathscr{V}$. The vector component of the total force on the element due to this surface force component will be

$$(F^s)_j = \int_{\mathscr{V}} (f^s)_j d^3 x.$$

From vector analysis, we recall that a volume integral on a given volume of a scalar can be transformed into a surface integral over the boundary of the volume if and only if the scalar is a divergence of a vector. If we are dealing with a vector $(f^s)_j$, it has to be a divergence of a tensor of rank two. We denote the latter by σ_{jk} and call it the *stress tensor*. Thus, we must have for $j = 1,2,3$

$$(f^s)_j = \left(\frac{\partial \sigma_{jk}}{\partial x_k}\right), \tag{1.19}$$

and the full surface force vector arising from the stress tensor is

$$(F^s)_j = \int_{\mathcal{V}} \left(\frac{\partial \sigma_{jk}}{\partial x_k}\right) d^3x = \oint_{\partial \mathcal{V}} \sigma_{jk} n_k dS, \tag{1.20}$$

where $\partial \mathcal{V}$ is the entire surface of the volume element and dS is a differential surface element. Equations (1.18) and (1.20) constitute thus the body and surface forces, correspondingly, acting on a fluid element. As we shall see, the integrands of these expressions will be used when the differential equations of motion are derived (see in Sect. 1.4.2).

1.3.2 Cauchy Theory of Stress and Its Physical Meaning in a Fluid

Here we briefly discuss the above defined stresses and their meaning in a fluid. The kind of reasoning we use was first proposed by A.-L. Cauchy, a nineteenth century prolific pioneer of mathematical analysis. We have seen that the definition of stress arises from the consideration of surface forces. We first imagine a very small, and thus flat, surface element δS, which may (or may not) be a part of some closed surface in a fluid. Assume that this element is centered on position $\mathbf{x}$ and let $\mathbf{n}$ be its unit surface normal vector determining the element's sense of direction. Let $\mathbf{T}(\mathbf{x}, \mathbf{n}, t)$ be the force per unit area, acting on the surface element by the fluid on the side *towards* which the normal $\mathbf{n}$ points. This situation is schematically depicted in Fig. 1.3. The surface force on the element is thus $\delta \mathbf{F}^s = \mathbf{T}\delta S$ and the vector $\mathbf{T}$, which depends on the position, time and *direction* of the surface element, that is, $\mathbf{n}$, is usually called the *stress or traction force*. One should not forget that its units are force per unit area. As we shall see, considering the components of $\mathbf{T}$ normal and

Fig. 1.3 A schematic depiction of the traction (stress) force $\mathbf{T}$ acting on a very small surface element δS with $\mathbf{n}$ being normal to the surface

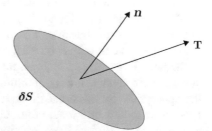

Fig. 1.4 The three
components of the stress
tensor $\sigma_{j1}, j = 1, 2, 3$, acting
on the surface element $\delta S \hat{\mathbf{x}}_1$

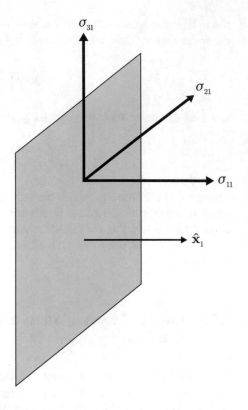

tangential to the surface, will be useful in understanding the physical meaning of the
stress tensor. We choose now to redefine the nine components of the stress tensor
σ_{jk}, consistently with Eq. (1.20), as the j component of the surface force (per unit
area), acting on a surface element δS, whose normal $\mathbf{n}$ is pointing in the k direction.
For the sake of a clear demonstration, let us choose the small surface element δS
so that its normal is $\mathbf{n} = \hat{\mathbf{x}}_1$. Figure 1.4 depicts this choice. In this particular case,
the stress σ_{11} is called the *normal* stress, as it is the component of the surface force
(per unit area) acting in a direction parallel to the normal. The components σ_{j1}, with
$j = 2, 3$ are called *shearing* stresses and they relate to surface forces per unit area
acting in a direction perpendicular to the normal, that is, parallel to the surface. We
hope that this figure, which depicts a very particular case, helps in understanding
the difference between normal and shearing stresses.

The definition of σ_{jk} found above allows us to clearly express the traction vector
in terms of the stress tensor using Cauchy's theory. In principle, if we could not use
it we would have to utilize, instead, a brute force evaluation of the surface integral.
In the latter case, the surface integral in Eq. (1.20) involves splitting $\partial \mathscr{V}$ into very
small surface elements, δS_α, summing over α to produce $\partial \mathscr{V}$, followed by taking
the appropriate infinitesimal limit. It is in this way that the integral would have
been decomposed into a sum and, correspondingly, each term in the sum would
contribute to the surface force, traction, expressed per unit area. However, Cauchy's

Fig. 1.5 An auxiliary
drawing for the proof of
formula (1.24). The *front
shaded area* is $\mathbf{n}\delta S$

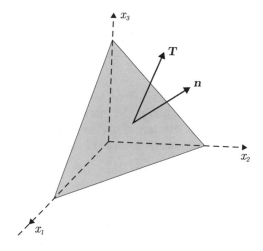

idea and construction is clearer, simpler, and more elegant, involving no calculus, since the underlying reasoning is purely geometrical. Following Cauchy, we imagine that our surface element δS is the triangular base (the large face) of a tetrahedron, whose "top" vertex is the axes origin, in this case depicted as the farthest vertex from the base triangle, and whose other three faces are the 90° triangles formed by a pair of axes and one side of the triangle δS (see Fig. 1.5). Now for a given instant, consider the fluid that occupies the interior of the tetrahedron. The proof involves a heuristic limiting process, wherein we examine what happens when the linear dimensions of the tetrahedron, e.g., its sides that reside on the axes, tend to zero. Still, consistently with our continuum approximation, we may not consider the tetrahedron to be smaller than what we have been calling a fluid particle. Now we want to find the total force acting on the tetrahedron and since it is a vector, we look at the jth component thereof. This surface force's jth component, exerted by the external to the tetrahedron fluid on the base of the tetrahedron, is clearly $T_j \delta S$. The j components of the forces acting on the sides that lie in the planes normal to $\hat{\mathbf{x}}_k$ are obtained by the relevant stress components $-\sigma_{jk}$. The minus is here because a particular $\hat{\mathbf{x}}_k$ is perpendicular to the triangle, in which the other two $\mathbf{x}_{j \neq k}$ lie, and thus points *into* the volume element. It has also to be multiplied by the area of the relevant 90° triangle. The latter is simply the projection of the base on the plane in question, that is, $n_k \delta S$. This argument is good for all the three faces, which lie in the planes spanned by the axes. We thus have that the jth component of the total surface force exerted by the ambient fluid of the tetrahedron, on it, is

$$(T_j - \sigma_{jk} n_k)\delta S. \tag{1.21}$$

This force plus any body force $\rho b_j \mathscr{V}$ will finally give the total force acting on the fluid inside the tetrahedron:

$$\delta F_j = (T_j - \sigma_{jk} n_k)\delta S + \rho b_j \mathscr{V}. \tag{1.22}$$

By virtue of Newton's second law, this force has to be equal to the mass $\rho \mathscr{V}$ times the acceleration a_j. Thus

$$(T_j - \sigma_{jk} n_k)\delta S + \rho b_j \mathscr{V} = \rho \mathscr{V} a_j. \tag{1.23}$$

Sending the linear dimensions of the tetrahedron to zero in a manner preserving the surface orientation **n** and assuming that the acceleration remains *finite* implies that $\mathscr{V} \sim L^3$ tends to zero faster than $\delta S \sim L^2$. This gives finally the important relation, which expresses the essence of Cauchy's stress theory

$$T_j = \sigma_{jk} n_k \tag{1.24}$$

and furnishes a clear relation between the traction (or stress force) vector and the stress *tensor*. Those who worry that the tetrahedron cannot really tend to zero, because of our remark before, should not forget that a fluid description as a continuum is an approximation, employing differentials that are larger than a fluid parcel, but much smaller than typical macroscopic length scales in the fluid.

It can be shown, using the fact that the total moment of surface force acting on a fluid element must be a surface integral, that the stress tensor is symmetric. The moment of the force $\mathbf{F}^s$ is $\mathbf{x} \times \mathbf{F}^s$ (a torque), where **x** is the point where the force is applied. The total moment of force on the element is thus an anti-symmetrical tensor of rank two (call its components M_{jk}). In component language the result is $M_{jk} = \int [(f^s)_j x_k - (f^s)_k x_j] d^3x$. Using the form of the surface force components per unit volume, in terms of the stress tensor components [as in formula (1.19)], we get

$$M_{jk} = \int \left(\frac{\partial \sigma_{jl}}{\partial x_l} x_k - \frac{\partial \sigma_{kl}}{\partial x_l} x_j \right) d^3x$$

$$= \int \frac{\partial}{\partial x_l} (\sigma_{jl} x_k - \sigma_{kl} x_j) d^3x - \int \left(\sigma_{jl} \frac{\partial x_k}{\partial x_l} - \sigma_{kl} \frac{\partial x_j}{\partial x_l} \right) d^3x.$$

Since the first integral on the right-hand side can be readily converted into a surface integral and the derivatives in the second integral are actually Kronecker deltas $(\partial x_m / \partial x_n) = \delta_{mn}$ we get

$$M_{jk} = \oint (\sigma_{jl} x_k - \sigma_{kl} x_j) n_l dS + \int (\sigma_{kj} - \sigma_{jk}) d^3x.$$

The requirement that this quantity must be a surface integral establishes that the volume integral part is zero and this, in turn, guarantees the symmetry of the stress tensor

$$\sigma_{ij} = \sigma_{ji}, \tag{1.25}$$

because this is valid for any domain of integration. This formal derivation is in accord with the fact that the Cauchy theory of stress was used here (the tacit assumption of which is isotropy of the fluid).

1.3.3   Some Mathematical Relations

In what follows, we shall give a number of mathematical relations that are helpful in the derivation and understanding of the equations of FD.

1.3.3.1   Euler's Formula

The first of these relations is here the so-called Euler's formula (there are many theorems and formulae bearing his name). It concerns the properties of continuous transformations as applied to fluid flow. In the Lagrangian picture, it is the continuous transformation, in terms of the parameter t, of the vector space of all $\mathbf{a}$ into the vector space of all $\mathbf{x}$. A well-known concept in the transformation theory of vector spaces is that of the *Jacobian*. In our case, the Jacobian is simply the determinant, J_t given in Eq. (1.3), with the index t reminding us that it is a continuous function of time. Euler's formula reads

$$\frac{DJ_t}{Dt} = J_t \nabla \cdot \mathbf{u}. \tag{1.26}$$

Note that this formula contains a Lagrangian, i.e., material, derivative as well as Eulerian derivatives of the Eulerian velocity field, that is,

$$\nabla \cdot \mathbf{u} = \left(\frac{\partial u_j}{\partial x_j} \right).$$

Euler's formula is heuristically obvious as the Jacobian is known to be related to the ratio of differential volume elements during the transformation, thus $J_t d^3 a = d^3 x$. Also, as is known from vector calculus, the relative rate of change of a differential volume element at a given time can also be expressed by $\nabla \cdot \mathbf{u}$, as this is the meaning of the divergence of velocity at a point where the element is located. These considerations are not rigorous enough to satisfy any mathematically oriented person, but we nevertheless state it here and it is proven, more rigorously, in Problem 1.5. Working out this problem may include some lengthy algebraic manipulations, but it is straightforward, especially if it is performed using the definitions we have presented and relevant vector quantities are decomposed into their constituent components.

1.3.3.2 Transport Theorems

Consider a fluid flow and concentrate on a finite *material*, i.e., moving with the fluid, volume $\mathscr{V}(t)$. Let $G(\mathbf{x},t)$ be some fluid property, expressed here as an Eulerian field. We wish to find the rate of change of the integral of G over $\mathscr{V}(t)$. We have

$$
\dot{I}_G \equiv \frac{d}{dt}\int_{\mathscr{V}(t)} G(\mathbf{x},t)\,d^3x = \frac{d}{dt}\int_{\mathscr{V}(0)} G(\mathbf{x},t)J_t\,d^3a = \int_{\mathscr{V}(0)} \left[\frac{\partial(GJ_t)}{\partial t}\right]_{\mathbf{a}} d^3a
$$

$$
= \int_{\mathscr{V}(0)} \frac{D}{Dt}(GJ_t)\,d^3a = \int_{\mathscr{V}(0)} \left(\frac{DG}{Dt}J_t + G\frac{DJ_t}{Dt}\right)d^3a. \tag{1.27}
$$

Some explanation is in order. The first term in the above equation includes a time derivative of the integrand and the limit of integration. As is often the case, we transform the integration variable, with the help of the appropriate Jacobian, which happens to be J_t, so that the limit of integration becomes time independent. Transforming this way makes the boundaries of the volume to appear fixed in time. Although one is allowed to do this, it will follow that the Jacobian of the transformation may be time dependent. Nonetheless, transforming into such coordinates (**a**) allows one to carry the differentiation into the integral, but we have to be careful, in this case, to recognize that it is the time derivative with **a** constant. This is nothing but the Lagrangian derivative. Continuing now the derivation, by using Euler's formula (1.26) and manipulations according to the rules of calculus, we get

$$
\dot{I}_G = \frac{dI_G}{dt} = \int_{\mathscr{V}(0)} \left[\frac{DG}{Dt} + G\nabla\cdot\mathbf{u}\right] J_t\,d^3a
$$

$$
= \int_{\mathscr{V}(t)} \left[\frac{DG}{Dt} + G\nabla\cdot\mathbf{u}\right] d^3x = \int_{\mathscr{V}(t)} \left[\frac{\partial G}{\partial t} + \nabla\cdot(G\mathbf{u})\right] d^3x. \tag{1.28}
$$

The two last expressions in formula (1.28) are two alternative forms of the *Reynolds transport theorem*, which gives the rate of change of the integral I_G, as defined in Eq. (1.27), i.e.,

$$
\frac{d}{dt}\int_{\mathscr{V}(t)} G(\mathbf{x},t)\,d^3x.
$$

We leave to Problem 1.6 the proof of the following third, and perhaps the most practical, form of the Reynolds transport theorem

$$
\frac{d}{dt}\int_{\mathscr{V}(t)} \rho(\mathbf{x},t)F(\mathbf{x},t)\,d^3x = \int_{\mathscr{V}(t)} \rho(\mathbf{x},t)\frac{DF}{Dt}d^3x, \tag{1.29}
$$

where $F(\mathbf{x},t)$ is any Eulerian field. O. Reynolds, after whom the above theorems are named, lived and acted in the nineteenth century. We shall see that the most important nondimensional number in FD bears his name and this is because of his fundamental contributions to FD.

Another useful transport theorem is the one dealing with the transport of *circulation*. Let $C(t)$ be a closed curve in the fluid. We consider it here being a material entity, that is, moving with the flow, and this is the meaning of making it a function of t. The *circulation* around C at time t is defined to be

$$\Gamma_C(t) \equiv \oint_{C(t)} \mathbf{u}(\mathbf{x},t) \cdot \mathbf{dl},$$

where $\mathbf{dl}$ is a differential line element along C and the integral is, of course, a line integral. The *circulation transport theorem* reads

$$\frac{d}{dt} \oint_{C(t)} \mathbf{u} \cdot \mathbf{dl} = \oint_{C(t)} \frac{D\mathbf{u}}{Dt} \cdot \mathbf{dl}. \tag{1.30}$$

The proof is straightforward, but a bit lengthy and technical. Consider the closed curve $C(t)$ and the curve $C(t+\delta t)$, which is the one carried by the fluid to time $t+\delta t$. We assume that δt is small (it will be eventually taken to zero as a limit), so that we may assume that the closed curve is only slightly distorted during the time interval δt (Fig. 1.6). We may write, by definition,

$$\dot{\Gamma}_C \equiv \frac{d}{dt} \oint_{C(t)} \mathbf{u} \cdot \mathbf{dl} = \lim_{\delta t \to 0} \frac{1}{\delta t} \left[\oint_{C(t+\delta t)} \mathbf{u}(\mathbf{x},t+\delta t) \cdot \mathbf{dl} - \oint_{C(t)} \mathbf{u}(\mathbf{x},t) \cdot \mathbf{dl} \right]. \tag{1.31}$$

If we call the curve $C(t+\delta t)$ by the name C', for convenience, we immediately notice that the (imaginary) surface composed of S,S', the surfaces formed by the bounding curves C and C' with the addition of the surface defined by the integral

Fig. 1.6 An auxiliary drawing for the proof of the circulation transport theorem (1.30)

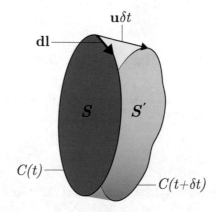

$\oint_{C(t)} \mathbf{dl} \times \mathbf{u}\delta t = -\oint_{C(t)} \mathbf{u} \times \mathbf{dl}\delta t$ form altogether a closed surface (care has to be taken, though, that the vector representing these surfaces points *outward* the closed surface). In addition, we may expand up to first order, i.e.,

$$\mathbf{u}(t+\delta t) = \mathbf{u}(t) + (\partial \mathbf{u}/\partial t)\delta t.$$

Thus

$$\dot{\Gamma}_C = \oint_{C(t)} \frac{\partial \mathbf{u}}{\partial t} \cdot \mathbf{dl} + \lim_{\delta t \to 0} \frac{1}{\delta t} \left[\oint_{C'} \mathbf{u}(\mathbf{x},t) \cdot \mathbf{dl} - \oint_{C(t)} \mathbf{u}(\mathbf{x},t) \cdot \mathbf{dl} \right].$$

By virtue of the Stokes theorem of vector analysis, the line integrals in the square parentheses of the previous expression may be transformed into surface integrals. If to the terms inside the square parentheses we also add and subtract the term $\oint_C (\nabla \times \mathbf{u}) \cdot \mathbf{u} \times \mathbf{dl}\,\delta t$, we end up with only the line integral $-\oint_C \mathbf{u} \times (\nabla \times \mathbf{u}) \cdot \mathbf{dl}$, after taking the limit $\delta t \to 0$. The addition of these terms turns the partial surface integral into a total surface integral of $\nabla \times \mathbf{u}$ over a closed surface which becomes identically zero. Remember that parts of a closed surface have directionality and therefore the minus in front of the integral over $S(t)$ turns into a plus. Invoking now the vector identity $\nabla(|\mathbf{u}|^2) = (\mathbf{u} \cdot \nabla)\mathbf{u} + \mathbf{u} \times (\nabla \times \mathbf{u})$ and remembering that a line integral over a closed loop of a gradient is zero, we finally obtain the statement of the theorem

$$\dot{\Gamma}(C) = \oint_{C(t)} \left[\frac{\partial \mathbf{u}}{\partial t} + (\mathbf{u} \cdot \nabla)\mathbf{u} \right] \cdot \mathbf{dl} = \oint_{C(t)} \frac{D\mathbf{u}}{Dt} \cdot \mathbf{dl}. \tag{1.32}$$

1.4 The Fluid Equations: Conservation Laws

The basic fluid equations are, as it often happens in physics, nothing but expressions of conservation laws. Since FD, at least in this book, is the Newtonian dynamics of a continuum, these equations are expressions of conservation of mass, momentum, and energy.

1.4.1 Mass Conservation

The field related to mass is *density*. Formally, it is defined at any point and at any time by the limiting process

$$\rho(\mathbf{x},t) = \lim_{\delta \mathscr{V} \to 0} \frac{\delta m}{\delta \mathscr{V}},$$

where it is understood that the mass δm and the volume $\delta \mathscr{V}$ containing it may be shrunk onto a point $\mathbf{x}$ at time t, without violating the continuum approximation.

Using the Reynolds transport theorem in the form (1.28), and choosing $G = \rho$, we exploit our assumption that the total mass of an element, which is a *material volume*, is constant, i.e.,

$$\frac{d}{dt}\int_{\mathscr{V}(t)} \rho(\mathbf{x},t)d^3x = 0,$$

and obtain at once the mathematical expression of mass conservation, called the *equation of continuity*,

$$\frac{\partial \rho}{\partial t} + \nabla \cdot (\rho \mathbf{u}) = 0, \tag{1.33}$$

recalling that the volume over which the integration is done is arbitrary. This equation is in a fully Eulerian form, but the second form of the Reynolds transport theorem gives

$$\frac{D\rho}{Dt} + \rho \nabla \cdot \mathbf{u} = 0, \tag{1.34}$$

which contains the Lagrangian derivative and, depending upon circumstances, this equation is often useful.

A simple volume integration should suffice in helping to understand physically that Eq. (1.33) indicates that the mass inside any volume can change only through the mass flux $(\rho\mathbf{u})$ into and out of that volume's boundary. This is the usual depiction found in many elementary textbooks on the subject. Equation (1.34) gives the compression rate of a fluid particle (density rate of change) during its motion. It follows that its relative compressibility is nothing but $-\nabla \cdot \mathbf{u}$. Flows in which $D\rho/Dt = 0$ is valid everywhere are called *incompressible*. The mathematical meaning is obvious: the density in incompressible flows is a material invariant. We shall discuss its meaning from a more physical approach later on. Here it will be enough to say that the fluid element retains the same density *along its motion*. This can happen if and only if $\nabla \cdot \mathbf{u} = 0$. In the case that the density of a fluid is constant in space and time, all flows of this fluid are incompressible.

Formula (1.34), together with Euler's formula (1.26), gives

$$\frac{D}{Dt}(\ln\rho + \ln J_t) = 0,$$

i.e., ρJ_t is a material invariant. Thus, we may set it equal to its value at the initial instant, when $J_0 = 1$, obtaining the *Lagrangian* version of the mass conservation equation

$$\rho(\mathbf{x},t)J_t = \rho_0, \tag{1.35}$$

where we have abbreviated $\rho(\mathbf{a},0) = \rho_0$. This immediately leads to the following incompressibility condition: $\rho = \rho_0$, which leads to $J_t = 1$. By virtue of Euler's formula one gets $\nabla \cdot \mathbf{u} = 0$. Both of the previous expressions mean that the flow preserves volume.

1.4.2   Momentum Conservation

We apply Newton's second law on a material volume, $\mathcal{V}(t)$, whose bounding surface is denoted by $\partial \mathcal{V}(t)$,

$$\frac{d}{dt} \int_{\mathcal{V}(t)} \rho \, \mathbf{u} \, d^3x = \int_{\mathcal{V}(t)} \rho \, \mathbf{b} \, d^3x + \oint_{\partial \mathcal{V}(t)} \mathbf{T}(\mathbf{n},\mathbf{x},t) dS,$$

where the body force and the surface force (traction) integrals have been included.

Now we exploit the Reynolds transport theorem in the form found in (1.29) on any component of the left hand side of this equation and use Eq. (1.24) for expressing the traction in terms of the stress. The previous equation written in component form gives for the kth component

$$\int_{\mathcal{V}(t)} \rho \, \frac{Du_k}{Dt} \, d^3x = \int_{\mathcal{V}(t)} \rho \, b_k \, d^3x + \oint_{\partial \mathcal{V}(t)} \sigma_{kj} n_j dS.$$

The last integral here is a surface integral over the closed surface embedding $\mathcal{V}(t)$. By virtue of the Gauss theorem, it can be replaced by a volume integral

$$\oint_{\partial \mathcal{V}(t)} \sigma_{kj} n_j dS = \int_{\mathcal{V}(t)} \frac{\sigma_{kj}}{\partial x_j} d^3x.$$

Since $\mathcal{V}(t)$ is arbitrary, the integrands must be equal, giving finally

$$\frac{Du_k}{Dt} = b_k + \frac{1}{\rho} \frac{\partial \sigma_{ki}}{\partial x_i}. \tag{1.36}$$

This is usually referred to as the *Cauchy equation for fluid motion*.

Now we can define more formally and thus precisely what we call in this book an *ideal fluid*. In an ideal fluid the stress tensor satisfies $\sigma_{ki} = -P\delta_{ki}$ with $P > 0$, that is, the stress tensor is diagonal and all its elements are equal to the negative of the Eulerian field, $P(\mathbf{x},t)$, which is called the *hydrostatic pressure*. The minus sign should by now be clear to the reader. The fact that the stress tensor is diagonal reflects the underlying fact that the force resulting from the hydrostatic pressure, which is nothing but force per unit area in the fluid, acts perpendicularly to a surface. The equality of all three components on the diagonal means *isotropy* of the hydrostatic pressure is usually referred to as *Pascal's law*. In the case of an ideal fluid (while there are no perfectly ideal fluids in nature, there are close approximations), and from Cauchy's equation, we get

$$\frac{D\mathbf{u}}{Dt} = \mathbf{b} - \frac{1}{\rho}\nabla P, \quad \text{or more explicitly} \quad \frac{\partial \mathbf{u}}{\partial t} + (\mathbf{u} \cdot \nabla)\mathbf{u} = \mathbf{b} - \frac{1}{\rho}\nabla P. \quad (1.37)$$

This is known as the *Euler equation* for fluid motion. It describes the flow of a fluid, in which nondiagonal terms in the stress tensor have been neglected. The physical meaning of this approximation will be made clear later on, when the Reynolds number is discussed. The Euler equation is the equation for a nondissipative motion of fluid. Dissipative effects will be added soon. Before that, it is useful defining here an important concept, namely, that of *hydrostatic equilibrium*. This situation occurs when the flow is not only stationary, that is, time independent, but the fluid is actually at rest, meaning that $\mathbf{u} = 0$ everywhere. Thus, hydrostatic equilibrium is just the balance of forces on the right-hand side of Eq. (1.37). The body force is a long range force, e.g., gravity, either self or external, magnetic force on ionized fluid, and more complicated forces. In the first case, hydrostatic equilibrium has an obvious meaning and is frequently found in astrophysics or geophysics, indicating, as we shall see in a number of problems, the balance of gravity and pressure gradient force.

The Euler equation of motion is written for Eulerian fields. If one wants to write the Lagrangian version of this equation for ideal fluids, then one has to translate the Eulerian coordinates $(\mathbf{x}, t)$ into the corresponding Lagrangian variables. This gives an equation for $\mathbf{x}(\mathbf{a}, t)$ which, for convenience, we write in component form. Starting with

$$\ddot{x}_i = b_i - \frac{1}{\rho}\frac{\partial P}{\partial x_i} \quad (1.38)$$

and using Eq. (1.2) and elementary calculus we get

$$\ddot{x}_i = b_i - \frac{1}{\rho}\mathbb{J}_{ik}^{-1}\frac{\partial P}{\partial a_k}, \quad (1.39)$$

where the functions x_i, b_i, ρ, and P have all to be expressed in Lagrangian variables $(\mathbf{a}, t)$. Remembering that $\mathbf{x}$ is the result of the flow transformation $\mathbf{x} = \Phi_t(\mathbf{a})$ [see Eq. (1.1)], this may be useful if a transformation of a solution, back to its Eulerian coordinates, is required. We repeat for the sake of clarity that Eq. (1.39) is the Lagrangian equation of motion, reflecting momentum conservation; it holds for an ideal fluid and is written in component form. This completes the two descriptions of the equation of motion for ideal fluids.

In a *nonideal* fluid, the off-diagonal components of the stress tensor σ_{ik} with $i \neq k$ cannot be neglected. This means that the "friction" of a given fluid layer with its adjacent layer plays a dynamical role. Thus, the stress tensor can be split in the following way

$$\sigma_{ik} = -P\delta_{ik} + \tau_{ik}, \tag{1.40}$$

where, as before, P is the hydrostatic pressure and τ_{ik} is called the *deviatoric* or *viscous* stress. We shall try here to motivate the choice for the form of the deviatoric stress, using an analogy with the theory of elasticity. To do so we have to establish a *constitutive* equation. Such an equation here can follow from a relation connecting the stress to the rate of deformation, which we have called $\mathscr{D}$ before. We shall obtain this relation using reasoning appropriate for a *Newtonian* fluid. A Newtonian fluid (we shall not deal with non-Newtonian fluids in this book) is defined by two properties:

1. The relation between the deviatoric stress and the deformation is *linear*, which is usually called the stress constitutive relation.
2. The fluid is *isotropic*.

Linearity of the constitutive relation can be expressed as

$$\tau_{ik} = K_{ikmn}\mathscr{D}_{mn}.$$

In a three-dimensional space, there are thus $81 = 3^4$ components of the tensor $\mathbb{K}$, which depend on the thermodynamic state of the material. However, owing to the isotropy and tensor index symmetries, the 81 parameters are reduced to 2! The symmetry of the stress and rate of deformation tensors ($\sigma_{ik} = \sigma_{ki}$ and $\mathscr{D}_{mn} = \mathscr{D}_{nm}$) require $K_{ikmn} = K_{kimn} = K_{iknm} = K_{kinm}$, so that we are left with 36 distinct values for the coefficients. Exploiting now the fact that the fluid is isotropic, we understand that a tensile stress in a certain direction must result in a deformation in the direction of that tension. A corollary of this observation is that the principal axes of the stress and of the deformation tensors coincide. The 36 coefficients reduce thus to only 2.[2] Omitting some technical details, like the splitting of a tensor into a traceless part and the rest, for example, we finally get

$$\sigma_{ik} = -P\delta_{ik} + 2\eta\left(\mathscr{D}_{ik} - \frac{1}{3}\delta_{ik}\mathscr{D}_{mm}\right) + \zeta\delta_{ik}\mathscr{D}_{mm}, \tag{1.41}$$

where the two surviving coefficients of the tensor $\mathbb{K}$, which we rename η and ζ are called the dynamic *viscosity* coefficients. The two are preceded by the adjectives *shear* and *second* or *bulk*. respectively. We shall discuss later in more detail the meaning of shear viscosity and of second viscosity from their physical perspective.

---

[2]These conclusions follow from elementary tensor analysis.

Using the definition of the rate of deformation $\mathcal{D}_{ik}$ (1.15) and that of the deviatoric (viscous) stress (1.40), we get the latter's form in terms of velocity derivatives and coefficients of viscosity:

$$\tau_{ik} = \eta \left[\left(\frac{\partial u_i}{\partial x_k} + \frac{\partial u_k}{\partial x_i} \right) - \frac{2}{3} \delta_{ik} \frac{\partial u_m}{\partial x_m} \right] + \zeta \delta_{ik} \frac{\partial u_m}{\partial x_m}. \tag{1.42}$$

The definitions $\eta \equiv \rho v$ and $\zeta \equiv \rho v_2$, using the *dynamic* shear and second (or bulk) viscosity coefficients, introduce v and v_2 which are the corresponding *kinematic* viscosity coefficients.

The momentum conservation equation, which includes also the viscous stress, can now be written using Cauchy's equation (1.36), with σ_{ki} fully incorporating the diagonal as well as the deviatoric parts of the stress

$$\frac{Du_k}{Dt} = b_k - \frac{1}{\rho} \frac{\partial P}{\partial x_k} + f_k^{\text{vis}}, \tag{1.43}$$

where f_k^{vis} is a shorthand notation for

$$f_k^{\text{vis}} \equiv \frac{1}{\rho} \frac{\partial \tau_{ik}}{\partial x_k},$$

as can be seen using Eq. (1.42). It is worth stressing that the above term is not a force. It just reflects the effect of viscous dissipation on the kth component of the motion and has the units of [force]/[mass]. We shall return to this equation in Sect. 1.4.4.

1.4.3 Energy Conservation

Internal energy, also known as thermal energy, is one of the thermodynamic variables of matter and in statistical mechanics we learn that it is actually the kinetic energy of the microscopic particles resulting from their random thermal motions and is intimately related to temperature. We shall discuss it in more detail in Sect. 1.5, devoted to the thermodynamics of a fluid. Here we shall be content with only accepting that internal energy is a well-defined field in the fluid. The local thermodynamic equilibrium assumption allows this, as we shall see in Sect. 1.5. Thus, we posit that the internal (thermal) energy, per unit mass, is an Eulerian field $e(\mathbf{x}, t)$, or alternatively that this quantity is well defined for a fluid parcel, in the Lagrangian description.

We now write, by inspection, a conservation equation for the sum of the kinetic energy of the fluid macroscopic motion plus the internal thermal energy, per unit mass, in a material fluid element $\mathcal{V}(t)$. It reads

$$\frac{d}{dt} \int_{\mathscr{V}(t)} \left(\frac{1}{2} |\mathbf{u}|^2 + e \right) \rho \, d^3x = \int_{\mathscr{V}(t)} \mathbf{u} \cdot \mathbf{b} \rho \, d^3x + \oint_{\partial \mathscr{V}(t)} \mathbf{u} \cdot \mathbf{T} n \, dS + \dot{Q}_{\mathrm{el}}, \qquad (1.44)$$

where the terms on the right-hand side involve $\mathbf{u} \cdot \mathbf{b}$, the power input of the body force integrated over the element, plus $\mathbf{u} \cdot \mathbf{T}$, that of the traction (stress force) which is naturally integrated over the closed *surface* of the moving element, plus the heat gain by the element $\mathscr{V}(t)$, from both total external heat inflow and internal heat production, per unit time. All these are marked together by $\dot{Q}_{\mathrm{el}}$. We denote by $\mathbf{n} \, dS$ the differential vector surface element of the moving volume's embedding boundary, where $\mathbf{n}$ is the outward normal at the position of the surface element. Using the Reynolds transport theorem (1.29), the relation $T_k = \sigma_{ki} n_i$, and assuming, for convenience, that the fluid does not possess any internal sources (or sinks) of heat, save dissipation due to viscosity, we may obtain the following energy conservation equation:

$$\rho \frac{D}{Dt} \left(e + \frac{1}{2} u_k u_k \right) = \rho u_k b_k + \frac{\partial}{\partial x_i} (u_k \sigma_{ki}) - \frac{\partial}{\partial x_k} q_k, \qquad (1.45)$$

where $\mathbf{q}$ is the outward directed heat flux through a point on the surface $\partial \mathscr{V}$ of the element. The direction of $\mathbf{n}$ is always outward and so we have $\mathbf{q} \cdot \mathbf{n} > 0$ and thus the last term, including the sign, represents the gain of heat energy by the element $\mathscr{V}(t)$, or in other words

$$\dot{Q}_{\mathrm{el}} = - \oint_{\partial \mathscr{V}} \mathbf{q} \cdot \mathbf{n} \, dS. \qquad (1.46)$$

If the surface heat flux happens to be, at some point of the surface, *into* the volume element, its contribution to the integral will be with a negative sign and a sufficiently large area of strong enough heat influx may lead to an overall energy gain by the element. We may now exploit the Cauchy equation (1.36) in order to get an equation for the internal thermal energy e alone. Multiplying Eq. (1.36) by u_k and summing over k we get

$$\rho \frac{D}{Dt} \left(\frac{1}{2} u_k u_k \right) = \rho u_k b_k + u_k \frac{\partial \sigma_{ki}}{\partial x_i}. \qquad (1.47)$$

Subtracting this relation from Eq. (1.45) and noting that by virtue of the freedom to interchange dummy indices and the stress tensor's symmetry, that is,

$$\sigma_{ik} \frac{\partial u_k}{\partial x_i} = \frac{1}{2} \sigma_{ik} \frac{\partial u_k}{\partial x_i} + \frac{1}{2} \sigma_{ki} \frac{\partial u_i}{\partial x_k} = \sigma_{ik} \mathscr{D}_{ik},$$

we finally obtain an equation for the material rate of change of the internal thermal energy

$$\rho \frac{De}{Dt} = \sigma_{ik} \mathcal{D}_{ik} - \frac{\partial}{\partial x_k} q_k, \tag{1.48}$$

where, as before, we do not allow any heat sources or sinks *inside* the volume element, save viscous dissipation which is given by $\sigma_{ik} \mathcal{D}_{ik}$ summed on both indices. This relationship will prove to be useful a little later on. We remark here that other forms of the energy equations exist as well, most of which will be given later in the text of this book.

1.4.4 Summary of the Fluid Dynamical Equations

We are now in the position in which we can write three of the fluid equations. Choosing the Eulerian notation, we have for the fields $\mathbf{u}(\mathbf{x}, t)$, $\rho(\mathbf{x}, t)$, and $e(\mathbf{x}, t)$ the following general equations, written here in component notation:

Mass conservation

$$\frac{\partial \rho}{\partial t} + \frac{\partial (\rho u_m)}{\partial x_m} = 0. \tag{1.49}$$

Momentum conservation

$$\rho \left(\frac{\partial u_k}{\partial t} + u_j \frac{\partial u_k}{\partial x_j} \right) = \rho b_k - \frac{\partial P}{\partial x_k} + \rho f_k^{\text{vis}}, \tag{1.50}$$

with

$$\rho f_k^{\text{vis}} = \frac{\partial \tau_{kj}}{\partial x_j} = \frac{\partial}{\partial x_k} \left[\eta \left(\frac{\partial u_k}{\partial x_j} + \frac{\partial u_j}{\partial x_k} \right) - \frac{2}{3} \delta_{ki} \frac{\partial u_m}{\partial x_m} \right] + \frac{\partial}{\partial x_k} \left(\zeta \frac{\partial u_m}{\partial x_m} \right).$$

Energy conservation

$$\rho \frac{De}{Dt} = \sigma_{ik} \mathcal{D}_{ik} - \frac{\partial}{\partial x_k} q_k. \tag{1.51}$$

Note that even with appropriate initial and boundary conditions, this is not a closed set of partial differential equations (PDEs). The function P is not specified, i.e., either given explicitly in terms of coordinates and time or expressed by some other unknown functions. The coefficients of viscosity are also not supplied. In addition, the body force $\mathbf{b}$ has to be given. Finally, we have to specify the physical source, and thus explicit form, of the heat flux $\mathbf{q}$ function. If the heat flux is the result of conduction, one can write a heat transport equation $\mathbf{q} = -\kappa \nabla T$, where T is the absolute temperature (also unknown) and κ is the thermal conductivity. In astrophysics the heat flux is often dominated by radiation, rather than conduction, and consequently the formula for $\mathbf{q}$ is more complicated. In the *heat diffusion* approximation, valid in most stellar interiors, the equation for heat transport is not too complex and may be found in any serious book on stellar structure, e.g., that of Kippenhahn & Weigert (reference [12] in the *Bibliographical Notes* of this chapter). If this approximation cannot be used, then some kind of radiative transport theory, which is beyond the scope of this book, is needed (see, e.g., reference [14]). Alternatively, a third mode of energy transport often exists in FD—*convection*. This is a very difficult problem, involving fluid motions, often turbulent, cf. Chap. 9. The simplest treatment of convection, the so-called *mixing length theory*, can also be found in books on stellar structure, like reference [12], which is in our opinion the best among many others. We should single out also Prianlik's book (reference [13]) for being concise, precise, and clear. In any case, additional specifications connecting thermodynamic variables to one another are needed to complete the fluid equations. For instance, one or several relations called *equations of state* are needed and except for simple gases and liquids they are very complex and are often taken from available results of lengthy computer calculations.

If the coefficients of viscosity are not explicit functions of position, the first two equations can be written in vectorial form, and therefore succinctly, as

$$\frac{\partial \rho}{\partial t} + \nabla \cdot (\rho \mathbf{u}) = 0, \tag{1.52}$$

$$\rho \left[\frac{\partial \mathbf{u}}{\partial t} + (\mathbf{u} \cdot \nabla) \mathbf{u} \right] = \rho \mathbf{b} - \nabla P + \eta \nabla^2 \mathbf{u} + \left(\zeta + \frac{1}{3} \eta \right) \nabla (\nabla \cdot \mathbf{u}). \tag{1.53}$$

The shear viscosity η is easily understood physically—it is frictional and results from the exchange of particles between one layer of fluid and an adjacent parallel one, moving at a different velocity. This interaction of fluid layers results in dissipation. The bulk viscosity ζ affects the equation only if the flow cannot be considered incompressible. Incompressible flows, which do not necessarily require that the density is constant throughout the fluid at all times, are often a good approximation for many flows (see below, in Chap. 2). The incompressibility condition ($\nabla \cdot \mathbf{u} = 0$) causes ζ to drop out from the equation of motion, as can be seen in Eq. (1.53), and one can then ignore the bulk viscosity. In the opposite case, when compressibility is important, the bulk viscosity remains and, as we can see in Eq. (1.50), its product with $-(\partial u_m / \partial x_m) \equiv -\nabla \cdot \mathbf{u}$, that is, compression, produces a term whose physical significance is to add resistance, formally adding to what we have called the hydrostatic pressure P. Now the question arises: is it correct to identify the hydrostatic pressure with the *thermodynamic* pressure (see next section), that is, the one which prevails in the fluid in local thermodynamic equilibrium? If this identification is made, then it becomes *formally* equivalent to setting the bulk viscosity to zero. The bulk or second viscosity is relevant only in significantly rarefied non-perfect gases, possessing internal degrees of freedom. Such degrees of freedom (perhaps rotational or vibrational motion modes of molecules) may be excited or damped when rapid changes of volume occur, i.e., when the timescale defined by $(\nabla \cdot \mathbf{u})^{-1}$ is much shorter than a typical flow time. The resultant coupling to the translational degrees of freedom may give rise to a frictional force opposing changes of volume. This is equivalent to an extra force per unit area, which has to be added to the hydrostatic pressure to yield the thermodynamic pressure. Symbolically we may write

$$P_{\text{thermo}} = P_{\text{hydro}} + \zeta (\nabla \cdot \mathbf{u}). \tag{1.54}$$

Most flows, with the exception of those in which local high energy phenomena occur in non-perfect gases, can be approximated by the condition in which the second viscosity can be ignored. However, one should keep in mind that in rapid changes of volume, this viscosity coefficient cannot be neglected. In some instances, like in very strong explosions, or perhaps in cosmology, it may play a leading role. It is also important to remark here that in some systems, astrophysical as well as certain physical experimental ones, fluids are often subject to strong radiation fields. These fields contribute to the stress tensor, but these terms are usually neglected. Radiative contributions to the stress tensor are addressed in reference [14] of the *Bibliographical Notes* found at the end of this chapter.

Equations (1.52)–(1.53) are known in the literature as the *Navier–Stokes* equations. The importance of these equations to fluid dynamics is hard to overestimate. As we shall see in this book, most of the fluid dynamical problems are approached by using them as a starting point. Often, this name is reserved only for incompressible flows, i.e., $\nabla \cdot \mathbf{u} = 0$, in which the latter equation replaces Eq. (1.52).

1.4.4.1   Mass and Momentum Equations Written in Tensor Conservation Form

It is sometimes useful and certainly more elegant to write the above equations in a tensor *conservation form*, i.e., relating the time derivatives of relevant functions to the divergences of appropriate fluxes. Mass conservation is written trivially as

$$\frac{\partial \rho}{\partial t} + \frac{\partial (\rho u_j)}{\partial x_j} = 0 \tag{1.55}$$

The momentum equation reads

$$\frac{\partial (\rho u_i)}{\partial t} + \frac{\partial \Pi_{ij}}{\partial x_j} = b_i, \tag{1.56}$$

where the momentum flux density tensor is defined as

$$\Pi_{ik} \equiv \rho u_i u_k + P\delta_{ik} - \tau_{ik}.$$

Proof of the equivalence of Eq. (1.56) with the Navier–Stokes equation is the subject of Problem 1.11.

1.4.5 Examples

We give below two very simple examples of solutions to the above FD equations.

1.4.5.1 Hydrostatic Plane-Parallel Atmosphere

As defined before, by hydrostatic equilibrium we mean a state of the fluid in which all the components of the velocity are zero, i.e., $\mathbf{u}(\mathbf{x},t) = 0$ at all locations and times. Consider now a one-dimensional plane-parallel equilibrium state, in which the pressure and density are functions of $x_3 = z$ only. Let $\mathbf{b} = -g\hat{\mathbf{z}}$ be the body force, with the constant $g > 0$. The physical meaning of this may be that the self-gravity of an atmosphere is negligible with respect to that of the body on which it resides and that we are considering a small enough region so that it may be approximated by a plane. We also assume that the system is shallow enough, so that g may be considered constant. The only surviving fluid dynamical equation is then

$$\frac{dP}{dz} = -\rho g. \tag{1.57}$$

As simple as this equation may look, we have to remark that when both P and ρ are allowed to vary, clearly some relation between them is needed, so that Eq. (1.57) can be solved. To give an example, assume that such a relation is given by

$$\frac{P(z)}{P_0} = \left[\frac{\rho(z)}{\rho_0}\right]^\gamma,$$

where P_0 and ρ_0 are constants equal to, say, the values of the pressure and density at the base of the atmosphere, $z = 0$. We require that the constant $\gamma \geq 1$. If γ is the adiabatic exponent, whose meaning will be recalled later in this chapter, it is a well-defined constant, $5/3$, for an ideal monoatomic gas. The atmosphere in which such a relation holds is called *adiabatic* (see below in Sect. 1.5.2). It follows that the differential equation for the density is

$$\frac{d\rho^\gamma}{dz} = -\frac{\rho_0^\gamma}{P_0} g\rho.$$

Straightforward integration of the above relationship gives the solution

$$\rho(z) = \rho_0 \left(1 - \frac{\gamma-1}{\gamma} \frac{g\rho_0}{P_0} z \right)^{\frac{1}{\gamma-1}},$$

for $\gamma \neq 1$. The validity of this equation breaks down at some z_1 where the density goes to zero, i.e., $\rho(z_1) = 0$. This occurs for $z_1 = [\gamma P_0]/[(\gamma-1)g\rho_0]$. For $\gamma = 1$ the equation is simpler and has the decaying exponential solution, appropriate for an isothermal case:

$$\rho(z) = \rho_0 \exp \left(-\frac{g\rho_0}{P_0} z \right),$$

in which the density goes to zero only as $z \to \infty$ (you may wish to compare this with an isothermal hydrostatic sphere solution, which will be given at the end of Sect. 1.6.1).

1.4.5.2 Uniformly Rotating Fluid of Constant Density

Let there be given a steady flow of a constant density fluid, which in Cartesian coordinates is $\mathbf{u}(x,y,z) = (-\Omega y, \Omega x, 0)$, where we denote by (x,y,z) the components of the position vector $\mathbf{x} = (x_1, x_2, x_3)$. Gravity is directed in the $-\hat{\mathbf{z}}$ direction and its force per unit mass is a constant g. Although the flow is given, the objective of this example is to find the surfaces of constant pressure.

Writing out the components of the Euler equations (1.37) for this flow we get

$$\frac{\partial P}{\partial x} = \Omega^2 \rho_0 x, \qquad \frac{\partial P}{\partial y} = \Omega^2 \rho_0 y, \qquad \frac{\partial P}{\partial z} = -\rho_0 g, \tag{1.58}$$

where the terms on the right-hand side of the first two equations represent the corresponding centripetal force associated with uniform rotation (see Chap. 5 for further details). Direct integration yields the following pressure distribution as a solution

$$P(x,y,z) = \frac{1}{2}\Omega^2 \rho_0 \left(x^2 + y^2 \right) - \rho_0 g z + \text{const.}$$

This means that the *isobaric* (of constant pressure) surfaces have the form

$$z(x,y) = \frac{\Omega^2}{2g}(x^2 + y^2) + \text{const.} \tag{1.59}$$

If the rotating fluid has a free surface, the shape of the surface will also have such a paraboloid of rotation surface, with the constant corresponding to the atmospheric pressure (imagine a vigorously stirred cup of liquid).

1.5 Thermodynamics of Fluid Motion

So far we have encountered only one or two aspects of thermal physics. In Sect. 1.4.3 the internal thermal energy, e, was introduced, and heuristically interpreted as the kinetic energy of the random velocities of microscopic particles. We also invoked a thermodynamic relation between P and ρ in an adiabatic atmosphere and mentioned the adiabatic exponent γ. Pressure and density were *defined* fluid dynamically, by using only mechanical considerations. Yet we should not lose sight of the fact that thermal physics is a significant and non-trivial branch of physics. Classical thermodynamics was the first to treat thermal phenomena, in a rather axiomatic and abstract way, largely detached from other branches of physics. In thermodynamics, e was simply a variable, or a state function. Eventually, statistical mechanics provided a deeper understanding of thermal phenomena and their physical meaning. When discussing the thermodynamics of fluid motion, we shall not restrict ourselves to one view or the other but, instead, we co-mingle both approaches. This should be permissible from a twenty-first century perspective of physics.

1.5.1 Local Thermodynamic Equilibrium

In classical thermodynamics one deals with states, unchanging in time, called *thermodynamic equilibria*, in which the system under consideration is uniform, i.e., it does not possess spatial gradients in temperature, density, specific energy, composition, or in any other thermodynamic variable. Global temporal changes to the system are assumed to occur at a rate which is considered very slow. To be more specific, one speaks of *quasi-static* changes, which approximates changes to the system as a sequence of permanent states. In other words, the slow changes we speak of are perceived as a series of thermodynamic equilibria. The approximation is good only if global changes occur on a timescale which is much longer than the relaxation timescale of the system towards thermodynamic equilibrium.

Fluid motions are sometimes not very slow and the above-mentioned thermodynamic variables may certainly vary in space. It turns out, however, that we may adopt the approximation that a *local* thermodynamic equilibrium (LTE) prevails, if the relevant variables' gradients are not too large and the changes are not too fast. Quantitatively, it means the following two conditions must be met for a given thermodynamic function $F(\mathbf{x},t)$: (1) the typical length scale of the thermodynamic function be much larger than the size scale of a fluid element, i.e., $\Lambda_F \equiv (|\nabla(\ln F)|)^{-1} \gg (\mathcal{V}_\varepsilon)^{1/3}$, where, as before, $\mathcal{V}_\varepsilon$ is the volume of the fluid element considered a fluid particle and (2) the global time changes are on scales much longer than the relaxation time towards thermodynamic equilibrium, i.e., $\tau_F \equiv (\partial \ln F / \partial t)^{-1} \ll \tau_{\text{relax}}$, where the relaxation time τ_{relax} is typically the time for a few collisions between the microscopic particles to take place and is, e.g., of the order of $\sim 10^{-5}$ s for terrestrial and planetary gases. It turns out that the LTE approximation works well if we have *linear* constitutive equations for the stress and for heat transport, as we saw in our discussion of Newtonian fluids. Fluids endowed with such linear constitutive equations are called *simple fluids*.

Thus, we may be content with the fact that the LTE is usually a good approximation and the laws of thermodynamics are obeyed locally, that is, at any $(\mathbf{x}, t)$, for the majority of flows we shall consider. To sum up the notation, sometimes repeating ourselves, the following list of the various thermodynamics functions, is considered: P (thermodynamic pressure), T (absolute temperature), ρ (mass density), e (specific internal thermal energy), $h = e + P/\rho$ (specific enthalpy), and finally s (specific entropy). The last variable (entropy) plays a key role in thermal physics, and we shall be discussing it below in considerable detail. As we noted earlier, in principle, the pressure listed here may not be the hydrostatic pressure as defined in our discussion of the diagonal elements of the stress tensor. We shall, however, distinguish between these two definitions of pressure, keeping in mind that only in extremely rapid processes may the bulk viscosity cause the thermodynamic pressure to differ from the hydrostatic one. In the context of the FD treated in this book, the bulk, or second, viscosity is neglected.

1.5.2 Equations of State and the Laws of Thermodynamics

The equation of state is clearly the property of the material in question. We do not intend to discuss the various conditions encountered in applications and derive the appropriate equation of state for them. We have previously mentioned that there are good sources for such information, for a wide range of densities and temperatures.

We shall explicitly treat here only the simple, but useful cases. It is customary to single out the following forms of equations of state (EOS). Each has its own name, and these are for a given constant chemical composition, (1) the *thermal* EOS: $P = P(\rho, T)$, (2) the *caloric* EOS: $e = e(\rho, T)$, and (3) the *potential* or *canonical* EOS: $e = e(\rho, s)$. Manipulation of thermodynamic identities and laws shows that one form of EOS can always be transformed into another, because any thermodynamic variable can always be expressed as a function of the remaining ones, which characterize the system. In the case of simple systems, any one thermodynamic variable can be expressed as a function of two other thermodynamic variables. More complicated gases (e.g., a mixture involving more than one type) requires more variables. We shall now give just two examples of equations of state, one applicable for dilute gases and the other for a liquid like water.

An *ideal*, or *perfect*, gas is a substance, in which the microscopic particles interact only through binary collisions between particles. It is straightforward to show, as be seen in virtually any undergraduate course or book on thermal physics, that such a gas obeys the following thermal EOS

$$P(\rho,T) = \frac{\mathscr{R}}{\mu}\rho T \equiv R\rho T, \tag{1.60}$$

where $\mathscr{R} = 8.314 \times 10^7$ erg/deg mol is the gas constant, $\mathscr{R} \equiv N_A k_B$, i.e., the gas constant is the product of the perhaps more familiar Avogadro number and the Boltzmann constant. μ is the mean molecular weight, that is, the mean mass of a microscopic particle in the gas in atomic mass units. The notation $R = \mathscr{R}/\mu$ is sometimes used, where R depends on the composition of the gas. For $\mu = 1$ (is there a gas having this value of μ, if only approximately?), $\mathscr{R} = R$.

Pure water at pressure P_0, temperature T_0, and having density ρ_0 obeys approximately a linear equation of state, if the changes from the reference values are small. We may write

$$\rho(P,T) = \rho_0\left[1 - \beta_T(T - T_0) + \beta_P(P - P_0)\right], \tag{1.61}$$

with $\beta_T \approx 2 \times 10^{-4}$ 1/K and $\beta_P \approx 4.1 \times 10^{-11}$ cm^2/dyn.

1.5.2.1   Some Thermodynamic Relations for an Ideal (Perfect) Gas

We find it useful to remind the reader about several of the most important thermodynamic quantities, which are often used, also in thermodynamics of fluid motion and give their explicit form for an ideal gas. The first of these is the *adiabatic sound speed*, c_s, defined by

$$c_s^2 = \left(\frac{\partial P}{\partial \rho}\right)_s, \tag{1.62}$$

where the partial derivative is taken at constant entropy. In FD, c_s is a field and we shall see, when discussing sound waves in compressible fluids, that this indeed is their phase velocity. Next we mention the so-called *adiabatic exponent*, γ:

$$\gamma = \left(\frac{\partial \ln P}{\partial \ln \rho}\right)_s. \tag{1.63}$$

Some basic relations satisfied by γ, in an ideal gas, as related to adiabatic derivatives of thermodynamic functions should also be mentioned. With $\upsilon \equiv 1/\rho$ being the specific volume we have

$$\left(\frac{\partial \ln T}{\partial \ln \upsilon}\right)_s = 1 - \gamma,$$

$$\left(\frac{\partial \ln P}{\partial \ln T}\right)_s = \frac{\gamma}{\gamma-1}. \tag{1.64}$$

For an ideal gas we have the result that γ is a constant, depending on the number of atoms in a gas molecule. For a monoatomic perfect gas $\gamma = 5/3$ and its more general value can be shown to be

$$\gamma = 1 + \frac{2}{f},$$

where f is the number of unfrozen, i.e., accessible, degrees of freedom of a given gas molecule. Thus, a monoatomic gas has $f = 3 \Rightarrow \gamma = 5/3$, while diatomic nitrogen, say, in the conditions of the Earth's atmosphere, has $f = 3$ (*translational*) + 2 (*rotational*) = 5 and thus $\gamma = 7/5 = 1.4$. The vibrational degrees of freedom are frozen, because their excitation requires a very high temperature.

Finally, a few words about specific heats of an ideal gas. These quantities, when defined at constant volume or pressure, indicate the amount of added heat (per unit mass) to the substance, per rise of $1\,°K$ in the temperature. So

$$c_V = \left(\frac{\partial q}{\partial T}\right)_V, \qquad c_P = \left(\frac{\partial q}{\partial T}\right)_P. \tag{1.65}$$

It is a simple matter to show that, using equilibrium reversible thermodynamic processes, i.e., $dq = Tds$, where q means here heat per unit mass, for a perfect gas

$$c_V = (\gamma - 1)^{-1}\frac{\mathscr{R}}{\mu}, \quad c_P = c_V + \frac{\mathscr{R}}{\mu}, \quad \text{and} \quad \frac{c_P}{c_V} = \gamma.$$

1.5.2.2 Some General Thermodynamic Identities

Before turning to the formulation of the two important laws of thermodynamics appropriate for fluid motion, we state the all-important *Gibbs equation* relating the aforementioned quasi-static changes of thermodynamical variables, where we choose to use $1/\rho$, instead of v for visual clarity:

$$Tds = de + Pd\left(\frac{1}{\rho}\right). \tag{1.66}$$

J.W. Gibbs (1839–1903), who lived in the USA, was one of the most important figures in thermal physics and the development of statistical mechanics. This equation has a number of equivalent forms which can be derived from (1.66) utilizing standard thermodynamic relations, such as those appearing in any elementary thermodynamics book. The most important of Gibbs equation equivalent forms, the *Gibbs relations*, are

$$dh = Tds + \left(\frac{1}{\rho}\right)dP, \quad df = -s\,dT - Pd\left(\frac{1}{\rho}\right), \quad dg = -s\,dT + \left(\frac{1}{\rho}\right)dP, \quad (1.67)$$

where $h = e + P/\rho$ is the *enthalpy*. $f \equiv e + P/\rho$ and $g \equiv h - Ts$ are the Helmholtz and Gibbs thermodynamic functions, respectively, all per unit mass.

Turning now to the *laws of thermodynamics*, we write them here in a form appropriate for FD, that is, considering a moving fluid particle as our thermodynamic system in LTE.

1.5.2.3 The First Law and Bernoulli's Formula

The first law of thermodynamics is in fact an energy conservation energy statement. As such, its form is similar to Eq. (1.51), and we rewrite it here for convenience:

$$\rho\frac{De}{Dt} = \sigma_{ik}\mathcal{D}_{ik} + \dot{\mathcal{Q}}, \qquad (1.68)$$

where $\dot{\mathcal{Q}}$ is the time rate of heat absorption by the fluid particle, per unit volume. In the case that this heat comes from outside, by the heat flux $\mathbf{q}$, we have $\dot{\mathcal{Q}} = -\partial q_k/\partial x_k$. If internal heat sources or sinks exist (e.g., chemical or nuclear reaction heating or neutrino losses), the rate of such heat gain, per unit volume, should be appropriately included in $\mathcal{Q}$. Note that we have not included, so far, body forces. They will be added in the generalization immediately below. One can easily identify in Eq. (1.68) the first law, as it is usually written in thermodynamics. The internal thermal energy rate of change, per unit volume, is equal to the rate of work done on the fluid per unit volume, including here the stress times deformation tensors double sum, plus the rate of heat absorption by a unit volume of the system.

Assume now that the body force is derivable from a static potential function, that is, $\mathbf{b}(\mathbf{x},t) = -\nabla\Phi(\mathbf{x})$. Very often this is the case, e.g., if the body force is a static external gravitational potential. Using now the energy conservation equation in our original form (1.45), with the general term $\dot{\mathcal{Q}}$ instead of minus the divergence of the heat flux, and with the body force included in terms of (minus) a potential gradient, we get

$$\rho\frac{D}{Dt}\left(e + \frac{1}{2}u_k u_k\right) = \frac{\partial}{\partial x_i}(u_k\sigma_{ki}) - \rho u_k\frac{\partial\Phi}{\partial x_k} + \dot{\mathcal{Q}}. \qquad (1.69)$$

Separating now the stress tensor in the usual way $\sigma_{ki} = -P\delta_{ki} + \tau_{ki}$, we find after some algebra that

$$\rho \frac{D}{Dt}\left(h + \frac{1}{2}u_k u_k + \Phi\right) = \frac{\partial P}{\partial t} + \frac{\partial}{\partial x_i}(u_k \tau_{ki}) + \dot{\mathcal{Q}}, \qquad (1.70)$$

remembering that $h = e + P/\rho$.

The *Bernoulli function* for a flow is now defined as the field

$$\mathscr{B}(\mathbf{x},t) \equiv h(\mathbf{x},t) + \frac{1}{2}|\mathbf{u}(\mathbf{x},t)|^2 + \Phi(\mathbf{x}), \qquad (1.71)$$

and thus we get the prototype, in fact the most general form of the unsteady *Bernoulli's equation*. The term Bernoulli's *formula* is also frequently used.

$$\rho \frac{D\mathscr{B}}{Dt} = \frac{\partial P}{\partial t} + \frac{\partial}{\partial x_i}(u_k \tau_{ki}) + \dot{\mathcal{Q}}. \qquad (1.72)$$

The expression as appearing here, especially in its steady state, is useful for several different types of flows, see below in Chap. 2. The above equation bears the name of D. Bernoulli, who lived in the eighteenth century and belonged to a large family of outstanding physicists and mathematicians.

1.5.2.4   The Second Law

The state function *entropy*, per unit mass, which we denote by $s(\mathbf{x},t)$, plays a special role in thermodynamics and in statistical mechanics. One of the often used simplifications are flows in which the specific entropy does not change along the flow of a fluid particle, that is,

$$\frac{Ds}{Dt} = 0. \qquad (1.73)$$

We call such flows *isentropic*. It applies when fluid elements do not acquire heat from external sources or reactions from within (chemical, nuclear, etc.). The case where the specific entropy is a constant everywhere and at all times within the fluid, i.e.,

$$\frac{\partial s}{\partial t} = \nabla s = 0, \qquad (1.74)$$

will be referred to as *homentropic*. It is not to be confused with the concept of adiabaticity. A flow is *adiabatic* if a fluid particle does not absorb any heat from the outside or from nondissipative source inside it. However, heat may be produced by the deviatoric stress working on the fluid element (*viscous dissipation*).

Thus, an adiabatic flow may be non-isentropic. However, if an adiabatic flow is also reversible (nondissipative) then it is clearly isentropic. These subtle details are often important.

The *second law* of thermodynamics can be formulated by the celebrated statement that the entropy of a closed system never decreases. Several other statements of this law have been given over the years, but all have been proven equivalent. We may translate it to conditions appropriate for a fluid element in the following way:

$$\frac{d}{dt}\int_{\mathcal{V}(t)} s\rho\, d^3x \geq -\oint_{\partial\mathcal{V}(t)} \Xi\cdot\mathbf{n}dS, \qquad (1.75)$$

where $\Xi(\mathbf{x},t)$ is the entropy flux vector, that is, the entropy crossing a unit area perpendicular to Ξ, per unit time. The statement thus reads that the total entropy inside the element can never decrease faster than the rate that the entropy leaves the element. The entropy loss here is through the element's closed surface.

Some interesting and important results follow if we use the result proved in Problem 1.13:

$$\frac{De}{Dt} - \frac{P}{\rho^2}\frac{D\rho}{Dt} = \frac{1}{\rho}\tau_{ik}\mathcal{D}_{ik} + \frac{1}{\rho}\dot{\mathcal{Q}}.$$

Applying on the left-hand side of this equation the Gibbs equation (1.66), one gets

$$T\frac{Ds}{Dt} = \frac{1}{\rho}\tau_{ik}\mathcal{D}_{ik} + \frac{1}{\rho}\dot{\mathcal{Q}}.$$

For the sake of definitiveness and simplicity we shall now abandon the more general notation $\dot{\mathcal{Q}}$ and replace it by $-\nabla\cdot\mathbf{q}$, that is, consider the non-adiabatic effects as coming only from external heat flux $\mathbf{q}$. A generalization for the case when there are point sources or sinks inside the element should not be too difficult. Thus we have

$$T\frac{Ds}{Dt} = \frac{1}{\rho}\left(\Psi - \nabla\cdot\mathbf{q}\right), \qquad (1.76)$$

where

$$\Psi \equiv \tau_{ik}\mathcal{D}_{ik} \qquad (1.77)$$

is called the *dissipation function*. It is not too difficult to show (try it) that this function must be nonnegative, i.e., $\Psi \geq 0$, otherwise there would be a serious problem in thermal physics. The above exercise can be done by developing the components and showing that it is a sum of squares multiplied by positive viscosity coefficients. Viscous work is thus being created by deviatoric stresses multiplied by deformation. To visualize that work is being done imagine deforming an elastic material by deviatoric stresses. This formulation is consistent with what is known

in thermodynamics as the *Clausius inequality* $Tds \geq \delta q$, where $\delta q = -\upsilon \nabla \cdot \mathbf{q}\, dt$ is
the heat transfer to a unit mass of the fluid particle in time dt. $\upsilon \equiv 1/\rho$, while T and
ds are for the fluid particle as well.

For the simplest case of heat transfer, namely, that of the *Fourier law of conduction*, we have $\mathbf{q} = -\kappa \nabla T$. J. Fourier was one of the most important mathematicians
and a very successful administrator in the Napoleonic era. His contributions will be
used frequently in this book. It is now only a matter of algebra to show that

$$-\frac{1}{T}\nabla \cdot \mathbf{q} = \kappa \left(\frac{\nabla T}{T}\right)^2 + \nabla \cdot \left(\kappa \frac{\nabla T}{T}\right).$$

Substituting this in Eq. (1.76) gives

$$\rho T \frac{Ds}{Dt} = \Psi + \frac{\kappa}{T}(\nabla T)^2 + T\nabla \cdot \left(\frac{\kappa \nabla T}{T}\right). \tag{1.78}$$

If we identify now $\Xi \equiv \mathbf{q}/T$ as the entropy flux vector, we can get an interesting
expression, consistent with the second law (see 1.75), since the right-hand side of
the following expression is nonnegative:

$$\rho \frac{Ds}{Dt} + \nabla \cdot \Xi \geq 0.$$

Moreover, we get explicitly the specific entropy production rate in the possibly
moving fluid particle for this case. For more general cases of internal heat point
sources and/or other heat transport laws, one can proceed in a similar way and get
also the entropy production rate in the fluid particle. Finally, we remark that in the
case when the pressure is stationary and the flow is adiabatic and nondissipative, that
is, isentropic, the unsteady Bernoulli's equation (1.72) implies that the Bernoulli
function is material invariant, that is, $D\mathscr{B}/Dt = 0$. This is one of the *Bernoulli
theorems* (see Chap. 2).

We conclude this rudimentary section on thermodynamics of fluid motion by
making an even shorter comment on statistical mechanics. As is well known,
statistical mechanics provides a mechanistic formulation of thermodynamics, by
means of statistical analysis. Since it is clearly beyond the scope of this book to
go into details of that subject, we shall only state that the important notion of
entropy was formulated by L. Boltzmann and others, as a state function proportional
to the logarithm of the number of different accessible microscopic states in
thermodynamic equilibrium, consistent with the macroscopic constraints defining
the state. The outrage caused by Boltzmann's important insight was very severe.
Possibly, it cost Boltzmann his life. As is often the case in science, Boltzmann's
triumph was only achieved posthumously. On his gravestone, the famed formula

expressing the central statement about the entropy is engraved. In this chapter's *Bibliographical Notes* we shall give some general references on the basics of thermodynamics and statistical mechanics.

1.6 Similarity and Self-Similarity in Fluid Dynamics

It turns out that several nondimensional numbers exist, embodying basic qualities of a fluid flow. We have already met the Knudsen number, (Kn), which was important in determining if a continuum description of a fluid is possible. In the upcoming pages of this book we will encounter many of these numbers. Here we introduce the *Reynolds number* (Re), an important quantity which played a key role in the discovery of the *law of similarity*. But before entering into the issue of similarity, or self-similarity of solutions, we start with stating that the value of the *Reynolds number* indicates some important details on the nature of a fluid flow.

In large fluid systems allowing large velocities, where the meaning of "large" here will be clear soon, we almost always encounter flows, in which viscosity, which is a physical property of the fluid, plays a negligible role. The meaning of this statement is best formulated using the Reynolds number, Re, to be introduced and discussed shortly. However, we found it necessary to give a short discussion of such circumstance, because it usually leads to turbulence, a detailed description of which is deferred to Chap. 9. Turbulence, which occurs often in astrophysical flows, mimics many of the characteristics of a flow with effective enhanced viscosity, and because of the lack of a reliable theory of turbulence, the effective viscosity approach is usually used, certainly in astrophysics.

The Reynolds number of a *flow* is probably the most well-known of the nondimensional numbers. As we shall see, Re, as well as other nondimensional numbers, go a long way in characterizing fluid dynamical flows. For the sake of simplicity, consider a constant density fluid in a steady flow ($\partial/\partial t = 0$), without any body forces $\mathbf{b} = 0$. The only parameter characterizing the flow is the ratio of the viscosity η to the constant density. This parameter $v = \eta/\rho$, which was already defined as kinematic viscosity, appears in the Navier–Stokes equation that is applicable to our simplified case [see (1.53)]:

$$(\mathbf{u} \cdot \nabla)\mathbf{u} = -\nabla\left(\frac{P}{\rho}\right) + v\nabla^2\mathbf{u}. \qquad (1.79)$$

We proceed by making Eq. (1.79) nondimensional. Let the typical value of the velocity in the particular flow we consider be U and a typical length scale L. These are obviously dimensional quantities (in the c.g.s. system $[U] = \text{cm/s}$, $[L] = \text{cm}$, and $[v] = \text{cm}^2/\text{s}$). The only nondimensional quantity that can be formed from the above three quantities is the combination UL/v. This combination is the *Reynolds number*, Re, so that

$$\text{Re} \equiv \frac{\rho\, UL}{\eta} = \frac{UL}{\nu}. \tag{1.80}$$

Thus in steady flows Re is important in assessing the relative value of the advective, usually called *inertial*, term to the viscous one, in the fluid equation of motion. We may consider the density as a nondimensional number, because it may be expressed in units of its constant value, in fact $\rho = 1$, but we shall keep it for the sake of transparency. There remains the issue of scaling the pressure. Here we choose to use its dynamic scaling ρU^2, noting, however, that other possibilities also exist, which involve viscosity. This will be discussed in Sect. 3.4.1, where the Stokes flow, having a very small Reynolds number, will be discussed. So the nondimensional version of Eq. (1.79) is

$$(\mathbf{u}\cdot\nabla)\,\mathbf{u} = -\nabla\left(\frac{P}{\rho}\right) + \frac{1}{\text{Re}}\nabla^2\mathbf{u}. \tag{1.81}$$

The Reynolds number of a flow is an indicator, as said before, of the nature of that flow. We may say that it measures the typical ratio of the advective, i.e., inertial effects in the flow to the viscous ones. In this way, Re^{-1} is a measure of the strength of the viscous effects. Note that as the Reynolds number includes also a length scale, it may be the case that viscous effects are negligible on some large length scale, but are pronounced on short scales. Similarly, the typical velocity size appears in Re, and thus, even though viscous effects may be unimportant for a particular flow, in which the velocity is high, those effects may be nonnegligible for low velocity flows having the same viscosity and length scale.

As we shall see in the chapters on linear and nonlinear instability (Chaps. 7, 8) and turbulence (Chap. 9), the Reynolds number plays a key role in rendering a flow unstable. In general, high Re implies instability. Typical astrophysical flows are endowed with extremely large Reynolds numbers, the crucial factor being the scale L. For example, a rotating star has the following order of magnitude relevant parameters $U \approx 10^5$ cm/s, $L \approx 10^{11}$ cm, and $\nu \approx 1\,\text{cm}^2/\text{s}$. This gives a formidable value for the Reynolds number $\text{Re} \approx 10^{16}$. The situation is similar in accretion disks and interstellar clouds or intergalactic flows. As hinted before, such a value of Re would imply turbulent flows, which are often approached by considering *eddy viscosity*, which with respect to transport plays the role of the usual *microscopic* viscosity that we have been discussing so far. This issue will be discussed in more detail in the chapter on turbulence (Chap. 9).

In problems of the type that is given by Eq. (1.81), we may restrict ourselves to using only nondimensional quantities, e.g., express velocities in units of U and lengths in units of L; it follows that the velocity function, obtained as a solution of the relevant problem, must have the following form:

$$\mathbf{u} = U\mathbf{f}(\mathbf{x}/L, \text{Re}),$$

where $\mathbf{f}$ is some vector function of its arguments. This is so because the only dimensionless parameter in the problem is the Reynolds number. The above formula shows that two flows, which are geometrically similar, including the boundaries, can be considered also physically similar, if the Reynolds number is the same. That is to say that one flow can be obtained from the other by changing the units of measurement, here for velocities and lengths. This *law of similarity*, which we have given here for a simple case, was discovered by O. Reynolds in 1883. The pressure distribution in the fluid, for this problem, can be written as

$$P = \rho \, U^2 f(\mathbf{x}/L, \mathrm{Re}),$$

choosing the above dynamic scaling for pressure. Here, too, f is some suitable function. Physical quantities, which are not functions of coordinates, can also be expressed in a way lending itself to similarity arguments. For example, if in the problem considered here a body of a particular shape is immersed in the flow, one can express the drag force on that body from dimensional numbers related to the flow and the body, i.e., ρ, L, and U, thus

$$F_D = \rho \, U^2 L^2 f(\mathrm{Re}).$$

Other flows of differing types are characterized by yet other dimensional units, and a similar procedure of finding nondimensional numbers composed of them is possible to determine similarity. Moreover, a flow may be *self-similar*, that is, when expressed and solved in nondimensional variables at a particular time we obtain a configuration that, at later times, will simply be an enlargement or rescaling of the previous configuration.

Before moving to some examples, we would like to single out a nondimensional number related to the dimensional quantity of thermal conductivity, κ. In addition to the Reynolds number, whose reciprocal value indicates the importance of viscosity, we define the *Prandtl number* $\mathrm{Pr} \equiv \nu/\chi$, where $\chi \equiv \kappa/(\rho c_P)$ is the thermometric conductivity. The Prandtl number is just a constant of the material and for most gases in conditions similar to our atmosphere it is of the order of unity. Pr obviously indicates in a nondimensional way the relative importance of the viscous dissipative processes to those originating from thermal conduction. L. Prandtl (1875–1953) was a very prominent fluid dynamicist and we shall undoubtedly see some of his contributions in this book. In liquids like water or alcohol Pr is of order 10, while in glycerine, say, it is much larger and in mercury much smaller.

Finally we mention briefly, for the sake of our effort at completeness, two more nondimensional numbers. The force of gravity is also important in a problem, a relevant nondimensional number known as the *Froude number*, and defined here (later on in Chap. 6, we shall define the square root of this as the Froude number) as

$$\mathrm{Fr} \equiv \frac{U^2}{Lg}. \tag{1.82}$$

Another number is indispensable for nonsteady flows. In such flows two independent nondimensional numbers can be constructed: the *Strouhal number*

$$\mathtt{St} \equiv \frac{U\tau}{L}, \tag{1.83}$$

where τ is a typical timescale of variations (oscillations, say) of the flow, and also the Reynolds number $\mathtt{Re}$. If there is no additional external force to induce the time changes in the flow, we must have the functional relation

$$\mathtt{St} = f(\mathtt{Re}).$$

We shall now give some examples of similarity and self-similar solutions in basic problems of astrophysical interest.

1.6.1   Similarity of Polytropic Stars Having the Same Index

The simplest model of a star constitutes of a *gas sphere*. If a pressure–density relation is supplied, one can solve the fluid dynamical equations without the need for dealing with the energy equations and thermodynamics. It turns out that in certain stars a relation of this kind is not a bad approximation and, in any case, omitting all the details of energy production and transport in stars greatly simplifies the problem and allows one to focus on some salient features of stellar structure. For historical reasons the assumed pressure–density relationship takes the form called a *polytropic* relation:

$$P = K\rho^{1+1/n}, \tag{1.84}$$

where the constant n is called the polytropic index and K is a constant. Note that it is not an equation of state, it is a geometrical relation for a specific configuration, but obviously if a fluid is a homentropic perfect gas, then a relation like (1.84) follows from the equation of state $P \propto \rho^{\gamma}$. If a polytropic relation of the kind (1.84) holds throughout, then static, spherically symmetric models of stars can be constructed and they are called *polytropes*. One writes a spherically symmetric hydrostatic equation, using r as the independent variable, in Eulerian form, as

$$\frac{dP}{dr} = -\frac{Gm(r)}{r^2}\rho(r), \qquad \frac{dm}{dr} = 4\pi r^2 \rho(r), \tag{1.85}$$

where now $m(r)$ is the mass inside the sphere of radius r. In Problem 1.12, we find this equation in its Lagrangian form:

$$\frac{dP}{dm} = -\frac{Gm}{4\pi r^4} \quad \text{and} \quad \frac{dr}{dm} = \left(4\pi r^2 \rho\right)^{-1}. \tag{1.86}$$

Differentiating the first of the equations found in (1.85) with respect to r, after some rearrangement, incorporating it with the second one yields a single, second-order ordinary differential equation (ODE):

$$\frac{1}{r^2} \frac{d}{dr} \left(\frac{r^2}{\rho} \frac{dP}{dr} \right) = -4\pi G\rho. \tag{1.87}$$

In what follows we nondimensionalize this equation and substitute into it the polytropic relation (1.84). This is done in the following order: first, we substitute for the density $\rho \equiv \rho_c \phi^n$, where ρ_c is the dimensional central density, which will also serve as a density unit, and where $\phi(r)$ is a dimensionless function. Then, choosing the length unit to be the quantity

$$\tilde{L} \equiv \sqrt{\frac{(n+1)K\rho_c^{(n-1)/n}}{4\pi G}}$$

greatly simplifies (1.87) and turns it into the more concise equation

$$\frac{1}{\xi^2} \frac{d}{d\xi} \left(\xi^2 \frac{d\phi}{d\xi} \right) = -\phi^n. \tag{1.88}$$

This nondimensional equation ($\xi \equiv r/\tilde{L}, \rho = \rho_c \phi^n$) is called the *Lane–Emden equation* and it is supplemented by the following boundary conditions:

$$\phi(0) = 0 \quad \text{and} \quad \frac{d\phi}{d\xi}(0) = 0. \tag{1.89}$$

The second of these conditions is a consequence from a regularity requirement, namely that the pressure field ought not to have spikes at the very center of the hydrostatic spherical model. Explicit analytical solutions $\phi(\xi)$ are known only for three values of the polytropic index $n = 0, 1$, and 5. For other indices it is a simple matter to numerically integrate the equations. For $n = 1$ the solution is

$$\phi(\xi) = \frac{\sin \xi}{\xi}.$$

The model extends up to $\xi = \pi$ only, where ϕ, and therefore the density, vanishes. We have chosen this problem to demonstrate the law of similarity. All $n = 1$ polytropes are similar to one another and their explicit dimensional solutions are determined by the choice of ρ_c (Fig. 1.7).

An isothermal gas sphere, despite being formally a polytrope with index $n = \infty$, does not lend itself to a treatment of the kind given above. It is an interesting case, in which it is impossible to invoke the law of similarity. In dimensional units the equation of hydrostatic equilibrium for an isothermal sphere reads

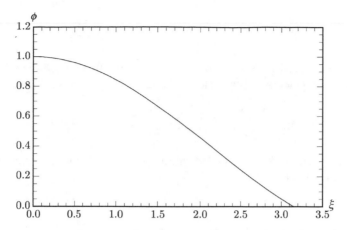

Fig. 1.7 Polytropic gas sphere of index $n = 1$. Similarity is achieved by multiplying the nondimensional density ϕ by the central density ρ_c and the nondimensional radius ξ by L; for details, see text

$$\frac{c^2}{\rho}\frac{d\rho}{dr} = -\frac{d\Phi}{dr}, \tag{1.90}$$

where $\Phi(r)$ is the gravitational potential, c_T is the isothermal sound speed, and we have used $dP = c_T^2 d\rho$ and then dropped the subscript T. The solution is

$$\rho(r) = \rho_0 \exp\left(-\frac{\Phi}{c^2}\right),$$

where ρ_0 is the density at the place we choose to be a zero of the potential. To get the explicit form of $\Phi(r)$ one needs to solve a Poisson-like equation with the potential appearing also on the right-hand side:

$$\frac{1}{r^2}\frac{d}{dr}\left(r^2\frac{d\Phi}{dr}\right) = 4\pi G\rho_0 \exp\left(-\frac{\Phi}{c^2}\right). \tag{1.91}$$

Equation (1.91) is a second-order ODE that does not have known analytical solutions, but solutions may be found numerically and have been extensively tabulated in the literature. Their nature depends, obviously, on the boundary conditions. For example, if we choose at $r = 0$

$$\Phi(0) = 1 \qquad \text{and} \qquad \frac{d\Phi}{dr}(0) = 0,$$

the solutions are regular and extend to infinity if an external bounding surface is not postulated. All solutions have the same asymptotic behavior for $r \to \infty$, namely,

$$\rho(r) \to \frac{c^2}{2\pi G} r^{-2} \qquad \text{for} \qquad r \gg \frac{c}{\sqrt{4\pi G \rho_0}}.$$

ρ_0 is the central density (in which we have chosen $\Phi(0) = 0$ to be the minimum of the gravitational potential). Note that particular limiting solutions

$$\rho(r) = \frac{c^2}{2\pi G} r^{-2} \qquad \text{and} \qquad \frac{d\Phi}{dr}(r) = 2c^2 r^{-1} \tag{1.92}$$

exactly satisfy Eqs. (1.90)–(1.91), but unfortunately the solution is singular at the center since $\rho \to \infty$ as $r \to 0$.

1.6.2 The Self-Similar Solution of a Collapsing Isothermal Sphere

To demonstrate a *self-similar* solution, we give here the collapsing isothermal sphere problem. In spherical symmetry, the Eulerian continuity equation is

$$\frac{\partial \rho}{\partial t} + \frac{1}{r^2} \frac{\partial}{\partial r} \left(r^2 \rho u \right) = 0.$$

The variable $m(r,t)$ (the mass interior to radius r), shown before in our discussion of polytropes, is clearly a Lagrangian variable in this one-dimensional case. In this problem we shall use it explicitly. The Lagrangian continuity, that is, mass conservation, equation in this one-dimensional spherical case takes the form

$$\frac{\partial m}{\partial r} = 4\pi r^2 \rho. \tag{1.93}$$

Now, combining these two equations, by multiplying the first by $4\pi r^2$ and substituting the second, yields

$$\frac{\partial}{\partial r} \left(\frac{\partial m}{\partial t} + u \frac{\partial m}{\partial r} \right) = 0.$$

It is clear that the above relation, whose meaning in a spherical flow translates to $\partial (Dm/Dt)/\partial r = 0$ requires that $Dm/Dt = f(t)$ for all r, where f is a suitable function. Physically, this means that the mass interior to any r is a universal function of t. Clearly, this is impossible for $r = 0$, if $f(t)$ itself is not zero. With this in mind, we now list the three equations to be solved. These include two statements of mass conservation, the first one is actually a relation between the functions m and ρ and is just Eq. (1.93). The other equations are

$$\frac{\partial m}{\partial t} + u\frac{\partial m}{\partial r} = 0 \tag{1.94}$$

and the equation of motion

$$\frac{\partial u}{\partial t} + u\frac{\partial m}{\partial r} = -\frac{c_T^2}{\rho}\frac{\partial \rho}{\partial r} - G\frac{m}{r^2}, \tag{1.95}$$

where c_T, as before, is the isothermal sound speed in which the identification $c_T^2 \equiv (\partial P/\partial \rho)_T$ has been made.

To proceed, it is necessary to form a dimensionless similarity variable, called here ξ, which is constructed out of the dimensional radius and time, together with the sound speed c_T, which we shall mark by c for economy of notation and clarity:

$$\xi \equiv \frac{r}{ct}. \tag{1.96}$$

The name *acoustic depth*, sometimes used for the variable ξ, has a revealing physical meaning. Now we postulate the following form of new dependent variables, guided by their nondimensional nature:

$$\delta(\xi) = 4\pi G t^2 \rho(r,t), \quad \mu(\xi) = \frac{Gm(r,t)}{c^3 t}, \quad v(\xi) = \frac{u(r,t)}{c}. \tag{1.97}$$

Substituting the dimensional functions u, ρ, and m from the last equation into Eqs. (1.93)–(1.95) gives

$$\mu + (v - \xi)\frac{d\mu}{d\xi} = 0, \quad \frac{d\mu}{d\xi} = \xi^2\delta,$$

which upon the elimination of $d\mu/d\xi$ finally gives

$$\mu(\xi) = \xi^2\delta(\xi)[\xi - v(\xi)]. \tag{1.98}$$

Some straightforward but lengthy algebra (see Problem 1.14) produces the final coupled nonlinear ODEs to be solved:

$$[(\xi - v)^2 - 1]\frac{dv}{d\xi} = \left[(\xi - v)\delta - \frac{2}{\xi}\right](\xi - v),$$

$$[(\xi - v)^2 - 1]\frac{d\ln\delta}{d\xi} = \left[\delta - \frac{2}{\xi}(\xi - v)\right](\xi - v). \tag{1.99}$$

As can be checked by direct substitution

$$\mu(\xi) = 2\xi, \qquad v(\xi) = 0, \qquad \delta(\xi) = \frac{2}{\xi^2}, \qquad (1.100)$$

is an analytical solution of the coupled ODEs (1.99). Using Eqs. (1.97), (1.100) translates into the corresponding dimensional solution,

$$\rho(r) = \frac{c^2}{2\pi G} r^{-2} \qquad \text{and} \qquad m(r) = \frac{2c^2}{G} r,$$

which is, clearly, the limiting *singular* (at the center) isothermal gas sphere solution found above.

Now, if we choose this singular solution to serve as the initial condition for our time-dependent problem, we find self-similar behavior for $t > 0$. A complication arises, however, for $\xi - v = 1$ and $\delta = 2$ using the nondimensional variables, since Eq. (1.99) contains critical points, i.e., the equations lose their meaning at these values of the functions. We recommend that the reader also considers whether or not the case $\xi - v = -1$ is possible and what may be its physical meaning. Special care, using Taylor expansions, is needed in order to pass the critical points smoothly. We encounter a very similar situation in the next chapter, when discussing the Bondi problem. It will be easier to explain the problem and its remedy there, even though the method to deal with critical points applies equally to the present case.

1.7   The Virial Theorem and Some of Its Consequences

We conclude this chapter with an important general global statement called the *virial theorem*, which results from the fluid equations of motion. This statement has played an important role, especially in astrophysics, in understanding physical principles pertaining to fluid masses. R. Clausius in 1860 pioneered the global approach embodied in the virial theorem, employing it on a system composed of many small discrete bodies. We shall present the derivation based on the work of S. Chandrasekhar (1910–1995), a prominent astrophysicist and recipient of a Nobel prize in physics, whose work encompasses many areas of astrophysics and general relativity.

1.7.1   General Derivation

Considering an ideal fluid, i.e., one in which the stress includes only the pressure, occupying a fixed volume in space, $\mathscr{V}$, we assume it to be self-gravitating with no other body forces acting. The surface of the above volume is denoted by $\partial \mathcal{V}$. The equation of motion is then

$$\rho \frac{Du_i}{Dt} = -\frac{\partial}{\partial x_i} P - \rho \frac{\partial}{\partial x_i} \Phi, \tag{1.101}$$

where the gravitational potential Φ is the solution of the Poisson equation:

$$\Phi(\mathbf{x}) = -G \int_{\mathcal{V}} \frac{\rho(\mathbf{x})}{|\mathbf{x} - \mathbf{x}'|} d^3 x'. \tag{1.102}$$

Actually, since we have not limited the fluid to be stationary, the functions describing it should be dependent on time as well. For simplicity of presentation we omit this notation, unless explicitly necessary.

Multiplying Eq. (1.101) by x_k and integrating over the entire volume gives

$$\int_{\mathcal{V}} x_k \rho \frac{Du_i}{Dt} d^3 x = -\int_{\mathcal{V}} x_k \frac{\partial P}{\partial x_i} d^3 x - \int_{\mathcal{V}} x_k \rho \frac{\partial \Phi}{\partial x_i} d^3 x. \tag{1.103}$$

Using now the mixed notation, introduced in Eq. (1.5), we may transform the left-hand side of the above equation to

$$\int_{\mathcal{V}} x_k \rho \frac{Du_i}{Dt} d^3 x = \int_{\mathcal{V}} x_k \rho \frac{D}{Dt} \left(\frac{Dx_i}{Dt}\right) d^3 x = \int_{\mathcal{V}} \rho \frac{D}{Dt} (x_k \dot{x}_i) d^3 x - \int_{\mathcal{V}} \rho \dot{x}_k \dot{x}_i d^3 x. \tag{1.104}$$

The last integral is related to the kinetic energy of the flow in the entire fluid. We define here the positive quantity as the *kinetic energy tensor* of the configuration, which is symmetric:

$$\mathscr{K}_{ik} = \frac{1}{2} \int_{\mathcal{V}} \rho u_k u_i d^3 x. \tag{1.105}$$

Thus we may summarize this part of our derivation by writing

$$\int_{\mathcal{V}} x_k \rho \frac{Du_i}{Dt} d^3 x = \int_{\mathcal{V}} \rho \frac{D}{Dt} (x_k \dot{x}_i) d^3 x - 2\mathscr{K}_{ik}. \tag{1.106}$$

Regarding the right-hand side of Eq. (1.103), we deal with its two terms separately. For the first term we have

$$-\int_{\mathcal{V}} x_k \frac{\partial P}{\partial x_i} d^3 x = -\int_{\mathcal{V}} \frac{\partial (x_k P)}{\partial x_i} d^3 x + \delta_{ki} \int_{\mathcal{V}} P d^3 x$$

$$= -\oint_{\partial \mathcal{V}} P x_k n_i dS + \delta_{ik}(\gamma - 1)U, \tag{1.107}$$

where U is the total internal energy and actually the above result follows the expression obtained in Problem 1.15. It is valid in this form for a perfect gas and $\partial \mathcal{V}$ is the surface of the fluid body, where in the integration the surface differential is written as dS. The second term on the right-hand side of equation (1.103), which we define as the tensor $\mathcal{W}_{ik}$, gives

$$\mathcal{W}_{ik} \equiv -\int_{\mathcal{V}} x_k \rho \frac{\partial \Phi}{\partial x_i} d^3x = G\int_{\mathcal{V}} x_k \rho(\mathbf{x}) \frac{\partial}{\partial x_i}\left[\int_{\mathcal{V}} \frac{\rho(\mathbf{x}')}{|\mathbf{x}-\mathbf{x}'|}d^3x'\right]$$

$$= -G\int_{\mathcal{V}}\int \rho(\mathbf{x})\rho(\mathbf{x}')\frac{x_k(x_i-x_i')}{|\mathbf{x}-\mathbf{x}'|^3}d^3x d^3x'. \quad (1.108)$$

(Can you see why?) It is a straightforward task to symmetrize this expression by noting that upon exchanging primed and unprimed independent variables the expression does not change as both types of variables are integrated over. Consequently, adding half of the expression to half of the same expression but with primed and unprimed variables interchanged yields the original expression. So we get a symmetrical expression for, what is now called, the *gravitational energy tensor*:

$$\mathcal{W}_{ik} = -\frac{1}{2}G\int_{\mathcal{V}}\int \rho(\mathbf{x})\rho(\mathbf{x}')\frac{(x_k-x_k')(x_i-x_i')}{|\mathbf{x}-\mathbf{x}'|^3}d^3x d^3x'.$$

Collecting all terms, we obtain the expression

$$\int_{\mathcal{V}} \rho \frac{D(x_k \dot{x}_i)}{Dt} d^3x = 2\mathcal{K}_{ik} + \delta_{ik}(\gamma-1)U + \mathcal{W}_{ik} - P_S \oint_{\partial \mathcal{V}} x_k n_i dS, \quad (1.109)$$

where we have assumed that the pressure on the surface of the configuration is constant and equals P_S. If we consider very large bodies of fluid (extending to "infinity," so to speak), it is reasonable to assume that the pressure $P_S \to 0$ and the surface term vanishes altogether. Otherwise, if the fluid body is immersed in pressure P_S this term has to be retained.

Assuming that the left-hand side of this equation is a symmetric tensor since the right-hand side, without the surface term, certainly is, by manipulating this left-hand side of the equation we can get to the following form:

$$\int_{\mathcal{V}} \rho \frac{D(x_k \dot{x}_i)}{Dt} d^3x = \frac{1}{2}\int_{\mathcal{V}} \frac{D(x_k \dot{x}_i + \dot{x}_k x_i)}{Dt} d^3x.$$

Finally, we may write a fairly general statement of the virial theorem as

$$\frac{1}{2}\frac{d^2}{dt^2}\mathcal{I}_{ik} = 2\mathcal{K}_{ik} + \delta_{ik}(\gamma-1)U + \mathcal{W}_{ik} - P_S \oint_{\partial \mathcal{V}} x_k n_i dS, \quad (1.110)$$

where we have defined the *moment of inertia tensor*,

$$\mathscr{I}_{ik} = \int_{\mathscr{V}} \rho x_i x_k d^3 x, \tag{1.111}$$

and used twice the Reynolds transport theorem. To sum up: the only assumptions, so far, were that the surface pressure has to vanish if the body extends practically "to infinity" and that the relation between the pressure and the specific internal thermal energy of the gas, of which the large fluid mass is composed, may be written as $P = (\gamma - 1)\rho e$ with γ constant.

1.7.2   Some Specific Consequences

General and elegant as the virial theorem statement in Eq. (1.110) may be, it is, as it stands, of a limited practical use. If one takes trace of that formula, i.e., puts $i = k$ and sums over the index, a procedure called, in the language of tensor theory, *contraction*, the following expression is obtained:

$$\ddot{I} = 2K + 3(\gamma - 1)U + W - 3P_S\mathscr{V}, \tag{1.112}$$

where I, K, U, W, and $\mathscr{V}$ are, respectively, the moments of inertia, kinetic energy, internal thermal energy (the factor 3 appears as a result of δ_{ii}), the gravitational energy, and the total volume of the configuration. The reemergence of the surface term here is for the sake of generality and we may keep it if the tensors are contracted, even if $P_S \neq 0$. The form of the surface term here follows trivially from

$$-P_S \oint_{\partial \mathscr{V}} x_k n_k dS = -P_S \int_{\mathscr{V}} \nabla \cdot \mathbf{x} d^3 x = -3P_S\mathscr{V}. \tag{1.113}$$

One particular example, the case of a static star, composed of a perfect gas, goes a long way to understanding the properties of such a system, using the virial theorem alone. We take the simplest case with $P_S = 0$ and because the star is static both $\ddot{I} = 0$ and $K = 0$ we have

$$3(\gamma - 1)U + W = 0 \qquad \Longrightarrow \qquad W = -3(\gamma - 1)U. \tag{1.114}$$

Now, the total energy of the star is $\mathscr{E} = U + W = (4 - 3\gamma)U$ and since U is a nonnegative quantity we have $\mathscr{E} \leq 0$, which indicates a gravitationally bound star if $\gamma > 4/3$. This is certainly the case for monoatomic perfect gas, for which $\gamma = 5/3$. This simple principle shows that a gravitationally bound star, which *loses energy* to space, because of radiation from the surface, must necessarily *heat up*, because if we define the energy loss as $\mathscr{L} = -d\mathscr{E}/dt$, it follows that

$$\mathscr{L} = (3\gamma - 4)\frac{dU}{dt}.$$

Now if $\mathscr{L} > 0$ and $\gamma > 4/3$, we must have $dU/dt > 0$, that is, the energy losing star must increase its thermal energy and thus heat up! This curious phenomenon, characteristic of gravitationally bound systems, is interpreted as resulting from the system having a negative specific heat(!). However, this is not the usual *thermodynamic specific heat*, which is always positive. Rather, we are referring here to what is called the *gravo-thermal specific heat*, which takes into account also the gravitational energy. It is easy to see that also

$$\mathscr{L} = \frac{3\gamma - 4}{3(\gamma - 1)}\frac{dW}{dt},$$

and this shows that if the gravitationally bound star loses energy, its negative gravitational potential energy decreases, which means that the star shrinks which, in turn, causes it to heat up. In fact, for the case of perfect monoatomic gas, $\gamma = 5/3$ and thus

$$\mathscr{L} = \frac{dU}{dt} = -\frac{1}{2}\frac{dW}{dt},$$

i.e., half of the energy liberated by contraction is radiated away, while the other half is used to heat up the star.

Another trivial example using the virial theorem is the case of a satellite of mass m_s in a circular orbit (at constant velocity v_s) around a central mass. In this case, we retain only the kinetic energy as the situation is not static, but certainly $\ddot{I} = 0$ and we have, denoting the kinetic energy in this case by $T = K = 1/2 m_s v_s^2/2$,

$$2T + W = 0 \qquad \longrightarrow \qquad W = -2T. \tag{1.115}$$

The total energy of the orbit is $\mathscr{E} = T + W = -T$. As is known from elementary mechanics, the total energy loss because of friction, say, diminishes $\mathscr{E}$ and thus increases T and the satellite speeds up by sinking into a lower orbit so that the gravitational energy becomes more negative. Also, the well-known relation $T = -W/2$ comes up naturally. The virial theorem is also often used to describe clusters of stars and galaxies. Problem 1.16 elaborates on one such example.

Problems

1.1. Estimate the Knudsen number for the following systems:

1. A cubical room $4 \times 4 \times 4$ meter filled with air, whose density is $1.2\,\mathrm{kg\,m^{-3}}$.

2. An interstellar neutral hydrogen cloud, whose size is ~ 1 parsec (pc), where $1\,\mathrm{pc} = 3.26$ light years (ly) $= 3.08567758 \times 10^{18}\,\mathrm{cm}$ and its density is $10^{-23}\,\mathrm{g\,cm^{-3}}$.
3. A spherical bulb of radius $5\,\mathrm{cm}$, in which the pressure is equal to $P = 0.01$ atm, at temperature $20\,^\circ\mathrm{C}$. Assume that the ideal gas approximation holds.
4. A glass of water.

On the basis of the order magnitude of Kn in these systems, determine in which the continuum description is a good one.

1.2.
The expressions in this problem are all nondimensional.

1. Given is a two-dimensional steady flow $\mathbf{u}(x,y) = x\hat{\mathbf{x}} - y\hat{\mathbf{y}}$. Note that the steadiness of the flow means that the Eulerian spatial coordinates at all points do not change in time. Find the Lagrangian coordinates of a particle, which at $t = 0$ was at position (x_0, y_0).
2. Examine now another simple two-dimensional flow $\mathbf{u}(x,y) = (y\hat{\mathbf{x}} - x\hat{\mathbf{y}})/(x^2 + y^2)$ where we shall formally exclude the point $x = y = 0$. Find the Lagrangian coordinates of the same particle as above. Also find the streamlines and describe the fluid motion on them.
3. Now consider the nonsteady two-dimensional flow $\mathbf{u}(x,y) = y\hat{\mathbf{x}} - (x + c\sin\omega t)\hat{\mathbf{y}}$ with c constant. Find the instantaneous streamlines, expressing them as a function of t. Plot the particle path through the point $x = 1, y = 2$, and also the streakline emanating from this point for $0 < t < 2\pi$, taking $c = 1$, $\omega = 0.75$.

1.3.
Prove that the vector $\hat{\mathbf{x}}_i \mathcal{O}_{ik} dx_k$, see Eq. (1.16), can be written as the second term of Eq. (1.17). Use the hints in the text.

1.4.
Let $\mathbb{S}_{ij}$ and $\mathbb{A}_{kl}$ be any symmetric and antisymmetric tensors, respectively. Prove that $\mathbb{S}_{nm}\mathbb{A}_{nm}$ is identically zero.

1.5.
Prove Euler's formula $DJ_t/Dt = J_t\nabla \cdot \mathbf{u}$. Use the hints in the text.

1.6.
Prove yet another (and useful!) form of the Reynolds transport theorem

$$\frac{d}{dt}\int_{\mathcal{V}(t)} \rho(\mathbf{x},t)F(\mathbf{x},t)\,d^3x = \int_{\mathcal{V}(t)} \rho(\mathbf{x},t)\frac{DF}{Dt}d^3x, \qquad (1.116)$$

where F is any Eulerian field. Hint: choose $G = \rho F$ in the previous versions of the Reynolds theorem.

1.7.

Suggest an alternative derivation of the mass conservation equation and the Euler equation, without the use of Reynolds transport theorem, i.e., by directly applying the appropriate conservation laws on a material volume element.

1.8.

A solid body of volume $\mathscr{V}$ and density ρ_b is immersed in a very large container of fluid, whose density is $\rho(\mathbf{x})$ and it is in hydrostatic equilibrium under the downward acting body force $\mathbf{b}$. Show that the net force on the body is

$$\mathbf{F}^{\text{net}} = \int_{\mathscr{V}} \rho_b \mathbf{b} \, d^3 x + \int_S \mathbf{T} \cdot d\mathbf{S} = \int_{\mathscr{V}} (\rho_b - \rho) \mathbf{b} \, d^3 x. \tag{1.117}$$

$\mathbf{T}$ is the traction and S is the surface of the solid body. It is perhaps time to remind here our notation $d\mathbf{S} \equiv \mathbf{n} dS$. The result is simply the *Archimedes principle*.

1.9.

Given the following nondimensional velocity field $\mathbf{u}(\mathbf{x},t) = (ax/t)\hat{\mathbf{x}}_2$, where a is constant, calculate the material acceleration $D\mathbf{u}/Dt$. Describe and interpret the result for $a = 1$.

1.10.

In this problem only nondimensional quantities are used. Consider the two-dimensional stagnation point flow $(u, \mathrm{v}) = (x, -y)$ with initial density $\rho_0(x, y) = x^2 + y^2$. Show that the solution for the density at later times is $\rho(x, y, t) = (xe^t)^2 + (ye^t)2 = x^2 e^{2t} + y^2 e^{2t}$. Describe qualitatively how the lines of constant density change with time.

1.11.

The Eulerian continuity equation is already written in *conservation form*, that is, the partial derivative of a physical quantity is expressed as (minus) the divergence of an appropriate flux. Among other things, the conservation form is appropriate for numerical calculations. Derive the conservation form of the momentum conservation equation for the case in which there are no body forces on the fluid:

$$\frac{\partial}{\partial t}(\rho u_i) = -\frac{\partial}{\partial x_j}(\rho u_i u_j + P\delta_{ij} - \tau_{ij}). \tag{1.118}$$

1.12.

In stellar hydrodynamics, assuming spherical symmetry, the Lagrangian description is usually used. The coordinates are t (time) and m—the mass interior to a sphere, whose radius at time t is $r(m,t)$, which is used as the dependent variable (the unknown function of the differential equation). Clearly m is a Lagrangian coordinate. Write the equation of motion for a spherical star, assuming ideal fluid and writing the pressure as $P(m,t)$. Complement this equation by a mass conservation equation, where you include also the function $\rho(m,t)$.

1.13.
Show that by splitting the stress tensor into a diagonal and deviatoric (τ_{ik}) part, the first law of thermodynamics, as in Eq. (1.68), can be written in the following way:

$$\rho \frac{De}{Dt} - \frac{P}{\rho^2} \frac{D\rho}{Dt} = \frac{1}{\rho} \tau_{ik} \mathscr{D}_{ik} + \frac{1}{\rho} \dot{\mathcal{Q}}. \tag{1.119}$$

1.14.
Use Eq. (1.98) to transform Eqs. (1.93)–(1.95) into the nondimensional pair of ODEs:

$$[(\xi - v)^2 - 1] \frac{dv}{d\xi} = \left[(\xi - v)\delta - \frac{2}{\xi} \right] (\xi - v),$$

$$[(\xi - v)^2 - 1] \frac{d \ln \delta}{d\xi} = \left[\delta - \frac{2}{\xi}(\xi - v) \right] (\xi - v). \tag{1.120}$$

1.15.
Let the total internal thermal energy of a fluid mass enclosed in volume $\mathscr{V}$ be $U = \int_{\mathscr{V}} \rho e d^3x$, where $e(\mathbf{x}, t)$ is the internal thermal energy per unit mass. Show that if the fluid is composed of an ideal gas,

$$U = \frac{1}{\gamma - 1} \int_{\mathscr{V}} P d^3x, \tag{1.121}$$

where γ is the adiabatic exponent.

1.16.
As early as in the 1930s, the respected astronomer F. Zwicky tried to estimate the total mass of a rich galaxy cluster, called the Coma cluster. He had at his disposition the velocity dispersion of the cluster, $\sigma_v = 1000\,\text{km/s}$, whose square may be perceived as an average kinetic energy (per unit mass) of a galaxy in the cluster. Assume that the cluster is gravitationally bound and its effective radius is $R_C = 5\,\text{Mpc}$. This radius is defined so as to give the gravitational energy of the cluster, per galaxy, to be equal to $W_g = GM_C/R_C$. Estimate the Coma cluster mass, using the virial theorem.
If it is known that the blue luminosity of the Coma cluster is $L_{bC} = 2 \times 10^{12} L_\odot$, estimate (M/L_{bC}) as compared to $M_\odot/L_\odot$. What can be the source of the discrepancy you have found? Note that in this problem we assume that there is no contribution to any of the quantities because of the uncertainty in the Hubble constant (i.e., we put $h = 1$).

Bibliographical Notes

General

There are numerous books on fluid dynamics. Among them we single out two: the classical text of the Landau series *Course of theoretical physics* and Thomson's book. We chose them from the multitude as corresponding closely our approach to the topics discussed in them. Our book includes several additional "modern" topics, mainly in nonlinear behavior and approximation methods.

1. L.D. Landau, E.M. Lifshitz, *Fluid Mechanics*, 2nd edn. (Elsevier, Amsterdam, 2004)
2. P.A. Thompson, *Compressible Fluid Dynamics*. Advanced Engineering Series (Rensselaer, Troy, 1988)
   The following introductory and rather short book includes a longer discussion on plasma physics than most FD books and focuses on astrophysical systems:
3. A.R. Choudhuri, *The Physics Of Fluids and Plasmas – An Introduction for Astrophysicists* (Cambridge University Press, Cambridge, 1998)
   We would like to also recommend, in general, the concise excellent textbook
4. D.J. Acheson, *Elementary Fluid Dynamics* (Oxford University Press, Oxford, 1990)

Section 1.1

The following article by Edward Spiegel from a 1987 summer school clearly and thoughtfully lays the groundwork for understanding the concept of a fluid particle and its relation to Eulerian description. The short book of Richard Meyer also clarifies the basis of Lagrangian and Eulerian descriptions:

5. E.A. Spiegel, in *Astrophysical Fluid Dynamics*. Les Houches, vol. XLVII, ed. by J.-P. Zahn, J. Zinn-Justin (North-Holland, Amsterdam, 1987)
6. R.E. Meyer, *Introduction to Mathematical Fluid Dynamics* (Wiley, London, 1971)

Section 1.2

The nomenclature of fluid kinematics is largely based on reference [5] and on

7. L.D. Landau, E.M. Lifshitz, *Theory of Elasticity*, Chap. 1, 3rd edn. (Elsevier, Amsterdam, 1986)

The meaning of the word *chaotic trajectory* is simply and well explained in the introductory text:

8. O. Regev, *Chaos and Complexity in Astrophysics* (Cambridge University Press, Cambridge, 2012)

Sections 1.3 and 1.4

Most general books (e.g., [1, 4]) give a good account of the dynamical equations. One may also benefit from the concise summary in

9. J.-L. Tassoul, *Theory of Rotating Stars*, Chap. 3 (Princeton University Press, Princeton, 1978)

Section 1.5

In the general books on FD, good discussion on the topic of this section can be found in the relevant parts of reference [1] and in Chap. 2 of reference [2].
The following books can serve as a very good introduction to thermodynamics and statistical mechanics, in general:

10. H.B. Callen, *Thermodynamics and An Introduction to Thermostatistics*, 2nd edn. (Wiley, London, 1985)
11. K. Huang, *Introduction to Statistical Physics* (CRC, Boca Raton, 2001)
    A good description of equations of state, basics of radiation transport, and point energy sources or sinks can be found in the following first two books on stellar structure. Serious treatment of radiative transport is given in the third book. All three are recommended:
12. R. Kippenhahn, A. Weigert, *Stellar Structure and Evolution* (Springer, Berlin, 1994)
13. D. Prialnik, *An Introduction to the Theory of Stellar Structure and Evolution*, 2nd edn. (Cambridge University Press, Cambridge, 2009)
14. D. Mihalas, B.W. Mihalas, *Foundations of Radiation Hydrodynamics* (Dover, New York, 1999)

Section 1.6

Similarity is discussed in detail in an entire book devoted to it:

15. L. Sedov, *Similarity and Dimensional Methods in Mechanics*, 10th edn. (CRC, Boca Raton, 1993)

A good description of polytropes can be found in references [12] and [13] and in

16. F.H. Shu, *The Physics of Astrophysics*, vol. II (University Science Books, Mill Valley, 1992)
    where the isothermal sphere receives special treatment, including a semi-quantitative description of its collapse. This collapse details are given in the following article:
17. F.H. Shu, Self-similar collapse of isothermal spheres and star formation. Astrophys. J. **214**, 488 (1977)

Section 1.7

An entire section of this classic book is devoted to an exact derivation of the virial theorem, in its most general form.

18. S. Chandrasekhar, *Hydrodynamic and Hydromagnetic Stability*, sect. 117 (Dover, New York, 1961)

Chapter 2
Restricted and Vortical Flows

Make things as simple as possible, but not simpler.

Albert Einstein (1879–1955)

2.1 Introduction

The fluid dynamical equations of motion are a formidable set of nonlinear PDEs
(partial differential equations). It seems hopeless to look for solutions of these
equations in any general case, if specific boundary and initial conditions are given.
We can, however, learn a lot about the physical properties of flows, i.e., solutions
of the equations by defining auxiliary functions and deriving theorems about flows
that are valid under special circumstances. That is the subject of this chapter. We
shall introduce and use here the *vorticity* field $\omega = \nabla \times \mathbf{u}$, whose importance in
FD is paramount and it will accompany the discussions along most of the book. It
is very useful to consider flows occurring under various *restricted conditions*, e.g.,
steady, inviscid, incompressible, barotropic, and *irrotational* or a combination of a
small number thereof. We include in this category flows with simplistic geometries
and boundary conditions, including those with manageable initial conditions. In this
chapter we choose a number of flows with explicitly defined restricted conditions
and we go on to examine what can be learned analytically about such flows. It is
important, in our view, to know the physical properties of relevant special flows and
understand them before attempting to numerically solve more complicated general
flows. We shall also introduce here some mathematical approximation methods
to help in the derivations and understanding of approximate analytical solutions.
Naturally, analytical approximations are best suited to equations that are simplified

© Springer Science+Business Media, LLC 2016
O. Regev et al., *Modern Fluid Dynamics for Physics and Astrophysics*,
Graduate Texts in Physics, DOI 10.1007/978-1-4939-3164-4_2

by special conditions. We shall try to accompany our general discussion of different restricted flows with a substantial number of examples and also give problems related to the various relevant flows and topics.

2.2 Vorticity Basics and the Crocco Theorem

We begin by introducing one of the most important vector fields in fluid dynamics, the *vorticity*, which is nominally defined by taking the $\nabla \times$ of the velocity field $\mathbf{u}(\mathbf{x}, t)$. We denote it in the customary way as ω, and it is being defined formally as

$$\omega(\mathbf{x}, t) \equiv \nabla \times \mathbf{u}(\mathbf{x}, t). \tag{2.1}$$

It is evident that a flow with a given velocity has a unique vorticity field, while the converse is not true. For a given vorticity field ω a whole family of velocity fields is possible, differing from each other by a gradient of any scalar function. For example, if for a given scalar function f we have two velocity fields $\mathbf{u}$ and $\mathbf{u}' = \mathbf{u} + \nabla f$, then both velocity fields produce the same vorticity field, see also Sect. 2.5.4. We add the concept of flow *helicity* $H \equiv (1/2)\mathbf{u} \cdot \omega$ which is sometimes useful. Note that in some definitions the factor of one half is omitted. Some elementary concepts from vector analysis like the definition of the $\nabla \times$ operator and the Stokes theorem immediately link the vorticity to the *local* rotation of the fluid. Indeed, it is possible to state that the vorticity at a point is the amount of circular motion of the fluid, per unit area, as one performs a limiting process to the point in question. The direction of the vorticity vector is assessed using the right-hand screw rule. These considerations can be made precise, of course, but rather than doing it here in an abstract formulation we prefer to give some simple examples.

Assume a known two-dimensional flow, given by $\mathbf{u}(x_1, x_2, t)$ in the Cartesian plane. Consider two perpendicular infinitesimal line elements in the plane so that the x_1 and x_2 axes are chosen along these elements and the elements originate at time t and point (x_1, x_2). Now, since

$$u_2(x_1 + dx_1, x_2, t) - u_2(x_1, x_2, t) = \frac{\partial u_2}{\partial x_1} dx_1,$$

$$-[u_1(x_1, x_2 + dx_2, t) - u_1(x_1, x_2, t)] = -\frac{\partial u_1}{\partial x_2} dx_2,$$

the partial derivatives on the right-hand side of these equations, including the minus sign in the second one, are the instantaneous angular velocities of the perpendicular line elements (in the anticlockwise direction). Their average is nothing else than half of the vorticity at the point (x_1, x_2), because

$$\frac{1}{2}\left(\frac{\partial u_2}{\partial x_1} - \frac{\partial u_1}{\partial x_2}\right) = \frac{1}{2}\omega_3, \tag{2.2}$$

where ω_3 is the vorticity component in the x_3 direction. Incidentally, this z-component of the vorticity is often called ζ, which should not be confused with the bulk viscosity defined before. The consideration summarized in Eq. (2.2) states more precisely what we have already said before: the vorticity serves as a measure of *local* rotation, but since it is a vector it can only do so with regard to anticlockwise circulation in a plane perpendicular to it. An important clarification should be made: vorticity is a local property while rotation is not, since a point cannot rotate. Given the velocity field of a fluid, one can determine the effects of vorticity on the fluid only in a very small open set. In two-dimensional flows, a fluid patch in which the fluid rotates with a constant angular velocity, $\dot{\theta}$, say, the perpendicular vorticity, is twice this value. It is also true that when vorticity is sufficiently large there is significant rotation in the fluid. We note here, in passing, that we shall use the indical notation, e.g., $\mathbf{x} = (x_1, x_2, x_3)$, $\mathbf{u} = (u_1, u_2, u_3)$, whenever necessary, e.g., when overall tensor notation is used, but in specific examples, we shall opt for the more natural $\mathbf{x} = (x, y, z)$ and $\mathbf{u} = (u, v, w)$.

The existence of *shear* in a flow context can be explained in the simplest way when it is meant to imply that the Eulerian Cartesian velocity component u_α depends on the coordinate β with $\beta \neq \alpha$, where α and β are x, y, or z. Consider now what is perhaps the simplest shear flow:

$$\mathbf{u}(x, y, z) = (-\lambda y, 0, 0)$$

with constant $\lambda > 0$. Its corresponding vorticity vector is easily calculated and shown to be constant

$$\omega = \lambda \hat{\mathbf{z}}, \tag{2.3}$$

that is, this flow, although not rotating at all (globally), has a nonzero vorticity λ in the z direction. Thus, we see that the vorticity is related to the local rotation rate only and not to any global property of the flow. Indeed, an argument similar to the one used before (exploiting perpendicular infinitesimal line elements) reveals that the fluid locally has a nonzero angular velocity (equal to $\lambda/2$) everywhere.

A more thorough discussion of vorticity per se will be given in Sect. 2.5, but we shall use the notion of vorticity in many places in this book and its physical significance will emerge as paramount. In particular, vorticity is important in formulating the useful theorem discussed next. Before ending this section, it will be useful to define that a vorticity free flow is called *irrotational*.

2.2.1 Crocco's Theorem

We conclude this section by deriving a useful relation valid for general flows and known as the *Crocco theorem*. In the form given here, the only special requirement is that any body force present be derivable from a potential, i.e., there exists a function Φ, so that $\mathbf{b} = -\nabla\Phi$. The significance of the theorem is that it includes explicitly the vorticity and ties it into the dynamics of the fluid.

By applying the vector identities

$$(\mathbf{u}\cdot\nabla)\mathbf{u} = (\nabla\times\mathbf{u})\times\mathbf{u}+\nabla\left(\frac{1}{2}u^2\right),$$

$$\nabla^2\mathbf{u} = \nabla(\nabla\cdot\mathbf{u})-\nabla\times(\nabla\times\mathbf{u}), \qquad (2.4)$$

where $u = |\mathbf{u}|$, the general compressible Navier–Stokes equation,

$$\frac{\partial\mathbf{u}}{\partial t}+\mathbf{u}\cdot\nabla\mathbf{u} = -\nabla\Phi-\frac{1}{\rho}\nabla P+\nu\nabla^2\mathbf{u}+\left(\nu_2+\frac{\nu}{3}\right)\nabla(\nabla\cdot\mathbf{u}),$$

where $\mathbf{b} = -\nabla\Phi$, and $\nu = \eta/\rho$ and $\nu_2 = \zeta/\rho$ are the *kinematic* viscosities, may be rewritten into the form

$$\frac{\partial\mathbf{u}}{\partial t}+\omega\times\mathbf{u} = -\nabla\left(\frac{1}{2}u^2+\Phi\right)-\frac{1}{\rho}\nabla P-\nu\nabla\times\omega+\left(\nu_2+\frac{4\nu}{3}\right)\nabla(\nabla\cdot\mathbf{u}). \qquad (2.5)$$

Now, using the first of the equivalent forms of the Gibbs equation (see Eq. (1.67)) and making the natural identification $d\mapsto\nabla$ yields $\rho^{-1}\nabla P = \nabla h-T\nabla s$. Then the term including the pressure gradient can be substituted from this thermodynamic relation and we finally get the viscous Crocco theorem, which is equivalent to (2.5):

$$\frac{\partial\mathbf{u}}{\partial t}+\omega\times\mathbf{u} = -\nabla\mathscr{B}+T\nabla s-\nu\nabla\times\omega+\left(\nu_2+\frac{4\nu}{3}\right)\nabla(\nabla\cdot\mathbf{u}), \qquad (2.6)$$

where $\mathscr{B} = \frac{1}{2}u^2+h+\Phi$ is the Bernoulli function, as defined before in (1.71). Some books that refer to Crocco's theorem often times refer to the inviscid form of the theorem as in (2.6), but with the viscosity coefficients equal to zero.

   In conclusion, we point out some deep physical content contained in Eq. (2.6). For simplicity, we assume incompressible flow, i.e., $\nabla\cdot\mathbf{u} = 0$, and assume that the viscosity coefficients are constant. Operating on (2.6) by $\nabla\times$ drops the gradient terms and leaves behind

$$\frac{\partial\omega}{\partial t} = \nabla\times(\mathbf{u}\times\omega)+\nabla\times(T\nabla s)-\nu\nabla\times(\nabla\times\omega). \qquad (2.7)$$

Now, using the relations

$$\nabla \times (\mathbf{u} \times \omega) = -(\mathbf{u} \cdot \nabla)\omega + (\omega \cdot \nabla)\mathbf{u},$$

which is correct for $\nabla \cdot \mathbf{u} = 0$ (our simplifying assumption) and $\nabla \cdot \omega = 0$ (an identity), and adding another identity

$$\nabla \times (\nabla \times \omega) = -\nabla^2 \omega,$$

we get the meaningful equation

$$\frac{D\omega}{Dt} = (\omega \cdot \nabla)\mathbf{u} + \nabla \times (T\nabla s) + \nu \nabla^2 \omega. \tag{2.8}$$

This equation indicates what are the possible sources of vorticity in a flow. Clearly, there is a viscous term ($\nu \nabla^2 \omega$), but it cannot produce globally any vorticity from zero value, in the interior of the fluid domain, and the same is true for the $(\omega \cdot \nabla)\mathbf{u}$ term. Rather, together with the time derivative term, the viscous term can cause vorticity diffusion or dispersion through the body of the fluid provided vorticity is produced either by viscosity through the boundaries and/or by the remaining term $\nabla \times (T\nabla s)$. This baroclinic (see below) forcing term can produce vorticity anywhere, even if there is no initial vorticity at all. The baroclinic forcing term is zero for a barotropic flow, properly defined in the next section. Our discussion thus far has been heuristic, but we shall revisit it with more mathematical rigor in the next section. We conclude this section by commenting that, as we shall see in Chap. 6, discontinuities, which are called *shocks* by physicists and *weak solutions* by mathematicians, can appear in flows. Physically, discontinuities are impossible, but we are already used to the fact that mathematical descriptions of physical systems may be just approximations. Discontinuities of this kind can create jumps in the Bernoulli function, for example. Such jumps, if they are not uniform, e.g., curved shock fronts, may also generate vorticity.

2.3   Some Basic Theorems and Results

In this section we shall derive several very important results, valid in general for barotropic flows, which will be discussed in considerable detail in Sect. 2.3.1. The important Kelvin's and Ertel's theorems hold on the condition that the flow is also inviscid and we shall discuss the meaning of this restriction. Next we shall discuss, in detail, the influential Bernoulli's theorem for which we shall clarify the conditions that lead to the different restricted statements made on the basis of the most general case. We feel that students would benefit from such a systematic organization of the various Bernoulli statements. We shall conclude this section with two important examples.

2.3.1   Barotropic Flows

In the previous chapter, we have set precise nomenclature to be used in this book to define homentropic, isentropic, and adiabatic flows. An important class of flows arises when one has at his or her disposal a *geometric* relation, i.e., it need not be a thermodynamic relation between the pressure and density. The relation has to be of the form $P = P(\rho)$ representing a stratification, so that surfaces of constant density (isopycnic) coincide with those of constant pressure (isobaric). If a relation like this exists the flow is called *barotropic*, with the fluid portion in which the relation holds termed *barotrope*. Polytropes, which we described in the previous chapter, are an example of barotropes. In this section, we give some general and meaningful results for barotropic flows. When a flow is not barotropic we call it *baroclinic*. In the latter case, there is a nonzero angle between the above-mentioned two families of surfaces. It is clear that in a homogeneous fluid, homentropic or isothermal flows are necessarily barotropic. This is due to the fact that two admissible equations of state forms, e.g., $P = P(\rho, s)$ and $P = P(\rho, T)$, equivalently reduce to a barotropic relation $P = P(\rho)$. The converse, however, need not be true. For example, a barotropic flow does not imply homentropy.

2.3.2   Inviscid Flows

We readily understand that no physical flow can be truly inviscid, that is, we cannot drop a priori the viscous terms from the relevant equations. However, it is also clear that for very large Reynolds numbers, provided we are not near boundaries, flows are essentially unaffected by omitting the viscous terms. Such an omission would be an approximation good up to $\mathscr{O}\left(\mathrm{Re}^{-1}\right)$, where the symbol $\mathscr{O}\left(.\right)$ means *the order of magnitude* and we use it interchangeably with $\sim$ in this book. The flow in this approximation is henceforth termed an *inviscid flow*. We also make note here that an isentropic flow must be necessarily inviscid and adiabatic, because the only terms that can create entropy are the dissipative ones, which by their very nature include viscosity. Of course adiabaticity and, therefore, isentropy, can also be broken by absorption of heat from the outside or by some internal thermo-reaction process within the fluid particle.

It is useful to transform what we have called the viscous Crocco theorem (2.6) into the inviscid, in the above explained sense, limit[1]

$$\frac{\partial \mathbf{u}}{\partial t} + \omega \times \mathbf{u} = -\nabla \mathscr{B} + T \nabla s. \tag{2.9}$$

---

[1] As we have already remarked some books refer to relation (2.9) as *the* Crocco theorem.

Operating with $\nabla \times$ on this equation, using the vector identity for the $\nabla \times$ of a vector product and remembering that $\nabla \cdot \omega = 0$ (recall why), we end up with the material derivative of the vorticity

$$\frac{D\omega}{Dt} = (\omega \cdot \nabla)\mathbf{u} - (\nabla \cdot \mathbf{u})\,\omega + \nabla T \times \nabla s. \tag{2.10}$$

Remembering another form of the Gibbs relation, following from the one already used above, after Eq. (2.5),

$$T\nabla s = \nabla h - \rho^{-1}\nabla P.$$

Equation (2.9) may then be reexpressed as

$$\frac{\partial \mathbf{u}}{\partial t} + \omega \times \mathbf{u} = -\nabla(\mathscr{B} - h) - \rho^{-1}\nabla P, \tag{2.11}$$

and after operating with $\nabla \times$ on it results in

$$\frac{D\omega}{Dt} = (\omega \cdot \nabla)\mathbf{u} - (\nabla \cdot \mathbf{u})\,\omega - \nabla(\rho^{-1}) \times \nabla P. \tag{2.12}$$

The last equation brings out again and perhaps more directly the fact that an initially zero vorticity can grow in an inviscid flow only due to the baroclinic forcing term $\nabla(\rho^{-1}) \times \nabla P$, now its nonvanishing being clearly and directly related to the lack of a barotropic relation like $P(\rho)$, which necessarily renders this term to be zero as surfaces of constant ρ are parallel to those of constant P. Thus, vorticity can grow as a result of baroclinicity and perhaps viscosity.

2.3.3   Ertel's Theorem, Potential Vorticity

Assume the existence of an inviscid flow and consider the quantity ω/ρ. Straight-forward algebraic manipulation gives

$$\begin{aligned}
\frac{D}{Dt}\left(\frac{\omega}{\rho}\right) &= \frac{1}{\rho}\frac{D\omega}{Dt} - \frac{\omega}{\rho^2}\frac{D\rho}{Dt} = \frac{1}{\rho}\left[(\omega \cdot \nabla)\mathbf{u} - \omega(\nabla \cdot \mathbf{u})\right] \\
&\quad + \frac{\omega}{\rho}\nabla \cdot \mathbf{u} - \frac{1}{\rho}\nabla(\rho^{-1}) \times \nabla P \\
&= \frac{1}{\rho}(\omega \cdot \nabla)\mathbf{u} - \frac{1}{\rho}\nabla(\rho^{-1}) \times \nabla P, \tag{2.13}
\end{aligned}$$

where we have used Eq. (2.12). For a barotropic flow, the above calculation gives

$$\frac{D}{Dt}\left(\frac{\omega}{\rho}\right) = \left(\frac{\omega}{\rho}\cdot\nabla\right)\mathbf{u}. \tag{2.14}$$

As basic as this equation may look, it is not useful in providing solutions to FD equations. It is, however, extremely useful in determining various general properties of the flow. In particular, the above equation leads to a noteworthy statement known as *Ertel's theorem*. In one of its formulations, it states that if some scalar function $F(\mathbf{x},t)$, say, is a material invariant of the flow, that is, $DF/Dt = 0$, then

$$\frac{D}{Dt}\left[\left(\frac{\omega}{\rho}\right)\cdot\nabla F\right] = 0; \tag{2.15}$$

in other words, the scalar product of the vorticity (divided by ρ) with the gradient of any material invariant F is also a material invariant. This scalar product has a general nature because of the freedom in the choice of F. It is often called the *potential vorticity* and denoted by Q,

$$Q \equiv \left(\frac{\omega}{\rho}\right)\cdot\nabla F, \tag{2.16}$$

where $F(\mathbf{x},t)$ is *any* material invariant. However see also Sect. 5.5, where the definition of potential vorticity, as used in geophysics, is given and discussed. Here in Eq. (2.16) the definition depends on the choice of the material invariant F. In the words of J. Pedlosky (a leading geophysical fluid dynamicist): "It is hard to exaggerate the importance of the potential vorticity conservation." Before discussing this matter in a little more detail, we give now a proof of the theorem in the version given above. We start by scalar multiplying Eq. (2.14) by ∇F and adding to both its sides a term which makes the left-hand side equal to the expression in the theorem (by virtue of a product derivative). Thus

$$\frac{D}{Dt}\left[\left(\frac{\omega}{\rho}\right)\cdot\nabla F\right] = \left[\left(\frac{\omega}{\rho}\cdot\nabla\right)\mathbf{u}\right]\cdot\nabla F + \frac{\omega}{\rho}\cdot\frac{D}{Dt}(\nabla F). \tag{2.17}$$

It is a straightforward matter (see Problem 2.2) to show that the last term in the above equation can actually be transformed into

$$\frac{\omega}{\rho}\cdot\nabla\left(\frac{DF}{Dt}\right) - \left[\left(\frac{\omega}{\rho}\cdot\nabla\right)\mathbf{u}\right]\cdot\nabla F, \tag{2.18}$$

which proves the theorem since, by assumption, $DF/Dt = 0$.

The significance of the potential vorticity concept and Ertel's theorem is best seen in rotating fluids (it will be discussed in some detail in Chap. 5), and this is probably the reason for its great importance in geophysics. Still it is instructive to notice here that the potential vorticity has also a purely physical significant meaning. This can be seen if we, for the sake of simplicity, consider a homentropic flow. Let us assume that **a** is the initial position of a fluid particle. Obviously, each of the components of **a** is a material invariant and therefore Ertel's theorem implies that

$$\frac{D}{Dt}\left(\frac{\omega}{\rho}\cdot\nabla a_j\right) = 0, \tag{2.19}$$

for $j = 1, 2, 3$. Note that a_j are a priori independent, by which we mean that ∇a_j point, in general, in different directions. By virtue of (2.19), it is clear that initially independent a_j remain so during the flow. Being independent, we can regard a_j also as curvilinear coordinates, and for the sake of clarity we may call them α_j. Since Eq. (2.19) holds, these are also Lagrangian coordinates and surfaces of constant α_j move with the flow. Let us consider $\nabla\alpha_j$ to be basis vectors, corresponding to these coordinates. Naturally, these basis vectors get moved and distorted by the flow. This effect also happens to ω/ρ; however Ertel's theorem guarantees that their dot product remains a material invariant. Now assume that $\mathbf{U} \equiv (U_1, U_2, U_3)$ are components of the velocity in that basis, i.e.,

$$\mathbf{u} = U_1\nabla\alpha_1 + U_2\nabla\alpha_2 + U_3\nabla\alpha_3. \tag{2.20}$$

From Problem 2.4 it follows that this leads to the relation

$$\left(\frac{\omega}{\rho}\cdot\nabla\right)\alpha = \nabla_\alpha \times \mathbf{U}, \tag{2.21}$$

where $\alpha = (\alpha_1, \alpha_2, \alpha_3)$ and $\nabla_\alpha = (\partial/\partial\alpha_1, \partial/\partial\alpha_2, \partial/\partial\alpha_3)$. This means in words that the materially invariant potential vorticity, as defined here, is nothing other than the $\nabla\times$ of the velocity in Lagrangian coordinates. The Ertel theorem can thus be written as

$$\frac{D}{Dt}(\nabla_\alpha \times \mathbf{U}) = 0,$$

i.e., the potential vorticity, including Lagrangian coordinates as the material invariant, is simply the ordinary vorticity expressed in Lagrangian coordinates!

2.3.4 Kelvin's Theorem, Circulation

A concept related to the vorticity in a flow is that of *circulation*. The vorticity is a function of position, while the circulation has to be defined over a given contour

C, which lies entirely inside a simply connected fluid domain. The circulation, commonly denoted by Γ, is defined as

$$\Gamma_C = \oint_C \mathbf{u} \cdot \mathbf{dl}.$$

The relation of velocity to vorticity is obvious. Stokes theorem of vector calculus relates the circulation along a contour to the surface integral of the normal vorticity over any open surface containing the same contour as its boundary: $\Gamma_C = \int_S \omega \cdot \mathbf{dS}$ where $\mathbf{dS}$ is the vector surface element of the bounded surface S, i.e., $\mathbf{n}\, dS$, where $\mathbf{n}$ is the unit outward normal to the surface. If you worry about which, out of the possible two, direction of a flat surface is positive, remember that the path, bounding the surface, has a positive direction (anticlockwise) and this relates to the surface's direction by the right-hand screw convention.

We have already seen in Eq. (1.30) a circulation transport theorem, that is, an expression giving the rate of change of the circulation over a contour materially *following* the flow. We repeat the statement here, for convenience:

$$\dot{\Gamma}_C \equiv \frac{d}{dt} \oint_{C(t)} \mathbf{u} \cdot \mathbf{dl} = \oint_{C(t)} \frac{D\mathbf{u}}{Dt} \cdot \mathbf{dl}. \tag{2.22}$$

In 1869, Lord Kelvin (W. Thomson, perhaps the British most prominent physicist of that period) noticed and published a very significant result about the material transport of circulation. In its most general form the statement of that result, known as *Kelvin's circulation theorem*, is: in barotropic and inviscid flows, in which all body forces are derivable from a potential, the velocity circulation over a contour materially moving with the flow is conserved, provided that the contours remain in a simply connected fluid domain. In mathematical terms the theorem can be formulated as

$$\dot{\Gamma}_{C(t)} \equiv \frac{d}{dt} \oint_{C(t)} \mathbf{u} \cdot \mathbf{dl} = 0, \tag{2.23}$$

provided that all the aforementioned conditions are met. Different proofs of the theorem exist in the literature and we shall give here one that is based on (1.30) and some vector algebra. In our opinion this is the mathematically clearest and most comprehensible proof, at least to us. First, we use the circulation transport theorem (2.22) and concentrate on the term on the right-hand side, transforming it into a surface integral by virtue of the Stokes theorem

$$\oint_{C(t)} \frac{D\mathbf{u}}{Dt} \cdot \mathbf{dl} = \int_{S(t)} \left(\nabla \times \frac{D\mathbf{u}}{Dt} \right) \cdot \mathbf{n}\, dS.$$

Now expanding the material derivative in the integrand on the right-hand side of this equation gives

$$\nabla \times \frac{D\mathbf{u}}{Dt} = \frac{\partial \omega}{\partial t} + \nabla \times [(\mathbf{u} \cdot \nabla)\mathbf{u}],$$

and remembering the vector identity given in the first of Eq. (2.4) we get

$$\nabla \times \frac{D\mathbf{u}}{Dt} = \frac{\partial \omega}{\partial t} + \nabla \times (\omega \times \mathbf{u}).$$

Finally, operating with $\nabla \times$ on Eq. (2.11), which is a corollary of the inviscid Crocco's theorem (2.9), and remembering that the flow is barotropic, one gets

$$\frac{\partial \omega}{\partial t} + \nabla \times (\omega \times \mathbf{u}) = 0. \tag{2.24}$$

This completes the proof as it implies

$$\int_{S(t)} \left(\nabla \times \frac{D\mathbf{u}}{Dt} \right) \cdot \mathbf{n} \, dS = 0. \tag{2.25}$$

Kelvin's circulation theorem has profound consequences. Among them is the observation that, provided the conditions of the theorem are met, a flow which is irrotational, i.e., $\omega = 0$ everywhere in a simply connected domain at some given time, will remain irrotational thereafter. This is also equivalent to the velocity circulation being zero and remaining so over every material contour which is internal to the fluid and *never* encompassing "holes," meaning that the contours stay around a simply connected fluid region. The statement $\omega = \nabla \times \mathbf{u} = 0$ implies that the velocity must be a gradient of some scalar function, called the *velocity potential*. Thus the knowledge of the velocity field in this case, which is an irrotational flow, is determined by one function only (instead of three), a fact that simplifies problems, where appropriate. We shall devote the whole of Sect. 2.4 in this chapter to irrotational flows, as well as the majority of Chap. 4. This kind of flow is referred to as *potential flow* and holds in each domain in which it is irrotational.

2.3.5 The Various Forms of Bernoulli's Theorem

Assuming that the body force is derivable from a potential Φ, and that the flow is inviscid, that is, all viscous terms are dropped, we remind the reader of two results derived previously. The first one is identical to (2.6), but with the viscous terms omitted,

$$\frac{\partial \mathbf{u}}{\partial t} + \omega \times \mathbf{u} = -\nabla \mathcal{B} + T\nabla s, \tag{2.26}$$

and the second one is the Bernoulli theorem (1.72) for the inviscid case

$$\frac{D\mathcal{B}}{Dt} = \frac{1}{\rho}\left(\frac{\partial P}{\partial t} + \dot{\mathcal{Q}}\right), \tag{2.27}$$

where $\dot{\mathcal{Q}}$ is the heat absorbed by the system per unit volume and $\mathcal{B}$ is the Bernoulli function, which we repeat here, for completeness:

$$\mathcal{B}(\mathbf{x},t) = h + \frac{1}{2}u^2 + \Phi. \tag{2.28}$$

Here h is the specific enthalpy, defined as $h = e + P/\rho$, $u = |\mathbf{u}|$, and Φ is the body force potential. So far, we have assumed only an *inviscid flow and that the body force is derivable from some potential*. These two assumptions will be valid in all the cases that we now enumerate:

1. The additional assumptions in this case is that the flow is *steady* and *adiabatic*. The latter means that $\dot{\mathcal{Q}} = 0$ is obeyed. Now since the flow is steady we can get from Eq. (2.27) $(\mathbf{u} \cdot \nabla)\mathcal{B} = 0$. This means that the Bernoulli function is constant on streamlines.
2. The additional assumption in this case is that the flow is *steady* and *the density is constant* $\rho = \rho_0$. Now, remembering one of the equivalent forms of the Gibbs equation (1.67), one gets for constant ρ the relation

$$T\nabla s = \nabla h - \nabla\left(\frac{P}{\rho_0}\right).$$

So, taking the dot product of (2.26) by the velocity $\mathbf{u}$ gives

$$\mathbf{u} \cdot \nabla\left(\mathcal{B} - h + \frac{P}{\rho_0}\right) = 0.$$

Thus, the modified Bernoulli's function $\mathcal{B}' \equiv \mathcal{B} - h + P/\rho_0$ is constant on streamlines.
3. Finally, we assume in this case that the flow velocity is derivable from a velocity potential, that is, $\mathbf{u} = \nabla\phi$. Given such an *irrotational flow*, Eq. (2.26) implies the following two separate cases:

   a. If the flow is also *homentropic*, i.e., $\nabla s = 0$, then $\nabla(\mathcal{B} + \partial_t\phi) = 0$, that is,

$$\mathcal{B} + \frac{\partial\phi}{\partial t} = f(t) \qquad \text{everywhere,}$$

   where $f(t)$ is a function of time, consistent with the initial condition.

b. If the flow is also of *constant density* then, similarly,

$$\mathscr{B}' + \frac{\partial \phi}{\partial t} = f(t) \qquad \text{everywhere.}$$

In these two cases if the flow is also steady, then $\mathscr{B}$ or $\mathscr{B}'$ is a constant everywhere.

2.3.6 Examples

Bernoulli's theorems, in their various forms, constitute powerful statements and allow one to find interesting properties of solutions, without actually solving the fluid equations. In what follows we shall give two important examples of this kind. Both examples are of formally *compressible* flows but appear in this book before discussing thoroughly the effects of compressibility in Chap. 6. However, that chapter will primarily discuss sound waves, shock waves, detonations, and similar extreme manifestations of compressibility. In the examples given below, compressibility is hardly manifested and they can be readily understood by the student at this point of the book. These important flows do not exhibit the above typically compressible and sometimes extreme phenomena.

2.3.6.1 The de Laval Nozzle

The origin of this problem is in jet engines, where a gas is expelled from a combustion chamber through a short channel of varying cross section. For simplicity, we shall consider the case in which that channel is horizontal and the flow is approximately one-dimensional (in the x direction, say), homentropic, and viscosity is neglected. The rationale for the assumption of a homentropic flow is in the speed of the process. The nondissipative fluid also has no time to exchange heat with its surroundings and there are no heat sources in the nozzle. The flow is treated as being inviscid owing to its relatively large Reynolds number.

In such circumstances, with the additional proviso that the flow be steady, although this is actually not needed for the following statement, but will be used later. We have from case 1 of the Bernoulli theorem, above, that the Bernoulli function is a streamline constant, because the *de Laval flow* is treated as one-dimensional, identical to each other and with parallel streamlines. Thus,

$$\mathscr{B} = \frac{1}{2}u^2 + h = \text{const}$$

along each streamline, where the force potential term has been dropped, because the flow is assumed horizontal. No body forces except for gravity are assumed to be present. The one-dimensionality of the flow allows the above equation to hold for all streamlines and so the equation can be written, under these assumptions:

$$u\,du + dh = u\,du + \frac{1}{\rho}dP = 0,$$

where we have used a thermodynamic identity given in (1.67) and the fact that $ds = 0$. Rewriting this in a slightly different form and adding into it the one-dimensional equation of continuity yields

$$u\,du + c_s^2 \frac{d\rho}{\rho} = 0, \tag{2.29}$$

$$\rho u A = \text{const}, \tag{2.30}$$

where $c_s = \sqrt{dP/d\rho}$ in this case is the adiabatic sound speed and A is the cross-sectional area of the nozzle. The equation of mass conservation results trivially from the flow being steady. Even though, as said above, it is somewhat early in this book to deal with supersonic flow, we think that the current problem may serve as a nice prelude to Chap. 6, without touching the salient concepts of compressible flow. In any case, it is possible to define now the important nondimensional number in this context, $M \equiv u/c_s$, the *Mach number*, whose importance is great in compressible flows (Chap. 6). Equation (2.29) can thus be rewritten as

$$\frac{d\rho}{\rho} = -M^2 \frac{du}{u}. \tag{2.31}$$

E. Mach (1838–1916), whose name was immortalized mainly by this number, was an important physicist and philosopher, noted for his contributions to both disciplines. As a philosopher of science, he was a major influence on logical positivism and through his criticism of Newton, a forerunner of Einstein's relativity.

   If the flow is very subsonic, then it is a good approximation to use the assumption that the fluid is incompressible. Significant compressibility does not occur for subsonic flows. Now taking the logarithmic derivative of Eq. (2.30) and substituting into it relationship (2.31) yields an equation having a number of significant physical implications:

$$\left(1 - M^2\right)\frac{du}{u} = -\frac{dA}{A}. \tag{2.32}$$

We shall discuss here only three observations following from this equation and leave some other results as problems. It is clear from Eq. (2.32) that for subsonic ($M < 1$) flows $du > 0 \Longrightarrow dA < 0$ and vice versa as u is assumed positive in this problem. This can be expected on the basis of daily experience: for example, higher (lower) speed of a steady gas flow in a narrower (wider) region of a horizontal pipe. As we shall explicitly see in Chap. 6, effects of compressibility can be neglected at subsonic speeds. The result for supersonic ($M > 1$) flow is more intriguing. Increasing the speed ($du > 0$) requires an increase ($dA > 0$) in the crosssectional area, because $(1 - M^2) < 0$. However, this seemingly strange result has actually a sound physical explanation. $M > 1$ and therefore from Eqs. (2.31) and (2.32) the density decreases faster than the area increases and to ensure mass conservation the area must increase. It is for this reason that exhaust parts of jet aircraft engines are designed to vary their cross-sectional area A which is small when the exhaust is subsonic, but opens up when it becomes supersonic. A caveat is in order here. It is generally impossible to achieve a smooth reverse transition, that is, from supersonic to subsonic flow. As we shall see in Chap. 6, a discontinuity (shock) is expected in such circumstances.

Perhaps the most interesting observation is that the flow can undergo a smooth (with du/dx finite) sonic transition, that is, M can be 1 only at the throat of the nozzle $dA = 0$. Thus, even at this level of approximation we can understand that if a subsonic flow with increasing velocity enters the nozzle, whose cross-sectional area decreases, it reaches the speed of sound at the narrowest point and continues increasing its speed to growing supersonic values, leaving the nozzle through a region in which the cross-sectional area increases in the flow direction. In addition to the obvious applications of de Laval nozzle theory to jet engines, this theory has also been used to provide idealized models of gas breakout from high pressure bubbles, confined by higher density medium and of astrophysical jets, whether from young stars or active galactic nuclei.

2.3.6.2   Spherical Accretion and Winds

In many astrophysical systems, a relatively compact mass M accretes gas from its surroundings or, conversely, expels out a gaseous wind. To make such a problem tractable analytically, we neglect the size of the central object, relatively to typical scales of the flow, see below, and assume spherical symmetry and steady flow. We shall also assume here that the flow is adiabatic and actually isentropic, because it is inviscid as well. This problem was first treated successfully by H. Bondi in the early 1950s. We see that the Bernoulli formula applies here as well and we have

$$\frac{1}{2}u^2 + h - \frac{GM}{r} = \text{const}, \tag{2.33}$$

$$4\pi r^2 \rho u = \text{const}, \tag{2.34}$$

where we chose u here as the radial component of the velocity $\mathbf{u}$. Equation (2.33), while formally valid only along a streamline, is actually valid everywhere as the streamlines are radial and the flow is spherically symmetric. Equation (2.34) is the mass conservation equation in this flow. The value of the constant can be negative or positive, constituting the steady *accretion rate* or the *mass loss rate*. In both cases, the absolute value of the constant is usually denoted by $\dot{M}$.

It is not difficult to see that the constant in Eq. (2.33) can be reasonably set to zero, when we consider inflow only. Limiting ourselves here to just accretion, we first integrate along the streamline from infinity to get the enthalpy

$$h = \int_{\rho_\infty}^{\rho} \frac{1}{\rho} dP. \tag{2.35}$$

Setting $\rho = \rho_\infty$ for $r \rightarrow \infty$, we get $h_\infty = 0$ and also $u_\infty = 0$, so we have the value of the Bernoulli constant in this case equal to zero. For isentropic flow $P \propto \rho^\gamma$, which allows one to explicitly integrate the equation for the enthalpy, which is

$$h(r) = \frac{\gamma}{\gamma - 1} \frac{P_\infty}{\rho_\infty} \left\{ \left[\frac{\rho(r)}{\rho_\infty} \right] - 1 \right\}. \tag{2.36}$$

$C_{T,\infty}^2 \equiv P_\infty/\rho_\infty$ is now defined for the sake of notational economy, but it has also a physical meaning: $C_{T\infty}$ is the isothermal sound speed at infinity. The actual adiabatic sound speed along the flow is variable and can be easily computed

$$c_s^2(r) \equiv \frac{dP}{d\rho} = \gamma C_{T,\infty}^2 \left[\frac{\rho(r)}{\rho_\infty} \right]^{\gamma-1}. \tag{2.37}$$

It is now advantageous to switch to nondimensional variables. First and foremost we notice that the typical length scale of the problem is the Bondi radius, given by

$$R_B \equiv \frac{GM}{C_{T,\infty}^2}. \tag{2.38}$$

Other natural units are also self-evident and can be used to scale the relevant variables thus:
$\xi \equiv r/R_B$; $V \equiv |u|/C_{T\infty}$; $D \equiv \rho/\rho_\infty$. Most central to the solutions that follow, we define a dimensionless accretion rate

$$\mu \equiv \frac{\dot{M}}{4\pi R_B^2 \rho_\infty C_{T,\infty}}. \tag{2.39}$$

The relevant Eqs. (2.33)–(2.34) can now be recast in the nondimensional form:

$$\xi^2 DV = \mu, \tag{2.40}$$

$$\frac{1}{2}V^2 + H(D) - \frac{1}{\xi} = 0, \tag{2.41}$$

with

$$H(D) = \frac{\gamma}{\gamma - 1} \left(D^{\gamma-1} - 1 \right).$$

(2.42)

There are different types of solutions for the velocity as a function of distance, depending on the mass accretion rate μ. It turns out that there exists a critical accretion rate, μ_c, which gives a *transonic* accretion solution, usually referred to as the *Bondi solution* but sometimes this name is reserved to isothermal flow only. To find μ_c we take the differential of both nondimensional equations (2.40)–(2.41) and eliminate dD between the two equations. This gives

$$\left[DH'(D) - V^2 \right] \frac{dV}{V} = \left[\xi^{-1} - 2DH'(D) \right] \frac{d\xi}{\xi}.$$

(2.43)

The point at which the factors multiplying the differentials are equal to zero is a *singular point* of this differential equation. This singular point gives rise to the so-called *critical transonic solution* which occurs at

$$\xi = \frac{1}{2DH'(D)} \qquad \text{and satisfies} \qquad V^2 = DH'(D).$$

(2.44)

Now, using Eq. (2.41) and the last two equalities we get the existence condition for the critical transonic solution

$$H(D) = \frac{3}{2} DH'(D),$$

(2.45)

and remembering the form of $H(D)$ from Eq. (2.42) we finally get the critical nondimensional density at the sonic point:

$$D = \left[\frac{2}{5 - 3\gamma} \right]^{\frac{1}{\gamma-1}} \qquad \text{for} \quad \gamma < \frac{5}{3}.$$

(2.46)

Special care has to be taken for $\gamma = 5/3$, because then the solution actually acquires the sonic point at the origin. With the help of Eqs. (2.44) and (2.40) we can obtain the critical accretion rate, which gives the transonic solution (it depends on γ). Some of its values are: $\mu_c(\gamma = 1.2) = 0.663$; $\mu_c(\gamma = 1.4) = 0.377$; $\mu_c(\gamma = 1.5) = 0.272$. Another line of reasoning is very useful for understanding the physical nature of the different solutions, including the outflow ones. Combining the momentum conservation equation for a spherical flow

$$u \frac{du}{dr} = -c_s^2 \frac{1}{\rho} \frac{d\rho}{dr} - \frac{GM}{r^2},$$

(2.47)

with the radial derivative of the logarithm of the mass conservation equation (2.33), in order to eliminate the terms containing ρ, gives

$$(u - c_s)\frac{1}{u}\frac{du}{dr} = \frac{2c_s^2}{r}\left[1 - \frac{GM}{2c_s r}\right]. \tag{2.48}$$

This equation clearly shows that there exists a certain radius of the flow, defined by

$$r_s \equiv \frac{GM}{2c_s^2}, \tag{2.49}$$

and at this r_s either u has an extremum or there is a sonic point there, i.e., $u(r_s) = c_s(r_s)$.

The transonic accretion solution is depicted by the inflowing hyperbola in Fig. 2.1, which sketches the physically acceptable solution of the inflowing velocity, in units of the sound speed, as a function of radius, in units of the sonic point radius. We notice that in addition to the transonic Bondi accretion solution, there also

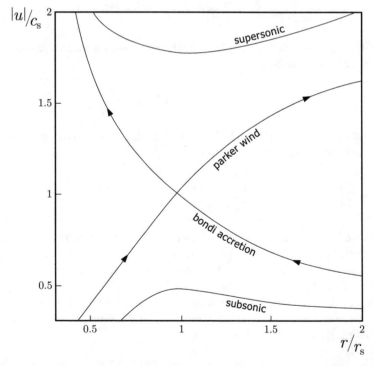

Fig. 2.1 The various types of solutions of the spheri-symmetrical accretion/ejection problem. The absolute value of the radial velocity in units of the local sound speed c_s (see text) is plotted as a function of the radius, in units of r_s (see text). The transonic solutions are the Bondi accretion and the Parker wind

exists a fully subsonic solution, which actually approaches a hydrostatic solution for $r \to 0$. There is also a fully supersonic solution, which appears in the upper part of the figure. While acceptable mathematically, it has no physical application. If the constant in the continuity equation is positive, nothing substantially changes but we get other types of solutions—outflows (winds). The *Parker wind* solution is depicted in the figure as another type of critical solution, an accelerating outflow, crossing the sonic point on the way out. In Problem 2.8, the case of isothermal flow is addressed in a case of the kind discussed above. Finally, we should remember that we have assumed a steady flow and this may force artificial, unphysical effects.

2.4   Potential (Irrotational) Flows

A frequently studied flow type is that of the above-introduced potential flow. In such flows, the velocity field is a gradient of the velocity potential, denoted by ϕ. We thus have the relation $\mathbf{u} = \nabla \phi$. It is sufficient and necessary for a flow to be a potential flow if it is *irrotational*, that is, $\nabla \times \mathbf{u} = 0$ in some simply connected region. By virtue of the Stokes theorem, this is equivalent to the integral characterization of a potential flow,

$$\oint_C \mathbf{u} \cdot \mathbf{dl} = 0, \tag{2.50}$$

over every closed contour C that lies *inside* the fluid domain. The integral on the left-hand side of this equation is the circulation of the velocity $\mathbf{u}$ on the contour C. The additional demand is that the domain be simply connected, zero circulation everywhere is equivalent to the flow being irrotational and hence also to its being a potential flow. One of the most important questions pertaining to irrotational flow is: under what condition(s) will such a flow persist, that is, when can one be confident that a potential flow remains as such for all times. This is important, since in an irrotational flow, the velocity, a three-dimensional vector, can be obtained from the scalar velocity potential. This reduces the number of dependent variables from three to one and so is bound to simplify the problem. Kelvin's theorem guarantees that a potential flow will persist, as long as the conditions for the theorem's validity hold in a simply connected region. In particular, a fluid initially at rest, $\mathbf{u}(\mathbf{x}, t = 0) = 0$, and satisfying the conditions of Kelvin's theorem, will remain irrotational. That baroclinicity, viscosity, or the non-potential nature of the body force can be a source of vorticity and thus destroys a potential flow, follows from operating with $\nabla \times$ on Eq. (2.5), which gives

$$\frac{D\omega}{Dt} = (\omega \cdot \nabla)\mathbf{u} - \omega \nabla \cdot \mathbf{u} - \nabla \times \mathbf{b} + \nabla(\rho^{-1}) \times \nabla P$$

$$-\nabla \times (\nu \nabla \times \omega) + \nabla \times \left[\left(\nu_2 + \frac{4\nu}{3} \right) \nabla(\nabla \cdot \mathbf{u}) \right], \tag{2.51}$$

where we have not assumed that the viscosity coefficients are constant. Now if $\omega = 0$ everywhere at a particular instant, vorticity can arise and grow only if the fluid is baroclinic ($\nabla \rho \times \nabla P \neq 0$), or the force is not derivable of the potential ($\mathbf{b} \neq \nabla f$), or there is a nonconstant finite viscosity. Vorticity can be created further if the flow is already rotational, i.e., the first two terms are non-zero. In this case, additional vorticity can be produced by its interactions with the velocity field and also by the above-mentioned effects.

The flow around a solid body immersed in a large fluid volume as compared to the body's dimensions is a classical case of potential flow. As an example, imagine a stationary fluid satisfying the conditions of Kelvin's theorem. Now suppose that the body starts gently, accelerating with respect to the stationary fluid until it acquires a constant velocity $\mathbf{U}$. In such circumstances, we may now look at the problem in the body's frame, centered at $\mathbf{x} = 0$, say, and assume an infinite fluid flow, now with a boundary condition at infinity written symbolically as $\mathbf{u}(\infty, t) = -\mathbf{U}$. The flow upstream is what we call *laminar*, defined as a streamline flow, when the fluid flows in layers, with no disruption between the layers. Before proceeding to deal with problems of this kind, a number of caveats are to be considered:

1. A streamline touching the body is not entirely inside the fluid. So, strictly speaking, one cannot claim the conservation of circulation along this streamline and the assumption of potential flow breaks down on the body's surface. In a region behind (downstream) the body, the fluid may be detached from the potential flow coming from the front (upstream). Such a phenomenon in a flow is called *separation*, which invariably results from the presence of tangential discontinuity and is, ultimately, related to the existence of some source of vorticity production in the near surface zone of the body. Under such circumstances in which separation has occurred, and mathematically speaking, we have here a breakdown of even the uniqueness of the solution. This irrotational formulation is therefore correct only for flows in which the viscosity coefficient is identically set to zero, indicating that viscous effects are mathematically unimportant near tangential discontinuities.

2. Related to our previous point: In any realistic physical situation involving a classical fluid, the viscosity cannot be strictly zero. Even though a tiny value of viscosity has a negligible effect almost everywhere in a fluid, it must have noticeable influence in a region susceptible to tangential discontinuities. These are places where close to each other fluid layers would appear to slide on one another or on the surface of a solid body. A small wake may then appear behind the body. We have already mentioned the fact that even though the Reynolds number may be very large on the scales of the flow, it may be of order unity for very small scales, like the ones encountered on the surface of the discontinuity as illustrated in the previous item, as well as on body surfaces. The latter results from the fact that an almost exactly inviscid fluid cannot slide in a direction parallel to the surface of an impenetrable body. That is true because the surface is truly rough on a small enough resolved scale. It possesses, therefore, a sticky topography at a sufficiently small enough scale,

no matter how insignificant the viscosity may be. Viscosity is the means by which kinetic energy flow dissipates. Heuristically, one may reason that if the fluid, having a given small viscosity, does not have small enough scales within the flow to dissipate energy, the flow will create such a scale, near those aforementioned solid sticky surfaces. This is the phenomenon of the emergence of viscous boundary layers. In general, a *boundary layer* is a very thin region separating two different flow regimes or a flow and its solid boundary, in which the flow variables change rapidly. We shall encounter boundary layers in several flows described in this book (see, e.g., Sect. 3.5) and discuss some special mathematical techniques to treat this phenomenon (see Sect. 3.5.1).

3. At a certain, still higher, Re, the boundary layer separates from the body and in the separated region vortices are symmetrically created on both sides of the body and are shed, creating a   *vortex street.* This is prominent around long cylinders perpendicular to the flow (two dimensions) and accounts for the "singing" of telephone or electric wires in the wind. The circulation in one street is the same as that in the parallel street, but in the opposite direction. Numerical experiments, real ones in the lab and observations, all confirm this finding, which was proposed by T. von Kármán (1881–1963), who is considered as the outstanding aerodynamic theoretician of the twentieth century. He also devised a simplistic analytical model of the street which we shall leave to the reader as guided Problem 2.21. It should be tackled only after learning Sect. 2.5.3.

4. If the flow speed is of the order of the speed of sound another complication occurs: a shock forms in front of the body or obstacle. Shocks will be discussed in detail in Chap. 6, but it should be clear that in practice potential flow past an obstacle has sense only for incompressible (see below) flows. In Sect. 2.4.1, we shall be more specific about what is exactly meant by incompressibility of a flow and when it can be expected to be a good enough approximation (Fig. 2.2).

In Fig. 2.3, one can see a von Kármán vortex street in the atmosphere (compare with Fig. 2.2d). The above caveats make potential flow a rarely valid approximation for flows encountered in nature, see, however, below. Nevertheless, there are certain types of flows in which despite these caveats, potential flow may be a good approximation to realistic conditions. For example, if the body in question is streamlined, the separated region may be negligible, consisting of a very narrow wake downstream from the body, and the boundary layer on the body disturbs the flow only very little. Another case is that of very small, compared to the body size, oscillations of a body immersed in a fluid. This case is the subject of Problem 2.9.

One may consider irrotational flows which are also incompressible (see below) or simply being of constant density. This leads to the problem of solving the *Laplace equation* $\nabla^2 \phi = 0$ for the velocity potential, under various boundary conditions. Indeed, the wealth of incompressible potential flows is significant (see below in Sect. 2.4.1), but it should not be forgotten that there are flows whose velocity fields are derivable from a potential but are not necessarily incompressible. In fact, as

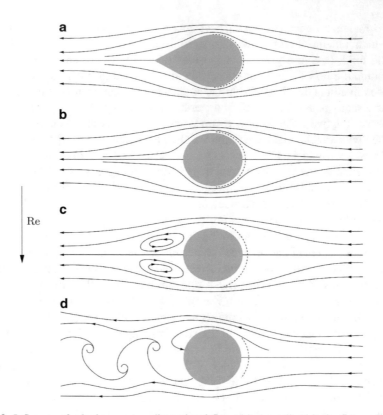

Fig. 2.2 Influence of a body on a two-dimensional flow: (**a**) streamlined body, (**b**) small wake, (**c**) flow reversal in laminar wake, (**d**) vortex street. Note that the bodies have cylindrical symmetry, in the sense that their profile is unchanged in the direction perpendicular to the page. Re increases downward

we shall see in the chapter on compressible flows (6), we often use the velocity potential, in describing the physics of sound waves. The Laplace equation is among the most important equations in mathematical physics. It is a prototype of elliptic equations, see below in Chap. 6. It is named after P.-S. Laplace, who was a prolific mathematician and astronomer and whose work was pivotal to the development of mathematical physics, astronomy, and statistics. He lived from approximately mid eighteenth century until the 1830s.

In the following example, we shall consider an inviscid steady isentropic potential flow, without body forces. Such a flow is governed by the equation of motion, where we replace ∇P by $c_s \nabla \rho$:

$$\rho(\mathbf{u} \cdot \nabla)\mathbf{u} + c_s \nabla \rho = 0. \tag{2.52}$$

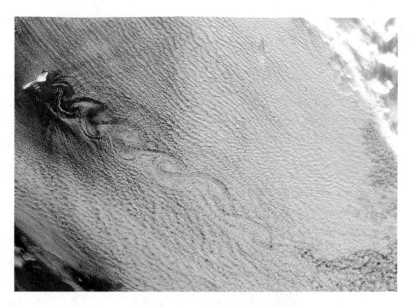

Fig. 2.3 A satellite photograph of the von Kármaán vortex street in the cloud pattern, caused by wind flowing around the Jan Mayen island, between Greenland and Norway (*Public Domain. Courtesy of MODIS Rapid Response Project at NASA/GSFC-http://rapidfire.sci.gsfc.nasa.gov*)

From the equation of continuity we have

$$\rho \nabla \cdot \mathbf{u} = -(\mathbf{u} \cdot \nabla)\rho. \tag{2.53}$$

Thus, multiplying Eq. (2.53) by $\mathbf{u}$ allows us to eliminate ρ between the two equations. The result is

$$c_s^2 \nabla \cdot \mathbf{u} - \mathbf{u} \cdot (\mathbf{u} \cdot \nabla)\mathbf{u} = 0. \tag{2.54}$$

Writing now $\mathbf{u} = \nabla \phi$ and denoting partial derivatives over α, say, by ∂_α for economy of notation, we get the equation

$$[c_s^2 - (\partial_x \phi)^2]\partial_{xx}\phi - [c_s^2 - (\partial_y \phi)^2]\partial_{yy}\phi - [c_s^2 - (\partial_z \phi)^2]\partial_{zz}\phi$$
$$- 2(\partial_x \phi \partial_y \phi \partial_{xy}\phi + \partial_y \phi \partial_z \phi \partial_{yz}\phi + \partial_z \phi \partial_x \phi \partial_{zx}\phi) = 0. \tag{2.55}$$

Throughout this book we use both ∂_α and $\frac{\partial}{\partial \alpha}$ for a partial derivative, interchangeably. The above equation does not seem to be easy to solve, but a particular flow case can be nicely treated. Imagine a flow created by a stream $\mathbf{u_1}$ coming from $x = -\infty$. Let $\mathbf{u} = \mathbf{u_1} + \mathbf{u}'$ be the velocity at a point, which is equal to the stream plus a small perturbation (allowing a small compressibility), assuming that potential flow is being preserved by the small perturbation, so that $\mathbf{u}' = \nabla \phi'$. The x-axis is taken in the direction of $\mathbf{u_1}$ and we have $\phi = xu_1 + \phi'$, so that ϕ' may be considered a small quantity. (We will revisit how to formally assess this smallness in Chap. 4.) So in the lowest order in this approximation equation (2.55) leads to

$$(1 - M_1)\partial_{xx}\phi' + \partial_{yy}\phi' + \partial_{zz}\phi' = 0, \tag{2.56}$$

with the Mach number $M_1 \equiv u_1/c_s$. At the limit $M_1 \ll 1$, we are very close to incompressible potential flow, i.e., Laplace's equation for ϕ', but for larger Mach numbers the departure is significant.

2.4.1 Incompressible Potential Flows

As we have already seen when discussing the mass conservation equation in form (1.34), a flow is considered incompressible when the density is a material invariant. Thus formally

$$\frac{D\rho}{Dt} = 0 \qquad \Longleftrightarrow \qquad \nabla \cdot \mathbf{u} = 0. \tag{2.57}$$

This obviously includes the case of any flow in a fluid whose density is, to a good approximation, a constant $\rho(\mathbf{x}, t) = \rho_0$. Incompressibility has also implications for flows which are not necessarily irrotational, but a potential incompressible flow allows for an attractive formulation:

$$\mathbf{u} = \nabla\phi \quad \& \quad \nabla \cdot \mathbf{u} = 0 \quad \Longrightarrow \quad \nabla^2\phi = 0. \tag{2.58}$$

That is, the velocity potential satisfies the Laplace equation. As readers may remember from their studies in electromagnetism, the electrostatic potential satisfies this equation in charge-free vacuous regions. Mathematically, this equation is of the elliptic PDE type which requires well-defined boundary condition in order for a solution to both exist and be unique. In the case of potential flow, the boundary conditions are usually given at infinity, or on the walls of a vessel containing the fluid, and on solid bodies immersed in the fluid. The obvious boundary condition on surfaces of solid bodies is the one stating that the perpendicular to the surface component of the velocity is zero and this readily translates to what is called *Neumann* boundary conditions

$$\frac{\partial\phi}{\partial n} = \hat{\mathbf{n}} \cdot \nabla\phi = 0 \tag{2.59}$$

on the bodies' surfaces. Here $\hat{\mathbf{n}}$ is a unit normal to these surfaces at the point in question. This is perhaps the place to mention, in passing, that the other type of boundary conditions, which have to do with the existence and solution of the Laplace equation, are named after *Dirichlet* and consist of the value of the potential on the boundary. In unbounded flows, the requirement usually refers to the manner in which the velocity approaches a constant value or zero at infinity, that is, very far as compared to any meaningful physical distance. For the sake of completeness we mention also *mixed boundary conditions*, which have different types of above mentioned conditions on different parts of the boundary. Before proceeding any further, we shall now mention the conditions under which a flow can be considered practically incompressible.

2.4.1.1   Physical Conditions for Incompressibility

When the fluid has constant density, every flow in it is incompressible, as mentioned before. We have also seen in Eq. (2.31), when discussing the example of the de Laval nozzle, that for small Mach numbers (subsonic flows), compressibility is insignificant. We repeat this important point here again—very subsonic flows cannot have any significant properties resulting from compressibility and, for all practical purposes, they may be considered incompressible. Speaking a little more precisely, the usual condition ($M < 1$) is not sufficient. Indeed, when the flow is unsteady, and we let l and τ be the length and timescale, respectively, over which the velocity changes appreciably, we may use estimates based on the Euler equation ($\partial u/\partial t \sim \nabla P/\rho$) and on the continuity equation, obtaining an additional condition for incompressibility:

$$\tau \gg l/c_s. \tag{2.60}$$

This condition has a clear meaning, namely, any changes in the flow variables must proceed on a much longer time than it takes to communicate their occurrence over a significant distance. Another way to say this is that disturbances propagate at the sound speed within the fluid but this happens extremely fast, the propagation is effectively instantaneous. This quality is another typical characteristic of incompressibility. In the preceding discussion we have ignored the body force term in Euler's equation. Is there any condition, related to this term, which may destroy the approximation of incompressibility? We defer this issue to Problem 2.10.

We may now consider some specific potential flow problems and examples, when the flow is incompressible. For simplicity we shall choose a vessel of a fluid having constant density, in addition to being irrotational. Since the governing equation is the Laplace equation, there exists a vast literature on its solution methods (potential theory). We mention some references to these sources in the *Bibliographical Notes* at the end of this chapter. Laplace's equation has application in many branches of physics, but we, naturally, address only fluid dynamical problems.

2.4.1.2   Two Simple Examples

1. Consider the potential $\phi(\mathbf{x}) = Vx$, where V is a constant. As can be easily seen this represents a uniform flow, having a constant velocity V in the x direction.

2. We move to the investigation of a steady flow in two dimensions, in polar
   coordinates (r, θ), around a circular cylinder of radius R, which perpendicularly
   cuts the plane and whose formally infinite axis aligns along the z-axis in three-
   dimensional cylindrical coordinates. The corresponding Laplace equation is

$$\frac{\partial^2 \phi}{\partial r^2} + \frac{1}{r^2} \frac{\partial^2 \phi}{\partial \theta^2} = 0. \tag{2.61}$$

The potential flow far away from the cylinder has a fixed velocity U in the
horizontal ($\theta = 0$) direction. On the surface of the impenetrable cylinder
we should have $u_r(a) = (\partial \phi / \partial r)_{r=a} = 0$. Among the solutions of the Laplace
equation, there are positive and negative powers of r (and $\ln r$, as well)
multiplied by $\sin \theta$ or $\cos \theta$. Exploiting the superposition principle valid for
the Laplace equation, which is linear, we readily find a solution satisfying the
equation and the boundary conditions:

$$\phi(r, \theta) = U \left(r + \frac{a^2}{r} \right) \cos \theta. \tag{2.62}$$

The solution here is easily verified by re-inserting back into (2.61). Is the solution
unique? If it is not, does this not violate one of the basic theorems on the existence
and uniqueness of Laplace's equation? Think about the necessary form of the
boundary conditions at infinity.

2.4.2   General Three-Dimensional Potential Flow Past a Solid Body

Consider potential flow with constant density, ideal fluid past an arbitrary solid body,
with the assumptions that there are no body forces. Let the velocity of the fluid
very far from the body be $\mathbf{U}$. As we have already mentioned, it is evident that the
problem is completely equivalent to the motion of the fluid when the same body
moves through it. It is just a question of changing the coordinate system so that
the fluid be at rest at infinity, that is, the transformation on the fluid velocity will
be $\mathbf{u} \mapsto \mathbf{u} - \mathbf{U}$. We shall be interested in the nature of the fluid velocity very far
away from the body. By that we mean that if the fluid extension, say L, is much
larger than the dimensions of the body, say l, then $\varepsilon \equiv l/L \ll 1$. We should get
a reasonably good approximation already at the lowest order in ε. The governing
equation is $\nabla^2 \phi = 0$ where ϕ is the fluid velocity potential.

Assume that at some instant the body starts to move and quickly acquires a
constant velocity, $\mathbf{U}$. As discussed before, in the body frame there exists a potential
steady flow of the fluid. Thus, we choose a point on the body as the origin and
work in a frame moving with the body. If the body's shape is such that there is a
streamline perpendicular to its surface, it is trivial to find the maximal pressure of

the flow. Indeed, since the conditions of case 3b of the Bernoulli theorems apply, we have

$$\frac{P}{\rho_0} + \frac{1}{2}u^2 = \text{const.} \tag{2.63}$$

The point at which the streamline impinges perpendicularly on the body we have $u = 0$ and thus the maximal pressure. So the maximal pressure in the flow is just $P_{\max} = P + \rho_0 u^2/2$ for all points on the streamline and actually here, see case 3b of the Bernoulli theorems, this relation is valid everywhere.

The easiest looking three-dimensional problem is for a spherical obstacle. However, we now consider a body of a general shape and try to formulate some general mathematical statements on the fluid velocity distribution far from the body. Once again, we have the Laplace equation $\nabla^2\phi = 0$, with the boundary condition of zero flow at infinity and some well-defined specific Neumann conditions which depend intimately upon the shape of the body at the body's surface. This guarantees a unique solution. The fluid is at rest and undisturbed very far from the body and mathematically this means that we want the potential and its derivatives to vanish at infinity. It should be stressed again that we look at the case of a moving body, with the coordinate center fixed on it, that is, the solution will be obtained in a moving frame.

It is known in potential theory (e.g., reference [3]) that up to second order in ε the most general solution in such a case is the expansion

$$\phi = -\frac{a}{r} + \mathbf{A} \cdot \nabla\left(\frac{1}{r}\right) + \cdots, \tag{2.64}$$

where the ellipsis indicates higher order terms, for which the acronym "HOT" will be often used throughout this book, where a is a scalar and $\mathbf{A}$ a vector, both coordinate independent. We show now that $a = 0$. Indeed, the lowest order potential term gives the following velocity with respect to the body:

$$\mathbf{u} = -a\nabla\left(\frac{1}{r}\right) = \frac{a}{r^2}\hat{\mathbf{r}}, \tag{2.65}$$

where $\hat{\mathbf{r}}$ is a unit vector in the radial direction. So the mass flux through a large spherical surface of radius R should be $f_R = 4\pi R^2 \rho a/R^2 = 4\pi\rho a$. But in an incompressible fluid the mass flux through any closed surface must be zero. Hence $a = 0$.

Thus, the leading behavior of the potential for large r is

$$\phi = \mathbf{A} \cdot \nabla\left(\frac{1}{r}\right) + \text{HOT} = -\mathbf{A} \cdot \hat{\mathbf{r}}\frac{1}{r^2} + \text{HOT}. \tag{2.66}$$

In what follows we concentrate on the far field, that is, drop the higher order term and get the velocity field in the lowest order of the small parameter defined as $\varepsilon \equiv l/L$. We remind the reader that we are dealing here with a general shape of the body, nevertheless, as we shall see the result is interesting, significant, and useful. Thus we have

$$\phi(r) = -\mathbf{A} \cdot \frac{\hat{\mathbf{r}}}{r^2} \quad \Longrightarrow \quad \mathbf{u}(r) = \nabla\phi = -(\mathbf{A} \cdot \nabla)\frac{\hat{\mathbf{r}}}{r^2}, \tag{2.67}$$

where $\mathbf{A}$ is dependent on the shape of the body and its velocity and can be determined only by solving the problem completely. The notation $\hat{\mathbf{r}}$ serves as the unit vector in the r direction, while $\hat{\mathbf{n}}$ denotes the unit normal to the body and the two coincide if the body is a sphere. At a large enough distance from the body, the two may also be perceived as the same since at large distances the body may be approximated by a point.

We now use an identity from vector calculus, see Eq. (A.16) in Appendix A of the book:

$$(\mathbf{A} \cdot \nabla)\left[\hat{\mathbf{r}}f(r)\right] = \frac{f(r)}{r}\left[\mathbf{A} - \hat{\mathbf{r}}(\mathbf{A} \cdot \hat{\mathbf{r}})\right] + (\mathbf{A} \cdot \hat{\mathbf{r}})\frac{\partial f}{\partial r}. \tag{2.68}$$

The formula holds for any constant vector $\mathbf{A}$ and a scalar function $f(r)$. Using this identity in formula (2.67), i.e., with $f(r) = 1/r^2$, gives immediately that the lowest order in ε, the behavior of the velocity, potential, and the velocity, in a potential flow around an arbitrary body:

$$\phi(r) = -\mathbf{A} \cdot \frac{\hat{\mathbf{r}}}{r^2}; \quad \mathbf{u}(r) = \frac{1}{r^3}\left[3\hat{\mathbf{r}}(\mathbf{A} \cdot \hat{\mathbf{r}}) - \mathbf{A}\right]. \tag{2.69}$$

We are confident that readers who have some training in potential theory will not find this formula alien. The vector $\mathbf{A}$ is related to the total momentum and energy of the fluid in its motion past the body. Equation (2.69) is a solution and because of the uniqueness theorem for solutions of Laplace's equation and since Neumann boundary conditions are satisfied on the boundary (convince yourself that the conditions at infinity suffice to complement the Neumann condition on the body), it is also the only solution of the problem. We should not forget, however, that this solution is correct in the frame moving with the body. Problem 2.13 is a demonstration that a general formula for the total kinetic energy of the fluid and its linear momentum may be computed from Eq. (2.69), when it is taken into account that we have calculated the velocity of the fluid with respect to the body.

The above problem shows that the total momentum of the fluid in this potential flow is

$$\mathbf{P} = 4\pi\rho\mathbf{A} - \rho\mathcal{V}_0\mathbf{U},$$

where $\mathscr{V}_0$ is the body's volume. It follows that the force the body exerts on the fluid is

$$\mathbf{F}_{b \to f} = \frac{d\mathbf{P}}{dt} = 0, \tag{2.70}$$

since all the factors in $\mathbf{P}$ are constant in time. It follows from Newton's third law ($\mathbf{F}_{f \to b} = -\mathbf{F}_{b \to f}$) that the fluid also does not exert any force on the body, i.e., the *drag*, defined as $\mathbf{F}_{f \to b} \cdot \hat{\mathbf{U}}$, is zero as well as the *lift*, defined as $\left(\mathbf{F}_{tot} - \mathbf{F}_{f \to b}\right) \cdot \hat{\mathbf{U}}$ with $\mathbf{F}_{tot} = d\mathbf{P}/dt$. This result is clearly contrary to the physical reasoning that the body must invest work in order to move at constant speed in a fluid. It has to push and displace the fluid out of its way and thus suffer back-reaction. This is a legendary conundrum having even acquired a special name—the *d'Alembert paradox*. The respected L. Prandtl tried to resolve the paradox by simply stating that a completely inviscid fluid does not exist in nature, thus an exactly potential flow around a body is an idealization. He thought that energy is dissipated in the boundary layer near the body and therefore the body must suffer a drag, i.e., work has to be done in order to move it at constant speed. Remarkably, the definite resolution of the paradox had to wait to the twenty-first century, when high resolution numerical calculation revealed that the zero-drag potential solution seems to be unstable and develops a time-dependent turbulent wake causing substantial drag. Thus the action happens, so it seems, at the rear of the body, where the boundary layer separates. This occurs no matter how small the viscosity is. The turbulent wake may acquire wavy irregular forms, allowing drag and lift. Several further elementary examples of potential flows in three dimensions are the subject of Problem 2.12.

2.4.3 Two-Dimensional Flows: Stream Function and Complex Potential

Two-dimensional potential flows of an incompressible fluid have little to do with real physical or astrophysical flows, but are important because of the possibility of elegant mathematical treatment and may also have significant applications in aeronautics. This is probably the cause that these types of flows are frequently referred to as classical aerofoil theory. We do not discuss this subject at great length as many books exist on just this one topic, but since some of the results are very basic and powerful and, as mentioned before, the mathematical techniques are very elegant, we devote some space to a comprehensive summary of the subject. We give in the *Bibliographical Notes* of this chapter some references to books that devote a substantial part of their discussion of two-dimensional, incompressible, potential flows to methods using complex potential and conformal mappings.

Any two-dimensional flow of an incompressible fluid, potential or not, allows for a definition of an important scalar function, the *stream function*. Assume that the flow is two-dimensional in Cartesian coordinates, $(x_1, x_2) \equiv (x, y)$ and similarly for the velocities $(u_1, u_2) \equiv (u, \mathsf{v})$. The two-dimensional incompressibility condition is

$$\nabla \cdot \mathbf{u} = \frac{\partial u}{\partial x} + \frac{\partial v}{\partial y} = 0. \tag{2.71}$$

If a function $\psi(x,y,t)$ is now defined, such that

$$u = \frac{\partial \psi}{\partial y} \quad \text{and} \quad v = -\frac{\partial \psi}{\partial x}, \tag{2.72}$$

the incompressibility condition is satisfied automatically. The streamlines of the flow in two dimensions are just one of the equalities of Eq. (1.12) and in our notation it reads

$$\frac{dx}{u} = \frac{dy}{v}. \tag{2.73}$$

Substitution of u and v from Eq. (2.72) into this streamline equation gives

$$\frac{\partial \psi}{\partial x} dx + \frac{\partial \psi}{\partial y} dy = 0.$$

But the left-hand side of this equation is nothing else than $d\psi$, therefore it is evident that the stream function is constant on streamlines. In other words, the streamlines of a two-dimensional, incompressible flow are lines of constant stream function, ψ.

Consider now two very close streamlines, denoted by ψ and $\psi + d\psi$, respectively, and let an arbitrary point $\mathbf{x}$ reside on the streamline corresponding to ψ. Let an arbitrary line element $d\mathbf{x} = (dx, dy)$ connect the point $\mathbf{x}$ on ψ to the point $\mathbf{x} + d\mathbf{x}$ on $\psi + d\psi$. Choosing a case in which both dx and dy are positive, we obtain an expression equal to the difference

$$(v dx - u dy) = -\frac{\partial \psi}{\partial x} dx - \frac{\partial \psi}{\partial y} dy = -d\psi, \tag{2.74}$$

showing that the volume rate of flow across the element $d\mathbf{x}$, between a pair of streamlines, is numerically equal to the difference between their ψ values. The sign of the stream function is such that facing the direction of motion ψ increases to the left. In addition, if we want in general two streamlines, separated by a finite distance, a simple integration procedure will approve the results which we obtained for differentially separated streamlines. We saw already that the stream function is constant on streamlines, and this can be verified by

$$\mathbf{u} \cdot \nabla \psi = \frac{\partial \psi}{\partial y} \frac{\partial \psi}{\partial x} - \frac{\partial \psi}{\partial x} \frac{\partial \psi}{\partial y} = 0. \tag{2.75}$$

A compact way of expressing the velocity vector from the stream function defined for a two-dimensional flow in the x–y plane, equivalent to Eq. (2.72), is

$$\mathbf{u} = \nabla \times (\psi \hat{\mathbf{z}}). \tag{2.76}$$

This equivalence is easy to check (do it!). A similar prescription is useful for polar coordinates in the plane as well, yielding

$$u_r = -\frac{1}{r}\frac{\partial \psi}{\partial \varphi} \quad \text{and} \quad u_\varphi = \frac{\partial \psi}{\partial r}. \tag{2.77}$$

An incompressible potential flow in two dimensions allows, as we have seen, the definition of the velocity potential $\phi(x,y)$ and stream function $\psi(x,y)$, here expressed in Cartesian coordinates for convenience. We have also seen that the velocity components u and v are equal, correspondingly, to

$$\frac{\partial \phi}{\partial x} = \frac{\partial \psi}{\partial y} \quad \text{and} \quad \frac{\partial \phi}{\partial y} = -\frac{\partial \psi}{\partial x}. \tag{2.78}$$

Using elementary complex function theory, we identify Eq. (2.78) as the *Cauchy–Riemann* conditions and they ensure, provided that all the derivatives are continuous, the existence of an analytic function of a complex variable:

$$w(z) = \phi(x,y) + i\psi(x,y), \tag{2.79}$$

where henceforth, in this section, $z \equiv x + iy$, that is, x and y define the complex plane and z is the appropriate complex number. $w(z)$ goes under the name *complex potential* and it is evident from the Cauchy–Riemann conditions that both the real and the imaginary parts of the complex potential, i.e., ϕ and ψ satisfy the Laplace equation. If one knows the complex potential, it is easy, knowing just the basics of complex function theory, to find the velocity, using

$$\frac{dw}{dz} = \frac{\partial \phi}{\partial x} + i\frac{\partial \psi}{\partial y} = u - i\mathrm{v}. \tag{2.80}$$

Thus $u = \Re(dw/dz)$, $\mathrm{v} = -\Im(dw/dz)$, and

$$|\mathbf{u}| = \sqrt{u^2 + \mathrm{v}^2} = \left|\frac{dw}{dz}\right|. \tag{2.81}$$

The complex potential method is very powerful in describing flows having the above-mentioned properties. One important observation is that every complex analytical function describes an acceptable two-dimensional incompressible potential (irrotational) flow. The method also allows us to solve analytically nontrivial problems, as well as derive some important theorems, as we shall see in this section and the associated problems at the end of the chapter.

2.4.3.1   Two-Dimensional Potential Flow Past a Circular Cylinder

We are now set to discuss the problem of a two-dimensional flow past a circular cylinder with radius a whose circular profile cuts the flow plane. The cylinder is assumed to be of mathematically infinite height, but a physical approximation $H_c \gg a$, where the height of the cylinder is H_c, will suffice. Since the complex potential method will be used, it should be remarked that a superposition of two complex potentials, representing two different flows, is itself a new complex potential and, therefore, represents a superposed flow.

For example, Problem 2.14 is an examination of a "dipole" or "doublet" flow, whose complex potential is $w_d(z) = -\mu/z$. Let us superpose this flow with another flow whose complex potential is $w_u(z) = Uz$ with U real constant. It is easy to verify that the real part of this complex potential is $\phi(x,y) = Ux$ and, thus, it represents a uniform flow in the x direction with constant velocity U. The aggregate complex potential is $w = w_u + w_d$

$$ w = Uz - \frac{\mu}{z} = \left(U - \frac{\mu}{x^2 + y^2} \right) x + i \left(U + \frac{\mu}{x^2 + y^2} \right) y. \qquad (2.82) $$

Thus the velocity potential is

$$ \phi(x,y) = \left(U - \frac{\mu}{x^2 + y^2} \right) x $$

and the stream function is

$$ \psi(x,y) = \left(U + \frac{\mu}{x^2 + y^2} \right) y. $$

As we discussed in Sect. 2.4.3, surfaces of constant stream function values are surfaces across which there is no mass flux. We then identify the locus where $\psi = 0$ (it is an arbitrary choice) and, accordingly, construct its shape. Looking at the above form for ψ and setting it to zero, we see very clearly that it consists of two branches $y = 0$ and $(x^2 + y^2)U = -\mu$. If $\mu = -Ua^2$ then the cylinder circumference will be a part of the streamline $\psi = 0$. In Fig. 2.4a one can see the streamlines of this problem, but it is also possible to add yet another component to the flow, viz. circulation around the cylinder. This is achieved by adding a complex potential

$$ w_c = i \frac{\Gamma}{2\pi} \ln z. $$

That such a complex potential represents circulation can be easily verified by the reader and will be discussed in Sect. 2.5.3. Γ is the strength of the circulation and the flow pattern depends on the parameter

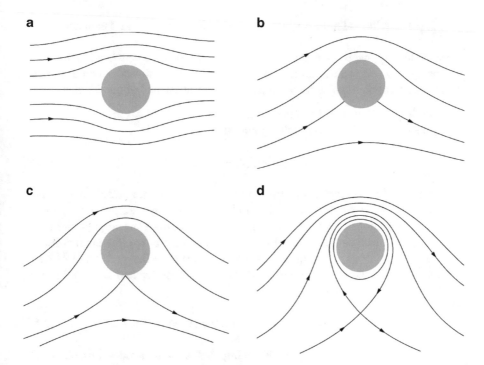

Fig. 2.4 Flow past a circular cylinder for different values of β representing circulation around the cylinder. For details see text: (**a**) $\beta = 0$, (**b**) $\beta < 2$, (**c**) $\beta = 2$, (**d**) $\beta > 2$

$$\beta = -\frac{\Gamma}{2\pi U a}.$$

The full complex potential (not forgetting that $\mu = -U a^2$, so as to ensure $\psi = 0$ on the surface of the cylinder) thus reads

$$w(z) = U\left(z + \frac{a^2}{z} - i\beta a \ln z\right). \tag{2.83}$$

Note that in Eq. (2.62) we have presented a solution to a problem that seems equivalent to this one, but it was not clear that it was the unique solution. Here we have several solutions with circulation around the cylinder. Why is uniqueness violated here?

2.4.3.2 The Milne–Thomson, Blasius and Kutta–Joukowski Theorems

In the following, we present some theorems valid in two-dimensional incompressible potential flows, whose proofs are examples of the complex potential method. Additional examples can be found in the problems at the end of the chapter.

1. *The Milne–Thomson theorem.*
   Consider the following complex potential

$$w(z) = f(z) + f^*(r_0^2/z^*),\tag{2.84}$$

where $f(z)$ is a complex function, whose singularities lie outside the circle of radius r_0 centered on the origin, i.e., in $|z| > r_0$. The superscript asterisk denotes complex conjugate. We now show that $w(z)$ cannot have any singularities inside the circle. If all singularities of $f(z)$ are outside the circle, the singularities of $f(r_0^2/z^*)$ are in $|r_0^2/z^*| > r_0$, that is, in $|z| < r_0$ (inside the circle) and therefore those of the second term in Eq. (2.84) are outside the circle. Next, on the circle itself $zz^* = r_0^2$ and thus

$$w(z) = f(z) + f^*(z) \quad \text{on the circle} \quad |z| = r_0,\tag{2.85}$$

so $w(z)$ is real on the circle, implying that $\psi = 0$ (a constant) on it. The circle $|z| = r_0$ is thus a streamline. This is the statement of the theorem. In Problem 2.16 we consider an application of this theorem.

2. *The Blasius theorem.*
   This theorem gives the forces acting on a body in a two-dimensional potential flow of constant density, whose complex potential is known: $w(z)$. We assume that the flow is stationary and it is uniform at infinity. It may thus be written symbolically $\mathbf{u}(x = \pm\infty, y, t) = U\hat{\mathbf{x}}$. The statement of the theorem is

$$F_x - iF_y = \frac{1}{2} i\rho_0 \oint_C \left(\frac{dw}{dz}\right)^2 dz,\tag{2.86}$$

where F_x and F_y are the x and y force components per unit length, respectively, acting on the body, whose boundary is the simple closed contour C. They are equal, respectively, to the drag and the lift experienced by the body. If we denote by ℓ the arc length along the contour, making an angle α, say, with the positive x direction and by considering the differential line element $d\ell = dz = dx + idy$, we readily get the differential forces on this differential segment of the body, per unit length, because in an inviscid flow there is only pressure force, perpendicular to the body. $dF_x = -Pdy$ and $dF_y = Pdx$, and so

$$dF_x - idF_y = -Pdy - iPdx = -iPdz^*.\tag{2.87}$$

Thus

$$F_x - iF_y = -i \oint_C P dz^* \text{ counterclockwise on contour } C \text{ as usual.} \qquad (2.88)$$

Now remembering case 3a of the Bernoulli's theorem for steady potential flow, we get the relation between pressure and velocity

$$P + \frac{1}{2}\rho_0(u^2 + v^2) = \mathscr{B}'\rho_0 = P_\infty + \frac{1}{2}\rho_0 U^2. \qquad (2.89)$$

Thus

$$P = P_\infty + \frac{1}{2}\rho_0 U^2 - \frac{1}{2}\rho_0(u + iv)(u - iv). \qquad (2.90)$$

When integrating over the contour C as in Eq. (2.88), the first two constant terms do not contribute. Also $u + iv = \sqrt{u^2 + v^2}e^{i\alpha}$ and $dz = |dz|e^{i\alpha}$ because on the contour dz is parallel to the velocity vector. So $(u + iv)dz^*$ is real and thus equal to $(u - iv)dz$. Thus the integral in Eq. (2.88) finally becomes

$$F_x - iF_y = \frac{1}{2}i\rho_0 \oint_C \left(\frac{dw}{dz}\right)^2 dz, \qquad (2.91)$$

where we have used

$$\frac{dw}{dz} = u - iv.$$

We remind the reader that this problem is two-dimensional and accordingly, due to the behavior of the solution of Laplace's equation far from the object, the result is different from the three-dimensional case which, as we saw, gives rise to the d'Alembert paradox.

3. *The Kutta–Joukowski theorem.*

   We consider again a steady potential flow of constant density in two dimensions, past a two-dimensional body, the cross section of which is a simple curve C. As before, we assume that the flow is stationary and is uniform at infinity, $\mathbf{u}(x = \pm\infty, y, t) = U\hat{\mathbf{x}}$. The statement of the theorem is that if the circulation around the body (around C) is Γ, then $F_x = 0$ and $F_y = -\rho_0 U\Gamma$. This fully inviscid case is an example where circulation induces lift on the body even though the drag remains zero. To prove this theorem we take dw/dz to be an analytic function in the entire flow range and, correspondingly, expand it in Laurent series around an origin point O, chosen to be inside the body. Thus we have

$$\frac{dw}{dz} = U + \frac{a_1}{z} + \frac{a_2}{z^2} + \text{HOT}. \qquad (2.92)$$

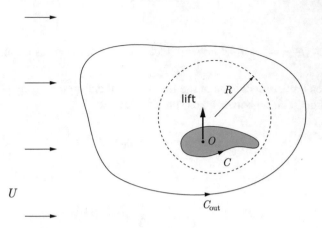

Fig. 2.5 Auxiliary sketch for the Kutta–Joukowski theorem

Inspection shows that the above form is consistent with the flow being of uniform of speed $U\hat{\mathbf{x}}$ at infinity. Now, we shall use the Blasius theorem, but before that we surround the body (whose cross section is C) by a larger circular contour C_b of radius R, which is the smallest circle containing the curve C. In Fig. 2.5 it is the unmarked dashed circle. Squaring the Laurent series (2.92) and substituting the result into that of Blasius, we find that only the $1/z$ term contributes, giving

$$F_x - iF_y = -2\pi\rho_0 U a_1. \tag{2.93}$$

Using the residue theorem we may easily determine a_1

$$2\pi i a_1 = \oint_{C_{\text{out}}} \frac{dw}{dz} dz \tag{2.94}$$

where C_{out} is an arbitrary curve *containing* the above dashed circle and all the contribution, after the Laurent expansion is used, comes from the point O. Cauchy theorem of complex function theory guarantees that without singularities in the simply connected domain, bounded by C_{out} and C (see Fig. 2.5) an integral of any analytical function over both of these contours (consisting together of two parts) gives zero. Thus, there is only a contribution to the integral on C from the interior of C. But C is a streamline thus the change of ψ along C is zero, while the change in ϕ is just the circulation Γ. Thus, $2\pi i a_1 = \Gamma$ and substituting it into formula (2.93) gives the desired result for the lift

$$F_y = -\rho_0 U \Gamma. \tag{2.95}$$

It is imperative to exercise care in determining what is the sign of Γ and thus the lift. As is usually the case, physical reasoning is superior to mathematical pedantry, and this case is no different. In general, if the shape of the body is such that it deflects the horizontal stream of fluid downward, as is clearly the case in Fig. 2.5, especially if we imagine a series of such bodies one above the other, then Negative (downward) momentum is imparted on the fluid and thus the body experiences a reaction force which is upward, i.e., positive lift $(F_y > 0)$. The opposite is true when the cross section of the body is such that its upper part is inclined, on the average, upward, with respect to the fluid stream. Do you recognize the changes in the aircraft wing during takeoff as opposed to landing or, similarly, with the shape of a sail during a tacking maneuver? We should, however, not forget that all the above results assume ideally irrotational flow everywhere around the body.

However as we have already remarked, boundary layers are found to be created in experiments and numerical simulations (see Chap. 3) of aerodynamic aerofoils immersed in fluids of minute viscosity. The boundary layer may separate and create a wake behind the body (as explained in the next chapter). Usually vortices are created in the wake and may be shed into the fluid. These vortices can have positive or negative vorticity (the sign being arbitrarily defined), with respect to the upper surface of the body being convex (like in Fig. 2.5) or concave. This has import in determining the circulation around C_b, because by Stokes theorem it is equal to the surface integral of the vorticity over the fluid area bounded by the path. If there are vortices there, then they must contribute (whether positively or negatively) to Γ.

2.5 Vortex Motion

Vortices are common and prominent features of fluid flows. They are observed on a variety of scales, in our atmosphere, e.g., hurricanes and tornadoes, in the Earth's oceans, rivers, and streams as well as on the giant Solar system planets, most prominently on Saturn and Jupiter, displaying its giant vortex. Their importance for both geophysics and astrophysics is doubtless. The above examples might suggest that vortices appear only on fluid bodies that are rotating, a topic which is the subject of the next chapter. But in fact, vortices also appear in nonrotating flows and also in laboratory scale flows. It is of theoretical interest to understand them and their motion and, in the process, to uncover their remarkable beauty.

2.5.1 Helmholtz–Vortex Theorems

Before formulating these theorems, which have their nineteenth century origin in the work of H. von Helmholtz and Lord Kelvin, it is imperative to define several terms:

- A *vortex line* is a curve having the same direction as the vorticity vector ω, at any particular time. Its definition is similar to a streamline, where we speak about curves having the same direction as the velocity vector. Formally, a vortex line, given parametrically by $\mathbf{x}(s)$, is the solution of

$$\frac{dx_1/ds}{\omega_1} = \frac{dx_2/ds}{\omega_2} = \frac{dx_3/ds}{\omega_3}, \tag{2.96}$$

which is an elementary concept from multidimensional calculus. This definition is essentially the same as Eq. (1.12), for the velocity, but here we use the auxiliary length parameter s.
- A *vortex tube* is a volume enclosed by all vortex lines that pass through some simple closed curve.
- The *strength* of a vortex tube is the circulation around its cross section.
- A *vortex ring* is a vortex tube that is closed on itself, i.e., it is the volume embedded in a collection of closed vortex lines that pass through a closed curve.
- A *vortex sheet* is a series of vortices, whose centers are very close to one another. Actually, the distances between them tend to zero along some curve.

Usually, only two Helmholtz vortex theorems are cited, but we shall give here four statements, the first two being the usual ones, summarizing most of the important vortex properties. Three of the statements will be proven here and the proof of the fourth is deferred to the problems. The assumptions needed for the Helmholtz theorems are identical to those of Kelvin's circulation theorem.

1. *Vortex lines are material invariants.*
   This means that they move with the fluid. In our opinion, the easiest proof of this theorem is to consider a vortex tube and a closed curve $C1$ encompassing an area $S1$ and lying on the *side surface* of the tube (see Fig. 2.6). Clearly, the circulation around $C1$ is zero as $\int_{S1} \omega \cdot \hat{\mathbf{n}}\, dS = 0$. After some time the curve $C1$ is materially transported to another curve $C2$, encompassing now an area $S2$. According to Kelvin's theorem, the circulation around $C2$ remains zero. But this is true for any time, i.e., any $S2$. Thus, there cannot be a component of vorticity normal to the flux tube's side surface and this means that $S2$, during its material motion, must remain on the same side surface of the vortex tube. So we have actually proven a bit more than the theorem says. We have shown that a particular vortex tube is material invariant. Effecting a limiting procedure, so that the cross section of the vortex tube tends to zero, proves the first theorem, as formulated.

Fig. 2.6 Auxiliary sketch for
the Helmholtz vortex
theorems. The *shaded
regions*, surrounded by the
paths *C*1 and *C*2 lie on the
side outer surface of this
vortex tube. *A*1 and *A*2 are
capping the tube at two
perpendicular to the vorticity
positions

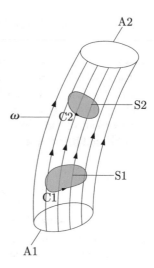

2. *The strength of a vortex tube is constant along its length.*

   The proof of this statement is straightforward. Consider a vortex tube, capped
   by two surfaces perpendicular to ω, $A1$ and $A2$, say. In this way, we create a
   volume embedded by the side surfaces of the tube and the two areas $A1, A2$.
   We call this volume $\mathscr{V}_{\text{tube}}$ and its bounding surface S_{tube}. Clearly, by the Gauss
   theorem

   $$\int_{S_{\text{tube}}} \omega \cdot \hat{\mathbf{n}} \, dS = \int_{\mathscr{V}_{\text{tube}}} \nabla \cdot (\nabla \times \mathbf{u}) d^3x = 0. \tag{2.97}$$

   Remembering the convention about the outward direction of the normal to
   a closed surface, calculating the strength of the tube we keep the same
   directionality of the surfaces $A1$ and $A2$ and we get using the Stokes integral
   theorem

   $$\int_{A1} \omega \cdot \hat{\mathbf{n}} \, dS = \int_{A2} \omega \cdot \hat{\mathbf{n}} \, dS \implies \oint_{C1} \mathbf{u} \cdot \mathbf{dl} = \oint_{C2} \mathbf{u} \cdot \mathbf{dl}. \tag{2.98}$$

   Since $A2$ is arbitrary, the theorem has actually been proved. A little reflection
   shows that this theorem, together with the first theorem, proves a little more,
   which is the content of the next theorem.
3. *The strength of a vortex tube remains constant in time.*
4. *A vortex tube cannot end within the fluid.*

   It can either end on the boundary or close on itself, creating a vortex ring. The
   proof of this statement is the subject of Problem 2.22.

2.5.2 Inviscid Two-Dimensional Vortex Equation of Motion

In severely restricted conditions, the equations governing vortex motion simplify considerably. In particular, consider a flow which is two-dimensional. Suppose that all fluid quantities are represented in the Cartesian plane. This is not necessary, of course, but merely convenient for our purposes here. Assume that all quantities are independent of the vertical z coordinate. Then assuming, as well, that the fluid is inviscid and barotropic equation (2.13) simplifies to

$$\frac{D}{Dt}\left(\frac{\zeta}{\rho}\right) = -\hat{\mathbf{z}}\cdot\left[\nabla\left(\frac{1}{\rho}\right)\times\nabla P\right]. \tag{2.99}$$

The symbol ζ is used here to denote the vertical component of the total vorticity, i.e., $\zeta \equiv \hat{\mathbf{z}}\cdot\boldsymbol{\omega}$. Under even more restricted conditions, wherein the fluid is assumed to be everywhere of constant density,

$$\frac{D\zeta}{Dt} = 0, \tag{2.100}$$

where we remember that the material derivative in such two-dimensional flows is

$$\frac{D}{Dt} = \frac{\partial}{\partial t} + u\frac{\partial}{\partial x} + \mathrm{v}\frac{\partial}{\partial y}. \tag{2.101}$$

It follows that (see also Problem 2.19) the usual stream-function formulation is appropriate and, as such,

$$u = \frac{\partial\psi}{\partial y}, \qquad \mathrm{v} = -\frac{\partial\psi}{\partial x}, \qquad \zeta \equiv \frac{\partial\mathrm{v}}{\partial x} - \frac{\partial u}{\partial y} = -\left(\frac{\partial^2\psi}{\partial x^2} + \frac{\partial^2\psi}{\partial y^2}\right). \tag{2.102}$$

The important consequence of Eq. (2.101) is the fact that, in two-dimensional Lagrangian flows, fluid particles conserve the vorticity with which they are initially endowed. This observation leads to an interesting result, which we discuss in the next section.

2.5.3 Hamiltonian Dynamics of Point Vortices

Vortex dynamics is a vast subject, which could not be seriously covered in a book like this; however, we shall mention in the *Bibliographical Notes* at least one comprehensive reference. We have chosen a few example two-dimensional flows to include in our concise discussion of vortices. One interesting example that has connections to statistical mechanics consists of a collection of point vortices. From Problem 2.19 it can be shown that Eq. (2.101) enables a *vorticity–stream function formulation* of an inviscid, two-dimensional incompressible flow, whose primary equation may be recast into the following form:

$$\frac{\partial \zeta}{\partial t} + J[\psi, \zeta] = 0, \quad \text{with} \quad \zeta = -\nabla^2 \psi, \tag{2.103}$$

where J is here the two-dimensional (x, y) Jacobian, as defined in Problem 2.19, and where ∇^2 is the two-dimensional Laplacian. The above equation can be rewritten as a single differential equation for the stream function

$$\frac{\partial (\nabla^2 \psi)}{\partial t} + J[\psi, \nabla^2 \psi] = 0, \tag{2.104}$$

where J is given by

$$J[\psi, \nabla^2 \psi] = \frac{\partial (\psi, \nabla^2 \psi)}{\partial (x, y)} = \partial_x \psi \partial_y (\nabla^2 \psi) - \partial_y \psi \partial_x (\nabla^2 \psi). \tag{2.105}$$

We should not forget that the stream function satisfies the two-dimensional Poisson equation, as can be seen, e.g., in the second expression in Eq. (2.103), and that the source term of that Poisson equation is the vorticity. Our focus shifts now to a vorticity field composed of only very tight vortices, actually approximated by point ones as the source term to the above-mentioned Poisson equation, thus

$$\zeta = \sum_{j=1}^{N} \Gamma_j \delta(\mathbf{x} - \mathbf{x}_j), \tag{2.106}$$

where Γ_k is the strength, i.e., the circulation around, of the k-th point vortex, which is located at a point $\mathbf{x}_k = (x_k, y_k)$ of the x–y plane. The function $\delta(x)$ is the Dirac delta function, which is not a new concept to the reader, who learned any undergraduate quantum mechanics and perhaps a graduate course on electromagnetism. The Dirac delta function's more formal definition, in particular the question how it can be mathematically perceived as a function and some of its properties, will be discussed in some detail later, in Sect. 4.1.3, in the context of waves. The sum is over all the point vortices, here assumed to be N, and they constitute the aggregate entirety of the vorticity field. The solution of the Poisson equation in two dimensions having point sources is well known and one may easily apply it here to get

$$\psi(\mathbf{x}) = \frac{1}{2\pi} \sum_{j=1}^{N} \Gamma_j \log |\mathbf{x} - \mathbf{x}_j|. \tag{2.107}$$

The fluid velocity at the vortex core of the ith vortex, $\mathbf{u}_i \equiv (u_i, v_i)$, is obtained by taking the derivative of the total stream function sans the contribution of the ith vortex. In other words, the velocity, in the Lagrangian description, of the fluid particle at the ith point vortex location is given by the induced velocity of the remaining $N-1$ point vortices, thus

$$
\frac{dx_i}{dt} = u_i = \frac{\partial \psi}{\partial y_i} = -\frac{1}{2\pi} \sum_{j \neq i}^{N} \Gamma_j \frac{(y_i - y_j)}{|\mathbf{x} - \mathbf{x}_j|^2},
$$

$$
\frac{dy_i}{dt} = v_i = -\frac{\partial \psi}{\partial x_i} = \frac{1}{2\pi} \sum_{j \neq i}^{N} \Gamma_j \frac{(x_i - x_j)}{|\mathbf{x} - \mathbf{x}_j|^2}. \tag{2.108}
$$

When examining Eq. (2.108), it is not too difficult to notice the intriguing fact, namely, that they have the form of Hamilton's equations, where the Hamiltonian function is given by

$$
\mathscr{H} = -\frac{1}{4\pi} \sum_{\substack{i,j \\ i \neq j}} \Gamma_i \Gamma_j \log |\mathbf{x}_i - \mathbf{x}_j|. \tag{2.109}
$$

The plane coordinate y_k is proportional to the canonically conjugate momentum of x_k. Actually, they have to be multiplied by $\sqrt{\Gamma_k}$. We have then the Hamilton equations

$$
\Gamma_i \frac{dx_i}{dt} = \frac{\partial \mathscr{H}}{\partial y_i} \quad \text{and} \quad \Gamma_i \frac{dy_i}{dt} = -\frac{\partial \mathscr{H}}{\partial x_i}, \tag{2.110}
$$

with this pair clearly equivalent to Eq. (2.108). It is surprising, at least to the authors of this book, to find such an unusual Hamiltonian system arising from FD. In any case, the question if this system is integrable or not is certainly of interest. Since its discovery, the system of N point vortices (and the limit $N \to \infty$) have been extensively studied. We mention only a few significant results: systems of two or three vortices are integrable, while that of $N = 4$ displays typical characteristics of Hamiltonian chaos. References to deeper studies of this problem are listed in the *Bibliographical Notes* of this chapter. Also, in Chap. 5 we examine similarities and implications of these ideas to point vortices in geostrophic flows (Problem 5.10).

2.5.4 The Velocity Field, Derived from a Given Vorticity Field

We mentioned at the outset of this chapter that problems involving vortex flows do not always have a unique solution owing to the fact that many different velocity fields can produce the same vorticity profile. It turns out that under appropriate restricted conditions uniqueness is possible. Consider a given vorticity field $\omega(\mathbf{x})$, with the proviso that very far from the origin, that is, for a large enough $r \equiv |\mathbf{x}|$, the vorticity decays to zero faster than $1/r^2$. We wish to claim that a solenoidal (a synonym for "divergence-free") velocity field given by

$$\mathbf{u}(\mathbf{x}) = \frac{1}{4\pi} \int_{\mathscr{V}_L} \frac{(\mathbf{x}' - \mathbf{x}) \times \omega(\mathbf{x}')}{|\mathbf{x} - \mathbf{x}'|^3} d^3 x', \tag{2.111}$$

where the volume $\mathscr{V}_L$ is large enough, so as to satisfy the requirement posed above. To show this, define

$$\mathbf{V}(\mathbf{x}) = \frac{1}{4\pi} \int_{\mathscr{V}_L} \frac{\omega}{|\mathbf{x} - \mathbf{x}'|} d^3 x', \tag{2.112}$$

Posit now $\mathbf{u} = \nabla \times \mathbf{V}$. Operating with $\nabla \times$ on (2.112) leads to

$$4\pi \mathbf{u}(\mathbf{x}) = \int_{\mathscr{V}_L} \left(\nabla_{\mathbf{x}} \frac{1}{|\mathbf{x} - \mathbf{x}'|} \right) \times \omega d^3 x' = -\int_{\mathscr{V}_L} \left(\nabla_{\mathbf{x}'} \frac{1}{|\mathbf{x} - \mathbf{x}'|} \right) \times \omega d^3 x'$$

$$= -\int_{\mathscr{V}_L} \left(\frac{\mathbf{x} - \mathbf{x}'}{|\mathbf{x} - \mathbf{x}'|^3} \right) \times \omega d^3 x', \tag{2.113}$$

which can be shown using rules of vector calculus, see Problem 2.23. Thus $\mathbf{u}$ as given in formula (2.111) satisfies $\nabla \times \mathbf{V}$, as required. The proof of its uniqueness follows, as usual in such problem, from assuming that there is some other $\mathbf{u}' = \nabla \times \mathbf{V}$. So $\mathbf{u} - \mathbf{u}'$ must also be solenoidal and vanish for a far enough boundary. But it must also be true that $\nabla \times (\mathbf{u} - \mathbf{u}') = 0$ and also both velocity vectors and therefore their difference are gradients of a scalar potential $\mathbf{u} - \mathbf{u}' = \nabla \phi$, so $\nabla^2 \phi = 0$ is implied. The only bounded solution for this potential satisfying all the above conditions is $\phi =$ const, and hence the desired uniqueness follows.

2.5.5 The Rankine Vortex

Consider a simple one-dimensional example of ideal fluid flow with a region of nonzero vorticity, called *forced vortex*, surrounded by a region with circular rotation but with zero vorticity—such a vortex is called *free*. Naturally, it should be possible to have compound vortices, e.g., forced up to some cylindrical radius and free above it. Such a compound vortex is known as the *Rankine vortex*.

A simple example of a Rankine vortex is what is called the bathtub vortex (for obvious reasons), but it should be remembered that the solution presented here is just a good approximation of a real vortex of this type. Consider a constant density fluid in a container and let the function $z = H(r)$ be the height of the free surface of the fluid, which we describe in cylindrical polar coordinates (r, φ, z). Implicit in the form of H is an axial symmetry assumption of the surface. The body force, due to gravity, has the form $\mathbf{b} = -g\hat{\mathbf{z}}$ and the external pressure is a constant P_0 say. Let us formally develop the velocity field of the Rankine vortex using a stream function formulation. We assume that the flow has no vertical velocity in steady state. In that event, the equations governing the motion of the fluid are two-dimensional and, hence, are given by Eqs. (2.101)–(2.102). Let a be the radius of the "sink" creating a tube of radius a above it. We assume that the time-independent vorticity contained inside this circular patch is also spatially constant and given by $\zeta(r, \varphi) = 2\Omega$. Thus we write

$$\zeta(r, \varphi) = \begin{cases} 2\Omega, & r < a, \\ 0, & r \geq a, \end{cases} \tag{2.114}$$

remembering that Ω is a constant. Since we assume no variations with respect to the polar coordinate, the vorticity equation is given by

$$\frac{1}{r}\frac{d}{dr}\left(r\frac{d\psi}{dr}\right) = -\zeta = -\begin{cases} 2\Omega, & r < a, \\ 0, & r \geq a. \end{cases} \tag{2.115}$$

For the sake of this formal development, let us assume that the velocity field is zero at $r = 0$. Thus the solution to the above equation for the stream function satisfying the condition that velocity field decays as $r \to \infty$ is given by

$$\psi = \begin{cases} A - \frac{1}{2}\Omega r^2, & r < a, \\ B\ln r, & r \geq a, \end{cases} \tag{2.116}$$

where A and B are unknown constants to be determined by applying a suitable matching procedure at the transition $r = a$. In particular, we require that the stream function and its first derivative with respect to r match at $r = a$. The latter condition is the same as requiring that the circular velocities match each other at the surface, and see below. Thus the conditions become

$$A - \frac{1}{2}\Omega a^2 = B\ln a, \qquad \Omega a = B/a. \tag{2.117}$$

The solution to the above shows that $B = \Omega a^2$ and $A = \Omega a^2 \left[\ln(ae^{1/2})\right]$.

In polar coordinates, the radial and azimuthal velocities are given in Eq. (2.77). Given the symmetry in the problem $\partial_\varphi \mapsto 0$, it comes as no surprise that the radial velocity is zero, $u_r = 0$, and that the azimuthal component of the velocity field is given by

$$u_\varphi = \begin{cases} \Omega r & r < a, \\ \Omega\left(a^2/r\right) & r \geq a. \end{cases} \tag{2.118}$$

Thus, here Ω is the angular velocity of the vortex core and it gives the solid body rotation rate of the tube. The factor 2Ω appearing in the designation of the vorticity of the patch is a common one and will reappear once again when we speak of Coriolis effects and the physics of fluid flows in globally rotating reference frames (Chap. 5). Note here the explicit meaning of the continuity of the radial gradient of ψ: without it the azimuthal velocity field u_φ would show a jump at $r = a$. Because the vorticity outside the vortex tube is zero, i.e., $\zeta = 0$ for $r \geq a$, the external flow is irrotational and we may use the appropriate version of Bernoulli's theorem, in particular, case 3b of the steady detailed in Sect. 2.3.5, to construct the pressure field externally:

$$P = P_0 - \frac{\rho\Omega^2 a^4}{2r^2} - \rho g z, \qquad (2.119)$$

for $z < H$, where H is the height of the free surface and we assume $H \to 0$ for $r \to \infty$. The free surface for $r \geq a$ can thus be expressed as

$$H(r) = -\frac{\Omega^2 a^4}{g} r^{-2}.$$

Inside the vortex core, we actually have very simple equations of motion

$$\frac{1}{\rho}\frac{\partial P}{\partial r} = \Omega^2 r, \qquad \frac{1}{\rho}\frac{\partial P}{\partial z} = -g, \qquad (2.120)$$

which can be integrated to give

$$\frac{1}{2}\Omega^2 a^2 - gz + C = P_c, \qquad (2.121)$$

where P_c is the pressure in the core. Now on the cylinder $r = a$ for $z < H$, we should require that $P_c = P$. This gives the value of the constant

$$C = \frac{P_0}{\rho} - \Omega^2 a^2$$

and substituting it in Eq. (2.121) we get

$$P_c = P_0 - \rho\Omega^2 a^2 \left(1 - \frac{r^2}{2a^2}\right) gz.$$

The expression for the surface follows

$$H(r) = \begin{cases} -\frac{a^4\Omega^2}{2} r^{-2} & \text{if } r \geq a, \\ \frac{a^2\Omega^2}{g}\left(\frac{r^2}{2a^2} - 1\right) & \text{if } r < a. \end{cases} \qquad (2.122)$$

Remark: Is du_φ/dr continuous at $r = a$? Is this allowed? What does it mean? Offer a physical rationalization of this feature by taking into account that all fluids in nature are viscous, if however slightly. We also urge the reader to tackle Problem 2.25, which is another important example.

2.5.6   Tumbling Kirchhoff–Kida Vortices

In Sect. 2.5.5, we discussed a simple case of the Rankine vortex, which is a circular patch of constant vorticity $\zeta = 2\Omega$. The flow in its interior rotates like a solid body ($u_\varphi \propto r$), while the exterior flow shows a $1/r$ drop off in u_φ. All streamlines are nested circles with a common origin. Such vortices are constructed assuming there is no slippage on the boundary of the vortex. Because the globally integrated vorticity field is nonzero, it comes as no surprise that the total circulation is a constant for any closed path containing the interior patch (of surface area S_p, say) outside of which there is no vorticity, that is, $\oint \mathbf{u} \cdot \mathbf{dl} = S_p \zeta$.

   In this final section of the chapter, we provide a more sophisticated calculation involving the response of an elliptical patch of vorticity. Imagine taking a circular vortex with constant interior vorticity like the Rankine vortex, say, and stretching it into an ellipse and then letting it go. This stretched elliptical patch will tumble in place with a constant frequency $\dot{\varphi}$. This setup is an example where an asymptotic approximation analytical solution can be constructed for a relatively complex arrangement. It is the purpose of this section to develop this solution by highlighting the techniques and procedures used for the calculation. Aside from providing a platform for developing deeper intuition on vortical flows, such vortices are often used as time-dependent test-case solutions to assess the performance of numerical experiments. Henceforth, we shall refer to this vortex solution as the Kirchhoff–Kida vortex (KK vortex for short). The original treatise on this problem is actually attributed to both H. von Helmholtz and G. Kirchhoff, while its importance was recognized and expanded upon, to include external strain and shear, only recently (in 1981) by S. Kida. It is an important construction in developing an understanding of what happens when vortices interact beyond the simpler circular assumption. It is also interesting from a mathematical point of view since it makes use of elliptical coordinates to construct the full answer. One of the purposes of this discussion is to serve as an introduction to this often overlooked, yet powerful, asymptotic method of solution.

   Thus consider a two-dimensional, inviscid, constant density fluid. The equations governing the vorticity of such a system is given in Eqs. (2.101)–(2.102). The KK vortex is the solution in this setting giving the behavior a patch of constant vorticity, $\zeta \equiv \Lambda = \text{const.}$ contained within an ellipse of major and minor axes of size a and b, respectively. Without loss of generality we assume $a > b$. The ellipse, in turn, tumbles in place with a constant frequency $\dot{\varphi}$ given by

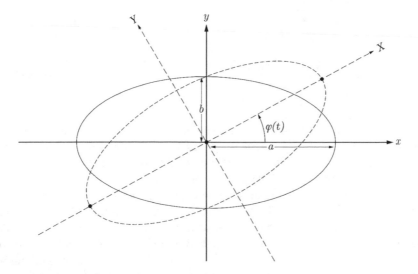

Fig. 2.7 Elliptic rotating patch, a diagram

$$\dot{\varphi} = \frac{\Lambda ab}{(a+b)^2}. \tag{2.123}$$

The discussion in the rest of this section will center on the derivation of the relationship in (2.123), which might perhaps seem innocuous enough; however as we shall see, its development is subtle. Let the laboratory frame coordinates be given by x and y. Now suppose that in a frame rotating with an angle as a function of time $\varphi(t)$, the patch of vorticity of interest appears fixed. Call these coordinates in the rotating frame X, Y (see the diagram in Fig. 2.7). The two coordinates relate to one another according to

$$X = x\cos\varphi(t) + y\sin\varphi(t), \quad Y = -x\sin\varphi(t) + y\cos\varphi(t). \tag{2.124}$$

While the points X and Y are fixed in the rotating frame, they appear to have a time dependence in the laboratory frame with corresponding time derivatives

$$\dot{X} = \dot{\varphi}Y, \quad \dot{Y} = -\dot{\varphi}X, \quad \dot{\varphi} \equiv \frac{d\varphi}{dt}, \tag{2.125}$$

where, as usual, over-dots denote ordinary derivatives with respect to time. However, do note that this coordinate transformation is area preserving as this will prove beneficial a little further on. In the rotating frame, the shape of the ellipse is fixed in time. We shall call the ellipse shape, S, by using the name surface function. In the laboratory coordinates, it is expressed by the relation $S(x, y, t) = 0$ and in this frame the third derivative in the above Equation is relevant:

$$
S(x,y,t) = \frac{X^2}{a^2} + \frac{Y^2}{b^2} - 1
$$

$$
= \frac{\left[x\cos\varphi(t) + y\sin\varphi(t)\right]^2}{a^2} + \frac{\left[-x\sin\varphi(t) + y\cos\varphi(t)\right]^2}{b^2} - 1 = 0.
$$

$$(2.126)$$

Our objective will be to follow the Lagrangian motion of the boundary over time.[2] The solution developed by Kirchhoff makes the important ansatz that boundary points of the patch rotate with constant rotational velocity, i.e., $\dot{\varphi}$ = const. We shall return to this during the matching procedures below.

In order to ensure that the developed solution is self-consistent, the solution procedure must go through the following three stages similar to what was done for the Rankine vortex. Here the method of *matched asymptotic expansions* will be used. In Chap. 3, this method will be applied to a simpler case and we encourage the reader (and teacher) to study the self-contained Sect. 3.5.1 before delving into the following lengthy calculation employing matched asymptotic expansions. As references to the technique we find the books of Nayfeh, Van Dyke, and Kevorkian and Cole (references [21, 22] and [23] of Chap. 4) to be the best. The above-mentioned three stages are:

1. Calculate the solution to the vorticity and stream function in the interior and, furthermore, have motion of the surface be consistent with the interior solution. We hereafter refer to this as the derivation of the *interior solution*.
2. Develop the solution to the stream function exterior to the rotating patch. This will entail calculating the stream function in elliptical coordinates. It is referred to as the derivation of the *exterior solution*.
3. Match the stream function and its derivative normal to the elliptical surface at the location of the surface of the elliptical vortex patch.

To give the different regions symbolic designations, we refer to the interior region as I and the exterior region as E. We symbolically reference the surface separating the interior from the exterior as ∂S, reminding ourselves that the equation describing the boundary of this surface is given by $S(x,y,t) = 0$ as expressed in Eq. (2.126).

1. *Interior solution.*
   The vorticity in the interior is constant and given by Λ. The equation for the stream function in region I is

$$
\zeta = \Lambda = -\left(\frac{\partial^2 \psi}{\partial x^2} + \frac{\partial^2 \psi}{\partial y^2} \right).
$$

$$(2.127)$$

---

[2]A more pedagogic discussion of the mathematical formulation of boundaries and how they are formally tracked will be given in Chap. 4, in and around the discussion surrounding equation (4.41).

However, for reasons which will become self-evident below, it will prove easier for us to proceed with the calculation in the X, Y coordinate frame. Now, because the coordinate transformation from $(x, y) \mapsto (X, Y)$ is area preserving, the Laplace operator also transforms preserving its form in both coordinate representations. This may be verified by applying the coordinate transformation in Eq. (2.124) directly to Eq. (2.127). The result becomes

$$\zeta = \Lambda = -\left(\frac{\partial^2 \psi}{\partial X^2} + \frac{\partial^2 \psi}{\partial Y^2}\right). \tag{2.128}$$

The above equation has the particular solution

$$\psi = -\tfrac{1}{2}\Lambda\left(AX^2 + BY^2\right), \tag{2.129}$$

with the, as yet, unspecified constants A and B. The solution in Eq. (2.129) is relatively general and, thus, in order to ensure that $-\nabla^2 \psi = \Lambda$, it follows that A and B must relate to each other according to

$$A + B = 1. \tag{2.130}$$

Now, u and v are the velocity components expressed in the (x, y) coordinate system; thus we know that since this is a two-dimensional incompressible fluid,

$$u = \frac{\partial \psi}{\partial y}, \qquad v = -\frac{\partial \psi}{\partial x}. \tag{2.131}$$

If we reference the corresponding velocities in the (X, Y) coordinate system by U and V, respectively, they are given by

$$U = \frac{\partial \psi}{\partial Y} = -BY, \qquad V = -\frac{\partial \psi}{\partial X} = AX. \tag{2.132}$$

We will make explicit use of the above shortly.
As we have posited, points on the boundary S always remain on the boundary S, even if material is moving along the boundary, and, as such, the boundary surface rotates with solid body rotation. This then means

$$\frac{DS}{Dt}\bigg|_{x,y \in \partial S} = 0, \qquad \Longrightarrow \qquad \frac{\partial S}{\partial t} + u\frac{\partial S}{\partial x} + v\frac{\partial S}{\partial y} = 0. \tag{2.133}$$

See also the future discussion on Eq. (4.41). We proceed by individually assessing the terms appearing in Eq. (2.133). By working in the transformed coordinate frame (X, Y) and making use of Eq. (2.125), it follows that

$$\frac{\partial S}{\partial t} = -\dot{\phi}(a^2 - b^2)XY. \tag{2.134}$$

The remaining two terms in Eq. (2.133) can be transformed from the (x,y) coordinate system to the (X,Y) coordinate system showing that the entire term form is exactly preserved, i.e.,

$$u\frac{\partial S}{\partial x} + v\frac{\partial S}{\partial y} = \frac{\partial \psi}{\partial y}\frac{\partial S}{\partial x} - \frac{\partial \psi}{\partial x}\frac{\partial S}{\partial y} = \frac{\partial \psi}{\partial Y}\frac{\partial S}{\partial X} - \frac{\partial \psi}{\partial X}\frac{\partial S}{\partial Y} = U\frac{\partial S}{\partial X} + V\frac{\partial S}{\partial Y}. \tag{2.135}$$

The above is none other than the statement that the Poisson bracket of two scalar functions is conserved in an area preserving coordinate transformation (see Problem 2.26). Thus it follows that

$$u\frac{\partial S}{\partial x} + v\frac{\partial S}{\partial y} = (Aa^2\Lambda - Bb^2\Lambda)XY. \tag{2.136}$$

Reassembling terms we find

$$\left.\frac{DS}{Dt}\right|_{x,y\in\partial S} = 0 = \left[Aa^2\Lambda - Bb^2\Lambda - \dot{\phi}(a^2 - b^2)\right]XY, \tag{2.137}$$

which is satisfied only if

$$\dot{\phi} = \Lambda\frac{Aa^2 - Bb^2}{a^2 - b^2}. \tag{2.138}$$

At this stage the solution is only partially complete. Equations (2.130) and (2.138) are two relationships for the three unknowns A, B, and $\dot{\phi}$. To proceed we must determine the solution in the external region E.

2. *Exterior solution.*

The exterior solution is one where the vorticity is zero. We have therefore to solve

$$\frac{\partial^2 \psi}{\partial x^2} + \frac{\partial^2 \psi}{\partial y^2} = 0. \tag{2.139}$$

While this looks straightforward in principle, the solution of this Laplace equation has to satisfy the condition that the velocity field decays to zero as the polar radius $r = \sqrt{x^2 + y^2} \to \infty$. The situation is further complicated by the requirement that whatever solution we develop in this exterior region, it must smoothly match the stream function of the interior solution on the elliptical boundary ∂S. To facilitate this, we should transform this problem into one in elliptical coordinates (see Fig. 2.8). Let us set up the exterior solution in terms of X and Y (going into this frame of reference does not affect the mathematical

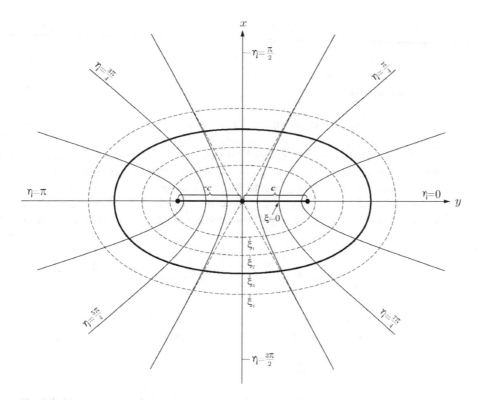

Fig. 2.8 Elliptical coordinates

form of the solution) and then transform into elliptical coordinates. We now briefly define elliptical coordinates (ξ, η) through

$$X \equiv c \cosh \xi \cos \eta, \qquad Y \equiv c \sinh \xi \sin \eta. \qquad (2.140)$$

The constant c will be chosen momentarily. Four important features should be observed at the outset, namely, that (see Problem 2.27)

- Constant values of η trace out pairs of hyperbolae
- Constant values of ξ trace out ellipses
- $\xi \to \infty$ is the same limit as $r \to \infty$ in polar coordinates
- The elliptical angle η asymptotes to the angle coordinate (φ) in polar coordinates as $r \to \infty$

Because in these coordinates constant values of ξ correspond to surfaces of ellipses, we can determine the value of c so as to uniquely define a family of ellipses containing the one of interest to us. Specifically, an ellipse family will share the same major to minor axis ratio and this is determined by choosing the correct value of c. Once c is chosen (and our ellipse family is determined), we can determine the value of $\xi = \xi_0$ that uniquely identifies the boundary associated with the vortex patch defined in Eq. (2.126). We start by assuming ξ_0 is known in order to determine c. Once c is determined, we can solve for ξ_0. Thus, we suppose that the coordinates X, Y corresponding to the boundary of the elliptical patch is

$$X \equiv c \cosh \xi_0 \cos \eta, \qquad Y \equiv c \sinh \xi_0 \sin \eta. \qquad (2.141)$$

In order for these values of X and Y to describe the coordinates of the vortex patch boundary, we should arrange X and Y found in Eq. (2.141) in such a way as to be able to directly compare it to the functional form for the boundary found in Eq. (2.126). Using trigonometric identities, we relate the coordinates of the boundary in Eq. (2.141) directly to the equation of the boundary in Eq. (2.126) and this gives

$$\frac{X^2}{c^2 \cosh^2 \xi_0} + \frac{Y^2}{c^2 \sinh^2 \xi_0} = 1 = \frac{X^2}{a^2} + \frac{Y^2}{b^2}. \qquad (2.142)$$

The above correspondence will be true only if the coefficients of the X^2 and Y^2 terms are equal to one another on either side of the equation, which means

$$c \cosh \xi_0 = a, \qquad c \sinh \xi_0 = b, \qquad \implies \qquad c^2 = a^2 - b^2. \qquad (2.143)$$

In this way, we have determined the value of c that selects the family of ellipses appropriate to our vortex patch. Now we can use this to determine the value of ξ_0 identifying the boundary of the vortex patch of interest to us here. This is done via

$$\cosh \xi_0 + \sinh \xi_0 = e^{\xi_0} = \frac{a+b}{c}, \qquad \implies \qquad \xi_0 = \log\left(\frac{a+b}{c}\right). \qquad (2.144)$$

Conveniently, the Laplace equation of the stream function expressed in elliptical coordinates is

$$\frac{\partial^2 \psi}{\partial \xi^2} + \frac{\partial^2 \psi}{\partial \eta^2} = 0. \qquad (2.145)$$

To see why this is so, see Problem 2.27. The general solution which shows decay as $|X|, |Y| \to \infty$, which is the same as $\xi \to \infty$, is given by an integer sum over n of individual solutions, i.e.,

$$\psi = A_{ext} + C_0 \xi + \sum_{n=1}^{\infty} e^{-n\xi} (C_n \cos n\eta + D_n \sin n\eta), \qquad (2.146)$$

with A_{ext}, C_n, D_n unknown coefficients determined from imposing matching conditions on the boundary ∂S. With hindsight (see below) we know that the only nonzero value of the coefficients are C_2 and A_{ext}; thus the external solution is given by

$$\psi(x, y \in E) = A_{ext} + C_2 e^{-2\xi} \cos 2\eta, \qquad (2.147)$$

where all the rest of the C_n are zero. The symbol $\psi(X, Y \in E)$ is the stream function for values of X, Y contained in the exterior region E. We have developed the exterior solution but what remains is the determination of the two unknown constants C_2 and A_{ext}. This is done next.

3. *Matching.*

We have solutions for the stream function in regions I and E as found in Eqs. (2.129) and (2.147) with a number of undetermined constants including $A, B,$ and $\dot{\phi}$ as well as C_2 and A_{ext}. A partial set of relationships exist between these constants found in Eqs. (2.130) and (2.138). However, to complete the solution we must uniquely ascertain the values of the five unknown constants in terms of $a, b,$ and Λ. This is done by matching the stream functions and their derivatives at the boundary ∂S. Matching of the solutions means enforcing, first, that the stream functions of both the inner and outer solutions match at the boundary of the ellipse and, second, that the gradients of the two solutions in the direction normal to the ellipse boundary, ∂S, match each other as well. The matching procedure must be done in a transparent way. The best strategy for this is to reexpress the solution in the inner region in terms of the elliptical coordinate system that was used to express the solution in the outer region. Once transformed in this way, direct comparisons of the stream functions of both regions, as well as their normal derivatives, can be done with just little complication.

Starting with the stream function in the interior zone as expressed in Eq. (2.129), we can transform it from the X, Y variables into the elliptical variables η, ξ by explicitly applying the transformation in Eq. (2.141). This yields

$$\psi(X, Y \in I) = -\tfrac{1}{2}\Lambda (AX^2 + BY^2) = -\tfrac{1}{2}\Lambda \left(c^2 A \cosh^2 \xi \cos^2 \eta + c^2 B \sinh^2 \xi \sin^2 \eta \right), \qquad (2.148)$$

where the symbol $\psi(X, Y \in I)$ is the stream function for values of X, Y contained in the interior region I. The convenient utility of transforming into these elliptical coordinates is that the vortex patch boundary, ∂S, is at $\xi = \xi_0$ (for all η). As such, the functional form of the stream function as one approaches this boundary from inside is easily represented by

$$\psi(X, Y \to \partial S^-) = -\tfrac{1}{2}\Lambda \left(c^2 A \cosh^2 \xi_0 \cos^2 \eta + c^2 B \sinh^2 \xi_0 \sin^2 \eta \right)$$

$$= -\tfrac{1}{4}\Lambda \left[a^2 A + b^2 B + (a^2 A - b^2 B) \cos 2\eta \right]$$

$$= -\tfrac{1}{4}\Lambda \left(a^2 A + b^2 B \right) - \tfrac{1}{4}\dot{\phi} c^2 \cos 2\eta, \qquad (2.149)$$

where the symbolic notation $\psi(X, Y \to \partial S^-)$ is meant to represent the value of the stream function as one approaches the boundary of the stream function from points contained inside the ellipse, hence the minus superscript appearing on ∂S. In arriving at the second line of the above equation, we have made use of the definitions of c and ξ_0 found in Eq. (2.143) while also making use of some basic trigonometric double-angle identities. In writing the last line, we have made explicit use of (2.138) to bring out the expression of $\dot{\varphi}$ in the coefficient of $\cos 2\eta$.

The value of the exterior stream function, Eq. (2.147), as one approaches the boundary ∂S from outside is

$$\psi\left[(\xi, \eta) \to \partial S^+\right] = \psi(\xi \to \xi_0^+, \eta) = A_{\text{ext}} + C_2 e^{-2\xi_0} \cos 2\eta, \qquad (2.150)$$

where we have used the same notational convention (i.e., the expression $\psi[\xi, \eta \to \partial S^+]$) to express the corresponding value of the stream function on the boundary ∂S as viewed from within the exterior region E. The two stream function solutions on the boundary, Eqs. (2.149) and (2.150), are equivalent if the constants match each other, i.e., $A_{\text{ext}} = -\frac{1}{4}\Lambda\left(a^2 A + b^2 B\right)$ and if the coefficients in front of the function $\cos 2\eta$ are equal. The second of these implies that

$$C_2 e^{-2\xi_0} = -\frac{1}{4}\dot{\varphi} c^2 \qquad \Longrightarrow \qquad C_2 = -\frac{\dot{\varphi}}{4}(a+b)^2. \qquad (2.151)$$

In getting the last expression above, we have made use of the alternate form of ξ_0 found in Eq. (2.144). Thus, the exterior solution may be reexpressed as

$$\psi\left[(X, Y) \in E\right] = -\frac{\dot{\varphi}}{4}(a+b)^2 e^{-2\xi} \cos 2\eta - \frac{1}{4}\Lambda\left(a^2 A + b^2 B\right). \qquad (2.152)$$

Note that had we retained all of the unknown coefficients (i.e., C_n and D_n) of the general external solution for the stream function ψ found in Eq. (2.146), then the step we have just executed would have demanded that they all be set to zero except, obviously, for A_{ext} and C_2. This justifies our earlier dropping of them in our lead up to Eq. (2.147).

Our next task involves matching the normal gradient of the interior and exterior solutions at the boundary ∂S. We note that this normal gradient represents the tangential flow around the vortex patch and by assessing this differentiability condition across the ellipse boundary we are assessing the circulation profile on it as well. The tangent flow is easily identified in elliptical coordinates because it corresponds to gradients of ψ along the direction of ξ evaluated at the surface $\xi = \xi_0$. Thus, the procedure is similar to what we just performed for matching the stream functions: take the partial derivative with respect to ξ of the interior stream function solution expressed in elliptical coordinates and evaluate it as one approaches the boundary from inside, i.e., as $\to \partial S^-$. The result of the last calculation should be set equal to the same partial derivative with respect to ξ of the exterior solution as evaluated as one approaches the boundary from outside (i.e., as $\to \partial S^+$). Thus the matching condition is nothing but

$$\frac{\partial \psi\left[(X,Y) \in \mathrm{I}\right]}{\partial \xi}\bigg|_{(X,Y) \to \partial S^-} = \frac{\partial \psi\left[(\xi,\eta) \in \mathrm{E}\right]}{\partial \xi}\bigg|_{(X,Y) \to \partial S^+}. \tag{2.153}$$

The matching procedure will require the identification of the coefficients appearing in front of the remaining trigonometric functions appearing in the above equation. We leave out the details and show the resulting form after some algebra (see Problem 2.28) to simplify into the following final condition:

$$\frac{1}{2}\dot{\varphi}(a+b)^2 e^{-2\xi_0} = \frac{\Lambda}{2}c^2(A-B)\cosh\xi_0 \sinh\xi_0. \tag{2.154}$$

We now have three conditions: Eqs. (2.130), (2.138), and (2.154) which, together with the relationship in Eq. (2.143), lead to a unique solution in which all constants of the problem can be expressed in terms of Λ, a, and b. We get therefore,

$$A = \frac{b}{a+b}, \qquad B = \frac{a}{a+b}, \tag{2.155}$$

and the value of $\dot{\varphi}$ as is required in (2.123),

$$\dot{\varphi} = \frac{\Lambda ab}{(a+b)^2}.$$

We have presented here a solution for an elliptical vortex patch. This solution is highly idealized in the sense that it is constructed of only a constant vorticity patch. Nevertheless, the Kida solution has been extended to incorporate effects of ambient flow shear and strain and such extensions can be studied using tools from dynamical systems theory. Such kind of calculation is presented in reference [15] in the *Bibliographical Notes* of this chapter.

Problems

2.1.
Show that in a flow of a fluid having a constant density, the following equation for the vorticity holds:

$$\frac{D\omega}{Dt} = (\omega \cdot \nabla)\mathbf{u} + \nu\nabla^2\omega. \tag{2.156}$$

What can you say about the evolution of vorticity in a two-dimensional flow, in the plane (x_1, x_2), say, in such a flow?

2.2.
Show that

$$\frac{\omega}{\rho} \cdot \frac{D}{Dt}(\nabla F) = \frac{\omega}{\rho} \cdot \nabla \left(\frac{DF}{Dt} \right) - \left[\left(\frac{\omega}{\rho} \cdot \nabla \right) \mathbf{u} \right] \cdot \nabla F, \qquad (2.157)$$

where we are writing out here the terms used in the proof of the Ertel theorem—in Eqs. (2.17) and (2.18). Hint: it seems that the easiest way to do it is by writing out all the vectors in their component form.

2.3.
Prove the Ertel theorem (2.15), but instead of the condition that F is a material invariant, it is a function of P and ρ only.

2.4.
Derive Eq. (2.21) from (2.20). Are there any restrictions on α_j needed for the proof?

2.5.
Show that in a steady, that is, $\partial_t = 0$, and an inviscid flow the following relation holds:

$$\omega \times \mathbf{u} = -\nabla \mathcal{B} + T \nabla s, \qquad (2.158)$$

where $\mathcal{B}$ is the Bernoulli function, defined in (1.71).

Assume now that the Bernoulli function is a constant (as it sometimes happens, see Sect. 2.3.5) and the flow is homentropic. Thus $\omega \times \mathbf{u} = 0$. The meaning of the case $\omega = 0$ is obvious and the case $\mathbf{u} = 0$ is trivial. The case where the vorticity and velocity are parallel, $\omega \| \mathbf{u}$, is referred to as *Beltrami* flow. Give an example of Beltrami flow, which is a swirl flow in a pipe. Hint: use of cylindrical coordinates simplifies the derivation.

2.6.
In the example of a uniformly rotating flow, worked out in the previous chapter, we got the result that isobaric surfaces are parabolas of revolution with a minimum at the center (see Eq. (1.59)). Now, with the Bernoulli theorems at our disposal we may reason in the following way: Bernoulli's theorem (case 2) gives

$$\frac{1}{2}u^2 + \frac{P}{\rho_0} + gz = \text{const}, \qquad (2.159)$$

leading to the following shape of surfaces of constant pressure:

$$z(x, y) = \text{const} - \frac{\Omega^2}{2g}(x^2 + y^2). \qquad (2.160)$$

This is in contradiction to Eq. (1.59) and seems wrong, the surface here has a maximum at the origin. Can you explain this?

2.7.
In our discussion of the de Laval nozzle we have concluded that a sonic point in the flow through the nozzle is in the throat. Is the converse also true, i.e., that at the throat there is always a sonic point? Is it possible that the flow through the de Laval nozzle is subsonic throughout its entirety? If yes, what will be the general shape of $A(x)$?

2.8.
Consider now the isothermal case of the accretion vs. wind problem, discussed in Sect. 2.3.6.2. Equation (2.48) now reads

$$(u - C)\frac{1}{u}\frac{du}{dr} = \frac{2C^2}{r}\left[1 - \frac{GM}{2Cr}\right], \qquad (2.161)$$

where C is the isothermal sound speed and is a constant. Discuss the Parker wind solution and show that such a solution starts at $r = 0$, accelerates through the sonic point at $r = r_s = (r_B)/2$, and goes out to infinity with ever increasing speed. Clearly then, in such an outflow problem, conditions at infinity cannot be specified. Rather than that, one can show that the density at the sonic point is given by $\rho_s = e^{3/2}\rho_\infty$ and the mass loss acquires then a critical value (express it in nondimensional units, as defined in the text).

2.9.
Show that inviscid adiabatic flow of a fluid with no body forces, around a body of linear dimension b executing co-linear oscillations of amplitude $a \ll b$ inside the fluid, is a potential flow.

2.10.
Assume that the absolute value of a body force is b. Show that for incompressibility to be a good approximation, the magnitude of the force has to be much smaller than some quantity related to the velocity change time and length scales. Interpret physically this result.

2.11.
Show that the correct solution of a potential, incompressible flow is the one whose kinetic energy is the smallest among all incompressible flows satisfying the given boundary conditions (as first shown by Lord Kelvin in 1849).

2.12.
We assume that a source at the origin of strength q in three dimensions satisfies $\nabla \cdot \mathbf{u} = q\delta(\mathbf{x})$, where $\delta\mathbf{x} = \delta(x)\delta(y)\delta(z)$ is the Dirac three-dimensional delta function, and the velocity, excluding the origin, is derivable from a potential $\mathbf{u} = \nabla\phi$. Show that the potential of this source flow is

$$\phi(\mathbf{x}) = -\frac{q}{4\pi}\frac{1}{r},$$

where r is the radial coordinate in spherical coordinates $r \equiv \sqrt{(x^2 + y^2 + z^2)}$. What would be the potential of a source q at $x = a/2, y = z = 0$, and a sink $-q$ at $z = -a/2, x = y = 0$ (a dipole). Find a simple expression for ϕ in this case in spherical coordinates for $r \gg a$.

2.13.
Using the general formula (2.69) show that the total kinetic energy and the total momentum of a constant density fluid in which a body of volume $\mathcal{V}_0$, having an arbitrary shape, is moving at velocity $\mathbf{U}$ and inducing a potential flow in the fluid are

$$E_{\text{kin}} = \frac{1}{2}\rho\left(4\pi\mathbf{A}\cdot\mathbf{U} - \mathcal{V}_0 U^2\right), \tag{2.162}$$

$$\mathbf{P} = \rho\left(4\pi\mathbf{A} - \mathcal{V}_0\mathbf{U}\right), \tag{2.163}$$

respectively, where $\mathbf{A}$ is the vector appearing in (2.69) and ρ is the fluid density.
Hints:

• Consider a large sphere of fluid containing the body and use the identity

$$u^2 = U^2 + (\mathbf{u}+\mathbf{U})\cdot(\mathbf{u}-\mathbf{U}).$$

• Show that the following angle integral, containing two constant vectors $\mathbf{B}$ and $\mathbf{C}$, is

$$\int(\mathbf{B}\cdot\hat{\mathbf{r}})(\mathbf{C}\cdot\hat{\mathbf{r}})d\Omega = \frac{4\pi}{3}\mathbf{B}\cdot\mathbf{C},$$

where $d\Omega$ is the solid angle and $\hat{\mathbf{r}}$ is the radial unit vector in spherical coordinates.

2.14.
(a) Consider the complex potential $w(z) = m\ln(z - z_0)$. Find the stream function and show that it represents a source in two dimensions located at $z_0 = x_0 + iy_0$. (b) If there is a sink of the same strength situated at $-z_0$, what is the complex potential? (c) Show that in the limit when the source and sink are very close to the origin, such a doublet (compare with a dipole) has a complex potential $w(z) = -\mu/z$, with μ complex. The strength of the doublet is $|\mu|$ and its axis points in the direction of the complex number is μ. Explain.

2.15.
Given the complex potential

$$w(z) = -i\frac{\Gamma}{2\pi}\log(z), \tag{2.164}$$

find the flow velocity. Hint: it will be easier to express it in circular polar coordinates, r and φ. Interpret it physically and give the physical meaning of the constant Γ.

2.16.

Consider a two-dimensional uniform, irrotational flow of fixed density having a speed $V\hat{x}$ at infinity, past a fixed cylinder of radius a. Show that the polar velocity components are

$$u_r = V\left(1 - \frac{a^2}{r^2}\right)\cos\varphi; \qquad u_\varphi = V\left(1 + \frac{a^2}{r^2}\right)\sin\varphi. \qquad (2.165)$$

Hint: use the Milne–Thomson theorem and assume that there is no circulation around the cylinder.

2.17.

Demonstrate the power of the complex potential method by showing that any analytic function of z that has an inverse (conformal mapping) generates a transformed complex potential and in this way streamlines are mapped to streamlines. Use the so-called Joukowski mapping $Z = z + c^2/z$ on a circle $z = ae^{i\varphi}$ with $a \geq c \geq 0$. What does the circle transform to? Using the results of the previous problem, sketch (do not perform all the calculations) the way of obtaining the solution to the transformed problem for the case in which the circulation along the body is 0.

2.18.

Let a two-dimensional harmonic (satisfying the Laplace equation) flow have a complex potential $f(z)$, analytic in the circle $|z| \leq R$. Show that if a circular cylinder of radius R is placed so that its circular cross section coincides with the circle $|z| \leq R$, the complex potential of the new flow is $w(z) = f(z) + f^*(R/z^2)$, where $*$ indicates complex conjugate. Can you think of a way to generalize this result to a three-dimensional case? The result is called Butler's sphere theorem.

2.19.

Consider a two-dimensional, dependent on x and y only, say, incompressible flow. Do not assume that the flow is steady and/or inviscid. Show that the following two relations hold

1.

$$\nabla^2\psi = -\zeta, \qquad (2.166)$$

where ψ is the stream function and $\zeta = \hat{z}\cdot\omega$, with $\hat{z} = \hat{x}\times\hat{y}$

2.

$$\frac{\partial\zeta}{\partial t} + J[\psi,\zeta] = \nu\nabla^2\zeta, \qquad (2.167)$$

where ν is the kinematic viscosity and we have used the shorthand notation $J[a,b] = \partial_x a\,\partial_y b - \partial_x b\,\partial_y a$.

This is called, as mentioned in the text, the *vorticity–stream function* formulation and is frequently useful in calculations of two-dimensional, incompressible, viscous flows. Note the ψ is determined up to a constant and one can show that the function is a constant on the boundary of the domain (so it is possible to choose it as 0 there).

2.20.
Show that in two dimensions, $\omega \cdot \nabla \mathbf{u} = 0$, i.e., ω is a scalar material invariant and specify under what conditions on the flow this is true.

2.21.
Derive, following von Kármán, the horizontal speed of two infinite parallel point vortex, spaced by δ, lines comprising a street, where the vertical distance between the lines is h and the vortices sequence on the lower line is shifted by $-\Delta/2$ with respect to the vortices on the upper line, using the following stages:

1. From Eqs. (2.106)–(2.107) for the vorticity and stream function of an N point vortex line with strengths Γ_i, find the complex velocity potential of the double line of point vortices (all separated by distance Δ on a line and all of the same strength $-\Gamma$)

$$
w_N = -\frac{i\Gamma}{2\pi} \sum_{k=-N}^{k=N} \log(z - k\Delta) = -\frac{i\Gamma}{2\pi} \log\left[\frac{\pi}{\Delta} z \prod_{k=1}^{N} \left(1 - \frac{1}{k^2\Delta^2}\right) z^2\right] + \text{const.}
$$

$$(2.168)$$

2. Use the identity

$$
\sin z \equiv z \prod_{k}^{\infty} \left(1 - \frac{1}{k^2\pi^2} z^2\right)
$$

$$(2.169)$$

to get at the limit $N \to \infty$

$$
w_\infty = \frac{i\Gamma}{2\pi} \log\left[\sin \frac{\pi}{a} z\right],
$$

$$(2.170)$$

and this can be repeated for the other line, as well.

3. After evaluating this limit, find the total velocity potential for a body moving at velocity $(U,0)$ and consider the velocity at the point $(0, h/2)$ for the system minus that of the vortex at this point. This should give, by symmetry, the speed of this vortex, relative to the body, V, say. You will have to evaluate a limit to get the result

$$
V = \frac{\Gamma}{2\Delta} \tanh\left(\pi \frac{h}{\delta}\right).
$$

$$(2.171)$$

2.22.
Prove the fourth Helmholtz theorem, as given in the text, regarding the possible termination of a vortex tube.

2.23.
Prove formula (2.113).

2.24.
The integral kernel

$$\frac{(\mathbf{y}-\mathbf{x}) \times \mathbf{J}(\mathbf{y})}{|\mathbf{x}-\mathbf{y}|^3}$$

used in formula (2.111) appears also in the *Biot–Savart* law in electromagnetic theory. What is then the vector $\mathbf{J}$? Explain.

2.25.
Consider a steady two-dimensional flow of an ideal fluid of constant density. Show that in this case a relation

$$\nabla^2 \psi = f(\psi), \tag{2.172}$$

where ψ is the stream function and f an arbitrary function, must exist. Take $f(\psi) = -k^2 \psi$ and try to solve the resulting equation for ψ in the disk $r \leq a$ whose velocity can match the velocity in $r > a$ that is the same as irrotational flow past a circular body of radius a. Hint: separating $\psi(r, \varphi) = B(r) \sin(\varphi)$ you will arrive at the ODE for B, Bessel equation. Take as solution the Bessel functions of order one. Can you explain the interpretation of the solution as a propagating vortex dipole?

2.26.
Suppose there are two functions $f(x, y)$ and $g(x, y)$ and, furthermore, suppose there exists a coordinate transformation given by Eq. (2.124), where $\phi(t)$ is a parameter. Show explicitly using the chain rule that the Poisson bracket is preserved in the transformation, meaning

$$\frac{\partial f}{\partial x}\frac{\partial g}{\partial y} - \frac{\partial g}{\partial x}\frac{\partial f}{\partial y} = \frac{\partial f}{\partial X}\frac{\partial g}{\partial Y} - \frac{\partial g}{\partial X}\frac{\partial f}{\partial Y}. \tag{2.173}$$

Argue that this is necessarily true for area preserving coordinate transformations.

2.27.
Starting with the transformation to elliptical coordinates (see Fig. 2.8 in the text)

$$X \equiv c \cosh \xi \cos \eta, \qquad Y \equiv c \sinh \xi \sin \eta.$$

(a) Prove that the locus of points for constant values $\xi = \xi_0$ trace out ellipses with major and minor axes a and b expressed through the formulae $a^2 = c^2 \cosh^2 \xi_0$ and $b^2 = c^2 \sinh^2 \xi_0$.
(b) Similarly, show that constant values of $\eta = \eta_0$ trace out hyperbolae satisfying the relationship $x^2/a^2 - y^2/b^2 = \pm 1$.

(c) Show that the elliptic angle η is equivalent to the angle coordinate in polar coordinates in the limit of very large ξ_0.

(d) Show that $0 \leq \eta < \pi$ uniquely corresponds to points in the upper two quadrants of the Cartesian plane, while $\pi \leq \eta < 2\pi$ corresponds to points in the lower two quadrants. Argue that all points $Y = 0$ and $X > c$ are identified with $\eta = 0$ and arbitrary ξ, while all points $Y = 0$ and $X < -c$ are associated with $\eta = \pi$ and arbitrary ξ.

(e) Show that the coordinate system has a coordinate representation ambiguity for values of $\xi = 0$. In particular, show that the chord along the $Y = 0$ axis connecting the points $\pm c$ cannot be uniquely assigned a value of η (two values of η coincide with the same point). Conclude that the coordinate system is only useful to represent points that do not contain this dangerous chord.

(f) Show that the Laplacian in these coordinates very easily converts from Cartesian coordinates:

$$\frac{\partial^2}{\partial X^2} + \frac{\partial^2}{\partial Y^2} \mapsto \frac{2}{c^2(\cosh 2\xi - \cos 2\eta)} \left(\frac{\partial^2}{\partial \xi^2} + \frac{\partial^2}{\partial \eta^2} \right) \qquad (2.174)$$

and conclude that the denominator in the last expression is not zero for all values of $\xi > 0$. Show this by explicitly putting in the coordinate transformation and working through all of the partial derivatives. Hint: work backwards starting from the above result.

2.28.
Complete the steps leading to (2.154).

Bibliographical Notes

Section 2.1

Here we focus on restricted flows, that is, on flows that an assumption (or set thereof) is simplifying the general equations. The general references on FD, mentioned in the beginning of the previous chapter, usually contain some material of special flows. In the next sections we add some specific bibliography on restricted flows, which may be quite useful as further reading.

Section 2.2

The vorticity concept is helpful in formulating the fluid dynamical equations in an alternative way, which is often more suitable to the understanding of certain physical aspects of the flow. In the following references, the reader can find a good discussion of such a formulation (due to mainly Crocco).

1. F.H. Shu, *The Physics of Astrophysics*, vol. II (University Science Books, Mill Valley, 1992)
2. P.A. Thompson, *Compressible Fluid Dynamics*. Advanced Engineering Series (Rensselaer, Troy, 1988)

Section 2.3

The Kelvin and Ertel theorems are discussed in

3. L.D. Landau, E.M. Lifshitz, *Fluid Mechanics*, 2nd edn. (Elsevier, Massachusetts, 2004)

Essentially all FD books, in particular, references [3] and [2], have some version(s) of the Bernoulli theorem. The following books, together with reference [1], include discussions and examples of application of this theorem to astrophysical flows:
4. C. Clarke, B. Carswell, *Principles of Astrophysical Fluid Dynamics* (Cambridge University Press, Cambridge, 2007)
5. S. Shore, *Astrophysical Hydrodynamics* (Wiley, New York, 2007) (a new edition of the 1993 Academic Press book)

Section 2.4

Many general books discuss irrotational flows; however, we point out some books on Laplace's equation in general, two of them ([9] and [10]) containing extensive chapters on the use of complex potential in two-dimensional flows.

6. P.M. Morse, H. Feshbach, *Methods of Theoretical Physics* (Feshbach Publishing, New York, 1981)
7. D.R. Bland, *Solutions of Laplace's Equation* (Routledge & Kegan Paul, London, 1961)
8. J.D. Jackson, *Classical Electrodynamics*, 3rd edn. (Wiley, New York, 1998)
9. C.S. Yih, *Fluid Mechanics* (West River Press, Ann Arbor, 1998)
10. P.K. Kundu, I.M. Cohen, *Fluid Mechanics*, 4th edn. (Elsevier, New York, 2008)

Section 2.5

Most books on FD contain discussions of vorticity and of vortices in different degrees of detail. More advanced aspects of vortices are discussed in considerable depth in the following references, which are recommended for further reading:

11. P.G. Saffman, *Vortex Dynamics* (Cambridge University Press, Cambridge, 1995)
12. H. Aref, Integrable, chaotic and turbulent vortex motion in two-dimensional flows. Ann. Rev. Fluid Mech. **15**, 345 (1983)
13. P.K. Newton, *The N-vortex Problem, Analytical Techniques* (Springer, New York, 2001)
14. Sir H. Lamb, *Hydrodynamics* (Dover, New York, 1945)
    The original papers of Kirchhoff are found in the reference to Lamb's book. The more modern study by S. Kida is found in his paper
15. S. Kida, Motion of an elliptic vortex in a uniform shear flow. J. Phys. Soc. Jpn. **50**, 3517 (1981)

Chapter 3
Viscous Flows

We fail!
But screw your courage to the sticking place,
And we'll not fail.

William Shakespeare (1564–1616); 'Macbeth' I, vii 59

3.1   Introduction

We hope it is both appreciated and understood that the nondimensional number indicating the importance of viscosity in a definite flow is the Reynolds number Re, as defined in Eq. (1.80). Viscous flows, for which Re is neither exceedingly large nor vanishingly small, are naturally more complicated than inviscid ones. Moreover, the class of solutions describing viscous flows is very different from those of the inviscid Euler equations. The Navier–Stokes equations are of the mixed kind, while the Euler equation is hyperbolic. Thus, it can be reasonably deduced that the limit $\mathrm{Re}^{-1} \to 0$ is a singular one. More about the singular limits is given in Sect. 3.5.1. In this chapter we offer a few examples of simple, analytically soluble viscous flows and discuss briefly their nature.

Perhaps the simplest example is the following. Assume a one-dimensional flow $\mathbf{u}(\mathbf{x},t) = u(y,t)\hat{\mathbf{x}}$, in which both the density and pressure are constant. It is easy to see that this is a solution of the two-dimensional Navier–Stokes equations, i.e., u satisfies

$$\frac{\partial u}{\partial t} = v \frac{\partial^2 u}{\partial y^2}, \tag{3.1}$$

where v is the kinematic viscosity. Keep in mind that the nonlinear advection term is automatically zero since there is no velocity component in the y direction while

© Springer Science+Business Media, LLC 2016

O. Regev et al., *Modern Fluid Dynamics for Physics and Astrophysics*,
Graduate Texts in Physics, DOI 10.1007/978-1-4939-3164-4_3

there is no dependency of the x-component of velocity in the x direction. The above expression is simply the well-known diffusion equation in which the quantity being diffused in the perpendicular (y) direction is the horizontal (x) momentum, per unit mass. So we readily understand that shear viscosity causes the transport of linear momentum, usually in a direction perpendicular to the momentum. Note that the bulk viscosity does not comply with this simple prescription but, as we have said before, we neglect the effects of bulk viscosity except when dealing with rarefied gases in extreme conditions. The approximate value of the kinematic viscosity coefficient, which depends on the temperature and is given here in the dimensional units (cm^2/s), is, e.g., 0.01 for water, 0.15 for air, and 7 for glycerine, at room temperature. Of primary interest is the *viscous time*, i.e., the typical timescale associated with momentum diffusion/transport over a length scale of interest L, but this depends on L, of course. The solution of equation (3.1) is well known. We do not deal with it here, because we do not want to get bogged down by the boundary conditions, but detailed solutions of this equation will be examined in Chap. 7. The general form of the viscous time, which can approximately be found from dimensional analysis, is $\tau_{\text{visc}} \sim L^2/\nu$. This will be useful in much of what follows. It would seem that all one needs to know is the value of τ_{visc} to assess the effects of viscosity on general flows, but this can be misleading. One interesting observation is that water rotating in a drinking glass ($L \sim 5$ cm) has a very long viscous time of the order of $\tau_{\text{visc}} \sim 2500$ s, suggesting that it should take about this long for the water to come to a standstill. But this is in contradiction with everyday experience as the swirling liquid in the glass actually comes to rest in under a minute. The solution to this puzzle lies in recognizing the emergence of boundary layers at the bottom of the glass and reevaluating the validity of the Helmholtz vortex theorems to such scenarios, the latter of which has already been mentioned. This particular problem of water swirling in a glass is further examined in Chap. 5.

We proceed with some of the aforementioned analytically tractable viscous flow solutions. The first two sections and their associated problems have become well-known paradigms of viscous flows because of the possibility of getting exact analytical solution for them. Unless otherwise stated, in all of the following examples we assume a steady flow of constant density.

3.2   Elementary Flows When the Governing Equations Are Linear

Some of these flows are exact solutions of the Navier–Stokes equations and contribute a lot to the understanding of various physical effects in viscous flows.

3.2.1 Plane Couette Flow

The idealized example is of two infinite parallel plates located at $y = 0$ and $y = h$, with a plane parallel flow between them whose properties are x and z independent. The flow is set in motion by the horizontal velocity of the sliding upper plate at constant velocity $U\hat{\mathbf{x}}$, while the lower plate is stationary. We assume that the fluid is incompressible and make the steady flow ansatz that

$$\mathbf{u}(\mathbf{x},t) = u(y)\hat{\mathbf{x}}. \tag{3.2}$$

Substitution of the above into the Navier–Stokes equations gives

$$0 = -\frac{1}{\rho}\frac{dP}{dy}\hat{\mathbf{y}} + v\frac{d^2u}{dy^2}\hat{\mathbf{x}}, \tag{3.3}$$

with the solution

$$P = \text{const} \quad \text{and} \quad \mathbf{u} = U\frac{y}{h}\hat{\mathbf{x}}, \tag{3.4}$$

which satisfies the boundary conditions of a viscous flow: parallel components of fluid velocity equal the boundary velocity, while normal components of the fluid velocity are zero. It is fairly easy to convince oneself that the upper plane exerts a force per unit area of $v\rho(U/h)\hat{\mathbf{x}}$ on the fluid, in other words, this is a horizontal stress, while the lower plate exerts on the fluid a force per unit area of the same magnitude but opposite in direction. Newton's third law gives the answer about the forces the fluid exerts on the plates.

3.2.2 Poiseuille Flow

We next consider the ubiquitous *pipe Poiseuille flow*. Planar Poiseuille flow exists as well—see the end of this section. Pipe Poiseuille flow takes place, as the name suggests, in cylindrical pipes and is often times used as a model for blood flow in a circular blood vessel, even though blood is not a simple viscous fluid but has a shear dependent kinematic viscosity (blood is sometimes known as a *shear-thinning* fluid since its kinematic viscosity weakens with increased shear); however, such generalized Newtonian or non-Newtonian fluids are not discussed in this book. We should remark, however, that the name of G.H.L. Hagen is also associated with pipe Poiseuille flow, since his research on such flows was done independently of J.L.M. Poiseuille and at about the same time.

Consider a finite circular pipe of constant cross section and radius R. Take the axis of the pipe to be the x-axis and assume that the fluid flowing through it is of constant density ρ_0 and constant kinematic viscosity coefficient v. Dictated by convenience, the two other directions may be represented by y and z, or by polar coordinates r and φ. We postulate a steady axi-symmetric flow in the pipe, whose direction is the $\hat{x}$ direction only, and the velocity along the x-axis is independent of x, i.e., the flow ansatz is $u_x \equiv u(r)$. Such a flow automatically satisfies the continuity equation while the y and z components of the Navier–Stokes equations give that the pressure is constant over the cross section of the pipe, i.e., $\partial P/\partial y = \partial P/\partial z = 0$. On the other hand, the x component of the Navier–Stokes equation is

$$0 = -\frac{1}{\rho_0}\frac{\partial P}{\partial x} + \nu\nabla^2 u. \tag{3.5}$$

In order not to violate the ansatz that u be independent of x, it must be true that the value of the axial pressure gradient, $\partial P/\partial x$, is independent of x. Given that the flow is axi-symmetric and that the pressure is independent of r and φ, it immediately follows that $\partial P/\partial x$ = constant. Consequently, the pressure gradient along the pipe must be $-\Delta P/L$, if ΔP is the pressure drop between the two ends of a pipe of length L, where the minus sign indicates forcing for $\Delta P < 0$.

The velocity distribution can thus be determined from an equation having the form $\nabla^2 u$ = constant and is best solved for in cylindrical coordinates. Rewriting the Laplacian in cylindrical coordinates and substituting the value of the constant so as to ensure that the boundary conditions on P be properly satisfied, we finally get

$$\frac{1}{r}\frac{d}{dr}\left(r\frac{du}{dr}\right) = -\frac{\Delta P}{\rho_0 \nu L}. \tag{3.6}$$

Integrating twice over the cylindrical radial coordinate, between the appropriate limits and imposing the condition that the velocity must remain finite at the center of the pipe plus the requirement that $u = 0$ at $r = R$ (no slip), we obtain the final parabolic velocity distribution over the pipe cross section

$$u(r) = \frac{\Delta P}{4\nu\rho_0 L}(R^2 - r^2). \tag{3.7}$$

This is the acclaimed pipe Poiseuille, sometimes called Poiseuille–Hagen flow (Fig. 3.1). Note that the amount of fluid passing per unit time through the pipe is according to (3.7)

$$Q = 2\pi\rho_0 \int_0^R u r dr = \frac{\pi}{8}\frac{\Delta P}{\nu L}R^4. \tag{3.8}$$

A prominent feature of this is that the discharge rate per unit pressure drop has a R^4 dependence. Thus, for example, to get the same amount of fluid discharge through a pipe with half the radius of the original one, the pressure has to increase 16-fold, or, conversely, if the pressure drop is the same only 1/16 of the fluid will pass through the pipe, per unit time. This indicates that clogging one's pipes may incur a heavy price. We leave to Problem 3.2 the solution for *plane* Poiseuille flow.

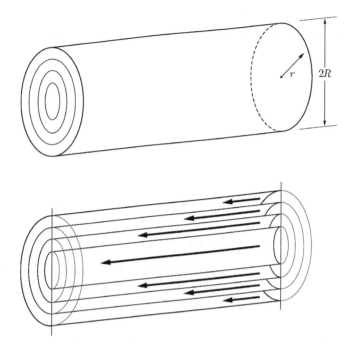

Fig. 3.1 A schematic depiction of pipe Poiseuille flow. For details see text

3.2.3   Flow on an Inclined Plane

We study now the flow of a viscous fluid down an inclined plane. Let the plane be perpendicular to the $\hat{\mathbf{z}}$ unit vector and thus coincide with the x–y plane of a Cartesian system. We consider only a flow in the $\hat{\mathbf{x}}$ direction of a fluid layer whose thickness is h. The gravitational force, per unit mass, $\mathbf{g}$ is directed at an angle θ to the direction $-\hat{\mathbf{z}}$ and it lies in the $x, -z$ plane, see Fig. 3.2. Because of symmetry considerations the velocity must be in the form $(u, v, w) = (u(z), 0, 0)$ and the pressure be a function of z alone. The fluid is forced down the inclined plane by the gravitational body force. The relevant equations are

$$\rho g \sin \theta + \eta \frac{d^2 u}{dz^2} = 0,$$

$$\frac{dP}{dz} + \rho g \cos \theta = 0, \tag{3.9}$$

where η is the coefficient of dynamic viscosity. Let the external pressure be P_0. Thus on the surface $z = h$ the stress has only a normal component (equal to the external pressure) $\sigma_{zz} = -P_0$. On the other hand, we assume that the ambient air above induces no tangential stresses. We recall that the form of the tangential stress

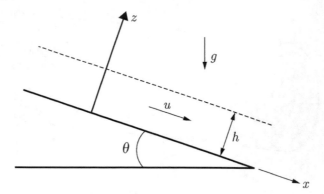

Fig. 3.2 Schematic illustration of the flow of a fluid layer down an inclined plane. For details see text

across the surface is, in this configuration, given by $\sigma_{xx} = v\,du/dz$. Thus, the second boundary condition is that $\sigma_{xx}(z = h) = 0$, which means to say that we require $du/dz = 0$ at $z = h$. Therefore it follows that

$$u = \frac{\rho g \sin\theta}{2\eta} z(2h - z) \quad \text{and} \quad P = P_0 + \rho g(H - z)\cos\theta. \tag{3.10}$$

We may calculate easily the total volume flux (per unit length in the y direction) to be

$$\int_0^h u\,dz = \frac{gh^3 \sin\theta}{3v}.$$

3.2.4 The Rayleigh Problem

This problem, named after Lord Rayleigh (J.W. Strutt), the well-known physicist and 1904 Nobel laureate, results from the following setup: viscous fluid of constant density occupies the $z > 0$ half space. There are no body forces and we assume that the $z = 0$ plane moves with a prescribed, time-dependent, velocity $U(t)\hat{\mathbf{x}}$. The problem is to solve for the fluid flow for all $z > 0$ with the assumption that the fluid velocity goes to zero as $z \to \infty$.

A brief inspection is enough to show that the resulting flow profile depends on the function $U(t)$. For illustration here, we assume that the boundary oscillates at some frequency ω, i.e., $U(t) = U_0 \cos\omega t$. The symmetry of the problem dictates that all dependent variables must be functions of only the vertical coordinate z and the time t, while the only nonzero component of the fluid velocity is in the x direction, e.g., since $\nabla \cdot \mathbf{u} = 0$, w must be a constant and because it is zero on the boundary, it must be zero everywhere. We have in general

$$\frac{\partial \mathbf{u}}{\partial t} = -\frac{1}{\rho}\nabla P + v\nabla^2 \mathbf{u}, \tag{3.11}$$

where v is the kinematic viscosity coefficient. The z component of the above equation is $\partial P/\partial z = 0$ and since all quantities are functions of z only, it must be true that $P = \text{const}$. Thus, we are left only with the diffusion equation in z for the velocity $u(z,t)$:

$$\frac{\partial u}{\partial t} = v\frac{\partial^2 u}{\partial z^2}, \qquad u(0,t) = U(t). \tag{3.12}$$

With the above $U(t)$ we make the ansatz $u(z,t) = \Re\left[Z(z)e^{i\omega t}\right]$, where Z is a complex function of the real variable z. The substitution gives

$$i\omega Z - v\frac{\partial^2 Z}{\partial z^2}, \qquad Z(0) = U_0. \tag{3.13}$$

The solution for the velocity satisfying the condition that $u(z \to \infty) = 0$ is (Fig. 3.3)

$$u(z,t) = \Re\left[U_0 e^{-(i+1)z/\delta}\, e^{i\omega t,}\right] = U_0 \cos\left(\omega t - \frac{z}{\delta}\right)e^{-z/\delta}, \tag{3.14}$$

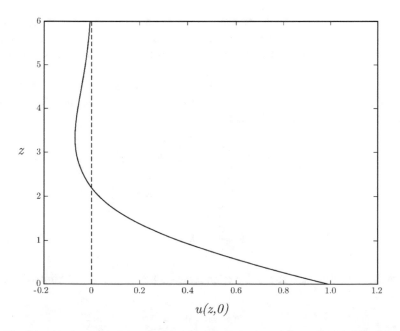

$u(z,0)$

Fig. 3.3 The velocity (horizontal axis) as a function of the distance from the oscillating boundary at time $t = 0 \mod 2\pi$. z is in units of $\sqrt{v/\omega}$

where $\delta \equiv \sqrt{(2\nu/\omega)}$ is the depth over which transverse fluid waves are attenuated, reminding the reader of the idea of *skin depth* in electromagnetism. As can be expected from physical intuition, the depth of penetration of the waves increases with viscosity and decreases with frequency.

3.3 Some Additional Viscous Flows

In a fair number of flows, the nonlinear Navier–Stokes equations are reducible to a single nonlinear ODE or a set thereof, simplifying the problem. Despite this simplification, the resulting ODE usually must be solved either numerically or via the use of some kind of similarity technique. There are cases, however, that the reduced nonlinear ODE is tractable analytically, either exactly or by approximation methods. We give now a number of such flows that we feel may have import in physical applications, some of which are also described in the collection of problems at the end of this section, e.g., problems 3.4 and 3.13.

3.3.1 Two-Dimensional Flow Towards a Stagnation Point

Consider a two-dimensional incompressible steady viscous flow in the domain $y \geq 0$ in which $y = 0$ is a stationary rigid boundary. The relevant Navier–Stokes equations are expressed in Cartesian coordinates $\mathbf{x} = (x, y)$, with a corresponding two-dimensional velocity field $\mathbf{u} = (u, \mathrm{v})$. The velocity field far from the rigid boundary is assumed to be that of straining flow, i.e., as $y \to \infty$ we assume that $u = \alpha x$ and $\mathrm{v} = -\alpha y$, where α is a positive constant. Thus, the Navier–Stokes equations for steady flow, in this case, are

$$u\frac{\partial u}{\partial x} + \mathrm{v}\frac{\partial u}{\partial y} = -\frac{\partial P}{\partial x} + \nu\nabla^2 u,$$

$$u\frac{\partial \mathrm{v}}{\partial x} + \mathrm{v}\frac{\partial \mathrm{v}}{\partial y} = -\frac{\partial P}{\partial y} + \nu\nabla^2 \mathrm{v},$$

$$\frac{\partial u}{\partial x} + \frac{\partial \mathrm{v}}{\partial y} = 0, \tag{3.15}$$

in which the Laplacian operator ∇^2 is understood here to be in two dimensions, i.e.,

$$\nabla^2 = \frac{\partial^2}{\partial x^2} + \frac{\partial^2}{\partial y^2}.$$

As remarked before, when a flow problem in question is dimensionally restricted like it is here, the Laplacian operator symbol (∇^2) should be understood to be restricted in dimension in the same way. In the above quoted equations we have set the density to a constant, $\rho = 1$. To proceed towards finding a solution, we make the following ansatz:

$$u = \alpha x f'(\xi), \qquad v = -\sqrt{v\alpha} f(\xi), \qquad (3.16)$$

where $\xi = y\sqrt{\alpha/v}$. Viscous problems in which we assume *rigid* boundary conditions are those in which the velocity field matches that of the boundary. In this particular case, the velocity must be zero at $y = 0$, i.e., $u = v = 0$ at $y = 0$ since the boundary at $y = 0$ does not move. As can be seen, the assumed velocity solution form found in Eq. (3.16) satisfies the boundary condition at $y = 0$, provided $f(0) = f'(0) = 0$. Furthermore, the solution takes the asymptotic form $u \sim \alpha x$, $v \sim -\alpha y$ very far from the boundary which is consistent with the far-field behavior assumed for the problem at the outset. Equation (3.16) is also an exact solution of equations (3.15) provided f satisfies the nonlinear ODE:

$$f''' + ff'' + 1 - f'^2 = 0, \qquad \text{with } f(0) = f'(0) = 0, \qquad f'(\infty) = 1. \qquad (3.17)$$

The last of the conditions means that $f(y \to \infty) \to \sqrt{\alpha/v}$ which automatically satisfies the boundary condition on the velocity field v in the far field. Here Eq. (3.17) must be solved numerically. The solution yields the flow in Fig. 3.4.

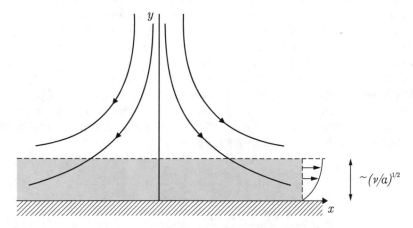

Fig. 3.4 Two-dimensional flow head on a wall, that far from the wall has the character of classic strained flow, i.e., $u \to \alpha x$ and $v \to -\alpha y$ as $y \gg \sqrt{v/\alpha}$

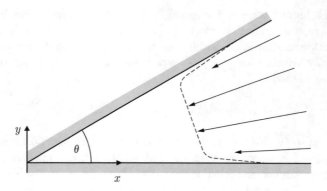

Fig. 3.5 The geometry and the schematic flow in a converging channel

In the layer near the wall there is a balance between viscous diffusion of vorticity from the wall and the advection of vorticity towards the wall by the flow. What can you say about the effect of the increase of ν?

3.3.2   Flow in a Converging Channel

We shall consider this problem for a relatively high Reynolds number and therefore we expect (see Sect. 3.5) the appearance of boundary layers near the walls. The geometry of the problem is shown in Fig. 3.5. The flow we consider is two dimensional and nothing depends on the coordinate z. We assume that at $x = y = 0$ there is an extremely narrow slit allowing fluid to exit (enter) when the flow is towards (out of) the converging channel. This problem was first solved by G. Hammel in 1917 and we essentially follow his reasoning while developing the solution. Switching now to cylindrical coordinates (r, φ, z) it is natural to assume, because of the independence on z, that the flow is purely radial, that is, to say, $u_\varphi = u_z = 0$ together with $u_r \equiv u(r, \varphi)$. The Navier–Stokes equations in cylindrical coordinates then give for this incompressible flow

$$u\frac{\partial u}{\partial r} = -\frac{1}{\rho}\frac{\partial P}{\partial r} + \nu\left(\frac{\partial^2 u}{\partial r^2} + \frac{1}{r^2}\frac{\partial^2 u}{\partial \varphi^2} + \frac{1}{r}\frac{\partial u}{\partial r} - \frac{u}{r^2}\right), \qquad (3.18a)$$

$$0 = -\frac{1}{\rho r}\frac{\partial P}{\partial \varphi} + \frac{2\nu}{r^2}\frac{\partial u}{\partial \varphi}, \qquad (3.18b)$$

$$0 = \frac{\partial(ru)}{\partial r}. \qquad (3.18c)$$

Now, since the last of these equations indicates that $ru(r, \varphi)$ is a function of φ only, it is advantageous to introduce a new function of φ

$$q(\varphi) \equiv \frac{ru}{6v} \qquad \Longrightarrow \qquad u = \frac{6vq(\varphi)}{r}.$$

From the second equation of the above set, (3.18b), we have

$$\frac{1}{\rho}\frac{\partial P}{\partial \varphi} = \frac{12v^2}{r^2}q', \tag{3.19}$$

where $q' = dq/d\varphi$. Integrating in φ introduces an arbitrary function $f(r)$, an integration "constant," and so, substituting the result into Eq. (3.18a) yields

$$q'' + 4q + 6q^2 = \frac{1}{6v^2}f'(r). \tag{3.20}$$

Both sides of this equation must be equal to a constant (do you understand why?) and we call this constant C. The solution for the pressure is thus

$$P(r,\varphi) = \rho\frac{6v^2}{r^2}\left[2q(\varphi) - \frac{C}{2}\right] + \text{const.}$$

and $q(\varphi)$ has to satisfy the ODE

$$q' + 4q + 6q^2 = C. \tag{3.21}$$

We may multiply this equation by q' and integrate once, yielding

$$\frac{1}{2}q'^2 + 2q^2 + 2q^3 - Cu + D = 0, \tag{3.22}$$

where D is yet another constant. It is clear the constants will be found when this equation is solved with *no*-slip boundary conditions, that is, with the velocity parallel to the boundary equalling zero. This condition has to be applied on both walls. This kind of situation indicates that the fluid meets what is called a *rigid boundary*: velocity zero in both normal and tangential directions. Can you deduce any flow properties without actually solving Eq. (3.22)?

3.4 Motion in Very Viscous Fluid

Even though most fluid flows encountered in nature are endowed with high Reynolds numbers, there exist several applications of flows having $Re \ll 1$. This can happen if the viscosity of the fluid is very high (e.g., as in honey or molasses) or the relevant length scale of the flow is tiny, e.g., flow in thin films, or best if both situations occur. Also, when small particles in a liquid (oceans) or a gas (heavily polluted air, a dusty proto-planetary disk with many embedded more sizeable solid bodies) are critical to important larger scale astrophysical and geophysical problems, they require a description using a very viscous, or equivalently a very low Re approach. We bring here a few examples of these *creeping* flows, commonly known as *Stokes flows*, concentrating on some of their interesting physical phenomena. This seems to be the right place to point out that in the nineteenth century G.G. Stokes's, contributions to fluid dynamics, optics, and mathematical physics were paramount. As we saw, the central equation of viscous FD bears also his name, together with C.-L. Navier, an engineer, who arrived at a similar equation, at approximately the same time, independently. From the outset, let us suppose the uniqueness of steady incompressible flow when Re ≪ 1. The straightforward lowest order assumption is to multiply the nondimensionalized Navier–Stokes equations by Re, followed by taking the limit Re → 0. The resulting set of equations should be a good approximation to a very viscous, very slow, creeping flow. This limiting form of the Navier–Stokes equation are called the *Stokes equations* and the resulting solutions are referred to as mentioned before—Stokes flows.

3.4.1   Stokes Flow and Its Properties

One should, however, be slightly more careful. The Stokes equations result from the steady incompressible Navier–Stokes equation. When we discussed the nondimensionalization of the Navier–Stokes equation, which naturally gives rise to the appearance of the Reynolds number, we adopted typical values for U and L (velocity and length scale). With the dynamic viscosity η assumed constant, the only parameter of a steady incompressible flow becomes the nondimensional number Re. Furthermore, we had deduced (see Sect. 1.6) that the natural units for the pressure are ρU^2 where ρ is a constant which we set, henceforth, $\rho = 1$. In this case, we obtain in the limit Re → 0 the nondimensional equations,

$$\nabla^2 \mathbf{u} = 0 \quad \text{and} \quad \nabla \cdot \mathbf{u} = 0, \tag{3.23}$$

since the inertial terms in the Navier–Stokes equation ($D\mathbf{u}/Dt$) are multiplied by Re and, hence, go to zero in the limit taken. Now, we leave it to the problems at the end of the chapter to deal with an example of a flow obeying equation (3.23), hoping the reader will be convinced that in some cases this equation is a less physically plausible choice than the following alternative to represent highly viscous flows. The reason that Eq. (3.23) may be unsatisfactory is that expelling the pressure at the outset is not physically justified since it may play an important role in shaping the resulting flow. In order to carry the pressure into the nondimensional formulation, we must scale it by $\eta U/L$, another combination having units of pressure, which yields

$$\nabla P - \nabla^2 \mathbf{u} = 0 \quad \text{and} \quad \nabla \cdot \mathbf{u} = 0, \tag{3.24}$$

where we remember that u and P, as appearing now, have been nondimensionalized by U and $\eta U/L$, respectively. Note that this pressure scaling has in it the values of viscosity, while the previous scaling was purely dynamic. With this "viscous scaling," so to speak, we obtain the *Stokes equations*. The above form found in Eq. (3.24) reflects the smallness of Re and, as a result, the inertial terms in the equations of motion are inconsequential, as we mentioned above.

Another physical scenario which leads to the form appearing in Eq. (3.24) is the case of a flow in a very thin layer, as in a layer between two parallel, very close walls, giving rise to a relatively very small typical flow length, and hence to Re $\ll 1$. Yet another case is a flow which involves a timescale different from L/U and, as a result, the time derivative in the equation cannot be dropped. Can you think of an example of such a flow? Returning our attention to Eq. (3.24), we also assume now that the flow governed by this equation takes place in a finite volume $\mathscr{V}$, bounded by the closed surface $\partial \mathscr{V}$ on which velocity boundary conditions of a viscous flow are satisfied, i.e., the no-slip conditions discussed earlier. We now list several properties of the Stokes equations, as posed here.

1. *Uniqueness.*

   Let the following boundary condition be specified, $\mathbf{u}(\mathbf{x}) = \mathbf{u}_s$ on $\partial \mathscr{V}$. For the sake of the solution's uniqueness proof, assume that there are two possible distinct solutions to (3.24), where we label the second one with superscript asterisk. Forming now the so-called difference flow, we get

$$ p \equiv P - P^*, \quad \text{and} \quad \mathbf{v} \equiv \mathbf{u} - \mathbf{u}^*, \tag{3.25} $$

   with $\mathbf{v} = 0$ on the boundary. Thus we obtain the Stokes equations expressed in terms of the difference variables. This is possible because the equations are linear. We chose to write Eq. (3.25) into its component forms

$$ 0 = -\frac{\partial p}{\partial x_i} + \frac{\partial^2 v_i}{\partial x_j^2}, \quad \frac{\partial v_i}{\partial x_i} = 0. \tag{3.26} $$

   Multiplying the first equation by v_i, integrating the result over $\mathscr{V}$, making use of the divergence statement, and finally using the Gauss divergence theorem results in

$$ 0 = -\oint_{\partial \mathscr{V}} p v_i n_i dS + \int_{\mathscr{V}} v_i \frac{\partial^2 v_i}{\partial x_j^2} d^3 x. \tag{3.27} $$

   The first term vanishes since $v_i = 0$ on $\partial \mathscr{V}$ and so

$$ \int_{\mathscr{V}} v_i \frac{\partial^2 v_i}{\partial x_j^2} d^3 x = \int_{\mathscr{V}} \left[\frac{\partial}{\partial x_j} \left(v_i \frac{\partial v_i}{\partial x_j} \right) - \left(\frac{\partial v_i}{\partial x_j} \right)^2 \right] d^3 x = 0. \tag{3.28} $$

Using the divergence theorem again and re-exploiting the fact that $v_i = 0$ on the boundary, we are left with

$$\int_{\mathcal{V}} \left(\frac{\partial v_i}{\partial x_j} \right)^2 d^3x = 0. \tag{3.29}$$

The above integrand has nine terms, all squares, thus it is positive or zero. The only way that the integral can vanish is if indeed all the squares of derivatives are zero and thus $\mathbf{v}$ is constant, but it is zero on the boundary, thus it is zero everywhere. So we have proven here that the solution to equations (3.24), if it exists, is also unique.

2. *Reversibility and difficulty of swimming.*

   Now we claim more. Let a particular vector function, say $\mathbf{f}$, be the boundary condition $\mathbf{u}_s$, of a unique solution to the problem, $\mathbf{u}$. The corresponding pressure field of the solution will be some spatially dependent scalar function P', say, plus some physically inconsequential constant P_0, that is to say, $P = P' + P_0$. Suppose that we then change the boundary condition to $\mathbf{u}_s = -\mathbf{f}$. By inspection of the Stokes equations, now $-\mathbf{u}$ will be a solution to the problem with a corresponding pressure distribution given by $P_0^* - P'$, with P_0^* being some other inconsequential constant pressure. We have proven that this is the only solution. Thus, it follows from the Stokes equations that reversing the boundary conditions leads to reversing the flow. This is the explanation of the schematic drawing in Fig. 3.6. The two small patches of two kinds of dyes are smeared out by the rotation, but return to their original position when the flow is reversed. Experimental verifications of this phenomenon are quite stunning, but one has to remember that not all fluid particles return, a small amount of them are sensitive enough to small disturbances caused by thermal molecular motion, which breaks this symmetry. The film "Low Reynolds Number Flow" in the NCFMF collection (web.mit.edu/hml/ncfmf.html) contains a beautiful demonstration by G.I. Taylor of the Stokes flow reversibility and much more material on the subject of this chapter. Sir G.I. Taylor (1886–1975) was a physicist and mathematician, and a major figure in FD and wave theory. The fact that creeping flows are reversible makes regular swimming in very viscous fluids very problematic. It would appear at first glance that swimming is achieved (like in fish) by repeated motions to and fro of the tail. Based on our above considerations, if this motion is time symmetric, then swimming forward is impossible. In order to achieve swimming forward, some breaking of the time symmetry of the above motion is called for. This is sometimes referred to as *the scallop theorem*. The following example is a consequence of this theorem.

3. *Swimming by time symmetry breaking.*

   Tiny swimmers can overcome the difficulties explained above and achieve forward motion. A biological swimmer, like a sperm cell executing helical motion, say, of its tail, is one form of action that is not time reversible. Another good example of what some microorganisms do to break the time symmetry is described in the following example:

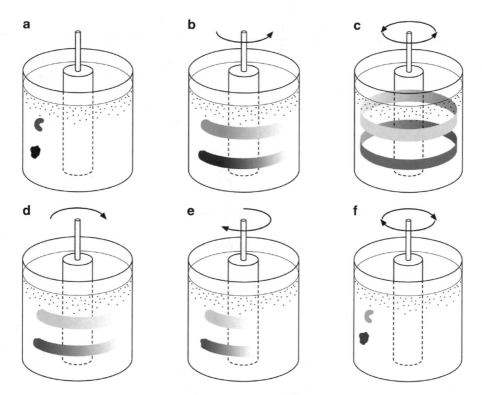

Fig. 3.6 The reversibility of Stokes flow: (**a**) initial dye patches, (**b**)-(**c**) spreading of the dye due to the rotation of the inner cylinder, (**d**)–(**f**) reversal of rotation causing the return to initial condition. During the motion some of the dye microscopic particles are lost to the unstained fluid due to molecular diffusion. This experiment has to be done very carefully because rapid accelerations violate the Stokes scaling

Consider a system that is a thin flexible sheet. Let (x_*, y_*) be the coordinates of any point on this sheet and let each of these points execute the following motion:

$$x_* = x, \qquad y_* = \alpha \sin(kx - \omega t), \tag{3.30}$$

that is, a wave travels in the x_* direction with *phase velocity* $V_p \equiv \omega/k$. We shall discuss the concept of phase velocity with physical insight and some mathematical rigor in the next chapter. The particles of the sheet exhibit oscillatory motion in the y_* direction:

$$\frac{dy_*}{dt} = -\omega\alpha\cos(kx_* - \omega t), \tag{3.31}$$

where k is in units of inverse length and ω is in inverse time units. Despite its oscillatory nature, this motion is not time reversible, the wave runs in the opposite direction if $t \mapsto -t$ and, moreover, it is possible to show that a steady flow component is also induced. In this particular case, we will be able to show this result analytically, using an approximation for the case $\alpha/\lambda = \alpha k/(2\pi) \ll 1$. ε is defined as $\varepsilon \equiv \alpha k$ and is thus also very small ($\ll 1$). As will be apparent shortly, the flow will have, to an ε^2 level of approximation, a steady flow component in the x direction,

$$V = 2\pi^2 \left(\frac{\alpha}{\lambda}\right)^2 V_{\mathrm{p}}, \tag{3.32}$$

when viewed from the frame of the fluid, which is on the average at rest. We get a steady motion of the sheet, that is swimming to the left.

Now, the vorticity-stream function representation of the flow in vector form is

$$\nabla \times \mathbf{u} = (0, 0, -\nabla^2 \psi), \tag{3.33}$$

and after operating with $\nabla\times$ on the dimensional version of the Stokes equations (3.24)

$$0 = -\nabla P + \eta \nabla^2 \mathbf{u}, \qquad \nabla \cdot \mathbf{u} = 0, \tag{3.34}$$

the *biharmonic equation* for the stream function follows

$$\nabla^4 \psi = 0, \tag{3.35}$$

with the condition that $\partial \psi/\partial y = 0$ and $\partial \psi/\partial x = \omega \alpha \cos(kx - \omega t) = 0$, both of them on the surface $y = \alpha \sin(kx - \omega t)$. Since this is our first encounter with the use of an asymptotic approximation method, we perform it explicitly using some elementary intermediate results found in problem (3.5). First, it is convenient to switch to nondimensional spatial variables:

$$kx - \omega t \mapsto x', \qquad ky \mapsto y', \qquad k\psi/\omega\alpha \mapsto \psi',$$

and we remark that the time variable is only a parameter here. Now, we perform the above substitutions and subsequently drop the primes from all expressions hereafter, i.e., $x' \mapsto x, y' \mapsto y$, and $\psi' \mapsto \psi$, where now it is understood that all of these variables are their nondimensional counterparts. This practice is common also outside this book and we are confident that the reader has encountered it before; however to avoid confusion, we shall be very explicit when variables are to be manipulated in this way. The nondimensionalization yields the biharmonic equation (3.35) with boundary conditions expressed in these nondimensional coordinates as

$$\frac{\partial \psi}{\partial y} = 0, \qquad \frac{\partial \psi}{\partial x} = \cos x, \qquad \text{on} \quad y = k\alpha \sin x.$$

The subtlety here is that the boundary conditions are applied on a surface that is not simple and flat but which has some structure to it ($\sim \sin x$). The situation before us can be analyzed if we assume $\varepsilon \equiv k\alpha << 1$, in which case the reader can show after doing Problem 3.5 that the boundary conditions found above can be expressed in the form of the following Taylor series expansions:

$$\left(\frac{\partial \psi}{\partial y}\right)_{y=0} + \varepsilon \sin x \left(\frac{\partial^2 \psi}{\partial y^2}\right)_{y=0} + \text{HOT} = 0,$$

$$\left(\frac{\partial \psi}{\partial x}\right)_{y=0} + \varepsilon \sin x \left(\frac{\partial^2 \psi}{\partial y \partial x}\right)_{y=0} + \text{HOT} = \cos x. \qquad (3.36)$$

Seeking an approximate solution we formally expand ψ asymptotically in powers of ε:

$$\psi = \psi_0 + \varepsilon \psi_1 + \text{HOT}. \qquad (3.37)$$

Substituting this expansion into the biharmonic equation, we get a series of problems for the different powers of ε. We shall list just the first two

$$\left(\frac{\partial^2 \psi_0}{\partial x^2} + \frac{\partial^2 \psi_0}{\partial y^2}\right)^2 = 0 \qquad \text{with} \qquad \frac{\partial \psi_0}{\partial y} = 0$$

$$\text{and} \quad \frac{\partial \psi_0}{\partial x} = \cos x \qquad \text{on} \qquad y = 0; \qquad (3.38)$$

$$\left(\frac{\partial^2 \psi_1}{\partial x^2} + \frac{\partial^2 \psi_1}{\partial y^2}\right)^2 = 0 \qquad \text{with}$$

$$\frac{\partial \psi_1}{\partial y} + \frac{\partial^2 \psi_0}{\partial x^2}\sin x = 0 \quad \text{and} \quad \frac{\partial \psi_1}{\partial x} + \frac{\partial^2 \psi_0}{\partial y \partial x}\sin x = 0, \quad \text{on } y = 0. \ (3.39)$$

Trying the solution for ψ_0 as

$$\psi_0(x,y) = \left[(A + By)e^{-y} + (C + Dy)e^{y}\right] \sin x, \qquad (3.40)$$

we must have $C = D = 0$ for physical reasons (the solution has to be bounded for $y \to \infty$). Using the boundary conditions for $y = 0$ we get the solution

$$\psi_0 = (1 + y)e^{-y} \sin x. \qquad (3.41)$$

We may substitute this solution directly into the boundary conditions for ψ_1 at $y = 0$, giving

$$\frac{\partial \psi_1}{\partial y} = \sin^2 x = (1 - \cos 2x)/2, \qquad \frac{\partial \psi_1}{\partial x} = 0. \tag{3.42}$$

Using similar considerations as we used in finding the solution to ψ_0, we readily find the solution:

$$\psi_1 = \frac{1}{2} y \left(1 - e^{-2y} \cos 2x \right). \tag{3.43}$$

Finally, returning to dimensional variables, i.e., $x \mapsto kx - \omega t$, $y \mapsto ky$, and $\psi \mapsto k\psi/(\omega\alpha)$, henceforth, considering x, y, ψ as dimensional, we may determine the x component of the fluid velocity, up to order ε^2, to be

$$u = \frac{\partial \psi}{\partial y} = -\varepsilon \omega y e^{-ky} \sin(kx - \omega t) + \varepsilon^2 V_p \left[\frac{1}{2} + \left(ky - \frac{1}{2} \right) e^{-2ky} \cos 2(kx - \omega t) \right]$$

$$+ \text{HOT}. \tag{3.44}$$

Do you see where the extra factor of ε comes from? This shows immediately that there is a non-oscillating (in fact, steady) component to the velocity, $\varepsilon^2 V_p/2$, which has a value stemming from the phase velocity V_p, which broke the time symmetry of problem (3.32).

4. *Steady motion of a spherical body at very low Re. Drag.*
   The Stokes equations admit an analytical solution for the motion of a spherical body of radius R in the presence of an oncoming stream whose velocity is a constant very far from the body. The problem can also be phrased as a sphere moving at constant velocity U through a fluid, which is at rest at infinity. The most convenient way of approaching this axially symmetric problem is by using the Stokes stream function which is examined in Problem 3.6. We set the problem up by orienting the symmetry axis to be along the direction of the incoming flow and position ourselves in the reference frame of the sphere. From this vantage point we assume that very far from the sphere the velocity field is a constant whose magnitude is equal to $-U$ and oriented parallel to the symmetry axis of the sphere. It is therefore best to express the stream function solutions in spherical polar coordinates (r, θ, φ). Although spherical and cylindrical coordinates will be discussed, in detail, in Chap. 5, we expect the reader to remind himself or herself of these important fundamental technical details and their properties.
   Consider the Stokes stream function $\Psi(r, \theta)$, with the axial symmetry of the problem reflected in the function's independence of the azimuthal coordinate angle φ, so that

$$u_r = \frac{1}{r^2 \sin\theta} \frac{\partial \Psi}{\partial r}, \quad u_\theta = -\frac{1}{r\sin\theta} \frac{\partial \Psi}{\partial \theta}, \quad u_\varphi = 0. \tag{3.45}$$

We now find

$$\nabla \times \mathbf{u} = \left(0, 0, -\frac{1}{r\sin\theta} \mathbb{D}_s^2 \Psi\right), \tag{3.46}$$

with the definition of the differential operator

$$\mathbb{D}_s^2 \equiv \frac{\partial^2}{\partial r^2} + \frac{\sin\theta}{r^2} \frac{\partial}{\partial\theta}\left(\frac{1}{\sin\theta} \frac{\partial}{\partial\theta}\right). \tag{3.47}$$

Now we write the creeping flow equations (3.24) in a slightly altered, but entirely equivalent, form

$$\nabla P = -\eta \nabla \times (\nabla \times \mathbf{u}), \tag{3.48}$$

where we have restored units to the expression and having made use of the divergence free condition $\nabla \cdot \mathbf{u} = 0$. This expression can be separated component by component in spherical coordinates, giving two equations, one for $\partial P/\partial r$ and the other for $\partial P/\partial\theta$. We leave it to the reader to write out these equations in full. Differentiating the first equation in θ, and second in r, and subtracting the result eliminates the pressure and produces a simple looking homogeneous equation for the Stokes stream function:

$$\mathbb{D}_s^2(\mathbb{D}_s^2)\Psi = 0. \tag{3.49}$$

We note here that this equation resembles a spherical analog biharmonic equation (3.35). To determine a unique solution, boundary conditions are needed: we require that the flow does not penetrate the surface of the sphere and, furthermore, there is no slip there. We saw before that this is nothing other than the typical requirement of viscous flow: both components of the fluid velocity must vanish on the surface of the stationary body, viz.,

$$\partial_r \Psi = \partial_\theta \Psi = 0 \quad \text{at} \quad r = R, \tag{3.50}$$

while we would like the fluid to be undisturbed at infinity, in other words, that the flow be equal to $-U$ in the direction oriented along the body's symmetry axis. For this to be the case, we suppose that the far field form of the stream function should be given by

$$\Psi \sim \frac{1}{2} U r^2 \sin^2\theta \quad \text{as} \quad r \to \infty. \tag{3.51}$$

Indeed, upon using Eq. (3.45), we obtain that $u_r \rightarrow U\cos\theta$ and $u_\theta = -U\sin\theta$, as needed. To solve Eq. (3.49) we try separation of variables. Inspired by the form of the Stokes stream function at infinity, we try the ansatz $\Psi = f(r)\sin^2\theta$. Substituting this into Eq. (3.49) confirms that the angular part has the correct form and the ansatz produces the following radial ODE for $f(r)$:

$$\frac{d^4f}{dr^4} - \frac{4}{r^2}\frac{d^2f}{dr^2} + \frac{4f}{r^4} = 0. \tag{3.52}$$

This is a linear, homogeneous, fourth order, nonconstant coefficient equation for $f(r)$. Inspecting the form of the equation, it is possible to attempt a polynomial solution. We substitute a trial solution r^λ for some unknown λ and we find that λ should satisfy

$$[(\lambda-2)(\lambda-3)-2][\lambda(\lambda-1)-2] = 0. \tag{3.53}$$

This fourth order equation has four distinct solutions: $\lambda = -1, 1, 2, 4$ indicating that there are no pathologies in the assumed form[1] and, therefore, we can write

$$f(r) = \frac{A}{r} + Br + Cr^2 + Er^4, \tag{3.54}$$

where A, B, C, E are constants yet to be determined. The condition at infinity requires $C = U/2$ and $E = 0$ while the conditions on the surface of the sphere require that $f(R) = f'(R) = 0$, which finally determines the solution

$$\Psi = \frac{1}{4}U\left(2r^2 + \frac{R^3}{r} - 3Rr\right)\sin^2\theta. \tag{3.55}$$

With the Stokes stream function found, it is possible to determine the fluid velocities as well as the drag force on the sphere, which has the classical Stokes value $F_D = 6\pi\eta UR$. Problem 3.7 is devoted to this calculation.

3.4.2 The Stokes "paradox" and the Reciprocal Theorem

Consider a Stokes flow around an infinitely long cylinder of radius R whose axis lies perpendicular to the flow. The flow in the far field is assumed to be of a constant velocity U, oriented along the y-axis. Thus aligning the x coordinate direction with

---

[1]Note that this is an example of an *indicial equation* and is a part of the procedure of developing ODE solutions using the *method of Frobenius*. A good treatment of this matter can be in books on ODEs (see the *Bibliographical Notes*).

the cylinder's axis, we require that as $r \gg 1$ the velocity field should take on the form $\mathbf{u} = U\hat{\mathbf{y}} = U\sin\varphi\,\hat{\mathbf{r}} + U\cos\varphi\,\hat{\boldsymbol{\varphi}}$. We have already seen above that after operating with $\nabla\times$ on the first equation of (3.24), we are led to the biharmonic equation for the stream function:

$$\nabla^4 \psi = 0. \tag{3.56}$$

Inspired again by the requirement that for $r \to \infty$ we have to get a uniform flow, we choose to separate variables in plane polar coordinates, thus $\psi(r,\varphi) = f(r)\sin\varphi$. This satisfies the differential equation (3.56) provided $f(r)$ is the solution of the equation

$$\left(\frac{d^2}{dr^2} + \frac{1}{r}\frac{d}{dr} - \frac{1}{r^2}\right)^2 f(r) = 0, \tag{3.57}$$

with the no-slip requirement at the surface of the cylinder, i.e., $\psi(R,\varphi) = (\partial\psi/\partial r)_{r=R} = 0$, while the unperturbed flow, condition at $r \to \infty$, enforces $f(r) \to Ur$ for very large r (do you understand why?). The most general solution satisfying Eq. (3.57) can be found through routine integration

$$f(r) = Ar^3 + Br\ln r + Cr + \frac{D}{r}, \tag{3.58}$$

with A, B, C, D yet to be determined constants. The far field requirement of a smooth matching onto a uniform flow forces us to choose $A = B = 0$. The rigid/no-slip conditions on the surface says that both $CR + D/R = 0$ and $C - D/R^2 = 0$. But these can be both satisfied only if $C = D = 0$ and *there is no satisfactory solution around the cylinder in unbounded fluid*. This is known as the *Stokes paradox* and its resolution involves some subtleties involving the retention of order Re terms, which ultimately leads to the important Oseen equation (see Problem 3.9).

This kind of difficulty arises also in other branches of physics that are governed by elliptic equations (which demand boundary conditions on the surface of the domain, or at infinity if the domain is infinite). Two-dimensional perpendicular cuts through infinite linear structures, like the cylinder in the last example, and trials to solve the problem in two dimensions (of the cut) fail to take into account that the infinite, in three dimensions, structure does not allow proper boundary conditions at the three-dimensional infinity. Therefore, logarithmic contributions appear, which diverge at large distances. This is similar to the calculation, e.g., of the electrical potential at a distance r of an infinite charged wire.

Consider now two different Stokes flows, i.e., both satisfying equations (3.24): $\mathbf{u}^{(1)} \neq \mathbf{u}^{(2)}$. The volumes in which these flows take place are identical in size and shape and we assume that there are no body forces, but we are not assuming that the fluids have the same dynamic viscosity η. One of the versions of the reciprocity quality (sometimes called a theorem, which was originally discussed by H. Lorentz, as early as 1906, but proved in its general form by H. Brenner only in 1963) is then

$$\eta^{(2)} \int_S dS_i \sigma_{ik}^{(1)} u_k^{(2)} = \eta^{(1)} \int_S dS_i \sigma_{ik}^{(2)} u_k^{(1)} dS_i. \tag{3.59}$$

We can arrive at the reciprocal expression (3.59) using our previous relations (1.15), (1.40), and (1.41) evaluating the product $\sigma_{ij}^{(1)} \mathscr{D}_{ij}^{(2)}$, which comes out to be $-P^{(1)} \delta_{ij} \mathscr{D}_{ij}^{(2)} + 2\eta^{(1)} \mathscr{D}_{ij}^{(1)} \mathscr{D}_{ij}^{(2)}$, where we have not forgotten the incompressibility of both fluids and it seems to be a proper place to remind the reader that the summation convention on two identical indices is used. The first term in the last expression drops out on account of incompressibility. Thus we are left with

$$\sigma_{ij}^{(1)} \mathscr{D}_{ij}^{(2)} = 2\eta^{(1)} \mathscr{D}_{ij}^{(1)} \mathscr{D}_{ij}^{(2)},$$
$$\sigma_{ij}^{(2)} \mathscr{D}_{ij}^{(1)} = 2\eta^{(2)} \mathscr{D}_{ij}^{(2)} \mathscr{D}_{ij}^{(1)}. \tag{3.60}$$

The second expression comes out upon interchanging the upper indices, which is obviously allowed. Now, since $\mathscr{D}_{ij}^{(1)} \mathscr{D}_{ij}^{(2)} = \mathscr{D}_{ij}^{(2)} \mathscr{D}_{ij}^{(1)}$ it follows that

$$\eta^{(2)} \sigma_{ij}^{(1)} \mathscr{D}_{ij}^{(2)} = \eta^{(1)} \sigma_{ij}^{(2)} \mathscr{D}_{ij}^{(1)}. \tag{3.61}$$

From this relation follows the reciprocity relation by a relatively simple calculation, which we leave to Problem 3.8. Reciprocity relationships may be useful in calculating the resistance of particles and concomitant pressure gradients when the particles (or drops) are immersed in a creeping flow of a fluid.

3.4.3   Viscosity of Suspensions

Fluids mixed with a large number of particles (in various phases, but of a different material than the fluid) are common in nature and industry. Understanding the behavior of suspensions is therefore very important and we shall devote this section to the problem of calculating the correction to the viscosity of a fluid whose dynamic viscosity coefficient, in the absence of suspended particles, is given to be η_0. Before proceeding, we must first elucidate some preliminary assumptions that are valid for the specific case we treat.

3.4.3.1   Definition of a Suspension and the Case Treated Here

We consider the mixture as a homogeneous suspension if the fine particles in question are randomly distributed in the fluid and if the phenomena we wish to study have a characteristic length scale that is much larger than the size of the particles. We call η the value of the fluid dynamic viscosity coefficient of the *suspension* and it will be some function of the basic fluid viscosity η_0 and, obviously, some properties of the suspensions themselves which we shall develop hereafter. The value of η can be calculated easily, as it turns out, if the suspension is *dilute*, that is, the total volume of the particles is much smaller than the volume of the fluid in which they reside. This case is particularly amenable to a calculation when the suspended particles are all of equal size and spherical. This calculation was first done by A. Einstein (1906).

3.4.3.2 Dilute Spherical Particles Case When the Unperturbed Flow Has Low Re

Consider first the effect of one particle when immersed in a fluid whose unperturbed flow has constant velocity gradients, in linear form as, say,

$$(u_0)_i = \alpha_{ik} x_k, \tag{3.62}$$

where α_{ij} is a constant symmetric tensor. The unperturbed pressure is assumed constant and is denoted by P_0. The fluid is assumed incompressible, which means that $\alpha_{ii} = 0$; thus the tensor defining the unperturbed flow is traceless because $\partial(u_0)_i/\partial x_i = 0$. Placing the spherical particle of radius a, say, at the origin we denote the new fluid velocity by $\mathbf{u} = \mathbf{u}_0 + \mathbf{u}_1$, where the perturbation $\mathbf{u}_1$ tends to zero far away from the origin. Nevertheless, the velocity perturbation, at least in the region of the particle, cannot be considered very small with respect to the flow in the absence of a particle. At $r = a$, where r is the spherical coordinate, we must have the boundary condition $\mathbf{u} = 0$, because of the symmetry of the problem. Now using Eq. (3.24), in view of the additional assumption that the flow is characterized by small Re, we may try the solution for $\mathbf{u}_1$ to be

$$\mathbf{u}_1 = \nabla \times \nabla \times [(\alpha \cdot \nabla)f], \quad P = \eta_0 \alpha_{ik} \frac{\partial^2 (\nabla^2 f)}{\partial x_i \partial x_k}, \tag{3.63}$$

where $(\alpha \cdot \nabla)f = \alpha_{ik}\partial f/\partial x_k$. It is a vector since α is a second order tensor, and we are also assuming that the function f is of the form $f(r) = Ar + B/r$. Expanding all these expressions and using the boundary conditions the following result emerges:

$$(u_1)_i = 2.5 \left(\frac{a^5}{r^4} - \frac{a^3}{r^2} \right) \alpha_{km} n_i n_k n_m - \frac{a^5}{r^4} \alpha_{ik} n_k, \tag{3.64}$$

with the pressure given by

$$P = -5\eta_0 \frac{a^3}{r^3} \alpha_{ik} n_i n_k. \tag{3.65}$$

$\mathbf{n}$ is a unit vector in the radius vector direction. The momentum flux density tensor is equal to the stress tensor because of the linear approximation of the velocity. Consider the volume average of the stress tensor

$$\overline{\sigma}_{ij} \equiv \frac{1}{\mathcal{V}} \int_{\mathcal{V}} \sigma_{ij} d^3 x, \tag{3.66}$$

where $\mathcal{V}$ is considered very large. We denote average (or mean) values in this book, interchangeably by an overline ($\bar{\cdot}$) or angle brackets ($\langle \cdot \rangle$). In each case its meaning is spelled out and in Chap. 9, where averaging plays a central role, it will be discussed in considerable detail. Define now the integral

$$I \equiv \frac{1}{\mathcal{V}} \int_{\mathcal{V}} \left[\sigma_{ik} - \eta_0 \left(\frac{\partial u_i}{\partial x_k} + \frac{\partial u_i}{\partial x_k} \right) + P \delta_{ik} \right]. \tag{3.67}$$

Evidently, the integrand is zero in the fluid (by definition), but it may have contribution from the volume of the spherical suspended particles. On the other hand

$$\overline{\sigma}_{ij} = \eta_0 \left(\frac{\overline{\partial u_i}}{\partial x_k} + \frac{\overline{\partial u_k}}{\partial x_i} \right) - \overline{P} \delta_{ik} + I \tag{3.68}$$

is an identity. Convince yourself why. Now it is reasonable to drop the average pressure, using the argument that it is a scalar and thus is given in terms of linear combinations of α_{ij}, subject to $\alpha_{ii} = 0$. It is also evident that $\partial(\sigma_{im} x_k)/\partial x_m = x_k(\sigma_{im})/\partial x_m + \sigma_{ik}$. This allows one to transform the volume integral I into a surface integral, and it has to be multiplied by n, denoting here the number density of the spherical particles, because it can be assumed to be done over each particle alone (due to the low density of the suspended spherical particles). The result is

$$I = n \oint [\sigma_{im} x_k dS_m - \eta_0 (u_i dS_k + u_k dS_i)]. \tag{3.69}$$

The mission now is to calculate I, but we may keep only terms $\propto 1/r^2$ and neglect $\propto 1/r^4$ in the solution for the velocity (3.64). This algebra finally leads to

$$I = 20\pi a^3 n \eta_0 \left(5\alpha_{lm} \overline{n_i n_k n_l n_m} - \alpha_{im} \overline{n_k n_m} \right).$$

An elementary exercise in tensor calculus gives $\overline{n_k n_m} = \delta_{km}$ as well as

$$\overline{n_i n_k n_l n_m} = \frac{1}{15} \left(\delta_{ik} \delta_{lm} + \delta_{ik} \delta_{lm} + \delta_{im} \delta_{kl} \right),$$

and thus the result for the average stress is

$$\overline{\sigma}_{ij} = \eta_0 \left(\frac{\overline{\partial u_i}}{\partial x_k} + \frac{\overline{\partial u_k}}{\partial x_i} \right) + \frac{20}{3} \pi \eta_0 \alpha_{ik} a^3 n. \tag{3.70}$$

Upon substituting for the unperturbed velocity from Eq. (3.62) into the first term, we get $2\eta_0\alpha_{ik}$ plus a first order small component that averages to zero. The correction to the viscosity resides entirely in the second term. Thus, the corrected viscosity works out to be

$$\eta = \eta_0(1+2.5\beta), \tag{3.71}$$

where $\beta = 4\pi a^3 n/3$, conventionally called the *volume fraction*, is effectively the ratio of the total volume of the suspended spheres to the total fluid volume. The calculations for nonspherical suspended particles becomes very involved even for spheroids. It is also much more difficult to treat non-dilute suspensions, in which the fluid coupling among nearby spheres cannot be neglected.

3.4.4 Hele–Shaw Flow

When a fluid is restricted to flow between two parallel flat plates (made of glass, say) a distance h apart and this distance is very small with respect to the expected typical length scale L of the flow in the plane of the plates, we may transparently consider the response of the fluid for the case of fixed density, steady flow, and free of body forces. This kind of flow is called *Hele–Shaw* and the space in which the flow proceeds is sometimes referred to a Hele–Shaw cell. There are a number of applications for such a flow, the most obvious is the case when a fluid sample is prepared for microscopic examination.

Choosing the plane of the plates as horizontal, that is, with the z-axis perpendicular to it, we nondimensionalize the velocities (u,v,w) by a typical horizontal velocity U, the lengths in the x–y plane by a typical length L, and the one in the z direction by h, the distance between the plates. With these scalings, the Navier–Stokes equations greatly simplify if we remember that $\varepsilon \equiv h/L \ll 1$. The density is scaled by its constant value and disappears from the equations. To obtain a meaningful balanced set of equations for the flow of a thin layer between two plates, we scale the pressure by $(\varepsilon^{-2}\eta U/L)$. The following components of the steady Navier–Stokes equation follow, with the functions scaled as explained above:

$$\left\{\frac{U^2}{L}\right\}(u\partial_x + \mathrm{v}\partial_y + w\partial_z)u = -\left\{\varepsilon^{-2}\eta\frac{U}{L^2}\right\}\partial_x P + \left\{\frac{U}{L^2}\right\}(\partial_{xx} + \partial_{yy} + \varepsilon^{-2}\partial_{zz})u, \tag{3.72}$$

$$\left\{\frac{U^2}{L}\right\}(u\partial_x + \mathrm{v}\partial_y + w\partial_z)\mathrm{v} = -\left\{\varepsilon^{-2}\eta\frac{U}{L^2}\right\}\partial_y P + \left\{\mathrm{v}\frac{U}{L^2}\right\}(\partial_{xx} + \partial_{yy} + \varepsilon^{-2}\partial_{zz})\mathrm{v}, \tag{3.73}$$

$$\left\{\varepsilon\frac{U^2}{L}\right\}(u\partial_x + \mathrm{v}\partial_y + w\partial_z)w = -\left\{\varepsilon^{-2}\eta\frac{U}{L^2}\right\}\partial_z P + \left\{\eta\frac{U}{L^2}\right\}(\partial_{xx} + \partial_{yy} + \varepsilon^{-2}\partial_{zz})w. \tag{3.74}$$

It is not difficult to convince oneself, dividing the equations by $\eta U/(\varepsilon^2 L)^2$, say, that in lowest order in ε the following nondimensional set of equations emerges:

$$-\frac{\partial P}{\partial x} + \frac{\partial^2 u}{\partial z^2} = 0,$$

$$-\frac{\partial P}{\partial y} + \frac{\partial^2 v}{\partial z^2} = 0,$$

$$-\frac{\partial P}{\partial z} = 0. \tag{3.75}$$

Using the boundary conditions $u = v = 0$ on $z = 0, 1$, setting w to be zero throughout, at this order of approximation, and exploiting the nondimensional continuity equation

$$\partial_x u + \partial_y v + \partial_z w = 0, \tag{3.76}$$

we readily obtain

$$u = -\frac{1}{2}\frac{\partial P}{\partial x}z(1-z),$$

$$v = -\frac{1}{2}\frac{\partial P}{\partial y}z(1-z), \tag{3.77}$$

and $P = P(x, y)$. Now, integrating the continuity equation over z, between $z = 0$ and $z = 1$ after substituting from Eq. (3.77) for the velocities, a two-dimensional Laplace's equation for the pressure, which is a function of x and y alone, follows:

$$\frac{1}{12}\nabla^2 P = 0. \tag{3.78}$$

It is also true that the vertically averaged two-dimensional velocity in a Hele–Shaw flow is

$$\bar{\mathbf{u}} = \int_0^1 \mathbf{u}\,dz = \frac{1}{12}\nabla P.$$

An interesting result can be observed, when a Hele–Shaw cell is only partially filled with fluid, as when it is injected or sucked (by a syringe, say) through a point in one of the walls (assumed to be the origin), as shown in Fig. 3.7. Assume that at the free surface Γ (sideways) there is constant pressure and also fixed surface tension leads to the conclusion that the radius of curvature of the boundary of the advancing or retreating fluid is constant (see figure). Let us assume that the boundary Γ is given by the function $b(x, y, t) = 0$. The z averaged boundary condition at the surface is thus

$$P = 0 \quad \text{and} \quad \frac{Db}{Dt} = 0 \quad \text{on } b = 0, \tag{3.79}$$

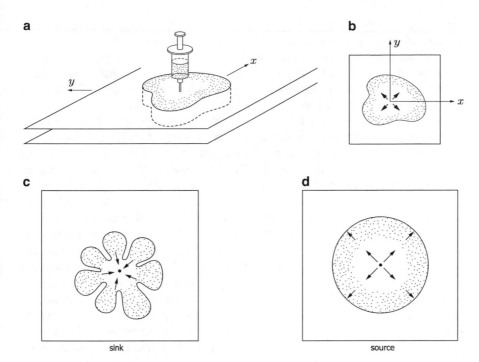

Fig. 3.7 Hele–Shaw cell

where

$$\frac{D}{Dt} = \frac{\partial}{\partial t} + \overline{\mathbf{u}} \cdot \nabla,$$

where the ∇ operator contains, of course, only derivatives on x and y. The model is reversible (replacing t by $-t$ and P by $-P$ leaves the equations unchanged). However, advancing fluid flow (as a result of injection) is stable, while a retreating interface, because of a sink, is observed to be unstable because of the related breaking of the interface to form "fingers" (see Fig. 3.7c). We leave further investigation of the problem, as well as related flows through porous media, as being outside the scope of this book.

3.5 Viscous Boundary Layers

Thus far, we have both intimated and explicitly shown that viscous effects may play an important role also in flows endowed with very high values of Re. This counterintuitive fact arises for reasons we have already discussed, involving the response due to the presence of both boundaries and/or obstacles in the flow, where the velocity of the fluid is forced to take on specific values (often zero). The viscous effects we speak of become manifested on very short spatial scales, to offset (dimensionally) the small viscosity. In the flow development it arises, generally, in the near-zone of the boundary or obstacle. Such regions go under the name of *boundary layers*. In a boundary layer a small length scale is created giving rise to a smaller *effective* Re in the region, despite its being large on the flow scale.

We wish to discuss boundary layers a little more thoroughly in this section, even though it is obvious that we shall be able to touch only upon the basics. Entire books have been written on the subject and we refer the reader to the relevant material in the *Bibliographical Notes* section at the end of this chapter. We would like, however, to state at the outset that we limit ourselves to laminar boundary layers, that is, those in which the flow is largely stable and appears ordered. We have already seen an example of this in our earlier discussion of two-dimensional flow towards a stagnation point (Sect. 3.3.1). This should be held up in contrast to unstable boundary layers, which present the phenomena of detachment and transition to a disordered flow on the way to turbulence. We have discussed briefly the phenomenon of boundary layer separation, but a serious discussion of turbulence must await Chap. 9. Regardless, the example of the *Blasius boundary layer* which we discuss below is well known to undergo a linear instability due to the emergence of a *dynamical* boundary layer, i.e., a boundary layer that appears in the flow only when some time-dependent perturbation is applied to the basic state. We shall see this examined further in Chap. 7. In order to fully appreciate this sequential progression towards flow complexity and turbulence, we must start with the basics, that is, mathematical boundary layers in ODE, treated by matched asymptotic expansions. Moving to physics, the particular Blasius flow basic state is a good place to start and is described in some detail in this section. P.R.H. Blasius, whose name was given to the boundary layer which became a seminal work, was the first student of L. Prandtl.

We should make clear what our attitude and perspective is with regards to the matter of fluid flows and how we think of them in relation to turbulence. Fluid flow, like most natural physical motions, develops in the direction of kinetic energy dissipation and entropy increase. If the viscosity is dominant $Re = \mathscr{O}(1)$, and viscous dissipation readily occurs. In high Re flows, confined by solid boundaries, e.g., walls, or including solid obstacles, the flow creates short scales near the boundaries in order to be able to dissipate in the boundary layers as explained above. Unbounded flows with very large Re, like many astrophysical flows, if they have no obstacles, can dissipate their kinetic energy only by becoming unstable and forming small-scale eddies in the flow. This is the way to turbulence which is endowed with high *effective* turbulent viscosity (see Chap. 9). As we shall see, this does not mean that bounded flows cannot become turbulent. We feel that an exposition of the subject of FD and, in particular, the development of turbulence is best based on a look at historical developments. At the beginning of the twentieth century, there were two types of well-studied flows: those of inviscid fluids, which under some additional simplifying assumption gave rise to beautiful analytical theory, e.g., that of potential flow, and those of very viscous, i.e., Stokes, or creeping flows, which are also well understood when subject to certain simplifying assumptions, as for flows on thin films. We have given a number of examples and problems on these two types of flows in this chapter. As is often the case, both types of flows can be present in a given physical setup and, as we explained at the beginning of this section, these two types of flows appear to be linked to each other in boundary layers, where the relevant length scales of the flow become very small in at least one spatial dimension of the flow. This was first noticed by L. Prandtl who, in his seminal paper of 1905, introduced and initiated the study of boundary layer theory by writing a set of appropriate equations. We shall not, for the reason on economy of space, detail these equations and direct the interested reader to reference [7] in the *Bibliographical Notes*, a voluminous classical work on boundary layers.

3.5.1 A Mathematical Digression

We shall make now a mathematical digression, discussing a rigorous way to perform boundary layer matching in a simple ODE, not related to any particular fluid problem. We intend the following example to be a guide for the student in constructing similarly approximate solutions to problems defying exact analytical solution. Consider the following two-point boundary value problem for the function $f(x)$:

$$\varepsilon \frac{d^2 f}{dx^2} + 2 \frac{df}{dx} + 2f(x) = 0, \qquad (3.80)$$

with the boundary conditions $f(0) = 0$, $f(1) = 1$, and where $1 \gg \varepsilon > 0$, i.e., ε is a small positive number. Notice that the solution to the problem posed in Eq. (3.80) can be exactly written down, as it is a second order linear constant coefficient ODE. We use this simple system to detail how an approximate solution is developed and compare it to the exact solution. In the description of the asymptotic approximation method, we shall proceed as if the exact solution to the problem is not known. We shall, naturally, exploit the fact that ε is a small number, in the construction of the approximate solution. As remarked in Sect. 2.5.6, that calculation should perhaps appear after this discussion, but constraints compelled us to include that topic in Chap. 2.

The first thing to notice, when one considers an approximate solution to Eq. (3.80), is that a simplistic, straightforward approach to a possible perturbative approximation technique will not work here because this is clearly a *singular perturbation* theory problem. This concept means that such a perturbative expansion fails, because $\varepsilon = 0$ changes the mathematical nature of the problem. In this particular instance, the singular nature of the problem arises because setting $\varepsilon = 0$ reduces the order of the ODE and it becomes impossible, in general, to simultaneously satisfy both of the required boundary conditions for $f(x)$. The technique that will be employed instead is especially suited for such problems and is perhaps the simplest example of boundary layer theory. The idea is to consider first an *outer* solution (for x large), in which the danger of singularity is absent, and later move on into an *inner* region (for $x \geq 0$ and small enough), where the apparent singularity is properly resolved. The procedure concludes by showing how one links the solutions found in the two separate regions.

1. *Outer solution.*

   Proceeding in a straightforward manner, we assume a regular perturbation series to represent the approximate solution:

   $$f(x) = f_0(x) + \varepsilon f_1(x) + \text{HOT}. \tag{3.81}$$

   We get, in lowest order in ε,

   $$\frac{df_0}{dx} + f_0(x) = 0, \tag{3.82}$$

   whose solution $f_0(x) = Ae^{-x}$ cannot satisfy, nontrivially, the $x = 0$ boundary condition. This is, thus, the outer solution and we shall, in this simple example, be content with only its lowest order, i.e., the $\mathcal{O}(1)$ term. It is reasonable that this solution satisfies the outer boundary condition $f_0(1) = 1$ and thus $A = e$ and the lowest order outer solution is

   $$f_0^{\text{out}}(x) = e^{1-x}. \tag{3.83}$$

   It is evident that this outer solution does not satisfy the boundary condition at $x = 0$, i.e., $f_0^{\text{out}}(0) = e \neq 0$.

2. *Inner solution.*

   The breakdown of the outer solution occurs very close to the left boundary, where the full solution must undergo a rapid variation, as is obvious from the inspection of (3.80). In other words, a boundary layer is reasonably assumed to exist near the left boundary. Thus we should look for a suitable inner solution. To capture it, one may stretch the x variable, with the help of ε, e.g., by defining a new spatial variable

   $$X \equiv \frac{x}{\varepsilon^a}, \tag{3.84}$$

   with $a > 0$ undetermined, as yet, for the sake of generality. Trying an expansion appropriate for the boundary layer region—the inner solution,

   $$F^{\text{in}}(X) \approx F_0(X) + \varepsilon^c F_1(X) + \text{HOT}, \tag{3.85}$$

   where we call this inner solution F^{in} and again allow for a positive power c of ε for the sake of generality. Note that here we understand that X is held fixed when $\varepsilon \to 0$.

Substituting the inner solution into the original equation and balancing the
various terms, in order to achieve consistency in the powers of ε, gives rise
to the determination that only $a = 1$ can give consistency. This is called the
distinguished limit for this problem. With this choice the lowest order boundary
layer equation is

$$\frac{d^2 F_0}{dX^2} + 2\frac{dF_0}{dX} = 0 \quad \text{for } 0 < X < \infty, \tag{3.86}$$

with the boundary condition $F_0(0) = 0$. This equation can be readily solved with
one undetermined integration constant (B) thus

$$F_0(X) = B(1 - e^{-2X}), \tag{3.87}$$

where we now add the superscript $F_0^{\text{in}}(X) \equiv F_0(X)$ for the sake of clarity.

3. *Matching.*

   The only, as yet, undetermined constant is B and it will be determined by the
   procedure of *matching* explained below. The important point to remember is
   that both the inner and the outer solutions actually are approximations of the
   same function (in different regions); therefore, we expect that there is a smooth
   transition between the two in the intermediate region. In this example as one
   comes out of the boundary layer, that is, $X \to \infty$, the inner expansion has to be
   equal to the value of the outer expansion as one comes into the boundary layer,
   that is, $x \to 0$. More formally, the matching condition is

$$\lim_{\varepsilon \to 0, \, x \text{ fixed}} F^{\text{in}}(X) = \lim_{x \to 0} f^{\text{out}}(x), \tag{3.88}$$

   or more simply, symbolically

$$(F^{\text{in}})^{\text{out}} = (f^{\text{out}})^{\text{in}}, \quad \text{as } \varepsilon \to 0. \tag{3.89}$$

   Imposing this kind of condition on Eqs. (3.87) and (3.83) immediately yields
   $B = e$, and the final result for the lowest order term of the outer expansion is

$$F_0^{\text{out}}(X) = e - e^{1-2X}, \tag{3.90}$$

   where it has to be remembered that $X = x/\varepsilon$.

4. *Composite solution.*

   The objective of the final step is to obtain the complete approximate solution,
   correctly composed of the inner and outer solutions. This can be readily achieved
   by adding the two solutions over the entire domain (the two solutions do not
   contribute anything outside the intermediate overlap region) and subtracting their
   common part (which has nonzero contribution only in the intermediate, matching
   region). In the present example this yields in lowest order

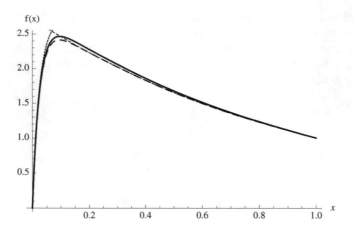

Fig. 3.8 Outer and inner solutions in lowest order (*dotted*) and the composite solution at this order (*dashed*) is compared with the exact solution (*solid*). In this example, $\varepsilon = 0.05$. To achieve an even better approximation the inclusion of HOT is called for

$$f \sim f_0(x) + F_0\left(\frac{x}{\varepsilon}\right) - f_0(0) \sim e^{1-x} - e^{1-2x/\varepsilon}. \tag{3.91}$$

The good quality of the boundary layer method approximation in this simple example can be seen clearly in Fig. 3.8, where the exact solution of (3.80) (solid line) is shown along with the outer and inner solutions (dotted) and the composite approximation (dashed). The value of ε here is 0.05 and further improvement of the approximation can be achieved, if the next order (too complicated to be discussed here) is included.

3.5.2   The Blasius Boundary Layer

We consider now the Blasius boundary layer problem and use this opportunity in order to show how to estimate the order of magnitude of the different equation terms relevant in this situation and exploit the information in a systematic procedure. Let there be a steady, viscous, two-dimensional flow taking place in the x–y plane with corresponding velocity components u and v. The flow proceeds in the x direction parallel to a semi-infinite plate, e.g., a strip $0 \le x \le L$, infinite in the y direction. L is assumed to be large so that it is much larger than the average width of the boundary layer whose size is denoted by $\bar{\delta}$. The boundary layer does not have to be of fixed thickness, and we shall see further on that it is, in fact, dependent upon one's position along x. The flow is arriving from the left and it is in the x direction. Its magnitude far away from the object is U_0. The boundary conditions to be used are

1. $u = v = 0$ for $y = 0$ and $0 < x < L$,

2. $u \to U_0$ for $y \to \infty$.

Considering only the first component of the equation of motion, i.e.,

$$u\frac{\partial u}{\partial x} + v\frac{\partial u}{\partial y} = -\frac{1}{\rho}\frac{\partial P}{\partial x} + v\left(\frac{\partial^2}{\partial x^2} + \frac{\partial^2}{\partial x^2}\right)u, \tag{3.92}$$

where v is the kinematic viscosity, we notice that if the equation is nondimensionalized, all that changes is $v \mapsto 1/\mathrm{Re}$. We shall now use the key assumption of the boundary layer approximation, namely, that in such boundary layers, variations in y are much more rapid than those in x. The boundary layer thickness is δ, so we estimate at once that, since

$$\frac{U_0}{\delta} \sim \left|\frac{\partial u}{\partial y}\right| \gg \left|\frac{\partial u}{\partial x}\right| \sim \frac{U_0}{L}, \tag{3.93}$$

where $\sim$ means, as usual, $\mathcal{O}$, that is "order of"" that the boundary layer thickness is much smaller than a typical scale of the main flow, i.e.,

$$\delta \ll L \quad \text{or} \quad \varepsilon \equiv \frac{\delta}{L} \ll 1. \tag{3.94}$$

Now, since we assume that in the boundary layer the viscous term is of the order of the inertial term, one may estimate

$$\frac{U_0^2}{L} \sim \left|u\frac{\partial u}{\partial x}\right| \sim \left|\frac{\partial^2 u}{\partial y^2}\right| \sim v\frac{U_0}{\delta^2}, \tag{3.95}$$

that is,

$$\delta^2 = v\frac{L}{U_0} \implies \frac{\delta}{L} \sim \mathrm{Re}^{-1/2} \implies \mathrm{Re} \sim \varepsilon^{-2}. \tag{3.96}$$

Another order of magnitude estimate that follows is

$$\left|\frac{\partial P}{\partial x}\right| \sim \rho\left|u\frac{\partial u}{\partial x}\right|. \tag{3.97}$$

Returning to the dimensional equations we get the boundary layer equations (Problem 3.10):

$$u\frac{\partial u}{\partial x} + v\frac{\partial u}{\partial y} = -\frac{1}{\rho}\frac{\partial P}{\partial x} + v\frac{\partial^2 u}{\partial y^2}, \quad 0 = -\frac{\partial P}{\partial y}, \quad \frac{\partial u}{\partial x} + \frac{\partial v}{\partial y} = 0. \tag{3.98}$$

This set of equations is not as simple as it may look and its solution depends, of course, on the boundary conditions employed (non-penetrating no-slip conditions

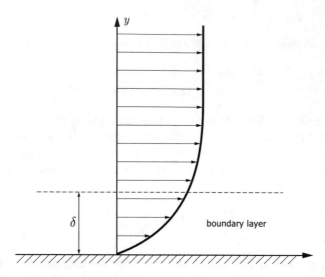

Fig. 3.9 Schematic drawing of a boundary layer forming on a plate, located at $y = 0$ in a flow proceeding parallel to the plate and above it. The x-axis in the figure plane direction, at the bottom, with $x = 0$ at the same position as $y = 0$ in the figure plane

as in our case) or even on more complicated bottom wall conditions. There also remains the question of how a boundary layer solution can be smoothly matched to the external flow. It is beyond the scope of the book to deal with these issues in detail but we shall give the proper physical and mathematical framework to better perceive and handle the problem posed.

We adopt the above boundary conditions and proceed towards developing an approximate equation that the basic flow state must satisfy for the case depicted in Fig. 3.9, where it is assumed that we have a flow over a flat wall coinciding with our strip on the $y = 0$ plane. As said before, well above the plate the flow is assumed inviscid and satisfies the irrotational version of the Bernoulli equation; in fact, it is a uniform flow with velocity $\mathbf{u} = U_0\hat{\mathbf{x}}$. Along the streamline just above the boundary layer, we thus have $P/\rho_0 + (1/2)U_0^2 = C$ (where ρ_0 is a constant density). It is also not difficult to see that all terms involving v are $\mathcal{O}(\varepsilon)$ so the pressure y derivative is $\mathcal{O}(\varepsilon^2)$ and along any streamline (they are parallel to the x-axis) the pressure is constant. Hence, up to order ε our (dimensional) equations reduce to

$$u\frac{\partial u}{\partial x} + v\frac{\partial u}{\partial y} = v\frac{\partial^2 u}{\partial y^2}. \tag{3.99}$$

Do you understand why the second equation was discarded? Now we shall try to obtain the stream function $\psi(x,y)$ for this problem from dimensional and similarity analysis. Defining $\mathrm{Re}_x = U_0x/v$ and $\xi \equiv yx^{-1}\mathrm{Re}_x^{1/2}$ and using similarity solution considerations, we make an educated guess of the following ansatz for the stream function

$$\psi(x,y) = \sqrt{vxU_0}f(\xi) \qquad (3.100)$$

and obtain that if, as usual, $u = \partial\psi/\partial x$ and $v = -\partial\psi/\partial y$ we will have $u/U_0 = f'(\xi)$ and the equation with its boundary conditions will be satisfied if only the function f satisfies the ODE

$$f''' + \frac{1}{2}ff'' = 0, \qquad (3.101)$$

with the boundary conditions $f(0) = f'(0) = 0$, $f(\xi \to \infty) = 1$. What is the meaning of the last condition? Equation (3.101) has to be solved numerically and from its solution one can reconstruct the boundary layer structure (see Problem 3.11).

3.5.3 Concluding Remarks on Boundary Layers

The body of research on mundane boundary layers tends to be large. This is chiefly so because of the engineering needed for designing solid objects (e.g., cars, rockets) that have to move through the air and partially or fully submerged ones in liquids (ships, submarines). Physical understanding must accompany the designing tasks. One of the important recent achievements of this research was the understanding of

Fig. 3.10 Outflow from a New Jersey lake, after hurricane Irene, is forced to narrow by a concrete bridge support. Flow separation is clearly evident at the sharp bend and even though the flow is not laminar, the photograph illustrates well this phenomenon. (*Author: Altac csi, licensed under the Creative Commons Attribution-Share Alike 3.0 Unported—http://creativecommons.org/licenses/by-sa/3.0/deed.en*)

the mechanisms leading to boundary layer *separation* (Fig. 3.10). It is beyond the scope of this book to describe these mechanisms in greater detail.

Problems

3.1.
Consider an incompressible, viscous fluid (having a constant kinematic viscosity v) at rest between two parallel rigid boundaries at $y = 0$ and $y = H$. Assume that at $t = 0$ the lower boundary ($y = 0$) starts to move in the x direction with velocity U and continues with this speed for all $t > 0$. Prove that the velocity distribution for $0 \leq y \leq H$ is

$$\mathbf{u}(y,t) = \left[\left(1 - \frac{y}{H} \right) - \frac{2U}{\pi} \sum_{j=1}^{\infty} \frac{1}{j} \exp\left(-\frac{j^2 \pi^2 vt}{H^2} \right) \sin\left(j\pi \frac{y}{H} \right) \right] \hat{\mathbf{x}}. \qquad (3.102)$$

Hint: Clearly, this problem is governed by the diffusion equation as in the Rayleigh problem but with no pressure driving. However, the boundary conditions are not homogeneous—there is an impulse of the lower boundary at $t = 0$. Thus, the mathematical technique has to use a general solution to the homogeneous problem plus a particular solution to the inhomogeneous one.

3.2.
Solve for the *plane Poiseuille* flow, that is, a flow between two parallel plates (at $y = 0$ and $y = H$) driven by a pressure difference in the x direction (that of the flow) being ΔP over length L, with $u_y = u_z = 0$.

3.3.
Solve the Rayleigh problem (see Sect. 3.2.4) for an impulsive start of the boundary motion, that is, $U(t) = 0$, for $t < 0$; $U(t) = U_0$, for $t \geq 0$ with $u = 0$ for all finite t at $z \to \infty$.

3.4.
Imagine a solid disk of a very large radius (ideally, infinite) rotating at the $z = 0$ plane with angular velocity Ω. Find a nonlinear set of ODEs, whose solution will describe the entrainment of the fluid, i.e., its flow for $z > 0$, where the fluid resides originally. Neglect any body forces, and assume a constant uniform inflow, $w = -W$, at very large z, as needed.

3.5.
Start from the biharmonic equation for the stream function (3.35) subject to the condition $\mathbf{u} = \mathbf{u}_*$ on the sheet (see the definition of the wave on the sheet through the equations for x_* and y_* in the text). Expand the x and y derivatives of ψ around $y = 0$ to second order in ε (see text). The result should be Eq. (3.36).

3.6.

A stream function ψ' expressed in a plane (r,z) automatically satisfies the incompressible condition. Imagine an axi-symmetric flow expressed in cylindrical coordinates (r, z, φ). According to (2.76) a representation $\mathbf{u} = \nabla \times (\psi' \hat{\varphi})$ is valid. However, ψ' does not satisfy the *physical* quality of being constant on streamlines. Show it. Examine, instead, the prescription for the velocity given by

$$\mathbf{u} = \nabla \times \left(\frac{\Psi}{r} \hat{\varphi} \right), \tag{3.103}$$

where Ψ is the *Stokes stream function*. Show that it guarantees incompressibility and is constant on streamlines and thus can serve as a bona fide stream function for axi-symmetric flows.

3.7.

From the Stokes stream function given in Problem 3.6 find the fluid velocities in the problem of a viscous fluid flowing from $-\infty$ (where its velocity is U) towards a spherical body of radius R. By integrating the pressure equations and finding the viscous stresses imparted on the surface of the sphere, show the sphere suffers a drag force of $F_D = 6\pi\eta RU$. Calculate the terminal velocity of a ball falling in a container of glycerine in terms of $\Delta\rho$, the difference between the density of the ball and the of the glycerine (in c.g.s.). Given are $R = 1\text{cm}$, $\eta_{gl} = 10$ c.g.s.

3.8.

1. Complete the proof of the reciprocal theorem (3.59) for Stokes flows, from the equality in the text (3.61). Hints: start from evaluating the following expression $\sigma_{ij}^{(1)} \mathscr{D}_{ij}^2$ and use tensor symmetries, dummy index interchange, and product of derivatives formula to arrive at appropriate expressions that are in the form of a divergence.
2. How can a body force be incorporated in the theorem to generalize it?
3. Can you give an example in which this reciprocal theorem is applied?

3.9.

The limit $\text{Re} \to 0$ was taken in the text, at lowest order, by setting the Reynolds number to zero and thus implicitly assuming that this is a regular limit, while in reality it is a *singular* one (see in Sect. 3.5.1). Explain the mathematical and physical meaning of this. Taking into account the fact that for $r \sim 1/\text{Re}$ (r in units relevant to the problem, e.g., the radius of the moving sphere in the Stokes problem) the neglect of the inertial terms is not justified. Oseen suggested to keep an approximate contribution of the inertial term. Suggest such a linear term, remembering that U is in the $\hat{\mathbf{y}}$ direction. Keep $\rho = 1$ and write the resulting Oseen equation. Show that it, in turn, leads to a modified equation for the stream function

$$\left[\mathrm{Re}\left(U\sin\varphi\frac{\partial}{\partial r}+\frac{U}{r}\cos\varphi\frac{\partial}{\partial\varphi}\right)-\nabla^2\right]\nabla^2\psi=0,\qquad\text{where}$$

$$\nabla^2=\frac{1}{r}\frac{\partial}{\partial r}\left(r\frac{\partial}{\partial r}\right)+\frac{1}{r^2}\frac{\partial^2}{\partial\varphi^2}.$$

(3.104)

3.10.
Derive the dimensional boundary layer set found in Eq. (3.98), using the procedure explained in the text.

3.11.
Numerically solve equation (3.101) and describe how to proceed in order to construct from it the Blasius boundary layer, as detailed in the text. Plot $f(\xi)$ and $f'(\xi)$, using $f(0)=f'(0)=0$ and $f'(\xi\to\infty)=1$ and find the asymptotic behavior of f for large ξ.

3.12.
The dimensional similarity analysis leading to the stream function ansatz (3.100) in the Blasius boundary leads to the understanding that

$$\xi=y\sqrt{\frac{U_0}{\nu x}}.$$

Using the results of the previous problem find the stream function as $y\to\infty$ up to $\mathcal{O}(\varepsilon)$.

3.13.
Determine the viscous flow in the setting of a jet emerging from the end of a narrow tube into an infinite space filled with the fluid. This problem goes under the name of a *submerged jet*. Hint: use polar coordinates (and the fact that the problem is axially symmetric) and justify the use of the ansatz $u_r(r,\varphi)=F(\varphi)/r$ and $u_\varphi(r,\varphi)=f(\varphi)/r$. Find the functions f and F from algebraic and differential manipulations of the equation of motion and complete the calculation by imposing physical conditions on the jet momentum.

Bibliographical Notes

Sections 3.1 and 3.2

Essentially all books on FD contain discussions of simple viscous flow. We recommend, in particular, the book of Landau and Lifshitz, reference [1] as listed in the *Bibliographical Notes* of Chap. 1, as well as the one of Acheson (here, [1]).

Section 3.3

1. C.S. Yih, *Fluid Mechanics* (West River Press, Ann Arbor, 1998) contains a wealth of examples of nontrivial viscous flows.

Section 3.4

Stokes flow is well explained in

1. D.J. Acheson, *Elementary Fluid Dynamics* (Clarendon Press, Oxford, 1990)
2. S. Childress, *Mechanics of Swimming and Flying* (Cambridge University Press, Cambridge, 1981)
3. J.M. Ottino, Sci. Am. **260**, 40 (1989)
   A very good discussion on the method of Frobenius and ODEs, in general:
4. G.F. Simmons, S. Kranz, *Differential Equations; Theory, Technique, Practice* (McGraw-Hill, New York, 2006)
5. E.L. Ince, *Ordinary Differential Equations* (Dover, New York, 2012). Suspensions are treated in the Landau and Lifshitz book and much more thoroughly in
6. J. Happel, H. Brenner, *Low Reynolds Number Hydrodynamics* (Martinus Nijhoff, Dordrecht, 1983)

Section 3.5

Boundary layers are treated in a variety of books, the most comprehensive is the classic

7. H. Schlichting, K. Gersten, *Boundary Layer Theory, 8th edn.* (Springer, Heidelberg, 2003)
   The following review may have special appeal to students interested in the subject:
8. J.P. Boyd, The Blasius function: computation before computers, the value of tricks, undergraduate projects and open research problems. SIAM Rev. **50**, 791 (2008)
   Asymptotic methods, sometimes useful in treating fluid boundary layers, are well described in a number of books, the best of which are, in our opinion, those by Nayfeh, Kevorkian and Cole and Van Dyke. They are explicitly referenced in Chap. 4.

Chapter 4
Linear and Nonlinear Incompressible Waves

Through the dear might of Him that walked the waves.

John Milton (1608–1674); 'Liquids'

Waves in fluids are ubiquitous, from the outward propagation of pond ripples, ocean waves crashing ashore, sound and shocks in the air, and so many others that to list them all would be impossible. Waves appear also in a multitude of other branches of physics. For this reason we start this chapter with a mathematical and physical description of waves and adopt a linear point of view. This is a more basic and significantly simpler approach to discuss mathematically than the general one. It is a starting point which is empirically motivated, a natural conceptual tool to bridge the gap between mathematical results and observed fluid phenomena. Many of the wave phenomena that we observe day-to-day reside in the so-called *linear* regime wherein the mathematical and conceptual machinery of waves and wave dynamics is captured by *small* (in meaning defined more precisely later on) departures from steady states.

Most of the notions related to linear waves, including when and where linearization works, are discussed in the main initial part of this chapter. Because of their strong link to a solid mathematical construction, we spend a considerable part of the discussion in this chapter developing the proper mathematical formalism necessary for producing reasonably rigorous results, not just in regard to waves but also for other calculations encountered throughout the book. We also introduce here how to use scaling arguments to develop *reduced* equations, like those of shallow water. In order to ease the task of introducing the above mathematical machinery and demonstrating its use in interpreting a large class of fluid phenomena, we develop the language mainly in the context of the simplest kind of dynamical fluid: the incompressible irrotational, i.e., potential flow in one spatial dimension.

© Springer Science+Business Media, LLC 2016
O. Regev et al., *Modern Fluid Dynamics for Physics and Astrophysics*,
Graduate Texts in Physics, DOI 10.1007/978-1-4939-3164-4_4

4.1 Waves, A Mathematical Primer

Before delving into how waves emerge as natural solutions of fluid equations, it is important to introduce some terminology. Suppose there exists what may be called a one-dimensional wave-train, embodied in the function $\eta(x,t)$, describing the departure of some quantity from equilibrium or steady state in one space plus a time dimension. Mathematically, we start with a simple wave-train, as depicted in Fig. 4.1, which may be *locally* approximated by the sinusoidal waveform

$$\eta(x,t) \approx A_k \cos(kx - \omega t).$$

If the *amplitude* $\equiv A_k$ is a constant, i.e., has a fixed numerical value in some units, not depending on k, despite the subscript, this wave train is strictly *periodic*. The functional form is then familiar and so we define and give meaning to the various quantities appearing. In any given fixed value of time t, we may assess a *wavelength* λ the nominal spatial (x) distance measure of successive crests or troughs in η, see Fig. 4.1. This is related to the *wavenumber* k appearing in the expression for η by

$$k = \frac{2\pi}{\lambda}.$$

The quantity ω represents an *angular frequency* of the maxima, say, of η, or, e.g., its throughs. In this sense we mean that in a fixed position in space, ω is a measure of the number of radians passed-through, per unit time by the peak, say, of η, which is moving to the right. If T denotes the *period* between two successive peaks, passing at said fixed position, then the peak passes through 2π radians, i.e., one cycle, in time T given by

$$T = \frac{2\pi}{\omega}.$$

Even though we shall use the angular frequency ω throughout, it should be mentioned, for the sake of completeness, that another measure of frequency is often

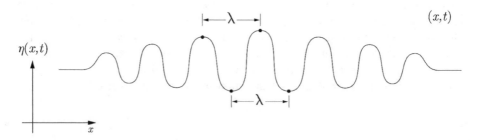

Fig. 4.1 A wave-train schematic diagram

used, namely, the number of *cycles* per second. It is denoted by f or v and the following obvious relations hold

$$f = v = \frac{\omega}{2\pi} = \frac{1}{T}.$$

The physical unit "cycles per second" has special name, honoring a prominent nineteenth-century physicist, H. Hertz, who was among the discoverers of electromagnetic radiation. The unit is called simply "Hertz," abbreviated in writing as Hz. An interesting feature appears in examining the waveform η: one can choose a moving reference frame, in which the sinusoidal pattern appears fixed in time. Let us define a Galilean reference frame boost V_p through the expression $X = x - V_p t$. Rewriting the waveform η in terms of the following new variable, X, i.e., $x \mapsto X + V_p t$ reveals explicit time independence for

$$V_p = \frac{\omega}{k},$$

that is,

$$\eta\left(kX + V_p kt - \omega t\right) = \eta\left(kX\right), \quad \text{if} \quad V_p = \frac{\omega}{k},$$

in other words, the wave-train pattern appears independent of time to an observer moving with the appropriate velocity $V_p \equiv \omega/k$, which is usually referred to as the *phase velocity*. Note that from the above expression for η, it follows also that $V_p = \lambda f$, whose meaning is that in the lab frame a distance of a wavelength is traversed by the wave-form in a time of one cycle. A wave in which the amplitude A_k can be perceived as a fixed parameter may be represented by the functional form $\eta \propto \cos(kx - \omega t)$ and is referred to as a *monochromatic* wave, meaning "of single color," i.e., of a single wavelength λ.

Sticking with our role as empirical observers, we could tabulate, from direct observation of carefully designed experiments, the frequency corresponding to a monochromatic wave of given wavelength. In general, one would find that for most given phenomena the frequency of a given waveform depends specifically on the waveform's wavelength and thus on the wavenumber so that $\omega = \omega(\lambda)$ and $\omega(k)$, where usually physicists prefer the latter form $\omega = \omega(k)$, which can be considered more useful. We shall see momentarily why. A familiar example is the relation for electromagnetic waves in vacuum $\omega = ck$, where c is the speed of light in vacuum, a universal constant. A qualitatively different example is the propagation of waves in deep water where the relation is given, as we shall see, by $\omega^2 = gk$ where g is the magnitude of the vertical component of the gravitational acceleration. The relationship $\omega(k)$, whether tabulated or expressed analytically, is important for understanding the properties of the wave in question. It is referred to as the *dispersion relation*. With $\omega(k)$ given, it implies that the special velocity $V_p = \omega(k)/k$ we referred to before as phase velocity of a given monochromatic waveform may depend upon the wavenumber (and thus wavelength) of the waveform. In the case of light propagating in vacuum, it is obvious that the phase speed in vacuum is independent of wavelength and, therefore, the speeds of propagation of the two waveforms of differing wavelength are the same, or else there would be news! But in the case of deep water waves, the phase speed depends upon wavenumber, i.e., $V_p \propto k^{-1/2}$, and for two given waveforms of slightly differing wavelengths and thus wavenumber, there emerges a gradual separation in their relative phases whenever

$$V_p = \frac{\omega}{k} \neq \frac{\partial \omega}{\partial k} \tag{4.1}$$

and, when this is the case, the waves are called *dispersive*, i.e., exhibiting dispersive phenomena (see the bottom panel of Fig. 4.2). The derivative on the rightmost side of this formula is a velocity, as is obvious in the equation. We shall show below that this velocity plays a particularly important role in dispersive waves, and it will become clearer as the theoretical underpinnings of these ideas are more fully developed along this chapter.

We can speak of linear monochromatic waves in more than one spatial dimension as well. Let the quantity η represent a phenomenon dependent on all three spatial dimensions, so that we may write its waveform as

$$\eta(\mathbf{x}, t) = \Re \left\{ A_k \exp\left[i(\mathbf{k} \cdot \mathbf{x} - \omega t) \right] \right\}, \tag{4.2}$$

where we use, as is frequently the habit for periodic phenomena, complex notation, with the physical quantity being the real part of the expression. The vector $\mathbf{k} = k_x \hat{\mathbf{x}} +$

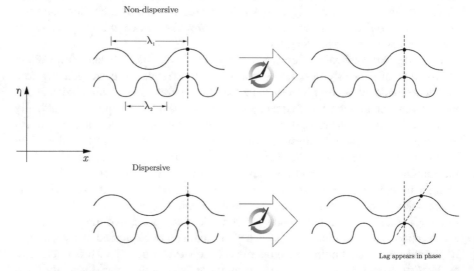

Fig. 4.2 Schematic depiction of both non-dispersive (*top panel*) and dispersive (*bottom panel*) waves

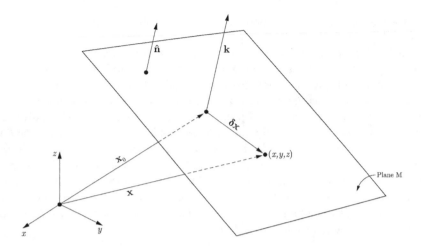

Fig. 4.3 Three dimensional wavefront and in a coordinate frame. The wave-vector **k** is perpendicular to the plane M with normal $\hat{\mathbf{n}}$

$k_y\hat{\mathbf{y}} + k_z\hat{\mathbf{z}}$ replaces now the familiar wavenumber and is called *wave-vector*. Its direction is $\hat{\mathbf{k}} \equiv \mathbf{k}/k$ where $k \equiv (k_x^2 + k_y^2 + k_z^2)^{1/2}$. The monochromatic waveform expressed in Eq. (4.2) would, for real values of A_k, take on very similar appearance as the one-dimensional spatial form, i.e., $\eta(\mathbf{x},t) = A_k\cos(\mathbf{k}\cdot\mathbf{x} - \omega t)$. Evidently, the three-dimensional waveform may be visualized as a collection of parallel planar fronts moving in the direction defined by the wave-vector **k**. To better connect the geometry of the physics to mathematics, we recall from elementary vector calculus that the set of points (x,y,z) constituting a plane M containing the point (x_0,y_0,z_0) is defined by its unit normal $\hat{\mathbf{n}}$ via $\delta x \cdot \hat{\mathbf{n}} = 0$, where

$$\delta x = \mathbf{x} - \mathbf{x}_0 = (x - x_0, y - y_0, z - z_0),$$

is a vector contained in the plane M (Fig. 4.3).

In light of the monochromatic wave's form, given in Eq. (4.2), we can see that the unit vector $\hat{\mathbf{n}}$ corresponds to the direction indicated by the wave-vector **k**. Thus at a given time t, there exists a plane M_t comprised of points $\mathbf{x}_t = (x_t, y_t, z_t)$ such that for all points on it $\mathbf{k}\cdot\mathbf{x}_t = \text{constant} = \omega t$. Furthermore, we can orient our perspective and choose a new coordinate system containing the direction defined by $\hat{\mathbf{k}}$ via $x' = \hat{\mathbf{k}}\cdot\mathbf{x}$ so that the three-dimensional waveform reduces to the familiar one-dimensional case, discussed earlier, that is, $\eta(\mathbf{x},t) = \eta(x,y,z,t) = \eta(x',t) = A_k\cos(kx' - \omega t)$. With this orientation in mind, it is easy to interpret the meaning of the three-dimensional monochromatic wave as a series of plane fronts propagating in the direction pointed to by $\hat{\mathbf{k}}$. Pretending that we are empirical observers having tabulated the dispersion relation for ω, we may very well find that it depends functionally upon the three components of the wave-vector **k**, that is, $\omega = \omega(\mathbf{k}) = \omega(k_x, k_y, k_z)$. It is possible, but by no means general, to find the tabulated dispersion relationship for some phenomenon to depend only upon the absolute value of the wave-vector, i.e., $\omega = \omega(|\mathbf{k}|) = \omega(k)$.

Finally, we introduce one concept specific to the relation of waves with their physical appearance. The motivation for the following definitions will become clear in an upcoming chapter but, for the sake of completeness, we define now the terms *transverse* and *longitudinal* waves. If a wave phenomenon is represented locally by a plane wave, then the direction of propagation of the wave, as we have previously defined, is $\hat{\mathbf{k}}$. Now if associated with the wave is a local velocity disturbance field given by the vector $\mathbf{u}$, then we say that the wave disturbance is transverse if $\hat{\mathbf{k}} \cdot \hat{\mathbf{u}} = 0$, which literally says that all velocity disturbance fields are perpendicular to the direction of motion of the wave. Longitudinal waves are those in which $\hat{\mathbf{k}} \times \hat{\mathbf{u}} = 0$ indicating that velocity disturbance vectors co-align with the direction of wave propagation. This definition carries over to disturbances in any field, e.g., $\phi(\mathbf{x}.t)$, propagating as waves in a definite direction.

4.1.1 One-Dimensional Linear Waves

We begin our discussion about waves, arising as mathematical solutions of actual equations, describing some physical phenomenon and show how the various wave properties described above follow naturally in these solutions. A simplest equation of this sort is a *first order, one-dimensional, unidirectional wave equation* with constant phase velocity V_p, given by

$$\frac{\partial \psi}{\partial t} + V_\mathrm{p} \frac{\partial \psi}{\partial x} = 0, \tag{4.3}$$

for some function $\psi(x,t)$. We show here that whatever the general solution, its functional dependence must possess the mathematical form[1]

$$\psi(x - V_\mathrm{p}t).$$

One way to show this is to simply take the form given above, directly substitute it into the governing equation (4.3), and verify by the differentiation chain-rule that indeed this is a solution of the equation. Another more revealing approach is to make a coordinate transformation and look for the solution ψ in terms of new variables defined by

$$X = X(x,t) = x - V_\mathrm{p}t$$

$$T = T(x,t) = t.$$

---

[1] We do not deal as yet with boundary and/or initial conditions which are necessary to find a particular solution of a PDE.

Clearly, X defines a (Galilean) reference frame moving with speed V_p to the right (increasing x). Substituting these new variables directly into Eq. (4.3) reduces the governing equation to the simple form

$$\frac{\partial \psi}{\partial T} = 0. \tag{4.4}$$

Thus, we see that whatever the general solution $\psi(X, T)$ is, we have now determined that it is independent of T and, as such,

$$\psi = \psi(X) = \psi(x - V_p t).$$

Any arbitrary function $\psi(X)$ satisfies the wave equation (4.3). In order to complete the formal solution to the wave equation (4.3), an initial condition must be given and we naturally suppose that the functional form of $\psi(x)$ is given at $t = 0$, i.e.

$$\psi(x, t = 0) = \psi_0(x).$$

The initial shape given at the initial time $\psi_0(x)$ remains fixed as viewed from the perspective of the observer moving with speed V_p in the positive x direction.[2] No specific boundary conditions are assumed and we look at the waves for $(-\infty < x < \infty)$.

These ideas may be extended to problems admitting two or more waves. For example, consider the second order partial differential wave equation

$$\frac{\partial^2 \psi}{\partial t^2} - V_p^2 \frac{\partial^2 \psi}{\partial x^2} = 0, \tag{4.5}$$

where V_p is, as before, constant. Introduce now the change of variables X and Y, given by

$$X = X(x, t) = x - V_p t,$$
$$Y = Y(x, t) = x + V_p t.$$

Comparing this to the form we defined earlier, it can be seen that whereas X represents a Galilean translation in a positive x direction and speed V_p, Y represents the same but in a negative x directions. Performing this change of variables in (4.5) transforms the second order PDE into

$$\frac{\partial^2 \psi}{\partial X \partial Y} = 0. \tag{4.6}$$

---

[2]Here we ignore any additive constants in the solution, determining that they are zero, owing to the initial condition.

This equation has the general solution

$$\psi(x,t) = f(X) + g(Y) = f(x - V_{\mathrm{p}}t) + g(x + V_{\mathrm{p}}t), \tag{4.7}$$

where f and g are each an arbitrary function, determined by initial conditions on the function ψ and its time derivative (see Problem 4.2). The interpretation is the same as before, however, with a minor twist: the solution to ψ is comprised of a linear sum of two solutions f and g where in f is a pattern that remains fixed in time in the frame of X while g remains similarly fixed in time in the frame of Y.

In principle, there could be many waves possible in a system. From a pure mathematical point of view, if we envision the following PDE in one space and time dimension

$$\frac{\partial^n \psi}{\partial t^n} + \alpha_{n-1} \frac{\partial^n \psi}{\partial t^{n-1} \partial x} + \cdots + \alpha_1 \frac{\partial^n \psi}{\partial t \partial x^{n-1}} + \alpha_0 \frac{\partial^n \psi}{\partial x^n} = 0,$$

then, if the constant coefficients $\{\alpha_j\}$ are such that the above equation may be factorized, that is, written as

$$\left(\frac{\partial}{\partial t} + V_{\mathrm{p}}^{(n)} \frac{\partial}{\partial x} \right) \left(\frac{\partial}{\partial t} + V_{\mathrm{p}}^{(n-1)} \frac{\partial}{\partial x} \right) \cdots \left(\frac{\partial}{\partial t} + V_{\mathrm{p}}^{(1)} \frac{\partial}{\partial x} \right) \psi = 0, \tag{4.8}$$

where the members of the set $\left\{ V_{\mathrm{p}}^{(j)} \right\}$ are all real and distinct. We see that the equation admits n-independent waves with each wave's phase velocity given by $V_{\mathrm{p}}^{(j)}$, respectively. The general solution can then be quite easily constructed by superposition and is given by

$$\psi(x,t) = \sum_{j=1}^{n} f^{(j)} \left(x - V_{\mathrm{p}}^{(j)} t \right),$$

where the functional forms $f^{(j)}$ are assessed from initial conditions: the spatial form of the function and of its $n - 1$ time derivatives at $t = 0$. The above form is but one of a myriad of wave equations and solutions. We shall return to this later in this chapter.

4.1.2   One-Dimensional Linear Wave Equation Reexamined as an Initial Value Problem

In general, linear wave equations may be more complicated than the simple example we began with, Eq. (4.3). For example, borrowing from quantum mechanics the Schrödinger wave equation, named after E. Schrödinger (1897–1961), one of that branch of physics founders, the equation has the form

$$\frac{\partial \psi}{\partial t} = i\beta \frac{\partial^2 \psi}{\partial x^2}, \tag{4.9}$$

where β is a parameter of the system. We may consider this as a purely mathematical example of a wave equation, having nothing to do with quantum mechanics, and solutions to this equation will involve, in general, a collection of waves with frequency $\omega(k)$, such that the waves, as we will see, are highly dispersive, i.e., each individual monochromatic wave will progress with a phase velocity that depends upon the wavenumber of the wave. We shall explore this character of dispersion in more detail in the next section. Before doing so, it will be instructive, however, if we introduce some more mathematical formalism and demonstrate more tools in the context of the simple wave equation (4.3). We will then use these to analyze the properties of other equations like the Schrödinger wave equation above.

It is important to realize that the relatively simple analysis we were able to apply in generating the general solution of the wave equation (4.3) is actually based upon a more systematic procedure that we can apply to linear wave equations of many kinds. Furthermore, the generic nature of the solution determined for Eq. (4.3) does not automatically carry over to the solution of more complicated wave equations like Eq. (4.9). It may be valid only if one considers the evolution of monochromatic disturbances involving an infinite (in x) periodic wave-train (e.g., a sine function for all x). In order to develop solutions of more sophisticated wave equations, like Eq. (4.9), we must appeal to more general techniques.

To demonstrate this procedure, we return to the first order wave equation (4.3) and analyze it in more formal terms as a general *initial value problem* on which we utilize both Laplace and Fourier transforms. It may seem odd that we are re-deriving the solution to Eq. (4.3), since we already know its solution, from the much simpler considerations in the previous section, to be $\psi_0(x - V_p t)$ with the function $\psi_0(x)$ being the initial condition at $t = 0$. However, it should be understood that through this simple demonstration, we show the proper procedure behind generating mathematically rigorous solutions to *general* linear differential equations, wave equations, and other ones.

Let us, as before, seek the solution of Eq. (4.3) with an initial condition $\psi(x, t = 0) = \psi_0(x)$. We further assume that the initial data function ψ_0 is continuous at x and that the integral of its absolute value square is bounded, i.e., $\int_{-\infty}^{\infty} |\psi_0|^2 dx < \infty$. As a general rule, initial value problems are handled by using *Laplace transforms*. The Laplace transform of a function $\psi(x, t)$ is defined as

$$\Psi(x, s) = \int_0^\infty e^{-st} \psi(x, t) dt,$$

where s is generally *complex*. Since the integral in the Laplace transform is done on the time variable at a fixed spatial position x, we shall drop x from the function arguments, without any danger of confusion. For the purpose of convergence, we suppose that the function ψ is locally integrable and the temporal behavior of ψ is at most exponential.[3] The inverse transform, occasionally referred to as the *Bromwich integral*, is given by

$$\psi(t) = \frac{1}{2\pi i} \int_{\sigma-i\infty}^{\sigma+i\infty} e^{st}\Psi(s)ds, \tag{4.10}$$

where the integration path in the complex plane lies along a line parallel to the imaginary s-axis and is shifted so as to run to the right of all poles of the function $\Psi(s)$, in order for the integral to converge. This is schematically depicted in Fig. 4.4a. The choice of sufficiently large $\sigma > 0$ achieves the latter requirement and we are setting the integration path to be the vertical line in the complex plane in the interval $(\sigma - i\infty, \sigma + i\infty)$, or more formally, the one between the endpoints $\sigma \mp iR$, with $R > 0$ real. Later on $\lim_{\to\infty} R$ is taken for the whole *closed* contour (see below). Ultimately the Bromwich integral will be evaluated using the *residue theorem* of complex function analysis, by completing this straight integration interval into a closed semicircular contour at a distance R from the origin with $R \to \infty$. The complete closed contour, along which the complex integral of $e^{st}\Psi(s)$ is evaluated, includes the infinite semicircular path C_∞, connecting its endpoints to those of the straight interval of the Bromwich integral. The encircled domain now contains the entire negative $\Re(s)$ axis and all of the $\Im(s)$ axis, as can be seen in Fig. 4.4b, where one has to imagine that the limit $R \to \infty$ has already been taken. The choice of the contour C_∞ has to be such that the integrand function be bounded (for physical reasons). In the figure we show an example, in which $\Psi(s)$ has three singularities, e.g., simple poles, which are obviously points, but are depicted in the figure in an artificially prominent manner. The construction shown in Fig. 4.4b serves sometimes in the proof of the important residue theorem, which we shall quote later, by letting the forth and back paths to approach each other until they ultimately merge and cancel out, leaving all poles surrounded by a small circular path, in a direction opposite to that of the big contour. The interested reader is encouraged to find a deeper treatment of this topic in, e.g., references [5] and [7] listed in the *Bibliographical Notes* to this chapter.

A general property of the Laplace transform has now to be mentioned: if one is taking this transform of a function $f(t) = dg/dt$, then putting this form directly into the Laplace transform definition and integrating once by parts reveals

$$F(s) = \int_0^\infty e^{-st} f(t)dt = \int_0^\infty e^{-st} \frac{dg(t)}{dt} dt = s \int_0^\infty e^{-st} g(t)dt - g(0) = sG(s) - g(0)$$

---

[3]This is a prerequisite for the use of Laplace transforms. In particular, it means that the long time behavior of ψ could have exponential divergences associated with it $\propto e^{at}$ for $a > 0$, but not *worse*. Functional behavior like (for instance) e^{at^2} cannot be formally handled by the given definition of the Laplace transform.

Fig. 4.4 Bromwich integrals and associated contour paths used to evaluate them. (**a**) The path of the integral to determine the inverse transform and (**b**) the contour path traversed in the complex plane to facilitate evaluation. Poles of the function being integrated are designated by an exaggerated, prominent, *dark circle* for the sake of visual clarity

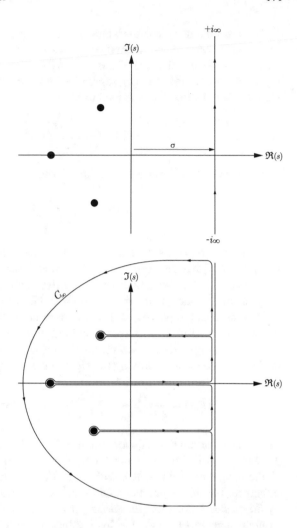

where $G(s)$ is the Laplace transform of $g(t)$ and $g(0)$ is the value of g at $t = 0$. The overall procedure in handling the solution of a wave equation involves thus the following steps. We take the Laplace transform of (4.3) yielding

$$s\Psi(x,s) + V_p \frac{\partial \Psi(x,s)}{\partial x} = \psi_0(x),$$

wherein $\Psi(x,s)$ is understood to be the Laplace transform of $\psi(x,t)$. Next, we perform a spatial *Fourier transform* to this equation by transforming it once more into

$$s\hat{\Psi}(k,s) + ikV_p\hat{\Psi}(k,s) = (s + ikV_p)\hat{\Psi}(k,s) = \hat{\psi}_0(k). \tag{4.11}$$

We are confident that the student has already been familiarized with the all important, at least to applied mathematics and physics, Fourier theorem, series, and transforms. To refresh some of this memory we define now the Fourier transform of a general spatial function $h(x)$, satisfying some suitable mathematical requirements. Denoted here by $\hat{h}(k)$, it is defined to be

$$\hat{h}(k) = \frac{1}{\sqrt{2\pi}} \int_{-\infty}^{\infty} e^{-ikx} h(x) dx, \tag{4.12}$$

where k is real. The inverse Fourier transform is given by

$$h(x) = \frac{1}{\sqrt{2\pi}} \int_{-\infty}^{\infty} e^{ikx} \hat{h}(k) dk. \tag{4.13}$$

An important remark should be made: if the function $h(x)$ is real, then its Fourier transform $\hat{h}(k)$ has the property that $\hat{h}(-k) = \hat{h}^*(k)$, where f^* denotes the complex conjugate of f. Further properties of Fourier analysis will appear in the following chapters of this book, and similarly reminded to the reader.

It is important to pause here keeping in mind that the function $\hat{\Psi}$ in Eq. (4.11) is the *double* transform: one Laplace and one Fourier, of the original function ψ for which we are seeking a solution, however $\hat{\psi}_0$ is just the Fourier transform of the initial condition $\psi_0(x)$. Solving the algebraic equation (4.11) for $\hat{\Psi}$ easily gives

$$\hat{\Psi}(k,s) = \frac{\hat{\psi}_0}{s + ikV_p} = \frac{\hat{\psi}_0}{s + i\omega(k)}, \qquad \text{because} \quad \omega(k) = kV_p.$$

The sum appearing in the denominator of the rightmost above expression is critical. Those values of s for which the denominator is zero, i.e., $s = -i\omega(k)$, are the poles of the integrand of Eq. (4.10), with an additional inverse Fourier transform of this integrand, but this is not essential here. The poles are the same and are identified by the dispersion relation, discussed at the beginning of the chapter. For our worked example here, $\omega(k) = kV_p$. Having a solution to $\hat{\Psi}$ we may take the inverse Laplace transform to construct $\hat{\psi}(k,t)$, i.e.,

$$\hat{\psi}(k,t) = \frac{1}{2\pi i} \int_{\sigma-i\infty}^{\sigma+i\infty} e^{st} \frac{\hat{\psi}_0(k)}{s + i\omega(k)} ds, \qquad \omega(k) = kV_p.$$

Since the poles of the integrand occur on the imaginary s axis, the location of the integration contour of the Bromwich integral can be set to $\sigma = 0^+$, meaning to say that the contour is placed infinitesimally to the right of the imaginary s axis, so that the poles found on the imaginary s-axis are located only slightly to the left of the vertical part of the contour, in contrast to the situation depicted in Fig. 4.4b, where the poles are a *finite* distance from the imaginary s axis, because $\sigma > 0$ and finite. Now, if we call C closed contour composed of the limits of integration of the Bromwich integral plus the infinite semicircular contour defined earlier and denoted by C_∞, then it follows from the residue theorem that

$$\int_C \left[\bullet \right] ds = \int_{\sigma-i\infty}^{\sigma+i\infty} \left[\bullet \right] ds + \int_{C_\infty} \left[\bullet \right] ds$$

$$= 2\pi i \sum_{j=1}^{N} \left\{ \text{Residues of } \left[\bullet \right] \right\}$$

where N is the total number of individual residues of the expression. This is schematically depicted in Fig. 4.4b. For our problem we would formally write

$$\int_{\sigma-i\infty}^{\sigma+i\infty} \frac{e^{st}}{2\pi i} \cdot \frac{\hat{\psi}_0(k)}{s+ikV_p} ds + \int_{C_\infty} \frac{e^{st}}{2\pi i} \cdot \frac{\hat{\psi}_0(k)}{s+ikV_p} ds$$

$$= 2\pi i \sum_j \hat{\psi}_0(k) \frac{e^{s_j t}}{2\pi i} = \hat{\psi}_0(k) e^{-ikV_p t}, \tag{4.14}$$

where s_j labels, in general, the poles in the integrand. Of course, in this case there is only the single pole located at $s_1 = -ikV_p$. Notice that for $t > 0$ the integral along the contour C_∞ is zero: because the semicircular contour extends out to infinite radius on the complex s plane by encompassing the $\Re(s) < 0$ axis, all contributions to the integral are killed off by the e^{st} term in the integrand. The solution for $\psi(x,t)$ may now be constructed by applying the inverse Fourier transform upon the expression $\hat{\psi}(k,t)$:

$$\psi(x,t) = \frac{1}{\sqrt{2\pi}} \int_{-\infty}^{\infty} \hat{\psi}(k,t) e^{ikx} dk$$

$$= \frac{1}{\sqrt{2\pi}} \int_{-\infty}^{\infty} \int_{\sigma-i\infty}^{\sigma+i\infty} \frac{e^{st}}{2\pi i} \cdot \frac{\hat{\psi}_0(k)}{s+ikV_p} e^{ikx} ds dk$$

$$= \frac{1}{\sqrt{2\pi}} \int_{-\infty}^{\infty} \hat{\psi}_0(k) e^{ik(x-V_p t)} dk.$$

To continue the calculation, one has to refer back to the definition of the inverse transform found in Eq. (4.13); it follows that the last line above is the inverse transform of the function $\hat{\psi}_0$ evaluated at the "parameter" $x - V_p t$ (instead of x). Thus the result of the last equation is simply

$$\psi(x,t) = \psi_0(x - V_p t). \tag{4.15}$$

At first glance it would seem that we have expended an unusual amount of energy to obtain an otherwise simple result, derived previously with equal simplicity.

However aside from defining a mathematical machinery, the procedure also offers insights that would not have been had otherwise. We expound on this in the following section.

4.1.3 Dispersion Relations, Phase and Group Velocities, Dirac Delta Function

Suppose one is given any linear, first order, one-dimensional wave equation, with constant coefficients, for a function ψ, e.g., the Schrödinger equation (4.9), mentioned before, and further suppose that the initial condition of ψ is given by the function $\psi_0(x)$. Then the general space-time solution $\psi(x,t)$ is given by

$$\psi(x,t) = \frac{1}{\sqrt{2\pi}} \int_{-\infty}^{\infty} \hat{\psi}_0(k) e^{ikx - i\omega(k)t} dk. \tag{4.16}$$

In general, we mean that one may "short-circuit" the Laplace transform stage of the process and, instead, express the solution of the system by evaluating Eq. (4.16) once the dispersion relation $\omega(k)$ is known, because we know that the Laplace transform involves an integral along a path, like we had before, of a function whose pole is at $s = -i\omega(k)$. A shortcut way, then, to determine the solution for a linear equation, without having to go through the Laplace transform route is to assume solutions of the wave equation to be a superposition (sum or integral) of the so-called *Fourier form* solutions, i.e.,

$$A_k e^{ikx - i\omega(k)t} \tag{4.17}$$

and insert this ansatz into the governing equation, utilizing the dispersion relation. We shall refer to an expression of the form (4.17) the *Fourier* or normal mode ansatz. A *normal mode* of an oscillating system is a pattern of motion, in which all parts of the system move sinusoidally with the same frequency and with a fixed phase relation.

In the example of the simple wave equation (4.3),

$$\left(-\omega + V_p k\right) A_k e^{ikx - i\omega t} = 0, \tag{4.18}$$

and a nontrivial solution (i.e., $A_k \neq 0$) arises when $\omega = \omega(k) = V_p k$, that is, ω is given by the above dispersion relation. In practice, when the dispersion relation is determined and provided that the initial condition is given, the spatio-temporal evolution of the system is determined wholly by evaluating the integral (4.16). Generally speaking for linear partial differential equations (not just limited to the wave equation), it is of central importance to ascertain the dispersion relation. Much of the analysis performed in this and also in a few subsequent chapters is organized around determining dispersion relations. Suppose that one is given the dispersion relation $\omega(k)$. As we have discussed above, a sinusoidal monochromatic wave in a medium with this dispersion relation $\omega(k)$, i.e.,

$$\psi = A_k \sin\left[kx - \omega(k)t\right] = A_k \sin\left\{k\left[x - \frac{\omega(k)}{k}t\right]\right\}, \tag{4.19}$$

will appear fixed in form to an observer who moves to the right with the phase speed $V_p \equiv \omega(k)/k$. Depending upon the problem, the required observer's frame speed may be a function of the wavenumber, k, because the dispersion relation need not to be of the simple form $\omega(k) \propto k$, like in electromagnetic waves in vacuum, but may be somewhat more complicated, e.g., $\omega(k) = V_p(k)k$, where the phase velocity is a nontrivial function of k. The mathematical underpinnings of this intuitive interpretation should be by now clear, given the discussion of the previous section.

Let us apply now the concepts and techniques that we have developed, to the Schrödinger equation (4.9). Assume, as before, that the initial condition for the function $\psi(x,t)$ is given as $\psi_0(x)$. We may find the dispersion relation by inserting the Fourier form (4.17) into the governing equation and find

$$A_k(-\omega + \beta k^2) = 0 \qquad \Longrightarrow \qquad \omega(k) = \beta k^2.$$

Observe the first departure from our previous discussion. Whereas the phase velocity of the pattern is given by

$$V_p = \frac{\omega}{k} = \beta k,$$

it is now apparent that it depends upon the disturbance wavenumber (thus wavelength). In this way, then, solutions of the Schrödinger equation are clearly dispersive.

Before continuing any further with this example of the Schrödinger equation, we take a little detour motivating a new important construction called the *Dirac delta function*. This concept was already mentioned in Chap. 2 but was treated in an heuristic way, see, e.g., Eq. (2.106). Here we would like to base Dirac's delta on a mathematically firmer ground. Consider a system that is wholly described by its Fourier form, and its initial ($t = 0$) condition is simply

$$\psi(x,0) = \frac{1}{\sqrt{2\pi}} A_{k_0} e^{ik_0 x}, \tag{4.20}$$

in which A_{k_0} is the amplitude. If we compare this expression to its equivalent definition via its inverse Fourier transform, the structure of the initial condition may be reverse-engineered, i.e.,

$$\psi(x,0) = \frac{1}{\sqrt{2\pi}} \int_{-\infty}^{\infty} \hat{\psi}_0(k)e^{ikx}dk$$

$$A_{k_0}e^{ik_0x} = \int_{-\infty}^{\infty} \hat{\psi}_0(k)e^{ikx}dk$$

where, on inspection, we see that the functional form on the left-hand side of the expression is the same as the form of the integrand, at $k = k_0$. So we may identify the initial data function $\hat{\psi}_0$ in terms of a special kind of function. Actually, it is usually perceived as a generalized function, $\delta(\xi)$, called the Dirac delta function, after P. A. M, Dirac, a leading theoretical physicist of the first half of the twentieth century, who provided much of the mathematical basis of quantum mechanics and was among the initiators of particle physics and quantum field theory. The following identification practically defines Dirac's delta function

$$\hat{\psi}_0(k) = A_{k_0}\delta(k - k_0) \tag{4.21}$$

and gives its basic property in the expression

$$\int_{-\infty}^{\infty} \delta(x - x_0)f(x)dx = f(x_0),$$

which says that the integral of the product of a function and the Dirac delta function is the function evaluated where the argument of δ is zero.[4] The following expressions clarify further its properties and those of its derivative

$$\frac{1}{\sqrt{2\pi}} \int_{-\infty}^{\infty} e^{ikx}dx = \delta(k),$$

$$\int_{-\infty}^{\infty} \delta'(x - x_0)f(x)dx = -f'(x_0)$$

where primes denote derivatives with respect to the argument of the function f. There exists also an associated function called the step or *Heaviside* function defined by

$$H(x) = \begin{cases} 0, x < 0; \\ 1, x > 0. \end{cases}$$

Throughout this book, we shall set its value at the discontinuity $H(0) \equiv 1/2$ and adopt its derivative with respect to x as the Dirac delta function

---

[4]There is considerable mathematical worry over the formalness of Dirac's δ as a true function per se. A useful conceptualization is to treat it as a *distribution*, familiar in probability theory.

$$\frac{dH}{dx} = \delta(x).$$

This is consistent with all of our previous definitions as $\int_{-\infty}^{\infty} \delta(x)dx = \int_{-\infty}^{\infty}(dH/dx)dx = H(\infty) - H(-\infty) = 1$.

Suppose now that we are interested in solving the Schrödinger equation with initial data $\psi(x,0) = \psi_0(x)$, as given in Eq. (4.20). In other words, we can say that this system is initiated with power in no other wavenumber than $k = k_0$. Its Fourier transform corresponds to a single Dirac delta function at k_0, i.e., $\hat{\psi}_0(k) = A_{k_0}\delta(k - k_0)$. Given the preceding analysis, the time evolution of ψ is given by

$$\psi(x,t) = A_{k_0}e^{ik_0(x-V_p t)} = A_{k_0}e^{ik_0(x-\beta k_0 t)}. \tag{4.22}$$

With the initial data given as a Dirac delta function at k_0, we find the time-dependent solution of the Schrödinger equation to be a rightward shift of the original pattern with phase velocity $V_p = \beta k_0$.

Now imagine a modified situation in which the initial data $\psi(x,0) = \psi_0(x)$ has a corresponding Fourier transform, $\hat{\psi}_0(k)$, that qualitatively resembles a Dirac delta function centered at k_0, but is not identical to it: it is an acceptable regular function, which is strongly peaked around $k = k_0$ and exhibits a strong decay on either side of $k = k_0$. A function fitting this prescription, and one which also may serve as a reasonable facsimile of a Dirac delta function, when $\Delta \to 0$, is the Gaussian

$$\varphi(k;k_0,\Delta) = \frac{1}{\sqrt{2\pi\Delta}}e^{-\frac{(k-k_0)^2}{2\Delta}}, \tag{4.23}$$

with the parameter Δ measuring the half-width at half maximum of φ. Note that the symbol φ here should not be confused with its role as the azimuthal angle variable in both polar and cylindrical coordinates, for which it was used in other chapters of the book. The Gaussian above has been normalized, so that its integral is $\int_{-\infty}^{\infty} \varphi(k)dk = 1$. The resemblance of φ to the Dirac delta function is most apparent in the limit where the half-width approaches infinitesimally small values, i.e., $\lim_{\Delta \to 0} \varphi(k;k_0,\Delta) \to \delta(k-k_0)$ (see Fig. 4.5). For the discussions found in this book, we shall always interpret the Dirac delta function as representing something like the Gaussian functional form found in Eq. (4.23) in the limit of an infinitesimally small width. In practice, the meaning of infinitesimally small width ($\Delta \to 0$) shall be understood as a width Δ that is much smaller than any relevant scale L of the system, i.e., $0 < \Delta \ll L$ and, as such, serve in approximations. The Dirac delta function, in addition to its general role in mathematical physics, is an important tool in facilitating analytical solutions to, e.g., fluid problems involving sharp interfaces, a situation we shall encounter in the upcoming pages.

An initial condition $\psi_0(x)$ whose Fourier transform is a strongly peaked function at some wavenumber k_0, like that expressed by φ of Eq. (4.23), qualitatively depicts a function in physical space that looks like the form found in Eq. (4.20), but with the added new feature of a spatial modulation of the envelope of its wave extrema.

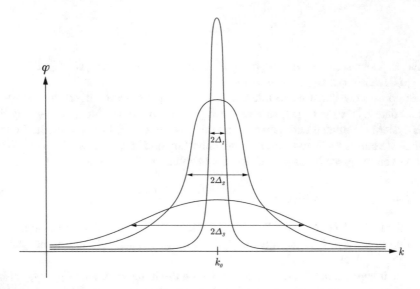

Fig. 4.5 Gaussian form with unit integral. It resembles a Dirac delta function in the limit where its width measure $\Delta \to 0$

Such a spatially modulated wave train is called a *wave-packet*. If the peak is narrow enough, i.e., Δ is suitably small, then it stands to reason that the waveform will show a correspondingly long length scale modulation of the basic periodic wave form with wavelength $\lambda_0 = 2\pi/k_0$. Indeed, as depicted in Fig. 4.6, the wave-train corresponding to the Gaussian form $\varphi(k)$ is the exponential form (4.20) with the constant amplitude A_{k_0} replaced by a spatially modulated function that we call $\Phi(x)$. In this example, that function is itself a Gaussian centered at $x = 0$ and decaying slowly (but surely) as $|x| \to \infty$. Speaking quantitatively, Φ is the Fourier transform of φ, as given above,

$$\psi_0(x) = \Phi(x)e^{ik_0x}, \qquad \Phi(x) \propto e^{-\frac{1}{2}\Delta x^2}. \tag{4.24}$$

These features naturally lead into a new concept called the *group velocity*, a name that fleshes out the evolutionary consequences of initial conditions that are not in the form of exact monochromatic waves represented by Dirac delta functions in wavenumber. Instead, we imagine an arguably more realistic waveform, initiated with a Gaussian profile in wavenumber space. The group velocity, as a concept, provides some information on the bulk evolutionary character of a collection of waves initiated together. For example, let us consider the fate of a slightly smeared monochromatic wave form given by Eq. (4.24). Writing the dispersion relation

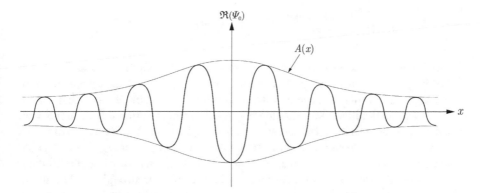

Fig. 4.6 The corresponding spatial form of $\psi_0(x)$ whose Fourier transform is φ given in the text. The envelope $A(x)$ is shown

$\omega(k) = \beta k^2$ as a Taylor expansion around the peak value of $k = k_0$ and retaining only terms up to second order

$$\omega(k) = \omega(k_0) + V_g(k - k_0) + \frac{1}{2} V_g'(k - k_0)^2,$$

in which the following identifications are made

$$V_g \equiv \frac{d\omega}{dk}\bigg|_{k_0}, \qquad V_g' = \frac{d^2\omega}{dk^2}\bigg|_{k_0}, \qquad (4.25)$$

we see that V_g, as given, defines the group velocity. In this particular example $V_g = 2\beta k_0$ and $V_g' = 2\beta$. Now we insert this rewritten form of the dispersion relation into the general wave solution Eq. (4.16) and perform some rudimentary algebraic manipulations to reexpress it in a more revealing way:

$$\psi(x,t) = \frac{1}{\sqrt{2\pi}} \int_{-\infty}^{\infty} \hat{\psi}_0(k) e^{ikx - i\omega(k)t} dk$$

$$= \frac{1}{\sqrt{2\pi}} \int_{-\infty}^{\infty} \hat{\psi}_0(k) e^{ik_0x + i(k-k_0)x - i[\omega(k_0) + V_g(k-k_0) + \frac{1}{2}V_g'(k-k_0)^2]t} dk.$$

This expression describes what is called a *wave-packet*. We have a superposition of waves with continuous wave numbers, whose amplitudes are expressed by the function $\hat{\psi}_0(k)$. The continuity of the superposition is reflected by the integral, a continuous sum, over k. If $\hat{\psi}_0(k)$ is indeed a very strongly peaked function around $k = k_0$, then we can define a new wavenumber variable $\delta k \equiv k - k_0$ to represent deviations from this peak wavelength. Inspection of $\varphi(k)$, as found in Eq. (4.23), shows that it is purely a function of δk and, therefore, so is the initial condition: $\hat{\psi}_0(k) = \hat{\psi}_0(\delta k)$. Thus

$$\psi(x,t) = \frac{e^{ik_0 x - i\omega(k_0)t}}{\sqrt{2\pi}} \int_{-\infty}^{\infty} \hat{\psi}_0(\delta k) e^{i\delta k \left(x - V_g t\right) - i\frac{1}{2}V_g' \delta k^2 t} \, d\delta k.$$

Up to this point, no approximation has been made in developing the solution to the Schrödinger equation with a general initial condition $\hat{\psi}_0(k)$. It is perhaps the right place to point out that we have used the Schrödinger equation, for no other reason than its being a familiar wave equation whose dispersion relation $\omega(k)$ is nontrivial, but the equation is linear. This allowed to meaningfully define V_g and distinguish it from V_p. The full solution is completed after evaluating the integral found in the above expression (see Problem 4.3). We can extract some insight without evaluating exactly the integral. Certain properties may be inferred for strongly peaked initial conditions $\hat{\psi}_0(\delta k)$, for example, like the form $\varphi(k)$ found in Eq. (4.23), by dropping the term proportional to δk^2 in the exponent. Formally speaking, dropping this term is acceptable only if one is interested in the evolution of the system for times significantly shorter than some maximum time $t_{\max} \equiv 1/(V_g' \cdot \Delta)$ (also verified in Problem 4.3). With this omission, the solution for these relatively short times is

$$\psi(x,t) \approx \Phi\left(x - V_g t\right) e^{ik_0 \left(x - V_p t\right)}, \quad \Phi\left(x - V_g t\right) \equiv \frac{1}{\sqrt{2\pi}} \int_{-\infty}^{\infty} \hat{\psi}_0(\delta k) e^{i\delta k \left(x - V_g t\right)} \, d\delta k.$$

$$(4.26)$$

The result can now be interpreted in the following way: the monochromatic wave k_0 with phase velocity $V_p = \omega(k_0)/k_0$ has as an amplitude envelope function Φ whose argument depends only upon the expression $x - V_g t$, i.e. $\Phi(x,t) \mapsto \Phi\left(x - V_g t\right)$. It indicates that in the frame moving rightward with the group velocity V_g, function Φ appears unchanged. One can envision this fixed shape Φ as defining an envelope function for the monochromatic wave in which the relatively high frequency action of the basic carrier wave (the basic wave with phase speed V_p) is amplitude localized around, for example, $x - V_g t = 0$ (see Fig. 4.7). Problem 4.3 examines the results of this interpretation against the exact solution of this Schrödinger system with the initial condition $\psi(x, t = 0) = \varphi$ given in Eq. (4.24). Although we will demonstrate this more explicitly in the upcoming sections, we should point out here that the group velocity typically is associated with the propagation of some physical quantity of the system under scrutiny. In FD applications, it will indicate the direction of propagation of the disturbance energy. Physical interpretation usually dictates that the quantity of interest will involve the square, or, in the complex function case, the absolute value squared, of the resulting wave solution, which may also carry information. If, for example, we interpret ψ as the wave-function of a quantum mechanical system, a physically meaningful and measurable quantity is the probability amplitude (in standard quantum mechanics interpretations) given by $|\psi|^2$. In what we have developed thus far this would be

$$|\psi|^2 \approx \left| \Phi\left(x - V_g t\right) e^{ik_0 \left(x - V_p t\right)} \right|^2 = \left| \Phi\left(x - V_g t\right) \right|^2.$$

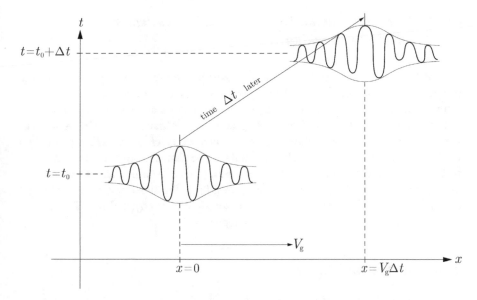

Fig. 4.7 The forward propagation of a mainly wavy initial condition with envelope function Φ. The bulk of the wave-train (where the amplitude in Φ is extremal) propagates rightward with group velocity V_g even though the individual wave pattern within exhibits a phase velocity V_p which is not always the same as V_g. In the figure, for display clarity, it is shown that the maxima of the modulated wave coincide with the maximum, of the envelope. This is, of course, true only if V_g/V_p is a rational fraction

Although we have introduced this concept through the specific solution of the Schrödinger equation, the conceptual framework of the group velocity carries over to any solution of any wave equation. The approximations we have used and the perspective we have developed here in interpreting the solution to the classical Schrödinger equation help in rationalizing the dynamics that emerge in a whole class of fluid problems. For example, a solution to a general wave equation may have a much more complicated dispersion relation $\omega(k)$. Nevertheless, together with an initial condition that is strongly peaked around some wavenumber k_0, or set of wavenumbers $k_0^{(j)}, j = 1, 2, \ldots, N$, where N is some finite integer, the concept, interpretation, and consequences of the group velocity will similarly carry over. We will see shortly that in certain wave propagation problems, the group velocity, evaluated in a multidimensional wave problem, will indicate the direction of energy propagation which is not always even in the same direction as the corresponding phase velocity.

4.1.4 One-Dimensional Unidirectional Nonlinear Waves and Their Breaking

We take a brief detour and introduce a way of intuitive thinking involving the progression of *nonlinear* waves. The approach grounds a certain amount of physical intuition into the notion of nonlinear wave distortion and relies on a perspective introduced in Chap. 1, namely, of the Lagrangian coordinates. From the mathematical point of view, we are talking about the *method of characteristics*, as applied here to incompressible waves. This method will be discussed in considerable detail in Chap. 6, before being applied to compressible flow. The mathematical solution expressed here, albeit in an implicit form, is simple. We show here how to connect the mathematical construction and derived solution to its physical manifestation in the hope that its intuitive power becomes apparent.

4.1.4.1 An Implicit Solution

Let the wave equation, for the function $\phi(x,t)$, be

$$\frac{\partial \phi}{\partial t} + V(\phi)\frac{\partial \phi}{\partial x} = 0, \tag{4.27}$$

in other words, the wave equation is characterized by a phase velocity V, suppressing the subscript "p" usually used to denote the phase velocity, which is a function of the wave-function itself ϕ. Suppose also that at some initial time $t = 0$, the function ϕ is initiated by some given function in space $\phi(x,0) = \phi_0(x)$. Assume now, if only for the sake of the following demonstration, that the functional form for the phase velocity is given by

$$V = V_0 + \phi.$$

The solution for the above system, valid, as we shall see, only for a limited time is

$$\phi(x,t) = \phi_0\left[x - (V_0 + \phi)t\right], \quad \text{for } t < t_b \quad \text{where} \quad t_b \equiv \min_x \left\{\left(-\frac{d\phi_0}{dx}\right)^{-1}\right\}, \tag{4.28}$$

and in the following discussion we prove it by showing how this solution is constructed. We stress here the curiosity that this solution is both implicit in its argument and, under certain conditions, is valid only up to a finite time, indicating a discontinuity (see below). We shall both prove this statement, discuss what it means, and show how it may be geometrically intuited from simple kinematic arguments. With V as given above and in accordance with standard linear wave theory, we may introduce a new variable

$$\xi \equiv x - V_0 t \qquad (4.29)$$

which represents a coordinate that moves with speed and direction indicated by V_0. In these new coordinates, the general nonlinear wave equation (4.27) is reduced to what is the celebrated *Burgers equation*:

$$\frac{\partial \phi}{\partial t} + \phi \frac{\partial \phi}{\partial \xi} = 0. \qquad (4.30)$$

The solution's form found in Eq. (4.28) comes from directly inputing the ansatz $\phi = F(\xi - \phi t)$ into Burgers's equation itself. It indicates that the quantity ϕ is some unspecified function F of an argument algebraically including itself. The curiosity of the implicit definition of the solution is handled, of course, via chain rule differentiation. Defining the variable $\Xi \equiv \xi - \phi t$

$$\frac{\partial \phi}{\partial t} = \frac{dF}{d\Xi} \cdot \left(-\phi - t \frac{\partial \phi}{\partial t} \right) \qquad \Longrightarrow \qquad \frac{\partial \phi}{\partial t} = \frac{-\phi \frac{dF}{d\Xi}}{1 + t \frac{d\phi}{d\Xi}},$$

$$\frac{\partial \phi}{\partial \xi} = \frac{dF}{d\Xi} \cdot \left(1 - t \frac{\partial \phi}{\partial \xi} \right) \qquad \Longrightarrow \qquad \frac{\partial \phi}{\partial \xi} = \frac{\frac{dF}{d\Xi}}{1 + t \frac{d\phi}{d\Xi}}.$$

Thus, putting this expression directly back into (4.30) we find the requisite cancellation, except for a troublesome feature, i.e.,

$$\frac{\partial \phi}{\partial t} + \phi \frac{\partial \phi}{\partial \xi} = \frac{-\phi \frac{dF}{d\Xi} + \phi \frac{dF}{d\Xi}}{1 + t \frac{dF}{d\Xi}} = \frac{(\phi - \phi) \cdot \frac{dF}{d\Xi}}{1 + t \frac{dF}{d\Xi}}.$$

While the numerator cancels exactly, indicating that this solves the system, there is a problem associated with the denominator if it ever crosses zero and the continued satisfaction with this solution cannot be ensured. With the function F identified with the initial distribution ϕ_0, we define the first moment this can happen as the time that the wave "breaks" and we will associate with it the symbol t_b defined by: $t_b \equiv \min \left(-1 / \frac{d\phi_0}{dx} \right)_{\forall x}$. In a more rigorous way, we say that at time t_b the solution develops a *weak discontinuity*. This appears again in our discussion of shock waves in Chap. 6. It is important to keep in mind the following subtlety: if we accept this solution as a valid one, even for a finite time, until t_b, then the functional form and evolution of the solution depends critically upon the functional form of the initial disturbance of ϕ, i.e., $\phi_0(x)$. Therefore, the solution $\phi = \phi_0(\xi - \phi t)$ and the breaking time t_b, if it is reached, is evaluated from the functional form of the initial distribution ϕ_0. One immediate consequence is clear: if $d\phi_0/dx$ is never negative, then there will be no possibility of wave-breaking. Below we shall sequentially develop the intuition that should help to better understand the meaning of this solution.

4.1.4.2   Characteristics: A Qualitative Demonstration for a Linear Wave

We shall now consider, for the sake of a qualitative understanding of the use of *characteristics* in an incompressible fluid, a linear example with the flow field set a priori. Consider a physical quantity expressed by the function $F(x,t)$, which at some initial time $t = 0$ is given by the spatial function $F(x,t) = F_0(x)$. Assume that this quantity is subject to a velocity field $V(x)$, which is spatially dependent, but fixed in time. This means that at $t = 0$, a fluid particle at a point x is endowed with a value of F equal to $F_0(x)$. Then the path $X(t)$, defined by the solution of the differential equation

$$\frac{dX}{dt} - V(X) = 0; \quad X(t = 0) = x_0, \tag{4.31}$$

on which the following is also satisfied

$$\left. \frac{dF_0}{dt} \right|_{x=X(t)} = 0, \tag{4.32}$$

will be such that, along it, the quantity F_0 will be constant. In other words, F_0, at a given starting position x_0, will remain the same along the path $x = X(t)$ given by the solution of the first order differential equation for $X(t)$ in which $x = x_0$ is the value of x at $t = 0$.

The physical consequence of this interpretation is illustrated in Fig. 4.8, depicting the fate of a tent-like localized pulse in F_0 initiated wholly in a region $x < 0$. The velocity field we consider is simple: $V(x) = V_-$ for $x < 0$ and $V(x) = V_+$ for $x > 0$ with $V_+ > V_- > 0$. Consider four points belonging to the initial profile labelled by their corresponding positions in x, e.g., $x_1 < x_2 < x_3 < x_4$. We further suppose that associated with these labelled points are four values of the quantity, which is to be advected F_i ($i = a,b,c,d$) such that $F_i = F_0(x_i)$. According to the flow dynamics, each of the labelled points will have their position readings kinematically evolve according to the prescription above, with their positions given by $x = X_i(t)$ with $X_i(t = 0) = x_i$. Along each path $X_i(t)$ the corresponding values of F_i are preserved. When the x value of each of these trajectories remains less than zero, each element labelled by i moves rightward with a velocity V_- in accordance with our established intuition. So while the entire pulse structure remains within $x < 0$, the evolution of the distribution F_0 moves with simple parallel pattern translation without distortion. However, the leading trajectory associated with label x_d reaches $x = 0$ before the others. Once it passes this point it will continue forward with the faster speed V_+. In this way distortion is experienced by the structure, as indicated by the illustration. Once the entirety of the pulse has passed beyond $x = 0$, its shape will still be recognizable but it will look stretched out. As is apparent from this depiction, the effective "wavelength" of the structure becomes longer. In this simple case, when the velocity field is given, it is easy to get the evolution of $F(x,t)$ following the flow,

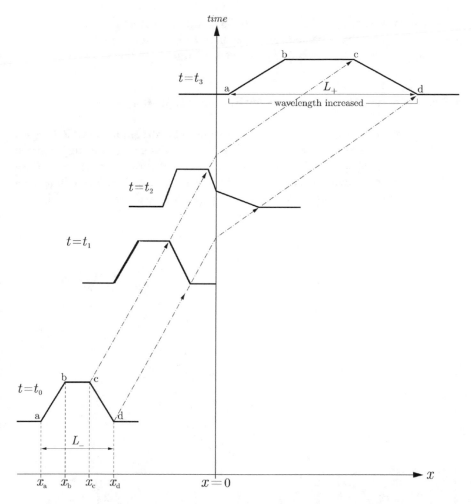

Fig. 4.8 Linear tent-like wave-form of the function F propagating rightward in a medium with two constant phase velocities on either side of $x = 0$. While totally contained in $x < 0$, the wave propagates rightward with constant velocity V_- and preserves its shape. However, as the waveform crosses $x = 0$ the phase velocity of the medium is a different constant $V_+ > V_-$ and the shape deforms, e.g., at $t = t_2$. Once the wave structure has completely crossed $x = 0$, here it happens at $t = t_3$, the wave has an effectively larger wavelength

which, as we have discussed in Chap. 1, is just

$$\frac{DF}{Dt} = \frac{\partial F}{\partial t} + V(x)\frac{\partial F}{\partial x} = 0, \tag{4.33}$$

for the initial distribution of $F|_{t=0} = F_0(x)$. The physical outcome of an equation like this, written in Eulerian form, may be understood within this intuitive framework. All that is now necessary is to appropriately interpret the meaning of the velocity term in the wave equation, i.e., $V(x)$ in (4.33). With this in mind we can now apply the same interpretative scheme to an equation like (4.27) and, in particular, develop a working mental image of how $V(\phi)$ dictates the time-evolution of the solution.

4.1.4.3 Characteristics: Analysis for Burgers's Equation

Building upon the tools of the last section, we examine the solution to the Burgers equation (4.30) from the perspective of characteristics. For the sake of simplicity we consider the case $V_0 = 0$, so that we have $\xi = x$, see (4.29), otherwise we would have a time-dependent horizontal shift to the right $x \mapsto x - V_0 t$. As before, we imagine a localized disturbance of $\phi(x, t = t_0)$ given by $\phi_0(x) > 0$, $\forall x$. By examining Fig. 4.9,

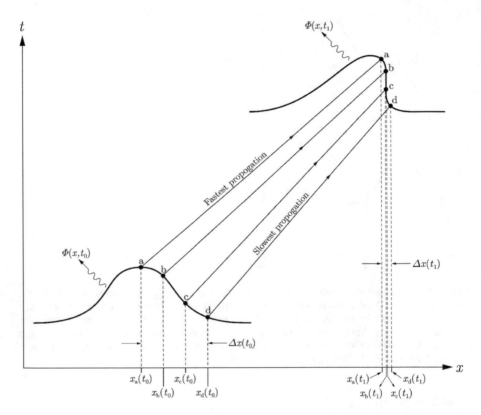

Fig. 4.9 One-dimensional nonlinear wave breaking schematic drawing. The shape of the wave ϕ is shown for two times. The speed of translation of the wave pattern is locally determined by the value of ϕ, which does not change as the wave moves and distorts its shape. Breaking (onset of weak discontinuity) is indicated at $t = t_1$ when the points $X_b = X_c = \tilde{X}_s$. Note that the shape of the wave distorts as it progresses forward in space and time. On the side of the wave structure in which $d\phi/dx < 0$, the two extreme ends of the labelled points are separated from one another by a distance $\Delta x = x_d - x_a$. Up to the beginning of wave breaking the value of Δx shrinks continuously

we similarly select four points on the profile ϕ at $t = t_0$ labelled by their x-coordinate positions (at $t = t_0$), x_i. Similarly we note the corresponding value of ϕ_0 at these four positions, ϕ_i, $i = a, b, c, d$. Now the key point to realize is the following: the velocity of the points on the structure is determined by the value of the structure at the initial time since in the Burgers equation the speed of the wave in ϕ is equal to the value of ϕ at the relevant point. Thus the initial disturbance $\phi(x, t = t_0) = \phi_0(x)$ determines how each individual fluid element of the initial disturbance moves forward in time. So, for instance, if we consider a fluid particle associated with the wave at point x_a and corresponding to it having the value of ϕ equal to $\phi_a = \phi_0(x_a)$, then the forward time trajectory of this point on the waveform is

$$X_a(t) = x_a + \phi_a(t - t_0).$$

As this "x_a" labelled element moves along, it also preserves its associated value of $\phi_0(x)$, equal to ϕ_a. In general it follows that the time evolution of each point on the wave, starting from (and labelled by) x_i, proceeds according to $X_i(t) = x_i + \phi_i(t - t_0)$.

As the figure indicates, since the elements of the disturbance travel rightward with differing speeds, there is the distinct possibility that two separated parts of the propagating disturbance may meet at some later time. But this is problematic, because it would imply that some point in space (call it $\tilde{X}_s$) at some specific time (call it t_1) would harbor two different values of $\phi_0(x)$. However the initial condition ϕ_0 is assumed to be a well-defined continuous function (we did not spell it out, because this is not a mathematics book). So this is an impossibility. Physically, the time at which this occurs indicates wave breaking. In reference to Fig. 4.9, for instance, let us consider the two adjacent elements of the original disturbance x_b and x_c. If such a breakdown point exists for a time t_1, then it means $X_b(t_1) = X_c(t_1) = \tilde{X}_s$. So

$$x_b + \phi_b(t_1 - t_0) = x_c + \phi_c(t_1 - t_0) \quad \Longrightarrow \quad (t_1 - t_0) = -\frac{x_c - x_b}{\phi_c - \phi_b} = -\frac{1}{\frac{\phi_0(x_c) - \phi_0(x_b)}{x_b - x_c}}$$

Notice that since $x_c - x_b > 0$, the value of $t_1 - t_0$ is greater than zero only if $\phi_0(x_c) - \phi_0(x_b) < 0$. As the figure indicates, the special point $\tilde{X}_s$ would correspond to two different values of ϕ_0. This means that there is a tendency to form a discontinuity if $\phi_0(x)$ is a decaying function of space. As this time is approached, the wave appears to steepen until a discontinuous feature appears, with singularity in the first derivative. The wave technically breaks for the earliest possible time which corresponds to the minimum value of all possible combinations of $t_1 - t_0$ in the expression above. Since we assume that the initial profile $\phi_0(x)$ is a continuous and differentiable function, it follows that the earliest possible instance for which breaking is possible occurs at a time Δt_b after t_0:

$$\Delta t_b = \min(t_1 - t_0) = \min_{\forall x} \left\{ -\left(\frac{d\phi_0}{dx}\right)^{-1} \right\}.$$

If $\forall x$, $d\phi_0/dx > 0$, then there is no danger of wave breaking. Physically speaking, when this multi-valued situation arises the system undergoes a process in which a *shock front* appears and propagates at some speed. This concept will be handled more formally, for compressible fluids, in Chap. 6. We should note here that in reality a fluid system characterized by the Burgers equation may have some viscosity. So as the propagating front steepens and the tendency of developing into a discontinuous jump is approached (a shock), the viscosity, however small, tends to smooth out the rough edges on the steepened side of the developing disturbance. In those cases one may find a propagation of a front which, over time, decays due to the viscosity.

In Chap. 6 we shall delve deeper into the mathematical theory of characteristics, in particular for handling fronts and shocks associated with compressible fluids, in which disturbances of any kind propagate at the speed of sound, complicating the dynamics. Here, in incompressible fluids, the speed of sound is nominally infinite, but the mathematical similarity to relevant equations facilitate the use of characteristics.

4.2 Gravity Waves on Water Surface as Irrotational Flows

The exposition of the propagation of linear waves over the surface of water in a constant gravitational field, treated as an irrotational flow, is an ideal setting to introduce procedures including linearization and its justification through basic scaling analysis, as well as development of linearized disturbances and the consequent wave motion. Near the end of this section, we will introduce a rudimentary mathematical modeling of surface tension and its possible role in such surface gravity waves. The overall discussion is quite detailed in order to introduce some general mathematical tools that will not only be used throughout the rest of this book, but also may have applications in other branches of physics.

4.2.1 Formulation

We examine the behavior of a constant density fluid configuration involving only two spatial dimensions: x for the horizontal and z vertical. Constant, antiparallel to $\hat{\mathbf{z}}$ the gravitational force per unit mass, g, is assumed to act on the system and this is a good approximation if the region we consider has a very small depth and height, as compared to the Earth's radius. We should also keep in mind that all of the following discussion generalizes if a second horizontal dimension y is included. The velocities corresponding to the Cartesian directions are (u, v, w). As stated above the fluid is incompressible, moreover we choose it to be constant density and it is assumed to be irrotational as well. This situation allows for potential flow, which was described and investigated in Chap. 2. In the present model, the fluid is ideally infinite in horizontal

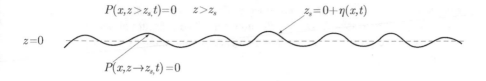

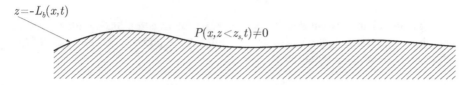

Fig. 4.10 Geometry of incompressible surface water waves. Note that we chose to mark the pressure above and on the surface as being equal to zero, while in real circumstances it is the atmospheric pressure P_{atm}. Will it make a difference in the waves equation of motion?

extent and in equilibrium its surface level is at $z = 0$. Let the surface of the disturbed fluid be $z = z_s$. Since initially we shall investigate the *linear* case of surface gravity waves, z_s will have to be small in some sense, which we will define more precisely momentarily. We allow for a more general case than just flat bottom and assume that the bottom of the body of water is located at $z = -L_b$. A schematic depiction of this system is given in Fig. 4.10.

In Sect. 2.4 we saw how an incompressible irrotational fluid is described by a velocity potential $\mathbf{u} = \nabla\phi$ where ϕ is a solution to the Laplace equation,

$$\nabla^2\phi = 0, \tag{4.34}$$

here, two-dimensional. We shall now call the vertical deviation to the water's surface equilibrium, which is at $z = 0$, due to a small disturbance $\eta(x,t)$, a function of position and time. The location of the surface of the fluid with the perturbation z_s is then

$$z_s = 0 + \eta(x,t),$$

or, in other words, we prefer to use η rather than z_s to denote a perturbation, in accord with widely accepted notation. Since the irrotational fluid of constant density is in a constant gravitational field, the evolution in terms of the velocity potential is described by the corresponding Bernoulli's equation. The various cases for the Bernoulli formula were discussed at length in Chap. 2, Sect. 2.3.5. Here we use the version called case 3b there. Thus

$$\frac{\partial\phi}{\partial t} + \frac{1}{2}\left(\nabla\phi\right)^2 + \frac{P}{\rho} + gz = f(t), \tag{4.35}$$

where we have used external constant gravitational acceleration g pointing downward and f is an arbitrary function of time only. The pressure above the surface is the atmospheric pressure and since the surface height changes by only a little, we may put

$$\frac{P}{\rho} = \frac{P_{\text{atm}}}{\rho_0} = \text{const} \equiv H_0, \tag{4.36}$$

since the fluid has constant density ρ_0 say. Now the freedom in $f(t)$ allows one to absorb $H_0 - f(t)$ in $\partial\phi(t)/\partial t$ changing nothing in the physics. So the result of the appropriate Bernoulli's equation is

$$\frac{\partial\phi}{\partial t} + \frac{1}{2}u^2 + g\eta = 0, \tag{4.37}$$

and this is correct at $z = \eta(x,t)$. We now linearize the surface condition by dropping u^2 from Eq. (4.37) and since on the free surface

$$\frac{\partial\eta}{\partial t} + u\frac{\partial\eta}{\partial x} = w \qquad \text{on} \quad z = \eta(x,t), \tag{4.38}$$

it yields $w(x,\eta,t) = \partial\eta/\partial t$. One can also Taylor expand w on the surface thus

$$w(x,\eta,t) = w(x,0,t) + \eta\frac{\partial w}{\partial z} + \text{HOT}.$$

Neglecting terms of higher order than the first gives:

$$w(x,0,t) = \frac{\partial\eta}{\partial t} \quad \Rightarrow \quad \frac{\partial\phi}{\partial z} = \frac{\partial\eta}{\partial t} \qquad \text{on} \quad z = 0. \tag{4.39}$$

Linearizing now the condition for the pressure, which is P_{atm} at $\eta(x,t)$ and using the equation of motion in the z direction and the possibility to express velocities as potential gradients, we finally get

$$\frac{\partial\phi}{\partial t} + g\eta = 0 \qquad \text{on} \quad z = 0. \tag{4.40}$$

To complete the setup of this problem we have to establish a *kinematic* condition, connecting the moving surface $z_s(x,t) = \eta(x,t)$ to the interior velocity field. As we discussed in Chap. 1, the motion of the surface is monitored by imagining the initial undisturbed surface's fluid particles labelled for the purpose of its Lagrangian description. It is almost a tautology to say that, in reference to the moving surface, the surface has zero velocity. Focusing for a moment on the function $Z(x,z,t) \equiv z - \eta(z,t)$, it is thus evident that its Lagrangian time derivative is zero on the surface. Thus the kinematic boundary condition for the surface is

$$\frac{DZ}{Dt} \equiv \frac{\partial Z}{\partial t} + u\frac{\partial Z}{\partial x} + w\frac{\partial Z}{\partial z} = 0 \text{ for } z = \eta(x,t), \qquad (4.41)$$

Using the above definition of the function Z we can identify

$$\partial_t Z = -\partial_t \eta, \qquad u\partial_x Z = -\partial_x \eta \qquad w\partial_z Z = w.$$

So the kinematic condition (4.41) evaluated at the *perturbed* surface $z = z_s = 0 + \eta(x,t)$, in accord with the above definition, is equivalent to Eq. (4.38):

$$\frac{\partial \eta}{\partial t} + u(x,\eta,t)\frac{\partial \eta}{\partial x} = w(x,\eta,t). \qquad (4.42)$$

This expression for the moving surface, as viewed from the laboratory frame, appears simple, however it proves to be quite challenging to handle in general circumstances. If the velocity fields are given a priori, then there is no trouble in following the moving surface. However, if the determination of the velocity fields dynamically depends upon the motion of the surface, then working with Eq. (4.42) involves some nuance. Below we show how to work with this in a relatively simple way for problems involving linearized disturbances. At the end of this chapter, we demonstrate how this equation can be handled in a nonlinear case—the example of solitons propagation in shallow or deep water.

To recap the situation for this incompressible irrotational fluid layer in a constant gravitational field: the velocity field in the interior fluid is given by the solution of the two-dimensional Laplace equation (4.34) for the velocity potential. The corresponding solution for the velocity must concurrently connect to the relationships governing the disturbed surface through Eqs. (4.38)–(4.42). This system forms a generalized nonlinear set of PDEs. To showcase how one goes about developing solutions of this system, we treat below the equations governing the linearized system in which the equilibrium state is a still fluid ($z_s = 0$) sitting atop a flat bottom boundary located at $z = -L_b$ ($L_b > 0$).

4.2.2   Linearization and Waves: Scaling and Normal Mode Analysis

From a mathematical point of view, linearization typically involves the following procedure:

1. Introduction of small disturbances about an equilibrium or mean/steady state, for all dependent variables.

2. Substitution of the dependent variables *plus* the disturbances into the governing equations of motion.
3. Bringing the equations into the most transparent form by algebraic manipulation and ignoring all disturbance quantities appearing that are quadratic products and higher.

It should be remarked that the first step can be performed also on states that are steady oscillatory. From the standpoint of a scaling analysis, linearization per se is justified if the discarded terms are significantly smaller in magnitude compared to the other terms that are kept. Clearly, assuming that one is dealing with continuous and differentiable functions, a disturbance can be introduced arbitrarily small to achieve this desired disparity. In practical situations, however, one would like to have some measure of the dependent variables' magnitude, typifying the system under consideration in order that one may use it as a comparison to the disturbances themselves. In other words, when we say *small* we have to provide a measure of small compared to what? The answer is that we are after an *approximation* up to the accuracy of quadratic divided by linear terms. Below we demonstrate this procedure.

Following the linearization procedure as enumerated above, we start with a fluid that is at rest. Since the velocity field is determined by the potential ϕ, the steady state potential is a constant in space and time, call it ϕ_0. Thus the potential is written as

$$\phi = \phi_0 + \phi'(x, z, t),$$

where the prime notation hereafter indicates that the said quantity is a disturbance. According to $u' = \nabla \phi'$ the corresponding velocity components are

$$u = 0 + u'(x, z, t) = \frac{\partial \phi'}{\partial x},$$

$$w = 0 + w'(x, z, t) = \frac{\partial \phi'}{\partial z}.$$

The velocity potential must satisfy Eq. (4.34) which amounts to

$$\frac{\partial^2 \phi'}{\partial x^2} + \frac{\partial^2 \phi'}{\partial z^2} = 0. \tag{4.43}$$

The solution of this equation requires boundary conditions in both the z and x directions. We will treat in this case the horizontal direction as unbounded. For the boundary conditions in z if we require that there is no normal flow at the bottom boundary $z + L_b = 0$, or $z = -L_b$, then since in this example L_b is a constant, the kinematic condition amounts to requiring that the vertical velocity disturbance w' be zero at this location:

$$w'(x, -L_b, t) = \left. \frac{\partial \phi'}{\partial z} \right|_{z=-L_b} = 0. \tag{4.44}$$

This example is limited and if we suppose, instead, that we have a general velocity vector $\mathbf{u}(x,y,z)$ and a bottom surface S_b, defined by $S_b(x,y,z) = L_b(x,y) + z = 0$, the explanation of what should be done in this case of a general bottom surface is not long. The requirement that the normal component of the velocity to the surface must be zero is expressed as

$$\left(\mathbf{u} \cdot \frac{\nabla S_b}{|\nabla S_b|} \right)_{S_b=0} = 0 \qquad \Longrightarrow \qquad \left(\mathbf{u} \cdot \frac{\nabla S_b}{|\nabla S_b|} \right)_{z=-L_b(x,y)} = 0.$$

If the functional form of the bottom surface shape S_b has no mathematical pathologies, then this condition simply becomes $(\mathbf{u} \cdot \nabla S_b)_{z=-L_b} = 0$. Since in the above example of a flat bottom $S_b = z + L_b = 0$ the no normal flow condition amounts to $w' = 0$ at the surface $S_b = 0$, i.e., $z = z_b = -L_b$. If the bottom has some topography to it, i.e., $L_b = L_b(x,t)$, then $S_b = z + L_b(x) = 0$ and we must have

$$w' + u' \frac{\partial L_b}{\partial x} = 0, \qquad \text{on } z = -L_b(x). \tag{4.45}$$

Summing up now the equations governing the disturbance terms of the velocity potentials in the present case, we write

$$\left. \frac{\partial \phi'}{\partial t} \right|_{z=\eta} = -\frac{1}{2} \left[\left(\frac{\partial \phi'}{\partial x} \right)^2 + \left(\frac{\partial \phi'}{\partial z} \right)^2 \right]_{z=\eta} - g\eta, \tag{4.46}$$

$$\frac{\partial \eta}{\partial t} = -\frac{\partial \eta}{\partial x} \cdot \left(\frac{\partial \phi'}{\partial x} \right)_{z=\eta} + \left(\frac{\partial \phi'}{\partial z} \right)_{z=\eta}, \tag{4.47}$$

where the second condition follows from Eq. (4.42). The Laplace equation (4.43), and the bottom boundary condition, which is Eq. (4.44) for a flat bottom, are also obviously needed. A striking illustration of gravity waves on Lake Superior is given in Fig. 4.11.

Linearization amounts, as explained above, to neglecting *products* of perturbations which, in this case, appear only in the first term on the right-hand side of each of the above two equations. The justification of this neglect is the following. The perturbation in the velocity potential ϕ' scales as $u_0 L_0$ with L_0 representing the horizontal and vertical scales of motion, here assumed to be of the same order of magnitude, and u_0 the disturbance velocity scale. Now, let η_0 represent the length scale of the surface perturbation. Then the relative scalings for the right-hand side of Eq. (4.46) are

$$\frac{1}{2} \left[\left(\frac{\partial \phi'}{\partial x} \right)^2 + \left(\frac{\partial \phi'}{\partial z} \right)^2 \right] = \mathcal{O}(u_0^2) \qquad \text{and} \qquad g\eta = \mathcal{O}(\eta_0 g),$$

Fig. 4.11 Gravity waves on Lake Superior. (*Public Domain. Author: NASA/Johnson Space Flight Center. Image cropped. Courtesy of NASA via Wikimedia commons - http://eol.jsc.nasa.gov/Info/ use.htm*)

while for Eq. (4.47) they are

$$\frac{\partial \eta}{\partial x} \cdot \left(\frac{\partial \phi'}{\partial x}\right) = \mathcal{O}\left(\frac{\eta_0}{L_0} u_0\right) \quad \text{and} \quad \left(\frac{\partial \phi'}{\partial z}\right) = \mathcal{O}(u_0). \tag{4.48}$$

Focusing first upon Eq. (4.48) we see that for the nonlinear term to be much smaller than the linear term, that is, making the latter *dominant* so that the following is necessary: $\eta_0/L_0 \ll 1$. In the preceding equation the same linear over nonlinear dominance holds if $u_0^2 \ll \eta_0 g$. These two strong inequalities are mutually compatible if

$$u_0^2 \ll g L_0 \implies u_0 \ll c_g, \quad \text{where} \quad c_g \equiv \sqrt{g L_0}.$$

The *gravity wave-speed* c_g, being a natural unit of velocity in this system, becomes the measure by which all other perturbation quantities will be assessed. From this scaling analysis we say that the system is in the linear regime if both

1. The disturbance velocities u_0 are much smaller than c_g.
2. The height disturbance η_0 is much smaller in scale compared to the typical horizontal and vertical scales L_0.

To complete this discussion, we now define a naturally occurring timescale $t_0 \equiv L_0/c_g$ wherein it follows that the time derivative terms on the left-hand sides of Eqs. (4.46) and (4.47) scale, respectively, as

$$\left.\frac{\partial \phi'}{\partial t}\right|_{z=\eta} = \mathcal{O}\left(\frac{c_g}{L_0}L_0 u_0\right) = \mathcal{O}\left(c_g u_0\right), \qquad \frac{\partial \eta}{\partial t} = \mathcal{O}\left(\frac{c_g}{L_0}\eta_0\right),$$

and, as such, in order for these terms to be balanced to the corresponding linear terms on the right-hand side of the same respective equations it must be that the height perturbation scaling $\eta_0 = \mathcal{O}\left(c_g u_0/g\right) = \mathcal{O}\left(L_0 u_0/c_g\right)$. Thus, the linearization of Eqs. (4.46) and (4.47) yields now simply

$$\left.\frac{\partial \phi'}{\partial t}\right|_{z=\eta} = -g\eta, \tag{4.49}$$

$$\frac{\partial \eta}{\partial t} = \left(\frac{\partial \phi'}{\partial z}\right)_{z=\eta}, \tag{4.50}$$

together with the linear equation (4.43) and the boundary condition in Eq. (4.44).

   The central PDE to be solved is the two-dimensional Laplace equation and so the natural way to construct a solution is by assuming Fourier forms for the dependent quantities:

$$\phi' = \hat{\phi}(z)e^{ik\left(x-V_p t\right)} + \text{c.c.}, \qquad \eta = \hat{\eta}e^{ik\left(x-V_p t\right)} + \text{c.c.}, \tag{4.51}$$

where the horizontal wavenumber is k and the corresponding phase velocity, as yet undetermined, is V_p. We are concerned with developing a dispersion relation for V_p and, as such, we are performing a normal mode analysis of the system. There are a number of subtleties associated with this procedure that will be addressed in later chapters. The notation "c.c" is a shorthand for the words *complex conjugate* and adding it to a complex quantity verifies that the sum is real, as all physical quantities should be. Another possibility to achieve the same is to take the real (or imaginary) part, when we are using complex numbers as physical quantities.

   With the Fourier ansatz, the Laplace equation (4.43) turns into

$$\frac{\partial^2 \hat{\phi}}{\partial \hat{z}^2} - k^2 \hat{\phi} = 0,$$

and, when subject to the boundary condition at $z = -L_b$, admits the solution

$$\hat{\phi} = A\cosh\left[k(z+L_b)\right],\qquad(4.52)$$

where A is an arbitrary amplitude. At this point we are ready to complete the solution by constructing solutions of (4.49)–(4.50), but before doing so we have to address one last important matter. Inspection indicates that the potential ϕ' and its derivatives must be evaluated at $z = \eta$. Given that η is small in the sense we have heretofore explained, it means that we can achieve the above goal by performing a Taylor expansion of ϕ' around $z = 0$, in other words

$$\phi'(x,z=\eta,t) = \phi'(x,0,t) + \left(\frac{\partial\phi'}{\partial z}\right)_{z=0}\eta + \frac{1}{2}\left(\frac{\partial^2\phi'}{\partial z^2}\right)_{z=0}\eta^2 + \text{HOT}$$

Examination of the expressions on the right-hand side indicates that the linear term is larger than the first nonlinear term by a factor of η_0/L_0 by the same scaling arguments developed earlier. In this way Eqs. (4.49)–(4.50) become approximately

$$\left.\frac{\partial\phi'}{\partial t}\right|_{z=0} \approx -g\eta,\qquad(4.53)$$

$$\frac{\partial\eta}{\partial t} \approx \left(\frac{\partial\phi'}{\partial z}\right)_{z=0},\qquad(4.54)$$

where the evaluation of the terms at $z = 0$ introduces an inaccuracy of the same order of magnitude as the other nonlinear terms that may be neglected in the above Taylor expansion of $\phi'(x,z=\eta,0)$. Substituting the above solutions, which are in Fourier form, from Eq. (4.51), together with the vertical structure equation (4.52) into the above equations we find the following algebraic relations:

$$-ikV_p A\cosh kL_b = g\hat{\eta},\qquad -ikV_p\hat{\eta} = kA\sinh kL_b.$$

This pair of equations can be solved for A and η; the non-trivial solution gives V_p, which with the identity $\omega = kV_p$, leads to

$$V_p^2 = \frac{g}{k}\tanh kL_b \qquad\Longleftrightarrow\qquad \omega^2 = gk\tanh kL_b,\qquad(4.55)$$

giving the explicit dispersion relation:

$$\omega(k) = \pm\sqrt{gk\tanh kL_b}.\qquad(4.56)$$

We note that two waves are admitted by this system, one left and one right propagating.

Of interest also are two limits of $\omega(k)$. The first, so-called deep water waves, is the one in which the horizontal extent of the disturbance is much smaller than the vertical depth of the fluid, thus $kL_b \gg 1$. In this case $\tanh kL_b \approx 1$ and the dispersion relation, with the corresponding phase velocity V_p, becomes

$$\omega^2 \approx gk, \implies \omega \approx \pm\sqrt{gk}, \quad \text{and} \quad V_p \approx \pm\sqrt{\frac{g}{k}}.$$

A curious feature of deep water waves is the property following from the above: as the wavelength is increased the oscillation frequency diminishes, yet the phase speed increases! Additionally, the group velocity V_g is

$$V_g = \frac{\partial \omega}{\partial k} \approx \pm\frac{1}{2}\sqrt{\frac{g}{k}} \approx \tfrac{1}{2}V_p,$$

indicating that V_g, which is the speed of the propagation of energy, equals half the phase velocity and is in the same direction as V_p. The other limit is opposite, namely, the horizontal extent of the disturbance is large compared to the vertical extent of the fluid. This physical setting is, in fact, the *shallow water* case. In this limit, wherein $kL_b \ll 1$, it follows that $\tanh kL_b \approx kL_b$ and

$$\omega^2 \approx gL_b k^2, \implies \omega \approx \pm k\sqrt{gL_b}, \quad \text{and} \quad V_p \approx \pm\sqrt{gL_b}.$$

This system is explored in more detail in the context of shallow water theory (Sect. 4.3).

Let us round out the discussion by briefly discussing the physical visualization of a surface water wave, asking the simple question: how does the position of a tracer on the surface, e.g., a very small buoy, approximating a fluid particle with no inertia, respond when a surface wave passes by. The vertical position of the surface is represented by the variable η, so let us say that we are witnessing a waveform for it, given by $\eta = A\cos\left[k(x - V_p t)\right]$, where A is an arbitrary small amplitude. Using the relations between the linear quantities established above, in this case explicitly $\partial_t \phi' = -g\eta$ (at $z = 0$), we find that the velocity potential at the surface $z = 0$ is given by $\phi' = [Ag/(kV_p)]\sin\left[k(x - V_p t)\right]$. We are after the Lagrangian coordinates of an element initially at point x at the surface ($z = 0$). Denote its horizontal Lagrangian position by χ' and the horizontal Eulerian velocity field at $z = 0$ by $u_0'(x,t) = \partial_x \phi'$. The appropriate relations between the Eulerian and Lagrangian descriptions, in the linear case, are

$$\left.\frac{\partial \chi'}{\partial t}\right|_{z=0} = \left.\frac{\partial \phi'}{\partial x}\right|_{z=0} = u'(x,z=0,t) = u_0'(x,t)$$

$$\implies \quad \chi' = -\frac{gA}{kV_p^2}\sin\left[k(x - V_p t)\right].$$

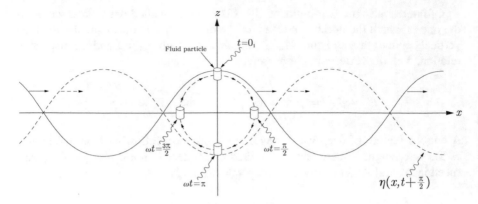

Fig. 4.12 Trajectory of an inertialess, very small surface buoy in the presence of a rightward moving surface wave given by $\eta = A \cos k(x - V_p t)$

We can use these expressions now to trace out the position of our tracer buoy. Let us follow, therefore, the position of the buoy which at rest is located at $z = x = 0$. The coordinate pair indicating its position would be written, parametrized by time, as $[x(t), z(t)]$:

$$[\chi_0, \eta_0] = \left[\chi'(x=0, t), \eta(x=0, t)\right] = A\left[\frac{g}{kV_p^2} \sin\left(kV_p t\right), \cos\left(kV_p t\right)\right].$$

The implication of the above is found by the form of the tracer (buoy) trajectory. It is schematically plotted in Fig. 4.12. This example tracer buoy executes elliptical motion as the rightward moving wave passes by, instead of the often erroneous notion that the buoy just bobs up and down only. The x and z coordinates of the buoy's position satisfy the relations of an elliptical curve.

$$\left(\frac{g^2}{k^2 V_p^4}\right) z^2 + x^2 = \frac{A^2 g^2}{k^2 V_p^4}.$$

4.2.3 The Effect of Surface Tension

Surface tension is a physical phenomenon appearing on the interface between two fluids or a fluid and vacuum. Microscopically it is easy to understand this heuristically, but the details require a deep understanding of the cohesion forces between the microscopic particles. Any microscopic particle in the bulk of the fluid experiences equal cohesion forces from all its ambience, i.e., its neighbors, assuming homogeneity, and is subject to thermal random motion. The film "Surface Tension

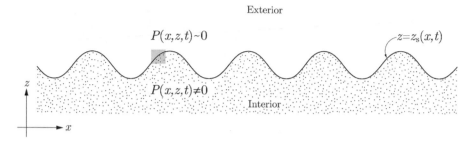

Fig. 4.13 Schematic surface wave train. Note the *shaded area*, on which we'll focus in the next figure

in Fluid Mechanics" in the site of NCFMF (web.mit.edu/hml/ncfmf.html) contains some nice demonstrations. Microscopic particles that lie on, or very close to a boundary with another fluid, experience different interaction with the material on their side of the interface from the interaction with the material on the other side of the interface. Since we have adopted at the outset a continuum approximation, we may say that to move a differential element of the surface, dS, a finite very small distance $\hat{n}\delta\xi$ in its perpendicular direction will, on the *average*, require an investment of mechanical work $dW = \delta\mathbf{F}\cdot\hat{n}\delta\xi dS$, where $\delta\mathbf{F}$ is the finite difference in the cohesion forces between the two sides. We may thus attribute to the interface surface element dS an energy dW. Defining $\gamma_{st} \equiv dW/dS$ we get a measure for energy *per unit area* that can be attributed to the interface per unit area at that point. γ_{st} is called *surface tension* and it is often interpreted as tangential surface force per unit length. The following elementary discussion will suffice for our purposes, enabling us to include surface tension effects in the physics of linear surface waves.

In Fig. 4.13, a long wave train is schematically depicted again but the purpose is to focus our attention to the shaded area, which is important for our upcoming explanation of the surface tension effect. The linearized Bernoulli's equation, without surface tension effects taken into account, helped us above to get an approximation to one-dimensional surface waves and their properties. Now our purpose is to supplant it with the contribution of surface tension. Focusing on the shaded region in Fig. 4.13 we sketch the forces per unit length, acting on the curve representing the water surface as projected on the plane $y =$ const., say, or any plane parallel to it, similarly to the what is depicted in the above mentioned shaded area. We obtain a one-dimensional curve and look at its short segment, whose x-coordinate is between x_1 and x_2. This segment, $\Delta x \equiv x_2 - x_1$, is shown at equilibrium in the lower part of Fig. 4.14. The forces per unit length acting on the segment are T on each side. When it is displaced upward by $\eta(x,t)$ the horizontal components of the forces are in equilibrium as there is no horizontal motion at this linear approximation. Thus we have $(T_2\cos\theta_2 - T_1\cos\theta_1)\Delta x = 0$. Now $\cos\theta = 1 - \theta^2/2 + \text{HOT}$ so for small angles linearization yields $T_2 \approx T_1 \approx T$. In the vertical direction, we have a net surface tension force acting on the segment

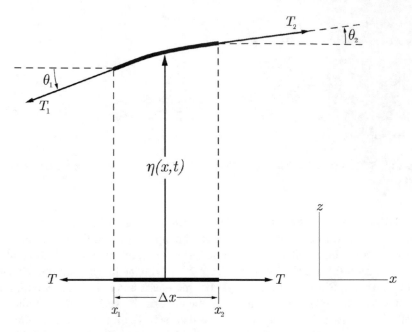

Fig. 4.14 Diagram to help calculate the effect of surface tension in one dimension

$$T_z = T_2 \sin\theta_2 - T_1 \sin\theta_1 \approx T(\tan\theta_2 - \tan\theta_1)$$

$$= T\left[\left(\frac{\partial\eta}{\partial x}\right)_{x_2} - \left(\frac{\partial\eta}{\partial x}\right)_{x_1}\right]\Delta x \approx T\frac{\partial^2\eta}{\partial x^2}\Delta x, \qquad (4.57)$$

where the sign $\approx$ denotes that linearization has been done. Thus the net upward force resulting from surface tension alone, per unit area on the fluid *surface* element, will be

$$F_z/\delta S = T\frac{\partial^2\eta}{\partial x^2}, \qquad (4.58)$$

and this has to be balanced by the pressure difference, to prevent vertical force on a massless body. At this point, we shall generalize the surface tension T to act on two-dimensional surfaces, by including some small "depth" in the perpendicular y direction and call it γ_{st}. Thus we write

$$\gamma_{st} = T, \qquad (4.59)$$

where the physical consistent dimensions of [force]/[length] or [energy]/[surface] are preserved. Thus

$$P_{\text{out}} - P_{\text{in}} = \gamma_{\text{st}} \frac{\partial^2 \eta}{\partial x^2} \quad \text{at} \quad z = \eta(x,t), \tag{4.60}$$

where there is no vertical force on a massless body. In Fig. 4.13, P_{out} is the exterior pressure, which, for convenience, we set it to be close to zero everywhere. P_{in} is the interior pressure, just below the surface, at the point and time of question. In any case, we should not forget that there is a constant in our disposal in the appropriate Bernoulli's equation (see the discussion before Eq. (4.40)). Now, the same procedure we did to linearize equation (4.37)–(4.40), but now with the extra surface tension term, gives

$$\frac{\partial \phi}{\partial t} + g\eta(x,t) - \frac{\gamma_{\text{st}}}{\rho} \frac{\partial^2 \eta}{\partial x^2} = 0 \quad \text{at} \quad z = 0. \tag{4.61}$$

Performing the same analysis as in the previous section leads to a modified dispersion relation for waves obeying (4.60); this amounts to solving (4.61) using (4.54) assuming, as before, solutions of the form $Ae^{ikx-i\omega t}$ + c.c. yields

$$\omega^2 = \left(gk + \frac{\gamma_{\text{st}}k^3}{\rho} \right) \tanh kL_b. \tag{4.62}$$

In the limit of large kL_b, known as *deep water*, the corresponding waves phase velocity is

$$V_{\text{p}} = \left(\frac{g}{k} + \frac{\gamma_{\text{st}}k}{\rho} \right)^{1/2}.$$

Further features of these waves are explored in Problem 4.7. One of the features uncovered is that a dimensional diagnostic parameter arises $q_c \equiv \gamma_{\text{st}}k/\rho$ which indicates the importance of surface tension in surface waves. With its help it is found that, e.g., for q_c very large, with respect to g/k, very short waves arise, that are new for us. They are called *capillary waves* and satisfy the following limit of their phase velocity:

$$\lim_{kL_b \to \infty} V_{\text{p}} \propto \sqrt{k},$$

while for the long wavelength disturbances we recover the familiar gravity wave phase velocity

$$\lim_{kL_b \to 0} V_{\text{p}} \propto 1/\sqrt{k},$$

where we showed only the dependence on the wavenumber. Other physical constants are contained in the constant of proportionality. It should be, by now, evident that the physical sense of extreme mathematical limits like 0 and ∞ actually mean

$\ll$ and $\gg$ than some typical value, determined a priori, of the relevant physical variable or algebraic combination thereof. These two limits also suggest that there is a minimum value of the phase velocity as a function of k. Calling this value of k k_m, it is straightforward to show that

$$k_m = \left(\frac{\rho g}{\gamma_{st}}\right)^{1/2}, \text{ where } V_m \equiv \left(\frac{4g\gamma_{st}}{\rho}\right)^{1/2}. \tag{4.63}$$

V_m is the minimum of V_p and it appears at the $k = k_m$ point. In Sect. 4.1.3, a somewhat heuristic argument was used to show that the group velocity measures the direction and speed in which energy propagates. A deeper discussion of this issue will be given in Sect. 4.4.3 as well as in Chap. 6, in which it will be shown for sound waves. Interesting features appear when examining V_g in the combined gravo-capilarity case. Again, looking at the case of deep water waves we find that

$$V_g = \frac{\partial \omega}{\partial k} = \frac{1}{2}V_p \frac{\left(g + 3\frac{\gamma_{st}k^2}{\rho}\right)}{\left(g + \frac{\gamma_{st}k^2}{\rho}\right)}. \tag{4.64}$$

Remarkably, at this special value of k_m it indicates that the phase and group velocities are the same, $V_g(k_m) = V_m$! This means that the speed corresponding to V_m defined in Eq. (4.63) will hold some significance which we shall see shortly. Incidentally, for water in the Earth's gravitational field the corresponding value of this special speed is approximately $V_m = 23.2$ cm/s. Generally speaking, then, one can refer to the surface response depending upon whether the phase speed exceeds or is under V_m: waves with wavenumber k in which $V_p(k) < V_m$ are known to belong to the so-called gravity branch, while waves whose phase velocities $V_p(k) > V_m$ are associated with the so-called capillary branch, see Fig. 4.15.

4.2.4 Surface Gravity Waves Induced by a Steady Flow Over a Corrugated Bed

We finalize our discussion on surface gravity waves by considering one example of a steady potential flow, in a geometry identical to the one dealt with in this entire section as depicted in Fig. 4.10. In the example we choose to abandon the assumption of a flat bottom and consider, instead, the resulting surface wave pattern generated by a steady stream flowing over a *corrugated* bed. Before proceeding we remind ourselves that the velocity potential for a pure undisturbed stream moving rightward in the x direction with constant speed U_0 is $\phi_0 = U_0 x + f_0(t)$, where $f_0(t)$ is a suitable function. It is known that $f_0(t)$ cannot affect the velocity, but we introduce it here, because for the full nonlinear Bernoulli's equation to hold, $f_0(t)$ has to assume particular form. Since we will be interested in the linearized case, this is

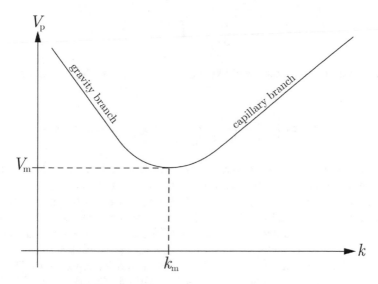

Fig. 4.15 Phase velocities for the mixed gravo-capillary wave problem

mentioned only to calm the mathematically worried. One expects that the presence of a bottom structure ought to influence the steady stream flow. We therefore write the total velocity potential as a sum of the steady stream flow plus ϕ' representing the contribution due to the obstacles, i.e., $\phi = \phi_0 + \phi'(x,z)$. The corrugated bed is represented with a spatially varying bottom topography z_b, $z_b(x) = -L_b + h_\varepsilon \zeta_b(x)$, where $L_b > 0$ is constant. For simplicity we assume $\zeta_b(x) = \cos k_0 x$ where k_0 corresponds to some intrinsic wavelength of the corrugation pattern $\lambda_0 = 2\pi/k_0$. The quantity h_ε is a reference height that we write as $h_\varepsilon = \varepsilon L_b$. We once again denote the surface of the fluid $z_s = \eta(x,t)$ and assume that η is small compared to L_b.

The full nonlinear equations of motion governing the problem are Eqs. (4.35) and (4.42) for the surface boundary conditions at $z = \eta$, the Laplace equation $\nabla^2 \phi = 0$ for the interior of the fluid $z_b < z < \eta$, and the bottom no-normal flow condition equation (4.45) evaluated at $z = z_b$. We are interested in steady state solutions for small values of η, ϕ' and ε so that approximate solutions may be developed by linearizing the equations governing the flow. Thus after substituting $\phi = \phi_0 + \phi'$ in the Bernoulli equation and linearizing, we find for a steady state ($\partial_t = 0$)

$$0 = -U_0 \left(\frac{\partial \phi'}{\partial x} \right)_{z=0} - g\eta. \tag{4.65}$$

Note that this was written for a point on the surface. Similarly, the kinematic condition at the upper surface becomes after linearization

$$U_0 \left(\frac{\partial \eta}{\partial x} \right) = \left(\frac{\partial \phi'}{\partial z} \right)_{z=0}. \tag{4.66}$$

The bottom no normal flow condition at $z = z_b = -L_b + h_\varepsilon \zeta_b(x)$ is, after linearization,

$$0 = \left[\mathbf{u} \cdot \nabla (z - z_b) \right]_{z=z_b} \approx \left(\frac{\partial \phi'}{\partial z} \right)_{z=-L_b} + h_\varepsilon k_0 U_0 \sin k_0 x = 0. \tag{4.67}$$

Note incidentally that what we have implied here is that the disturbance velocity scale is of the order of $h_\varepsilon k_0 U_0$ and for this to be significantly smaller than the steady stream flow U_0, we must have that $h_\varepsilon k_0 \ll 1$, i.e., that the corrugation wavelength cannot be too small, explicitly $\lambda_0 \gg \varepsilon L_b$. Inspection of the governing equations suggests the solution ansatz we should adopt. The solution of the two-dimensional Laplace equation for the interior of the fluid may be expressed easily after observing the form of the function indicated by the bottom condition equation (4.67). Trying the ansatz in the Laplace equation, which is periodic in the x direction $k_0 > 0$ given by

$$\phi' = \hat{\phi}(k_0, z) \sin k_0 x,$$

has the following general solution for $\hat{\phi}(k, z)$:

$$\hat{\phi}(k_0, z) = A_{k_0} e^{-k_0 z} + B_{k_0} e^{k_0 z},$$

where A_{k_0} and B_{k_0} are constants. The boundary conditions imply the following solution for the velocity potential disturbance

$$\phi' = \left\{ -h_\varepsilon U_0 \sinh \left[k_0 (z + L_b) \right] + A_{k_0} \cosh \left[k_0 (z + L_b) \right] \right\} \sin k_0 x, \tag{4.68}$$

satisfying the no normal flow condition at $z = z_b$. The extra term containing the arbitrary constant A_{k_0} must be retained since it both solves the Laplace equation and satisfies the bottom boundary condition equation (4.67). It remains for us to determine A_{k_0}, and for this we must turn to the upper boundary condition. Given that the x functional form of ϕ' is proportional to $\sin k_0 x$, inspection of Eqs. (4.65) and (4.66) guides us into proposing the ansatz form for the solution of η to be

$$\eta(x) = \bar{\eta} \cos k_0 x, \tag{4.69}$$

where $\bar{\eta}$ is a second unknown amplitude. Inserting Eqs. (4.68) and (4.69) into Eqs. (4.65) and (4.66), then removing the common functional forms, produces the following two equations for two unknowns A_{k_0} and $\bar{\eta}$:

$$h_\varepsilon U_0^2 k_0 \left[\sinh k_0 L_b + A_{k_0} \cosh k_0 L_b\right] - g\bar{\eta} = 0,$$

$$-\bar{\eta} + h_\varepsilon \left[\cosh k_0 L_b + A_{k_0} \sinh k_0 L_b\right] = 0.$$

The solution yields an expression for $\bar{\eta}$:

$$\bar{\eta} = \frac{h_\varepsilon \operatorname{sech} k_0 L_b}{1 - \dfrac{g}{k_0 U_0^2} \tanh k_0 L_b}. \tag{4.70}$$

It is interesting to note that there is a qualitative change in the features of the resulting solution depending upon the speed of the flow. Crests of the surface features appear immediately over troughs of the bottom bed whenever

$$\frac{g}{k_0} \tanh k_0 L_b > U_0^2,$$

while surface troughs coincide with troughs of the bottom bed otherwise. Recalling from our previous discussion that the phase velocity of deep water gravity waves for disturbances of wavenumber k_0 is given by $V_p = \sqrt{g/k_0}$, we readily notice that the qualitative change in the solution we just developed occurs whenever the stream flow exceeds the natural gravity wave phase velocity V_p. Finally, and most notably, we see that our analysis breaks down for specific values of the corrugation wavenumber corresponding to $k_0 = k_g \equiv g/U_0^2$ for the deep water limit.

4.3   Shallow Water Equations

Shallow water theory should be obviously valid when the fluid layer horizontal extension is much larger than its depth (the water is shallow). Its consequences perhaps do not necessarily belong to this chapter, even though waves in shallow water are certainly a physical possibility. The reason we have decided to place it here is twofold. First, the setup is similar to the problems we have been dealing with in this chapter, so the nomenclature and concepts are similar. Second, as it turns out, shallow water's *nonlinear* behavior has a formal similarity to one-dimensional gas dynamics, which will be explained in Chap. 6 and there we will also treat shallow water problems that are analogous to one-dimensional compressible gas phenomena. Both shallow water equations and one-dimensional gas dynamics are nonlinear, however, they allow the application of the method of characteristics to seek analytical solutions. This concept and the method are important in both the theory of PDEs and in FD. We have already used some basic concepts and techniques of the method in Sect. 4.1.4, where they were employed in understanding the breaking of nonlinear waves in incompressible fluids. As mentioned there, the issue will be discussed at considerable length in Chap. 6 and several problems will be devoted to it. In the discussion of surface water waves at the end of Sect. 4.2.2 we found that

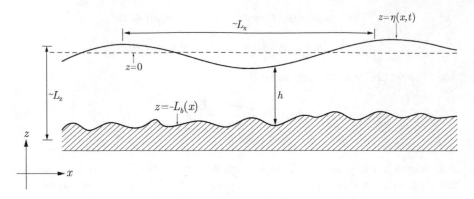

Fig. 4.16 Schematic drawing of a shallow water system

the phase velocity of surface water waves becomes independent of the horizontal wavenumber of the disturbance k, when the layer depth scale (we shall call it L_z) is much smaller than the horizontal disturbance scale $L_x \sim 2\pi/k$. The situation of interest is depicted in Fig. 4.16. Let $h(x, y, t)$ represent the vertical thickness of the layer and let h_0 indicate the static layer thickness of the fluid when there is no bottom topography so that L_b and h_0 are the same. The shallow water limit for surface water waves is the statement that the phase velocity $V_p \to c_g = \sqrt{gL_b} = \sqrt{gh_0}$ provided $kL_z \sim L_z/L_x \ll 1$. The symbol c_g is henceforth reserved to indicate the surface gravity wave speed in shallow water. Recalling our discussion in Sect. 4.2.2, water wave disturbances are in the linear regime if fluid velocities are much smaller than c_g.

The aim is to reformulate the equations of motion, in this shallow water limit, in order to develop a simpler set of *reduced* equations that are valid even if the fluid disturbances have velocities that are large and comparable to the surface gravity wave speed c_g. The procedure we employ is a classic scaling analysis. This tactic seeks to derive equations for phenomena characterized by very specific length and timescales. The spatio-temporal scalings used are always motivated by observations of one form or the other. Like in our discussion of boundary layers in Chap. 3, scaling analyses yield useful simplifications when there exist small parameters of the system. We detail the procedure next.

4.3.1   Derivation via Scaling Analysis

As explained above, a basic consideration in the scaling analysis of a shallow water problem exploits the fact that there exists a small parameter, which here is the aspect ratio $L_z/L_x \ll 1$. This implies that the corresponding timescales of interest are much longer than the vertical free fall time. Furthermore the central feature of shallow water is that dynamic motions are hydrostatic—an admittedly curious statement which will be explained shortly. Both of these features become self-evident during the course of the following analysis.

To begin, consider again the example of linearized surface gravity disturbances whose horizontal wavelength is measured by L_x. For shallow water gravity wave-speed c_g, the corresponding timescale T, typifying the wave's period is given by $T \sim L_x/c_g \approx L_x/\sqrt{gL_z}$. Recall that this means that the fluid locally bobs up and down on this timescale T. If vertical free fall time over the extent L_z is denoted by $T_{\mathrm{ff}} \sim \sqrt{L_z/g}$, the implication is that the above two timescales ought to be very different in shallow water as indicated by the ratio $T_{\mathrm{ff}}/T = L_z/L_x \ll 1$. The next observation we make is that of the state of the fluid in equilibrium. When there is no flow, i.e., $\mathbf{u} = 0$, the steady state is described by the equation for vertical hydrostatic equilibrium

$$\frac{\partial P}{\partial z} = -\rho g.$$

For a constant density fluid the steady configuration for a situation where the top boundary faces vacuum at $z = 0$ is expressed by $P = -\rho g z$. It is important to note that the scale of P in the interior, e.g., down to a level of $z \sim -L_z$, is given by $\rho g L_z \approx \rho c_g^2$.

The approach taken is to analyze the equations of motion without making the irrotational flow assumption a priori. We then apply the aforementioned scaling observations to the equations themselves and then simplify them by dropping terms in the equations that are small compared to the others. The equations are in two-dimensions (x, z) and come out as

$$\frac{\partial u}{\partial t} + u\frac{\partial u}{\partial x} + w\frac{\partial u}{\partial z} = -\frac{1}{\rho}\frac{\partial P}{\partial x} \tag{4.71}$$

$$\frac{\partial w}{\partial t} + u\frac{\partial w}{\partial x} + w\frac{\partial w}{\partial z} = -\frac{1}{\rho}\frac{\partial P}{\partial z} - g \tag{4.72}$$

$$\frac{\partial u}{\partial x} + \frac{\partial w}{\partial z} = 0. \tag{4.73}$$

We assume that all of the horizontal spatial dependencies can be estimated by the scale L_x and similarly by L_z for vertical scales. The above-mentioned disparities characteristic of shallow water obviously have to hold. The shallowness of the fluid layer is measured by the parameter $\varepsilon \equiv L_z/L_x$. The *shallow water limit* thus translates into the condition $\varepsilon \ll 1$.

Let us start with the incompressibility equation (4.73). Assume that the typical horizontal and vertical velocity scales are U_x and U_z, respectively. We do not yet know individually what these scales should be but they will be borne out of the analysis shortly. What we do establish here, instead, is what must be the relative scaling between these two velocities. If one of the terms in the incompressibility equation is of lower order than the other, a trivial state is obtained. Thus, the two terms comprising it have to be of the same order of magnitude. The magnitude of each term appearing is, respectively, $\partial u/\partial x = \mathcal{O}\left(U_x/L_x\right)$ and $\partial w/\partial z = \mathcal{O}\left(U_z/L_z\right)$.

If these two terms are to be of the same order of magnitude, then $U_z/U_x \sim L_z/L_x$. Thus in the shallow water limit $U_z/U_x = \mathcal{O}(\varepsilon)$. Armed with this relative scaling for $w \sim U_z \sim \varepsilon U_x$, we are now prepared to simplify the vertical momentum equation. We have said at the outset that we are interested in deriving equations describing dynamics of disturbances whose velocities start approaching that of the natural gravity wave phase velocity. So we examine the consequences of setting U_x to be $\sim c_g$. As it was already estimated, the timescales of interest are $T \approx L_x/c_g$. So a scaling analysis of the inertial terms on the left-hand side of (4.71) can now be performed. The scalings for each of the terms appearing are $\partial_t w \sim \dot{U}_z/T \sim \varepsilon c_g^2/L_x$, $u \partial_x w \sim (U_x)U_z/L_x \sim \varepsilon c_g^2/L_x$, and $w \partial_z w \sim U_z^2/L_z \sim \varepsilon c_g^2/L_x$. Thus all three terms are of the same order of magnitude and, consequently, we have that the inertial terms on the left-hand side of the equation collectively have the scaling $Dw/Dt \sim U_z/T = \varepsilon c_g^2/L_x$. The pressure term on the right-hand side of Eq. (4.71) is scaled according to $(1/\rho)\partial P/\partial z \sim c_g^2/L_z$. It follows that the inertial terms on the left-hand side of the vertical momentum equation are smaller than the hydrostatic terms by a factor of ε^3. Remarkably, it means that vertical hydrostatics is a dynamic feature in shallow water. This is the meaning of the *dynamic hydrostatic equilibrium*, referred to before. Thus

$$\frac{\partial P}{\partial z} = -\rho g.$$

Let the surface layer of the fluid dynamically vary and be given by the function $\eta(x,t)$. The solution of the vertical momentum equation is really the solution of the hydrostatic equation—but now it is time-dependent:

$$P(x,z,t) = \rho g \left[\eta(x,t) - z \right]. \tag{4.74}$$

The physical meaning of this relation is that the pressure in the fluid always adjusts so as to induce local vertical hydrostatic equilibrium, as the surface position of the fluid layer (η) varies with time.

With Eq. (4.74) in hand, we can analyze the horizontal momentum equation. We do so by directly inserting the solution for the pressure into Eq. (4.71) to find

$$\frac{Du}{Dt} \equiv \frac{\partial u}{\partial t} + u\frac{\partial u}{\partial x} + w\frac{\partial u}{\partial z} = -\frac{\partial(g\eta)}{\partial x}.$$

Now observe that the right-hand side of this expression is independent of the vertical coordinate. Thus if we assume that the horizontal velocity is independent of z as well, then the above statement reduces to the important equation

$$\frac{\partial u}{\partial t} + u\frac{\partial u}{\partial x} = -\frac{\partial(g\eta)}{\partial x}. \tag{4.75}$$

The assumed vertical independence of u has a reasonable way to justify it. Note that the horizontal momentum balance equation without this assumption really says that the Lagrangian rate of change of the horizontal velocity is independent of z. Thus, if u starts off as independent of z it will remain so for all times after. In any case, the assumption of $\partial_z u = 0$ has to be assessed on the basis of whether or not it leads to a contradiction and/or a physically unreasonable prediction. So far a problem of the sort is absent. Now, because the horizontal velocity u is independent of z, we see from inspection of the incompressibility equation (4.73) that the vertical velocity must bear a linear vertical coordinate dependence, i.e.,

$$
\frac{\partial w}{\partial z} = -\frac{\partial u}{\partial x}, \qquad \Longrightarrow \qquad w(x,z,t) = -z\frac{\partial u}{\partial x} + W_0(x,t),
$$

where $W_0(x,t)$ is an, as yet, unknown function independent of z. In order to write the equations with some more generality, we suppose that the bottom of our fluid layer has some time independent topography located at a horizontally dependent depth $z = -L_b(x)$. As we discussed in Sect. 4.2.2, the imposition of no-normal flow condition at some level $z = -L_b(x)$ means

$$
w + u\frac{\partial L_b}{\partial x} = 0, \qquad \text{on } z = -L_b(x).
$$

Substituting it in the PDE for w, written just above the last equation we have

$$
L_b\frac{\partial u}{\partial x} + W_0 + u\frac{\partial L_b}{\partial x} = 0,
$$

not forgetting our assumption that u is independent of z. Solving for $W_0(x,t)$ is trivial and this, in turn, yields the explicit solution for the vertical velocity

$$
w(x,z,t) = -(L_b + z)\frac{\partial u}{\partial x} - u\frac{\partial L_b}{\partial x} \tag{4.76}
$$

Finally, an equation describing the evolution of the surface is needed—but we already know that this is developed using the kinematic condition for the upper moving boundary S, i.e., the surface defined by the relationship $S(x,z,t) = z - \eta(x,t) = 0$, thus

$$
\left.\frac{dS}{dt}\right|_{z=\eta(x,t)} = 0, \qquad \Longrightarrow \qquad \frac{\partial \eta}{\partial t} + u\frac{\partial \eta}{\partial x} = w(x,z=L_b,t) = -(L_b+\eta)\frac{\partial u}{\partial x} - u\frac{\partial L_b}{\partial x}.
$$

Provided (as we have assumed here) that the bottom is not time-dependent, we can rewrite the above expression into a more revealing form (defining the total layer depth $h \equiv \eta + L_b$). The equation for the evolution of η turns then effectively into an equation for the layer-width

$$\frac{\partial h}{\partial t} + u\frac{\partial h}{\partial x} + h\frac{\partial u}{\partial x} = 0. \tag{4.77}$$

The shallow water momentum equation (4.75) expressed in terms of h is

$$\frac{\partial u}{\partial t} + u\frac{\partial u}{\partial x} = -\frac{\partial(gh)}{\partial x} + \frac{\partial(gL_b)}{\partial x}. \tag{4.78}$$

The set of Eqs. (4.77)–(4.78) are known as the *shallow water equations* for one horizontal dimension with bottom topography embodied in a "forcing" term $\partial(gL_b)/\partial x$ representing how the bottom boundary is shaped. The procedure we have implemented can be generalized to include a second horizontal dimension y on equal footing. In that case, there will also be an equation for the evolution of the other horizontal velocity v (see Problem 4.12).

Before proceeding to analyze the shallow water set of equations, we call to attention that the equation for the layer thickness (4.77) is, in fact, the one-dimensional continuity equation in disguise. Inspection quickly shows that Eq. (4.77) may immediately be rewritten as

$$\frac{\partial h}{\partial t} + \frac{\partial(hu)}{\partial x} = 0,$$

which has the form of a continuity equation. Thus, the layer depth h can be equivocated with an effective compressible density. This interpretation is quite accurate upon reflection of the fact that because the fluid of the shallow water itself is of constant density, vertical variations of the thickness means that the total vertically integrated mass of the layer can be a function of horizontal position and time. In this way, the thickness h may be more intuitively interpreted as a surface density. The similarity between the treatment of the solutions of this system and one-dimensional compressible gas dynamics by the method of characteristics is related to this observation.

Before continuing we pause to reflect upon what we have achieved here. We have taken a system of equations describing layer dynamics and effectively reduced the dimension of the system. Equations described by the coordinates (x, z, t) have been turned into equations in terms of (x, t), instead. The shallow water equations are an *asymptotic reduction* of the original equations of motion achieved using scaling arguments. The derivation of such reduced equations is common in FD treatments for several reasons, the most important of them being that the physical intuition that can be gleaned from such simpler models can be invaluable in interpreting and understanding the results of more complicated systems—both in experiments and in numerical computations.

4.4   Atmospheric Waves in the Boussinesq Approximation

Thus far, we have considered in this chapter problems in which the flow is characterized by constant density. Density changes occurred only at discontinuities separating fluid regions of constant density. But in reality density variations are present continuously throughout a fluid medium and, when significant, they should be taken into account. A general treatment of density variations must include a formal mathematical treatment of sound waves and shocks, and this is the subject of Chap. 6. There are however phenomena in which a fluid supports density fluctuations, while the flow remains *largely* incompressible. In this section, we explore and motivate the setup of the *Boussinesq fluid*, i.e., a model fluid whose flow fields are incompressible and whose density variations are dynamically significant to the fluid only when coupled to external forces. The model bears the name of J. V. Boussinesq, a mathematician and physicist, who lived close to the turn of the nineteenth and twentieth centuries. Sometimes the name of A. Oberbeck is also attached to Boussinesq's whenever this approximation or model is mentioned.

Such a fluid model might seem arbitrary, but it is motivated by vast amount of experimental and observational results. There exists a wide collection of fluid flows involving density fluctuations that affect the fluid flow and, yet, do not generate dynamically significant *acoustic* disturbances. The latter notion obviously refers to sound and we must wait until Chap. 6 for its proper definition and elucidation. It is however useful to develop a model framework in which the dynamical influence of density variations can be included, without generating acoustic phenomena— the model Boussinesq fluid fits the bill. Examples of fluid phenomena amenable to this kind of model include buoyant convection, internal gravity waves, dynamics of nematic fluids, atmospheric dynamics affected by moisture, and so on.

4.4.1   The Boussinesq Approximation and Corresponding Equations

One of the earliest lessons from classical thermodynamics is that any thermodynamic quantity may be written as a function of other two (or more, if the fluid is not homogeneous in its chemical composition) other thermodynamic quantities. We have dealt in detail with the thermodynamics of fluids in the first chapter of this book, in Sect. 1.5. Equations of state were among the topics discussed there and we wish to quote here the result for an ideal (perfect) gas. Its equation of state, i.e., Eq. (1.60) valid for a homogeneous composition, giving rise to a mean molecular weight μ, expressed in atomic weight units, was written in the form $P = \mathscr{R}\rho T/\mu$, where $\mathscr{R}$ is the gas constant. Of course, we are free to express also ρ as a function of

P and T in this and other, as well, equations of state. We might choose to express the density as a function of pressure and specific entropy instead, symbolically written

$$\rho = \rho(P,T) \qquad \text{or} \qquad \rho = \rho(P,s).$$

At the core of the Boussinesq approximation is the following, empirically motivated, rationale: considering a fluid element of a particular length scale L, embedded in a much larger global system, we posit that all dynamics on the global length scale occurs on a timescale that is long, as compared to the sound crossing time of the fluid element (L/c_s). Since acoustic waves play the role of establishing mechanical equilibrium in a fluid, it follows that the pressure inside the fluid element equilibrates very quickly, indeed approximately instantaneously, to the pressure found in the environment, precisely due to the short sound propagation timescales within the element. We may say that the situation is *quasi-static*. In other words, the equilibration takes the form of a fluid element's interior pressure matching its external pressure and, for all practical purposes, we assume this to be the case unless we are forced into abandoning this position for a good reason, as explained in the following sentences. There is a limit to this view since, mathematically speaking, there may exist fluid elements whose length scales are large enough, so that their sound crossing timescales are similar to the dynamical times, typifying the global system. It is at this scale that such a model picture breaks down and full compressibility should be taken into account. Henceforth, we assume that such conditions are never met for the present system of interest. In this way, however, it is important to be mindful of the scales characterizing the system of interest and to be aware of when the fundamental assumptions behind a model begins to breakdown.

We saw in Sect. 1.5.2, and it is actually a linearization procedure, or Taylor expansion in two variables, retaining only first order terms, that variations of the density may be written in the following form

$$\rho' \equiv \rho - \rho_0 = \left(\frac{\partial\rho}{\partial s}\right)_P (s - s_0) + \left(\frac{\partial\rho}{\partial P}\right)_s (P - P_0),$$

or expressed in a similar form to Eq. (1.61), in which the free variables are T,P, instead of s,P, as here:

$$\rho(P,s) = \rho_0 \left[1 - \beta_s(s - s_0) + \beta_P(P - P_0)\right],$$

$$\beta_P \equiv \frac{1}{\rho_0}\left(\frac{\partial\rho}{\partial P}\right)_s, \beta_s \equiv -\frac{1}{\rho_0}\left(\frac{\partial\rho}{\partial s}\right)_P.$$

Since in this approximation $|\beta_P(P - P_0)|$ is dwarfed by the magnitude of $|\beta_s(s - s_0)|$, we may approximate this form of the equation of state as

$$\rho' = -\beta_s\rho_0 s', \qquad \Longleftrightarrow \qquad \rho = \rho_0\left[1 - \beta_s s'(x,y,t)\right]. \tag{4.79}$$

The symbol β_s is known as the coefficient of thermal/entropic expansion at constant pressure. The curious sign convention is historical as the coefficient used to be referenced and related to what was called the "coefficient of volumetric expansion at constant pressure" which, of course, is the inverse of an increase in density. In general, this can be a complicated function of the fluid's local pressure and entropy, P_0 and s_0, but for our purposes here we shall approximate it a by constant. We recall that $s' \equiv s - s_0$ and we further note that s_0 is some base entropy profile/state, which might have a time independent spatial structure to it. This is different from what we suppose for the background mean density ρ_0 assumed to have a uniform character to it. Variations to this idea, where the mean density ρ_0 may show spatial dependence, yet the dynamics is treated in a quasi-incompressible way, are referred to and implemented as the *anelastic approximation*. In that case instead of $\nabla \cdot \mathbf{u} = 0$ the equation relating the three components of velocity is $\nabla \cdot (\rho \mathbf{u}) = 0$. Incidentally, S.Chandresekhar treated in his book referenced in the *Bibliographical Notes* this description differently but the mathematical end result is the same.

As we noted in Sect. 1.5.2, despite $|\beta_P P'| \ll |\beta_s s'|$, relative variations of the density due to entropy fluctuations under terrestrial conditions can still be quite small, on order of 10^{-3} or smaller. In the absence of any other external force, this weakness of the density variations implies that the flow is nearly incompressible for all practical purposes. However, when the density field couples to body forces like gravity, magnetic, or electric force fields, whose magnitudes are large enough, variations in density become the means by which these forces can affect the fluid and, as such, should be retained in the dynamical analysis. *Buoyancy*, resulting from variations in density (ρ'), is a fundamental phenomenon and traditionally it is mainly used in meteorology, because to describe how density fluctuations due to coupling to gravity affect fluid motions *à la* Archimedes is perhaps the most common example of an effect of this kind. The perspective this implies, and that which forms the basis of the Boussinesq approximation, is that density variations in the fluid are mostly zero, the flow is therefore largely incompressible, and that the only influence density variations have upon the fluid's momentum budget is when they couple to the fluid via some external force. One important consequence: treating the velocity field as incompressible effectively filters out sound waves from the analysis. The Oberbeck–Boussinesq approximation can be more formally stated as:

> The dynamics of a fluid may be treated as incompressible except when there is dynamical coupling to other external forces. In those cases, density variations of fluid particles are considered to occur at constant pressure. On the spot variations in density accordingly respond to variations in other thermodynamic quantities, e.g., entropy/temperature, salinity content, etc. *but not* to pressure variations.

The motivation for the generality implied by the last sentence above comes from noting that the buoyancy is both a function of the fluid particle's temperature (dependent on entropy content) and other thermodynamically measured quantities like the amount of salt carried by a fluid element. This is obviously applicable to the oceans, while for atmospheres we may think of it as various "pollutants." Since salt diffuses, any local changes in the salt content of a fluid element will correspondingly effect its density. If we assign the symbol C to indicate here salt concentration, then variations in the density (at constant pressure) would be expressed as

$$\rho' = -\beta_s \rho_0 (s - s_0) + \beta_C \rho_0 (C - C_0), \quad \beta_C \equiv \frac{1}{\rho_0} \left(\frac{\partial \rho}{\partial C} \right)_{P,s}, \quad \beta_s \equiv -\frac{1}{\rho_0} \left(\frac{\partial \rho}{\partial s} \right)_{P,C},$$

where we have extended the definition β_s to be dependent on the density variations at both constant pressure and salt concentration. β_C is similarly understood to be the expansion coefficient with respect to salt concentration at constant pressure and entropy. This extension procedure can be generalized to include two or more variables that effect the thermodynamic state of a fluid element. We emphasize that the Boussinesq approximation is still, in the end, an approximation. There have been several attempts at grounding it in other more mathematically rigorous arguments but its "proof" still remains outstanding. However, the empirical basis for its continual use is based upon decades of usage and verification via careful testing and observation, both experimental and numerical.

We give now an example of applying the Boussinessq approximation to a general inviscid plane-parallel atmosphere, acted upon by a body force, per unit mass, **b**. The equations for such a configuration are

$$\frac{Du}{Dt} = -\frac{1}{\rho} \frac{\partial p}{\partial x} + b_x, \tag{4.80}$$

$$\frac{Dv}{Dt} = -\frac{1}{\rho} \frac{\partial p}{\partial y} + b_y, \tag{4.81}$$

$$\frac{Dw}{Dt} = -\frac{1}{\rho} \frac{\partial p}{\partial z} + b_z, \tag{4.82}$$

$$\frac{\partial u}{\partial x} + \frac{\partial v}{\partial y} + \frac{\partial w}{\partial z} = 0, \tag{4.83}$$

in which $\mathbf{b} = (b_x, b_y, b_z)$ indicates the aforementioned general applied body force, per unit mass. The density is related to other quantities via the equation of state

$$\rho = \rho_0 \left(1 - \beta_s s' + \beta_C C' + \cdots \right), \quad s' = s - s_0, \quad C' = C - C_0. \tag{4.84}$$

For instance, if all that were relevant to the dynamics were variations in the entropy, then we supplement Eqs. (4.80)–(4.84) with an equation for the entropy $s = s_0 + s'$, i.e.,

$$\frac{Ds}{Dt} = \frac{\partial s'}{\partial t} + \mathbf{u} \cdot \nabla \left(s_0 + s' \right) = Q_s, \tag{4.85}$$

where Q_s is a source/sink rate of entropy, per unit mass into/out of the system and can be expressed as $Q_s = Q/\rho c_V T$, where Q is the familiar heat source/sink rate per unit volume and with c_V being the specific heat per unit mass at constant volume. The reader may find a short summary of this thermodynamical relation and more in Sect. 1.5.2.

In some studies, like in Rayleigh–Bénard convection, the evolution of the temperature is followed instead of (4.85). For all practical purposes involving incompressible flows, it makes no mathematical difference if one follows the evolution of the temperature or the entropy. We could have just as easily written the Boussinesq approximation in terms of the evolution of a mean background temperature and associated temperature fluctuations. The results found in either case are functionally equivalent. The temperature and entropy forms are being used interchangeably. We note that publications using the Boussinesq approximation written by scientists working in the old Soviet bloc, for instance, in the treatment of Landau & Lifschitz's book, cited in the *Bibliographical Notes*, the thermodynamical influence upon flow dynamics in terms of entropy fluctuations is preferred. Treatments in terms of temperature variations was preferred by scientists in the West. Additionally, in the parlance of meteorology, entropy is often referred to as *potential temperature* and is often assigned the symbol θ or Θ. It is traditional in such studies to speak of temperature fluctuations when they really represent entropy fluctuations.

4.4.2   3-D Waves in Plane-Parallel Atmosphere

The general Boussinesq formalism of the previous section will now be applied to a more specific case of a plane-parallel atmosphere. Consider such a system, composed of a constant density fluid in a vertical gravitational field, that is, $b_x = b_y = 0$ and $b_z = -g$. In addition, this atmosphere is actually a horizontal slab between $z = 0$ and $z = h_0$. It is otherwise infinite in extent in both horizontal directions x and y. We suppose that the bottom of the atmosphere is bounded by a rigid plate. We perturb now Eqs. (4.80)–(4.85), using the Boussinesq approximation in which density fluctuations are due solely to fluctuations in the entropy. We consider perturbations, denoted, as usual, by prime about steady states marked with the subscript "0." In the steady state we have $\rho' = 0, s' = 0, u = v = w = 0$, meaning that the basic state is in vertical hydrostatic equilibrium

$$\frac{\partial P_0}{\partial z} = -\rho_0 g.$$

Assuming that the fluid is exposed to an open vacuum at $z = h_0 = $ const., the solution for the pressure field P_0 is already familiar to us and is given by

$$P_0 = \rho_0 g(h_0 - z) \quad \text{for} \quad 0 < z < h_0.$$

We further assume in our setup the unperturbed entropy state $s_0(z)$ is a function of the vertical coordinate. For tractability we take the simple dependence

$$s_0(z) = s_{00} + \beta z, \tag{4.86}$$

with s_{00} a constant base value and where β is also a constant, measuring the rate of change of the atmosphere's mean entropy with height. For our purposes, we restrict attention to cases where $\beta > 0$, i.e., basic state entropy profiles that increase with height. The opposite case $\beta < 0$ is deferred to Chap. 7 to discuss this situation.

Adding now the small perturbations to the dependent variables like $P = P_0 + P'$ into Eqs. (4.80)–(4.85) and linearizing, results in the following set of equations, the first of which includes the perturbed pressure terms in the vertical momentum equation

$$-\frac{1}{\rho}\frac{\partial P}{\partial z} - g \approx -\frac{1}{\rho_0}\frac{\partial P_0}{\partial z} - \frac{1}{\rho_0}\frac{\partial P'}{\partial z} + \frac{\rho'}{\rho_0^2}\frac{\partial P_0}{\partial z} - g \implies = -\frac{1}{\rho_0}\frac{\partial P'}{\partial z} - g\frac{\rho'}{\rho_0}, \tag{4.87}$$

$$\frac{\partial u'}{\partial t} = -\frac{1}{\rho_0}\frac{\partial P'}{\partial x}, \tag{4.88}$$

$$\frac{\partial v'}{\partial t} = -\frac{1}{\rho_0}\frac{\partial P'}{\partial y}, \tag{4.89}$$

$$\frac{\partial w'}{\partial t} = -\frac{1}{\rho_0}\frac{\partial P'}{\partial z} + g\beta_s s', \tag{4.90}$$

$$0 = \frac{\partial u'}{\partial x} + \frac{\partial v'}{\partial y} + \frac{\partial w'}{\partial z}, \tag{4.91}$$

$$\frac{\partial s'}{\partial t} + \beta w' = 0. \tag{4.92}$$

Equations (4.84) with $C = 0$, i.e.,variation of ρ only because of entropy variation and (4.85) with, in addition, no entropy sources or sinks assumed in the system, that is, $Q_s = 0$, were all exploited in obtaining the above linearized set.

We may now combine these equations into a single one for the vertical vorticity. Remembering that the vertical component of vorticity is $\zeta \equiv \omega \cdot \hat{\mathbf{z}}$, we may first form the vertical component of the *vorticity perturbation* $\zeta' \equiv \partial_x v' - \partial_y u'$ and derive an equation for it by operating on Eq. (4.89) by $\partial/\partial x$ and subtracting from it the result of operating on Eq. (4.88) by $\partial/\partial y$ leaving simply

$$\frac{\partial \zeta'}{\partial t} = 0. \tag{4.93}$$

The last equality simply means that the vertical vorticity perturbation ζ' is time independent. Similarly, we can form the horizontal divergence of the first two Eqs. (4.88)–(4.89) and make use of Eq. (4.91) once again to find

$$-\frac{\partial}{\partial t}\frac{\partial w'}{\partial z} = -\frac{1}{\rho_0}\nabla^2 P',\qquad (4.94)$$

where we use the usual notation for a Laplacian, but mean here the horizontal Laplacian only, i.e., $\nabla^2 = \partial_x^2 + \partial_y^2$, hoping that this will not lead to confusion. Operating on Eq. (4.90) with ∇^2 and subtracting from it the result of operating upon the intermediate Eq. (4.94) with $\partial/\partial z$ leaves

$$\frac{\partial}{\partial t}\left(\frac{\partial^2}{\partial z^2} + \nabla^2\right)w' = g\beta_s\nabla^2 s'.$$

Lastly, we operate on the above with $\partial/\partial t$ and replace the term on the right-hand side above with the corresponding relation found in Eq. (4.92), to find the final equation for w'

$$\frac{\partial^2}{\partial t^2}\left(\frac{\partial^2 w'}{\partial z^2} + \nabla^2 w'\right) + g\beta_s\beta\nabla^2 w' = 0.\qquad (4.95)$$

We are interested in the wavelike solutions of this system. For illustration, consider the atmosphere to be bounded by an impenetrable lid at $z = h_0$. The physical meaning of this assumption will be explained at the end of the calculation. This adds to the no normal flow boundary condition at $z = 0$ the same condition at $z = h_0$, i.e., $w = 0$ at $z = 0$ & h_0. Choosing a natural Fourier form ansatz for the solution

$$w'(x,y,z,t) = \hat{w}(z)\exp\left(ik_x x + ik_y y - i\omega t\right) + \text{c.c.},$$

we deduce the following ordinary differential equation for $\hat{w}(z)$, which will be solved as two point boundary ODE problem

$$\frac{d^2\hat{w}}{dz^2} + \mu^2\hat{w} = 0,\qquad \mu^2 = K^2\left(\frac{g\beta_s\beta}{\omega^2} - 1\right),\qquad (4.96)$$

where $K^2 \equiv k_x^2 + k_y^2$. General solutions for $\hat{w}$ are linear combinations of $\sin\mu z$ and $\cos\mu z$. We select the former as it satisfies the boundary condition at $z = 0$. By writing

$$\hat{w} = A\sin\mu z,$$

the second boundary condition at $z = h_0$ enforces

$$\hat{w}(h_0) = A\sin\mu h_0 = 0,\quad \Longrightarrow\quad \mu h_0 = n\pi,\quad \Longrightarrow\quad \omega^2 = \frac{g\beta_s\beta h_0^2 K^2}{n^2\pi^2 + h_0^2 K^2}.$$

It follows that

$$\omega_n = \pm h_0 K \left(\frac{g \beta_s \beta}{n^2 \pi^2 + h_0^2 K^2} \right)^{1/2} \tag{4.97}$$

This dispersion relation is actually a solution of an eigenvalue problem and holds for the angular frequencies ω_n with $n = 1, 2, \ldots$. It depicts the properties of *internal gravity waves*. These waves are transverse on account of the incompressibility condition, that is, $\nabla \cdot \mathbf{u}' = 0$, because from it necessarily follows $\mathbf{k} \cdot \hat{\mathbf{u}}' = 0$. They are disturbances akin to surface gravity waves discussed in the previous sections except that they propagate in the interior of the fluid as opposed to be disturbances that are restricted to the top of the fluid. They come about because of the weak compressibility effects introduced via the Boussinesq approximation and reflect, ultimately, the stratification in the atmosphere. The no vertical flow condition at the top of the model atmospheric layer, which we chose in our illustrative example, has filtered out the surface gravity waves we developed at length in Sect. 4.2.2. If one relaxes this condition and, instead, imposes a zero total pressure condition on the top of the moving boundary, then one may see how surface waves are retained (Problem 4.16). In any event, in the next chapter we show how these internal gravity wave structures persist in fluid problems involving rotation. In that case, these modes will be generalized and referred to as *inertia-gravity* waves.

4.4.3   Energy Propagation in Internal Gravity Waves

In Sect. 4.1.3, we introduced the mathematical definition of the group velocity and offered a loose motivation for why it is interpreted as indicating the direction and speed of energy propagation associated with the wave disturbance. In this section, this identification is made explicit using the example of internal gravity waves. One begins by constructing an energy quantity out of the linearized disturbance equations by

1. Multiplying Eqs. (4.88)–(4.90), respectively, by u', v', w',
2. Multiplying Eq. (4.92) by $g \beta_{ss}'/\beta$,
3. Adding the four resulting equations together.

This yields the following relation for the energy, per unit volume, of the perturbations, here waves:

$$\frac{\partial \mathcal{E}'}{\partial t} + \nabla \cdot (\mathbf{u}' P') = 0, \qquad \mathcal{E}' \equiv \rho_0 \frac{u'^2 + v'^2 + w'^2}{2} + \frac{\rho_0 g \beta_s}{\beta} \frac{s'^2}{2}, \tag{4.98}$$

where, in deriving the above expression, explicit use was made of flow incompressibility, Eq. (4.91). The energy of the disturbances is composed of the kinetic energy portion ($\sim \mathbf{u}'^2/2$) and a part that represents a potential energy in the form of buoyancy since $\beta_s^2 s'^2$ corresponds to ρ'^2/ρ_0^2. The goal is to identify the direction of the wave energy flow, or of the flux. To find the energy flow we use some fundamental concepts regarding it: if a pressure, say P', is acting on a surface whose very small area is δS with normal unit vector $\hat{\mathbf{n}}$, i.e., $\delta \mathbf{S} \equiv \delta S \hat{\mathbf{n}}$, then we can calculate the work done across that surface as being $\delta S \Delta x P'$, where Δx is a length along the normal vector $\hat{\mathbf{n}}$ of the surface on which the pressure acts. If that surface is being worked upon by a velocity field $\mathbf{u}'$, then the rate of working being done across the surface is $P' \mathbf{u}' \cdot \delta \mathbf{S}$. From this simple construction we can understand the meaning of the expression $P' \mathbf{u}'$ appearing in Eq. (4.98).

The group velocity in multiple dimensions must be a vector, which is mathematically defined by the expression

$$\mathbf{V}_g \equiv \frac{\partial \omega}{\partial k_x} \hat{\mathbf{x}} + \frac{\partial \omega}{\partial k_y} \hat{\mathbf{y}} + \frac{\partial \omega}{\partial k_z} \hat{\mathbf{z}}. \tag{4.99}$$

In the following discussion, we demonstrate how the group velocity indicates the propagation direction of the mechanical power associated with $P' \mathbf{u}'$. To illustrate how this comes about we re-examine Eq. (4.95) for the case of three-dimensional simple plane wave solutions, without any boundary conditions on z, i.e., we assume that

$$w' \sim \hat{w} \exp\left(i k_x x + i k_y y + i k_z z - i \omega t \right) + \text{c.c.},$$

in which k_z, along with k_x and k_y, are real constants. The shorthand form $\mathbf{k} \equiv k_x \hat{\mathbf{x}} + k_y \hat{\mathbf{y}} + k_z \hat{\mathbf{z}}$ allows the three-dimensional wave ansatz to be written more compactly as

$$w' = \hat{w} \exp\left(i \mathbf{k} \cdot \mathbf{x} - i \omega t \right) + \text{c.c.} \tag{4.100}$$

Note that in this case, and unlike the boundary value problem we described earlier, $\hat{w} = \text{const}$. As we discussed earlier in this chapter, basic vector calculus informs us that $\mathbf{k}$ is the normal to the surface described by $\mathbf{k} \cdot \mathbf{x} - \omega t = \text{const}$. These surfaces, as parametrized by t are ones of constant phase/argument of the wave solution. Inserting the explicit three-dimensional wave ansatz into Eq. (4.95) yields the dispersion relation

$$\omega = \pm \left(\frac{g \beta \beta_s K^2}{k_z^2 + K^2} \right)^{1/2}, \tag{4.101}$$

where, as before, $K^2 \equiv k_x^2 + k_y^2$. Before proceeding we first note that because of (4.91) it follows that $\mathbf{k} \cdot \mathbf{u}' = 0$. Given the wave ansatz we have assumed, it indicates that the phase velocity and thus the wave propagation is in the direction of $\mathbf{k}$ and it is orthogonal to the velocity field fluctuations, reminding us again that these (and all) incompressible wave motions are *transverse*, similar in this aspect to freely propagating electromagnetic waves in a vacuum.

Consider now the average energy content of a simple monochromatic wave with wavevector given by $\mathbf{k} = (k_x, k_y, k_z)$. Everything that follows from the forthcoming discussion generalizes to a general wave disturbance composed of many superimposed waves. We shall use here the angular brackets to denote both spatial and temporal averages. To a given $\mathbf{k}$ there correspond three lengths $L_{x,y,z}$ which here are exactly the wavelength in each of the three directions: $L_x = 2\pi/k_x$, $L_y = 2\pi/k_y$, and $L_z = 2\pi/k_z$. T is, as usual, one temporal cycle of the wave $T = 2\pi/\omega$. For any given perturbation $f'(x,y,z,t)$, which behaves like a wave in three dimensions, similar to (4.100), the average is understood to be

$$\langle f' \rangle \equiv \frac{1}{L_x L_y L_z T} \int_0^{L_x} \int_0^{L_y} \int_0^{L_z} \int_0^{T} f'(x,y,z,t) \, dx \, dy \, dz \, dt. \tag{4.102}$$

Inserting now such a wave ansatz for all the components of the velocity plus the pressure and entropy perturbations into the governing linearized equations of motion (4.88)–(4.92) reveals the following relation between the hatted quantities

$$-\omega\hat{u} = -k_x\hat{P}/\rho_0, \qquad -\omega\hat{v} = -k_y\hat{P}/\rho_0,$$

$$-\omega\hat{w} = -k_z\hat{P}/\rho_0 - ig\beta_s\hat{s}, \qquad -\omega\hat{s} = i\beta\hat{w}. \tag{4.103}$$

We use these to now assess various averages. If we choose to express all quantities in terms of $\hat{P}$, then we find, for instance, that

$$\langle P'u' \rangle = \left\langle \left(\hat{P}e^{i\mathbf{k}\cdot\mathbf{x}-i\omega t} + \text{c.c} \right) \left(\hat{u}e^{i\mathbf{k}\cdot\mathbf{x}-i\omega t} + \text{c.c} \right) \right\rangle$$

$$= \left\langle \left(\hat{P}e^{i\mathbf{k}\cdot\mathbf{x}-i\omega t} + \text{c.c} \right) \left(-\frac{k_x\hat{P}}{\omega\rho_0}e^{i\mathbf{k}\cdot\mathbf{x}-i\omega t} + \text{c.c} \right) \right\rangle$$

$$= \frac{2k_x}{\omega\rho_0}|\hat{P}|^2. \tag{4.104}$$

In going from the first to the second line in the above derivation, we have made use of the first equality in Eq. (4.103). A similar procedure may be performed to determine $\langle P'v' \rangle$ and $\langle Pw' \rangle$ (see Problem 4.15). In the same vein we may calculate the average of the energy content $\langle \mathscr{E}' \rangle = \langle \mathscr{E}'_{kin} \rangle + \langle \mathscr{E}'_{pot} \rangle$ where

$$\langle \mathscr{E}'_{kin} \rangle \equiv \left\langle \frac{\rho_0}{2}\left(u'^2 + v'^2 + w'^2 \right) \right\rangle,$$

$$\langle \mathscr{E}'_{pot} \rangle \equiv \left\langle \frac{\rho_0}{2}\frac{g\beta_s}{\beta}s'^2 \right\rangle \quad \Longrightarrow \quad \langle \mathscr{E}' \rangle = 2\frac{K^2(K^2 + k_z^2)}{\omega^2 k_z^2}\frac{|\hat{p}|^2}{\rho_0}. \tag{4.105}$$

A part of Problem 4.15 is the proof that $\langle \mathscr{E}'_{kin} \rangle = \langle \mathscr{E}'_{pot} \rangle$ indicating that on average the amount of energy in the kinetic portion of the wave is equal to the amount present in the potential-like part of the energy budget, conforming to physical intuition (harmonic oscillator). After some further manipulation it follows that

$$\langle P'\mathbf{u}'\rangle = \langle P'u'\rangle\hat{\mathbf{x}} + \langle P'v'\rangle\hat{\mathbf{y}} + \langle P'w'\rangle\hat{\mathbf{z}} = \langle\mathscr{E}'\rangle\frac{\partial\omega}{\partial k_x}\hat{\mathbf{x}} + \langle\mathscr{E}'\rangle\frac{\partial\omega}{\partial k_y}\hat{\mathbf{y}} + \langle\mathscr{E}'\rangle\frac{\partial\omega}{\partial k_z}\hat{\mathbf{z}} = \langle\mathscr{E}'\rangle\mathbf{V}_g,$$

$$(4.106)$$

establishing the relationship we sought to show.

One of the more remarkable outcomes of this is the following fact: in this case the group velocity points in a direction *orthogonal* to the phase velocity! As we noted earlier, the direction of propagation of the wavefronts is the same as **k**. The phase velocity is $V_p = \omega/|\mathbf{k}|$ and is therefore expressed vectorially by

$$\mathbf{V}_p \equiv V_p\hat{\mathbf{k}}.$$

Given the form of the dispersion relation found in Eq. (4.101), some algebraic manipulation makes it easy to show that

$$\frac{\partial\omega}{\partial k_x} = \frac{k_z^2}{K^2}\frac{\omega k_x}{k}, \qquad \frac{\partial\omega}{\partial k_y} = \frac{k_z^2}{K^2}\frac{\omega k_y}{k}, \qquad \frac{\partial\omega}{\partial k_z} = \frac{k_z\omega}{k},$$

with the obvious notation $k = |\mathbf{k}|$. Thus

$$\mathbf{V}_p \cdot \mathbf{V}_g = V_p\frac{k_x^2}{K^2}\frac{\omega k_z^2}{k} + V_p\frac{k_y^2}{K^2}\frac{\omega k_z^2}{k} - V_p\frac{k_z^2\omega}{k} = 0,$$

remembering that $K = \sqrt{(k_x^2 + k_y^2)}$. Imagine now for the moment a source of gravity waves in an atmosphere. Then if the wavefront moves with an oblique angle upward then the disturbance delivers mechanical energy in a direction $\pi/2$ degrees rotated downward. There are many experiments in the laboratory as well as observations of mesoscale disturbances in the atmosphere that exhibit this very phenomenon. The curiosity is that common intuition associates energy propagation in the same direction as the phase fronts, like electromagnetic waves in vacuum or air, but we have shown here that the situation in fluids may be subtler.

4.5 Solitons in Shallow Water

As the story has famously been retold for almost two centuries, J. S. Russell accounted for the discovery (perhaps, rather, the first scientifically documented observation of the sort) of a "wave of translation" in his 1844 publication entitled "Report on waves." Today the object he observed is more commonly referred to as a *soliton*. The story is summarized in the following quote from de Jager's *On the origin of the Korteweg-de Vries equation*, 2011:

> It was in the year of 1834 that the Scottish naval architect followed on horseback a towboat, pulled by a pair of horses along the Union Canal, connecting Edinburgh and Glasgow. However the boat was suddenly stopped in its speed - presumably by some obstacle - but

not the mass of water, which it had put in motion. Our engineer perceived a very peculiar phenomenon: a nice round and smooth wave - a well defined heap of water - loosened itself from the stern and moved off in forward direction without changing its form with a speed of about eight miles an hour and about thirty feet long and one or two feet in height. He followed the wave on his horse and after a chase of one or two miles he lost the heap of water in the windings of the channel.

It is said that Russell remarked on the question how the localized wave structure maintained its coherence and experienced no dispersive effects during its propagation. Russell went on to design experiments to study and reproduce the long-time coherence of such water wave structures. After repeated examinations of their propagation speeds and stability, Russell went on to challenge the mathematical community "to give an a priori demonstration a posteriori" of the long-term robustness of these phenomena. The ensuing history of this challenge piqued the interest of many respected fluid dynamicists who contributed to a mathematical theory of solitons including, but not limited to, Boussinesq [18], Lord Rayleigh [19], and Korteweg and de Vries [20].

In dimensional form the Korteweg-de Vries (KdV) equation is

$$\frac{\partial \eta}{\partial t} + \sqrt{gL}\,\frac{\partial \eta}{\partial x} + \frac{3}{2}\frac{\sqrt{gL}}{L}\eta\frac{\partial \eta}{\partial x} + \frac{1}{6}L^2\sqrt{gL}\,\frac{\partial^3 \eta}{\partial x^3} = 0, \qquad (4.107)$$

where η denotes the vertical position of the fluid surface. We should state here, for the record, that the soliton-like solutions of this equation become increasingly valid both when the disturbance scale is much smaller than the depth of the fluid L_z, i.e., $\mathscr{O}(\eta) \ll \mathscr{O}(L_z)$, and when the horizontal scale L_x of the disturbance is much larger than the depth, that is, $\mathscr{O}(L_z) \ll \mathscr{O}(L_x)$. The exact relative measure of these various disparities will be explained further below. We emphasize here two points:

1. Even though the KdV equation lacks in its generality, for instance because it only describes the motion of unidirectionally propagating disturbances, it compensates by being rigorously derivable through standard perturbation series procedures. It is for this reason that we take the time in Sect. 4.5.1 to carefully develop the calculation as it showcases many of the techniques that get employed again in the development of nonlinear solutions of unstable systems in Chap. 8.

2. It would appear that the solitary wave structure described by Russell falls within the shallow water limit since $L_z \ll L_x$ and that, consequently, it should be treated using the equations of shallow water described in Sect. 4.3. Actually, the correct description of the soliton requires the retention of linear dispersive terms which are absent in the model shallow water equations.

4.5.1 An Asymptotic Derivation of the KdV Equation

In order to make progress towards an asymptotic development of any nonlinear system, one is guided by the following strategy: First, certain scalings measuring various fluid quantities of the system have to be set down, preferably informed by observations. Second, nondimensional quantities have to be identified and it has to be found which of these quantities can be made small, again based on empirical implications. Finally, size disparities have to be exploited in a perturbation series expansion of the equations of motion. When these relations are a priori chosen correctly, the perturbation series can be safely truncated at some low order expansion through the invocation of a kind of consistency criterion rooted in an existence condition for some higher order solution of the series expansion as it will become evident in the procedure below. In this way, the equation we seek will naturally "fall-out" of the analysis as an existence condition of one sort or another. The aforementioned relationships between the various nondimensional quantities that appear are usually guided either by some physical insight or bounds imposed by the parameters of the problem. In practice, however, establishing the relationships between these nondimensional quantities usually requires repeated trial and error attempts until the right proportions are achieved. The unspoken motto in asymptotic analysis is simply "when it works you know it works." Practice and experience figure prominently in this process and there are no substitutes. The derivation of the KdV equation is not only typical of this rule, but its presentation also serves to show that when the right relative choices of amplitudes and length scale disparities are made, a beautiful self-encapsulating simplification can be distilled from otherwise more complicated sets of equations.

The detailed asymptotic procedure begins in this case with the fundamental equations of the physical system developed in Sect. 4.2, which are the Laplace equation describing two-dimensional irrotational flow in the interior of the fluid-equation (4.43), the Bernoulli equation which relates dynamical quantities on the surface, and the kinematic condition describing the moving upper boundary- Eqs. (4.46)–(4.47). As before, the setting is described by a solid horizontal boundary at the position $z = -L_z < 0$ with the undisturbed surface of the fluid at $z = 0$. We recall that both of the equations mentioned above are expressions evaluated at the position of the moving surface. This means that the velocity potential function ϕ is evaluated at $z = \eta(x,t)$ which depends upon the function ϕ as well. As mentioned at the start of this section, we assume that the depth of the fluid is shallow compared to the horizontal length scales of the system. To formalize this we suppose then that disturbances are characterized by a horizontal length scale L_x and that the vertical scales are controlled by the scale L_z. Most importantly, we must say something about the velocity scales characterizing disturbances. We are guided by the one scale known for this system: as we indicated at the end of Sect. 4.2.2 the one and only natural speed in the shallow limit, given by $\sqrt{gL_z}$. Because the velocity potential is given in units of speed times length scale, we suppose that it is characterized by $\delta L_z \sqrt{gL_z}$ where δ here is a new nondimensional parameter, measuring the relative amplitude of the horizontal velocity disturbance as compared to $\sqrt{gL_z}$. The value of δ is not yet determined, but will be ascertained below. Again from the linear theory of Sect. 4.2.2, we found that in the shallow limit the wave-speed $\sqrt{gL_z}$ is accompanied by a timescale equal to $L_x/\sqrt{gL_z}$, resulting from the frequency for a disturbance of length scale L_x. This timescale is a natural one for the basic wave structure but we will later argue that there shall also be a need for a second, so-called, long timescale for the system. Finally, we also assume that the height fluctuations η are scaled by the vertical scale L_z multiplied by another unknown nondimensional small parameter, ε, which shall be assessed in comparison with the other quantities shortly. In summary, then, we make the following scaling replacements:

$$t \mapsto \frac{L_x}{\sqrt{gL_z}}T \qquad \Longrightarrow \qquad \frac{\partial}{\partial t} \mapsto \frac{\sqrt{gL_z}}{L_x}\frac{\partial}{\partial T},$$

$$x \mapsto L_x X \qquad \Longrightarrow \qquad \frac{\partial}{\partial x} \mapsto \frac{1}{L_x}\frac{\partial}{\partial X},$$

$$z \mapsto L_z Z \qquad \Longrightarrow \qquad \frac{\partial}{\partial z} \mapsto \frac{1}{L_z}\frac{\partial}{\partial Z},$$

$$\eta \mapsto \varepsilon L_z H \qquad , \qquad \phi \mapsto \delta L_z \sqrt{gL_z}\,\Phi,$$

in which $X, Z, T, H(X,T)$ and $\Phi(X,Z,T)$ are the corresponding nondimensionalized expressions of the variables of the system. Note that T denotes here a time variable and not an oscillation period or other physical variables it has been used for so far in the book. We have also explicitly depicted how all differential operators transform under these scalings. Note that in these variables the bottom boundary is at $Z = -1$. Inserting these into the governing equations (4.43), (4.46), and (4.47) we find

$$\left(\frac{L_z^2}{L_x^2}\right)\frac{\partial^2 \Phi}{\partial X^2} + \frac{\partial^2 \Phi}{\partial Z^2} = 0; \qquad -1 < Y < \varepsilon H, \text{(4.108)}$$

$$\frac{\partial \Phi}{\partial T} + \frac{1}{2}\delta\frac{L_x}{L_z}\left[\frac{L_z^2}{L_x^2}\left(\frac{\partial \Phi}{\partial X}\right)^2 + \left(\frac{\partial \Phi}{\partial Z}\right)^2\right] + \frac{\varepsilon}{\delta}\frac{L_x}{L_z}H = 0; \qquad \text{at } Y = \varepsilon H, \quad \text{(4.109)}$$

$$\frac{\varepsilon}{\delta}\frac{\partial H}{\partial T} + \varepsilon\frac{L_z}{L_x}\left(\frac{\partial \Phi}{\partial X}\right)\left(\frac{\partial H}{\partial X}\right) = \frac{\partial \Phi}{\partial Z}; \qquad \text{at } Z = \varepsilon H, \text{(4.110)}$$

together with the boundary condition $\partial \Phi/\partial Z = 0$ at $Z = -1$.

Because this system is shallow in the sense we have been using the phrase, thus $L_z/L_x \ll 1$. We want the height deviations to also be small compared to the mean state, therefore we expect that $\varepsilon \ll 1$. The key to the expansion scheme developed hereafter is to judiciously choose how L_z/L_x and ε relate to each other. We propose here, essentially as a guess, that the relationship between them is

$$\frac{L_z^2}{L_x^2} \sim \varepsilon,$$

or $L_z/L_x \sim \varepsilon^{1/2}$. With this in hand we can, with hindsight, note that in any perturbation expansion of the Bernoulli equation (4.109) we should like some sort of basic balance, in lowest order of the expansion parameter ε, between the time derivative term and the height variable H and the meaning of this will be explained a bit further below. Thus we want the coefficient in front of H in Eq. (4.109) to be also of $\mathcal{O}(1)$, from this it follows that

$$\frac{\varepsilon}{\delta} \frac{L_x}{L_z} \sim 1, \qquad \Longrightarrow \qquad \delta \sim \varepsilon \frac{L_x}{L_z} \sim \varepsilon^{1/2}.$$

Returning to the story of Russell's discovery of the propagating solitary wave in the rushing channel flow, we suppose that we are best served to transform the equations into a reference frame moving in tandem with the overall fluid structure. Because the surface water waves can move in either direction in X, we choose to follow a wave that moves in the positive direction, keeping in mind that we could have just as easily chosen to follow a wave moving in the negative direction. With the adopted scalings it would mean, therefore, that a natural variable to invoke would be a quantity moving with this wave which we define as the nondimensional ξ given simply by

$$\xi \equiv X - T.$$

Once in this moving frame we suppose that an additional long, or slow, timescale is needed. This is called a *multiple timescale analysis* and we shall delve deeper into this strategy and its implementation in Chap. 8. We propose this to be characterized by a timescale $\tau \sim \varepsilon T$, i.e., it is $1/\varepsilon$ times slower than the timescale T.

We may now express our solution ansatz as

$$\Phi = \Phi(X - T, Z, \tau) \qquad \Longrightarrow \qquad \Phi = \Phi(\xi, Z, \tau),$$
$$H = H(X - T, \tau) \qquad \Longrightarrow \qquad H = H(\xi, \tau),$$

where the new long time variable $\tau \equiv \varepsilon T$. Before carrying on any further, it is critical to note that all derivative operators appearing in Eqs. (4.108)–(4.110) must correspondingly be rewritten to reflect the last choice of the length scale and multiple timescales. We thus have to replace $\partial/\partial X$ and $\partial/\partial T$ in our equations by

$$\frac{\partial}{\partial X} \mapsto \frac{\partial}{\partial \xi}, \qquad \frac{\partial}{\partial T} \mapsto -\frac{\partial}{\partial \xi} + \varepsilon \frac{\partial}{\partial \tau}.$$

Finally, we insert the entirety of the above scalings relationships and solution assumptions into Eqs. (4.108)–(4.110) and after sorting out the details we find

$$\varepsilon \frac{\partial^2 \Phi}{\partial \xi^2} + \frac{\partial^2 \Phi}{\partial Z^2} = 0; \qquad -1 < Z < \varepsilon H, \quad (4.111)$$

$$-\frac{\partial \Phi}{\partial \xi} + \varepsilon \frac{\partial \Phi}{\partial \tau} + \frac{1}{2}\left[\varepsilon\left(\frac{\partial \Phi}{\partial \xi}\right)^2 + \left(\frac{\partial \Phi}{\partial Z}\right)^2\right] + H = 0; \qquad \text{at } Z = \varepsilon H, \quad (4.112)$$

$$\varepsilon\left[-\frac{\partial H}{\partial \xi} + \varepsilon \frac{\partial H}{\partial \tau} + \varepsilon\left(\frac{\partial \Phi}{\partial \xi}\right)\left(\frac{\partial H}{\partial \xi}\right)\right] = \frac{\partial \Phi}{\partial Z}; \qquad \text{at } Z = \varepsilon H, \quad (4.113)$$

again together with $\partial \Phi/\partial Z = 0$ at $Z = -1$.

As presented, Eqs. (4.111)–(4.113) are now ready to be solved through perturbation series expansions in powers of ε. At this point we have to adopt a strategy to address solutions of this mathematical system. First, we propose the expansion scheme

$$\Phi = \Phi_0(\xi, Z, \tau) + \varepsilon \Phi_1(\xi, Z, \tau) + \varepsilon^2 \Phi_2(\xi, Z, \tau) + \text{HOT}, \qquad (4.114)$$

$$H = H_0(\xi, \tau) + \varepsilon H_1(\xi, \tau) + \text{HOT}. \qquad (4.115)$$

Secondly, we observe that the Bernoulli and kinematic equations (4.112)–(4.113) are valid at the position $Z = \varepsilon H$. This means that the velocity potential and its corresponding partial derivatives must also be evaluated at $Z = \varepsilon H$. In order to evaluate these quantities at $Z = \varepsilon H$ we should go through these two preliminary stages: first develop the series solution for Φ in the fluid interior, i.e., for $-1 < Z < \varepsilon H$ by solving Eq. (4.111) order by order in ε followed by a second stage of taking the resulting solution for Φ, valid in the interior, and correspondingly evaluating it and its derivatives at $Z = \varepsilon H$. In effect this amounts to a secondary expansion of the solution found in Eq. (4.114), once determined after solving Eq. (4.111) out to some order in ε. In practice the latter will necessitate solving the equation out to either first or second order in powers of ε. We unambiguously depict this procedure below.

Guided by this strategy, our tactic now is to solve Eq. (4.111) order by order as proposed, subject to the boundary condition that $\partial \Phi/\partial Z = 0$ at $Z = -1$. Inserting the expansion found in Eq. (4.114) into Eq. (4.111) and sorting out terms in powers of ε we find that to lowest order

$$\frac{\partial^2 \Phi_0}{\partial Z^2} = 0,$$

which means that the lowest order solution of the velocity potential is $\Phi_0 = \Phi_0(\xi, \tau)$, i.e., an as yet undetermined function of ξ and τ with no Z-dependence. It will be this function that will be a solution to the KdV equation mentioned at the beginning of this section. This solution Φ_0 automatically satisfies the no normal flow condition at $Z = -1$. The next order equation is

$$\frac{\partial^2 \Phi_0}{\partial \xi^2} + \frac{\partial^2 \Phi_1}{\partial Z^2} = 0.$$

Solutions to this equation subject to the boundary condition at $Z = -1$ is

$$\Phi_1(\xi, Z, \tau) = -\frac{1}{2}(Z+1)^2 \frac{\partial^2 \Phi_0}{\partial \xi^2} + \tilde{\phi}_1(\xi, \tau), \qquad (4.116)$$

where $\tilde{\phi}_1$ is a similarly undetermined function of ξ and τ only. Knowledge of the exact form of this function will not be necessary, but this will become self-evident a little later on. At the next order of Eq. (4.111) we find

$$\frac{\partial^2 \Phi_1}{\partial \xi^2} + \frac{\partial^2 \Phi_2}{\partial Z^2} = 0$$

and this too similarly has solution given by

$$\Phi_2(\xi, Z, \tau) = \frac{1}{24}(Z+1)^4 \frac{\partial^4 \Phi_0}{\partial \xi^4} - \frac{1}{2}(Z+1)^2 \frac{\partial^2 \tilde{\phi}_1}{\partial \xi^2} + \tilde{\phi}_2(\xi, \tau), \qquad (4.117)$$

where $\tilde{\phi}_2$ is a similar function to $\tilde{\phi}_1$.

Now we may move onto the second part of the calculation strategy: inspection of Eqs. (4.112)–(4.113) shows that we must evaluate the quantities $\partial \Phi / \partial \tau$, $\partial \Phi / \partial \xi$, and $\partial \Phi / \partial Z$ at the position $Z = \varepsilon H = \varepsilon H_0 + \varepsilon^2 H_1^2 + \cdots$. In practice this means evaluating these functions around $Z = 0$ in a Taylor series expansion in powers of εH (see Problem 4.17):

$$\left. \frac{\partial \Phi}{\partial \tau} \right|_{Z=\varepsilon H} = \frac{\partial \Phi_0}{\partial \tau} + \mathcal{O}(\varepsilon), \qquad (4.118)$$

$$\left. \frac{\partial \Phi}{\partial \xi} \right|_{Z=\varepsilon H} = \frac{\partial \Phi_0}{\partial \xi} + \varepsilon \left(-\frac{1}{2} \frac{\partial^3 \Phi_0}{\partial \xi^3} + \frac{\partial \tilde{\phi}_1}{\partial \xi} \right) + \mathcal{O}(\varepsilon^2), \qquad (4.119)$$

$$\left. \frac{\partial \Phi}{\partial Z} \right|_{Z=\varepsilon H} = -\varepsilon \frac{\partial^2 \Phi_0}{\partial \xi^2} + \varepsilon^2 \left[-H_0 \frac{\partial^2 \Phi_0}{\partial \xi^2} + \frac{1}{6} \frac{\partial^2 \Phi_0}{\partial \xi^2} - \frac{\partial^2 \tilde{\phi}_1}{\partial \xi^2} \right] + \mathcal{O}(\varepsilon^3). \quad (4.120)$$

Now we may proceed towards completing the derivation of the equation governing the dynamics. At lowest order of the kinematic equation (4.113) we find simply

$$\frac{\partial \Phi_0}{\partial Z} = 0,$$

but this is an already satisfied condition so we may now move on to $\mathcal{O}(1)$ terms of the Bernoulli equation (4.112)

$$-\frac{\partial \Phi_0}{\partial \xi} + H_0 = 0, \qquad \Longrightarrow \qquad H_0 = \frac{\partial \Phi_0}{\partial \xi}, \tag{4.121}$$

which is basically a diagnostic relationship between the horizontal derivative of the velocity potential and the lowest order height deviation. At $\mathcal{O}(\varepsilon)$ in the expansion of Eq. (4.113) we find

$$\frac{\partial^2 \Phi_0}{\partial \xi^2} - \frac{\partial H_0}{\partial \xi} = 0. \tag{4.122}$$

But, on account of the relationship between H_0 and Φ_0 determined in Eq. (4.121), the above equation is automatically satisfied. $\mathcal{O}(\varepsilon)$ terms of Eq. (4.112) and $\mathcal{O}(\varepsilon^2)$ terms of Eq. (4.113) are, after some rearranging,

$$-\frac{1}{2}\frac{\partial^3 \Phi_0}{\partial \xi^3} + \frac{\partial \Phi_0}{\partial \tau} + \frac{1}{2}\left(\frac{\partial \Phi_0}{\partial \xi}\right)^2 = -H_1 + \frac{\partial \tilde{\phi}_1}{\partial \xi}, \tag{4.123}$$

$$\frac{\partial H_0}{\partial \tau} + \frac{\partial H_0}{\partial \xi}\frac{\partial \Phi_0}{\partial \xi} + H_0\frac{\partial^2 \Phi_0}{\partial \xi^2} - \frac{1}{6}\frac{\partial^2 \Phi_0}{\partial \xi^2} = \frac{\partial H_1}{\partial \xi} - \frac{\partial^2 \tilde{\phi}_1}{\partial \xi^2}. \tag{4.124}$$

The above would appear to imply that we have an underdetermined system: two equations for three unknown variables Φ_0, H_1, and $\tilde{\phi}_1$. As we alluded to earlier in this discussion, the right-hand sides of both Eqs. (4.123)–(4.124) actually represent a condition for consistency of the developed solutions, often referred to as a *solvability condition*. This notion will be further discussed and detailed in Chap. 8, where also a good relevant reference [6] will be given in the *Bibliographical Notes* of that chapter. Here it suffices to see how this consistency condition is enforced by operating by $\partial/\partial \xi$ on Eq. (4.123) and adding the resulting equation directly to Eq. (4.124). The procedure formally eliminates both of the unknown functions H_1 and $\tilde{\phi}_1$ from the two equations leaving finally the nondimensional Korteweg-de Vries equation:

$$\frac{\partial H_0}{\partial \tau} + \frac{1}{6}\frac{\partial^3 H_0}{\partial \xi^3} + \frac{3}{2}H_0\frac{\partial H_0}{\partial \xi} = 0, \tag{4.125}$$

where we have made use of the relationship $H_0 = \partial \Phi_0/\partial \xi$ in writing the above equation. We note here that the consistency condition embodied in Eq. (4.125) must be satisfied if a unique solution for H_1 and $\tilde{\phi}_1$ is to exist. This is the same idea we referred to at the beginning of this discussion.

Finally, restoring the dimensional variables into Eq. (4.125) and utilizing the fact that

$$\frac{\partial}{\partial t} \mapsto \varepsilon \frac{\sqrt{gL_z}}{L_x}\frac{\partial}{\partial \tau} - \sqrt{gL_z}\frac{\partial}{\partial x},$$

recovers the dimensional version of the KdV equation (4.107).

4.5.2 Linear Theory Reanalyzed and Some Exact Solutions

A test of the robustness of a reduced theoretical model like the KdV equation (4.107) is to see whether or not it captures the correct aspects of the linear theory, derived for the original set of equations as examined in Sect. 4.2.2. In that case we found the general dispersion relation, which included left and right propagating waves, in Eq. (4.55). Focusing on the waves propagating rightwards, that is, selecting the negative root of that expression and examining only waves having long horizontal scale, that is, assuming $kL_z \equiv kL \ll 1$, we may expand the dispersion relation (4.56) as

$$\omega = -\sqrt{gk\tanh kL} \approx -\sqrt{gL}\left(k - \frac{1}{6}k^2L^2 + \cdots\right).$$

Indeed, if we linearize Eq. (4.107), that is, drop the nonlinear, in η, term and insert in it a normal mode solution of the form $\eta = \hat{\eta}\exp(ikx - i\omega t) + \text{c.c.}$, we find that the KdV equation demands that for this substitution to be a solution

$$-\omega + \sqrt{gL}k - \frac{1}{6}\sqrt{gL}L^2k^3 = 0,$$

which is consistent with the expansion, as expected.

The crowning achievement of the reduced equation is that it has a nonlinear solution that is analytically tractable. The so-called solitary wave solution is an isolated soliton: a structure which is locally concentrated so that the disturbance amplitude decays to zero, as the horizontal extents get very large. Moreover, the theory also predicts the relations between the wave height, η, and the speed with which it propagates. Let us use again the definition of $c_g(= \sqrt{gL})$ that was encountered in Sect. 4.3 and turn our attention to a nonlinear traveling wave solution of Eq. (4.107). We check the consequences of this solution to have the form

$$\eta(x,t) = \eta(x - Vt), \tag{4.126}$$

where V is the constant rightward traveling wave speed, yet to be determined. We shall now recycle former variables and define a new Galilean transformation

$\xi \equiv x - Vt$, paying attention that ξ is now dimensional, and substitute the ansatz (4.126) into Eq. (4.107) to get

$$\left(1 - \frac{V}{c_g}\right)\frac{\partial \eta}{\partial \xi} + \frac{3}{2L}\eta\frac{\partial \eta}{\partial \xi} + \frac{1}{6}L^2\frac{\partial^3 \eta}{\partial \xi^3} = \frac{\partial}{\partial \xi}\left[\left(1 - \frac{V}{c_g}\right)\eta + \frac{3}{4L}\eta^2 + \frac{1}{6}L^2\frac{\partial^2 \eta}{\partial \xi^2}\right] = 0.$$

(4.127)

We are looking for *localized* solutions, that is disturbances whose structure is such that η $\partial\eta/\partial\xi$ and $\partial^2\eta/\partial\xi^2 \to 0$ as $\xi \to \pm\infty$. Equating the expression contained within the square brackets of the second equality to zero—this is the condition for the demanded localization as explained above—followed by multiplying it by $\partial\eta/\partial\xi$ results in

$$0 = \left(1 - \frac{V}{c_g}\right)\eta\frac{\partial \eta}{\partial \xi} + \frac{3}{4L}\eta^2\frac{\partial \eta}{\partial \xi} + \frac{1}{6}L^2\frac{\partial \eta}{\partial \xi}\frac{\partial^2 \eta}{\partial \xi^2}$$

$$= \frac{1}{L}\frac{\partial}{\partial \xi}\left[\frac{L^3}{3}\left(\frac{\partial \eta}{\partial \xi}\right)^2 - (h - \eta)\eta^2,\right]$$

where $h \equiv 2L(V/c_g - 1)$. Using similar reasoning as before, that all functions and their derivatives should decay to zero as $\xi \to \pm\infty$, the expression contained within the brackets of the last line in the above equation should also be equal to zero. Thus the task is reduced now to the determination of the solution to

$$\frac{L^3}{3}\left(\frac{\partial \eta}{\partial \xi}\right)^2 = (h - \eta)\eta^2.$$

(4.128)

Formally speaking, one may take the square root of both sides and integrate the equation using the method of quadratures. In general, such procedures lead to elliptic equations. Here the integration steps leading to Eq. (4.128) have the general form of

$$\frac{L^3}{3}\left(\frac{\partial \eta}{\partial \xi}\right)^2 = (h - \eta)\eta^2 + A\eta + B,$$

where A and B are constants. These constants must be real if the solutions of this equation are to be bounded. When these conditions on A, B are met, the type of periodic nonlinear structures to emerge are called *cnoidal waves*. These can be expressed in terms of Jacobi elliptic functions. However in the case that the domain is infinite, as it is here, the solution form is simple. We guess once more an ansatz having the functional form

$$\eta(\xi) = h\,\text{sech}^2(\gamma\xi),$$

where $\text{sech}(x) \equiv 1/\cosh(x)$ and where γ is an unknown constant. Substituting this ansatz into Eq. (4.128) and equating to zero like powers of the sech function that appear after the algebra has been sorted out leads to the identification that $\gamma = 3h/4L^3$ and that

$$\eta(x,t) = h \, \mathrm{sech}^2 \left[\left(\frac{3h}{4L^3} \right)^{1/2} (x - Vt) \right] \qquad \text{with} \qquad V \equiv c_0 \left(1 + \frac{h}{2L} \right). \tag{4.129}$$

This solution consists of a *single soliton*. h is here a *parameter* of the problem and it is apparent how the soliton propagation speed V is clearly a function of it, i.e., $V(h)$. Sometimes problems like this are referred to as *nonlinear eigenvalue problems* since there are perhaps one or more parameters of the system that are free and certain other quantities appearing in the problem will depend upon them. Problems involving, the so-called integrable PDEs supporting nonlinear wave solutions are a rich area of activity, as there are many problems in a variety of other contexts that exploit solution techniques as this—many involve problems of pattern formation theory— which we will address further in Chap. 8.

It is important to be aware of the differences between the form of the KdV equation (4.107) and one of its many solutions expressed in Eq. (4.129), and that of the equation form and solutions of the Burgers equation (4.30). Moving into a reference frame defined by c_g and doing away with the dispersive term in the KdV equation recovers the essential form of Burgers's equation. So what is interesting is the following: if one had initiated a solution of the Burgers equation with an initial soliton form as detailed in Eq. (4.129) with $t = 0$, then after some time $t = t_s = \min -1/f' > 0$, parts of the wave disturbance would begin to overtake itself and the solution would, technically speaking, wave-break. With the inclusion of the dispersive term, the tendency to wave-break is resisted. In a sense, what is happening is that as wave steepening occurs, the amplitude contained in high wavenumbers steadily grows. According to the dispersion relation, high wavenumber disturbances propagate forward with faster speeds that there corresponding longer wavelength brethren. Thus, the steepening tendency that would lead to wave breaking is effectively neutralized by the dispersive stretching of the front due to the energy pumped into high wavenumber components of the structure. We shall round out the discussion here with a reference to an interesting phenomenon on more complicated solutions of Eq. (4.107) involving two interacting solitons. Observing the form of the propagation speed V, we can see that the larger the amplitude of the individual soliton the faster it will propagate. Thus, we imagine a situation in which two solitons are initiated somewhat further apart but that the first one, centered at say $x = 0$, is initiated with a height h_1 which is larger than the amplitude h_2 of the second soliton, initiated somewhere at $x > 0$. Interpreting the solution indicates that the left soliton (number 1) has a speed V_1 which is faster than the right one's speed (number 2) V_2. After a certain amount of time the first soliton will overtake the second one. The remarkable feature of their interaction is that they pass through each other unaltered (see Fig. 4.17)!

Of course, it does not mean they have not interacted: proof of their mutual influence can be evinced in observing the arrival times (say at some measuring station $x = X_0$) of the two solitons long after they have interacted: the larger

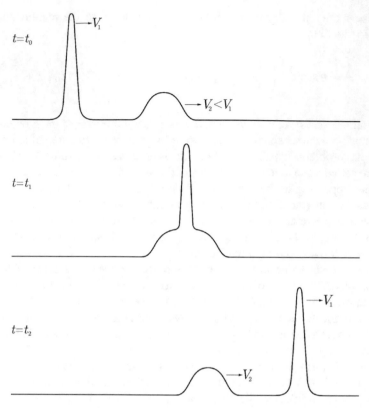

Fig. 4.17 Multiple soliton pulses. The larger amplitude pulse, starting from behind, overtakes the smaller pulse—passing through it and emerging with both structures unaltered

amplitude wave will arrive a bit sooner than expected while the smaller amplitude wave will arrive a bit later than expected. The waves will travel at their prescribed speeds $V_{1,2}$ when they are far from each other but during their interaction phase the larger amplitude wave speeds up a little while the lower amplitude wave correspondingly slows down a bit. Still, this does not conform to the usual intuition built upon interaction by superposition of waves. The feature of no interaction is typical for the above mentioned integrable nonlinear PDEs.

Problems

4.1.
Consider two monochromatic waves having close wave numbers k and $k + \delta k$, respectively. Suppose that the medium has a general dispersion relation given by

$\omega(k)$. Prove the assertion that these two nearly similar wavelength waves have differing phase velocities if the dispersive condition equation (4.1) is satisfied.

4.2.

Consider a second order wave equation in an infinite one-dimensional medium:

$$\frac{\partial^2 \psi}{\partial t^2} - V^2 \frac{\partial^2 \psi}{\partial x^2} = 0,$$

where $V > 0$ is a constant. Show that if the initial conditions of this problem are $\psi(x,0) = f(x)$ and $\partial_t \psi(x)\big|_{t=0} = g(x) = dh/dx$ for some functions f, g, and h, then the solution of this system for $t > 0$ is

$$\psi(x,t) = F(x+Vt) + G(x-Vt), \qquad F \equiv \frac{1}{2}\left(f + \frac{1}{V}h\right), \quad G \equiv \frac{1}{2}\left(f - \frac{1}{V}h\right).$$

4.3.

Develop the full solution to the Schrödinger equation (4.9) with initial condition given in Eq. (4.24).

1. Show that

$$\left|\psi(x,t)\right|^2 = \frac{\exp\left[-\frac{\Delta\left(x-V_g t\right)^2}{1+4t^2\beta^2\Delta^2}\right]}{2\pi\sqrt{1+4t^2\beta^2\Delta^2}}$$

   with $V_g = 2\beta k_0$.
2. Show that $\int_{-\infty}^{\infty} |\psi(x,t)|^2 dx$ is a constant.
3. Working directly with Eq. (4.9) prove that this previous result is consistent. That is, show without explicitly solving Eq. (4.9) that

$$\frac{d}{dt}\int_{-\infty}^{\infty} |\psi(x,t)|^2 dx = 0,$$

   for initial conditions that decay to zero as $x \to \pm\infty$.
4. Taylor expand the solution for values of $t \ll 1/(\beta\Delta)$ and show that the result recovers the approximate form motivated in Sect. 4.1.3.
5. Sketch the solution form of $\left|\psi(x,t)\right|^2$ for a few increasing values of time. Is the qualitative discussion developed in Sect. 4.1.3 consistent with the exact solution? How do they differ and what is the meaning of the difference?

4.4.

Argue that the function

$$\tilde{H} = \frac{1}{2} + \frac{1}{2}\tanh\left(\frac{x}{\Delta}\right)$$

is a reasonable continuous facsimile of a Heaviside function in the limit where Δ gets arbitrarily small. What would be, therefore, the corresponding Dirac delta function facsimile?

4.5.
Argue that a one-dimensional linear wave equation, like Eq. (4.33), with a prescribed continuous and differentiable function $V(x)$, can never show wave-breaking.

4.6.
How does the trajectory of a surface buoy on a fluid supporting capillary-gravity surface waves compare to when there are no surface tension effects? For the sketches assume that the surface wave is rightward and that the functional form for the surface position is given by $\eta = A \cos k(x - V_p t)$ where V_p is the appropriate phase velocity for surface capillary-gravity waves.

4.7.
Argue that in the limit of $kL_b \gg 1$ that the group velocity of gravo-capillary waves is given by

$$V_g = \frac{\partial \omega}{\partial k} = \sqrt{\frac{g}{4k} \frac{1 + 3q_c}{(1 + q_c)^{1/2}}}, \quad \text{with} \quad q_c \equiv \frac{Tk^2}{g\rho}.$$

Find the point on the k axis, k_m at which V_p of these waves change its derivative with respect to k and evaluate $V_m = V(k_m)$. What are the corresponding expressions when $kL_b \ll 1$?

4.8.
Examine the properties of the flow over a corrugated bed developed in Sect. 4.2.4. In particular calculate the mass flow flux, that is, the rate of fluid mass flowing across a unit surface perpendicular to the background mean flow.

4.9.
Recalculate the corrugated bed problem examined in Sect. 4.2.4 with the inclusion of surface tension.

4.10.
Develop the quasi-steady solution of the corrugated bed problem examined in Sect. 4.2.4 for a situation where the bottom bed surface oscillates with angular frequency ω_0 and is represented by the function $z_b = -L + h_\varepsilon \sin(\omega_0 t) \cos(k_0 x)$. How do the surface crests orient with the bottom troughs as a function of ω_0? (Hint: proceed by writing in $\sin(\omega_0 t)$ in exponential form and then separately solve boundary value problem for each of the responses proportional to $e^{\pm i\omega t}$. Note that there will appear statio-temporal phase lags in the response of the surface—so be sure to look out for that.)

4.11.
Recalculate the corrugated bed problem examined in Sect. 4.2.4 in two spatial dimensions. Assume, as before, that the mean flow is in the $\hat{x}$ direction but that the bottom topography has the functional form $z_b = -L + h_\varepsilon \cos(k_0 x) \cos(l_0 y)$. Develop the properties of the steady surface pattern that emerges.

4.12.
Repeat the derivation of the shallow water equations performed in Sect. 4.3.1 assuming the presence of a second horizontal dimension y with a horizontal length scale L_y. Denoting v as the corresponding velocity in that direction and by assuming that L_y is on the same order of magnitude as L_x show that for a thin shallow layer the corresponding equations are

$$\frac{\partial h}{\partial t} + u\frac{\partial h}{\partial x} + v\frac{\partial h}{\partial y} = -h\left(\frac{\partial u}{\partial x} + \frac{\partial v}{\partial y}\right), \tag{4.130}$$

$$\frac{\partial u}{\partial t} + u\frac{\partial u}{\partial x} + v\frac{\partial u}{\partial y} = -\frac{\partial (gh)}{\partial x} + \frac{\partial (gL_b)}{\partial x}, \tag{4.131}$$

$$\frac{\partial v}{\partial t} + u\frac{\partial v}{\partial x} + v\frac{\partial v}{\partial y} = -\frac{\partial (gh)}{\partial y} + \frac{\partial (gL_b)}{\partial y}, \tag{4.132}$$

where the bottom topography is steady and two-dimensional, $L_b = L_b(x, y)$. Furthermore argue that h can be interpreted as a surface density and that Eq. (4.130) is its corresponding continuity equation.

4.13.
Imagine we lived in a physical setting in which the downward pointing acceleration due to gravity was proportional to height, i.e., $g = \Omega_0^2 z$ where Ω_0 is a constant coefficient in units of inverse time. Assuming a constant density incompressible fluid whose horizontal dimension is much larger than the vertical, follow the arguments of Sect. 4.3.1 and develop the analogous equations of shallow water. Assume that the bottom of the atmosphere is at $z = 0$ (i.e., no bottom topography). Therefore derive

$$\frac{\partial h}{\partial t} + u\frac{\partial h}{\partial x} + h\frac{\partial u}{\partial x} = 0$$

$$\frac{\partial u}{\partial t} + u\frac{\partial u}{\partial x} = -\Omega_0^2 h\frac{\partial h}{\partial x},$$

where h is the layer width. Demonstrate that these equations, when written in characteristic form, indicate that the evolution of disturbances of this system is described by the solutions of a pair of uncoupled Burgers's equations, i.e.,

$$\left[\frac{\partial}{\partial t} + (u \pm \Omega_0 h)\frac{\partial}{\partial x}\right](u \pm \Omega_0 h) = 0.$$

4.14.

Derive the equations of shallow water in the same setting as in Problem 4.13 but with the difference that the bottom of the fluid layer is at a value of $z = Z_b > 0$.

4.15.

With reference to the group velocity calculation outlined in Sect. 4.4.3 prove the various intermediate steps leading to the final outcome:

1. Show that

$$\langle P'u' \rangle = \frac{2k}{\omega \rho_0} |\hat{p}|^2, \qquad \langle P'v' \rangle = \frac{2\ell}{\omega \rho_0} |\hat{p}|^2, \qquad \langle P'w' \rangle = -\frac{2K^2}{\omega \mu \rho_0} |\hat{p}|^2.$$

2. Show that the assertion $\langle \mathscr{E}'_{kin} \rangle = \langle \mathscr{E}'_{pot} \rangle$ is true and then prove that the expression for $\langle \mathscr{E}' \rangle$ found in (4.105) is correct.
3. Show finally that the following statement is true,

$$\langle P'\mathbf{u}' \rangle = \mathbf{V}_g \langle \mathscr{E}' \rangle.$$

4.16.

Repeat the calculation of Sect. 4.4.2 but, instead of a no-vertical flow condition at the boundary $z = h_0$, implement the condition that the total pressure is zero on the moving boundary $z = h_0 + \eta$, with $|\eta| \ll h_0$. Assume $w' | \propto |\hat{w}(z)e^{ik_x x + ik_y y - i\omega t} + \text{c.c.}$ solutions.

1. Show that the boundary condition amounts to the requirement

$$\left[\omega^2 \frac{d\hat{w}}{dz} - g(k_x^2 + k_y^2)\hat{w} \right]_{z=h_0} = 0$$

   on $z = h_0$.
2. Show that the corresponding characteristic equation relating ω to other properties of the system is

$$\tan(\mu h_0) = \frac{\mu}{g} \frac{\omega^2}{k_x^2 + k_y^2}$$

   where μ is as given in Eq. (4.96).
3. For the above relationship establish the conditions between k, μ, g, and vertical overtone number n for which

$$\frac{\mu}{g} \frac{\omega^2}{k_x^2 + k_y^2} \ll 1, \qquad \text{and,} \qquad \frac{\mu}{g} \frac{\omega^2}{k_x^2 + k_y^2} \gg 1.$$

   What is the corresponding dispersion relation for ω in those respective cases?
4. Under which conditions does one see surface gravity waves most prominently?

4.17.
Given the series solution Φ out to order ε^2, as expressed in Eq. (4.114), and using the functions given in (4.116) and Eq. (4.117), derive the expressions found in Eqs. (4.118)–(4.120).

4.18.
Assuming the single soliton solution of the KdV equation found in (4.129) is valid, based on de Jager's account of Russell's original soliton observation, estimate the average depth of the Union Canal along the stretch in which Russell chased his soliton. Is h/L_z sufficiently small?

Bibliographical Notes

Section 4.1

The general properties and description of waves is a subject that has been well developed over the years. There are a wide variety of discussions, in various mathematical degree of depth. Lighthill's book is entirely devoted to fluid waves and is an excellent further reading to this chapter that was more inspired, in the choice of its topics by the book of Acheson and in its style by Whitham's book.

1. J. Lighthill, *Waves in Fluids* (Cambridge University Press, Cambridge, 1978)
2. D.J. Acheson, *Elementary Fluid Dynamics* (Oxford University Press, Oxford, 1990)
3. G.B. Whitham, *Linear and Nonlinear Waves* (Wiley, New York, 1974)
   The mathematical formalism for the solution of linear equations, including an extensive discussion on Dirac delta functions, may be found in the classic text by Jackson:
4. J.D. Jackson, *Classical Electrodynamics*, 3rd edn. (Wiley, New York, 1998)
   An intermediate discussion on complex integration can be found in
5. R.V. Churchill, J.W. Brown, *Complex Variables and Applications*, 4th edn. (McGraw Hill, New York, 1984)
6. R. Courant, E. Hilbert, *Methods of mathematical physics*, vol. II (Interscience, New York, 1962)
7. P.M. Morse, H. Feshbach, *Methods of Theoretical Physics*, vol. 1 (Feshbach Publishing, Minneapolis, 1981)

Section 4.2

Recommended further reading pertaining to the development of disturbances in irrotational flows is Whitham's book [3]. Other texts in which the formalism for these kinds of disturbances is discussed are

8. L.D. Landau, E.M. Lifshitz, *Fluid Mechanics*, 2nd edn. (Elsevier, New York, 2004)
9. P.K. Kundu, I.M. Cohen, *Fluid Mechanics* 4th edn. (Elsevier, New York, 2008)
   For further reading and references on the foundations of surface tension and capillarity, we recommend the discussion found in Lamb's work
10. Sir H. Lamb, *Hydrodynamics* (Dover, New York, 1945)

Section 4.3

The physical motivation behind the derivation of shallow water equations was guided in large part by the exposition found in the classic text by J. Pedlosky (which focuses on the more general case of a rotating atmosphere—we shall deal with it in the next chapter).

11. J. Pedlosky, *Geophysical Fluid Dynamics*, 2nd edn. (Springer, New York, 1987)
12. G.K. Vallis, *Atmospheric and Oceanic Fluid Dynamics* (Cambridge University Press, Cambridge, 2006)
    Some useful discussion on the basics of the methods of characteristics as applied to water waves can be found in references [2] and [3]. A more general perspective can be found in
13. F.H. Shu, *The Physics of Astrophysics*, vol. II (University Science Books, Mill Valley, 1992) This topic gets developed further and deeper in Chap. 6.

Section 4.4

Weak compressibility is used prominently in most texts describing the flows of oceans and atmospheres, thus the books by Vallis [12] and Pedlosky [11] are very useful as references for the Boussinesq approximation. Furthermore, the detailed discussion by Chandrasekhar has a good discussion of its use in a whole variety of problems in plane-parallel atmospheres.

14. S. Chandrasekhar, *Hydrodynamic and Hydromagnetic Stability* (Dover, New York, 1981)

Section 4.5

There are many discussions on the derivation of the Korteweg-de Vries equation. We particularly focus on the extended discussion pertaining to solitons (in general) as being solutions to a wide variety of integrable equations (like the Sine-Gordon and nonlinear Schrödinger equations). A good, and often overlooked textbook is the one by Infeld and Rowlands. The historical discussion is taken from a recent lively discussion by E. M. de Jager. Other perspectives and discussions are also listed including some of the original papers

15. E. Infeld, G. Rowlands, *Nonlinear Waves, Solitons and Chaos* (Cambridge University Press, Cambridge, 1990)
16. E.M. de Jager, On the origin of the Korteweg-de Vries equation. arXiv: math/0602661v1
17. P.G. Drazin, Solitons, *London Mathematical Society Lecture Note Series*, vol. 85 (Cambridge University Press, Cambridge 1983)
18. J. Boussinesq, Théorie de l' intumescence liquide appeleé 'onde solitaire' ou 'de translation', se propageant dans un canal rectangulaire. C. R. Acad. Sci. Paris **72**, 755 (1871)
19. Lord Rayleigh, (J.W. Strutt), On waves. Phil. mag. **1**, 257 (1876)
20. D.J. Korteweg, G. de Vries, On the change of form of long waves advancing in a rectangular canal, and on a new type of long stationary waves. Phil. Mag. **39**, 422 (1895)

There are a number of books available that explicitly showcase many of the techniques used in the perturbation calculation, leading to the derivation of the KdV equation. Of the more transparent is the discussion in the work by Nayfeh. Others good texts with expanded discussions include the classic texts of Van Dyke and Kevorkian and Cole and others:

21. A.H. Nayfeh, *Perturbation Methods* (Wiley Classics, New York, 2000)
22. M. Van Dyke, *Perturbation Methods in Fluid Mechanics*, annotated edition (Parabolic Press, Stanford, 1975)
23. J. Kevorkian, J.D. Cole, *Perturbation Methods in Applied Mathematics* (Springer, New York, 1981)

Chapter 5
Rotating Flows

There is a sumptuous variety about the New England
weather that compels the stranger's admiration - and regret.
The weather is always doing something there; always attending
strictly to business; always getting up new designs and trying
them on the people to see how will they go

Mark Twain (1835–1910), in a speech to the American Geographical Society

The most significant rotating flows are found in nature. The atmospheres and oceans
of the Earth are fluids residing on a rotating planet and, as it turns out, this has
profound influence on these geophysical flows. In the recent decade or two, clever
techniques have revealed that planetary system abound in the Galaxy, orbiting many
stars. We should also not forget the planets of our own Solar system and their
satellites. As a rule, valid almost always, these objects rotate around their axes,
which may also change their orientations, albeit on a much slower timescale than
that of the rotation of the object. This rotation is dynamically important to the
fluids on the planet surfaces. Because terrestrial planets have solid surfaces with
complicated topographies and the interaction between liquid oceans and seas with
gaseous atmospheres is not simple, the problem of unravelling and understanding
the surface fluids motion is formidable. This is even true for the nearby objects of
our planetary system that we can observe relatively well. Extra-solar system planets
are, at this time, not resolved well enough to directly study the fluid layers that may
blanket their surfaces. One crude estimate based on everyday experience, especially
apparent to anyone who has traveled by air or sea, seems to be valid: the fluid masses
(air, water) on Earth are, on average, locked to the planet's rotation. Winds and ocean
currents appear to happen with respect to that.

Rotating fluids, like vigorously mixed tea cups or rotating buckets of water, are
also important to be understood. To this end, we have to study such systems in
controlled experiments in the laboratory. As is well known from studies of classical
mechanics, when the equation of motion of a particle is written in a uniformly

© Springer Science+Business Media, LLC 2016   241
O. Regev et al., *Modern Fluid Dynamics for Physics and Astrophysics*,
Graduate Texts in Physics, DOI 10.1007/978-1-4939-3164-4_5

rotating frame of reference, two additional terms appear: a *Coriolis* term and a *centrifugal* term. Reminding ourselves of analytical dynamics, these two terms, when concerning a particle of mass m, appear as forces. While the centrifugal force in a frame rotating with the constant vector angular velocity, Ω, is directed outward, $-2m \times \Omega \times (\Omega \times \mathbf{x})$, the basis for expressing the Coriolis force is subtler; it depends on the velocity, $\mathbf{u}_{rel}$, of the particle in the rotating frame and is $-2m\Omega \times \mathbf{u}_{rel}$. These forces are often referred to as being *fictitious*. In this sense, the effects they predict arise due to the introduction of a rotating coordinate system that is not entirely natural to the dynamics of a particle it is meant to describe. It is not hard to see why this moniker is used. Suppose we have a situation in which the particle moves without any force acting on it. Then in the nonrotating, laboratory frame, the particle obviously moves in a straight line. However when viewed from a coordinate reference frame rotating with a constant rotation, the particle appears to execute a complicated curved trajectory governed by the influence of these "forces." We say that the rotating frame introduces curvature terms. In this case, these terms cause apparent forces introduced to the observer in the rotating frame and are entirely artifacts of the coordinate transformation itself. The curved trajectory is really just the way a rotating coordinate frame represents the trajectory of a particle moving in a straight line in the laboratory frame. As we shall see, in the fluid dynamical equations, the centrifugal term can be usually absorbed in the pressure gradient term. The Coriolis term, however, remains and has far-reaching importance in rotating fluids. In a number of systems examined in this chapter we shall introduce some nontrivial mathematical techniques, which include asymptotic methods. It is our intention to use various topics from the general area of rotating fluids as a tutorial platform to demonstrate how these tools get used.

It is worth mentioning here the ubiquity of *disks*, or disk-like structures, as astrophysical objects. From the galactic plane in spiral galaxies, through accretion disks in central galactic supermassive black holes to the leftover disks after star formation, in which planets may form, and finally the relatively small accretion disks in various close binary systems, disks are found in many cosmic objects, over a very large scale range. Their existence is obviously the manifestation of one of the fundamental laws of nature, namely, angular momentum conservation in gas flow in the gravitational field of relatively compact object. We shall discuss a particular example of a thin accretion disk in Sect. 5.6.

5.1   Fundamentals

In this section we shall be concerned with the effects of rotation on relatively simple fluid configurations. The main dynamical feature that comes into play is the effect resulting from the influence of the Coriolis force, which we shall abbreviate as just the *Coriolis effect*. We repeat that the rotating fluid equations are expressed in a rotating frame of reference, as defined relative to an inertial frame that is often called the laboratory frame. We shall examine this effect and its manifestation in a small number of examples. We start by explicitly showing how the equations of motion appear when one views the dynamics in a globally uniformly rotating reference frame and go on to introduce three new nondimensional numbers: the Rossby number, the Burger number, and the Ekman number. V.W. Ekman (1874–1954) was an oceanographer and C-G.A. Rossby (1898–1957) a meteorologist. Their research contributions were paramount to our understanding of the oceans and the atmosphere.

5.1.1 Fluid Equations of Motion in a Uniformly Rotating Reference Frame

The previous discussion of particle dynamics in a rotating system sets the stage for the corresponding equations of motion for fluid flow. If the system is undergoing a solid body rotation around some arbitrary direction $\hat{n}$, the angular velocity is a vector Ω pointing in that direction. The momentum balance equation (1.53) rewritten in a frame determined by the above rotation is

$$\rho \left[\frac{\partial \mathbf{u}}{\partial t} + (\mathbf{u} \cdot \nabla) \mathbf{u} \right] + 2\rho \Omega \times \mathbf{u} = \rho \mathbf{b} - \nabla P + \rho \Omega \times (\Omega \times \mathbf{r}) + \eta \nabla^2 \mathbf{u} + \frac{1}{3} \eta \nabla (\nabla \cdot \mathbf{u}),$$

(5.1)

where the velocity is in the rotating frame and, for the sake of upcoming notational convenience, $\mathbf{r}$ is here the radius vector equal to $\mathbf{x} \equiv x\hat{x} + y\hat{y} + z\hat{z}$. Note that the bulk viscosity has been neglected for simplicity. The physical meaning of this has already been explained in Chap. 1. The Coriolis force per unit volume is explicitly written into the left-hand side of the above equation: $2\rho \Omega \times \mathbf{u}$. The equation of continuity (1.52) and the second law of thermodynamics (1.76) remain unchanged. Can you rationalize why the dissipation function Ψ is not altered in Eq. (1.76)? In the following we explicitly state the equations of motion in various coordinate systems. Two such coordinate systems are depicted, for reference, in Fig. 5.1.

5.1.1.1 Cylindrical Coordinates

The equations of motion in cylindrical coordinates (r, φ, z) are written assuming that the global rotation is oriented along the $\hat{z}$ axis, i.e., $\Omega = \Omega_0 \hat{z}$:

$$\frac{\partial u_r}{\partial t} + \mathbf{u} \cdot \nabla u_r - \frac{u_\varphi^2}{r} - 2\Omega_0 u_\varphi = -\frac{1}{\rho} \frac{\partial p}{\partial r} + b_r + \Omega_0^2 r + v \left(\nabla^2 u_r - \frac{2}{r^2} \frac{\partial u_\varphi}{\partial \varphi} - \frac{u_r}{r^2} \right),$$

(5.2)

$$\frac{\partial u_\varphi}{\partial t} + \mathbf{u} \cdot \nabla u_\varphi + \frac{u_\varphi u_r}{r} + 2\Omega_0 u_r = -\frac{1}{r\rho} \frac{\partial p}{\partial \varphi} + b_\varphi + v \left(\nabla^2 u_\varphi + \frac{2}{r^2} \frac{\partial u_r}{\partial \varphi} - \frac{u_\varphi}{r^2} \right),$$

(5.3)

$$\frac{\partial u_z}{\partial t} + \mathbf{u} \cdot \nabla u_z = -\frac{1}{\rho} \frac{\partial p}{\partial z} + b_z + v \nabla^2 u_z,$$

(5.4)

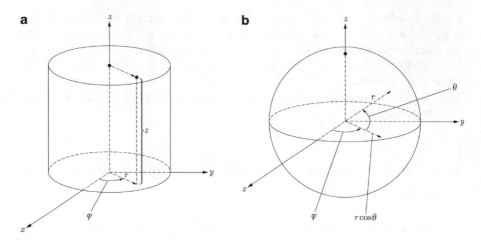

Fig. 5.1 A schematic depiction of the two main curvilinear coordinate systems: (**a**) cylindrical coordinates and (**b**) spherical coordinates

where explicitly

$$\mathbf{u} \cdot \nabla = u\frac{\partial}{\partial r} + \frac{u_\varphi}{r}\frac{\partial}{\partial \varphi} + u_z\frac{\partial}{\partial z}, \qquad \nabla^2 = \frac{\partial^2}{\partial r^2} + \frac{1}{r}\frac{\partial}{\partial r} + \frac{1}{r^2}\frac{\partial^2}{\partial \varphi^2} + \frac{\partial^2}{\partial z^2}.$$

The equation of mass conservation is

$$\frac{\partial \rho}{\partial t} + \mathbf{u} \cdot \nabla \rho + \rho\left(\frac{1}{r}\frac{\partial r u_r}{\partial r} + \frac{1}{r}\frac{\partial u_\varphi}{\partial \varphi} + \frac{\partial u_z}{\partial z}\right) = 0 \tag{5.5}$$

and the thermodynamic equation is

$$\rho T\left(\frac{\partial s}{\partial t} + \mathbf{u} \cdot \nabla s\right) = (\text{heat sources} - \text{sinks})/\text{unit volume}. \tag{5.6}$$

It should be understood that all quantities are measured with respect to the rotating frame, including the azimuthal coordinate φ. In the limit where $\Omega_0 \to 0$, the system reduces to the usual set of equations as viewed in the laboratory frame. Note that u_φ^2/r and $u_\varphi u_r/r$ found, respectively, in Eqs. (5.2) and (5.3) are the previously mentioned curvature terms. When the dynamics of interest is constrained to within a *thin gap*, i.e., a small radial extent, then the curvature terms are neglected because their influence is minor, especially when compared to the influence of the Coriolis effect, i.e., the terms $-2\Omega_0 u_\varphi$ and $2\Omega_0 u_r$. This approximation will be discussed in the next section and is particularly valid if the rotation is strong.

   In circumstances where the flow is treated strictly as two-dimensional, it is often more revealing to describe the dynamics in terms of the vorticity, not unlike the formulations in Sect. 2.2.1. In particular, if the flow is incompressible *and* the dynamics is independent of the $\hat{z}$ direction and that there are no external forces present and the vertical velocity is zero $w = 0$, then the equation for the flow is a simplification of equation (2.8), in which only the vertical component of the vorticity is tracked ($\omega_z \equiv \zeta$), i.e.,

$$\frac{D\zeta}{Dt} = \hat{\mathbf{z}} \cdot [\nabla \times (T\nabla s)] + \nu\nabla^2\zeta, \tag{5.7}$$

where T (temperature) and s (specific entropy) are also assumed to be independent of z. In this chapter we shall use the Lagrangian time derivative also in two-dimensional cases:

$$\frac{D}{Dt} = \frac{\partial}{\partial t} + u_r\frac{\partial}{\partial r} + u_\varphi\frac{1}{r}\frac{\partial}{\partial\varphi}. \tag{5.8}$$

Thus the material time derivative has special form in cylindrical coordinates. This is true for other curvilinear coordinate systems as well, so the reader is asked to be able to distinguish between different coordinate systems and the number of dimensions, before he or she writes out explicitly the terms of D/Dt. Notice also that in Eq. (5.7) the Coriolis term $2\Omega_0$ is explicitly absent. It drops out of the expression because of incompressibility,

$$\frac{1}{r}\frac{\partial(ru_r)}{\partial r} + \frac{1}{r}\frac{\partial u_\varphi}{\partial\varphi} = 0. \tag{5.9}$$

Specifically, when the left-hand side of Eq. (2.8) has the term $2\Omega \times \mathbf{u}$, operating with $\nabla\times$ on it gives

$$\nabla \times (2\Omega \times \mathbf{u}) = \hat{\mathbf{z}}\left[\frac{1}{r}\frac{\partial(ru_r)}{\partial r} + \frac{1}{r}\frac{\partial u_\varphi}{\partial\varphi}\right], \tag{5.10}$$

where the condition is that the incompressible flow is strictly two-dimensional. In cylindrical coordinates we have a stream function $\psi(r,\varphi)$ giving

$$u_r = -\frac{1}{r}\frac{\partial\psi}{\partial\varphi}, \quad u_\varphi = \frac{\partial\psi}{\partial r}, \tag{5.11}$$

which automatically satisfies Eq. (5.9), when substituted there, verifying incompressibility. Thus we can write

$$\zeta = \frac{1}{r}\frac{\partial(ru_\varphi)}{\partial r} - \frac{1}{r}\frac{\partial u_r}{\partial\varphi} = \frac{1}{r}\frac{\partial}{\partial r}\left(r\frac{\partial\psi}{\partial r}\right) + \frac{1}{r^2}\frac{\partial^2\psi}{\partial\varphi^2}, \tag{5.12}$$

which reveals the vorticity-stream function formulation for this two-dimensional problem in polar coordinates (r,φ).

How can the centrifugal terms affect the dynamics in a two-dimensional flow if they are explicitly absent from the vorticity evolution equation (5.7)? The answer follows from the observation that the centrifugal terms are connected to the pressure gradients in the flow. Since the flow is assumed to be two-dimensional and incompressible, one can administer the horizontal divergence of the horizontal momentum equations (5.2–5.3) and make use of two-dimensional incompressibility. This procedure results in the following diagnostic equation relating the flow fields to the Laplacian of the pressure field:

$$\nabla \cdot (\mathbf{u} \cdot \nabla \mathbf{u}) + 2\Omega_0 \zeta = \frac{1}{\rho}\nabla^2 P. \tag{5.13}$$

From the perspective of linear theory, the nonlinear term, the first on the left-hand side of the above equation, may be overlooked, and we can plainly the role of the centrifugal terms, appearing in a Poisson equation for the pressure. Thus, the centrifugal terms mitigate the shape of the resulting pressure field even if the flow is unsteady. We address this also in our discussion of the Taylor–Proudman theorem below.

5.1.1.2   Small Cartesian Segment

Our interest here is in the dynamics with respect to a small section of the fluid, rotating with it. Imagine a rectangular box-like segment centered at r_0. As defined above we envision that this segment rotates around the cylindrical $z = 0$ axis with rotation rate Ω_0. By writing $r \approx r_0 + x$ and $r\varphi \approx y$ where x, y, and z are small compared to r_0, the equations of motion in cylindrical geometry, i.e., Eqs. (5.2–5.4) may be approximated by dropping all curvature terms. Identifying $u_r \mapsto u$, $u_\varphi \mapsto v$ and $u_z \mapsto w$ we see that the equations simplify into

$$\frac{Du}{Dt} - 2\Omega_0 v = -\frac{1}{\rho}\frac{\partial P}{\partial x} + b_x + \Omega_0^2 x + \nu\nabla^2 u, \tag{5.14}$$

$$\frac{Dv}{Dt} + 2\Omega_0 u = -\frac{1}{\rho}\frac{\partial P}{\partial y} + b_y + \Omega_0^2 y + \nu\nabla^2 v, \tag{5.15}$$

$$\frac{Dw}{Dt} = -\frac{1}{\rho}\frac{\partial P}{\partial z} + b_z + \nu\nabla^2 w, \tag{5.16}$$

with the mass continuity and thermodynamic equations

$$\frac{D\rho}{Dt} + \rho\left(\frac{\partial u}{\partial x} + \frac{\partial v}{\partial y} + \frac{\partial w}{\partial z}\right) = 0, \quad \rho T \frac{Ds}{Dt} = (\text{entropy sources} - \text{sinks})/\text{unit volume},$$

$$\tag{5.17}$$

in which D/Dt and ∇^2 take on their usual forms appropriate for Cartesian coordinate geometry, i.e.,

$$\frac{D}{Dt} = \frac{\partial}{\partial t} + u\frac{\partial}{\partial x} + v\frac{\partial}{\partial y} + w\frac{\partial}{\partial z}, \quad \nabla^2 = \frac{\partial^2}{\partial x^2} + \frac{\partial^2}{\partial y^2} + \frac{\partial^2}{\partial z^2}.$$

Thus, the equation as expressed in the small segment is naturally in Cartesian coordinates. It is common to come across the expression *axi-symmetric* in reference to this Cartesian segment approximation of the equations of motion in a rotating system. It is meant to describe motions in which variations with respect to y are zero, which indicates the absence of variations with respect to φ in the equations of motion in cylindrical geometry. The utility of these equations is apparent: in many respects they are simpler than the full equations of motion in cylindrical geometry. Yet, they preserve enough of the physics associated with Coriolis effects to make them a useful platform for developing intuition on rotating systems. As we described in the discussion of cylindrical geometry, when the flow is incompressible and two-dimensional, i.e., $\nabla \cdot \mathbf{u} = 0$, no z-dependencies, and with $w = 0$, we find further simplification by adopting the vorticity-stream function representation of the equations of motion. In this two-dimensional formulation, ζ is the sole component of vorticity, the $\hat{z}$ one. D/Dt and ∇^2 are the usual Cartesian operators given above, but with $w = 0$ and $\partial/\partial z \mapsto 0$. Thus

$$\zeta = \frac{\partial v}{\partial x} - \frac{\partial u}{\partial y} = -\nabla^2 \psi, \qquad u = \frac{\partial \psi}{\partial y}, \qquad v = -\frac{\partial \psi}{\partial x}, \tag{5.18}$$

which is familiar from before, e.g., Sect. 2.5.2.

5.1.1.3 Spherical Coordinates

The equations of motion in spherical coordinates are presented here, for completeness. However for the sake of brevity, we choose the inviscid limit; see the *Bibliographical Notes* for a reference which includes all the terms. The coordinates are: r as the spherical radius coordinate, φ as the azimuthal angle, and θ as the polar coordinate. Since, as was explained in the beginning of this chapter, planet atmospheric/oceanic applications are the main ones, we define the polar angle to be measured from the equatorial plane so that $-\pi/2 < \theta < \pi/2$. In meteorological/planetary studies, the azimuthal and polar angles are often referred to as *the latitude* and *longitude*, respectively. The uniform rotation vector is aligned with the polar axis and has magnitude given by Ω_0. If we identify u_φ, u_θ, and u_r to be the azimuthal, polar, and radial velocities, respectively, then we have the following inviscid FD equations in a spherical coordinate rotating frame:

$$\frac{Du_\varphi}{Dt} - 2\Omega_0 \left(u_\theta \sin\theta - u_r \cos\theta \right) - \frac{u_\varphi u_r - u_\varphi u_\theta \tan\theta}{r} = -\frac{1}{\rho} \frac{1}{r\cos\theta} \frac{\partial P}{\partial \varphi} + b_\varphi \tag{5.19}$$

$$\frac{Du_\theta}{Dt} + 2\Omega_0 u_\varphi \sin\theta + \frac{u_r u_\theta + u_\varphi^2 \tan\theta}{r} = -\frac{1}{\rho} \frac{1}{r} \frac{\partial P}{\partial \theta} + b_\theta - \Omega_0^2 r \cos\theta \sin\theta, \tag{5.20}$$

$$\frac{Du_r}{Dt} - \frac{u_\theta^2 + u_\varphi^2}{r} - 2\Omega_0 u_\varphi \cos\theta = -\frac{1}{\rho} \frac{\partial P}{\partial r} + b_r + \Omega_0^2 r \cos^2\theta. \tag{5.21}$$

The Lagrangian time derivative in polar coordinates is

$$\frac{D}{Dt} = \frac{\partial}{\partial t} + \frac{u_\varphi}{r\cos\varphi}\frac{\partial}{\partial\varphi} + \frac{u_\theta}{r}\frac{\partial}{\partial\theta} + u_r\frac{\partial}{\partial r}. \tag{5.22}$$

The equation of continuity and the thermodynamic equation are

$$\frac{\partial\rho}{\partial t} + \underbrace{\frac{1}{r\cos\theta}\frac{\partial(\rho u_\varphi)}{\partial\varphi} + \frac{1}{r\cos\theta}\frac{\partial(\rho u_\theta\cos\theta)}{\partial\theta} + \frac{1}{r^2}\frac{\partial(r^2\rho u_r)}{\partial r}}_{\equiv\nabla\cdot(\rho\mathbf{u})} = 0, \tag{5.23}$$

$$\rho T\frac{Ds}{Dt} = (\text{entropy sources} - \text{sinks})/\text{unit volume}. \tag{5.24}$$

Note that, in comparison with our previous discussion in cylindrical coordinates, the form of the Coriolis effect in spherical coordinates is more complicated, as it gets expressed in all three coordinate directions

$$2\Omega_0\,\hat{\mathbf{z}}\times\mathbf{u} = 2\Omega_0\underbrace{\left(\cos\theta\,\hat{\boldsymbol{\theta}} + \sin\theta\,\hat{\mathbf{r}}\right)}_{=\hat{\mathbf{z}}}\times\left(u_\varphi\hat{\boldsymbol{\varphi}} + u_\theta\hat{\boldsymbol{\theta}} + u_r\hat{\mathbf{r}}\right)$$

$$= -2\Omega_0\left(u_\theta\sin\theta - u_r\cos\theta\right)\hat{\boldsymbol{\varphi}} + 2\Omega_0 u_\varphi\sin\theta\,\hat{\boldsymbol{\theta}} + -2\Omega_0 u_\varphi\cos\theta\,\hat{\mathbf{r}}, \tag{5.25}$$

where $\hat{\boldsymbol{\varphi}}, \hat{\boldsymbol{\theta}}, \hat{\mathbf{r}}$ are the unit vectors in the azimuthal, polar, and radial directions. Additionally, we have explicitly retained the centripetal acceleration terms in writing out the momentum balance equations above. In analyses which treat the density as a constant, the centripetal terms get absorbed into an effective pressure expression $P + (\rho/2)\Omega_0^2 r^2\cos^2\theta$. Consequently the centripetal terms are usually absent in other formulations of these equations in a rotating reference frame because it is usually understood in those contexts that they have been implicitly absorbed into the pressure.

5.1.2   Rossby, Burger, and Ekman Numbers

Rotating fluid systems are characterized by several nondimensional numbers indicating the relative importance of rotation, as captured in the Coriolis force, as compared to other physical effects. Incidentally, this force is named after G.-G. Coriolis a scientist who was first to publish in 1835 the full mathematical formulation of the effect, even though the qualitative knowledge about it was known before that. Here we present the most significant dimensionless quantities for rotating astrophysical and geophysical flows, the Rossby, Ekman, and Burger numbers.

Rossby Number This number, denoted by Ro, is the single most important quantity characterizing the dynamics in large scale rotating flows and thus in astrophysical and geophysical systems. It is defined as

$$\text{Ro} \equiv \frac{U}{2\Omega L} \tag{5.26}$$

where Ω is the global rotation rate, be it Ω_0 or $\Omega_0 \sin \varphi_0$ where φ_0 is a latitude of interest; U is a typical velocity scale and L is a horizontal length scale. For example, if the rotation axis of vortex is aligned with the effective global rotation of the system, for example, in a storm over the ocean on the earth, L/U is a rough measure of the turnaround timescale of the vortex while Ω^{-1} is of the order of the frame rotation timescale. The Rossby number is actually a measure of the ratio of timescales: that of flows in the rotating frame, $\tau_{\text{flow}} = U/L$ to the rotation timescale $\tau_{\text{rot}} = \Omega^{-1}$. In this way

$$\text{Ro} \approx \frac{\tau_{\text{flow}}}{\tau_{\text{rot}}}, \tag{5.27}$$

meaning that a *small* Ro flow is *strongly* affected by rotation, or the Coriolis effect. This generally means that such flows are primarily *geostrostrophic*, an important concept we will explain and use below in this chapter. Present assessments are that at mid to high latitudes of planetary atmospheres we have for Mars and the Earth Ro ~ 0.1 while for Jupiter, Saturn, Neptune, and Uranus it is ~ 0.1–0.01.

Burger Number Denoted by Bu and defined as

$$\text{Bu} = \frac{\sqrt{gH}}{2\Omega L}, \tag{5.28}$$

and is particularly important and relevant to geophysical fluid dynamics (GFD). An example is the following: a shallow fluid layer of height H in an external gravitational field g pointed downward in the same direction of the projected global rotation at a given latitude, taken here to be $\Omega\hat{z}$. In our previous discussion of shallow water, surface gravity waves propagate roughly at a speed $\approx \sqrt{gH}$, where H is a typical depth. If one is interested in dynamics occurring on a length scale L, then one may view Bu as measuring the ratio of the rotation timescale, $\tau_{\text{rot}} = \Omega^{-1}$, to the horizontal propagation timescale of a surface gravity wave across length L, $\tau_{\text{gw}} = L/\sqrt{gH}$,

$$\text{Bu} \approx \frac{\tau_{\text{rot}}}{\tau_{\text{gw}}}. \tag{5.29}$$

This is a loose definition and, in usual problems of interest, specific versions of Bu will appear depending upon the buoyancy of the medium. This usually means replacing the value of g with an effective value of g_{eff}, where g_{eff} is associated with buoyant oscillations. For example, this effective gravity can be a consequence of a vertical layer separated by two fluids of differing densities ($\rho_\pm$). In that case, the effective gravitational acceleration would be $g_{\text{eff}} = 2g(\rho_- - \rho_+)/(\rho_- + \rho_+)$. In other cases involving the vertical gradient of entropy, the expression g appearing in the definition of Bu gets replaced by $N^2 H$, where N is the Brunt–Väisälä frequency, defined by

$$N^2 = \frac{g}{s_0} \frac{ds_0}{dz}, \tag{5.30}$$

where s_0 is the horizontal mean of vertically varying entropy profile. H is identified as the vertical pressure scale-height given (roughly) by c_s^2/g, in which c_s^2 is the square of the typical sound speed. In that case Bu $= NH/2\Omega L$. A more detailed discussion of the Brunt–Väisälä frequency will be found in Chap. 7.

Ekman Number This number, denoted by E, concerns the relative importance of Coriolis effects and viscosity. It is defined as

$$E \equiv \frac{\nu}{2\Omega H^2}. \tag{5.31}$$

Keeping in line with our previous examples, consider now a fluid of vertical scale H and kinematic viscosity ν. As before, the global rotation vector is still considered aligned in the vertical direction. We have already encountered the timescale associated with viscous diffusion to be $\tau_{\text{vis}} \sim H^2/\nu$. Thus E is defined as the ratio of rotation time, τ_{rot}, and the viscous timescale, for a length scale H, or

$$E \approx \frac{\tau_{\text{rot}}}{\tau_{\text{vis}}}. \tag{5.32}$$

It is important to realize a subtle distinction compared to the previous examples: just because the Ekman number E may be a small does not necessarily mean that viscosity is unimportant. It just means that it is unimportant on the length scale H that was chosen. In turn, it means that there always exists an appropriately chosen length scale in which E ~ 1. In this way, it is similar to the Reynolds number. In the example of the Ekman layer that will be examined below, we shall see that the interesting dynamics will emerge on appropriately chosen length scales so that E ≈ 1.

Finally, we notice that the Reynolds number is simply related to these quantities according to

$$Re = \frac{Ro}{E}.$$

5.2 Some Rotating Flow Paradigms

A number of steady flow configurations that have rotation as a major feature are examined here. Starting with a discussion of the important Taylor–Proudman theorem and the idea of geostrophy and giving some examples, we move on to the Taylor–Couette flow. Next, we examine the simple Ekman layer, in which a steady wind interacts with a boundary layer in a rotating frame and show the curious behavior of the twisting of the mean wind field as one approaches the boundary.

5.2.1 Taylor–Proudman Theorem and Geostrophic Flow

In perhaps the simplest of possible configurations, we consider a fluid resting, in a rotating frame, on a flat plane at $z = 0$, and, neglecting gravity, consider the resulting steady flow where both the Rossby and Ekman numbers are so small that we assume they are zero. Keeping with the equations appropriate for rotating flows in a Cartesian geometry, we assume that $\boldsymbol{\Omega} = \Omega_0 \hat{z}$ and no other external body forces are present, $\mathbf{b} = 0$. This flow is not made rotating by any body force and writing the appropriate equations we find that

$$-2\Omega_0 v = -\frac{1}{\rho}\frac{\partial P}{\partial x}, \qquad 2\Omega_0 u = -\frac{1}{\rho}\frac{\partial P}{\partial y}, \qquad 0 = -\frac{1}{\rho}\frac{\partial P}{\partial z}. \tag{5.33}$$

Note that inertial nonlinear terms on the left-hand sides of the equations of motions are zero because of the $\mathrm{Ro} = 0$ assumption and this is instrumental for the result. Assume that the fluid is of constant density, obviously satisfying the condition of incompressibility $\nabla \cdot \mathbf{u} = 0$, which written out in velocity components gives in Cartesian coordinates

$$\frac{\partial u}{\partial x} + \frac{\partial v}{\partial y} + \frac{\partial w}{\partial z} = 0. \tag{5.34}$$

One can take the partial derivative with respect to z of the x-momentum equation (the first of (5.33)) and, likewise, take the partial derivative with respect to x of the z-momentum equation (the third of (5.33)) and find that

$$-2\Omega_0 \frac{\partial v}{\partial z} = 0, \tag{5.35}$$

indicating that v has no vertical variation with respect to z. The result that follows from a similar set of operations on the y and z-momentum equations, i.e., the second and third of (5.33), is the conclusion that u is also independent of z. Operating in similar appropriate way on the horizontal components: y derivative of the first equation subtracted from x derivative of the second equation of (5.33) and making use of the constancy of ρ shows that

$$2\Omega_0 \left(\frac{\partial u}{\partial x} + \frac{\partial v}{\partial y} \right) = \frac{1}{\rho} \left(\frac{\partial^2 P}{\partial x \partial y} - \frac{\partial^2 P}{\partial y \partial x} \right) = 0. \tag{5.36}$$

This result used in the incompressibility condition predicts that vertical velocity must be a constant since the vertical stretching term is zero, i.e.,

$$\frac{\partial w}{\partial z} = 0. \tag{5.37}$$

The vertical velocity is, consequently, set to zero, instead of some vertically independent function of the horizontal coordinates, because the presence of an impenetrable boundary at $z = 0$ means that w must be zero there and since the above equation means that there is no variation of w with respect to z, it follows that w must be zero everywhere.

Thus we have proved here an instance of the *Taylor–Proudman* theorem. Its strong meaning is that any flow under these stated conditions has properties that are *constant on vertical cylinders*, as there is no variation of the horizontal velocities with respect to z. This theorem includes the name of the revered G.I. Taylor and is perhaps the most significant result in rotating fluids theory. In plain words it states that strong rotation of a fluid causes it to behave as an essentially two-dimensional system. That is why two-dimensional flows and their properties are often treated in FD with the intention to apply the results to strongly rotating fluids. The notion of *geostrophic balance* is relevant to settings beyond the Cartesian one we have examined. Geostrophy is the explicit balance between the Coriolis effect and pressure gradient. The vectorial equation expressing geostrophy is

$$2\boldsymbol{\Omega} \times \mathbf{u} = -\frac{1}{\rho} \nabla P, \tag{5.38}$$

and this balance is most pronounced under conditions of strong rotation, commonly found in the atmosphere of our rapidly rotating home planet, the Earth. Thus the prefix "geo" originating in the Greek word for Earth. To be clear, geostrophic balance means that the Coriolis terms are more influential than the curvature terms and this becomes increasingly true if the value of Ro characterizing the system becomes small. As we have mentioned above a number of solar system planets, notably Mars have Ro value similar to Earth, while the value of this number for the outer giant planets, notably Jupiter, is even smaller. We stress again that a small value of Ro means that the global rotation time of the system is much shorter than the typical turnaround times associated with any velocity structure, as measured in the rotating reference frame. We revisit this again later in the chapter.

The Taylor–Proudman theorem, in a coordinate free general formulation, follows from the geostrophic balance equation (5.38), the only assumption being constant density. Indeed, assuming that the unit vector in the direction of the rotation axis is $\hat{\mathbf{z}}$, say, and operating with $\nabla \times$ on the equation gives

$$(\hat{\mathbf{z}} \cdot \nabla) \mathbf{u} = 0, \tag{5.39}$$

which means that **u** cannot be dependent on the $\hat{\mathbf{z}}$ direction. We may give an example of a very simple geostrophic flow which is y independent, that is, $\partial/\partial y \mapsto 0$, and in which the global pressure varies along x and is balanced by a form dependent on the velocity field v. If the pressure gradient, brought about by whatever means, is a constant, then the result is a mean constant flow for v:

$$v = \frac{1}{2\Omega_0 \rho} \frac{\partial P}{\partial x} = \text{const.} \tag{5.40}$$

If, along the x direction the pressure field exhibits, instead, a quadratic profile, i.e., $P = p_{00} + p_{01} \rho (x/L)^2$ with p_{00} and p_{01} constants, then the result will be a linear shear in the velocity field

$$v = \frac{p_{01}}{2\Omega_0 L^2} x. \tag{5.41}$$

The consequence of geostrophic flow is that pressure contours describe flow streamlines. First, since this flow is two-dimensional and incompressible by Eqs. (5.36) and (5.37), it can be described in terms of a stream function ψ in the usual way. Similarly, calculating the horizontal divergence of equation (5.38) results in

$$2\Omega_0 \left(\frac{\partial v}{\partial x} - \frac{\partial u}{\partial y} \right) = 2\Omega_0 \zeta = \frac{1}{\rho} \nabla^2 P, \tag{5.42}$$

where ζ is the vertical vorticity of the flow. Because we have determined that the flow is independent of the vertical coordinate z, the implication is that vorticity is traced by the horizontal Laplacian of the pressure field by the correspondence

$$\zeta = -\nabla^2 \psi = \nabla^2 \frac{P}{2\Omega_0 \rho}, \tag{5.43}$$

because ρ is assumed constant. Thus the isobaric contours are a proxy for the flow's stream function level curves which are streamlines. We have in geostrophic flow

$$\text{isobaric contours} \quad \Longleftrightarrow \quad \text{flow streamlines.} \tag{5.44}$$

This gives an indication as to why the Coriolis effect appears to not directly affect the evolution of two-dimensional flows according to the evolution equation for the vertical vorticity, as can be found in Eq. (5.7) and a general two-dimensional vorticity equation in a cartesian box (derive it!). In truth, its influence appears in the manifestation of the flow in concert with pressure contours. It plays a role in the diagnostic interpretation of the resulting dynamics by relating the flow streamlines to those of equal pressure (Fig. 5.2).

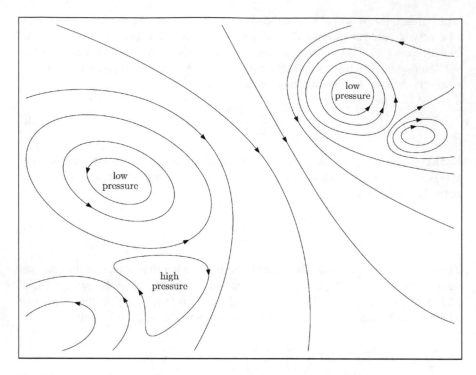

Fig. 5.2 Geostrophic flow diagram. Streamlines and isobaric contours are coincident in geostrophic flow. Flow streamlines are shown on the assumption $2\Omega_0 > 0$. Flow directions are reversed for $2\Omega_0 < 0$. Low pressure locations correspond anti-clockwise motion (high pressure $\leftrightarrow$ clockwise flow). The lines are meant to continue and ultimately close beyond the figure boundary

5.2.2 Taylor–Couette Flow

An important configuration that has experimental relevance and physical interest is a flow between two coaxial long cylinder surfaces whose axis coincides with the z-axis. Assume that a constant density incompressible fluid fills the gap between the cylinders and that the system is in a constant gravitational field of magnitude g pointing along $-\hat{\mathbf{z}}$. To handle the resulting steady flow, we then appeal to the equations of motion in cylindrical coordinates as found in Sect. 5.1.1.1. For the sake of this illustration we use a nonrotating frame, that is, set $\Omega_0 = 0$, and have only the vertical component of the body force $b_z = -g$. We seek solutions in which the only non-zero component of velocity is the azimuthal one. Furthermore, we assume that the purely azimuthally moving fluid flows over an impenetrable plane at $z = 0$ providing stress-free boundary conditions for the flow u_φ, which, for this calculation, means that u_φ shows no vertical variation. Other boundary conditions include no-slip and impenetrability at the cylinder's vertical walls. We consider an idealized case of essentially infinite cylinders, meaning that in some nondimensional units $z_{\text{top}} \gg 1$. The equations of steady motion are

$$-\frac{u_\varphi^2}{r} = -\frac{1}{\rho}\frac{\partial P}{\partial r} \tag{5.45}$$

$$0 = v\frac{\partial}{\partial r}\left[\frac{1}{r}\frac{\partial}{\partial r}(ru_\varphi)\right] \tag{5.46}$$

$$0 = -\frac{1}{\rho}\frac{\partial P}{\partial z} - g \tag{5.47}$$

The equation of hydrostatic balance (5.47) may be integrated with respect to z to reveal

$$P(r,z) = -\rho gz + p(r), \tag{5.48}$$

where $p(r)$ is an unknown function of r determined by the solution to Eq. (5.45),

$$\frac{dp}{dr} = \rho\frac{u_\varphi^2}{r}. \tag{5.49}$$

This problem is very similar to the Rankine vortex examined in Sect. 2.5.5 and we can, in principle, recover those results here. Instead we consider steady viscous solutions. In particular, the viscous stress divergence expression equation (5.46) can be rewritten into a more compact form, allowing for general solutions for u_φ to be readily stated,

$$u_\varphi = Ar + \frac{B}{r}, \tag{5.50}$$

for constants A and B which become determined once specific boundary conditions and settings are imposed. The Taylor–Couette (TC) flow is the azimuthal velocity profile that arises when one considers the flow contained between rotating coaxial cylinders with the additional proviso that the top of the relatively long cylinder is capped but the cap does not impose any stress. The upper and lower caps are therefore intended to obtain a flow in an idealized infinite cylinder. Unlike the Rankine vortex, there is here no free surface to worry about and so the basic state solution is much simpler. If the rotating concentric cylinders have inner and outer radii given by $R_{1,2}$, respectively, together with corresponding rotation rates $\Omega_{1,2}$, then the general solution of the velocity field (5.50) becomes

$$u_\varphi(r) = \frac{R_2^2}{R_2^2 - R_1^2}\left[\left(\Omega_2 - \Omega_1\frac{R_1^2}{R_2^2}\right)r + \frac{R_1^2}{r}\left(\Omega_1 - \Omega_2\right)\right]. \tag{5.51}$$

The corresponding radial pressure gradient follows from inserting this solution for u_φ into Eq. (5.49) and integrating the result.

We shall not consider here the laboratory TC problem, where the cylinders extend to $z = z_{top} < \infty$ since a variety of different boundary conditions may be applied at the top, giving rise to a host of separate different problems, some of them very complex. In any case, the solution presented here serves as basis for considering more complicated TC flows arising from special boundary conditions.

5.2.3 Simple Ekman Layer

We discuss here a rotating fluid structure referred to as the *Ekman layer*, but do not consider in detail the geophysical case, wherein the Ekman layer in the ocean is driven by wind stress. Rather, we focus on an idealized, simpler, case. Consider a flow in a rotating Cartesian frame, limited by a fixed, in that frame, bottom planar rigid plate at $z = 0$, so that all dynamics occurs for $z > 0$. The frame rotation vector is in the vertical, $\hat{z}$, direction having a constant rate Ω_0. The kinematic coefficient of viscosity is v, and the fluid has constant density. Suppose that the Rossby number is much less than the Ekman number, i.e., Ro $\ll$ E. We imagine that there is a large scale constant velocity field in the $\hat{y}$ direction, brought about by a global uniform pressure gradient in the $\hat{x}$ direction, similarly to the description in Sect. 5.2.1. We are interested in finding the velocities in this configuration. With the presence of viscosity, we expect that the horizontal velocity components should have some vertical dependence. We shall assume an extreme idealized such case and seek solutions for u and v that vary *only in the vertical direction*. The equations of motion in this case are

$$-2\Omega_0 v = -\frac{1}{\rho}\frac{\partial P}{\partial x} + v\frac{\partial^2 u}{\partial z^2}, \qquad 2\Omega_0 u = v\frac{\partial^2 v}{\partial z^2}. \tag{5.52}$$

Because the bottom plate should impose rigid boundary conditions, we have at its location $u = v = 0$ at $z = 0$, while we assume that the velocity field relaxes to the ambient configuration as $z \to \infty$, i.e., $u = 0$ and $v = v_0$, where

$$v_0 \equiv \frac{1}{2\Omega_0}\frac{\partial P}{\partial x} = \text{const.}$$

We can think of v_0 as a kind of geostrophic wind, since the value it achieves is a result of geostrophic balance. Since the global pressure gradient along x is constant, the equations governing the steady flow may be simplified to

$$-2\Omega_0(v - v_0) = v\frac{d^2 u}{dz^2}; \qquad 2\Omega_0 u = v\frac{d^2 v}{dz^2}. \tag{5.53}$$

Combining the above two equation into a single one for u reveals

$$u = -\frac{v^2}{4\Omega_0^2}\frac{d^4 u}{dz^4}. \tag{5.54}$$

This equation has four exponential solutions $e^{\kappa z}$, where $\kappa = \pm(1 \pm i)/H_0$ in which $H_0 \equiv \sqrt{2\Omega_0/v}$. The two solutions which would induce exponential growth as $z \to \infty$ are rejected on physical grounds. Thus an acceptable solution for u is

$$u = A e^{-z/H_0} e^{iz/H_0} + B e^{-z/H_0} e^{-iz/H_0}, \tag{5.55}$$

with unknown constants A and B. The above form for u is arranged so that $u = 0$ at $z = 0$, leaving us with

$$u(z) = A e^{-z/H_0} \sin\left(\frac{z}{H_0}\right),$$

in which A is undetermined. With this form for u in hand, we extract the functional form for v from the first of the equations in (5.53) and imposing the boundary condition on v see that it relaxes to the geostrophic wind state as $z \to \infty$. This results in determining the value of A and revealing the final solution form to be

$$v(z) = v_0\left[1 - e^{-z/H_0}\cos\left(\frac{z}{H_0}\right)\right]; \qquad u(z) = v_0 e^{-z/H_0}\sin\left(\frac{z}{H_0}\right). \tag{5.56}$$

Even before visualizing the solutions we can already see that the exponential drop-off on scale H_0 means that everything of interest is taking place in the region of extent H_0. Relating this back to the definition of the Ekman number in Eq. (5.31), we see that choosing $H = \mathcal{O}(H_0)$ means that on the scale H_0, $E = \mathcal{O}(1)$.

A close inspection of the velocity field solution in Eq. (5.56) shows something interesting: while the x component of the velocity field $\mathbf{u}$ decays to zero in the two limits $z \to 0, z \to \infty$, it actually takes on significantly non-zero values near $z = \mathcal{O}(H_0)$. This is qualitatively shown in Fig. 5.3, depicting the so-called Ekman spiral, showing the total velocity vector of the wind field as a function of distance away from the $z = 0$ plane. The spiral traces the envelope of the set of vectors for all $z > 0$. Equally remarkable is the existence of z values for which the resultant velocity vector is greater than $|v_0|$. In other words, it can be shown with little effort that the speed of the horizontal flow

$$|\mathbf{u}| = \sqrt{v^2 + u^2}$$

exceeds $|v_0|$ for values of z in the vicinity of H_0. Thus, even though the plate removes kinetic energy from the flow, there are places above the plate that the redistribution of momentum brought about by viscosity induces a speed that is actually faster than the ambient geostrophic wind speed characterizing the flow as $z \to \infty$.

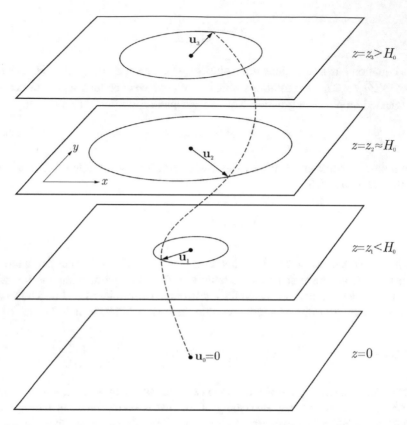

Fig. 5.3 Local visualization of Ekman flow. Three successive layers are shown, representing the distance from the rigid bottom boundary at $z = 0$. The spiral of the flow is shown by the dashed line tracing out the tip of the velocity vector as one moves upward. In this depiction, the mean geostrophic wind points in the y direction and is shown being asymptotically approached for $z \gg H_0$. The value of the velocity vector is greatest for $z \approx H_0$ exceeding the speed of the geostrophic wind. The greater magnitude of the speed for z near H_0 is exaggerated in the drawing

5.3 Linear Dynamics of Spin-Down

We examine here, in some detail, the Ekman spin-down problem and explicitly calculate the spin-down rate of a rotating column of fluid when interacting with solid surfaces through the fluid viscosity. This particular example makes use of matched asymptotic expansions and thus time is taken to carefully develop this formulation of the problem. The previous section dealing with simple Ekman flow demonstrates what happens when rotation and viscosity are both present, showing clearly that a flow profile shows a rotation of its velocity vector as a function of distance from a bounding surface. The present problem is no longer simple: we consider the fate of a rotating incompressible fluid between *two* rigid plates. In this discussion it is assumed that gravity plays no role in the ensuing dynamics. We envision a constant density fluid sandwiched between two horizontal impenetrable plates, parallel to the $x - y$ plane, separated from one another by a vertical distance L. Let the plates and the fluid between them rotate in unison with rotation vector $\Omega = \Omega_0 \hat{z}$. At some instant of time, we imagine that the plates stop their rotation abruptly. The question is: What happens to the rotating flow and on what timescale?

The flow immediately near a plate will match the motion of the plate effectively instantaneously showing a tendency to develop a vertical structure of the kind illustrated in the previous section. Now we have *two* plates and thus the flow just below the upper plate will show a mirror symmetric behavior. It is clear that the flow in the interior, between the two plates, will eventually come to rest. Our task is to decide, basing the answer on a calculation, what will be the timescale of the spin-down. Before delving into the calculation, notice that there are clearly *two* timescales in the problem: one associated with rotation and estimated by $\tau_{\text{rot}} \sim (2\Omega_0)^{-1}$ and the other due to viscosity. It is natural to estimate that the viscous timescale for the bulk of the fluid should be dictated by the fluid's length scale, L, say, and this means that the corresponding timescale is $\tau_{\text{visc}} \sim L^2/\nu$. As a simple example, for a rotating cylindrical vessel of water, whose typical dimension is $L \sim 10$ cm, one would estimate τ_{visc} to be slightly under an hour, given the value of water's viscosity $\nu \approx 0.01$ cm^2/s. However, experience shows that, in fact, it takes less than a minute for the water to come to a complete stop, after the vessel had abruptly stopped. One might argue that this timescale should depend upon the side walls of the vessel, but the timescales become insensitive to the presence of the walls once the horizontal dimensions of the vessel get sufficiently large, e.g., already for an aspect ratio of less than about a half the timescales become independent of the walls. It turns out that the spin-down timescale for this problem is the *geometric mean* of the rotation time and the viscous time, i.e., $\sim \sqrt{\tau_{\text{visc}} \tau_{\text{rot}}} \sim L/\sqrt{\Omega_0 \nu}$.

The aim in this section is to develop a relatively straightforward matched asymptotic expansion procedure, in which this peculiar timescale uniquely emerges as the correct one on which spin-down occurs. We begin this by turning this problem around to solve for the dynamics of the reverse problem, i.e., that of the spin-up. We will examine the response of the interior flow, which had been initially at rest for $t < 0$, to a change in the rotation rate of the bounding plates, having gone abruptly from no rotation to $\Omega_0 \hat{z}$ in the lab frame at time $t = 0$. We then decide to look at the problem in a frame rotating exactly as the above rotation, with respect to the lab, which we imagined starting abruptly at $t = 0$. The above will be *the rotating reference frame* for this problem. In this reference frame, the fluid's state after the rotation of the plates had started is that of a spinning column rotating on its axis, having constant rotation angular velocity $\Omega = -\Omega_0 \hat{z}$. We consider the velocity within the two boundary layer zones near the plate boundaries and, separately, in the bulk interior of the fluid, and ultimately link the two descriptions together. We chose to use Cartesian coordinates and adopt equations (5.14–5.16) together with the assumption of incompressibility. We start by formulating this problem in terms of small perturbations about a steady state, as viewed in the rotating frame, getting the following *linearized* equations of motion:

$$\frac{\partial u'}{\partial t} - 2\Omega_0 v' = -\frac{1}{\rho}\frac{\partial P'}{\partial x} + v\left(\frac{\partial^2 u'}{\partial x^2} + \frac{\partial^2 u'}{\partial y^2} + \frac{\partial^2 u'}{\partial z^2}\right), \qquad (5.57)$$

$$\frac{\partial v'}{\partial t} + 2\Omega_0 u' = -\frac{1}{\rho}\frac{\partial P'}{\partial y} + v\left(\frac{\partial^2 v'}{\partial x^2} + \frac{\partial^2 v'}{\partial y^2} + \frac{\partial^2 v'}{\partial z^2}\right), \qquad (5.58)$$

$$\frac{\partial w'}{\partial t} = -\frac{1}{\rho}\frac{\partial P'}{\partial z} + v\left(\frac{\partial^2 w'}{\partial x^2} + \frac{\partial^2 w'}{\partial y^2} + \frac{\partial^2 w'}{\partial z^2}\right), \qquad (5.59)$$

together with the assumption of incompressible flow

$$\frac{\partial u'}{\partial x} + \frac{\partial v'}{\partial y} + \frac{\partial w'}{\partial z} = 0. \qquad (5.60)$$

The matched asymptotic expansion procedure that we specifically detail here for this problem proceeds through the following four steps:

1. By assuming appropriate scalings for all the variables in the relevant equations here (5.57–5.60) make the equations nondimensional, so that they may be analyzed more transparently. At this point a small parameter should emerge, and in this case it is reasonable that it will be some positive power of the Ekman number. This is the crucial step in any asymptotic study, because the adjectives large and small have meaning only for nondimensional physical quantities. In addition, it is at this stage that one asserts initial guesses as to how quantities relate to one another. It is often true that if an incorrect scaling is assumed, it may result in a failure to determine bounded solutions, or any possible balancing of physically essential terms, in any of the following three subsequent stages.

2. Solve the resulting nondimensional equations of motion in the bulk of the rotating system, relatively far from the plate boundaries, as a perturbation series expansion. This will be called the *outer* or *exterior* solution as this is common nomenclature in boundary layer theory—we are out of the boundary layer. We have introduced this terminology previously, in the simplest case of boundary layer emergence in Sect. 3.5.1.

3. Solve the resulting nondimensional equations in the boundary layer zones near the two bounding plates at $z = 0$ and $z = 1$, respectively. The governing equations in the boundary layer should usually be expressed in stretched coordinates. In this case, the stretching will occur in the vertical coordinate. It is common, as we have seen in Sect. 3.5.1, to refer to such reexpressed equations as the *inner* equations, and to their solution as the inner solution.

4. Perform a matching procedure between the two inner solutions, i.e., those valid in the two boundary layers, onto the outer solution, which is valid in the bulk interior. Once the correct matching is made, the resulting equation can then be solved to recover the behavior we asserted to be the case at the beginning of this discussion.

5.3.1   Scalings and Nondimensionalization

The aim of this section is to rewrite Eqs. (5.57–5.60) in a way that will facilitate a perturbation series analysis. As we have done in previous examples, we must make certain assumptions about the relative scalings of various quantities. We will state them here and then formulate the equations of motion accordingly. Assume that the vertical *and* horizontal scales of the system are characterized by a length scale L and choose L to actually be the separation of the vertical boundaries. We also know from our previous experience with the simpler Ekman layer, worked out in the previous section, that there exists a boundary layer of scale $H_0 \sim \sqrt{\nu/2\Omega_0}$. The ratio of the boundary layer scale to L is easily deduced, using the above boundary layer scale definition: $H_0 \sim \sqrt{\nu/2\Omega_0} = L\sqrt{\nu/2\Omega_0 L^2} = L\mathrm{E}^{1/2}$. In other words, the ratio H_0/L is approximately the square-root of the Ekman number E. If the viscosity is sufficiently unimportant, then the Ekman number should indeed be small. We decide that E will be *the* small parameter of this system, as $\mathrm{E} \ll 1$.

Similarly to our analysis of the simple Ekman flow, we expect the dynamics within the bulk interior of the system to be in geostrophic balance. Suppose therefore that both horizontal velocities are scaled by a speed $\tilde{u}$. Then, if geostrophic balance is to be the dominant feature in this flow, then in order for $2\Omega_0 \hat{\mathbf{z}} \times \mathbf{u}' \sim (1/\rho)\nabla P'$, where ρ is a constant density, the pressures ought to have the approximate scaling of $\sim 2\Omega_0 \tilde{u}L\rho$. We rely upon observation to guide us in choosing the correct timescale characterizing spin-down. Coriolis effects have a timescale that is $\tau_{\mathrm{rot}} \sim 1/(2\Omega_0)$, while the corresponding viscous timescale, as determined based upon the system's large scale dimension L, is $\tau_{\mathrm{visc}} \sim L^2/\nu$. That much we have already proposed at the outset. As everyday experience with a well-stirred cup of liquid suggests, the fluid spin-down is faster than the viscous timescale but longer than the rotation timescale. We posit here, mainly as a guess, but hinted by the simple Ekman flow of the previous section (and later verified owing to the self-consistency of the final answer) that the relevant timescale is the geometric mean of the viscous and rotation times. This special timescale, τ_{E}, happens to be, as we saw $\sim 1/(2\Omega_0)\mathrm{E}^{1/2}$, which means to say that the timescale of interest is a factor $1/\sqrt{\mathrm{E}}$ longer than the Coriolis/rotation timescale of the system τ_{rot}. Thinking of one's experience with a rotating cup of fluid, and barring large amplitude vertical perturbations, if the cup is steadily rotating, then the vertical velocities of the rotating fluid ought to be very small compared to the corresponding horizontal circular velocities. Thus, we guess that the vertical velocities are smaller than the horizontal ones by a factor of $\mathrm{E}^{1/2}$ as well. In other words, we shall adopt the scaling that $w' \sim \mathrm{E}^{1/2}\tilde{u}$. In the following we reexpress all dependent and independent variables of this system in terms of these proposed relative scalings. Thus we write

$$x \mapsto Lx, \qquad y \mapsto Ly, \qquad z \mapsto Lz, \qquad t \mapsto 1/(2\Omega_0\mathrm{E}^{1/2})t,$$

$$u' \mapsto \tilde{u}u, \qquad v' \mapsto \tilde{u}v, \qquad w' \mapsto \mathrm{E}^{1/2}\tilde{u}w, \qquad P' \mapsto 2\Omega_0\tilde{u}L\rho\,\Pi, \quad (5.61)$$

with the understanding from here on that x, y, z, u, v, w, Π, t are nondimensional quantities related to the dimensional counterparts according to (5.61). Note now that we have deliberately chosen to *retain the same characters* (x, y, z, t) for the nondimensional space and time variables, primarily for aesthetic reasons, which also economizes the notation. The linearized equations of motion (5.57–5.60) now appear rewritten in terms of the nondimensional variables as

$$E^{1/2}\frac{\partial u}{\partial t} - v = -\frac{\partial \Pi}{\partial x} + E\left(\frac{\partial^2 u}{\partial x^2} + \frac{\partial^2 u}{\partial y^2} + \frac{\partial^2 u}{\partial z^2}\right), \qquad (5.62)$$

$$E^{1/2}\frac{\partial v}{\partial t} + u = -\frac{\partial \Pi}{\partial y} + E\left(\frac{\partial^2 v}{\partial x^2} + \frac{\partial^2 v}{\partial y^2} + \frac{\partial^2 v}{\partial z^2}\right), \qquad (5.63)$$

$$E\frac{\partial w}{\partial t} = -\frac{\partial \Pi}{\partial z} + E^{3/2}\left(\frac{\partial^2 w}{\partial x^2} + \frac{\partial^2 w}{\partial y^2} + \frac{\partial^2 w}{\partial z^2}\right), \qquad (5.64)$$

and

$$\frac{\partial u}{\partial x} + \frac{\partial v}{\partial y} + E^{1/2}\frac{\partial w}{\partial z} = 0. \qquad (5.65)$$

These Eqs. (5.62–5.65) are the ones that will be analyzed in the upcoming three sections. When the analysis is complete and solutions have been found, the results will be recast in terms of their original dimensional quantities as defined in (5.61). As we have assumed before, the bounding plates are vertically separated by the actual distance L, thus the vertical boundaries are located at $z = 0, 1$, in this nondimensional form. On the bounding plates we require rigid boundary conditions, i.e., the condition $u = v = w = 0$.

5.3.2   The Behavior in the Bulk-Interior: The Exterior Solution

We examine now the system in the interior part. Expecting it to be primarily characterized by geostrophic balance, but with some additional features which will become evident below, we assume a two term series expansion for all of the fluid quantities, except for the vertical velocity, for which we retain only one term. Thus we adopt the following perturbation series expansion in powers of the Ekman number

$$u(x, y, z, t) = u_1(x, y, z, t) + E^{1/2}u_2(x, y, z, t) + \text{HOT};$$

$$v(x, y, z, t) = v_1(x, y, z, t) + E^{1/2}v_2(x, y, z, t) + \text{HOT};$$

$$w(x, y, z, t) = w_1(x, y, z, t) + \text{HOT};$$

$$\Pi(x, y, z, t) = \Pi_1(x, y, z, t) + E^{1/2}\Pi_2(x, y, z, t) + \text{HOT}. \qquad (5.66)$$

We insert the series expansion given in Eq. (5.66) into the governing equations (5.62–5.65) and sort through the resulting set of equations, order by order, in powers of $E^{1/2}$. Thus at lowest order we have the set

$$v_1 = \frac{\partial \Pi_1}{\partial x}; \qquad u_1 = -\frac{\partial \Pi_1}{\partial y}; \qquad 0 = -\frac{\partial \Pi_1}{\partial z}; \qquad \frac{\partial u_1}{\partial x} + \frac{\partial v_1}{\partial y} = 0. \qquad (5.67)$$

The third of the equations above shows that the lowest order pressure is independent of the vertical coordinate, i.e., $\Pi_1 = \Pi_1(x,y,t)$. The first two equations are just a restatement of geostrophic balance, while the last equation is a consistency condition for geostrophic flow, i.e., geostrophic balance and leading order horizontal incompressibility appear together. The independence of Π_1 on z leads to the conclusion, based on the last set of equations, that the horizontal velocities are also z-independent: $u_1 = u_1(x,y,t)$, $v_1 = v_1(x,y,t)$.

Moving now to the next order equations we find

$$\frac{\partial u_1}{\partial t} - v_2 = -\frac{\partial \Pi_2}{\partial x}, \qquad \frac{\partial v_1}{\partial t} + u_2 = -\frac{\partial \Pi_2}{\partial y}, \qquad (5.68)$$

$$0 = -\frac{\partial \Pi_2}{\partial z}, \qquad \frac{\partial u_2}{\partial x} + \frac{\partial v_2}{\partial y} + \frac{\partial w_1}{\partial z} = 0. \qquad (5.69)$$

Similar to before, we see from the first of the equations in (5.69) that the next order pressure perturbation is also independent of height. We could go on to develop a solution to this order, but we will instead be satisfied with the following. Operating on the second of the equations in (5.68) by $\partial/\partial x$ and subtracting from it the result of operating on the first of the equations in (5.68) by $\partial/\partial y$ results in

$$\frac{\partial}{\partial t}\left(\frac{\partial v_1}{\partial x} - \frac{\partial u_1}{\partial y}\right) = -\frac{\partial u_2}{\partial x} - \frac{\partial v_2}{\partial y} = \frac{\partial w_1}{\partial z}. \qquad (5.70)$$

In reaching the last of the equalities in the above equation, we have made use of (5.69). Note that Eq. (5.70) is actually an evolution equation for the lowest order vertical vorticity $\zeta_1 \equiv \partial v_1 \partial x - \partial u_1/\partial y$. It is equal to the vertical stretching of the flow. In order to get a closed equation, we must develop an expression for the vertical stretching term $\partial w_1/\partial z$ in terms of the lowest order vertical vorticity ζ_1. This will be determined in the boundary layer zone in the upcoming subsection.

However, before proceeding let us pause and note the physical information contained in the approximate evolution equation for the lowest order vorticity found in Eq. (5.70). As opposed to the circumstances examined in classical geostrophic flow and the Taylor–Proudman theorem, we explicitly see here how the perturbation vertical vorticity term can be altered by vertical stretching embodied in the source term $\partial w_1/\partial z$. It is possible to show that the vertical velocity at the top of the column is positive and at the bottom it is negative. This gives $\partial w_1/\partial z > 0$ which, in turn, results in the local vorticity increase according to Eq. (5.70). Imagining for the moment that we are following a spinning cylinder, since the fluid is incompressible, a vertical stretch means that the corresponding radius of the cylinder has to shrink. This is in accord with Helmholtz's vortex theorems, see Sect. 2.5.1. The physical picture now becomes clear: as the column stretches its interior flow rate spins up in order to conserve the column's total angular momentum. Thus the spin response to stretch described here is a generic feature of many nearly two-dimensional flows. We will re-encounter this stretching dynamics when we examine geophysical flows. Problem 5.7 details an example setting showcasing the effect of vertical stretching upon the vorticity of a column in a rotating shallow water model with bottom topography.

5.3.3   Boundary Layers: One-Term Interior Solutions Near $z = 0, 1$

The flows within the two boundary layers are similar to the flow discussed in the simple Ekman layer detailed in Sect. 5.2.3. All fluid quantities excluding pressure vary strongly as one approaches the bounding plate. However when moving away from the boundaries, the fluid quantities have to gradually match onto the outer geostrophic flow. In the case here, the effective geostrophic wind will be the leading order vorticity ζ_1 in the exterior region of the flow as developed in Sect. 5.3.2.

We now move on to develop the boundary layer solution near the plate at $z = 0$. Later, by symmetry arguments, we will use the solution in this zone to obtain the solution appropriate for the boundary at $z = 1$. As discussed before, we expect the boundary layer zone to be a factor $\mathrm{E}^{1/2}$ smaller than the vertical scale of the system. It is on this scale that all physical quantities will vary vertically. Thus we rescale the equations of motion by introducing a new, inner stretched vertical coordinate Z

$$z \equiv \mathrm{E}^{1/2}Z; \qquad \frac{\partial}{\partial z} \equiv \frac{1}{\mathrm{E}^{1/2}}\frac{\partial}{\partial Z}. \tag{5.71}$$

The goal now is to explicitly show how the vertical derivatives transform in terms of this new, stretched scaling. Equations (5.62–5.65), now modified are given by

$$\mathrm{E}^{1/2}\frac{\partial u}{\partial t} - \mathrm{v} = -\frac{\partial \Pi}{\partial x} + \frac{\partial^2 u}{\partial Z^2} + \mathrm{E}\left(\frac{\partial^2 u}{\partial x^2} + \frac{\partial^2 u}{\partial y^2}\right), \tag{5.72}$$

$$\mathrm{E}^{1/2}\frac{\partial \mathrm{v}}{\partial t} + u = -\frac{\partial \Pi}{\partial y} + \frac{\partial^2 \mathrm{v}}{\partial Z^2} + \mathrm{E}\left(\frac{\partial^2 \mathrm{v}}{\partial x^2} + \frac{\partial^2 \mathrm{v}}{\partial y^2}\right), \tag{5.73}$$

$$\mathrm{E}^{3/2}\frac{\partial w}{\partial t} = -\frac{\partial \Pi}{\partial Z} + \mathrm{E}\frac{\partial^2 w}{\partial Z^2} + \mathrm{E}^2\left(\frac{\partial^2 w}{\partial x^2} + \frac{\partial^2 w}{\partial y^2}\right), \tag{5.74}$$

and

$$\frac{\partial u}{\partial x} + \frac{\partial v}{\partial y} + \frac{\partial w}{\partial Z} = 0. \qquad (5.75)$$

In this case we propose the following one-term series expansions. In order to distinguish the resulting perturbation series terms from the ones appropriate for the exterior solution, we write all the dependent variables in capital letters, except for the pressure which we mark by an overlying tilde:

$$u(x,y,z,t) = U_1(x,y,Z,t) + \text{HOT}, \qquad v(x,y,z,t) = V_1(x,y,Z,t) + \text{HOT},$$
$$w(x,y,z,t) = W_1(x,y,Z,t) + \text{HOT}, \qquad \Pi(x,y,z,t) = \tilde{\Pi}_1(x,y,Z,t) + \text{HOT}.$$

$$(5.76)$$

Substituting these expansions in Eqs. (5.72–5.75), we can generate a corresponding set of equations to solve, once again, order by order in powers of $E^{1/2}$. The lowest order terms of (5.74) and (5.75) are, respectively,

$$\frac{\partial \tilde{\Pi}_1}{\partial Z} = 0 \quad \text{and} \quad \frac{\partial U_1}{\partial x} + \frac{\partial V_1}{\partial y} + \frac{\partial W_1}{\partial Z} = 0. \qquad (5.77)$$

The first equation above confirms the suggestion made at the outset of this section that the pressure shows no vertical variation in the boundary layer zone *as well*. Therefore, it stands to reason that in order for the lowest order pressure perturbations in both regions to match onto each other it must be that $\tilde{\Pi}_1 = \Pi_1(x,y,t)$, in other words, the lowest order perturbation function in both the boundary layer and in the exterior zone are the same function. We make explicit use of that in writing out the leading order terms of Eqs. (5.72) and (5.73), namely,

$$-V_1 = -\frac{\partial \Pi_1}{\partial x} + \frac{\partial^2 U_1}{\partial Z^2}, \qquad \Longrightarrow \qquad -\left[V_1 - v_1(x,y,t)\right] = \frac{\partial^2 U_1}{\partial Z^2}, \quad (5.78)$$

$$U_1 = -\frac{\partial \Pi_1}{\partial y} + \frac{\partial^2 V_1}{\partial Z^2}, \qquad \Longrightarrow \qquad \left[U_1 - u_1(x,y,t)\right] = \frac{\partial^2 V_1}{\partial Z^2}, \quad (5.79)$$

where we have made use of the relationships between Π_1, u_1, and v_1 found in Eq. (5.67). Solutions to the above two equations can be found using similar mathematical operations as in the simple Ekman problem in the previous section. The results are

$$U_1 = u_1 - e^{-Z}\left(u_1 \cos Z + v_1 \sin Z\right), \qquad V_1 = v_1 - e^{-Z}\left(v_1 \cos Z + u_1 \sin Z\right).$$

$$(5.80)$$

The functions in (5.80) both solve Eqs. (5.78–5.79) and have been constructed with the property of matching onto the leading order exterior solutions u_1 and v_1 when the inner variable $Z = z/\sqrt{E}$ becomes very large, i.e., for $Z \to \infty, U_1 \to u_1$ and $V_1 \to v_1$.

Our aim now is to construct the solution to the vertical velocity perturbation. To do this, it proves itself useful to assess the horizontal divergence of the flow and make explicit use of the second relation in (5.77):

$$\frac{\partial W_1}{\partial Z} = -\left(\frac{\partial U_1}{\partial x} + \frac{\partial V_1}{\partial y}\right) = \underbrace{\left(\frac{\partial v_1}{\partial x} - \frac{\partial u_1}{\partial y}\right)}_{=\zeta_1(x,y,t)} e^{-Z}\sin Z + \underbrace{\left(\frac{\partial u_1}{\partial x} + \frac{\partial v_1}{\partial y}\right)}_{=0}\left(1 - e^{-Z}\cos Z\right)$$

$$= \zeta_1 e^{-Z}\sin Z. \tag{5.81}$$

It is possible to construct an explicit solution for the vertical velocity assuming that it is zero at $z = 0$ by integrating the above expression. The result is

$$w_1 = W_1(x,y,Z,t) = \frac{1}{2}\zeta_1\left[\left(1 - e^{-Z}\cos Z\right) + e^{-Z}\sin Z\right], \tag{5.82}$$

showing that as $Z \gg 1$ the vertical velocity limits to

$$W_1(Z \gg 1) = \left(\frac{1}{2}\right)\zeta_1(x,y,t). \tag{5.83}$$

The boundary layer solution near the top boundary is symmetrical to the one developed here near $z = 0$. We recycle the symbol Z but this time use it to stretch the vertical coordinate at $z = 1$, according to $(1 - z) = \mathrm{E}^{1/2}Z$. It therefore follows that the leading order boundary layer solution for the horizontal velocities near $z = 1$ is

$$U_1 = u_1 - e^{-Z}\left(u_1\cos Z + v_1\sin Z\right), \qquad V_1 = v_1 - e^{-Z}\left(v_1\cos Z + u_1\sin Z\right) \tag{5.84}$$

which is the same functional form as the solution near the $z = 0$ boundary. The leading order vertical velocity in the zone $z \approx 1$ is

$$w_1 = W_1(x,y,Z,t) = -\frac{1}{2}\zeta_1\left[\left(1 - e^{-Z}\cos Z\right) + e^{-Z}\sin Z\right], \tag{5.85}$$

where the minus sign appearing in front of ζ_1 reflects the symmetry and the recycled use of Z, mentioned above. Similarly to the other boundary, as one moves sufficiently far from the boundary, the vertical velocity takes on the form

$$W_1(x,y,Z,t) = -\frac{1}{2}\zeta_1(x,y,t). \tag{5.86}$$

Since we are interested in the vertical velocity and how it affects the exterior solution, let us restore the expressions in terms of the original non-stretched vertical coordinate, i.e., rewrite the above expressions transforming from Z to z. Thus, the vertical velocity near $z = 0$ is the leading order expression

$$w_1(x,y,z,t) = \frac{1}{2}\zeta_1\left[\left(1 - e^{-z/\mathrm{E}^{1/2}}\cos\frac{z}{\mathrm{E}^{1/2}}\right) + e^{-z/\mathrm{E}^{1/2}}\sin\frac{z}{\mathrm{E}^{1/2}}\right], \qquad (5.87)$$

while near the $z = 1$ boundary the leading order vertical velocity has the form

$$w_1(x,y,z,t) = \frac{1}{2}\zeta_1\left[\left(1 - e^{-(1-z)/\mathrm{E}^{1/2}}\cos\frac{1-z}{\mathrm{E}^{1/2}}\right) + e^{-(1-z)/\mathrm{E}^{1/2}}\sin\frac{1-z}{\mathrm{E}^{1/2}}\right].$$
$$(5.88)$$

These last two equations will be used again shortly.

5.3.4   Matching and Remarks

We now aim to complete the solution by matching the results of the three regions. This will make use of the equation governing the evolution of the bulk interior, Eq. (5.70), and the functional forms of the vertical velocity coming out of each of the respective boundary layers (5.87) and (5.88). In this particular application of the spin-down problem, matching will be straightforward because of the following steps. We can vertically integrate equation (5.70) along the extent of the vertical scale of the interior flow, but not delimit the integration explicitly on the boundaries. We represent this by

$$\int_\varepsilon^{1-\varepsilon}\frac{\partial\zeta_1}{\partial t}dz = \int_\varepsilon^{1-\varepsilon}\frac{\partial w_1}{\partial z}dz = \left[w_1(x,y,1-\varepsilon,t) - w_1(x,y,\varepsilon,t)\right], \quad (5.89)$$

where $\varepsilon > 0$ represents a smidgen sized distance *outside* the boundary layer zones. This region is often times referred to as the *overlap region* and we will show below how to access it in terms of the small parameter E. The subtle idea of the overlap region is that it should be larger than the boundary layer zone but reasonably small compared to the global scale of the system *and* that it maintains its existence as the main small parameter of the system goes to zero. Assume that the overlap regions are characterized by the small parameter E through the power law $\varepsilon = \mathrm{E}^\alpha$, where α, a positive constant, is as yet unknown. In this particular example, we input the value of the overlap region into the boundary layer solution for the leading order vertical velocity solution near $z = 0$, Eq. (5.87),

$$w_1(x,y,\varepsilon,t) = \frac{1}{2}\zeta_1\left[\left(1 - e^{-\mathrm{E}^{\alpha-1/2}}\cos\mathrm{E}^{\alpha-1/2}\right) + e^{-\mathrm{E}^{\alpha-1/2}}\sin\mathrm{E}^{\alpha-1/2}\right], \qquad (5.90)$$

and similarly for the leading order vertical velocity solution near $z = 1$, Eq. (5.88)

$$w_1(x,y,1-\varepsilon,t) = -\frac{1}{2}\zeta_1\left[\left(1 - e^{-\mathrm{E}^{\alpha-1/2}}\cos\mathrm{E}^{\alpha-1/2}\right) + e^{-\mathrm{E}^{\alpha-1/2}}\sin\mathrm{E}^{\alpha-1/2}\right].$$
$$(5.91)$$

We seek to have smooth matching of the solutions in the overlap region in the limit where $E \ll 1$. Looking at the mathematical form of the above expressions, we want the exponential terms to decay precipitously as $E \to 0$ and we see that this is achieved provided α satisfies the constraint $\alpha < 1/2$. Thus, the overlap region exists and the solutions have uniform asymptotic agreement, provided this constraint on α is met. We now explicitly substitute (5.90) and (5.91) into (5.89) and find that so long as $E \ll 1$ and $\alpha < 1/2$ we have the simple result

$$\frac{\partial \zeta_1}{\partial t} = -\zeta_1 + \mathscr{O}\left(E^{1/2-\alpha}, e^{-E^{\alpha-1/2}}\right). \tag{5.92}$$

Now it has emerged why $\alpha < 1/2$, for otherwise the error terms would grow exponentially as $E \to 0$. Dropping the error terms and restoring all dimensions to the quantities appearing, we get *to leading order*

$$\frac{\partial \zeta'}{\partial t} = -\frac{2\Omega_0 H_0}{L}\zeta' = -\frac{\sqrt{2\Omega_0 \nu}}{L}\zeta', \qquad \zeta' = \frac{\partial v'}{\partial x} - \frac{\partial u'}{\partial y}. \tag{5.93}$$

Recall that we were originally interested in the problem of spin-down but that, instead, we moved into the rotating frame. A column that is still in the laboratory frame has a vorticity equal to $-2\Omega_0$ in the rotating frame. Correspondingly, a spinning column with vorticity $2\Omega_0$ in the laboratory frame has zero vorticity in the rotating frame. Thus the solution to the vorticity equation (5.93) is

$$\zeta' = -2\Omega_0 \left(1 - e^{-t/\tau_0}\right), \qquad \tau_0 = \frac{L}{2\Omega_0 H_0} = \frac{L}{\sqrt{2\Omega_0 \nu}}, \tag{5.94}$$

describing the vorticity in the rotating frame. As $t \to \infty$ we have that $\zeta' \to -2\Omega_0$ which, in the laboratory frame, means no spin of the column. Most importantly we get *fairly* rigorously, if asymptotic methods of applied mathematics may be described in this way, that the relevant spin-down time is, in fact, $\tau_{s-d} = \tau_0 \approx L/\sqrt{2\Omega\nu}$. Finally we remark that now we see more clearly what we have alluded to in the beginning of the chapter. We mentioned that, for the instance of a rotating planet, the flow of air on a spinning planet effectively locks onto the motion of the ground below it. It is primarily through the mechanism of spin-up that this happens. The link that brings this idea to that of the scale of a planetary atmosphere like the Earth's is the viscosity coefficient. However, the actual observed spin-up timescale τ_0 for the Earth atmosphere is much shorter than what would be calculated using the values of simple molecular viscosity for a gas like air. What in fact occurs is that the effective viscosity coefficient is much higher due to turbulence, discussed in Chap. 9.

5.4 Linearized Dynamics: Inertial and Rossby Waves

In this section, we introduce the effects of stratification upon linearized disturbances including the development of inertia-gravity waves and Rossby waves. Using scaling arguments we motivate the development of a simpler set of equations, known as the quasi-geostrophic (QG) equations, describing the dynamical motion in the atmospheres of moderately fast rotating planets. The steps leading to the QG equations have strong parallels to the procedure involved in the Ekman spin-down problem in Sect. 5.3. We also examine how the QG model reveals an interesting feature called thermal wind. Inertial and Rossby waves are ubiquitous features of all rotating fluid environments. The *Rossby wave* plays a prominent role in the evolution of large-scale flows in a variety of settings, including those of planetary atmospheres and accretion disks. Inertial waves are also very important as they play several roles in the communication of vortex disturbances in atmospheres. They are most often considered in terms of *inertia-gravity waves*. We offer here a rudimentary analysis of these phenomena and refer the reader to the variety of texts on the topic listed at the end of the section, in the *Bibliographical Notes*, for further reading.

5.4.1 Introduction to Inertia-Gravity Waves

Consider a plane-parallel, inviscid Boussinesq fluid atmosphere with a constant vertical gravitational field, much like what was considered in Sect. 4.4.2. Consider this plane-parallel atmosphere to be part of a globally rotating system, so that the equations of motion relative to a small rotating Cartesian box, found in Sect. 5.1.1.2, are appropriate. Let the steady state of the atmosphere be one of rest with the steady profiles for the pressure and entropy found in Eqs. (4.4.2) and (4.86), respectively. Perturbations of this mean state, now with the inclusion of the Coriolis effect, lead to a modification of the equation set into (4.88–4.92), namely,

$$\frac{\partial u'}{\partial t} - 2\Omega_0 v' = -\frac{1}{\rho_0}\frac{\partial P'}{\partial x}, \qquad \frac{\partial v'}{\partial t} + 2\Omega_0 u' = -\frac{1}{\rho_0}\frac{\partial P'}{\partial y}, \qquad \frac{\partial w'}{\partial t} = -\frac{1}{\rho_0}\frac{\partial P'}{\partial z} + g\beta_s s',$$

$$(5.95)$$

and

$$0 = \frac{\partial u'}{\partial x} + \frac{\partial v'}{\partial y} + \frac{\partial w'}{\partial z}, \qquad \frac{\partial s'}{\partial t} + \beta_s w' = 0, \tag{5.96}$$

where we recall that β_s is known as the coefficient of thermal/entropic expansion at constant pressure, see Sect. 4.4.1. Our aim is to derive a single equation for any one of the variables shown. Similarly to previous practice, we identify ζ' with the vertical vorticity and recall that it is equal to $\partial_x v' - \partial_y u'$. Operating with $\nabla \times$ in two dimensions on the two horizontal momentum equations results in

$$\frac{\partial \zeta'}{\partial t} + 2\Omega_0 \left(\frac{\partial u'}{\partial x} + \frac{\partial v'}{\partial y} \right) = 0, \qquad \Longrightarrow \qquad \frac{\partial \zeta'}{\partial t} = 2\Omega_0 \frac{\partial w'}{\partial z}. \tag{5.97}$$

Notice the reappearance of the physical result we found in the Ekman spin-down problem, examined in Sect. 5.3: the perturbation vertical vorticity increases when the local fluid is vertically stretched. The next step necessitates taking the divergence of the three momentum equations (5.95) and making use of the incompressibility condition, revealing the relationship between ζ' P' and s':

$$-2\Omega_0 \zeta' = -\rho_0^{-1} \nabla^2 P' + g\beta_s \frac{\partial s'}{\partial z}. \tag{5.98}$$

Note immediately how the vertical vorticity has the same diagnostic form as in geostrophic balance when the Boussinesq term, the last one in Eq. (5.98), is absent. What is left then is just an expression of geostrophic balance. It says that the essential quality of geostrophic balance is true for perturbations as well. In fact, we will see that the inclusion of the entropy perturbations will motivate the definition of the *potential vorticity*, a quantity we have already seen in Chap. 2 and will encounter again in the next section, in which its geophysical form will be given and shown to be a special case of our broader definition in Sect. 2.3.3. Next, we form an evolution equation for $\partial_z w'$, by taking the two-dimensional, horizontal divergence of the two horizontal momentum equations and, once again, making use of the incompressibility condition. The result is similar to Eq. (4.94), except that now there is an additional term, which we repeat here for the sake of clarity:

$$-\frac{\partial}{\partial t} \frac{\partial w}{\partial z} - 2\Omega_0 \zeta' = -\frac{1}{\rho_0} \nabla^2 P'. \tag{5.99}$$

Here and throughout this problem we use two-dimensional (horizontal) operators involving ∇. In particular the Laplacian operator is

$$\nabla^2 \equiv \frac{\partial^2}{\partial x^2} + \frac{\partial^2}{\partial y^2},$$

where we avoid the often used subscript H, for horizontal, hoping that this will not cause confusion.

Following the same procedure as detailed in Sect. 4.4.2, we find the following equation for the vertical perturbation velocity

$$\frac{\partial^2}{\partial t^2} \nabla^2 w' + 4\Omega_0^2 \frac{\partial^2 w'}{\partial z^2} + g\beta_s \beta \nabla^2 w' = 0, \tag{5.100}$$

where β is related to vertical change of average entropy as in Sect. 4.4.2. Equation (5.100) is the linear partial differential equation governing linearized inertia-gravity waves. We can examine the frequency response in exactly the same way as we did in Sect. 4.4.2, by assuming that the atmosphere is capped at two heights $z = 0, h_0$. Therefore, assuming solutions of the form $\propto A(\mu, k_x, k_y) \sin(\mu z) \exp(ik_x + ik_y - i\omega t) + \text{c.c.}$, in which $\mu = n\pi/h_0$ we find the following dispersion relation

$$\omega^2 = \frac{\beta g \beta_s h_0^2 K^2 + n^2 \pi^2 4\Omega_0^2}{n^2 \pi^2 + h_0^2 K^2}, \quad \text{where} \quad K^2 \equiv k_x^2 + k_y^2. \tag{5.101}$$

The limit of $\Omega_0 \to 0$, i.e., no rotation, reproduces the internal gravity wave result from the previous chapter (see below). When internal variations of entropy are suppressed, i.e., $\beta \to 0$, then the result is the dispersion relation for *inertial waves*

$$\omega^2 = 4\Omega_0^2 \frac{n^2 \pi^2}{n^2 \pi^2 + h_0^2 K^2}. \tag{5.102}$$

We clearly see that in the absence of Coriolis effects, this whole family of inertial waves (5.102) vanishes. Otherwise, the frequency spectrum is located just under the maximum value of $2\Omega_0$, that is,

$$\omega^2(k_x, k_y, n) \leq 4\Omega_0^2.$$

In the case where stable gravity waves are supported, i.e., $\beta > 0$, the total disturbance energy is identical to that derived for the case where there are no Coriolis effects. In other words, the total energy is the same as found in Eq. (4.98). This means that all of the physical properties we discussed, regarding the energy propagation of disturbances, carry over to the case when Coriolis effects are included. Most prominently, the propagation of perturbation energy is always in a direction *perpendicular* to the phase velocity. An interesting variation of this situation, in which there is a global background shear of the $\hat{y}$ directed velocity along x, is detailed in Problems 5.4 and 5.5.

5.4.2 Introduction to Rossby Waves

Rossby waves are important because they play a leading role in the stabilization and destabilization of global velocity fields. In short, Rossby waves are wavelike disturbances in environments where there exists large-scale variation of the vorticity or potential vorticity. In planetary examples, often it is simply the latitude dependence of the effective frame rotation that causes vorticity variation, more on this in the next section. Treated as linear, Rossby waves are most easily analyzed for flows which are steady in time in the absence of perturbations. We detail below some example circumstances in which one might expect Rossby waves to arise

1. A localized shear or jet layer.
2. A classical shallow water system in which the bottom topography exhibits variation. These are called *topographic waves*. If one were to imagine Taylor columns, then the vorticity contained in a given column would have shown variation in its vorticity *per surface area* if the column were to travel over a bottom surface having varying height. See Problem 5.7 for an examination of this from the vantage point of the shallow water equations.
3. Planetary atmosphere in which the planetary rotation, as projected with respect to the ground at some latitude, shows variation as one moves in latitude, i.e., the latitude projected planetary vorticity $2\Omega_0 \sin\varphi$ is not constant with latitude as seen in Sect. 5.1.1.3.

The connection between variations of projected planetary rotation and the conservation of potential vorticity in the flow will be formulated mathematically for a small section of a rotating planet's atmosphere and will be discussed in terms of QG in Sect. 5.5. Let us begin by considering two-dimensional equations in a Cartesian geometry, i.e., the equations of motion found in Sect. 5.1.1.2 where there is no variation in the direction coincident with the background rotation $\hat{z}$. Let the global rotation show some weak variation in the y direction and let it be represented by $\Omega_0 = \Omega_{00} + (1/2)\beta_0 y$, where β_0 is a constant. Assume that the fluid is incompressible. Introduction of perturbations and linearization gives in this setting

$$\frac{\partial u'}{\partial t} - 2\Omega_0 v' = -\frac{1}{\rho_0}\frac{\partial P'}{\partial x}, \qquad \frac{\partial v'}{\partial t} + 2\Omega_0 u' = -\frac{1}{\rho_0}\frac{\partial P'}{\partial y}, \qquad \frac{\partial u'}{\partial x} + \frac{\partial v'}{\partial y} = 0. \tag{5.103}$$

Now, one can operate with $\nabla\times$ in two dimensions, on the first two equations considering them as two components of a vector. This will give a difference between the appropriate horizontal derivatives. A similar operation has been performed earlier in this chapter. We express now the result in terms of the perturbation stream function ψ' by making use of $u' = -\partial_y \psi'$ and $v' = \partial_x \psi'$ to show that

$$\frac{\partial}{\partial t}\left(\frac{\partial v'}{\partial x} - \frac{\partial u'}{\partial y}\right) + 2\beta_0 v' = 0, \qquad \Longrightarrow \qquad \frac{\partial}{\partial t}\underbrace{\left(\frac{\partial^2\psi'}{\partial x^2} + \frac{\partial^2\psi'}{\partial y^2}\right)}_{\text{vertical vorticity perturbation}} + \beta_0\frac{\partial\psi'}{\partial x} = 0.$$

$$\tag{5.104}$$

Assumption of normal mode solutions of the form $\propto \exp(ik_x + ik_y - i\omega t)$ shows here that ω is given by

$$\omega(k) = -\frac{\beta_0 k_x}{k_x^2 + k_y^2}. \tag{5.105}$$

To better understand the meaning of this dispersion relation, we recast this *linear* result in a form of an appropriate Lagrangian derivative. Notice that the linearization of

$$\frac{D}{Dt}\left(\frac{\partial v'}{\partial x} - \frac{\partial u'}{\partial y} + 2\Omega_0\right) = \left(\frac{\partial}{\partial t} + u'\frac{\partial}{\partial x} + v'\frac{\partial}{\partial y}\right)\left(\frac{\partial v'}{\partial x} - \frac{\partial u'}{\partial y} + 2\Omega_{00} + 2\beta_0 y\right),$$
(5.106)

reduces to the same perturbation equation (5.104). This observation allows us to interpret the result more clearly: the Rossby wave is one in which the total vorticity ζ_{tot} is conserved throughout the course of the fluid response. The total or absolute vorticity is made of two parts: a relative vorticity and the background vorticity, $2\Omega_0$. Thus

$$\zeta_{tot} \equiv \underbrace{\frac{\partial v'}{\partial x} - \frac{\partial u'}{\partial y}}_{\text{relative vorticity}} + 2\Omega_0,$$
(5.107)

and this language is used frequently in the context of GFD studies. In this sense, a given fluid parcel preserves the total vorticity ζ_{tot} it was endowed with at some initial time. If a given patch of, say, constant vorticity $\Delta\zeta$ moves into the positive y direction, and if $\beta > 0$, then total vorticity contained within the patch will be preserved throughout the course of the motion. Since $\Delta\zeta + 2\Omega_0 = \Delta\zeta + 2\Omega_{00} + \beta_0 y$, it follows that the value of $\Delta\zeta$ has to decrease. We note that the expression

$$\frac{D\zeta_{tot}}{Dt} = 0$$
(5.108)

is not approximate but is, in fact, exactly true for two-dimensional nonlinear disturbances with a spatially varying background rotation rate Ω_0 (see Problem 5.6). Thus the approximate Lagrangian preservation of the total vorticity in the linear case is consistent with its corresponding nonlinear conversation.

5.5 A Geophysical Example: Quasi-Geostrophy

Atmospheric flow is of great interest not just to meteorology and weather prediction, but also to the flows encountered in other planetary atmospheres and in accretion disks. Precise modeling of an atmosphere includes some complex algebra and geometry and a significant number of contributing physical effects, making it a daunting task. Despite the challenges it presents, a hundred years of theoretical and observational work exists to guide one's understanding. As such, we urge readers who are interested in this subject to consult the several relevant and excellent texts listed in the *Bibliographical Notes* of this chapter. We shall concentrate only upon a very narrow aspect of this vast topic, namely, that of the derivation of a theoretical model, namely, QG, that is both tractable and qualitatively accurate. The QG model captures many known large-scale atmospheric phenomena and is of theoretical interest because the equations can be analyzed with reasonable mathematical rigor. We devote this section to developing *one of many* formal derivations of the

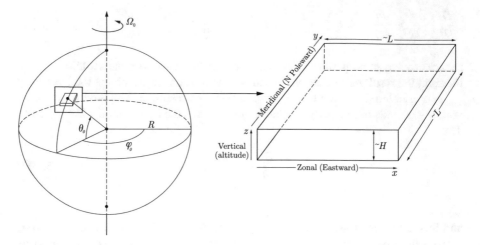

Fig. 5.4 A small synoptic scale section of a planetary atmosphere

QG equations. The presentation will follow a systematic derivation in terms of appropriate scaling analysis including detailed description of the assumptions made.

5.5.1   Physical Assumptions

Planetary atmospheres are considered to be thin fluid layers lying atop a rotating approximately spherical surface, possibly solid or liquid or a combination of both. The dynamics occurring in this spherical shell of fluid has the character of a familiar atmospheric flow because the Rossby number Ro is small, indicating that planetary rotation is the dominant agent dictating the dynamics of the atmosphere (through geostrophic flow). Let us therefore specify a number of quantities as viewed in the local rotating frame of a Cartesian section of an atmosphere at latitude θ_0, followed by a list of assumptions and, finally, present the equations from which we shall derive the QG approximation. Let U be the horizontal velocity scale which occurs on horizontal length scales of L. Let H denote the vertical scale, e.g., the pressure scale-height of the atmosphere, as measured from the ground which is at radius $\approx R$ from the planet's center. Assuming that the atmospheric gas, to a good approximation, obeys a perfect gas equation of state, it follows from (1.60) that $H \sim (\mathscr{R}/\mu)(T/g)$ where T is the temperature. Additional assumptions, referring also to Fig. 5.4, are:

1. The planet rotates along its polar axis with an angular velocity Ω_0. At some given latitude θ_0, the projected rotation rate is $\Omega_{00} = \Omega_0 \sin \theta_0$.
2. The planet's body is approximated by a sphere, so that the magnitude of the inwardly pointing radial component of the planet's gravity vector is given by g, which is constant.

3. The planetary atmosphere is a fast rotator. Thus the Rossby number, which here
   will be defined in terms of the projected rotation rate at latitude θ_0, i.e., Ro $=$
   $U/2\Omega_{00}L \ll 1$ so long as one stays clear of the equator.
4. Spatio-temporal variations of density are dynamically significant only when
   coupled to gravity. The fluid response is otherwise incompressible, that is, the
   Boussinesq approximation is applicable.
5. The following scalings are relevant

$$\frac{L}{R} \sim \frac{H}{L} = \mathcal{O}\left(\text{Ro}\right).$$

6. Small latitudinal variations $(\theta - \theta_0 \ll 1)$ of the projected rotation rate are
   written as

$$\Omega_0 \sin \theta \approx \Omega_{00} + (\theta - \theta_0) \cos \theta_0.$$

For the remainder of the derivation we use nondimensional equations of motion.
Thus we write

$$r - R_0 = Hz, \qquad \theta - \theta_0 = \frac{L}{R}y, \qquad \varphi - \varphi_0 = \frac{L}{R}x,$$

where the nondimensional coordinates' directions are:

  (i) x is eastward, i.e., in the azimuthal or zonal direction,
 (ii) y is north-poleward, i.e., in the meridional or polar direction,
(iii) z is upward, i.e., in the vertical direction or altitude.

A typical timescale, being the turnaround time of the planetary vortex $(\sim L/U)$,
will be used to scale time $t \mapsto t(U/L) = t/\Omega_0 \text{Ro}^{-1}$. The azimuthal and polar
velocities will be scaled by U and the vertical, that is, radial from the vantage
point external to the spherical planet, velocity by $HU/L \approx \mathcal{O}(\text{Ro } U)$. The reason
for this latter choice will become apparent below, upon examination of the equation
of continuity. However, it is important to keep in mind that because of the thinness
of the atmosphere, it is reasonable that the vertical velocities on the planet should
be small, or else the smallness of the aspect ratio would not be retained. Thus we
have the scalings

$$u_\varphi \mapsto Uu, \qquad u_\theta \mapsto Uv, \qquad u_r \mapsto \frac{H}{L}Uw, \qquad (5.109)$$

in which u, v, w are the nondimensionalized velocity variables in the east-west,
north-south, and vertical directions, respectively. Because of the Boussinessq
assumption, when the density appears not coupled to gravity, the latter will be
written as ρ_{00} and when coupled to gravity, we write it as a sum of a stably stratified
mean component $\overline{\rho}(z)$ plus a dynamically varying component so that

$$\underbrace{\rho \mapsto \rho_{00}}_{\text{not coupled to gravity}} \quad , \quad \underbrace{\rho \mapsto \rho_{00}\left[\overline{\rho}(z)+\rho'(x,y,z,t)\right]}_{\text{when coupled to gravity}},$$

where $\overline{\rho}$ and ρ' are nondimensional. Finally, we assess the pressure which should be scaled by $\rho_{00}gH$, since the atmosphere is hydrostatic in the vertical direction (see below). Similarly to density, we write the pressure to be expressed as a sum of a stratified, time independent, mean state $\overline{P}(z)$ and a dynamically varying piece Π. Thus

$$P \mapsto \rho_{00}gH\left[\overline{P}(z)+\Pi(x,y,z,t)\right].$$

Incorporating these assumptions and partial expansions into the momentum equations in (5.19–5.21) together with the continuity equation (5.23), followed by working out the algebraic details (see Problem 5.8), shows that the equations asymptotically simplify to the following set

$$\text{Ro}\frac{Du}{Dt} - \left(1+\text{Ro}\beta y\right)v + \frac{\text{Bu}^2}{\text{Ro}}\frac{\partial \Pi}{\partial x} = \mathscr{O}\left(\text{Ro}^2\right), \tag{5.110}$$

$$\text{Ro}\frac{Dv}{Dt} + \left(1+\text{Ro}\beta y\right)u + \frac{\text{Bu}^2}{\text{Ro}}\frac{\partial \Pi}{\partial y} = \mathscr{O}\left(\text{Ro}^2\right), \tag{5.111}$$

$$\frac{\text{Bu}^2}{\text{Ro}}\left(\frac{\partial \Pi}{\partial z}+\rho'\right) = \mathscr{O}\left(\text{Ro}^2\right), \tag{5.112}$$

where now $\beta \equiv \cot\theta_0$. Using the definition of the Burger number, we have $\text{Bu}^2 = gH\,/4\Omega_{00}^2 L^2$. The mean vertical variation of the pressure is based on hydrostatic equilibrium,

$$\frac{\partial \overline{\Pi}}{\partial z} = -\overline{\rho}+\mathscr{O}\left(\text{Ro}^2\right). \tag{5.113}$$

The Lagrangian time derivative operator takes on the familiar form

$$\frac{D}{Dt} = \frac{\partial}{\partial t}+u\frac{\partial}{\partial x}+v\frac{\partial}{\partial y}+w\frac{\partial}{\partial z}. \tag{5.114}$$

The incompressibility condition simplifies to

$$\frac{\partial u}{\partial x}+\frac{\partial v}{\partial y}+\frac{\partial w}{\partial z} = \mathscr{O}\left(\text{Ro}\right), \tag{5.115}$$

while, because of the incompressibility assumption, the continuity equation simplifies to the material derivative of the density profile

$$\frac{D\rho'}{Dt} + w\frac{\partial\bar{\rho}}{\partial z} = \mathcal{O}\left(\text{Ro},\rho'\right). \tag{5.116}$$

Inspection of the pressure gradient terms shows that they are scaled by Bu^2/Ro. For typical planetary conditions Bu is a number of $\mathcal{O}(1)$. For instance, on the Earth if we assume a synoptic scale, the exact meaning of this term being "of main weather features," of $L \sim 1000$ km, a vertical scale-height H of approximately 8.5 km, T is assumed to be 300 Kelvin at mid-latitudes and the mean molecular weight μ is ~ 28, representing a nitrogen and oxygen molecules' mixture. Thus $\text{Bu} \approx 2 - 3$, depending upon where on the Earth one is. Although there is some considerable variation from planet to planet, the typical value of Bu is generally $\mathcal{O}(1)$ and we shall, henceforth, treat it as a free parameter of the system.

Our aim is to apply an expansion approach to recover the familiar geostrophic balance we have seen several times in the discussion in this chapter. With Bu considered to be $\mathcal{O}(1)$, inspection of equations (5.110)–(5.112) shows that a leading order geostrophic/hydrostatic balance is possible only if Π and ρ' are $\mathcal{O}(\text{Ro})$. In such a case Eq. (5.116) shows that in order for the terms to sensibly balance, the vertical velocity should be even smaller than we have originally supposed, i.e., $w = \mathcal{O}(\text{Ro})$ too. We pause here to note that it means that vertical velocity variations are weak, being an $\mathcal{O}(\text{Ro}^2)$ smaller than the horizontal velocity scales U, consistent with the observations of the Earth's atmosphere and of other planets. Analysis of (5.112) for $\mathcal{O}(1)$ values of Bu^2, together with $\text{Ro} \ll 1$, indicates that hydrostasis is a consistent assumption. This would appear to be an odd statement, but we have encountered this already once before, in our derivation of the shallow water equations in Chap. 4. It means to say that while the system exhibits dynamics and local density fluctuations appear, the vertical pressure gradient always adjusts so as to achieve hydrostatic balance with whatever density and thus buoyancy force it "sees" at any instant. The adjustment occurs on a timescale that is very short compared to the ones assumed for the dynamics, $\approx L/U$. By construction, the dynamics timescales of interest are much longer than the vertical sound crossing time. For example, the turnaround time for a large tropical storm is on the order of several days while the vertical sound travel time over the extent of a vertical scale-height is on the order of 5–10 min.

5.5.2 Expansion Procedure and Derivation

Proceeding now mathematically, following the above conclusions, we solve Eqs. (5.110–5.112) and Eqs. (5.115–5.116) as a perturbation series expansion as follows:

$$u = u_0 + \text{Ro}\, u_1 + \text{HOT}, \qquad v = v_0 + \text{Ro}\, v_1 + \text{HOT}, \qquad w = \text{Ro}\, w_1 + \text{HOT},$$

$$\rho' = \text{Ro}\,\rho_1' + \text{HOT}, \qquad \Pi = \text{Ro}\,\Pi_1 + \text{Ro}^2\,\Pi_2 + \text{HOT}. \tag{5.117}$$

At lowest order Eqs. (5.110)–(5.112) become

$$-v_0 = -Bu^2 \frac{\partial \Pi_1}{\partial x}, \qquad u_0 = -Bu^2 \frac{\partial \Pi_1}{\partial y}, \qquad \rho_1 = -\frac{\partial \Pi_1}{\partial z}, \qquad (5.118)$$

where the first two equations are the expression of geostrophic flow, while the last one is hydrostatic equilibrium. The density equation becomes

$$\frac{D\rho_1'}{Dt} + w_1 \frac{\partial \overline{\rho}}{\partial z} = 0, \quad \text{with} \quad \frac{D}{Dt} = \frac{\partial}{\partial t} + u_0 \frac{\partial}{\partial x} + v_0 \frac{\partial}{\partial y}, \qquad (5.119)$$

where we note the absence of the vertical advection term, acting on the dynamical part of ρ' as it is weaker by a factor of $\sim$ Ro, in comparison with the other terms. However the vertical advection of the mean state remains. The incompressibility condition gives

$$\frac{\partial u_0}{\partial x} + \frac{\partial v_0}{\partial y} = 0, \qquad (5.120)$$

which is automatically satisfied owing to the geostrophically balanced state, recovered in Eq. (5.118).

At the next order we begin with Eq. (5.115) finding

$$\frac{\partial u_1}{\partial x} + \frac{\partial v_1}{\partial y} = -\frac{\partial w_1}{\partial z}. \qquad (5.121)$$

This means that these solutions exhibit weak vertical stretching. Therefore, we expect to see some facet of this in the evolution of the vorticity and, indeed, it will be evident a little further below. The situation encountered here is analogous to that in the Ekman spin-up/down calculation earlier. Equations (5.110) and (5.111) at this order are

$$\frac{Du_0}{Dt} - \beta y \, v_0 = v_1 - Bu^2 \frac{\partial \Pi_2}{\partial x}, \qquad (5.122)$$

$$\frac{Dv_0}{Dt} + \beta y \, u_0 = -u_1 - Bu^2 \frac{\partial \Pi_2}{\partial y}. \qquad (5.123)$$

Operating on Eq. (5.123) by $\partial/\partial x$ and subtracting from it the result of operating on (5.122) by $\partial/\partial y$, followed by making explicit use of Eq. (5.120) results in

$$\frac{D}{Dt}\left(\frac{\partial v_0}{\partial x} - \frac{\partial u_0}{\partial y}\right) + \beta u_0 = -\frac{\partial u_1}{\partial x} - \frac{\partial v_1}{\partial y} = \frac{\partial w_1}{\partial z} = -\frac{\partial}{\partial z}\left[\left(\frac{\partial \overline{\rho}}{\partial z}\right)^{-1} \frac{D\rho_1}{Dt}\right],$$

$$(5.124)$$

where we have made use of Eqs. (5.119) and (5.121).

Observe now what we previously alluded to: the rate of change of the vertical vorticity is proportional to the vertical stretching of the flow. Because of the ordering, the vertical stretching, though weak, is given as the material derivative of the density perturbation. It means that a spinning column instantly responds to the state of the density within the column. The vertical stretching is given by the vertical gradient of the corresponding pressure expression found in the equation. In this way, the dynamics describes the material invariance of some *total* quantity. To see this, we replace the expressions for u_0, ρ_1, and v_0 in terms of Π_1, as found in Eq. (5.118), and find

$$\frac{D\mathscr{Q}}{Dt} \equiv \frac{D}{Dt}\left[\frac{\partial^2 \Pi_1}{\partial x^2} + \frac{\partial^2 \Pi_1}{\partial y^2} + 2\Omega_{00} + \beta y - \frac{1}{Bu^2}\frac{\partial}{\partial z}\left(\frac{\partial \overline{\rho}}{\partial z}\right)^{-1}\frac{\partial \Pi_1}{\partial z}\right] = 0.$$
(5.125)

The above is intended to *define* $\mathscr{Q}$, which is comprised of the usual fluid vorticity $(\partial_x^2 + \partial_y^2)\Pi_1$ associated with the basic geostrophic flow, plus the background latitudinally varying vorticity $(2\Omega_0 = 2\Omega_{00} + \beta y)$ as viewed in the rotating frame at latitude θ_0. Most importantly $\mathscr{Q}$ is conserved by moving fluid particles.

5.5.3 Conservation of Potential Vorticity and Some Remarks

We restore all units to our variables and write the result in the conservative form

$$\frac{D\mathscr{Q}}{Dt} = \frac{D}{Dt}\left(q + 2\Omega_0\right) = 0, \qquad u = -\frac{\partial \psi}{\partial y}, \qquad v = \frac{\partial \psi}{\partial x}, \qquad \frac{\rho'}{\rho_{00}} = -\frac{2\Omega_{00}}{g}\frac{\partial \psi}{\partial z},$$
(5.126)

in which the effective stream function ψ is just the dynamical pressure perturbation $(P' \leftrightarrow \rho_{00}2\Omega_{00}\psi)$ in disguise—we shall refer to this further below. The definition of q must await a formula, which will come shortly, i.e., Eq. (5.128).

It is a somewhat unfortunate fact that there are two parallel conventions for the definition of the stream function in the FD literature. Throughout this book we have used the more traditional definition, e.g., that as defined in (2.101) and (2.102). However, in meteorology the convention is to have the stream function defined as the negative of the definition used in this book, as a comparison of (2.101) and Eq. (5.126) readily shows. Similarly, the corresponding velocity fields also have the signs in their definitions switched as well: see Eqs. (2.102) and (5.126). The overall results are, of course, unaffected by this but this can be a source of confusion for the beginning student. Additionally, we shall give here a definition of a very important concept in GFD, named here *potential vorticity* and denoted by $\mathscr{Q}$, which is unfortunately different from the usual FD definition, given in Chap. 2 of a quantity having the same *name*, given before. We chose this symbol to prevent confusion, because an ordinary Q has already been used in this book to denote several other quantities. We point out that in any given flow, there is a family of quantities that satisfy a conservation equivalent to that of Chap. 2; any of these may be called a potential vorticity. In Sect. 2.3.3 we have proven the Ertel theorem

$$\frac{D}{Dt}\left[\left(\frac{\omega}{\rho}\right) \cdot \nabla F\right] = 0, \tag{5.127}$$

that is valid for any material invariant function $F(\mathbf{x}, t)$. It was presented there as the potential vorticity conservation, and that concept, denoted by Q there, was defined in a more general way than here. Therefore it will be henceforth called Q_{gen}:

$$Q_{gen} \equiv \left(\frac{\omega}{\rho}\right) \cdot \nabla F,$$

where F, as said before, is *any* material invariant function. Clearly, $\mathscr{Q}$ as defined here is a particular case of Q_{gen} and seems well suited for GFD. $\mathscr{Q} = q + 2\Omega_0$ is given as a sum of the relative potential vorticity q plus the vorticity of the background state $(2\Omega_0)$, which is also a function of position, respectively:

$$q \equiv \frac{\partial^2 \psi}{\partial x^2} + \frac{\partial^2 \psi}{\partial y^2} + \frac{\partial}{\partial z}\frac{4\Omega_{00}^2}{N^2}\frac{\partial \psi}{\partial z}, \qquad \text{and} \qquad 2\Omega_0 = 2\Omega_{00}\left(1 + \cot\theta_0 \frac{y}{R}\right), \tag{5.128}$$

where the *stratification frequency* N^2, also known as the Brunt–Väisälä frequency,[1] is here used to express the gradient of the vertical background density field,

$$N^2 = -g\frac{1}{\rho_{00}}\frac{\partial \overline{\rho}}{\partial z}. \tag{5.129}$$

Note that earlier in our definition of the Burger number, we spoke of N^2 in terms of the vertical entropy gradient. It is common practice to have the Brunt–Väisälä frequency defined in different ways depending upon the context of the problem examined. In all instances, they represent the dynamical response of buoyancy oscillations. The vertical velocity is diagnostically determined from the equation of mass continuity and, as we saw in the derivation, is written in nondimensional form as found in Eq. (5.119). In dimensional form it is given by

$$w = -\left(\frac{\partial \overline{\rho}}{\partial z}\right)^{-1}\frac{D\rho'}{Dt} = -\frac{2\Omega_{00}}{N^2}\frac{D}{Dt}\left(\frac{\partial \psi}{\partial z}\right). \tag{5.130}$$

It is now time to define two additional geophysical concepts and the concomitant notation. We begin with the f-plane, which comes from the traditional practice to use f instead of $2\Omega_0$. So the f-plane is one in which $2\Omega_0$ is a constant. When θ_0 is near the midlatitudes and linear variations in f are retained, the setting is referred to

---

[1] Some researches call that quantity N, instead of N^2.

as the β-*plane*. The elegant feature of the asymptotic result in Eq. (5.126) is that we have identified a quantity that is exactly conserved by moving fluid particles. With this in hand, many conservation properties follow. This formulation also allows for an analysis that is both tractable and very revealing. We urge readers to examine the texts on GFD to explore the vast power of this simplified set of equations. We have provided here a mathematical foundation for them.

It is common to find the expression *balanced* flow in the GFD literature and this indicates flow fields which are in geostrophic balance, as they are here in this QG approximation. Much work focuses on assessing the relative influence of so-called ageostrophic effects, which include, as their name indicates, velocity components that are not geostrophic. For example, internal inertia-gravity waves are, by design, filtered out of the QG equations. The horizontal velocity fields associated with inertia-gravity wave motions are markedly not in geostrophic balance and, as such, are regarded ageostrophic. The question of when should inertia-gravity waves be necessarily retained in the analysis is not an easy one to answer, if one is seeking a general rule. However, consider that QG dynamics occurs on timescales that are usually 5–10 times longer than the local rotation times and that inertia-gravity waves periods are generally no longer than a rotation time, as can be seen in their dispersion relations in Eq. (5.101). So unless the circumstances involve values of Bu that are not $\mathcal{O}(1)$, but rather either $\ll 1$ or $\gg 1$, the timescale disparity of inertial-gravity waves and that of planetary vortex motions indicates that the influence of the former upon the latter is likely weak. However, if the vortex timescales of interest *are* similar enough to that of inertial-gravity waves, then one can no longer assume independence of one class of disturbances from the other and the use of QG equations should be considered with caution and only, at best, serve as a guide. For example, one possible outcome when a system of this kind is encountered is the destruction of a vortex induced by the inertia-gravity waves. The relative process in such a case is the *elliptical instability*, which we do not intend to discuss.

Despite these caveats the spirit and utility of quasi-geostrophy is apparent. At the beginning of the chapter, we introduced the Taylor–Proudman theorem and its consequences. Geostrophic flow is one of the most prominent of these, as discussed in Sect. 5.2.1 and displayed in Fig. 5.2. Nowhere in that description is there any prescription as to how the geostrophic flow evolves. QG is a descriptive prescription for situations such as when geostrophic rotating columns are manifest in a planetary atmosphere. Under certain simplifying assumptions these geostrophic columns may be examined relatively easily, e.g., revealing Hamiltonian structure when treated as two-dimensional point vortices like those examined in Sect. 2.5.3 (see Problem 5.10). Historically speaking, geophysical (i.e., meteorological, oceano-graphic) interest is the main origin of these ideas but because of their fascinating mathematical properties, they are the subject of much study in theoretical FD and applied mathematics.

5.5.4 Planetary Waves

An atmosphere at rest, as viewed in the rotating frame of the planet, naturally provides a good base state to examine the linearized perturbation of the QG equations. The resulting motions are often called *planetary waves*, and they are a generalization of the Rossby waves we discussed in Sect. 5.4.2. For the sake of illustration, suppose that the layer of interest is embedded between solid walls located at $z = 0$ and $z = H$ so that the perturbation of the vertical velocity is zero there. We assume that N^2 is constant, equal to N_0^2. Thus, we assume linearized Fourier perturbations of the form $\psi = \psi'(z)e^{ik_x x + ik_y y - i\omega t} + \text{c.c.}$ that are periodic on horizontal length scales of L and introduce them into Eqs. (5.126–5.128) and Eq. (5.130), in which we find, after some rearrangements,

$$-\omega\left(\frac{4\Omega_{00}^2}{N_0^2}\frac{d^2\psi'}{dz^2} - K^2\psi'\right) + \beta_0 k_x\psi' = 0, \quad \beta_0 \equiv 2\Omega_{00}\frac{1}{L}\frac{L}{R}\cos\theta_0, \quad K^2 \equiv k_x^2 + k_y^2,$$

(5.131)

with the boundary condition that $w = 0$ at the two surfaces:

$$-\omega\frac{d\psi'}{dz} = 0, \qquad \text{at } z = 0, H.$$

(5.132)

Inspection of the equation for ψ' (5.131) indicates that its solution is composed of sines and cosines. The boundary conditions therefore select the following

$$\psi' = A_n \cos\frac{n\pi z}{H},$$

(5.133)

where n is any integer ≥ 0 identifying different vertical overtones. For consistency with Eq. (5.131) ω has to satisfy

$$\omega\left(\frac{4\Omega_{00}^2}{N_0^2}\frac{n^2\pi^2}{H^2} + K^2\right) + \beta_0 k_x = 0,$$

(5.134)

which yields the dispersion relation

$$\omega(K) = \frac{-\beta_0 k_x}{K^2 + \left(\dfrac{1}{\text{Bu}^2}\dfrac{H_0^2}{H^2}\right)\dfrac{n^2\pi^2}{L^2}} = -2\Omega_{00}\cos\theta_0\left(\frac{L}{R}\right)\left[\frac{\text{Bu}^2 \cdot k_x L}{\text{Bu}^2 \cdot K^2 L^2 + n^2\pi^2(H_0^2/H^2)}\right],$$

(5.135)

with $H_0^2 \equiv g/N_0^2$. The above dispersion relation for planetary waves is the familiar one, found in many textbooks. The right-hand side of the second equality shows ω rewritten in a way which highlights the scalings of various terms. First and foremost, we see that the time scalings are consistent with the assumptions by which we motivated the derivation of the equations of QG: for H_0/H, Bu and KL all of $\mathcal{O}(1)$ a high frequency, corresponding to *slow time* can be introduced. This frequency scaling goes as $\mathcal{O}(\Omega_{00}L/R)$, and since L/R was assumed to be equal to $\mathcal{O}(\text{Ro})$, we see that the above high frequency scales as the Rossby number times the planet's rotation rate, locally projected. In addition, since the assumptions of the problem were so stated that the dynamics has Bu $= \mathcal{O}(1)$, the pressure scale-height H should be of the same order of magnitude as the distance between the boundaries of the problem, H_0. Thus unless one is considering extreme or pathological values of the problem, for instance $k_y = n = 0$ or $k_x L \ll 1$, then the inverse timescale of the planetary wave response is bounded from above by Ω_0 Ro.

The phase pattern of planetary waves always propagates westward, i.e., in the negative x direction. On the other hand, the energy associated with the Rossby wave propagates eastward if the zonal wavelength k_x is sufficiently large

$$\frac{\partial \omega}{\partial k_x} = \beta_0 \frac{k_x^2 - k_y^2 - k_z^2}{(k_x^2 + k_z^2 + k_z^2)^2}, \qquad k_z^2 \equiv \frac{1}{\text{Bu}^2} \frac{H_0^2}{H^2} \frac{n^2 \pi^2}{L^2}, \tag{5.136}$$

with similar considerations for the group velocities in the other directions.

Distinction is made between the different vertical overtones comprising planetary waves. The expression *barotropic waves* is used to designate $n = 0$ modes. This is used to indicate that the structure of the perturbations is uniform in the vertical direction. Indeed, in this case, the vertical velocity is zero and the motion is always geostrophic. The exactly two-dimensional motion recovers then the simple Rossby wave example we discussed in Sect. 5.4.2. For all values $n \geq 1$, the modes are referred to as *baroclinic waves* and this indicates that the structure of the dynamic pressure field has some vertical dependence. There is also a mathematical reason for calling them baroclinic. Recall that the general meaning of baroclinicity is defined to mean that isobars do not coincide with isopycnic surfaces, that is, when $\nabla P \times \nabla \rho \neq 0$. Supposing that the stream function, being a proxy for the pressure perturbation, is given by $\psi' = \cos(n\pi z/L)\sin(k_x x)\sin(k_y y)\cos(\omega t)$, then it follows that the dynamic density perturbation satisfies $\rho' \propto -n\sin(n\pi z/L)\sin(k_x x)\sin(k_y y)\cos(\omega t)$. Examining, for instance, the y component of the vector product of the density and pressure gradients shows that

$$\frac{\partial \rho'}{\partial x}\frac{\partial \psi'}{\partial z} - \frac{\partial \psi'}{\partial x}\frac{\partial \rho'}{\partial z} \sim \frac{kn^2}{2}\sin(2k_x)\sin^2(k_y y)\cos^2(\omega t) \neq 0, \tag{5.137}$$

except for isolated points in the domain. This is sufficient for baroclinicity. Low order baroclinic modes are important for planetary dynamics as they are the ones that traditionally give rise to weather in the midlatitudes. As it turns out, typical *thermal wind* profiles, which are zonal jet flows, i.e., flows of air that move in the east-west direction, with vertical shear, are unstable and it is these baroclinic modes that carry the instability. This effect is referred to as *baroclinic instability*.

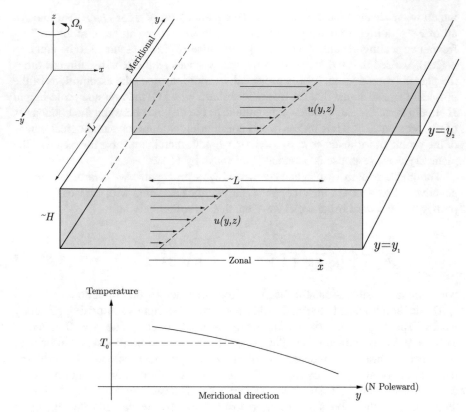

Fig. 5.5 Thermal wind profile in a representative section of a planet at latitude. A generalized zonal flow $u(y,z)$ is depicted. The graph below depicts the sense of the mean meridional gradient of temperature (density variation $\propto$-temperature variation)

5.5.5   Thermal Wind

We close this section with a short discussion of thermal wind (Fig. 5.5). It turns out that in the QG model, one can generate a configuration supporting the simplest vertically sheared flow. Consider now the diagnostic relationships between the pressure and the other relevant fluid variables: velocities u, v and the dynamic density variable ρ' detailed in Eq. (5.126). Taken as a mathematical system on its own merit, suppose that the dynamic density field possesses a latitudinal (y-directed) gradient. We write this as

$$\rho'_0 = \rho_{00}\left(1 + \delta y\right),$$

where if we are assuming dynamics on length scales of $\sim L$, then $\delta L \ll 1$.

This meridional density gradient could be due to a weak equator to pole temperature gradient. Recall that in the Boussinesq approximation $\rho' \approx -\beta_T \theta'$ where θ' is here the temperature perturbation and β_T is the familiar coefficient of thermal expansion at constant pressure. Taking the partial derivative with respect to the vertical coordinate z we get

$$\frac{\partial u_0}{\partial z} = -\frac{\partial}{\partial z}\frac{\partial \psi}{\partial y} = \frac{g}{2\Omega_{00}}\frac{1}{\rho_{00}}\frac{\partial \rho_0'}{\partial y} = \frac{g\delta}{2\Omega_{00}}. \tag{5.138}$$

Thus the zonal (east-west) mean velocity field possesses a vertical shear. Assuming that the ground provides some amount of frictional resistance as described for Ekman layers and spin-down in Sect. 5.3, we write the resulting solution for u, subject to the condition that u is zero at the ground $z = 0$, as

$$u_0 = \frac{g\delta}{2\Omega_{00}}z = \Lambda_{\text{tw}}z., \qquad \Lambda_{\text{tw}} \equiv \frac{g\delta}{2\Omega_{00}}. \tag{5.139}$$

The zonal flow u_0 is called the *thermal wind*. The flow at midlatitudes, known as the jet stream on the Earth, is intimately related to the thermal wind relationship. Because the equator to pole gradient of the temperature is negative, by the relationship $\rho' \propto -\theta'$ we deduce that δ must be positive. In turn, this means that $\Lambda_{\text{tw}} > 0$ and, consequently, the thermal wind *increases in the eastward direction* as one goes up in altitude. Thus the presence of an equator to pole variation in temperature competes with the tendency for air to remain still (and rotate with the rotating ground below) according to our intuition based on our discussion on the Ekman spin-down problem detailed in Sect. 5.3, in this chapter.

As we alluded to earlier, the appearance of weather arises from an instability associated with baroclinic planetary waves. J.G. Charney in 1947 and E. Eady a couple of years later published two of the earliest and simplest models of baroclinic instability. They analyzed the QG equations in the f-plane setting and showed the existence of instability when the basic state contains a thermal wind. The details of these models may be found in the *Bibliographic notes* in reference [7]. Figure 5.6 displays recent relevant simulations of global winds. The basic onset of the weather generating baroclinic instability is captured by the thermal wind model, but it does so by dispensing with other physical effects, including the latitudinal variation of the planetary rotation which affects the quantification of the resulting length scales from growth rates of the instability and the effective drag induced upon the flowing air due to Ekman spin-down. Since that time more sophisticated calculations have been performed, relaxing many of the assumptions made by Charney and Eady, leading to more accurate results conforming to observational data. For further details consult the references on geophysical flows found at the *Bibliographical Notes* at the end of the chapter.

Fig. 5.6 Visualization of global winds from a GEOS-5 simulation using 10-km resolution. Surface winds (0–40 m/s) are shown in white and trace features including Atlantic and Pacific cyclones. Upper-level winds (250 hectopascals) are colored by speed (0–175 m/s), with red indicating faster. The famous Jet stream is clearly visible (*Public Domain. Author:W. Puttnam/ NASA Goddard Space Flight Center. Courtesy of NASA -http://eol.jsc.nasa.gov/Info/use.html*)

5.6   An Astrophysical Example: Local Structure of Steady Thin Accretion Disks

Accretion disks are ubiquitous astrophysical objects formed by a gas, endowed with angular momentum with respect to a compact central object of mass M_* onto which it is falling, due to the object's Newtonian gravitational field. Accretion disks around *white dwarfs*[2] in cataclysmic variables and those around a freshly formed star, e.g., of the T-Tauri family, are classical examples. The central object is compact relatively to the extension of the fluid being gravitationally attracted to it. We shall limit ourselves in this example to *non-magnetic* accretion disks, that is, to those flows for which we may assume that the magnetic field is dynamically unimportant. Such disks may be found in close binary stars, when one of the stars overflows its Roche lobe, i.e., effectively, starts to lose mass to its more compact companion, in newly formed stars, which are still surrounded by the remains of the matter they were formed of, or even on much larger scales, when, e.g., a compact galactic nucleus attracts the gas of its neighborhood. In many cases of this kind, accretion gives rise to spectacular observable effects, as the disk and sometimes the central object emit prominent radiation. Indeed the first ideas about the existence of accretion disks were put forward in the beginning of the 1970s, in efforts to model the relatively recently observed compact galactic X-ray sources (Fig. 5.7).

---

[2] Very compact stars, their mass is similar to that of the Sun while their radius is approximately that of the Earth. Only neutron stars and black holes are more compact.

Fig. 5.7 Artist conception of an accretion disk around a white dwarf in a close binary. This particular system is WZ Saggitae, a cataclysmic dwarf nova, consisting of a white dwarf accreting from a low mass normal stellar companion (*Public domain. Author: P. Marenfeld and NOAO/AURA/NSF. Courtesy of NSF*)

The main subject of this book and its scope enforce space limitations preventing us from a fuller discussion on the formation of accretion disks and on the radiation they emit. This material can be found in a large number of existing books of which we shall mention, in the *Bibliographical Notes*, the best to our taste. Instead, we shall focus here on the description of disk models as accretion flows, after they have achieved cylindrically symmetric steady state. Thus to obtain the relevant equations for an accretion disk, we shall, in what follows, use Eqs. (5.2–5.4) and the continuity equation (5.5) in a nonrotating, that is, inertial, frame and with the body force, **b**, assumed to be derived only from the central object's gravitational potential, that is

$$\mathbf{b} = -\nabla \Phi(r,z); \quad \text{with} \quad \Phi(r,z) = -\frac{GM_*}{\sqrt{r^2 + z^2}}, \tag{5.140}$$

where G is the universal gravitational constant. In addition, we postulate, as is usual, that in a flow modeling an accretion disk, these equations have the following simplifying properties:

1. Steady state, that is, $\partial/\partial t = 0$,
2. Axial symmetry, that is, $\partial/\partial \varphi = 0$,
3. The flow appears viscous, despite its enormous Reynolds number, as defined using microscopic viscosity being enormous. In this statement, we mean that it is assumed that some kind of enhanced, probably turbulent, viscosity is present. Otherwise the gas will just orbit the central object.

We thus rewrite Eqs. (5.2–5.4) and the continuity equation (5.5) in an *inertial* cylindrical frame, and also replace the equation for the φ component of velocity using $u_\varphi \equiv r\Omega$, which defines the angular frequency Ω, by an equation for the angular momentum per unit mass, $ru_\varphi = r^2\Omega$. Ω is often referred to, incorrectly, as angular velocity but its units, radians/second, reveal that it is angular *frequency*.

The resulting set of equations reads

$$u_r\frac{\partial u_r}{\partial r} + u_z\frac{\partial u_r}{\partial z} - \Omega^2 r = -\frac{1}{\rho}\frac{\partial P}{\partial r} - \Omega_K^2(r)r\left(1+\frac{z^2}{r^2}\right)^{-3/2} + \frac{1}{\rho}f_r^{\text{visc}} \quad (5.141)$$

$$\rho u_r\frac{1}{r^2}\frac{\partial}{\partial r}\left(r^2\Omega\right) + \rho u_z\frac{\partial\Omega}{\partial z} = \frac{1}{r^3}\frac{\partial}{\partial r}\left(\nu_t\rho r^3\frac{\partial\Omega}{\partial r}\right) + \frac{\partial}{\partial z}\left(\nu_t\rho\frac{\partial\Omega}{\partial z}\right), \quad (5.142)$$

$$u_r\frac{\partial u_z}{\partial r} + u_z\frac{\partial u_z}{\partial z} = -\frac{1}{\rho}\frac{\partial P}{\partial z} - \Omega_K^2(r)z\left(1+\frac{z^2}{r^2}\right)^{-3/2} + \frac{1}{\rho}f_z^{\text{visc}}, \quad (5.143)$$

$$\frac{1}{r}\frac{\partial}{\partial r}(r\rho u_r) + \frac{\partial}{\partial z}(\rho u_z) = 0, \quad (5.144)$$

where $\Omega_K = GM_*/r^{3/2}$ is the Keplerian angular velocity at $r, z = 0$ induced by the central object. We posit the existence of an enhanced effective kinematic viscosity coefficient ν_{eff}, many orders of magnitude larger than the usual microscopic viscosity, which in this astrophysical flow should have a negligible effect. From now on we shall assume that ν_{eff} arises due to turbulence, and hence the notation $\nu_{\text{eff}} = \nu_t$. Thus, the above equation set is for appropriately averaged quantities in a turbulent flow (see Chap. 9). We also have to write out explicitly the contribution of the viscosity to the equations of motion

$$f_r^{\text{visc}} \equiv \frac{2}{r}\frac{\partial}{\partial r}\left(\nu_t\rho\frac{\partial u_r}{\partial r}\right) - 2\frac{\nu_t\rho u_r}{r^2} + \frac{\partial}{\partial z}\left[\nu_t\rho\left(\frac{\partial u_r}{\partial z} + \frac{\partial u_z}{\partial r}\right)\right] \quad (5.145)$$

$$f_z^{\text{visc}} \equiv 2\frac{\partial}{\partial z}\left(\nu_t\rho\frac{\partial u_z}{\partial z}\right) + \frac{1}{r}\frac{\partial}{\partial r}\left[r\nu_t\rho\left(\frac{\partial u_r}{\partial z} + \frac{\partial u_z}{\partial r}\right)\right], \quad (5.146)$$

which despite their units are *not* forces but dissipative terms. To bring out the typical properties of an accretion disk and in particular the *disparity of length scales* in the horizontal and vertical directions, we shall now nondimensionalize these equations by scaling all the physical variables, Ξ say, by their typical values, denoted by an upper tilde ($\tilde{\Xi}$). This is so, except for a number of selected quantities, among them the vertical (in z direction) scale-height, which is the typical disk half thickness and will be denoted by $\tilde{h}$, the radial variable $\tilde{r}$ which equals the radius of the accreting star, the angular rotation velocity Ω which will be scaled by its *Keplerian* value in the radial point $\tilde{r}$. This is done in the anticipation that this will be its typical value, thus $\tilde{\Omega} = \tilde{\Omega}_K \equiv \sqrt{GM_*/\tilde{r}^3}$. Finally, the radial and vertical velocities (u_r and u_z) will be scaled by the typical sound speed $\tilde{c}_s$ and the pressure by $\tilde{\rho}\tilde{c}_s^2$. Before effecting the scalings we note that in an accretion disk the azimuthal flow is highly supersonic, that is, $\tilde{u}_\varphi = \tilde{r}\tilde{\Omega} \gg \tilde{c}_s$. This result, which can be easily verified for real conditions in various astrophysical accretion disks is usually taken to mean that the disk is *cold* and it introduces a small, nondimensional parameter into the problem

$$\varepsilon \equiv \frac{\tilde{c}_s}{\tilde{\Omega}_K \tilde{r}} \ll 1. \tag{5.147}$$

We express the above postulated disparity of scales by writing $\tilde{h} = \varepsilon\tilde{r}$ and proceed to nondimensionalize the equations. The result up to second order in ε is

$$\varepsilon^2 u_r \frac{\partial u_r}{\partial r} + \varepsilon u_z \frac{\partial u_r}{\partial z} - \Omega^2 r = -\varepsilon^2 \frac{1}{\rho}\frac{\partial P}{\partial r} - \frac{1}{r^2} + \varepsilon^2 \frac{3z^2}{2r^4} + \varepsilon \frac{\partial}{\partial z}\left(v_t\rho\frac{\partial u_r}{\partial z}\right), \tag{5.148}$$

$$\varepsilon\rho u_r \frac{1}{r^2}\frac{\partial}{\partial r}\left(r^2\Omega\right) + \rho u_z \frac{\partial\Omega}{\partial z} = \varepsilon^2 \frac{1}{r^3}\frac{\partial}{\partial r}\left(v_t\rho r^3 \frac{\partial\Omega}{\partial r}\right), \tag{5.149}$$

$$\varepsilon u_r \frac{\partial u_z}{\partial r} + u_z \frac{\partial u_z}{\partial z} = -\frac{1}{\rho}\frac{\partial P}{\partial z} - \frac{z}{r^3} + \varepsilon^2 \frac{3z^3}{2r^5} + 2\varepsilon \frac{\partial}{\partial z}\left(v_t\rho\frac{\partial u_z}{\partial z}\right), \tag{5.150}$$

$$\varepsilon\frac{1}{r}\frac{\partial}{\partial r}(r\rho u_r) + \frac{\partial}{\partial z}(\rho u_z) = 0. \tag{5.151}$$

The units chosen for the viscosity are $\tilde{v}_t = \tilde{c}_s\tilde{h}$, anticipating the parametrization of this quantity, as a turbulent viscosity (see below). Notice that we have not included an energy equation. The reason for this is that in what follows we shall present only two *steady* accretion disk models, the first of which will be the vertically averaged model, based on the idea of N.I. Shakura and R.A. Sunyaev from 1973. The reference to the original paper can be found in several books in the *Bibliographical Notes*. The model is actually a lowest order approximation, in which the energy balance is straightforward. The second model will be based on an asymptotic expansion up to $\mathcal{O}(\varepsilon^2)$ and can be performed only if a relation $P(\rho)$ is given, e.g., a polytrope, and this allows us to circumvent the energy equation.

5.6.1   The Vertically Averaged Steady Model

This model actually derives from the lowest order terms of Eqs. (5.148–5.151) and the neglect of vertical motions $u_z = 0$ in view of the thinness of the disk. Thus we have

$$\Omega(r) = r^{-3/2}, \tag{5.152}$$

$$r\rho \frac{u_r}{\varepsilon}\frac{\partial}{\partial r}\left(r^2\Omega\right) = \frac{\partial}{\partial r}\left(v_t\rho r^3 \frac{\partial\Omega}{\partial r}\right), \tag{5.153}$$

$$\frac{\partial P}{\partial z} = -\frac{z^3}{r} \qquad (5.154)$$

$$\frac{\partial}{\partial r}(r\rho u_r) = 0. \qquad (5.155)$$

Several important comments should now be made. First and foremost it follows from (5.152) that the vertically averaged model's angular rotational velocity is Keplerian for all r. Thus it may seem that this rotating flow would not allow accretion. However, Eq. (5.153) can only hold if u_r is of $\mathscr{O}(\varepsilon)$. This means that the accretion demanding $u_r < 0$ is an $\mathscr{O}(\varepsilon)$ effect. To remember that u_r is of $\mathscr{O}(\varepsilon)$ we write $u_r = \varepsilon u_1$. Then we substitute it into Eq. (5.155) to find $r\rho u_1 = f_1(z)$, with the index 1 as a reminder that f was an $\mathscr{O}(\varepsilon)$ quantity. Now integration over z will give the total mass accretion rate, provided a factor 2π is inserted before. With $u_r < 0$, the result is

$$r\int_{-\infty}^{\infty}(\rho u_1)dz = -\dot{M}_1, \qquad (5.156)$$

where $\dot{M}_1$ is constant, representing the accretion rate, and the meaning of the index 1 should be self-evident. The factor 2π which is necessary here is absorbed in the units of $\dot{M}_1$. In the popular, by now, idea of parametrization of the turbulent viscosity, employing an unknown constant parameter $\alpha > 0$, Shakura and Sunyaev proposed to use

$$|\tau_{\varphi r}| \approx \alpha P, \qquad (5.157)$$

which approximately equates the absolute value of one component (φ, r) of the viscous stress tensor with αP, where P is the pressure, and neglects all other components of this tensor. α, the single unknown constant, was supposed to contain all our ignorance on the enhanced disk viscosity. In practice, when models of disks based on this prescription were constructed and the continuum radiation emanating from them was compared to disk observations, α was fitted to a constant, usually considerably less than 1. The prescription given above for parametrizing the viscosity is equivalent to setting $v_t = \alpha c_s h$. Can you show this equivalence with possible factors of $\mathscr{O}(1)$ being absorbed in a redefined new α? We shall refer to (5.157) or the equivalent explicit expression for v_t as the *ShS ansatz*. We have to include now the lowest order in ε term from the energy equation. The Shakura–Sunyaev vertically averaged model will follow. We write, in nondimensional units:

$$v_t \rho^2 \left(\frac{\partial \Omega}{\partial r}\right)^2 \approx \lambda \frac{\partial F^z}{\partial z}, \qquad (5.158)$$

where λ is a nondimensional constant, usually of $\mathscr{O}(1)$, and F^z is the vertical radiative flux out of the disk. Defining now the *surface density* of the disk as

$$\Sigma(r) = \int_{-\infty}^{\infty} \rho(r,z)dz \equiv 2\overline{\rho(r)}h(r), \tag{5.159}$$

where here also the averaged density $\overline{\rho(r)}$ is defined with the help of the disk half-thickness. It crudely follows from (5.154) to be

$$h(r) \approx \frac{c_s}{\Omega} = T^{1/2}r^{3/2}, \tag{5.160}$$

using the nondimensional relation between the sound speed and temperature. Integrating equations (5.158) and (5.153) now over z, we get as results, respectively,

$$\alpha v_t \Sigma r^2 \left(\frac{\partial \Omega}{\partial r}\right)^2 \approx 2\lambda F^z \tag{5.161}$$

and

$$-M_1 \frac{\partial}{\partial r}\left(r^2 \Omega\right) \approx \alpha \frac{\partial}{\partial r}\left(v_t \Sigma r^3 \frac{\partial \Omega}{\partial r}\right). \tag{5.162}$$

Perpetuating the habit of vertical averaging in disks, which introduces gross approximations, we use the appropriate EOS, and get $F^z \approx 0.5T^4/(\kappa\Sigma)$, with $\kappa = \rho T^{-3.5}$ as decided. It follows that

$$F^z \approx T^{7.5}h/\Sigma^2 \approx T^8 r^{3/2}/\Sigma^2. \tag{5.163}$$

Using now the Keplerian relation for $\Omega(r)$ and the ShS ansatz, Eq. (5.161) is brought to the form

$$T^7 \approx \left(\frac{9\alpha}{8\lambda}\right)\Sigma^3 r^{-3}. \tag{5.164}$$

Note that our units for v_t were $\tilde{c}_s \tilde{h}$ and thus the ShS ansatz brings out the parameter α. Integration of (5.162) requires an integration constant which has to be determined by the actual position where Ω achieves a maximum in a real disk, the *zero-torque* radius, from where Ω has to descend to its inner boundary value, which is presumably bound to happen for r close enough to $r = 1$. We assume here that this position is at $r = 1$ and the angular velocity maximum is its Keplerian value at $r = 1$. This inconsistency is assumed to hold up to $\mathcal{O}(\varepsilon)^2$ and we are careful not to take the results seriously as r becomes very close to 1. A number of algebraic manipulations finally give

$$T \approx A\alpha^{-1.5}\dot{M}_1^{3/10}r^{-3/4}(1 - r^{-1/2})^{3/10}, \tag{5.165}$$

$$\Sigma \approx B\alpha^{-4.5}\dot{M}_1^{7/10}r^{-3/4}(1 - r^{-1/2})^{7/10}, \tag{5.166}$$

with A and B appropriate nondimensional constants, depending on the constants introduced along the derivation. As hinted above, this solution ceases to be valid too close to the central object boundary and care has to be taken to treat that region in an appropriate way. Currently there is no general fully or even approximately analytical prescription for that. In any case, we shall not discuss here the problem of the boundary layer between the disk and the accreting object, always assuming that we are interested in points far enough away from the central object. Other variables can be found using the above derived relations, e.g., the inflow velocity is

$$u_1 \approx -\dot{M}_1/(r\Sigma), \tag{5.167}$$

etc. Note that we have used conditions appropriate for Kramers opacity and gas pressure dominance. Other cases can be treated in a similar way as well. In any case, we have worked in nondimensional units and the transformation to physical units may not be completely trivial. In what follows we shall give the relevant dimensional formulae.

For a typical disc in regimes as discussed here,[3] the dimensional results are

$$T \approx 1.4 \times 10^4 \alpha^{-1/5} \dot{M}_{16}^{3/10} R_{10}^{-3/4} m_1^{1/4} f^{6/5} \, \text{K}, \tag{5.168}$$

$$\Sigma = 5.2 \alpha^{-4.5} \dot{M}_{16}^{7/10} R_{10}^{-3/4} m_1^{1/4} f^{14/5} \, \text{g cm}^{-2}, \tag{5.169}$$

$$u_r \approx -2.7 \times 10^4 \alpha^{4/5} \dot{M}_{16}^{3/10} R_{10}^{-1/4} m_1^{-1/4} f^{-14/5} \, \text{cm s}^{-1}, \tag{5.170}$$

etc., where R_{10} is the radius in units of 10^{10} cm, $\dot{M}_{16}$ the mass accretion rate in units of 10^{16} g s^{-1} (within a reasonable range for cataclysmic variables) and with

$$f \equiv \left[1 - \left(\frac{R_*}{R} \right)^{1/2} \right]^{1/4}, \tag{5.171}$$

where R is the cylindrical radial distance, and R_* the central object's radius. Note the appearance of the factor m_1, which is the mass of the central object, expressed in solar masses ($m_1 \equiv M_*/M_\odot$). This factor did not appear in our previous nondimensional equations, because the mass of the central object entered only through the Keplerian angular velocity. Other quantities for the disk can be obtained from the above dimensional results, which are written for a mean molecular weight of $\mu = 0.615$, which is appropriate for a fully ionized gas having cosmic abundances. Finally we would like to justify our use of $\approx$ instead of $=$ in the equations. This stems from the *vertical averaging* procedure which holds well only if the disk is extremely thin. Otherwise we would estimate that the crude averaging causes inaccuracies up to factor $\mathcal{O}\left(\varepsilon^{1/2}\right)$ and this estimate is nothing but an educated guess.

---

[3] Recall: ideal gas pressure dominant, Kramers law of opacity.

5.6.2 The Asymptotic Polytropic Model with Vertical Structure

In this section we use the scaled equations (5.148–5.151). Specifically the scaling of P and ρ follow the barotropic relation $P(\rho)$, which actually will be an explicit polytropic one, as will be seen immediately below; thus, $\tilde{P} = P(\tilde{\rho})$. To be able to solve for the vertical structure one has the option to close the equation set, without the need of an energy equation. A natural option is that of a *polytrope*, that is assuming a specific form, which gives a polytrope of index n:

$$P(\rho) = \rho^{1+1/n}, \qquad c_s^2 = \frac{dP}{d\rho} = \left(1 + \frac{1}{n}\right)\frac{P}{\rho} = \left(1 + \frac{1}{n}\right)\rho^{1/n}. \qquad (5.172)$$

The asymptotic expansion up to $\mathcal{O}\left(\varepsilon^2\right)$ was first done in 1997 by W. Kluźniak and D. Kita and we shall try to summarize here their work. In what follows, we shall make use of c_s as the dependent variable, instead of the pressure P. Consequently, the pressure gradient terms will be replaced in the following way

$$\frac{1}{\rho}\frac{\partial P}{\partial r} = n\frac{\partial c_s^2}{\partial r}, \frac{1}{\rho}\frac{\partial P}{\partial z} = n\frac{\partial c_s^2}{\partial z}. \qquad (5.173)$$

In addition, for the sake of convenience, we make the replacements $u_r \mapsto u$ and $u_z \mapsto v$ and finally we opt for the following expression for the *dynamic viscosity*, following from the ShS prescription

$$\eta_t = \frac{2}{3}\alpha\frac{P}{\Omega_K} = \frac{2}{3}\alpha\frac{\rho c_s^2}{\Omega_K (1+1/n)}. \qquad (5.174)$$

As discussed before, this is equivalent up to $\mathcal{O}(1)$ multiplicative constant to the choice of $\nu_t = \alpha c_s h$ for the kinematic viscosity because the multiplicative constant may be absorbed in a "new" α parameter. The units of the *dynamic* viscosity coefficient are $\tilde{\rho}\tilde{c}_s\tilde{h}$.

Formally, we should now expand all quantities in asymptotic series in ε and substitute it into the appropriate equations. It has been found in the previous subsection that the leading order of u is ε and, as we can see from the scaled continuity equation (5.155), v is $\mathcal{O}\left(\varepsilon^2\right)$ in leading order. It can also be shown that the leading orders of Ω, c_s^2 and ρ should all be odd for the purpose of balancing the various orders of the equations. Thus in the expansions only the following terms contribute

$$\Omega(r,z) = \Omega_0(r) + \varepsilon^2\Omega_2(r,z) + \mathcal{O}\left(\varepsilon^4\right), \qquad (5.175)$$

$$u(r,z) = \varepsilon u_1(r,z) + \mathcal{O}\left(\varepsilon^3\right), \qquad (5.176)$$

$$v(r,z) = \mathcal{O}\left(\varepsilon^2\right), \qquad (5.177)$$

$$c_s^2 = c_{s0}^2(r,z) + \mathcal{O}\left(\varepsilon^3\right), \qquad (5.178)$$

$$\rho(r,z) = \rho_0(r,z) + \mathcal{O}\left(\varepsilon^3\right). \tag{5.179}$$

Substituting these expansions into Eqs. (5.152–5.155) and separating the different orders we notice that $\mathcal{O}(1)$ equations can be straightforwardly solved to yield solutions well-known in the literature. These are

$$c_{s0}^2(r,z) = \frac{h^2(r) - z^2}{2nr^3}, \quad \Omega_0(r) = \Omega_K(r) = r^{-3/2}, \quad \rho_0(r,z) = \left(\frac{nc_{s0}^2}{n+1}\right)^n, \tag{5.180}$$

where $h(r)$ the disk half-thickness will be determined by the solution of the next nontrivial order, the second. It is assumed that at $z = \pm h(r)$, the disk upper and lower boundary, the lowest order density ρ_0, and the sound speed c_{s0} vanish and remain zero for any $|z| > h(r)$. Also, the following relations hold:

$$\frac{P_0}{\rho_0} = \frac{nc_{s0}^2}{n+1}, \quad \frac{1}{\rho_0}\frac{\partial P_0}{\partial z} = -\frac{z}{r^3}. \tag{5.181}$$

Note that for the sake of convenience we have not expanded the viscosity coefficient in powers of ε. It will be taken as composed of zero order quantities only, thus, following Eq. (5.174) and using (5.181) we have for the dynamic viscosity coefficient

$$\eta_t = \frac{2}{3}\alpha P_0 r^{3/2}. \tag{5.182}$$

Before turning to the next nontrivial order we fix the polytropic index $n = 3/2$, a reasonable choice giving $P \propto \rho^{5/3}$, in order to simplify the form of the formulae. Equation of $\mathcal{O}(\varepsilon)$ does not give any additional nontrivial information, while $\mathcal{O}(\varepsilon^2)$ gives rise, after a considerable amount of algebra, to solutions for $u_1(r,z)$ and $v_2(r,z)$, as well as $h(r)$. It turns out that $h(r)$—note that similarly to the viscosity coefficient it is not expanded and taken to be equal to its lowest order—depends on the constant mass accretion rate. The latter can be easily related to u, ρ, and h by vertically integrating the continuity equation. It turns out that if one does this, paying attention to correct orders, it turns out that the mass accretion rate is an $\mathcal{O}(\varepsilon)$ quantity, given by

$$\dot{M}_1 = -2\pi \int_{-h}^{h} dz(r\rho_0 u_1) = \text{const}. \tag{5.183}$$

The constant $\dot{M}_1$ is an important parameter of the problem and serves as a constraint in the evaluation of the solutions, e.g., of $h(r)$. The latter can be derived in an asymptotically meaningful and valid way only if r is significantly distant from the above-mentioned *zero torque* radius, which we denote here by r_0. We also leave the issue of the difficulty encountered in the region between r_0 and the actual accreting object because, as indicated above, the boundary layer problem has not yet been successfully solved in any general way and numerical simulations of various cases suggest different results. The understanding is, at best, only qualitative. Thus we assume that r is significantly *larger* than r_0. Similarly we do not have detailed outer boundary condition, only the total mass influx constraint. Even though a solution can be found under this condition, it is not clear to what extent it is valid close to the disk edge. Consequently, we assume that r is also sufficiently *smaller* than the disk edge.

It can be shown, consult the *Bibliographical Notes* for references, that there is a value of α below which a certain amount of *back-flow* in the meridional flow pattern exists. This means that a stagnation radius exists, beyond which the radial velocity near the disk mid-plane is directed outwards. Finally we quote here the result for $h(r)$, for the polytropic index $n = 3/2$:

$$h(r) = (2\Lambda)^{1/6} r \left(1 - \sqrt{\frac{r_0}{r}} \right)^{1/6}, \qquad (5.184)$$

where

$$\Lambda \equiv \frac{0.73}{\Gamma(5/2)} \dot{M}_1 \alpha^{-1}.$$

It is not clear if the back-flows resulting from this analytical procedure do indeed exist in thin accretion disks and what may be their observational significance. We have decided to sketch the asymptotic procedure, not to achieve an explanation of an observation of some particular object. Rather, we wanted to show the power of advanced analytical techniques, which however limited they may be are completely transparent to the physical assumptions and properties of the model. This cannot usually be said about full-fledged numerical simulation.

5.6.3   Summary

Astrophysical accretion disks are still not well understood, despite their ubiquity. A lot of knowledge has accumulated since the first suggestions of their existence, most of it through crude approximations, guided by observations, which went a long way towards the elucidation of the formation of disks when matter possessing angular momentum is gravitationally attracted by a relatively compact massive object. The main difficulty is that the basic processes take place in different directions. This is in contrast to stars, which are much better understood, at least in their steady states. In accretion discs, which are best described in cylindrical coordinates (r, z, φ), the angular momentum transfer, indispensable for accretion, proceeds in the r direction, radiation emanating from the disk material heated by dissipation is transferred through the disk mainly in the z direction, while the main energy source resulting from the gravitational pull of the central object is being expressed by usually supersonic rotation in the φ direction with close to Keplerian shear. In addition, there exists the issue of the enhanced viscosity, since molecular viscosity cannot even come close to enabling accretion, as it is deduced from observed total accretion luminosity. It seems that the consensus is that accretion disks are turbulent and hence possess turbulent viscosity. The writers of this book have not yet seen convincing physical or mathematical description why a linear instability is needed to cause turbulence in a supersonically rotating very large $\mathrm{Re} > 10^{10}$ sheared cosmic body. More surprisingly, even in more simple and controlled laboratory shear flows, it is known that linear theory fails to explain the development of turbulence as can be studied in, e.g., references [14, 15, 16] in the *Bibliographical Notes*. Thus perhaps all the controversies regarding the magnetic source of the instability, which were at the center stage of the research effort for the past 20 years or so, are not the heart of the matter at all.

Problems

5.1.

An interesting model problem which is a variation of the simple Ekman layer is the one in which a flat layer of liquid, taken to be a model of an ocean, lies beneath a wind field exerting tangential stresses on the liquid layer. Assume the liquid is described in Cartesian geometry and the global rotation vector is aligned with the vertical direction, i.e., $\Omega = \Omega_0 \hat{z}$. Suppose that in the deep ocean (as $z \to -\infty$) the horizontal velocity field is given by $u = u_0$ and $v = v_0$. Assume that there is no vertical velocity. Consequently, the equations of motion for the interior of the fluid are similar to those developed in Sect. 5.2.3

$$-2\Omega_0(v - v_0) = \nu \frac{\partial^2 u}{\partial z^2}, \qquad 2\Omega_0(u - u_0) = \nu \frac{\partial^2 v}{\partial z^2}. \tag{5.185}$$

The boundary condition at $z = 0$ is set by applying given tangential stresses across the normal to the surface. Suppose therefore that the tangential stresses induced by the wind are $\tau_{xz} = \tau_x$ and $\tau_{yz} = \tau_y$ where τ_x and τ_y are constants. Matching of the stresses across the boundary means imposing the boundary condition

$$\tau_x = \rho \nu \frac{\partial u}{\partial z}, \qquad \tau_y = \rho \nu \frac{\partial v}{\partial z}, \qquad \text{at} \quad z = 0. \tag{5.186}$$

Show that the solutions for u and v are

$$u - u_0 = \frac{\sqrt{2}}{2\Omega_0 H} e^{z/H} \left[\tau_x \cos\left(\frac{z}{H} - \frac{\pi}{4}\right) - \tau_y \sin\left(\frac{z}{H} - \frac{\pi}{4}\right) \right], \tag{5.187}$$

$$v - v_0 = \frac{\sqrt{2}}{2\Omega_0 H} e^{z/H} \left[\tau_x \sin\left(\frac{z}{H} - \frac{\pi}{4}\right) + \tau_y \cos\left(\frac{z}{H} - \frac{\pi}{4}\right) \right]. \quad (5.188)$$

Keeping in mind that the Ekman layer depth H is proportional to $\sqrt{v}$ what is the relationship between the deep ocean current and the surface current as $v \to 0$? Does this surprise you? (See the book of Cushman-Roison and Beckers, given in the *Bibliographical Notes*, for further reference.)

5.2.
Re-examine the previous problem but, instead of the assumptions there, assume the following: the liquid below has density ρ_- and that the air above has density ρ_+. Assume also that the stresses that drive the heavier liquid below are derived from an airflow above. Suppose that the airflow above has a velocity field $u = u_0^{(+)}$ and $v = v_0^{(+)}$ as $z \to \infty$. Suppose the liquid below, similarly, has a current $u = u_0^{(-)}$ and $v = 0$ as $z \to -\infty$. Solve the flow equations in each separate region, $z > 0$ and $z < 0$, i.e.,

$$-2\Omega_0 (v - v_0^{(\pm)}) = v \frac{\partial^2 u}{\partial z^2}, \qquad 2\Omega_0 (u - u_0^{(\pm)}) = v \frac{\partial^2 v}{\partial z^2}, \quad (5.189)$$

assuming that the viscosities of the two fluids are equal. Complete the solution by matching the stresses and velocities across the surface $z = 0$:

$$\rho_- v \frac{\partial u}{\partial z}\bigg|_{z \to 0^-} = \rho_+ v \frac{\partial u}{\partial z}\bigg|_{z \to 0^+} \quad (5.190)$$

and similarly for v. Plot the solutions using appropriate computer software.

5.3.
Show that the total perturbation energy for linearized disturbances of an inviscid plane-parallel Boussinesq atmosphere in a rotating frame is identical to that when the atmosphere is not rotating. In other words, show that the total energy is still given by (4.98). Similarly, prove the assertion made that the group velocity propagates in a direction perpendicular to the direction of phase propagation.

5.4.
Consider an inviscid plane-parallel atmosphere in a rotating frame Ω_0. Assume that the equations governing three-dimensional disturbances are given by those found in Sect. 5.1.1.2. Suppose that there exists a global uniform background shear in the problem in which the mean velocity along the $\hat{y}$ direction is given by $v_0 = \Lambda x$. Examine the axi-symmetric perturbations of this system, i.e., those in which $\partial_y = 0$, in a periodic domain in z and x. Show that the total perturbation energy $\mathscr{E}'$ is given by

$$\mathscr{E}' = \rho_0 \frac{u'^2 + v'^2(2\Omega_0 + \Lambda)/2\Omega_0 + w'^2}{2} + \frac{\rho_0 g \alpha}{\beta} \frac{s'^2}{2}, \quad (5.191)$$

and is conserved.

5.5.

Examine the dispersion relation for the previous problem. Assume normal mode perturbations of the form $\sim \exp(k_x x + k_z z - i\omega t)$. Show that ω has the form

$$\omega^2 = \frac{2\Omega_0(2\Omega_0 + \Lambda)k_z^2 + g\alpha\beta k_x^2}{k_x^2 + k_z^2}. \tag{5.192}$$

Is the group velocity, indicating the propagation of the total disturbance energy, still perpendicular to the direction of the phase velocity $\mathbf{V}_p = \omega\mathbf{k}$? Repeat the procedure executed in the analysis regarding the direction of energy propagation in the simpler Boussinesq atmosphere found in Sect. 4.4.3. Determine that the group velocity is indeed given by $\mathbf{V}_g \equiv (\partial\omega/\partial k_x)\hat{\mathbf{x}} + (\partial\omega/\partial k_z)\hat{\mathbf{z}}$. Also, what does it mean if $\Lambda = -2\Omega_0$?

5.6.

Starting from the equations of motion for an inviscid, incompressible, two-dimensional fluid exhibiting variations in x and y only, with a global rotation vector $\Omega = (\Omega_{00} + \beta_0 y/2)\hat{\mathbf{z}}$ where Ω_{00} and β_0 are constants, show that the expression

$$\frac{D\zeta_{tot}}{Dt} = 0, \quad \text{with} \quad \omega_{tot} \equiv \frac{\partial v}{\partial x} - \frac{\partial u}{\partial y} + 2\Omega_{00} + \beta_0 y, \tag{5.193}$$

where ζ is the z component of the vorticity, is correct. How is this related to Crocco's theorem? Consider why and how this is another example of a special restricted flow, like those we have discussed in Chap. 2.

5.7.

We examine here the properties of *topographic waves* in the context of shallow water systems. To begin, it is not difficult to see how a shallow water system (Sect. 4.3) generalizes to include Coriolis effects leading to coupled equation set

$$\frac{Du}{Dt} - 2\Omega_0 v = -\frac{\partial(gh)}{\partial x}, \quad \frac{Dv}{Dt} + 2\Omega_0 u = -\frac{\partial(gh)}{\partial y}, \quad \frac{Dh}{Dt} = -h\left(\frac{\partial u}{\partial x} + \frac{\partial v}{\partial y}\right), \tag{5.194}$$

where Ω_0 and g are constants and where h is the layer height of the constant density fluid.

(a) Show that there exists a material invariant of the system which is like potential vorticity in which

$$\frac{D}{Dt}\left(\frac{2\Omega_0 + \zeta}{h}\right) = 0, \quad \zeta \equiv \frac{\partial v}{\partial x} - \frac{\partial u}{\partial y}. \tag{5.195}$$

Argue that the implication of this is depicted in Fig. 5.8 wherein if a given vortex column drifts in the direction of y where the layer thickness h changes, then the vorticity of the column must change in proportion to leave the quantity $\Xi \equiv (2\Omega_0 + \zeta)/h$ unchanged. Vertical variations in h means stretching of the vortex column and, consequently, spin-up or spin-down.

(b) Consider dynamics which is slow in comparison with gravity waves (see also next problem). This is done by approximating the velocity fields by dropping the inertial terms. Furthermore, approximate the mean height as $h = h_0 + \gamma y + h'(x,y,t)$ where h' is small in comparison with the mean height h_0 and where $\gamma \ll 1$. With

$$2\Omega_0 v \approx \frac{\partial (gh)}{\partial x}, \quad 2\Omega_0 u \approx -\frac{\partial (gh)}{\partial y}, \tag{5.196}$$

show that (5.195) may be approximately rewritten as

$$\left(\frac{\partial}{\partial t} + u'\frac{\partial}{\partial x} + v'\frac{\partial}{\partial y}\right)\left(\frac{\partial^2 h'}{\partial x^2} + \frac{\partial^2 h'}{\partial y^2} - \frac{4\Omega_0^2}{gh_0}h' - \frac{4\Omega_0^2}{gh_0}\gamma y\right) \approx 0, \tag{5.197}$$

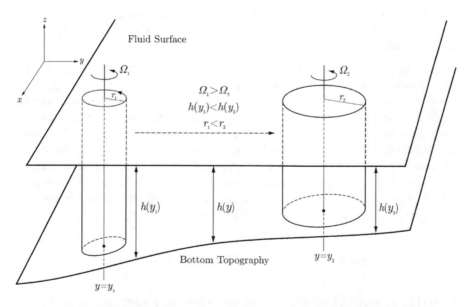

Fig. 5.8 Schematic depicting variations of bottom topography. Vortex column of cylindrical radius r_1 and height $h(y_1)$ with rotation scale Ω_1 is shown. If column drifts rightward into a shallower zone, then the column becomes fatter and spins slower

where

$$v' \approx \frac{g}{2\Omega_0}\frac{\partial h'}{\partial x}, \qquad u' \approx -\frac{g}{2\Omega_0}\left(\frac{\partial h'}{\partial x}+\gamma\right). \tag{5.198}$$

Note that $\overline{U} \equiv -g\gamma/2\Omega_0$ is a constant drift velocity.

(c) Introduce linear periodic perturbations of the form $h' = \hat{h}e^{ik_x x + ik_y - i\omega t}$ into this set. Show that the frequency ω satisfies

$$\omega = k_x \overline{U} - \frac{2\Omega_0 \gamma k_x}{k_x^2 + k_y^2 + 4\Omega_0^2/(gh_0)}, \tag{5.199}$$

and conclude that this form is mathematically similar to other Rossby wave dispersion relationships.

(d) Rationalize the meaning of these results by considering that as a vortex column moves in the *y-direction*, it becomes vertically stretched. Because the basic shallow water fluid is incompressible, as the column vertically stretches it must respond by spin-up to conserve the total potential vorticity which is another way of saying that the column must retain its total angular momentum.

5.8.
Beginning with the assumptions 1–6 laid out in the beginning of Sect. 5.5.1, perform a scaling analysis of (5.19–5.21) and (5.23) and show that they lead to the intermediate set (5.110–5.112) and (5.115–5.116).

5.9. In the discussion of planetary waves (Sect. 5.5.4), we assumed there that the atmosphere lies between rigid walls located at $z = 0, H$, which is an unrealistic boundary condition for such flows. As before, assume N^2 is a constant. Suppose now that the boundary condition for the top of the atmosphere is such that the total pressure is fixed on the moving boundary at $z = H$. In the linearized limit, this means that the Lagrangian pressure perturbation is zero at $z = H$. This is the same kind of boundary condition as used in the problem of surface water waves in Chap. 4. Supposing that $\eta'(x,y,t)$ represents perturbations of the initially level surface at $z = H$, argue that the boundary condition enforcing fixed pressure perturbation on the moving boundary is given, after linearization, by

$$P' - \rho_{00}g\eta' = \frac{\partial}{\partial t}\left(\frac{\partial \psi'}{\partial z}+\frac{N_0^2}{g}\psi'\right) = 0, \qquad \text{at} \qquad z = H.$$

Enforcing no normal flow at $z = 0$ and assuming the Fourier normal mode form $e^{ik_x x + ik_y y - i\omega t}$, show that the planetary wave frequency is

$$\omega = -\frac{\beta_0 k_x}{k_x^2 + k_y^2 + k_z^2 4\Omega_{00}^2/N_0^2}, \tag{5.200}$$

in which k_z is the solution of

$$\tan k_z H - \left(\frac{H}{H_0}\right) \frac{1}{k_z H} = 0, \qquad H_0 \equiv \frac{g}{N_0^2}. \qquad (5.201)$$

Argue that there are no barotropic modes for these boundary conditions. Graphically depict the solution of the condition for k_z. Argue that high overtone solutions are $k_z(n) = n\pi/H_0$, for $n \gg 1$.

5.10.
An analysis of the equations of quasi-geostrophy (5.128) can be simplified by considering a *single-layer* kind of model. The idea is to examine the behavior of atmospheric Taylor–Proudman columns which are vortex structures that are vertically uniform. However, the distinguishing feature of the quasi-geostrophic model is that it preserves in its description the physics of vertical stretching (e.g., see Problem 5.7). In order to keep this physical effect in a meaningful reduction, the second order partial derivative term with respect to z in Eq. (5.128) must somehow be retained.

Assume that the vertical scale of the layer is L_z and, furthermore assume that Ω_{00} and N are positive constants and that the analysis is restricted to the f-plane. Consequently, the vertical stretching effect is retained in *an effective two-dimensional reduction* by rewriting the stretching term as

$$\frac{\partial}{\partial z} \frac{4\Omega_{00}}{N^2} \frac{\partial \psi}{\partial z} \mapsto -\frac{1}{L_D^2}\psi, \qquad \frac{1}{L_D^2} \equiv \frac{8\Omega_{00}^2}{N^2 L_z^2}.$$

The quantity L_D is called the Rossby deformation radius. Thus the single layer geostrophic model equation set is

$$\frac{Dq}{Dt} = 0, \qquad q = \frac{\partial^2 \psi}{\partial x^2} + \frac{\partial^2 \psi}{\partial y^2} - \frac{\psi}{L_D^2}, \qquad u = -\frac{\partial \psi}{\partial y}, \qquad v = \frac{\partial \psi}{\partial x}. \qquad (5.202)$$

1. Show that the discretization of the vertical stretching term is consistent with a second order correct finite difference scheme for ∂_z^2.
2. Following the ideas discussed in Sect. 2.5.3, assume that the single layer is populated by a collection of N geostrophic point vortices labelled by the index j where each point vortex has strength Γ_j. Thus

$$\zeta = \sum_{j=1}^{N} \Gamma_j \delta(\mathbf{x} - \mathbf{x}_j)$$

wherein $\mathbf{x}_j = (x_j, y_j)$ is the x and y time dependent position of vortex j. Show that the total stream function and corresponding velocity fields are given by

$$\psi = \frac{1}{2\pi} \sum_{j=1}^{N} \Gamma_j K_0 \left(\frac{|\mathbf{x} - \mathbf{x}_j|}{L_D} \right) \tag{5.203}$$

together with

$$\frac{dx_i}{dt} = \frac{1}{2\pi} \sum_{j \neq i}^{N} \frac{\Gamma_j}{L_D} K_1 \left(\frac{|\mathbf{x} - \mathbf{x}_j|}{L_D} \right) \frac{(y_i - y_j)}{|\mathbf{x} - \mathbf{x}_j|},$$

$$\frac{dy_i}{dt} = -\frac{1}{2\pi} \sum_{j \neq i}^{N} \frac{\Gamma_j}{L_D} K_1 \left(\frac{|\mathbf{x} - \mathbf{x}_j|}{L_D} \right) \frac{(x_i - x_j)}{|\mathbf{x} - \mathbf{x}_j|} \tag{5.204}$$

where $K_\nu(x)$ is the modified Bessel function or order ν.

3. By examining the asymptotic behavior of the modified Bessel function K_ν, argue that geostrophic vortices do not appreciably influence each other if they are separated by a distance that is much greater than L_D.

4. Show that geostrophic vortices possess Hamiltonian structure. Namely, show that the equations of motion for the position of the vortex collection may be written as

$$\mathcal{H} = \frac{1}{4\pi} \sum_{\substack{i,j \\ i \neq j}} \Gamma_i \Gamma_j K_0 \left(\frac{|\mathbf{x}_i - \mathbf{x}_j|}{L_D} \right), \qquad \Gamma_i \frac{dx_i}{dt} = \frac{\partial \mathcal{H}}{\partial y_i}, \qquad \Gamma_i \frac{dy_i}{dt} = -\frac{\partial \mathcal{H}}{\partial x_i},$$

$$\tag{5.205}$$

and thereby showing explicitly the Hamiltonian structure.

Bibliographical Notes

General

It is not common to find books entirely devoted to rotating flows but we list a few to our liking:

1. H.P. Greenspan, *The Theory of Rotating Fluids* (Breukelen Press, Brookline, 1990)
2. P.R.N. Childs, *Rotating Flow* (Elsevier, Amsterdam, 2011)

Section 5.1

The equations of motion in curvilinear coordinates can be found in many texts. The following text includes very valuable appendices on coordinate forms and encompasses a thorough discussion of rotating self-gravitating masses and stars.

3. J.-L. Tassoul, *The Theory of Rotating Stars* (Princeton University Press, Princeton, 1978)

Section 5.2

The basic flows described in this section can be found in many texts. We find the following book very clear in its presentation of these topics.

4. P.K. Kundu, I.M. Cohen, *Fluid Mechanics*, 4th edn. (Academic Press, New York, 2008)

Section 5.3

The matched asymptotic expansion as a strategy in constructing a proper boundary layer solution can be found in

5. M. Van Dyke, *Perturbation Methods in Fluid Mechanics* (Parabolic Press, Stanford, 1975)

in addition to the books by A.H. Nayfeh and J.Kevorkian and J.D. Cole, referenced in the previous chapter.

Sections 5.4 and 5.5

There are several well-known books on GFD. Of the ones of current popular use are

6. G.K. Vallis, *Atmospheric and Oceanic Fluid Dynamics: Fundamentals and Large-Scale Circulation* (Cambridge University Press, Cambridge, 2006)
7. J. Pedlosky, *Geophysical Fluid Dynamics* (Springer, New York, 1987)
   An up-to-date account of geophysical fluid dynamics with a well-balanced discussion between numerical methods and theory is
8. B. Cushman-Roisin, J.-M. Beckers, *Geophysical Fluid Dynamics: Physical and Numerical Aspects* (Academic Press (Elsevier), Waltham, 2011)
   A finely succinct discussion on GFD encompassing both ocean and atmospheric dynamics is

9. J.C. McWilliams, *Fundamentals of Geophysical Fluid Dynamics* (Cambridge University Press, Cambridge, 2006)

Section 5.6

There are several well-known books including a discussion of accretion disks FD. We shall not cite here the original articles, which can be found as references in the books:

10. J. Frank, A. King, D.Raine, *Accretion Power in Astrophysics, 3rd edn.* (Cambridge University Press, Cambridge, 2002)
11. C.J. Clarke, R.F. Carswell, *Astrophysical Fluid Dynamics* (Cambridge University Press, Cambridge, 2007)
12. S.N. Shore, *Astrophysical Hydrodynamics* (Wiley-VCH, New York, 2007)
    Since a typical accretion disk is thin, starting from 1984 there were several attempts to develop semi-analytical theoretical models, using asymptotic expansions. For a fairly recent work in this line see this paper, where the early asymptotic works of Regev and of Kluźniak and Kita are referenced,
13. P. Rebusco, O.M. Umurhan, W. Klu'zniak, O. Regev, Global transient dynamics of 3-D hydrodynamical disturbances in thin accretion disks. Phys. Fluids **21**, 076601 (2009)
    For some other prominent works on turbulence developing without linear instability not in the case of accretion disks, see, e.g.,
14. F. Waleffe, Hydrodynamic stability and turbulence: beyond transients to a self-sustaining process. Stud. Appl. Math. **95**, 319 (1995)
15. S.J. Chapman, Subcritical transition in channel flows. J. Fluid Mech. **451**, 35 (2002)
16. T. Mullin, Experimental studies of transition to turbulence in a pipe. Ann. Rev. Fluid Mech. **43**, 1 (2011)

Chapter 6
Effects of Compressibility

And I will harden Pharoah's heart,
and multiply my signs and my wonders
in the land of Egypt.

Exodus 7, 3

6.1 Introduction

All fluids in nature are compressible, that is, if enough pressure is applied on a macroscopic fluid element, its volume decreases. Gases are usually much more easily compressed than other fluids and this is probably the reason that the regime of compressible flows is usually known as *gas dynamics*. We have, however, discussed in Chap. 2 the conditions under which the assumption of incompressibility of a particular *flow* is a good approximation and because it simplifies the equations considerably, we could consider there a number of important incompressible flows. The explicit conditions were summarized in Sect. 2.4.1, even though various considerations regarding this issue appear in other places of Chap. 2 as well. Moreover, it may happen that these explicit conditions for incompressibility are not met, even approximately, but it is still useful to assume incompressibility. Such is the case when the physical phenomena under study do not rely at all on the fluid property of compressibility.

In this chapter we shall consider, in contrast, flows for which compressibility is a crucial ingredient of the flow in question, that is, we shall examine physical phenomena which exist *because* of compressibility. For example, if we wish to study the physics of a phenomenon whose typical timescale is τ, say, and compression is

© Springer Science+Business Media, LLC 2016
O. Regev et al., *Modern Fluid Dynamics for Physics and Astrophysics*,
Graduate Texts in Physics, DOI 10.1007/978-1-4939-3164-4_6

instrumental in causing this phenomenon, we may describe it by the approximate size of the following nondimensional ratio, where ρ is a typical density of the fluid involved

$$\Delta_c \approx \frac{\tau}{\rho} \frac{D\rho}{Dt}. \tag{6.1}$$

If $\Delta_c \gg 1$, it is transparently physically understood that the process in question is faster than typical compression time and hence it is not reasonable to ignore compressibility. As we have already discussed in Chap. 2, compressibility can by no means be neglected, without losing the paramount effects of the flow, when the flow is *supersonic*, that is, the Mach number exceeds unity somewhere in the flow. So the Mach number is the concise and important diagnostic quantity. Defining Δ_c was an attempt to give, perhaps, a more physical formulation. All this is not to say that only in supersonic flows compressibility must be taken into account. In any case, in compressible flow the *sound speed* plays an important role. In particular, the value of the Mach number is proportional to the speed at which the quality of compressibility spreads out in the fluid. As an example, consider the motion of solid body, e.g., a wall in a fluid, so that its motion is supersonic. It would compress the fluid on the one side, while on the other side it cannot decompress, because the obstacle moves supersonically, while the decompression can proceed only at a sound speed. In approximations, which allow incompressibility, the sound speed c_s is nominally infinite, or at least much larger than any velocity scale in the flow. We saw in Chap. 4 a discussion of gravity waves on incompressible shallow water, defining a velocity of linear waves on the surface. As we shall see in Sect. 6.4.5 of this chapter, there is a remarkable analogy between these waves and one-dimensional compressible gas dynamics. The nondimensional number which applies in the case of shallow water gravity waves is the *Froude* number, mentioned previously in Sect. 1.6. In the present context we shall slightly modify the form of the Froude number as follows:

$$\mathrm{Fr} \equiv \frac{U}{\sqrt{gh}}, \tag{6.2}$$

with U being a typical fluid velocity and h the water depth, where now $c_g = \sqrt{gh}$ plays the role of sound speed in the above analogy. It is also easy to see that Fr for waves on shallow water plays a similar role to the Mach number, M, in the analogous compressible gas dynamics, which we shall discuss in detail below.

6.1.1 A Historical Note

We start this chapter on a historical note, assuming that the reader of the book is generally familiar with basic concepts related to fluid compressibility (e.g., sound waves, shocks), but is perhaps less familiar with the development of the ideas related to supersonic flows, and especially supersonic flight. These phenomena will be explained in more depth later in the course of the chapter.

The least violent manifestations of compressibility are small amplitude sound waves. Their study goes under the name of *acoustics*. Luminaries of no less a caliber than Pythagoras, Euclid, Aristotle, G. Galilei, R. Boyle, I. Newton, and Lord Rayleigh found time to deal with the phenomenon of sound, as its importance was self-evident, because it had connection with one of the human senses. The history of acoustics contains, almost solely, the phenomenon of sound waves in the air. Pythagoras produced sound waves by plucking a single string at constant tension, tied at both ends. He changed the length of the string and discovered that by halving the length of the string the musical nature of the note changes by an octave, i.e., the frequency of the sound just doubles. Pythagoras had, undoubtedly, a musical ear. He was fascinated by this phenomenon and it was a factor in his conviction about the connection of physics to music, which gave rise to his notion of "harmony of the spheres," an expression which probably inspired its use by Shakespeare and Kepler. Both Aristotle and Euclid tried to establish a quantitative understanding of sound propagation, but their progress was not significant, save perhaps the finding, not of Aristotle, but of his successor Theophrastus, that the speed of sound is independent of frequency, in contradiction to what Aristotle had stated. Galileo did a thorough study of vibrating strings, summarizing it in a self-contained theoretical framework. He also supported the idea that air is needed for the propagation of sound, like Aristotle had reasoned. Despite some controversy on this matter, which was largely based on faulty experiments, it was Robert Boyle who decided the issue. Theoretical physical acoustics began with Newton, who found that the speed of sound in a gas is $c_s \sim \sqrt{P/\rho}$, and culminated in the monumental work of Lord Rayleigh, published in 1877.

The study of nonlinear phenomena, developing from gas dynamics, started late by comparison with the above phenomena. The reason for this was undoubtedly the fact that the nonlinear effects appear when the compression is very rapid, that is, the flow is supersonic, usually caused by the motion of bodies at supersonic speeds in the air, say, or explosions, as a result of rapid combustion, which may occur in the gas. *Shock waves* are typical phenomena in such cases. Their research emerged from different disciplines and historically did not evolve along a direct path to the present state. From the historical point of view, the modern shock wave physics can be divided to the following developments:

1. The research of the ballistics of the early rifles (muskets) and guns in the mid eighth century, by Robins and Hutton, led them to the conclusion that the finite rapid disturbances created in these circumstances differ considerably from infinitesimal acoustic waves, as formalized by Euler. It was clear that the method of characteristics, developed for hyperbolic equations by G. Monge (1746–1818), a great mathematician and founder of the famed *Ecole Polytechnique*, can be used for theoretical understanding of nonlinear waves as well, as we have tasted in Chap. 4.

2. During first serious measurements of the velocity of sound, using a gun, it was found that very loud sounds, like the sound of the shot, propagate supersonically. The first attempts to develop shock wave theory were made by Poisson, Airy, Challis, and finally by the prominent and influential mathematician B. Riemann in 1859 and Rankine in 1869.

3. The classical period of experimental studies were conducted by Regnault in 1863, who showed that shock waves need supersonic motion as was later confirmed by laboratory studies of explosions by Mach and Sommer. Various experimental techniques were developed in the later decades of the nineteenth century by Mach, Salcher, and Boys. It is worth mentioning that this was also the time in which Reynolds introduced, not at all related to shock study, the ubiquitous and important nondimensional number, bearing his name.

4. Between 1870 and 1890, Rankine and independently Hugoniot discovered their invaluable contributions to shock theory. Additional systematic experimental and theoretical studies led to the discoveries of de Laval, Meyer, and Prandtl and to the development of the wind tunnel.

5. The thirties of the twentieth century was the period of high speed aviation and wind tunnel testing. A number of unique transonic and supersonic wind tunnels were set in operation.

6. During the Second World War and early post-war years the subject naturally blossomed. Supersonic rockets as well as the development and use of nuclear weapons brought a flurry of research by the scientists of the Manhattan project and later by Soviet scientists, headed by the respected Ya.B. Zel'dovich (1914–1987), who contributed to almost all problems in the field, as discussed in his seminal book with Raizer, see [4] in the *Bibliographical Notes*, and to other fields as well.

7. The study of hypersonic flight and theoretical calculations of extremely powerful explosions continue to the present day. In addition to classified research, the relatively new field of astrophysical gas-dynamics has provided a huge observational "laboratory" for the research of nonlinear sound waves, explosions, jets, and shock waves in general.

6.1.2   Overview of This Chapter

The field of compressible fluid dynamics, that is, fluid phenomena that are driven by compressibility, is vast. We cannot present all of it, not even a significant part. Our choice of topics is obviously subjective. We have tried to discuss those topics that are general and, in our view, important and skip detailed and technical issues. We shall discuss first the phenomenon of sound, in the approximation that the amplitude of the disturbances, propagating as linear waves, is relatively very small. After discussing sound waves and reviewing for the reader some rudiments of linear wave motion, we shall skip the large body of geometrical acoustics research, relying on the general properties of linear waves from Chap. 4. From the topic of physical acoustics, we shall single out for brief discussion only the momentum and energy of sound waves and some basics of the physics of the emission and attenuation of sound. The latter topic will culminate in a formal general solution of a sound wave propagating from a given source.

The method of characteristics has been gradually developed by mathematicians, starting from Cauchy and Monge, for the purpose of understanding, categorizing, and solving linear and quasilinear PDEs. The ideas are by no means trivial, repeating them here in a different context than in Chap. 2 and delving into them more deeply can be beneficial. The mathematical side of the topic, which matured during the time of B. Riemann, is outside the scope of this book, but we shall discuss some of its applications to the equations of gas dynamics. It is our hope that this will contribute to the understanding how sound waves, when they grow beyond linearity, may steepen and cause the loss of a continuous solution, in addition to developing other interesting continuous phenomena, such as rarefaction waves. In physics the discontinuities are called *shock waves* while in mathematics they are called *weak solutions*. As it turns out, the above-mentioned mathematical analogy between compressible gas dynamics equations in one-dimension and a particular formulation stemming from the shallow water equations [(4.77)–(4.78)] enables one to obtain interesting results related to the behavior of water when, e.g., a dam in a channel breaks or a wall is accelerated into it and a *hydraulic jump* is formed. We shall mention the first topic very briefly and expound on the second problem a little more, because of its analogy to shock formation in gas dynamics. We shall discuss normal shocks at considerable length and introduce the celebrated Rankine–Hugoniot relations, named after the scientists mentioned in Sect. 6.1.1, but not discuss in any detail oblique shocks, as we feel that they do not introduce new essential physical ingredients. We shall devote, however, a detailed problem to radiating shocks. Energetic phenomena like a strong explosion and detonation waves will be discussed next. This topic, which was developed when the arms race was at its peak, serves today as a theoretical basis for cosmic explosions, such as supernovae. The problems at the end of this chapter contain several important topics and we urge the reader to devote time to their solutions.

6.2 Sound

The phenomenon of sound is well known from our everyday experience through one of our primary senses, that is of hearing over some frequency range. The word *acoustics* itself originates from the Greek verb for hearing. In this section, we shall consider only the physical properties of sound and limit ourselves mainly to the acoustic approximation, that is, concentrate on relatively small perturbations in an underlying fluid flow, which we shall take usually to be a uniform state of rest, for simplicity. The by now familiar notion of linearization, that is, keeping in the equations only terms which are of first order in the perturbation, has been introduced in the beginning of Chap. 4 and we shall make use of it in acoustics. An additional element of the acoustic approximation that we shall use is the neglect of viscosity and this, in turn, is correct whenever the Reynolds number is extremely large. In this case

$$\text{Re} \sim \frac{\lambda c_s}{\nu} \sim \frac{\lambda}{\ell} \gg 1, \tag{6.3}$$

where ℓ is the mean free path of microscopic particles and we have approximated the thermal velocity by c_s, the sound speed (see below), and λ, the length scale in question, is the sound wavelength. Under standard conditions this is an excellent approximation and we shall always remark if a certain phenomenon can be expected when viscosity becomes important. In what follows we shall generally consider the unperturbed state of fluid as a uniform state of rest, unless explicitly stated otherwise. To make this more general, we actually accept a situation in which the fluid is in uniform motion and then make a Galilean transformation to that frame of reference. This allows us to concentrate on the study of sound per se. Our unperturbed fluid is assumed inviscid homentropic and devoid of body forces, therefore if it starts from rest (which is an irrotational flow) it will continue to be in a state of irrotational flow, $\nabla \times \mathbf{u} = 0$ at all times, due to Kelvin's theorem. This is the motivation for most of our study of sound to be done for *potential* flow, that is, for the case that the existence of a scalar velocity potential, $\phi(\mathbf{c}, t)$ satisfying $\mathbf{u} = \nabla \phi$ is guaranteed. In fact, the assumption of a barotropic flow (instead of homentropic) will suffice for that, but we prefer the assumption of homentropy for various reasons, e.g., the definition of adiabatic sound speed.

6.2.1 The Acoustic Wave Equation

We have already set up above the uniform, homentropic, unperturbed state of rest of a fluid with density ρ_0 and pressure P_0 say. The wave equation is obtained most easily, as hinted above, for an *inviscid fluid* in a *homentropic flow* and we shall make these approximations at the outset. We consider now a small perturbation, in the sense explained above, on the reference undisturbed state. Thus we have, in the perturbed state

$$P = P_0 + P', \quad \rho = \rho_0 + \rho', \quad \mathbf{u} = \mathbf{0} + \mathbf{u}, \tag{6.4}$$

where prime denotes a perturbation and the smallness of the perturbations may be explicitly expressed as

$$\frac{P'}{P_0} \ll 1, \quad \frac{\rho'}{\rho_0} \ll 1 \tag{6.5}$$

and in addition we have dropped the prime from $\mathbf{u}'$. Here we immediately note that in homentropic conditions we have, up to first order

$$P' = c_{s,0}^2 \rho' \tag{6.6}$$

where $c_{s,0}^2 \equiv (\partial P_0/\partial \rho_0)_s$ is the adiabatic sound speed squared, in the unperturbed medium, and we shall henceforth drop the 0 from its subscript, for economy of notation. Substituting the relations (6.4) into the continuity equation (1.33) and the Euler equation (1.37), dropping the derivatives of constant terms and neglecting second order terms because of the assumption of linearization, we get

$$\frac{\partial}{\partial t}\rho' + \rho_0 \nabla \cdot \mathbf{u} = 0, \tag{6.7}$$

$$\rho_0 \frac{\partial \mathbf{u}}{\partial t} + \nabla(P') = 0. \tag{6.8}$$

Note that $(\mathbf{u} \cdot \nabla)\mathbf{u}$ has been dropped, as it is second order in $\mathbf{u}$. In our following discussion it will be easy to understand that the interpretation of the velocity perturbation smallness velocity perturbation is its being small with respect to the sound speed in the unperturbed medium. Substituting now for ρ' from Eq. (6.6) into (6.7) gives

$$\frac{\partial P'}{\partial t} + \rho_0 c_s^2 \nabla \cdot \mathbf{u} = 0. \tag{6.9}$$

As explained above, we can, within our assumptions, safely use $\mathbf{u} = \nabla \phi$. Now, inserting this into Eq. (6.8) gives

$$\nabla \left(\rho_0 \frac{\partial \phi}{\partial t} + P' \right) = 0, \tag{6.10}$$

an equation with a trivial integral, which is an arbitrary function of time $C(t)$:

$$\rho_0 \frac{\partial \phi}{\partial t} + P' = C(t). \tag{6.11}$$

To simplify matters we may modify $\phi - C \mapsto \phi$ and this will not change the velocity. Physical considerations suggest that if the disturbance region is limited, like between two planes in a plane wave (see next section), then outside that region, the fluid is undisturbed, that is, the density and pressure are constant in space and time. In front of the disturbance this would suggest that the choice of the new $\phi = 0$ is reasonable. Behind the passing disturbance we still have $P' = 0$ and $\rho' = 0$ dictated by physical conditions. But we may choose the new ϕ to be a constant in space and time as well and consequently have $P' = -\rho(\partial \phi/\partial t)$ and $\rho' = -(\rho c_s^2)(\partial \phi/\partial t)$, as needed.

Substituting thus for P' from the above into Eq. (6.9) and using again $\mathbf{u} = \nabla \phi$ finally yields the classical scalar wave equation for the velocity potential

$$\frac{\partial^2 \phi}{\partial t^2} - c_s^2 \nabla^2 \phi = 0. \tag{6.12}$$

Since all the relevant physical variables of the flow are expressible, as we have seen, as spatial or time derivatives of ϕ, the velocity potential wave embodies in itself all the various aspects of a linear plane sound wave. In particular, taking the gradient of this wave equation guarantees that also each of the velocity components satisfies an identical equation. Note that this is a linear equation and when c_s is a constant, any superposition of solutions is also a solution. In Chap. 4, this equation was studied thoroughly and one observation we mention at the outset is that the constant c_s is simply the *phase* velocity of the wave described by Eq. (6.12), viz. $V_p = c_s$. We comment, for the sake of completeness, that sound or acoustic waves in a medium are sometimes referred to as acoustic *vibrations* in the medium. Before proceeding we shall make a slight change of notation, for convenience. The unperturbed quantities will be denoted with index zero (as before), but we shall drop the prime from the perturbation. Thus, e.g., the unperturbed density will remain ρ_0 but in the perturbation we make the change $\rho' \mapsto \rho$ and similarly for the other relevant perturbations. Taking the time derivative of equation (6.7) and using in it $\mathbf{u} = \nabla\phi$ yields

$$\frac{\partial^2 \rho}{\partial t^2} + \rho_0 \nabla^2 \left(\frac{\partial \phi}{\partial t} \right) = 0. \qquad (6.13)$$

Putting now $\rho_0(\partial\phi/\partial t) = -P = -c_s^2 \rho$ we obtain a wave equation,

$$\frac{\partial^2 \rho}{\partial t^2} - c_s^2 \nabla^2 \rho = 0, \qquad (6.14)$$

identical to (6.12), but for ρ and since P is merely proportional to ρ, a similar wave equation, but with different boundary conditions, holds for P as well.

Even though the wave equation and its solution were discussed in Chap. 4, we conclude this discussion by noting an important family of solutions to Eq. (6.12), that of plane traveling waves, whose form is $\phi(\mathbf{x}, t) = F(\hat{\mathbf{n}} \cdot \mathbf{x} \pm c_s t)$, where F is an arbitrary, smooth enough, function. $\hat{\mathbf{n}}$ is a unit normal in some direction and the $\pm$ refer to two possibilities of the direction of motion (see Problem 6.3).

6.2.1.1 The Poisson Formula

The problem we wish to address now is a general one—assume we know the initial condition of the velocity potential in a given spatial domain and wish to obtain a solution of the acoustic equation, that is to solve the initial value problem. It will naturally be expressed in a formal way, as the purpose is to get a closed form expression of the velocity potential at *any* time, if it is known at time $t = 0$, say. Density and pressure perturbation distribution will follow from that of the velocity potential. The final formula, which bears the name of S.D. Poisson (1781–1840), a prominent mathematician and physicist, known for, among other things, his contribution to probability theory (rumor has it that his interest stemmed from being a compulsive gambler), will hold in an infinite fluid, with the proviso that the sound waves tend to zero at infinity. Let $\phi(\mathbf{x},t)$ and $\phi_\delta(\mathbf{x},t)$ be *any* two different solutions of the wave equation which tend to zero at infinity. We cannot claim uniqueness of a solution to the wave equation, without specifying both initial and appropriate boundary conditions. The reason why we denote one of the solutions with this particular index will be apparent soon. Consider the time derivative of the following integral, taken over entire space

$$I \equiv I_1 + I_2 \equiv \int \left(\phi \frac{\partial \phi_\delta}{\partial t} - \phi_\delta \frac{\partial \phi}{\partial t} \right) d^3x. \tag{6.15}$$

It can be written as

$$\frac{dI}{dt} = \int \left(\phi \frac{\partial^2 \phi_\delta}{\partial t^2} - \phi_\delta \frac{\partial^2 \phi}{\partial t^2} \right) d^3x = c_s^2 \int (\phi \nabla^2 \phi_\delta - \phi_\delta \nabla^2 \phi) d^3x. \tag{6.16}$$

The second equality is on account of the fact that both ϕ and ϕ_δ satisfy the acoustic wave equation. Using standard methods of vector analysis this can be easily shown to vanish, that is, $dI/dt = 0 \Rightarrow I = \text{const.}$ (see Problem 6.1). Consider now an arbitrary point in space O, whose Cartesian coordinates are $\mathbf{x} = (x,y,z)$. We choose now the function ϕ_δ, which is a solution of the acoustic equation, to be a spherical wave in the form of a spike (delta function) incoming into a point O, and reaching it at the instant t_O. We denote by r the radial polar coordinate emerging from O. This incoming spherical wave can thus be written as

$$\phi_\delta(r,t) = \frac{1}{r} \delta[r - c_s(t_O - t)], \tag{6.17}$$

where $\delta(\xi)$ is the Dirac delta function. We shall calculate now the integral (6.15) over all space for two instants $t = 0$ and $t = t_O$ and equate them. This will yield the wanted Poisson formula.

To this end we use the explicit form of $\phi_\delta(r,t)$ as given above. First we evaluate the first term in the integral, for any time t

$$I_1(t) = 4\pi \int \phi \frac{\partial \phi_\delta}{\partial t} r^2 dr. \tag{6.18}$$

Now, noting that

$$\frac{\partial}{\partial t}\{\delta[r - c_s(t - t_O)]\} = c_s \frac{\partial}{\partial r}\{\delta[r - c_s(t - t_O)]\} \tag{6.19}$$

we obtain for I_1 at $t = t_O$

$$I_1(t_O) = 4\pi c_s \int r\phi \delta'(r) dr = -4\pi c_s \int (r\phi)' \delta(r) dr = -4\pi c_s \phi(r=0, t=t_O)$$

$$= -4\pi c_s \phi(x,y,z,t_O), \tag{6.20}$$

where (x, y, z) are the Cartesian coordinates of point O. It is not difficult to see that the second contribution to the integral (6.15) vanishes at $t = t_O$ since

$$I_2(t) = -4\pi \int \phi_\delta \frac{\partial \phi}{\partial t} r^2 dr, \tag{6.21}$$

where the integral is evaluated in spherical polar coordinates. Now, for $t = t_O$, $\phi_\delta = \delta(r)/r$ and thus the integrand contains $(\partial \phi/\partial t) r \delta(r)$ and thus integrates out to zero. The derivative of ϕ at $r = 0$ is assumed to be well behaved.

To sum up, the integral at $t = t_O$ has the value $I = -4\pi c_s \phi(x, y, z, t_O)$. Let us calculate now I for $t = 0$. Because I is a constant, its value must be equal to the one found above. Since the form of ϕ_δ is as given in Eq. (6.17), we may write

$$\frac{\partial \phi_\delta}{\partial t} = -\frac{\partial \phi_\delta}{\partial t_O} \tag{6.22}$$

and substitute it into the initial definition of I (6.15), not forgetting that the integral is calculated for $t = 0$. Thus

$$I = -\int \phi_{t=0} \left(\frac{\partial \phi_\delta}{\partial t_O} \right)_{t=0} d^3x = -\frac{\partial}{\partial t_O} \int (\phi \, \phi_\delta)_{t=0} d^3x - \int \left(\frac{\partial \phi}{\partial t} \phi_\delta \right)_{t=0} d^3x. \tag{6.23}$$

Writing now the differential volume element for the integral in polar spherical coordinates, $d^3x = 4\pi r^2 dr d\Omega$, where $d\Omega$ is the element of solid angle, the first integral in the above expression transforms to

$$\int (\phi \, \phi_\delta)_{t=0} d^3x = \int \phi_{t=0} r \delta(r - c_s t_O) dr d\Omega = c_s t_O \int \phi_{[t=0, r=c_s t_O]} d\Omega, \tag{6.24}$$

where we have used the explicit form of ϕ_δ and the properties of Dirac's delta function. The second integral in the expression (6.23) is transformed in a very similar way, and the final result for I, calculated for $t = 0$ is

$$I = -\frac{\partial}{\partial t_O} \left(c_s t_O \int \phi_{[t=0, r=c_s t_O]} d\Omega \right) - c_s t_O \int \left(\frac{\partial \phi}{\partial t} \right)_{[t=0, r=c t_O]} d\Omega. \tag{6.25}$$

Equating this expression for I with the one found for $t = t_O$ finally gives the Poisson formula

$$\phi(x, y, z, t) = \frac{1}{4\pi} \left\{ \frac{\partial}{\partial t} \left(t \int \phi_{[t=0, r=ct]} d\Omega \right) + t \int \left(\frac{\partial \phi}{\partial t} \right)_{[t=0, r=ct]} d\Omega \right\}. \tag{6.26}$$

Note that we have dropped the subscript O from t, because actually the expression is correct for any time. Explicitly, we have used the case $t = t_O$ in the derivation.

The importance of Poisson's formula is self-evident. It gives the spatial distribution of the velocity potential, appropriate for sound waves, at any time, if only the distribution of this potential and its time derivative, or alternatively, the density or pressure distribution, to which the potential time derivative is proportional, is known at some initial time $t = 0$. This is a typical situation for an initial value problem, also called in mathematics a *Cauchy problem* of a hyperbolic equation, which the wave equation is an example of. The Poisson formula (6.26) shows that the value of the velocity potential at any point O at any time t is determined by the value of the potential and its time derivative at time $t = 0$ on the surface of a sphere centered on $O(x, y, z)$ and whose radius is $R = c_s t$.

6.2.2 Plane Sound Waves

Recall now, from, e.g., Chap. 4 that plane waves, which, as we have seen, propagate in a straight line, have a constant value of the wave function on planes perpendicular to the direction on motion. We consider, without limiting the generality, plane wave solutions of equation (6.12) propagating in the $\pm\hat{\mathbf{x}}$ direction. We remind the reader that for the economy of notation, the subscript 0 has been dropped from the sound speed and other unperturbed quantities. We also denote as usual $u \equiv \mathbf{u} \cdot \hat{\mathbf{x}}$. One observation is immediately apparent: in a wave propagating in the x direction, completely arbitrary as we have chosen it for convenience, the fluid velocity u has only a component *parallel* to the direction of propagation. This means that at least plane sound waves are *longitudinal*, which is exceptional among the waves the reader has studied so far, which have been typically transverse as in the discussion of incompressible waves in Chap. 4. To be clear we refer to the example discussed in Sect. 4.4 on internal gravity waves: if a plane front has a wave-vector pointing in the $\hat{\mathbf{k}}$ direction and if $\mathbf{u}'$ is the instantaneous velocity of the fluid, then because the medium there was incompressible it follows that $\hat{\mathbf{k}} \cdot \hat{\mathbf{u}}' = 0$ and therefore this kind of wave disturbance is said to be transverse. On the other hand, wave disturbances for flows that are irrotational (like sound waves, here) must be that $\hat{\mathbf{k}} \times \hat{\mathbf{u}}' = 0$ and are, therefore, longitudinal in our usage of the word. One additional important matter of nomenclature is now in order. A wave propagating in a single straight direction will be called here a *simple* or a *progressive* wave. Thus $f(x, t) = q_1(x - c_s t) + q_2(x - c_s t)$ and $g(\hat{\mathbf{n}} \cdot \mathbf{x} + c_s t)$, for any $\hat{\mathbf{n}}$ are simple waves, but $h(x, t) = \phi(x + c_s t) - \phi(x - c_s t)$ is not.

As we have seen, the general traveling wave solution of equation (6.12), describing plane waves propagating along the x direction, actually consists of two families of solutions, each one being a simple wave

$$\phi(x, t) = \phi(x \mp c_s t), \tag{6.27}$$

where the function ϕ is an arbitrary one of its arguments $\xi \equiv x - c_s t$ or $\chi \equiv x + c_s t$. The relations between ϕ with its derivatives and the other physically relevant functions in a sound wave are summarized here, for convenience–see also reference [2]:

$$u = \nabla \phi; \quad P = -\rho \frac{\partial \phi}{\partial t}; \quad \rho = \frac{1}{c_s^2} P. \tag{6.28}$$

Noting that in a progressive traveling plane wave, in the x direction, say, there is a trivial mathematical relation between the x and time derivatives

$$\frac{\partial u(x - c_s t)}{\partial x} = \frac{1}{c_s} \frac{\partial u(x - c_s t)}{\partial t}, \tag{6.29}$$

and using the one-dimensional version of the linearized continuity equation (6.7)

$$\frac{\partial \rho}{\partial t} = -\rho_0 \frac{\partial u}{\partial x}. \tag{6.30}$$

We find the following relation for u in a plane traveling wave

$$u = c_s \frac{\rho}{\rho_0} = \frac{1}{\rho c_s} P. \tag{6.31}$$

Note that for the single wave propagating in the $-\hat{\mathbf{x}}$ direction, such a relation holds for a negative velocity. It is thus interesting to see, and this is the way it should be, that u is parallel to the direction of wave motion, as we have already seen. A sound wave is longitudinal, in waves propagating in the positive x direction, at a given instant, and the sign of the velocity is in the positive x direction, implying compression. But it is the other way around for the reverse part of such waves, i.e., the ones which are propagating in the negative x direction, the velocity is negative and rarefaction.

From Eq. (6.31) we obtain an important estimate of the fluid velocity in terms of the sound speed

$$\frac{u}{c_s} = \frac{\rho}{\rho_0}. \tag{6.32}$$

This is consistent with what we have already discussed, significant compressibility can be expected for a supersonic flow (Mach numbers over 1), but it also shows that actually the acoustic approximation requires $|u| \ll c_s$. We may mention also the relation between the temperature oscillation T' and the velocity. Evidently, dropping the prime, but preserving the meaning of the quantity as a perturbation, $T = (\partial T / \partial P)_s P$ to first order in the disturbances. Using now the thermodynamic identity

$$\left(\frac{\partial T}{\partial P} \right)_s = \frac{T}{c_P} \left(\frac{\partial v}{\partial T} \right)_P, \tag{6.33}$$

where $v \equiv 1/\rho$ is the specific volume, and from formula (6.32), but with P instead of $c_s\rho$, we obtain

$$T = \beta \frac{c_s}{c_P} T_0 u, \tag{6.34}$$

where

$$\beta \equiv \frac{1}{v}\left(\frac{\partial v}{\partial T}\right)_P \tag{6.35}$$

is the coefficient of thermal expansion.

Finally, we comment on the value of sound speed in a perfect gas. Using the equation of state of such a gas, $P = (\mathscr{R}/\mu)\rho T$, cf. Eq. (1.60), it is possible to show (see Problem 6.6) that the sound speed in perfect gas depends on the temperature alone, and for a given chemical composition and density it is proportional to $\sqrt{T}$. It should be perhaps stressed that the sound speed is *not* a constant for a particular fluid, but, as should be obvious from the above discussion, it is a state function. To get the feeling for the order of magnitude of c_s under normal conditions, we quote the value of the sound speed in dry air at $0\,°C$, if approximately considered as a perfect gas. It is $c_s \approx 330\,\mathrm{m/s}$.

6.2.2.1 Example: Sound Waves in a Stratified Medium

So far we have dealt with sound waves as small disturbances propagating in a uniform medium. We are interested now to see what can be said about sound waves which propagate in a nonuniform background and perhaps the easiest example for such a background is a plane-parallel density stratified isothermal atmosphere. We have seen a similar example of an *adiabatic* atmosphere in the discussion following Eq. (1.57). Thus the hydrostatic equilibrium equation in the z direction, say, reads

$$\frac{\partial P_0}{\partial z} = -g\rho_0(z), \tag{6.36}$$

where the density stratification is in the z direction and the body force per unit mass is an external constant gravitational acceleration $\mathbf{b} = -g\hat{\mathbf{z}}$. All the other symbols have their usual meaning. To solve for this equilibrium, we need an equation of state and assume that it is one of a perfect monoatomic gas $P_0 = R\rho_0 T_0$, where $R = \mathscr{R}/\mu$ is the gas constant divided by the mean molecular weight and T_0 is a constant temperature. We readily get the solution

$$\rho_0 = \rho_s e^{-z/H_z} \quad \text{and} \quad P_0 = P_s e^{-z/H_z} \tag{6.37}$$

where $H_z \equiv RT_0/g$ is here the constant vertical *scale-height*, satisfying

$$H_z^{-1} = -\frac{1}{\rho_0}\frac{d\rho_0}{dz} = -\frac{1}{P_0}\frac{dP_0}{dz}, \tag{6.38}$$

and all the thermodynamic symbols have their usual meaning. As is evident here, ρ_s is the bottom ($z = 0$) boundary value, $\rho_s = \rho_0(z = 0)$, which is assumed to be known and is indispensable for the solution. The lower boundary value pressure follows from Eq. (6.37) and is $P_0(z) = P_s$.

The relevant dynamical equations for this case have to include the body force, given above, as well as the fact that the flow is inviscid and compressible. In those equations we add to all the dependent variables, which satisfy the above static solution, a small perturbation, denoted by prime. Next, the equations are linearized, where certain relations between the perturbations are taken into account as well—e.g., we assume the perturbations to be adiabatic, because they are done on a timescale much shorter than any thermal process time, so that the pressure perturbation and the density perturbation, for example, are related to each other thus $P' = c_s^2 \rho'$ with

$$c_s^2 = \left(\frac{\partial P}{\partial \rho}\right)_s = \gamma R T_0 = \gamma \frac{P_0}{\rho_0}, \tag{6.39}$$

where the derivative is performed at constant entropy and the subscript zero has been dropped. As we already know, c_s is the sound speed in the unperturbed medium, and in this case, similarly to a uniform medium, it is a constant.

During the linearization we also neglect the y-variations of the variables, for the sake of simplicity, being content with a problem in two spatial dimensions, which is bound to capture the physical essentials. It is then possible, with the help of suitable algebraic manipulation, to obtain for this two-dimensional problem a fourth order partial differential equation (see Problem 6.7) for the function $\mathscr{L}(x, z, t) \equiv w\sqrt{\rho_0/\rho_s}$, where $w = \mathbf{u} \cdot \hat{\mathbf{z}}$, consistently with our usual notation:

$$\frac{\partial^4 \mathscr{L}}{\partial t^4} - c_s^2\frac{\partial^2}{\partial t^2}\left(\nabla^2 - \frac{1}{4H_z^2}\right) - c_s^2 N^2\frac{\partial^2 \mathscr{L}}{\partial x^2} = 0. \tag{6.40}$$

Here N^2 is the Brunt–Väisälä frequency, also called the buoyancy frequency, known from Chap. 5 and discussed in more detail in Chap. 7. The ∇^2 operates in the x and z directions only, but this is not marked by any additional symbol, for notational convenience. We express here the buoyancy frequency as

$$N^2 = -\frac{g}{\rho_0}\frac{d\rho_0}{dz}.$$

To proceed, we try an ansatz of a wave propagating in the (x, z) plane

$$\mathscr{L}(x, z, t) = \tilde{\mathscr{L}}\exp\left[i(k_x x + k_z z - \omega t)\right] + \text{c.c.}, \tag{6.41}$$

which when substituted in (6.40) leads to the following quartic equation for ω

$$\omega^4 - c_s^2\omega^2K^2 + c_s^2N^2k_x^2 = 0, \tag{6.42}$$

where $K^2 = k_x^2 + k_z^2 + 1/(4H_z^2)$ and this symbol denotes an expression somewhat different from how it was used in the last two chapters. The solution of the quartic gives a dispersion relation in the following form

$$\omega^2 = \frac{1}{2}c_s^2K^2\left[1 \pm \left(1 - \frac{4N^2k_x^2}{c_s^2K^4}\right)^{1/2}\right]. \tag{6.43}$$

The $\pm$ guarantees that there are two qualitatively different branches of roots, corresponding to acoustic waves arising from the plus sign, and internal gravity waves, which were found approximately in Chap. 4 when the Boussinesq approximation was used. Here, the fully compressible equations were used and therefore the properties of the internal gravity waves can be completely and exactly uncovered, however we are interested in this example only in some important qualitative properties of sound waves. To learn about the properties of the acoustic waves per se, the two branches have to be well separated and for that it is necessary that (see Problem 6.8) *either* $k_xH \gg 1$ *or* $k_zH \gg 1$ and then, as is apparent from the problem, we can get the approximate dispersion relation for the acoustics

$$\omega_a^2 \approx c_s^2\left(k_x^2 + k_z^2 + \frac{1}{4H_z^2}\right). \tag{6.44}$$

The important finding here is that there is a minimal acoustic frequency, given approximately by $\omega_{a,\min} \approx c_s/(2H_z)$, and no waves, however long, can possess a lower frequency. The limiting frequency tacitly (because of the approximation made here) depends on the direction of propagation and waves of this kind cannot propagate as they suffer from a king of "total internal reflection," so to speak. It is also possible to show that a wave whose wave-vector is $\mathbf{k} = (k_x, k_z)$ has a different direction of its phase and group velocity (show it!). Finally, an additional significant finding is the existence, in this case, of the so-called Lamb waves (see Problem 6.9), which propagate horizontally ($k_z = 0$), their curious property being that their velocity perturbation amplitude increases with height while the pressure perturbation behaves in the opposite way.

6.2.3 Spherical Sound Waves

In the case treated so far, we have assumed that sound waves are small perturbations on a uniform homentropic fluid, that is, no dissipative processes like viscosity or heat conduction were allowed. In the case of plane waves, this guarantees that neither the amplitude nor the form of the wave change as the wave propagates to infinity. In a *spherical wave*, that is, a wave outgoing from a center in all directions, this is not the case. The reason is no more than geometric—the wave has to spread over an ever-enlarging sphere and, as we shall show later on, the principal idea here is that the energy contained in a spherical wave is conserved as the wave spreads out, but the energy flux changes. In contrast to this, there exist also spherical *incoming* waves, which converge into a central point, coming from all directions. During the spherical spreading out, the given energy passes through a surface ever increasing in r, and so the *energy flux* diminishes gradually to zero. To stress again, this happens without any dissipative process.

Using the wave equation for ρ, for example (6.14), but expressing the Laplacian in spherical polar coordinates and including the assumption of spherical symmetry, i.e., the radial coordinate r is the only spatial coordinate left and we readily get

$$\frac{\partial^2 \phi}{\partial t^2} - \frac{c_s^2}{r^2} \frac{\partial}{\partial r} \left[r^2 \frac{\partial \phi}{\partial r} \right] = 0, \tag{6.45}$$

where the subscript 0 of the sound speed has been dropped. Perhaps surprisingly this equation can be cast into a Cartesian-like form, i.e., similar to (6.12), but for the function $r\phi$, where r is the radial spatial coordinate

$$\frac{\partial^2}{\partial t^2}(r\phi) - c_s^2 \frac{\partial^2}{\partial r^2}(r\phi) = 0. \tag{6.46}$$

We thus can write a general solution for the velocity potential

$$\phi(r,t) = \frac{F_1(r - c_s t)}{r} + \frac{F_2(r + c_s t)}{r}, \tag{6.47}$$

where F_1 and F_2 are smooth arbitrary functions. However, in this case we obviously do not have plane waves, but rather a spherical *outgoing* wave F_1/r and an *incoming* wave F_2/r. This perhaps is the place to see that for very large r, one may consider the r in the derivative approximately constant, since $r \pm c_s t$ causes F_1 and F_2 to vary significantly for an almost constant r. This hints that in the far field ($r \to \infty$) spherical phase fronts approach plane parallel fronts and, as can be expected, spherical waves are well approximated by plane waves. We shall now find the relation between the other physical variables of the wave and ϕ for the case of a spherical wave. Even though the procedure is similar to the case of plane waves, we find it useful to repeat it, to make sure that the expression of the equations in spherical coordinates does not change these relations, like it did for the solution of the wave equation itself. Thus, linearizing the equation of spherically symmetric motion gives, after dropping the zero index from unperturbed quantities,

$$\frac{\partial u_r}{\partial t} = -\frac{c_s^2}{\rho} \frac{\partial \rho}{\partial r}, \tag{6.48}$$

where u_r is the radial velocity. Using $u_r = \nabla\phi = \partial\phi/\partial r$ the last equation gives

$$\frac{\partial}{\partial r}\left[\rho + \frac{\rho_0}{c_s^2}\frac{\partial\phi}{\partial t}\right] = 0. \tag{6.49}$$

This equation can be integrated on r, and the expression in the square brackets becomes an arbitrary function of time. Considering for a moment only the outgoing wave, replacing F_1 by f for convenience, we have $\phi(r,t) = f(r - c_st)/r$. Now we assume that the disturbance is of a limited radial width. Far out, in regions where the disturbance has not yet passed, $\phi = f/r$ must be zero, because the fluid is in its initial uniform and static distribution. In the region behind the disturbance we have to force the value of ϕ to relax to a constant (in space and time). But this is possible only if $f(r - c_st)$ is identically zero behind the disturbance, otherwise $\phi = f/r$ cannot be a constant, as it diverges for $r = 0$. Thus we get the following expressions for the density and pressure perturbation, valid in a spherical wave as well,

$$\rho = -\frac{\rho_0}{c_s^2}\frac{\partial\phi}{\partial t}; \quad P = c_s^2\rho, \tag{6.50}$$

with the additional proviso that if the wave is a pulse limited in space, i.e., consisting of an outgoing spherical shell, the velocity potential ϕ must be zero in front and behind the shell.

Using the outgoing solution for the velocity potential

$$\phi(r - c_st) = \frac{1}{r}[f(r - c_st)], \tag{6.51}$$

we obtain for the compression and the velocity of the wave

$$\rho = -\frac{\rho_0}{c_s^2}\frac{\partial\phi}{\partial t} = \frac{\rho_0}{c_s}\frac{f'(r - c_st)}{r}, \tag{6.52}$$

$$u_r = \frac{\partial\phi}{\partial r} = \frac{f'(r - c_st)}{r} - \frac{f(r - c_st)}{r^2}, \tag{6.53}$$

where the prime denotes a derivative with respect to the whole argument. If the extension of the compression pulse(s) is small, with respect to r, it is a good approximation to assume that the *shape* of the compression disturbance does not change as the wave spreads out, while the amplitude, however, decreases as $1/r$. As we shall see in Sect. 6.2.4, the energy, per unit volume, carried by a sound wave is proportional to ρ^2. The volume of disturbance whose width is Δr, say, is $4\pi r^2\Delta r$ and thus the energy carried by the pulse is proportional to $r^2\rho^2$. Thus the factor $1/r$ in the compression disturbance, which is responsible for the decrease of the amplitude (squared) as $1/r^2$, guarantees energy conservation. Another important difference between a plane wave and a spherical one can be found when one examines the expression for the pressure disturbance

$$P = -\rho_0 \frac{\partial \phi}{\partial t}, \tag{6.54}$$

from which it follows that if we consider a shell-like disturbance and we pick a r_0 outside it, we may integrate the previous equation over all times, at this point, giving

$$\int_{-\infty}^{\infty} P dt = -\rho_0 [\phi(r_0, +\infty) - \phi(r_0, -\infty)] = 0, \tag{6.55}$$

because for $t \to -\infty$ the disturbance has not yet arrived at r_0, while for $t \to \infty$ it has already passed. In both cases the velocity potential is zero. This shows that the compression disturbance in a spherical wave cannot have only a positive part and must have also a part of rarefaction, so as to compensate for the compression and render the previous integral equal to zero. In our general discussion following Eq. (6.11) we have anticipated that in a plane wave a relation like (6.55) cannot be guaranteed, because ϕ may be equal to a non-zero constant behind the disturbance, even though it is zero ahead of it. Thus, a compression only or rarefaction only may be possible solely in a plane wave. The observation given here explains in a satisfactory manner a problem that might have been caused by fluid outflow in an outgoing spherical wave. It is certainly unphysical that an outgoing spherical wave gives rise to total fluid volume outflow. Indeed, for large r we may neglect the second term in the radial velocity according to Eq. (6.53). In the outgoing spherical wave the volume outflow rate for large r is

$$Q = 4\pi r^2 u_r \approx 4\pi r f'(r - c_s t) \propto r^2 \rho. \tag{6.56}$$

With r growing, outflows are balanced by inflows, since, as we showed, ρ must change sign and thus there is no danger that an outgoing spherical wave will cause persistent fluid outflow. This formula for Q is obtained taking into account Eq. (6.52) and $P = c_s \rho$ remembering also that ρ_0 and c_s are considered constant in this setting. It is also important to understand the behavior of the spherical wave solution at the point $r = 0$ (origin). This form of possible impediment must always be considered when spherical coordinates are used in physics. In our case we have to distinguish between two possibilities according to whether there is a fluid source or there is no such source at the origin. Lack of source does not cause any problems. When there is a source and there is a danger of a singularity, it is convenient to specify the possible central fluid source with the help of a time-dependent function $Q(t)$ which denotes the source strength by specifying the fluid rate of volume injection, in units of volume/time. The source strength may then conveniently be expressed as

$$Q(t) = 4\pi \lim_{r \to 0} \left(r^2 \frac{\partial \phi}{\partial r} \right). \tag{6.57}$$

We shall conclude the discussion of spherical waves by singling out monochromatic waves because, as mentioned before, any linear wave can be constructed using a sum (or integral) of monochromatic waves and so any general case can follow from it. Consider a monochromatic spherical wave. The only way to avoid singularity at the origin in this case, where there is no source at $r = 0$, is to consider *standing waves* (waves which actually do not propagate and possess nodes at fixed spatial locations). An appropriate wave of this sort has the form

$$\phi(r,t) = Ae^{-i\omega t}\frac{\sin k_r r}{r} + \text{c.c.} \qquad (6.58)$$

where it is understood that the physical solution is the real part of this expression and $k_r \equiv (\omega/c_s)$ is the wavenumber, which naturally is the r (and only) component of the wavevector. Now, this expression satisfies the spherical wave equation and is clearly not singular at $r = 0$.

6.2.4   Energy and Momentum Transport in Acoustic Waves

As is usual in physics, any wave motion carries with it energy and momentum as we have discussed this in Chap. 4. Also sound waves carry energy and momentum. We shall expound now on this topic and for simplicity consider only the nondissipative case, that is, neglect heat conduction and viscous stresses. In Chap. 4, we have already dealt with the energy and its propagation in three-dimensional internal waves in an atmosphere subject to the Boussinesq approximation (Sect. 4.4). Here we take into account the full effect of compressibility, but examine only a one-dimensional simple wave. We recommend that the student reminds himself or herself of the material in Sect. 4.4 before proceeding here.

Following the mechanical analogue, we may write the instantaneous fluid energy flux density vector (rate of working per unit volume, per unit area) in a sound wave as follows:

$$\Phi_{\mathscr{E}}(t) = P\mathbf{u} = (P_0 + P')\mathbf{u}. \qquad (6.59)$$

We shall consider here a simple, one-dimensional, acoustic wave, remembering that the meaning of "simple" in this context is containing only a component moving in a fixed direction. We can render the above quantity more physically meaningful if it is a time average over some time τ, say, giving in this way the time-independent average energy flux density

$$\langle \Phi_{\mathscr{E}} \rangle = \hat{\mathbf{n}}\frac{1}{\tau}\int_0^\tau \left(P_0 + c_{s,0}^2\rho\right) c_{s,0}\frac{\rho}{\rho_0}dt, \qquad (6.60)$$

where $\hat{\mathbf{n}} = \hat{\mathbf{u}}$ is the direction of propagation of our simple wave and we have retained, for the time being, the subscript 0, which denotes the undisturbed medium values. Also, we denote here time averages by $\langle \cdot \rangle$. The above average energy flux density is sometimes also called the *sound intensity*. The physical interpretation is quite simple, the expression in the integrand is the total pressure times the velocity, when use has been made of Eq. (6.32).

For strictly periodic waves, it is actually possible to choose the time interval τ over which the average is taken, as an integral number of periods, i.e., $\tau = nT = 2\pi n/\omega$, where ω is the frequency of the periodic wave, T the period, and n is an integer, as was done in Chap. 4 in the section on atmospheric waves. In the present case the first term integrates out to zero, as it is a constant multiplied by a periodic function. In any case, even if the exact period of the wave oscillation is unknown or unspecified, it is always possible to choose τ to be very large, with respect to any physically meaningful variability time of ρ, say, and then the integral of the first term, which contains an oscillating term ρ, becomes ultimately negligible with respect to the second one, in which the integrand contains the positive definite ρ^2. Thus we will have

$$\langle \Phi_{\mathscr{E}} \rangle = \hat{\mathbf{n}} c_{s,0}^3 \frac{\langle \rho^2 \rangle}{\rho_0} \tag{6.61}$$

We turn now to the momentum flux. As is usually the case in simple waves, we may easily show (see Problem 6.11) that the time averaged momentum flux density is

$$\langle \Phi_p \rangle = \hat{\mathbf{n}} \frac{1}{\tau} \int_0^\tau (\rho_0 + \rho) \, c_{s,0}^2 \frac{(\rho)^2}{\rho_0^2} \, dt. \tag{6.62}$$

Retaining only quantities up to second order in the disturbance one gets

$$\langle \Phi_p \rangle = \hat{\mathbf{n}} c_{s,0}^2 \frac{\langle \rho^2 \rangle}{\rho_0}, \tag{6.63}$$

And comparing with the average energy flux this gives immediately

$$\langle \Phi_{\mathscr{E}} \rangle = c_s \langle \Phi_p \rangle, \tag{6.64}$$

where the zero subscript has been dropped from the unperturbed value of velocity of sound. It is useful to understand that the above results could be anticipated and this may clarify their physical meaning.

Before turning to statements about balance and conservation of acoustic energy, we shall devote a short paragraph to the discussion of some accepted units for sound intensity and express in these units a number of commonly encountered example sounds. The sound intensity, similar to intensities of other phenomena, to which our senses respond logarithmically, is measured using a logarithmic scale of a ratio between the intensity and a reference value. The reference intensity, in this case, is set to the *threshold of hearing* (the weakest audible sound by an average healthy person) at the frequency of 1000 Hz. It is $|\Phi_{\mathscr{E}}^{\text{ref}}| = 10^{-12}$ W/m². Thus the sound intensity level expressed in *decibels* (dB) is

$$\Delta_{\text{dB}} \equiv 10 \log_{10} \left(\frac{\langle |\Phi_{\mathscr{E}}| \rangle}{|\Phi_{\mathscr{E}}^{\text{ref}}|} \right). \tag{6.65}$$

We now conclude the discussion of measures for sound intensity by citing several numbers, some of them surprising. It turns out, for example, that the human ear responds to sound intensity between 0 and 140 dB. The sound of intensity 0 dB is barely audible by those who hear well, while sounds close to 140 dB are bound to cause ear damage, still the 14 orders of magnitude ear sensitivity is remarkable. As is well known, the human eye is a very sensitive receiver of light, effectively able to detect a single photon, in the peak of the eye's spectral sensitivity. Here we add that the human ear is also a remarkable sound detector. In the language of sound engineering the gain of the human ear is very high. It should be mentioned, however, that sounds audible to humans fall into a limited spectral range ($\approx$20–20,000 Hz), but the sensitivity close to the middle of the spectrum is formidable. Another manifestation of that fact is the observation that at 0 dB the relative compression has the minute value of $\rho/\rho_0 \sim 10^{-20}$ while at 140 dB it is merely $\sim 10^{-6}$. In addition to the fantastic ability of the ear to detect minute perturbations in the air, we are confident that the acoustic approximation is certainly valid over the whole human range of hearing. Finally some numerical examples—a normal conversation: 50–60 dB, a diesel train at 100 ft: 80 dB, a jet takeoff at 1000 yd or a nearby jackhammer: 100 dB.

6.2.4.1 Acoustic Energy Conservation

In this section, we shall deal with the balance and conservation of acoustic energy, basing our discussion on the FD energy conservation equation. For convenience we choose the form (1.69), dropping however some terms due to the assumptions that we deal with an inviscid and adiabatic flow and that there are no body forces. What is left of that equation is thus

$$\rho \frac{D}{Dt} \left(e + \frac{1}{2} u^2 \right) = -\nabla \cdot (P\mathbf{u}), \tag{6.66}$$

where $u^2 \equiv \mathbf{u} \cdot \mathbf{u}$, e is the specific (per unit mass) internal energy and because the flow is inviscid, only the single term on the right-hand side survives.

Wishing to use this equation for a sound wave we write, as usual, $\rho = \rho_0 + \rho'$, $P = P_0 + P'$ with $P' = c_{s,0}^2 \rho'$ and $e = e_0 + e'$. Also, again, we shall drop the prime from the perturbations, when possible, that is, when there is no danger of confusion. Two comments should be immediately made—the first is that keeping terms of only first order in the acoustic disturbance would cause the loss of the kinetic energy contribution and thus not give the expected physically correct balance equation. Consequently, we have to keep terms up to second order in the disturbance. The second comment is actually a question. How to deal with e, that is, what should be the variables in which this quantity is best expanded, so as to make the calculation as efficient as possible? It turns out, after some trials, that expanding $e' = e - e_0$ in $v' = v - v_0$, where $v \equiv 1/\rho$ is a good enough choice. Thus we write

$$e' = \left(\frac{\partial e}{\partial v}\right)_s v' + \frac{1}{2}\left(\frac{\partial^2 e}{\partial v^2}\right)_s (v')^2 + \text{HOT}. \tag{6.67}$$

Now, through the use of thermodynamic identities and an expansion of $v - v_0$ in terms of $P' = P - P_0$ we get $(\partial e/\partial v)_s = -P$ and

$$v - v_0 = \left(\frac{\partial v}{\partial P}\right)_s P' + \text{HOT}. \tag{6.68}$$

This gives the following expression for the specific fluid energy

$$e + \frac{1}{2}u^2 = e_0 - P_0(v - v_0) + \frac{(P')^2}{2\rho_0^2 c_{s,0}^2} + \frac{1}{2}u^2 + \text{HOT}, \tag{6.69}$$

where we use, in addition to the substitution from the above thermodynamic identity and the expansion, the following self-evident expressions

$$\left(\frac{\partial P}{\partial v}\right)_s = -\rho^2 c_s^2 \quad \text{and} \quad \left(\frac{\partial v}{\partial P}\right)_s = \left(\frac{\partial P}{\partial v}\right)_s^{-1}. \tag{6.70}$$

Substituting the expression for the internal plus kinetic specific (per unit mass) energy from formula (6.69) in the fluid dynamical energy equation (6.66), using the above expression for the sound speed and retaining only terms up to second order in the acoustic disturbance yields

$$\rho\frac{D}{Dt}\left[e_0 - P_0(v - v_0) + \frac{(P')^2}{2\rho_0^2 c_{s,0}^2} + \frac{1}{2}u^2\right] = -\nabla \cdot (\mathbf{u}P). \tag{6.71}$$

We now write $P = P_0 + P'$ on the right-hand side of this equation and remember that all quantities with a zero subscript are constant in time and space and therefore should drop from all derivatives. We then get

$$\rho\frac{D}{Dt}\left[\frac{(P')^2}{2\rho_0^2 c_{s,0}^2} + \frac{1}{2}u^2\right] + \nabla \cdot (\mathbf{u}P') = P_0\left(\rho\frac{Dv}{Dt} - \nabla \cdot \mathbf{u}\right). \tag{6.72}$$

It is not difficult to convince oneself that the quantity in the parentheses on the right-hand side of this equation is zero, because of its being the continuity equation, expressed in a somewhat unusual form. If, in addition, we replace the convective derivative by a usual partial time derivative and the ρ in front of the equation by ρ_0 (consistently with the ordering, up to the second order in the acoustic disturbances), we finally get the acoustic energy equation:

$$\frac{\partial \mathscr{E}}{\partial t} + \nabla \cdot (\mathbf{u} P') = 0, \qquad (6.73)$$

where the instantaneous acoustic energy density (per unit volume) is

$$\mathscr{E} \equiv \frac{(P')^2}{2 \rho_0 c_{s,0}^2} + \frac{1}{2} \rho_0 u^2 \qquad (6.74)$$

and $(\mathbf{u} P')$ is the instantaneous acoustic energy flux density, in agreement with the *time averaged equation* (6.59), in which the part $P_0 \mathbf{u}$ is not related to the acoustic energy flux.

It should be remarked here that for simple waves, shown in Problem 6.12, the first term in the expression for instantaneous energy density in Eq. (6.74) is equal to the second term. This is not true in general, of course; however when we are dealing with small amplitude time-oscillations (like in acoustics), it is well known from mechanics that the time average of the first term, representing the potential energy of the oscillation, is actually equal to the time average of the second term—the kinetic energy of the oscillation. For simple waves, Eq. (6.73) acquires a particularly elegant form (see Problem 6.12)

$$\frac{\partial \mathscr{E}}{\partial t} + \nabla \cdot (\mathscr{E} \mathbf{c}_{s,0}) = 0, \qquad (6.75)$$

where $\mathbf{c}_{s,0}$ is the vector sound velocity, whose magnitude is the sound speed, and whose direction is the direction of propagation of the simple wave. Is this equation correct also for acoustic periodic waves, when we perceive $\mathscr{E}$ as a suitably defined average?

6.2.5 Normal Modes of Acoustic Vibrations

So far we have considered the propagation of sound waves without taking into account any boundary conditions. Mathematically, solutions of the wave equation which are propagating simple waves, say, are correct only if the medium is infinite, at least in the direction of propagation. However, since we have considered a uniform medium, the assumption that it is infinite held in each direction (equivalent to isotropy). As we have seen, in infinite media waves with any frequency can be propagated. Physically this is clearly an approximation, being satisfactory as long as the typical scale of the sound wave, l_{sw}, which is equal to the wavelength, $\lambda = 2\pi c_s / \omega$ for a single, fixed frequency (ω) wave, is much smaller than the size of the system ($\sim L$, say), i.e., $l_{sw} \ll L$. If this is not the case, we must take into account boundary conditions when solving the wave equation. Consider a fluid in a vessel of finite dimensions. We shall discuss here *free vibrations* of the fluid, which means that there is no force causing these oscillations. As is well known from the elementary problem of an elastic finite string tied at both ends, not all frequencies of vibrations are allowed. Mathematically, the solution of the wave equation with boundary conditions, such that the disturbance is zero at the boundaries, imposed on it, allows only a series of discrete definite frequencies. The wave equation separates and the spatial part becomes an eigenvalue problem with a discrete set of eigenvalues and eigenfunctions. The eigenfrequencies corresponding to the allowed wavelength are also called *characteristic frequencies* and the solutions *normal modes*. A simple order of magnitude estimate indicates that the longest allowed wavelength should be of the order of twice the vessel size, $\sim L$ say. Thus, the order of magnitude of the lowest characteristic frequency can be estimated

$$\lambda_1 \sim 2L, \qquad \omega_1 \sim \pi \frac{c_s}{L}. \tag{6.76}$$

Considering the 3-D wave equation for the velocity potential

$$\frac{\partial^2 \phi}{\partial t^2} = c_s^2 \nabla^2 \phi, \tag{6.77}$$

where we suppress, for simplicity, the subscript 0 from the sound speed; we stress again that in this problem there are definite boundary conditions on the walls of the vessel, specifically $\phi = 0$. Guided by our knowledge of similar problems with boundary conditions, we try to separate Eq. (6.77) into spatial and temporal parts, done using the technique of separation of variables. After writing $\phi(\mathbf{x},t) = \phi_{\mathrm{sp}}(\mathbf{x})\phi_{\mathrm{tm}}(t)$ and substituting it into the wave equation we eventually get

$$\frac{1}{\phi_{\mathrm{tm}}(t)} \frac{\partial^2 \phi_{\mathrm{tm}}}{\partial t^2} = \frac{c_s^2}{\phi_{\mathrm{sp}}(\mathbf{x})} \nabla^2 \phi_{\mathrm{sp}}(\mathbf{x}). \tag{6.78}$$

Since the left side of the above equation is a function of time only, and the right side depends only on space, both sides must be equal to a mutual *negative* constant. The negativity of the separation constant is dictated by our will to have a physically meaningful solution. A positive constant would allow exponentially growing or decaying in time (hence non-physical) solutions. We shall call the negative constant $-\omega^2$ and obtain the following pair of equations

$$\frac{\partial^2 \phi_{\mathrm{tm}}}{\partial t^2} = -\omega^2 \phi_{\mathrm{tm}}(t), \tag{6.79}$$

$$\nabla^2 \phi_{\mathrm{sp}}(\mathbf{x}) = -\frac{\omega^2}{c_s^2} \phi_{\mathrm{sp}}(\mathbf{x}). \tag{6.80}$$

Dealing with the temporal equation first it is easy to write its general solution

$$\phi_{\mathrm{tm}}(t) = C\cos(\omega t + \alpha), \tag{6.81}$$

with C and α being constants, determined by two initial conditions on the function and its time derivative, say. We choose to write the solution of the spatial equation as a spatial plane wave, having a slightly different, albeit equivalent choice of constants and form

$$\phi_{sp}(\mathbf{x}) = A\cos\left(\frac{\omega}{c_s}\hat{\mathbf{n}}\cdot\mathbf{x}\right) + B\sin\left(\frac{\omega}{c_s}\hat{\mathbf{n}}\cdot\mathbf{x}\right), \tag{6.82}$$

where A and B are constants to be determined by the boundary conditions and $\hat{\mathbf{n}}$ is a unit vector perpendicular to the equal phase planes of the wave. It is impossible to say that $\hat{\mathbf{n}}$ is the direction of propagation because the wave $\phi(\mathbf{x},t) = \phi_{sp}(\mathbf{x})\phi_{tm}(t)$ does not propagate! A short reflection reveals that at any point $\hat{\mathbf{n}}$ is perpendicular to the planes of equal phase. Combing the spatial and temporal portions of the solution we have, in general,

$$\phi(\mathbf{x},t) = \left[A\cos\left(\frac{\omega}{c_s}\hat{\mathbf{n}}\cdot\mathbf{x}\right) + B\sin\left(\frac{\omega}{c_s}\hat{\mathbf{n}}\cdot\mathbf{x}\right)\right] \times \cos(\omega t + \alpha),$$

where the constant C has been absorbed in A and B. We see already that the extent of the oscillation varies according to the position. It is thus a *standing wave*, sometimes called also *stationary*. To simplify the following calculation we choose $\hat{\mathbf{n}}$ to be in the x direction (this does not limit the generality) and define $k \equiv \omega/c_s$. We then impose the two spatial boundary conditions

$$u_x(0) = \frac{\partial\phi_{sp}}{\partial x}(0) = 0, \tag{6.83}$$

$$u_x(L) = \frac{\partial\phi_{sp}}{\partial x}(L) = 0, \tag{6.84}$$

and using Eq. (6.82), with $\hat{\mathbf{n}}\cdot\mathbf{x} = x$, we get

$$B = 0 \quad\text{and}\quad -\frac{\omega}{c_s}A\sin(kL) = 0. \tag{6.85}$$

Now both A and $\omega \neq 0$, otherwise there is no wave at all, and therefore only the following (eigen)values for k are allowed and these lead to the characteristic frequencies

$$k_n = n\frac{\pi}{L} \quad\Longrightarrow\quad \omega_n = n\frac{\pi c_s}{L}, \tag{6.86}$$

for $n = 1, 2, 3\ldots$ Thus the full solution in this case for a specific normal mode $n = m$ is

$$\phi_m(x,t) = A_m\sin\left(\frac{m\pi}{L}x\right)\cos\left(c_s\frac{m\pi}{L}t + \alpha_m\right), \tag{6.87}$$

where A_m and α_m can be determined from two initial conditions. Actually any linear combination of normal modes is an acceptable solution and the amplitudes and phases of the individual modes come out of the initial conditions. This is nothing but an elementary example of solving a PDE by Fourier series, examples of which we have already seen in Chap. 4 when discussing atmospheric waves.

Instead of trying to describe the structure of a standing plane harmonic wave, as the one in Eq. (6.87) which would be awkward, let us go back to an example of a particular standing wave, e.g.,

$$\phi = A \cos\left(\frac{\omega}{c_s} x\right) \cos(\omega t). \tag{6.88}$$

Clearly, this is a particular choice composed of the expressions as in (6.81) and (6.82), where we put $\hat{\mathbf{n}} \cdot \mathbf{x} = x$, $B = 0$, $C = 1$, and $\alpha = 0$. Having the expression for ϕ we can also write the corresponding expression for the velocity and for the pressure perturbation

$$u = \frac{\partial \phi}{\partial x} = -A \frac{\omega}{c_s} \sin\left(\frac{\omega}{c_s} x\right) \cos(\omega t), \tag{6.89}$$

$$P = -\rho_0 \frac{\partial \phi}{\partial t} = \rho \omega A \cos\left(\frac{\omega}{c_s} x\right) \sin(\omega t). \tag{6.90}$$

At points $x = 0, \pi c_s/\omega, 2\pi c_s/\omega \ldots j\pi c_s/\omega \ldots$, which are a fixed distance $\lambda/2$ from each other, the velocity is always zero. These are the *nodes* of the velocity standing wave. Midway between the nodes are points at which the velocity amplitude is the largest—these are called *antinodes*. It is evident that the pressure and density perturbations, as well as the velocity potential, have their nodes and antinodes reversed with respect to the velocity. The frequency of the sound wave and its harmonics actually determines the musical note that is heard by the human ear. It is therefore self-evident that the manufacturers of musical instruments do take into account the physics of characteristic sound frequencies of cavities and resonators (a cavity with a small aperture). A meaningful description of this topic is clearly beyond the scope of this book.

6.2.5.1   Example

Consider a cavity having the form of a parallelepiped with sides a, $2a$, and $3a$. We wish to find the characteristic acoustic frequencies of a fluid, contained in such a cavity. It is clear the choice of axes (x, y, z) is arbitrary since the answer cannot depend on coordinates. Thus we may decide that the fluid occupies $0 \le x \le a$, $0 \le y \le 2a$, $0 \le z \le 3a$, say. The normal modes of the velocity are 3-dimensional standing waves, with the velocity components normal to the walls of the vessel vanishing. Now, the spatial part of the velocity potential satisfies Eq. (6.80), which we try to separate in the spatial coordinates and thus assume an ansatz $\phi_{sp}(\mathbf{x}) = A\cos(kx)\cos(qy)\cos(pz)$. Clearly, this is a solution provided

$$k^2 + q^2 + p^2 = \frac{\omega^2}{c_s^2}. \tag{6.91}$$

The boundary conditions at $x = 0$, $y = 0$, and $z = 0$, i.e., the vanishing of the normal velocity components are satisfied by construction. Demanding now a similar boundary condition at $x = a$, $y = 2a$, and $z = 3a$ forces

$$k = \frac{n\pi}{a}, \qquad q = \frac{m\pi}{2a}, \qquad p = \frac{l\pi}{3a}, \tag{6.92}$$

where n, m, l are any integers. This gives the condition for ω to be a characteristic frequency

$$\omega_{n,m,l}^2 = \frac{c_s^2 \pi^2}{a^2}\left(n^2 + \frac{m^2}{4} + \frac{l^2}{9}\right). \tag{6.93}$$

This may remind some of counting the states in a box in elementary statistical quantum mechanics. Indeed mathematically, the process is identical. In the consideration up to this point we generally used an extended flow as the initial condition. We shall next consider also the case of a point sound source.

6.2.6 The Emission and Attenuation of Sound

The creation of a sound wave in an otherwise quiet and uniform medium can be achieved by inserting a sound *source* in the fluid. There are different possibilities for the form of such sound sources and under an appropriate approximation they may take the form of *point sources* emitting spherical waves or *plane sources* emitting plane waves. More complicated sources, for example, the fluttering of a hummingbird wing, may also generate a unique pattern of sound. These more complicated cases will be discussed here only briefly, if at all. In contrast to plane sound waves in a fluid medium under the assumption of isentropy, sound waves may be attenuated by geometrical effects as well as by dissipative absorption by the fluid. We shall discuss later several basic mechanisms for that.

6.2.6.1 Sound Emission by Oscillating Bodies

The simplest sound source is an oscillating body, immersed in a uniform fluid. In general, the body can have any form and execute any kind of oscillations. We shall consider only oscillations that produce velocities of the body's boundary which are much smaller than the sound speed, that is, $u_{osc} \ll c_s$. Since in an oscillation typically $u_{osc} = a\omega$, where a is the amplitude of the oscillation and ω the frequency, a sound wave emitted by such vibration must have a frequency ω and therefore a wavelength $\lambda = 2\pi c_s / \omega$, thus, the above strong inequality translates to $a \ll \lambda$. This should hold always and, in addition, the amplitude a has to be much smaller than the oscillating body in question, otherwise the potential flow assumption may be destroyed near the body (can you explain why?). In order to define the two limiting cases, on which we shall expound below, we imagine a body of any shape oscillating in any manner inside a quiet uniform compressible fluid. The basic equation is again the acoustic wave equation, and for our purposes it will be sufficient to choose the relevant function to be the velocity potential $\phi(\mathbf{x}, t)$:

$$\frac{\partial^2 \phi}{\partial t^2} - c_s^2 \nabla^2 \phi = 0, \tag{6.94}$$

where we do not forget that c_s is the sound speed in the unperturbed medium (i.e., $c_{s,0}$). In this discussion, we must take into account also the boundary condition on the surface of the body, which provides the sound source. The condition is that

$$(\hat{\mathbf{n}} \cdot \nabla \phi)_s = u_n, \tag{6.95}$$

where $\hat{\mathbf{n}}$ is the outer normal to the body and so u_n is the normal component of the oscillating body's outer surface velocity and thus also of the fluid immediately adjacent to the surface. We may turn now to the two limiting cases, where in both the typical dimension of the oscillating body is denoted by ℓ and the emitted sound wavelength as λ_{em}. The limiting cases are defined by comparing the body size to the sound wavelength.

- In the first case, we assume that the frequency is so large that the wavelength of the emitted wave $\lambda_{em} = 2\pi c_s / \omega \ll \ell$. This can be obviously always achieved for a large enough ω. The way to proceed is to divide the surface of the oscillating body to portions so small that they may be approximated as planar, but they are still large compared to λ_{em}, that is, each segment emits a plane wave. Remembering that the average energy flux density of a sound wave is as given by Eq. (6.61) and that in a plane wave the velocity (which is necessarily normal to the planes) satisfies Eq. (6.31), we get that any surface element dS of the body emits the following average acoustic energy flux density

$$\langle |\Phi_{\mathcal{E}}| \rangle = c_s \rho \langle u_n^2 \rangle. \tag{6.96}$$

Therefore the total power (energy/time) emitted by the source of the acoustic wave is

$$\mathscr{P} = c_s \rho \oint \langle u_n^2 \rangle dS. \tag{6.97}$$

- The other limiting case, namely, $\lambda_{em} \gg \ell$, is slightly more complicated. It is easy to see that in the immediate vicinity of the body, up to distance of a wavelength from it, say, the term containing the second time derivative in the wave equation is of $\mathcal{O}\left(\omega^2 \phi / c_s^2\right) \sim \phi / \lambda_{em}^2$, while the spatial derivative term is $\mathcal{O}\left(\phi / \ell^2\right)$. Thus, in this case we can neglect that first term in the wave equation and have the Laplace equation only, as long as we are in the vicinity of the body, closer than $\lambda_{em} = 2\pi c_s / \omega$ to it. The Laplace equation is valid for potential flow of an incompressible fluid, and so we may approximate the flow near the body as such a flow. This, of course, does not allow the presence of acoustic waves, however these are possible at large distances from the body.

We cannot write the solution of the Laplace equation in any general form within a distance of the order of the body size ℓ, as it depends on the actual shape of the body. However for distances $d \gg \ell$, but still $d \ll \lambda_{em}$, the Laplace equation is a good approximation and thus we have

$$\nabla^2 \phi = 0, \tag{6.98}$$

with the boundary condition than $\partial \phi / \partial n = u_n$, that is, the normal component of the velocity on the body and also $|\nabla \phi| = \mathbf{u}$ far away from the body (formally at infinity) both vanish. We have already solved such a problem, when describing incompressible potential flow in Sect. 2.4.1. We used what is called a *multipole expansion*, retaining only the first two terms, finding here

$$\phi = -\frac{a}{r} + \mathbf{A} \cdot \nabla \left(\frac{1}{r}\right) + \text{HOT}. \tag{6.99}$$

We remind the reader that the acronym HOT indicates "higher order terms," that scalar a and vector $\mathbf{A}$ are both constant in space, and r is the distance from any point in the body and we stress again that it has to be understood that this approximation is valid far from the body, in which case we can restrict ourselves to only two terms, those which decrease least rapidly with increasing r. We have shown in Sect. 2.4.1 that $a = 0$ unless the fluid volume within the radius a from the origin is not fixed. Consequently, the first (monopole) term had been dropped for that problem. In our case we may envisage a situation in which the body pulsates and as such changes its volume. We shall thus consider separately the case in which the volume of the body changes (case 1 below) and does not change (case 2).

1. Assume that the fluid volume inside a sphere of radius r changes because of a change in the volume of the body, which is responsible for the sound emission. In such a case we cannot guarantee $a = 0$ and consequently the first term is the dominant one, far enough from the origin. Indeed, denote the volume of the body, which is changing in time, by $\mathcal{V}_s(t)$, where the subscript stands for the word "source." It is not difficult to convince

oneself that for $\ell \ll r \ll \lambda_{em}$ the total volume of the fluid outflow through a surface of radius r is equal to the body volume change $d\mathscr{V}_s/dt \equiv \dot{\mathscr{V}}$. This is achieved if we assume that

$$\phi = \frac{-\dot{\mathscr{V}}(t)}{4\pi r}, \quad \text{since this gives} \quad u_r = \frac{\partial\phi}{\partial r} = \frac{\dot{\mathscr{V}}(t)}{4\pi r^2}. \quad (6.100)$$

As long as $r = a$ is large enough compared to ℓ, but still small enough compared to λ_{em}, we may retain only the monopole contribution $a \neq 0$. At distances $r \gg \lambda_{em}$, that is, in the far field wave region, ϕ has to have a form appropriate for an outgoing spherical wave, that is,

$$\phi(r,t) = -\frac{f(t-r/c_s)}{r}. \quad (6.101)$$

The discussion preceding this formula and the wish to preserve continuity of physical variables brings us to the conclusion that the emitted wave has to have the following form of the velocity potential if only $r \gg \ell$, that is, considering the emitting body as a point source

$$\phi(r,t) = -\frac{\dot{\mathscr{V}}(t-r/c_s)}{4\pi r}. \quad (6.102)$$

We are now interested in the behavior of this spherical wave in the true far field, that is, for $r \gg \lambda_{em}$. To obtain the velocity we differentiate the above potential with respect to r, as the problem is spherically symmetric in the domain in question therefore the velocity is radial. While performing the derivative we use the fact that r is much larger than the wavelength, which is of the order of the length scale of variation of the numerator, and consequently the function in the denominator varies slowly and we may approximate the derivative faithfully by performing it on the numerator only. So we get

$$u_r = \frac{\ddot{\mathscr{V}}(t-r/c_s)}{4\pi c_s r}, \quad (6.103)$$

because $\partial\dot{\mathscr{V}}(t-r/c_s)/\partial r = -\ddot{\mathscr{V}}(t-r/c_s)/c_s$. The average total power radiated in this case is

$$\mathscr{P} = \rho c_s \oint \langle u_r^2 \rangle dS = \frac{\rho}{16\pi^2 c_s} \oint \frac{\langle \ddot{\mathscr{V}}^2 \rangle}{r^2} dS, \quad (6.104)$$

where the closed surface over which the integration is taken has any form, so long as the source is inside it. Choosing the surface to be a sphere of radius r immediately gives the total sound power radiated by the point source in question:

$$\mathscr{P} = \frac{\rho}{4\pi c_s} \langle \ddot{\mathscr{V}}^2 \rangle. \tag{6.105}$$

2. Consider now the case in which the sound emitting body does not change its volume, e.g., it performs small oscillation to and fro in the fluid, a case which may include the complicated example of a flapping hummingbird wing. As explained above, the leading term is then the second (dipole) term, since $a = 0$. Thus we have the following expression coming from Eq. (6.99), valid near enough to the body:

$$\phi = \mathbf{A} \cdot \nabla \left(\frac{1}{r} \right) = \nabla \cdot \left[\frac{1}{r} \mathbf{A}(t) \right]. \tag{6.106}$$

At distances $r \gg \ell$ we deduce, as in the previous case, that the expression actually is $\phi = \nabla \cdot [\mathbf{A}(t - r/c_s)/r]$, which is a solution of the wave equation. The operator $\nabla \cdot$ is applied here to differentiate only the numerator, the same approximation as before for $r \gg \lambda_{em}$, giving finally

$$\phi(r,t) = -\frac{1}{c_s r} \dot{\mathbf{A}}(t - r/c_s) \cdot \hat{\mathbf{n}}, \tag{6.107}$$

where $\hat{\mathbf{n}}$ is the direction of emission. The velocity is obtained by $\mathbf{u} = \nabla \phi$ and it is a simple exercise in vector algebra to show that

$$\mathbf{u} = \frac{1}{c_s^2 r} \hat{\mathbf{n}}(\hat{\mathbf{n}} \cdot \ddot{\mathbf{A}}), \tag{6.108}$$

where we have used $\nabla(t - r/c_s) = -(1/c_s)\nabla r = -\hat{\mathbf{n}}/c_s$. As we already know, the total average acoustic wave flux density is obtained by the time average of u^2 multiplied by ρc_s. In this (dipole) case it gives

$$\langle |\Phi_{\mathscr{E}}| \rangle = \frac{\rho(\hat{\mathbf{n}} \cdot \ddot{\mathbf{A}})^2}{c_s^3 r^2}. \tag{6.109}$$

Typically of dipole emission we find proportionality of the flux to the cosine squared between the direction of emission and the vector $\ddot{\mathbf{A}}$. The total emission power can be obtained, again, by integrating over the surface of a large sphere. Choosing the polar axis in spherical coordinates to be in the direction of $\ddot{\mathbf{A}}$ this gives

$$\mathscr{P} = \frac{\rho}{c_s^3} \oint \frac{\langle (\hat{\mathbf{n}} \cdot \ddot{\mathbf{A}})^2 \rangle}{r^2} dS = \frac{4\pi\rho}{3c_s^3} \langle \ddot{\mathbf{A}}^2 \rangle. \tag{6.110}$$

It has to be understood that similarly to the problem of potential incompressible flow (see Sect. 2.4.1), $\mathbf{A}$ is dependent on the body shape and linearly dependent on its velocity. In both cases given here, it should be clear that if there is a single frequency oscillation of the emitting body, the second time derivative for a given velocity amplitude is proportional to ω^2. This gives the emitted power to behave like $\propto \omega^4$. In the case that the oscillation amplitude is given, the velocity amplitude itself behaves like ω^2 and thus the power is $\propto \omega^6$.

There is a simpler approach to the problem of sound emission. For example, one can deal with only the far field form of the wave. In such a case the source is taken as a point, with velocity potential like that of an ordinary spherical wave

$$\phi_s(r,t) = \frac{A}{r} e^{i(\omega t - kr)}, \tag{6.111}$$

with $k \equiv \omega/c_s$, so as to satisfy the wave equation. This is considered to be a monopole emission and a dipole emission is nothing else but two such sources at distance d, say, apart. They may be in phase or out of phase, in the latter case the second source's complex amplitude is written in a way that takes cares of the phase difference. An example of dipole sound radiation, calculated in this approach, may be found in Problem 6.16.

6.2.6.2 Sound Generation by Musical Instruments Ending with a Horn

Sound radiation by tubes of wind instruments depends on the flaring of the bell at the end of the tube. For example, the trumpet's very curved bell is responsible for emission of higher harmonics than the clarinet's soft shaped bell. In principle, the sharper the curvature of the bell, the brighter the sound. To give this statement a quantitative meaning it has to be defined in mathematical terms. The function $r(x)$ expressing the radius of the round horn shaped region of the instrument, as a function of linear distance, x, along the tube can be used to find curvature. The sharpness of the latter for a horn's bell can be quantified by the size of the second derivative d^2r/dx^2. In order not to complicate the problem and find the exact sound wave structure along the instrument, we shall not take into account the boundary condition at the beginning of the tube. We shall focus on a simple progressive wave in the tube and try to find out what frequencies (or wavelengths) we can expect emanating from a horn with a given d^2r/dx^2 of its bell.

The appropriate equation for this problem is known as the *Webster* equation and it reads

$$\frac{1}{S}\frac{\partial}{\partial x}\left(S\frac{\partial P}{\partial x}\right) = \frac{1}{c_s^2}\frac{\partial^2 P}{\partial t^2}, \tag{6.112}$$

where $S = \pi r^2$ is the cross sectional area of the horn. Define now $\Psi(x,t) \equiv P(x,t)S$. Taking, as an approximation valid in a tube with $r(x) =$const., $\Psi \propto e^{i(kx-\omega t)}$, it is possible to get by a few algebraic manipulations the equation

$$\frac{\partial^2 \Psi}{\partial x^2} - V(x)\Psi = -\omega^2 c_s^2 \Psi, \tag{6.113}$$

where

$$V(x) = \frac{1}{r}\frac{d^2 r}{dx^2}. \tag{6.114}$$

We remember that in our approximation we have taken an exact harmonic wave and therefore there is a single ω. But the whole idea is that because of the bell function, we actually have to allow for more general eigenvalues, that is, we have to solve a time independent Schrödingier equation, like (6.113) with a special potential determined by the curvature of the horn, that is (6.114) and general eigenvalues. Now, $V(x) > 0$ for horns flaring out, so actually we have a potential well, which ends at some x_{end}, the edge of the horn. We thus write

$$\frac{\partial^2 \Psi_n}{\partial x^2} - V(x)\Psi_n = E_n(\omega)\Psi_n, \tag{6.115}$$

where we have explicitly allowed for the eigenvalues to depend on the frequency of the simple wave in the tube. This is then a similar case to a quantum harmonic oscillator, having a zero point energy, a number of "bound state" energy levels and because the well ends at a finite x and thus height, there is some energy tunneling out and this is the radiation energy (depending on the frequency).[1] Unfortunately, we have to take recourse to numerical calculations of the zero point eigenvalue, its overtones, and the radiated sound. All this depends on the basic frequency being input to the instrument and on the horn function $r(x)$.

6.2.6.3 Sound Generation by Fluid Flow: General Treatment

The acoustic approximation consists of examining the physical development of small disturbances of an otherwise static and uniform but compressible fluid. In what follows we shall derive again the wave equation, but keep some of the neglected terms, in particular the velocity ones, which we shall perceive as a source to that equation. Consider the Euler equation, written in tensor conservation form, based on Eq. (1.56)

$$\frac{\partial(\rho u_i)}{\partial t} + \frac{\partial \Pi_{ij}}{\partial x_j} = 0. \tag{6.116}$$

---

[1] Our late colleague Robert Buchler, who contributed much to the theory of stellar pulsation and played the flute, used these kinds of arguments in explaining certain models of pulsating stars.

The momentum flux density tensor acquires here the form $\Pi_{ij} = \rho u_i u_j + P\delta_{ij}$, because we are considering inviscid flow, i.e., $\sigma'_{i,j} = 0$. Expanding now $P = P_0 + P'$, where P' is small, we can split $\Pi_{ij} = \mathcal{L}_{ij} + \mathcal{N}_{ij}$, that is, the momentum flux density tensor has a linear part $\mathcal{L}_{ij} = P'\delta_{ij}$ and a nonlinear part $\mathcal{N}_{ij} = \rho u_i u_j$. Adding now and subtracting from Π_{ij} a term $c_s^2 \rho' \delta_{ij}$ we get that the linear part is $\mathcal{L}_{ij} = c_s^2 \rho' \delta_{ij}$ while the nonlinear one is

$$\mathcal{N}_{ij} = \rho u_i u_j + (P' - c_s^2 \rho')\delta_{ij}. \tag{6.117}$$

Substituting now the linear plus nonlinear form of the tensor Π_{ij}, as above, $\Pi_{ij} = \mathcal{L}_{ij} + \mathcal{N}_{ij}$, into the tensor form of the Euler equation (6.116) gives

$$\frac{\partial}{\partial t}(\rho u_i) + c_s^2 \frac{\partial}{\partial x_i}\rho' = -\frac{\mathcal{N}_{ij}}{\partial x_j}. \tag{6.118}$$

Differentiating this equation with respect to x_i and using the equation of continuity for ρ', that is,

$$\frac{\partial}{\partial t}\rho' + \frac{\partial}{\partial x_i}(\rho u_i) = 0, \tag{6.119}$$

we obtain finally the inhomogeneous wave equation for the density perturbation $\rho' \mapsto \rho$:

$$\left(\frac{\partial^2}{\partial t^2} - c_s^2 \nabla^2\right)\rho = \mathcal{S} \tag{6.120}$$

where the source is given here by

$$\mathcal{S} = \frac{\partial^2 \mathcal{N}_{ij}}{\partial x_i \partial x_j}, \tag{6.121}$$

in which the summations on i and j should not be forgotten. The solution of equation (6.120) follows a standard, but perhaps not trivial, procedure of finding the appropriate Green's function, as is undoubtedly known by the reader who had studied electromagnetism on a graduate level. Thus, the closed form solution is here

$$\rho(\mathbf{x},t) = \frac{1}{4\pi c_s^2} \int \frac{\mathcal{S}(\mathbf{x}', t - |\mathbf{x} - \mathbf{x}'|/c_s)}{|\mathbf{x} - \mathbf{x}'|} d^3 x'. \tag{6.122}$$

6.2.6.4   Attenuation and Absorption of Sound

The isentropic acoustic waves do not lose energy and as such are obviously an idealization. In practice, fluids invariably include dissipative effects, however small as compared to the bulk of the acoustic energy in sound waves. In addition to the obvious geometrical effects of wave spreading, as is apparent in spherical waves due to the factor $1/r$ in the wave function and $1/r^2$ in energy flux, viscosity dissipates fluid motion to heat, and wave motion is no exception, while thermal conductivity spreads this heat spatially. The last process is especially prominent near walls of architectural spaces, but we will not consider that problem here. The discussion here will be limited to a short summary of the processes that attenuate and absorb sound energy in unbounded spaces. We will not discuss other possible processes of attenuation due to scattering by dust or fog, nor to *turbulent* scattering. Turbulent flow regions scatter sound by the random motions of smallish eddies, which are typical in turbulence (see Chap. 9). This is a difficult and not yet fully understood topic. Here, suffice it to mention one example: the rumbling, relatively long, sound of a thunder, following a short lightning, in some distant storm. The lightning presumably induces rapid turbulent flow into the region whose density is rapidly and locally diminished by the lightning.

The first effect we shall discuss here is the purely geometrical spatial spreading of the wave. This is not difficult to calculate for a spherical wave, which we shall consider here for simplicity

$$\mathscr{P}(r,t) = \frac{1}{r} f(r - c_s t), \tag{6.123}$$

where we use here for the first time the *condensation* $\mathscr{P} \equiv (\rho - \rho_0)/\rho_0$, with ρ_0 being the unperturbed density and $\rho = \rho_0 + \rho'$. The average energy flux density can be written for a spherical wave thus

$$\langle \Phi_e \rangle = \frac{1}{r^2} \Phi_0 \tag{6.124}$$

where the flux is, in general, a vector quantity, its direction being that of the group velocity. In our spherical wave we have used its absolute value and Φ_0 is a constant, if only the geometrical effect is taken into account. It is now easy to calculate the attenuation of sound energy, just because of the $1/r^2$ effect. If we increase the distance from the source by a factor of 2, for example, we get for the ratio of the average fluxes at the two points $r_2 = 2r_1$, the following $\langle \Phi_2 \rangle / \langle \Phi_1 \rangle = (r_1/r_2)^2 = 1/4$. Thus the attenuation in decibels

$$\delta\Delta = 10 \left(\log_{10} \frac{\langle \Phi_1 \rangle}{\Phi_{\text{ref}}} - \log_{10} \frac{\langle \Phi_2 \rangle}{\Phi_{\text{ref}}} \right) = 10 \log_{10} 4 \approx 6 \, \text{dB}. \tag{6.125}$$

We now turn to the problem of plane traveling waves whose slow energy loss, manifested by decrease of amplitude as they progress, is brought about by dissipative effects. Dropping the subscript e and considering Φ to be the already averaged quantity, we write the balance of energy upon its passage through a volume $\mathscr{V}$

$$S(\Phi_1 - \Phi_2) = \left\langle \int_{\mathscr{V}} \rho \frac{De}{Dt} d^3x \right\rangle + \left\langle \int_{\mathscr{V}} P\rho \frac{Dv}{Dt} d^3x \right\rangle, \tag{6.126}$$

where on the right-hand side we have time averages of the increase in internal energy plus the work done on expansion in the volume $\mathscr{V}$ upon passage of the wave through it. S is the surface area of the projection of the volume on a plane perpendicular to the wave propagation direction and $v = 1/\rho$. Φ_1 and Φ_2 are, respectively, the energy fluxes entering the volume and exiting it on the other side. We have average energy flux loss, which goes to dissipative entropy increase in the volume. The above can be written in a differential form, exploiting also the Gibbs equation (1.66) we then get, assuming wave propagation in the $\hat{\mathbf{x}}$ direction,

$$-\frac{d\Phi}{dx} = \left\langle \rho T \frac{Ds}{Dt} \right\rangle. \tag{6.127}$$

In Chap. 1 we have derived the formula (1.78), when discussing the second law of thermodynamics in the context of fluid motion. It is convenient to repeat it here for immediate use

$$\rho T \frac{Ds}{Dt} = \Psi + \frac{\kappa}{T}(\nabla T)^2 + T\nabla \cdot \left(\frac{\kappa \nabla T}{T}\right), \tag{6.128}$$

where the dissipation function Ψ is given by $\sigma'_{ik}\mathscr{D}_{ik}$. The former is the viscous stress tensor while the latter is the deformation rate tensor

$$\mathscr{D}_{ik} = \frac{1}{2}\left(\frac{\partial u_i}{\partial x_k} + \frac{\partial u_k}{\partial x_i}\right) \tag{6.129}$$

It is not difficult to see that with $\Psi = 2\eta \mathscr{D}_{ik}\mathscr{D}_{ik} + (\zeta - \frac{2}{3}\eta)(\mathscr{D}_{jj})^2$ and collecting all the terms from Eq. (6.128) (we limit ourselves to a progressive plane wave and remember that $\hat{\mathbf{x}}$ is the direction of propagation) we have for the dissipation function

$$\Psi = \left(\frac{4}{3}\eta + \zeta\right)\left(\frac{\partial u}{\partial x}\right)^2, \tag{6.130}$$

where the dynamic coefficients of viscosity shows up. Inserting this into Eq. (6.128) and time averaging we get

$$-\frac{d\Phi}{dt} = \left(\zeta + \frac{4}{3}\eta\right)\left\langle \left(\frac{\partial u}{\partial x}\right)^2 \right\rangle - \left\langle \frac{\kappa}{T}(\nabla T)^2 \right\rangle, \tag{6.131}$$

owing to the fact that the time average of the last term in Eq. (6.128) is zero, as follows from Problem 6.17. The appearance of the viscosity coefficient η is physically obvious. Viscosity is a mechanism of dissipation in the fluid and therefore it must have a role in the process of attenuation of sound.

Next, we deal with the second term on the right-hand side of equation (6.131). Taylor expanding $T(\rho, s)$ up to the first order of the condensation, $\mathscr{P}$

$$T = T_0 + \left(\frac{\partial T}{\partial \rho}\right)_s \rho \mathscr{P} = T_0 - \left(\frac{\partial T}{\partial v}\right)_s v \mathscr{P}. \tag{6.132}$$

Now, in this one-dimensional case $\nabla T = \hat{\mathbf{x}}(\partial T/\partial x)$ so using the above Taylor expansion and the thermodynamic identity

$$\left(\frac{\partial T}{\partial v}\right)_s^2 = (\gamma - 1)\frac{c^2 T}{v^2 c_P}, \tag{6.133}$$

where c_P is the specific heat at constant pressure (can you derive it for a perfect gas with constant specific heats?), the term of equation (6.131) we are evaluating, for a harmonic wave, that is, say $\mathscr{P}(x,t) = \mathscr{P}_0 \sin(kx - kct)$, becomes

$$\frac{\kappa}{T}\left\langle \left(\frac{\partial T}{\partial x}\right)^2 \right\rangle = \frac{(\gamma - 1)\kappa c^2}{2 c_P} k^2 \mathscr{P}_0^2. \tag{6.134}$$

Here k is the wavenumber and $\mathscr{P}_0$ the amplitude of the above harmonic wave. Since in a traveling one-dimensional sound wave $u = c\mathscr{P}$, Eq. (6.127) with the use of ,(6.128), (6.130), and (6.134) thus becomes

$$-\frac{d\Phi}{dx} = -\frac{1}{2}k^2 u_0^2 \eta c^2 \left(\frac{4}{3} + \frac{\zeta}{\eta} + \frac{\gamma - 1}{\mathrm{Pr}}\right), \tag{6.135}$$

where we remind the reader that the Prandtl number relates coefficients of viscosity and thermal conductivity. Here we use the dynamic viscosity coefficient and so $\mathrm{Pr} = (\eta c_P)/\kappa$.

Using a conventional definition for the attenuation relation,

$$\frac{d \ln \Phi}{dx} = -2\alpha \quad \Rightarrow \quad \Phi = \Phi_0 e^{-2\alpha x}, \tag{6.136}$$

it is easy to understand that α is defined for the attenuation of the wave amplitude, since the energy goes like the amplitude squared. This section contains all the equations needed for the calculation of the attenuation coefficient α for specific cases with known values of all required thermodynamic quantities.

6.3 Properties of Compressible Flows

We have already asserted several times in this book that effects of compressibility become important in flows in which the velocity $u = |\mathbf{u}|$ is large enough, so as to be comparable to the sound speed. This is expressed by the value of the Mach number

($M \equiv u/c_s$). Note that both u and c_s are *local* quantities and so the dimensionless Mach number may actually depend on position and if the flow is not steady then also on time. However in most flows, M is an increasing function of u (see below). As remarked before, compressible flows with $M = \mathscr{O}(1)$, or more, predominantly occur in gases and hence we talk of gas dynamics. When the motions in a flow are sonic or supersonic, the Reynolds number is usually very large, because the coefficient of microscopic kinematic viscosity is of the order $v \sim lc_s$, where here l denotes the mean free path of the microscopic particles comprising the gas. Thus $Re = Lu/(lc_s) \sim L/l$, where L is the size of the system and must be $\gg l$ in all cases except in very rarefied gases, when the continuum description fails. In most atmospheric and astrophysical systems, the Reynolds number is huge and if we expect to observe dissipative effects, like the results of viscosity or heat conductivity, we need to invoke some greatly enhanced, "macroscopic" coefficients of viscosity and conductivity and, as we shall see in Chap. 9, this may occur naturally when instabilities lead to great complexity of the flow which is referred to as *turbulent*.

In gas dynamics there is an important distinction between flows, which also determines their different behavior. The demarcation is between *subsonic* and *supersonic* flows. Supersonic flows may exhibit specific behavior which was not described hitherto, in this book. However in Chap. 4, when surface wave breaking in water or other constant density fluid was discussed, formal similarities to supersonic flow were encountered mathematically through the method of characteristics. We shall devote most of this section to general properties of supersonic flows and the rest of this chapter to specific prominent phenomena occurring in them. For simplicity we shall assume henceforth that the gas motion is steady.

6.3.1 Propagation of Disturbances, the Mach Cone

Consider, for a moment, a steady, uniform, one-dimensional supersonic flow with velocity u. We create a small perturbation, a pulse, say, in the density imposed on the basic flow at a certain place and time. The sound wave thus created will travel in both directions, forward and backward, with the speed of sound of the undisturbed medium c_s, relative to the fluid. Thus in the absolute, or an appropriate fixed inertial frame, the speed of the disturbance will be

$$v_{\text{disturb}} = u \pm c_s. \tag{6.137}$$

In the case $u > c_s$ both waves move downstream, that is, in supersonic flow disturbances cannot propagate upstream, as is obvious. This means that the upstream region is outside the *region of influence* of the disturbance. In the subsonic case ($u < c_s$) however, both the upstream and downstream regions are within the influence of the event, which here was the creation of a small density pulse.

The concept of region of influence is perhaps best demonstrated when we imagine a sound point source emitting pulses of sound at regular intervals of time.

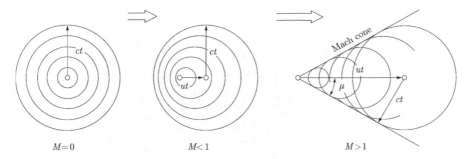

$M=0$ $M<1$ $M>1$

Fig. 6.1 Stationary sound source in uniform flow (to the *right*) of velocity u. The sound source emits very short pulses, which are illustrated by the circular wavefronts. The three panels from left to right show the situation in the three cases of M as denoted in the figure. The *rightmost panel* case (M > 1), i.e., supersonic motion, allows a geometrical construction of the Mach cone

The source itself is moving to the left, say, with constant velocity u, while the gas medium is stationary and an observer, actually listener, is stationary in the source frame. Looking at the problem in the source frame, we have the air moving to the right with uniform velocity u. The listener is observing the spherical wavefronts in the special frame in which the source is stationary, and is doing so assuming the medium is three dimensional. We distinguish between three cases $u = 0$ (left panel of Fig. 6.1), $u \lesssim c_s$ (central panel), and a supersonic flow $u > c_s$ (right panel). The first case is trivial and we do not discuss it at all. The second case is slightly more interesting, but still it does not carry much significance, save the distortion of the front array. If one is stationary in the frame of the fluid, this second case has an observer, assumed moving with the fluid, with the source receding to the left and thus is consistent with what is well known as the Doppler effect—the observer sees a lower frequency of pulses. In the third case the Mach cone, whose axis is ut, is constructed by allowing a time interval t to pass. Now we draw a perpendicular line from the source to the line drawn as the envelope of the wavefronts. The length of the line is ct. An observer hears no sound unless he or she is *in* the Mach cone. Outside the Mach cone he or she is outside the region of influence. When the Mach cone reaches and crosses the position of the observer he or she starts hearing the sound. The phenomenon may be familiar from observing a supersonic aircraft passing at a distance. The relation satisfied by the cone angle (called the *Mach angle*) can be easily derived by inspection of the triangle appearing in the rightmost panel of Fig. 6.1, that is:

$$\sin \mu = M^{-1}. \qquad (6.138)$$

Moreover, the familiar phenomenon of a sonic *boom* can be easily understood with the help of this construction. When an object such as a military aircraft is moving supersonically, it creates what is called an N wave, brought about by the nose and ended by the tail of the aircraft. The boom is experienced when there is a sudden change in pressure, therefore an N-wave causes two booms—one when the initial

Fig. 6.2 A photograph, using a special (so-called Schlieren or stereoscopy) technique of a bullet, taken by Mach himself in 1888, demonstrating very well the cause of the double boom. [*Public Domain (more than 70 years of copyright have expired). Courtesy of Ernst Mach*]

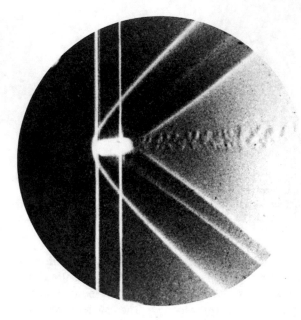

pressure rise from the nose hits the listener position, and another when the tail passes and the pressure suddenly returns to normal. There may be more complications brought about by various kinds of sonic booms, but we shall not dwell on these details and be content with the explanation of the phenomenon's basics.

In photographs of steady supersonic flows, which use special techniques to display Mach surfaces and lines, two-dimensional projections of Mach cones or their parts, the latter are visible emanating from points of a local disturbance, e.g., rough spots of supersonic projectiles of rockets (Fig. 6.2). In general, the nature of compressible flows depends on the Mach number. For $M < 1$ and, in particular, if the flow speed is low enough so that $M \lesssim 0.3$ the flows can be very accurately treated as incompressible. In contrast, for $M > 1$ new phenomena appear, unknown from incompressible flows, among them Mach cones and shock waves, the latter of which we shall describe in considerable detail in the remainder of this chapter. For the case $M \approx 1$ a special mathematical treatment is required and we shall not deal with this case here. The case $M \gg 1$ is referred to as *hypersonic* flow. Again, new physical phenomena related to ionization and viscosity modification must come into play. As we have already stressed, the main difference between subsonic and supersonic flow is in the limited range of influence in the supersonic case, as opposed to the unlimited such region in subsonic flow. We shall try to manifest it when discussing potential flow in the two cases in Sect. 6.3.2, but before that we wish to justify a statement we have made before, that the Mach number increases with flow velocity. We shall do this for isentropic flow. One of the Bernoulli theorems—when the flow is isentropic and inviscid—gives that the Bernoulli function is constant along streamlines, see Eq. (2.27) and the text following it (case 1). Thus treating now a situation when there is no body force, we have

$$\mathscr{B} = h + \frac{1}{2}u^2 = \text{const.} \quad \text{on streamlines.} \tag{6.139}$$

Remembering also the Gibbs equality, which for isentropy gives $dh = dP/\rho$, we may write that along streamlines

$$u\,du + \upsilon\,dP = 0 \tag{6.140}$$

holds, where $\upsilon = 1/\rho$ is the specific volume, as usual. There will also be a corresponding change in the sound speed squared along the streamline, viz.

$$dc_s^2 = \left[\left(\frac{\partial P}{\partial c_s^2} \right)_s \right]^{-1} dP. \tag{6.141}$$

On the other hand, we may write

$$c_s^2 = \left(\frac{\partial P}{\partial \rho} \right)_s = -\upsilon^2 \left(\frac{\partial P}{\partial \upsilon} \right)_s = -\frac{\upsilon^2}{(\partial \upsilon/\partial P)_s}, \tag{6.142}$$

and take the P derivative of this expression, keeping s constant. This yields

$$\left(\frac{\partial c_s^2}{\partial P} \right)_s = \frac{2}{\rho}(\Gamma - 1), \tag{6.143}$$

where the quantity Γ, defined as

$$\Gamma \equiv \frac{c_s^4}{2\upsilon^3} \left(\frac{\partial^2 \upsilon}{\partial P^2} \right)_s, \tag{6.144}$$

is called the *fundamental gasdynamic derivative*.

Using now the relations (6.141) and (6.143) in (6.140) we obtain

$$u\,du + \frac{c_s dc_s}{\Gamma - 1} = 0. \tag{6.145}$$

On the other hand, from the definition of the mach number, $\text{M} = u/c_s$, it follows that

$$\frac{d\text{M}}{\text{M}} = \frac{du}{u} - \frac{dc_s}{c_s}, \tag{6.146}$$

and using this together with the previous equation to eliminate c_s and its differential gives

$$\frac{du}{u} = \frac{d\text{M}/\text{M}}{1 + (\Gamma - 1)\text{M}^2}. \tag{6.147}$$

This proves that the Mach number increases monotonically with the fluid velocity u, provided $\Gamma \geq 1$. This condition is satisfied for most normal fluids. In Problem 6.19 it is found for a perfect gas.

6.3.2   Compressible Potential Flow

Assume an isentropic irrotational motion, that is, potential flow, satisfying by definition $\mathbf{u} = \nabla\phi$ and all the other conditions for such a motion; but in contrast to our discussion in Chap. 2, we allow now for the fluid to be compressible. For convenience we rewrite here the Euler equation (1.37) without body forces, that is, with $\mathbf{b} = \mathbf{0}$:

$$\frac{\partial \mathbf{u}}{\partial t} + (\mathbf{u} \cdot \nabla)\mathbf{u} = -\frac{1}{\rho}\nabla P.$$

Now, forming the scalar product of Euler's equation with $\mathbf{u}$ yields

$$\frac{1}{2}\left[\frac{\partial(u^2)}{\partial t} + \mathbf{u} \cdot \nabla(u^2)\right] = -\frac{1}{\rho}\mathbf{u} \cdot \nabla P. \tag{6.148}$$

We shall use the identity $\mathbf{u} \times \omega = \frac{1}{2}\nabla(u^2) - (\mathbf{u} \cdot \nabla)\mathbf{u}$ and exploit the condition $\omega = 0$ in this potential flow. Also, the right-hand side of equation (6.148) can be transformed to read

$$\frac{1}{\rho}\mathbf{u} \cdot \nabla P = \frac{1}{\rho}\left(\frac{DP}{Dt} - \frac{\partial P}{\partial t}\right). \tag{6.149}$$

Using now $dP = c_s^2 d\rho$ and the continuity equation this expression becomes

$$\frac{1}{\rho}\mathbf{u} \cdot \nabla P = -c_s^2\nabla \cdot \mathbf{u} - \frac{1}{\rho}\frac{\partial P}{\partial t}. \tag{6.150}$$

The scalar product of the Euler equation with $\mathbf{u}$ can thus be summarized as

$$\frac{1}{2}\frac{\partial u^2}{\partial t} + \frac{1}{2}\mathbf{u} \cdot \nabla u^2 - c_s^2\nabla \cdot \mathbf{u} = \frac{1}{\rho}\frac{\partial P}{\partial t}. \tag{6.151}$$

Taking now the partial time derivative of the Euler equation after expressing the velocity as the gradient of the velocity potential gives

$$\frac{\partial^2}{\partial t^2}\nabla\phi + \frac{1}{2}\frac{\partial}{\partial t}\nabla(\nabla\phi)^2 + \frac{\partial}{\partial t}\left(\frac{1}{\rho}\nabla P\right) = 0. \tag{6.152}$$

To proceed we need to use the identity (prove it!)

$$\frac{\partial}{\partial t}[f(P)\nabla P] = \nabla \left[f(P)\frac{\partial P}{\partial t}\right], \tag{6.153}$$

which, when used in the last term of equation (6.152) can bring that equation to the form

$$\nabla \left[\frac{\partial^2 \phi}{\partial t^2} + \frac{1}{2}\frac{\partial}{\partial t}(\nabla\phi)^2 + \frac{1}{\rho}\frac{\partial P}{\partial t}\right] = 0. \tag{6.154}$$

We now see that we have actually derived, in a somewhat alternative way, case 3b of the Bernoulli formula, as listed in Sect. 2.3.5, but without any body force. Indeed, the quantity in square bracket is an arbitrary function of time including a constant. In any case, it can be absorbed into the velocity potential, redefining it, but not changing anything in the physics. Thus we may write

$$\frac{\partial^2 \phi}{\partial t^2} + \frac{1}{2}\frac{\partial}{\partial t}(\nabla\phi)^2 + \frac{1}{\rho}\frac{\partial P}{\partial t} = 0. \tag{6.155}$$

Integration of this equation over time gives

$$\frac{\partial \phi}{\partial t} + \frac{1}{2}(\nabla\phi)^2 + \int \frac{dP}{\rho} = \text{constant}, \tag{6.156}$$

where this constant in time may depend on position.

We shall conclude this discussion of compressible potential flow by deriving a general equation, which follows from Eq. (6.155) substituted into Eq. (6.151). After expressing the flow velocity by $\mathbf{u} = \nabla\phi$ everywhere that it appears, the following interesting equation is uncovered:

$$\frac{\partial^2 \phi}{\partial t^2} + \frac{\partial}{\partial t}(\nabla\phi)^2 + \frac{1}{2}(\nabla\phi \cdot \nabla)(\nabla\phi)^2 - c_s^2\nabla^2\phi = 0, \tag{6.157}$$

where the sound speed can be found in terms of the velocity potential by using Eq. (6.156) and the relation $dP = c_s^2 d\rho$. This equation has two archetypal special cases. For $c_s \to \infty$, which is equivalent to incompressibility, we indeed obtain the incompressible potential flow, in which the velocity potential satisfies the Laplace equation $\nabla^2\phi = 0$. This is the archetype of an *elliptic* PDE. Equations of this type describe the end states of diffusive processes. The other case is obtained for finite speed of sound and for small amplitude motions, so that terms $(\nabla\phi)^2$ can be neglected. We obtain the wave equation, $\partial_t^2 \phi - c_s^2\nabla^2\phi = 0$, which is the archetype of a *hyperbolic* PDE. Hyperbolic equations are associated, as we have seen, with wave phenomena. We have already discussed at length, in Chap. 4, various cases of fluid dynamical waves. Here we have seen the acoustic waves example and will deal, later on, with the interesting phenomenon of shock waves and wave breaking in gravity waves on shallow water, which arise when the wave amplitude is not necessarily small. Although we shall not be able to give here the full mathematical theory of PDE classification and properties, we shall mention some of it below, in particular the idea of characteristics and their possible use.

6.3.3   Isentropic Flow of a Compressible Perfect Gas

The Bernoulli equation, valid for fluid flows fulfilling the appropriate conditions, presented where we have summarized the various Bernoulli's formulae in Sect. 2.3.5 asserts, for isentropic flows and when there are no body forces, that

$$\mathscr{B} = \frac{1}{2}u^2 + h \tag{6.158}$$

is constant along streamlines. In the case of potential flow, the constant is the same for all streamlines. If there is a stagnation point, i.e., $u = 0$ on a certain streamline, or anywhere in the potential flow we may write

$$\frac{1}{2}u^2 + h = h_0, \tag{6.159}$$

where h_0 is called the stagnation enthalpy. We now chose to discuss only this case of a perfect gas, for simplicity and transparency. The following equation of state then holds:

$$P\upsilon = \frac{P}{\rho} = \frac{\mathscr{R}}{\mu}T, \tag{6.160}$$

where $\mathscr{R}$ is the gas constant ($= 8.314 \times 10^7$ erg/deg mol) and μ is the mean molecular weight.

The velocity of sound, as we have seen, is:

$$c_s^2 = \gamma\frac{\mathscr{R}}{\mu}T = \gamma\frac{P}{\rho}, \tag{6.161}$$

where $\gamma = c_P/c_V$, i.e., is the ratio of specific heats, also known as the adiabatic exponent, note that $\gamma > 1$ always. For monoatomic gases it is equal to $5/3$ and for diatomic gases at not too extreme temperatures, it is $7/3$.

The specific internal energy, up to an insignificant constant, can be written as:

$$e = c_V T = \frac{P}{\rho(\gamma - 1)} = \frac{c_s^2}{\gamma(\gamma - 1)}. \tag{6.162}$$

The specific enthalpy is obtained using $c_P - c_V = \mathcal{R}/\mu$:

$$h = c_P T = \gamma \frac{P}{\rho(\gamma - 1)} = \frac{c_s^2}{\gamma - 1}. \tag{6.163}$$

The specific entropy is:

$$s = c_V \log \left(P \rho^{-\gamma} \right). \tag{6.164}$$

Using now the specific enthalpy equation for the flowing gas and substituting it in Eq. (6.159) yields

$$c_P T + \frac{1}{2} u^2 = c_P T_0, \tag{6.165}$$

where T_0 is the stagnation temperature. It is not difficult to deduce from the list of relations for a perfect gas the following two relations

$$c_s^2 = \gamma R T \quad \text{and} \quad c_P = \frac{\gamma R}{\gamma - 1}, \tag{6.166}$$

where as before in this book, $\mathcal{R}/\mu = R$, for convenience. When these are substituted back into Eq. (6.165) we get the important relation

$$c_s^2 + (\gamma - 1) \frac{u^2}{2} = c_0^2, \tag{6.167}$$

where c_0 is the sound speed at the stagnation point. From now on we shall drop the subscript s from the adiabatic sound speed, but remember that it is calculated with s constant. It is useful to introduce now the concept of the *sonic condition*, in essence a location at which $M = 1$. In one-dimensional flows this is usually one or very few points, in which the flow changes from subsonic to supersonic or vice versa. The value of a variable at a sonic point will be denoted by an asterisk subscript. Thus we have, from the last equation, remembering that at a sonic point necessarily $u = c = c_*$,

$$c_*^2 = \frac{2}{\gamma + 1} c_0^2 \implies c_* = \sqrt{\frac{2}{\gamma + 1}} c_0. \tag{6.168}$$

This allows Eq. (6.167) to be rewritten as

$$c_s^2 + \frac{\gamma - 1}{2} u^2 = \frac{\gamma + 1}{2} c_*^2 \tag{6.169}$$

or yet in another way, by dividing it through by c^2, which gives

$$1 + \frac{\gamma - 1}{2} M^2 = \left(\frac{T_0}{T} \right). \tag{6.170}$$

Setting $M = 1$ in this equation and making use of the adiabatic thermodynamic relations, which are valid here, i.e., $P \propto \rho^\gamma$; $T \propto \rho^{\gamma-1}$, we immediately get

$$T_* = T_0 \left(\frac{2}{\gamma + 1} \right); \quad \rho_* = \rho_0 \left(\frac{2}{\gamma + 1} \right)^{1/(\gamma-1)}; \quad P_* = P_0 \left(\frac{2}{\gamma + 1} \right)^{\gamma/(\gamma-1)}. \tag{6.171}$$

Before giving a list of formulae, which can be derived from the above relations using only simple algebra, we would like to discuss two concepts, which have a particular physical meaning. The first one is the *limit* or *maximum* flow velocity. Imagine that a large reservoir is filled with gas, whose velocity is zero. If the reservoir is evacuated into vacuum by opening a small nozzle and assuming one-dimensional flow, a steady flow is established, in which the pressure drops from its value in the reservoir, P_0, to zero. Using Eq. (6.170) and one of the adiabatic conditions yields

$$\left(\frac{P_0}{P} \right)^{(\gamma-1)/\gamma} = 1 + \frac{\gamma - 1}{2} M^2. \tag{6.172}$$

As $P \to 0$, formally $M \to \infty$. The flow velocity remains finite, however, because in vacuum $c = 0$. Substituting $c = 0$ in Eq. (6.167) gives the maximal velocity that may be attained in this kind of flow:

$$u_{max} = c_0 \sqrt{\frac{2}{\gamma - 1}}. \tag{6.173}$$

The second concept is the *normalized* Mach number. Formally, it is defined as $M_* \equiv u/c_*$. Note that it is not the Mach number at the sonic condition, which is by definition $M = 1$. Since c_* is a constant for any given stagnation condition, M_* can be regarded as a *normalized* flow velocity. It is easy to show, using Eqs. (6.170)–(6.171), that

$$M_* = \frac{M}{\left[\frac{2}{\gamma+1} + \frac{\gamma-1}{\gamma+1} M^2 \right]^{1/2}}. \tag{6.174}$$

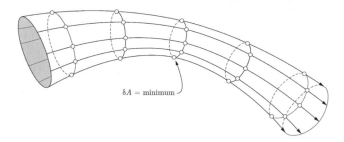

$\delta A = \text{minimum}$

Fig. 6.3 A schematic view of a stream tube, defined by a surface composed of streamlines. A minimal cross-sectional area is indicated

An abstract surface can then be perceived as formed by a collection of streamlines encompassing a volume, whose form is one of a stream tube (see Fig. 6.3). It is shown in Problem 6.20 that the mass flux density ρu can be written as

$$\rho u = \frac{\rho_* u_* M}{[2/(\gamma+1)+M^2(\gamma-1)/(\gamma+1)]^{(\gamma+1)/[2(\gamma-1)]}}. \tag{6.175}$$

The mass flux through the stream tube is $\rho u \delta A$, where δA is its cross-sectional area. If we consider δA to be small enough, so that all properties of the fluid are approximately uniform across any cross sectional area, mass conservation demands $\rho u \delta A = \text{const}$. If the mass flux density, ρu, is considered to be a function of M with all other quantities constant for given stagnation conditions, it is possible to find its extremum by differentiating with respect to M and equating the derivative to zero. The result of this simple exercise is that the flux density attains a maximum for $M = 1$. This means that the mass flux density is maximal at the sonic condition and the corollary from mass conservation is that the stream tube cross section is minimal at the sonic point. It should be remarked that the converse of these statements is obviously not true (can you think of a counterexample?). In terms of M_* this reads, using the relation between M and M_* (6.174),

$$\rho u = \rho_* u_* \left(\frac{\gamma+1}{2} - \frac{\gamma-1}{2} M_*^2 \right)^{1/(\gamma-1)}, \tag{6.176}$$

which is shown in Fig. 6.4. The maximal normalized flux density occurs for $M_* = 1$, since it is clear from the relation between the usual and normalized Mach numbers that if the first is equal to 1 then so is the other. When examining Fig. 6.4, it is apparent that for low normalized Mach numbers the curve rises linearly, corresponding to incompressible flow. However, at the sonic point the curve turns towards its maximum and for large normalized Mach numbers the mass flux density decreases. It is interesting that for the limit $M \to \infty$, the normalized Mach number tends to a finite number

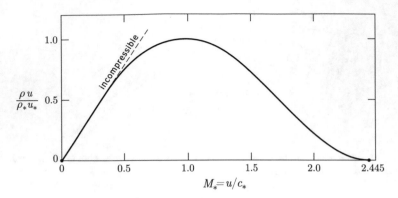

Fig. 6.4 Normalized mass flux density as a function of M_* for a perfect gas with adiabatic exponent $\gamma = 1.4$

$$M_* \rightarrow \sqrt{\frac{\gamma+1}{\gamma-1}}, \tag{6.177}$$

which for a perfect gas of $\gamma = 1.4$ is equal to 2.445. The mass flux at this value of the normalized Mach number vanishes as the density vanishes.

Summarizing the value of the thermodynamic variables in the steady isentropic flow of a perfect gas, in terms of the velocity and of the thermodynamic variables at the sonic point, we write

$$T = T_* \left(\frac{\gamma+1}{2} - \frac{\gamma-1}{2} \frac{u^2}{c_*^2} \right), \tag{6.178}$$

$$P = P_* \left(\frac{\gamma+1}{2} - \frac{\gamma-1}{2} \frac{u^2}{c_*^2} \right)^{\gamma/\gamma-1}, \tag{6.179}$$

$$\rho = \rho_* \left(\frac{\gamma+1}{2} - \frac{\gamma-1}{2} \frac{u^2}{c_*^2} \right)^{1/\gamma-1}, \tag{6.180}$$

$$u^2 = \frac{2\gamma}{\gamma-1} \frac{P_*}{\rho_*} \left[1 - \left(\frac{\rho}{\rho_*} \right)^{\gamma-1} \right]. \tag{6.181}$$

We should remark that in flows of the type discussed here surfaces of discontinuity may occur. We shall devote a full section to this phenomenon, but wish to stress here, that if s cannot remain constant along a streamline because of a discontinuity, not all our results hold in such a case. We shall discuss the conditions at a discontinuity in detail in the upcoming sections, dealing with shock waves.

6.4   One-dimensional Gas Dynamics

The salient phenomena of gas dynamics, that is, flows of a compressible fluid, happen when the flow possesses a velocity in the vicinity, or even above, the sound speed. In the previous section, we have already acquainted the reader with the basic phenomena and nomenclature of gas dynamics. Our exposition there was limited however to special circumstances, e.g., the flows studied were either steady, or isentropic, or irrotational, or involving a perfect gas. Sometimes we have assumed several of these special cases together. Still, it is our opinion that absorbing the essence of nontrivial phenomena, especially in the physical sciences, is best done by starting with simplistic approaches and then coming back to the subject on a more general level. In this section, we shall do just that for the subject of one-dimensional gas dynamics. Although we shall not treat the subject and its aspects in their full generality, the limiting assumptions will be kept at a necessary optimum between technical complexity and triviality which, we believe, may result in good understanding.

6.4.1   Characteristics

In our treatment of this subject, we shall assume one-dimensional gas motion and *homentropic* conditions, i.e., $s = $ const. throughout the fluid, unless explicitly noted otherwise. The concept of *characteristics* was introduced in Chap. 4 in our discussion of nonlinear water waves (Sect. 4.1.4). Because of the mathematical similarity of the governing equations, this tool is important for gas dynamics and we revisit it here, delving deeper into the foundations of the method. We note that the framework of characteristics originates in mathematics and plays a role in the theory of partial differential equations. In the types of equations relevant to one-dimensional gas dynamics, characteristics are curves whose primary importance is in determining the necessary boundary and initial conditions for the existence of a unique solution to a given PDE. Their use is most effective for so-called quasilinear (see below) and linear PDEs. We shall expand here upon our presentation in Chap. 4 and discuss first, briefly, the more mathematical aspects of characteristics, and later on we shall introduce this concept from a more heuristic and physical perspective.

6.4.1.1   Rudiments of the Mathematical Basis of Characteristics

A PDE is called *quasilinear* if the coefficients of the highest partial derivatives are functions of, at most, the lower order derivatives including the function itself, and of the independent variables. We shall give here a short account of the types of quasilinear PDEs of second order and of boundary conditions and see the connection thereof with the characteristic curves.

Starting with the general quasilinear second order PDE for a function $\Psi(t,x)$:

$$A\frac{\partial^2 \Psi}{\partial t^2} + 2B\frac{\partial^2 \Psi}{\partial t \partial x} + C\frac{\partial^2 \Psi}{\partial x^2} = D, \qquad (6.182)$$

where A, B, C, and D are functions of $t, x, \Psi, \partial \Psi/\partial t$, and $\partial \Psi/\partial x$, we are looking for a solution of equation (6.182) with given compatible (see below) $\Psi, \partial \Psi/\partial t, \partial \Psi/\partial x$ on a curve, Γ say, in the $t - x$ plane. This is a typical example of what is called a *Cauchy* or, initial value, problem. Thus we prescribe the Cauchy data on the above curve, which is given parametrically by $t = f(s)$, $x = g(s)$:

$$\Psi = h(s), \quad \frac{d\Psi}{dt} = q(s), \quad \frac{d\Psi}{dx} = \varphi(s). \qquad (6.183)$$

The above Cauchy problem must satisfy the compatibility conditions on the initial curve Γ:

$$h'(s) = q(s)f'(s) + \varphi(s)g'(s) \qquad (6.184)$$

among the Cauchy data. Thus, the three functions constituting the data are not all independent and at most two are arbitrary. Compatibility conditions of this type must hold also for the derivatives (first and second) of the function Ψ. If $\Psi(t,x)$ is a solution of equation (6.182) and both $q(s) = d\Psi/dt$ and $\varphi(s) = d\Psi/dx$ satisfy compatibility conditions similar to those for $h(s)$, we obtain the following three linear equations

$$A\frac{\partial^2 \Psi}{\partial t^2} + 2B\frac{\partial^2 \Psi}{\partial t \partial x} + C\frac{\partial^2 \Psi}{\partial x^2} = D,$$

$$f'(s)\frac{\partial^2 \Psi}{\partial t^2} + g'(s)\frac{\partial^2 \Psi}{\partial t \partial x} = q'(s),$$

$$f'(s)\frac{\partial^2 \Psi}{\partial t \partial x} + g'(s)\frac{\partial^2 \Psi}{\partial x^2} = \varphi'(s) \qquad (6.185)$$

for the partial derivatives of Ψ, which are valid along the curve Γ. A unique solution for the second order partial derivatives of Ψ is guaranteed, *unless*

$$\Delta \equiv \begin{vmatrix} A & 2B & C \\ f' & g' & 0 \\ 0 & f' & g' \end{vmatrix} = A(g')^2 - 2Bf'g' + c(f')^2 = 0. \qquad (6.186)$$

The initial curve Γ on which $\Delta = 0$ is called a *characteristic* for this differential equation and data. In the case described here, with the initial curve Γ being a characteristic, Eqs. (6.185) are inconsistent, thus a Cauchy problem with the initial condition prescribed on a characteristic generally has no solution. Conversely, when the curve on which the Cauchy data are prescribed is *not* a characteristic, the second derivatives of Ψ are consistently and uniquely determined on that curve. Moreover, higher derivatives can also be found, and the function itself can be formally written in the neighborhood of any point (t_0, x_0). The rigorous requirements, embodied in the Cauchy–Kowalevski theorem, are beyond the scope of this discussion.

What is important and relevant is the fact that the last equality allows one to write an equation satisfied by the characteristics

$$A dx^2 - 2B dt\, dx + C dt^2 = 0, \tag{6.187}$$

which can be solved for dx/dt thus

$$\frac{dx}{dt} = \frac{B \pm \sqrt{B^2 - AC}}{A}. \tag{6.188}$$

This relation is an ODE for the characteristic curve $x(t)$ provided A, B, and C are known functions of x and y, a case in which either a given solution $\Psi(x,t)$ is considered or the original PDE (6.182) is linear. In any case, the formal similarity to the classification of curves gives rise to the classification of second order quasilinear PDE. Equation (6.182) is called *hyperbolic* when $AC - B^2 < 0$, *elliptic* when the opposite is true, and *parabolic* in the limiting case $AC - B^2 = 0$. Regarding characteristic curves, it is immediately clear that a hyperbolic equation generally possesses two families of characteristics, on account of the $\pm$, while in the case of parabolic equations, only one family can be expected. Since we limit ourselves here to real equations, it is obvious that elliptic equations do not have characteristic curves at all. Even though it is not within the scope of this book to explain the meaning of this, we nevertheless remark that a clean classification of the above kind is possible only in the case of linear equations, but often even then there are different regions where the same equation may be of different type. In nonlinear equations the class or type of the equation can depend not only on the equation but also on the solution. Thus even from the mathematical point of view, the topic is murkier than it first appears.

6.4.1.2   Characteristic Curves for Compressible Gas Dynamics

The idea of characteristics, which is applicable also to incompressible water waves, was discussed through a linear example in our discussion of shallow water waves in Chap. 4. There we introduced the idea of characteristic curves at several stages of our discussion following our initial heuristic motivation of them in Sect. 4.1.4. The situation is analogous in problems involving one-dimensional gas dynamics, because of the similarity of the governing dynamical equations. We shall, therefore, expand on these ideas and explore them more deeply than we did there. In our discussion of acoustics we dealt with the wave equation for the velocity potential, say, in one spatial dimension, which we repeat here for convenience

$$\frac{\partial^2 \phi}{\partial t^2} - c_s^2 \frac{\partial^2 \phi}{\partial x^2} = 0, \tag{6.189}$$

where c_s is the sound speed in the undisturbed medium. This linear wave equation, as we have seen, possesses a general solution which can be initialized at t_0, say, by a given initial local disturbance and its time derivative at some point x_0, say. One can write this solution as

$$\phi(x,t) = f_1(x + c_s t) + f_2(x - c_s t), \tag{6.190}$$

where c_s is the wave speed equal to sound speed and f_1 and f_2 are functions, which are consistent with the initial condition. Even though we may need to use characteristics for nonlinear (not very small amplitude) waves in compressible fluids, we have decided to explain this mathematical technique on acoustic waves, which are linear. As we are considering a one-dimensional case, it is clear from the general solution that an initial disturbance, here in the velocity potential, will be divided into two parts. One part will propagate with velocity $-c_s$, i.e., to the left with respect to the x-axis, while the other will propagate in the $+x$-axis direction with velocity c_s. The division between the right and left propagating disturbance is completely determined, as mentioned before, by the initial condition. If the fluid is not stationary, but moves itself, as a whole, with a constant velocity u, say, then nothing essential changes—the above-mentioned disturbances propagate with velocities $u - c_s$ and $u + c_s$ with respect to the original inertial rest frame, respectively. This applies to a point x_0, at which a disturbance is created at time t_0, therefore *two* directions of propagation, starting from that point, are possible in the $t - x$ plane. Thus the two differential equations of the perturbation path in that plane

$$\frac{dx}{dt} = u + c_s,$$

$$\frac{dx}{dt} = u - c_s, \tag{6.191}$$

actually describe two families of positive $\mathscr{C}_+$ and negative $\mathscr{C}_-$ curves, which are nothing but the characteristics of the acoustic equation (6.189) in this case.

If we consider a more general case, in which the fluid fields, including the velocity and sound speed, are *not* constant, the above-mentioned mental picture of two sets of straight and parallel characteristics breaks down. The equations of the characteristics (6.191) continue to hold, though. The two different families are approximately straight and parallel to each other within a family, only in a close neighborhood of (t_0, x_0), because u and c_s are no longer constant. On larger scales the characteristics in the (x, t) plane may be curved as is schematically depicted in Fig. 6.5. Even though the above described attitude towards the nature of characteristics and their meaning is rather heuristic, we shall see that the qualitative and sometimes even quantitative information about the problem at hand is very useful, often superior to what can be deduced from mathematically rigorous

Fig. 6.5 Schematic drawing
of two sets of characteristics
of the acoustic equation in the
homentropic case

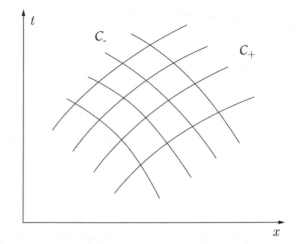

considerations. The property of the characteristic curves as the propagation direction
of small disturbances in the (x,t) plane, in one-dimensional flows, is by no means
the only important quality of these curves as we noted also in our discussion of
shallow water waves in Chap. 4. It is interesting to find, using the equations of gas
dynamics, how other relevant flow quantities change *along* the characteristics. As
we shall soon see, not only small disturbances but also specific combinations of the
flow fields are propagated along the characteristics.

Consider a given curve $x = f(t)$ in the time-space plane. Elementary mathematical considerations give rise to the derivative of any function $F(x,t)$ along that curve.
We have

$$\left(\frac{dF}{dt}\right)_f = \frac{\partial F}{\partial t} + \frac{\partial F}{\partial x}f', \tag{6.192}$$

where $f'(x) = dx/dt$. We shall now try to write the equations of gas dynamics so
that they contain derivatives of the above type, along a curve, but the curve is a
characteristic. In homentropic conditions we obviously have

$$\frac{D\rho}{Dt} = \frac{1}{c_s^2}\frac{DP}{Dt}, \tag{6.193}$$

where c_s is the adiabatic sound speed. On the other hand, the continuity equation
can be written:

$$\frac{D\rho}{Dt} = -\rho\frac{\partial u}{\partial x}. \tag{6.194}$$

Equating the two and multiplying through by c_s/ρ gives

$$c_s \frac{\partial u}{\partial x} + \frac{1}{\rho c_s}\left(\frac{\partial P}{\partial t} + u\frac{\partial P}{\partial x}\right) = 0. \tag{6.195}$$

The equation of motion is

$$\frac{\partial u}{\partial t} + u\frac{\partial u}{\partial x} + \frac{1}{\rho}\frac{\partial P}{\partial x} = 0. \tag{6.196}$$

Adding the last two equations we obtain our objective: derivatives only along the $\mathscr{C}_+$ characteristics:

$$\left[\frac{\partial u}{\partial t} + (u+c_s)\frac{\partial u}{\partial x}\right] + \frac{1}{\rho c_s}\left[\frac{\partial P}{\partial t} + (u+c_s)\frac{\partial P}{\partial x}\right] = 0. \tag{6.197}$$

Analogously, subtracting Eq. (6.195) from (6.196), we find an equation containing only derivatives along the other characteristics

$$\left[\frac{\partial u}{\partial t} + (u-c_s)\frac{\partial u}{\partial x}\right] + \frac{1}{\rho c_s}\left[\frac{\partial P}{\partial t} + (u-c_s)\frac{\partial P}{\partial x}\right] = 0. \tag{6.198}$$

Thus the gas dynamics equations can be written as

$$du + \frac{1}{\rho c_s}dP = 0 \quad \text{along } \mathscr{C}_+ \quad \text{i.e. } \frac{dx}{dt} = u + c_s,$$

$$du - \frac{1}{\rho c_s}dP = 0 \quad \text{along } \mathscr{C}_- \quad \text{i.e. } \frac{dx}{dt} = u - c_s. \tag{6.199}$$

Taking into account that we are dealing here with a homentropic flow, the condition of isentropy $Ds/Dt = 0$ follows. Its meaning is that the entropy is conserved along streamlines, which are given by $dx/dt = u$. Thus the streamlines may also be considered as a kind of trivial characteristics for isentropic flow and their family usually adopts the name $\mathscr{C}_0$. Considering for the moment the Lagrangian version of fluid mass conservation, Eq. (1.35), we write

$$\rho(\mathbf{x}, t)J_t = \rho_0, \tag{6.200}$$

where J_t is the time-dependent determinant of the Jacobian, characterizing the transformation from Lagrangian ($\mathbf{X}$) to Eulerian ($\mathbf{x}$) spatial coordinates. In our case we deal with only one dimension so that Eq. (6.200) yields just $\rho dX = \rho_0 dx$. Thus if one imagines the abstract plane (X, t), that is, initial positions, as Lagrangian coordinates, versus time, one would have to imagine the equations of the characteristics, which in these coordinates read

$$\mathscr{C}_+ : \frac{dX}{dt} = c_s\frac{\rho}{\rho_0}; \qquad \mathscr{C}_- : \frac{dX}{dt} = -c_s\frac{\rho}{\rho_0}; \qquad \mathscr{C}_0 : \frac{dX}{dt} = 0. \tag{6.201}$$

Frequently the gas dynamics equations written along characteristics, which may be perceived as alternative, natural coordinates for the particular equation, are especially convenient for numerical calculation.

6.4.2   Riemann Invariants, the Domains of Determination and of Influence

We shall maintain our assumptions of the previous section, specifically the ones of the flow being isentropic and one-dimensional, and try to achieve a deeper look into the equations as formulated along characteristics. First, it is evident that the constant entropy disappears altogether from the equations. The flow can thus be completely described by the velocity $u(x,t)$ and any *one* of the thermodynamic functions, P, ρ, or c_s. These variables, however, are uniquely related to each other at any one point by purely thermodynamic expressions, e.g., $P = P(\rho)$, $c_s = c_s(\rho)$, $c_s^2 = dP/d\rho$.

As it will turn out, the quantities $du \pm (\rho c_s)^{-1} dP$, which are equal to zero along the characteristics, where the $\pm$ correspond to the $\mathscr{C}_\pm$ families, respectively, are actually total differentials of the expressions J_+ and J_-, which are called *Riemann invariants*. They are

$$J_+ = u + \int \frac{dP}{\rho c_s} = u + \int c_s \frac{d\rho}{\rho}, \tag{6.202}$$

$$J_- = u - \int \frac{dP}{\rho c_s} = u - \int c_s \frac{d\rho}{\rho}. \tag{6.203}$$

Thus because $dJ_+ = 0$ along $\mathscr{C}_+$ characteristics, J_+ is constant, or invariant on them. So is the case with J_- along the $\mathscr{C}_-$ characteristics, and hence the indicative name *invariants*. The integrals are expressible in terms of thermodynamic quantities by purely thermodynamic considerations. The result depends, of course, on the properties of the fluid in question. The easiest and most useful example is, as usual, a perfect gas with constant specific heats. The Riemann invariants are then given by the functional form

$$J_\pm = u \pm \frac{2}{\gamma - 1} c_s. \tag{6.204}$$

In any case, the two Riemann invariants are constant along the corresponding families of characteristics and, as we have already stated, the characteristic curves are a kind of natural coordinates for the problem, and we conclude that the Riemann invariants are the natural variables. We are dealing with one-dimensional flows and thus our arena is the two-dimensional space-time (x,t) and we are reminded therefore that characteristics are curves in this plane. Note that in contrast to the diagrams showing characteristic curves, in the time-space diagram of a single dimension gas dynamics problem, here we prefer to use space-time, i.e., the plane

$x - t$. This, obviously, has no meaningful consequences. In addition, hometropic flows are fully described by just two variables: the velocity plus one of the thermodynamic functions. Consequently, we may expect that only two continuously infinite sets, as the Riemann invariants, can suffice to serve as dependent variables. Indeed, Eqs. [(6.202)–(6.203)] are formally invertible, when the thermodynamic properties of the fluid are known, that is, u and c_s, say, are expressible as a unique functions of J_+ and J_-. Again it is easy to express this explicitly for the case of a perfect gas. From Eq. (6.204) it follows that

$$u = \frac{1}{2}(J_+ + J_-); \qquad c_s = \frac{\gamma - 1}{4}(J_+ - J_-), \qquad (6.205)$$

and it is not a difficult exercise to get for the specific perfect gas in question also P, ρ, and T. It is possible to also obtain, from Eq. (6.204), the explicit equations of the characteristics for this case, expressed only in terms of the Riemann invariants

$$\mathscr{C}_+ : \quad \frac{dx}{dt} = \frac{\gamma + 1}{4}J_+ + \frac{3 - \gamma}{4}J_- = H_+(J_+, J_-),$$

$$\mathscr{C}_- : \quad \frac{dx}{dt} = \frac{\gamma + 1}{4}J_- + \frac{3 - \gamma}{4}J_+ = H_-(J_+, J_-), \qquad (6.206)$$

where $H_\pm$ are functions of only the Riemann invariants. In these equations we have explicitly given the form of the functions $H_\pm$, as evaluated for a perfect gas. Even though we have given again the perfect gas as an example, the evaluation of $H_\pm$ is possible, at least in principle, for any fluid, provided there is enough information on its thermodynamic properties. We conclude the discussion of Riemann invariants and their meaning by their connection to causality in compressible FD. Consider J_+, for example. This quantity is constant along any specified characteristic curve of the $\mathscr{C}_+$ family. Thus the only way that the slope of the characteristics can change is by the variation along the $+$ characteristic of the other invariant J_-, which is constant along the members of the $-$ characteristics. The flow equations, when written in the characteristic form, allow one to understand the causal connection between events in a gas dynamical flow. To demonstrate it in an easy and transparent way, we consider any one-dimensional homentropic flow and observe the behavior in the (x, t) plane. In Fig. 6.6, we illustrate the situation when an initial condition at $t = 0$ is specified for $x_1 \le x \le x_2$, i.e., on the segment $\overline{AB}$. This initial condition is most naturally expressed, as we have mentioned before, by the distribution of $J_+(x, 0)$ and $J_-(x, 0)$, which is equivalent to the distribution of physical variables, but may be more "telling." In the figure a number of $+$ and $-$ characteristics are sketched. It is obvious that an exact drawing of these characteristics may be achieved only after solving the particular problem, but we are dealing here with qualitative considerations. Let us consider an arbitrary general point of the flow $P(x, t)$. The values of the flow variables or Riemann invariants at P are determined only by the values of the two Riemann invariants at the initial points of the two characteristics, from the two families which meet at P. Thus we have at P: $J_+(x, t) = J_+(x_1, 0)$

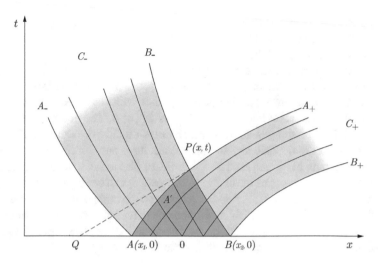

Fig. 6.6 The time-space diagram showing characteristic curves. The region of determination of P at $t = 0$, the initial condition segment $\overline{AB}$ is shown. The domain of determination and influence of that segment are the *darkly* and *lightly shaded regions*, respectively. For details, see text

and $J_-(x,t) = J_-(x_2,0)$. Again, we can be explicit for a perfect gas with constant specific heats and solve these equations for u and c, thus

$$u(x,t) = \frac{1}{2}(u_A + u_B) + \frac{1}{\gamma - 1}(c_A - c_B),$$

$$c_s(x,t) = \frac{1}{2}(c_A + c_B) + \frac{\gamma - 1}{4}(u_A - u_B), \tag{6.207}$$

where $v_A = v(x_1, 0)$ and $v_B = v(x_2, 0)$ for both $v = u$ or c.

The clarification of a possible misunderstanding is in order. The point P is the intersection of two characteristics from opposite families. Thus the conditions at that point, as in any point in the domain ABP, can be determined uniquely from the two Riemann invariants. This is the reason that the darkly shaded region ABP is referred to as the domain of *determination* of $\overline{AB}$. However, the location of the point depends on the slope of the characteristics and that is determined by the appropriate Riemann invariant of the characteristic curves of the other family, which cross the $\mathscr{C}_+$ characteristic, say, along its entire segment AP. These depend on the conditions at all the initial conditions between A and B. For example, the slope of the $+$ characteristic connecting A and P depends also on the value of the Riemann invariant of $J_-(O)$, which propagates from the intermediate point on the initial condition segment to A'. It can, however, be said that the state of the gas at point P is uniquely and completely dependent by the initial conditions between x_1 (point A) and x_2 (point B). Moreover this is the only initial segment which influences the condition at P! This is obvious because any change of the conditions at Q, say, will not have time to arrive at position x at time t. In the figure it is shown that if Q is to influence P it has to arrive at P on time, along the dashed curve, say, but then this $+$ characteristic, coming from Q, must cross other characteristics of the same family, which originate at $\overline{AQ}$, and this is not allowed.

This must not be confused with the domain of *influence* of $\overline{AB}$. This is the region which is the union of that delineated by the B_+ characteristic initiated at B and the A_+ characteristic starting at A, together with the region between the A_- characteristic initiated at A and the B_- one, starting at B. The points on the segment $\overline{AB}$ cannot influence the gas at subsequent times outside the domain of influence (lightly shaded region), since any signal will not have enough time to reach a point outside this domain of influence. This causality of phenomena in gas dynamics may be reminiscent of light cones in relativity, however it is true in gas dynamics only if characteristics of the same family do not intersect each other, since this would lead to non-uniqueness of the flow variables. Such occurrences give rise to the phenomenon of shock waves, which is allowed in gas dynamics but requires special treatment. We shall discuss shock waves in depth later in this chapter.

6.4.3   Nonlinear Simple Waves and the Steepening of Such Waves

The phenomenon of wave steepening is very important for understanding the spontaneous formation of surfaces of discontinuity in a compressible gas flow. We have defined simple waves in acoustics but the existence of such waves is not limited to small disturbances which result in the viability of a linear approximation. Thus consider a gas occupying an unbounded region and assume initial conditions such that one of the Riemann invariants, $J_-(x,0)$ say, is constant for all x. This is possible because we have at our disposal initial distributions of two variables, but the Riemann invariants are equivalent to any two such variables. For a perfect gas the relation (6.204) certainly guarantees it. The $\mathscr{C}_-$ characteristics originate at all points of the x-axis, and since J_- is constant on this axis, this constant value is carried along all the negative characteristics emanating from the x-axis. Now imagine that the gas occupies only a half space and is bounded on the left by a piston moving to the right according to the function $x_L = p_L(t)$. This left piston does not influence the $\mathscr{C}_-$ characteristics, which originate at the x-axis for $x > x_L$ and thus the value of J_- remains equal to the same constant, as determined before. The piston sends to the future, so to speak, only characteristics belonging to the $\mathscr{C}_+$ family. Remembering that the slopes of both families of the characteristics can be described, according to Eqs. (6.206), as $dx/dt = H_\pm(J_+,J_-)$ and that in the entire region in front of the piston J_- is equal to the same constant, it is easy to understand that each of the $+$ characteristics remains *straight* as it is crossed at all points by those $\mathscr{C}_-$ characteristics. Now, to be sure, the $\mathscr{C}_+$ characteristics have each a constant J_+ along them, but the constant is not the same for different characteristics! Thus $H_+(J_+,J_-)$ is actually a function of J_+ and J_- is a true constant. It is therefore possible to integrate and obtain the equation describing the straight line where each $\mathscr{C}_+$ characteristic is

$$x = H_+(J_+, J_-)t + K(J_+). \tag{6.208}$$

This is the equation of the characteristic on which a specific value J_+ is constant. Remembering that J_- is constant everywhere and K is a constant of integration, depending obviously on J_+. To write everything in the form of the velocity and a thermodynamic variable, we have to supplement Eq. (6.208) with the fact the J_- is a constant and can be written using Eq. (6.203)

$$J_- = u - \int \frac{dP}{\rho c_s} = \text{const.} \tag{6.209}$$

It follows that all the thermodynamic variables are, in this case, functions of the velocity u alone, and we do not forget that the fluid is homentropic. Thus Eq. (6.208) may be written as

$$x = [u + c_s(u)]t + K(u). \tag{6.210}$$

Can you explain why? This equation determines u implicitly, as a function of the coordinate x and time. Given values of u, $c(u)$ and thermodynamic variables are carried along x with the velocity $u + c(u)$ so we have actually that u, for example, has the form of a traveling wave, to the right:

$$u = f\{x - [u + c_s(u)]t\}. \tag{6.211}$$

Similar consideration can lead to a simple nonlinear wave traveling to the left and being distorted in an analogous way.

An acoustic wave, i.e., one that is limited to very small amplitudes, allowing one to neglect second order in the disturbance, is frozen, so to speak, in its shape, when it moves. The situation is different in the case discussed here, as it is apparent in Eq. (6.211) and demonstrated in Fig. 6.7. We shall discuss this mainly qualitatively, but it is possible to also use explicit formulae if we know the thermodynamic properties of the gas. Consider the initial profile $u(x,0)$, shown in the upper part of the figure. There are three points on the x-axis at which $u = 0$: A_0, C_0, and E_0. The slope of the characteristics starting at these points is all equal to $c_s(0)$, which we may simply call c_0 because this is the stagnation value of the sound speed. In contrast to this, the points marked B_0 and D_0 correspond to the maximal amplitude of the flow speed, where the former has a negative velocity and the latter positive. Thus the slopes of the corresponding characteristics, in the (x,t) plane, dx/dt, are

$$B_0: \text{ slope} = -v_{max} + c_s(-v_{max}),$$

$$D_0: \text{ slope} = v_{max} + c_s(v_{max}), \tag{6.212}$$

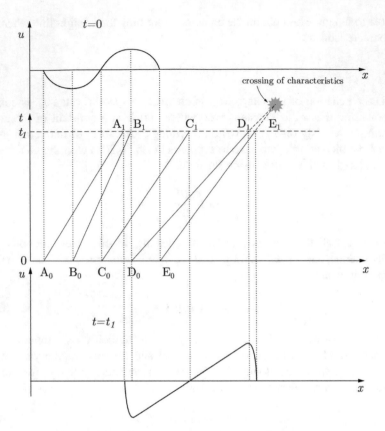

Fig. 6.7 A time-space diagram showing the characteristics corresponding to a nonlinear simple wave between $t = 0$ and $t = 1$ is shown. A diagrammatic drawing of the waveform is shown above and because some characteristics are more slanted than others, expressing higher communication speed, the wave gets steepened and distorted at t_1 to just before breaking. For details, see text and also see the discussion in Sect. 4.1.4

where $v_{max} \equiv |u_{max}| \geq 0$. Note that the slope is constant for each case because as we have explained J_- is constant everywhere and so the slope of the different $\mathscr{C}_+$ characteristics is different for each, but because of the constancy of the corresponding J_- on each characteristics the slopes are constant. However the slope of the B_0 characteristic is smaller than c_0. By the same token, the slope of the characteristic emerging from D_0 is larger than the one coming out from E_0. Thus at time t_1 we have, as can be seen on the lower part of the figure, a steepened form of the wave. The physics behind the steepening is that the wave crests travel faster than the nodes of the wave and the valleys correspondingly lower. This steepening is bound to ultimately cause a breaking of the waves and the breakdown of physical meaning of this reasoning. That happens when two characteristics of the same family actually cross (see Fig. 6.7). Mathematically it gives rise to non-uniqueness, as far as values of the function are concerned—a situation that is unphysical. In practice, the solution develops a kind of "pathology"—it turns out the function stays continuous and its derivative becomes discontinuous first, a situation referred to as a *weak discontinuity*. It was first encountered in our discussion of unidirectional nonlinear waves in Sect. 4.1.4. The fluid reacts by smoothing somewhat the profile, as the viscosity cannot in reality be zero, just very small. The narrow region of the weak discontinuity propagates with the sound speed. Pushing this reasoning further we get a situation in which the function's overshoot is prevented by viscous and heat conduction effects, a discontinuity in the function ensues. Such a discontinuity is referred to as *shock wave* (see Sect. 6.5). The high density and pressure region behind the shock continues to push the discontinuity so that it accelerates and ultimately reaches supersonic speeds, relative to the undisturbed medium, whose sound speed is c_0. With the stretching of the compression to more and more fluid, and without any source of energy and momentum input, the shock wave has to be degraded and dissipated. A persistent shock wave, as we shall see, requires continuous input of energy, e.g., a supersonically moving body in the gas.

6.4.4   Rarefaction Waves

Rarefaction waves appear sometimes as natural phenomena as well as in physical applications, and therefore we shall discuss some of their properties. In contrast to compression waves, like the one discussed before created by a piston moving forward, rarefaction waves, as their name indicates, are waves in which the density disturbance is below the undisturbed fluid density. Such a wave can be created, for example by a receding piston. In Fig. 6.8 we show a situation of this kind, where the left "wall" is actually a piston, whose motion is described by a function $x = x_{\text{piston}}(t)$, which is detailed in the next sentence. At $t \leq 0$, before the piston begins to move, we have a half space ($x > 0$) occupied by stationary uniform compressible fluid, whose thermodynamic functions are P_0, ρ_0, and c_0, say. The piston starts receding with some acceleration, so that initially $\ddot{x}_{\text{piston}} \neq 0$, but from time $t = t_1$, it continues to move to the left at constant speed, $V > 0$, say, so that $x_{\text{piston}} = -Vt$ +const. The gas, occupying the region to the right of the piston, will now be described using the method of characteristics. Figure 6.8 summarizes the outcome. In region I, on the right side of the $\mathscr{C}_+$ characteristic $\overline{OA}$, the gas remains undisturbed even after the piston starts to move. The characteristics are straight lines with slopes c_0 and $-c_0$ ($\mathscr{C}_-$ family). Any possible small disturbance, before the piston had started to move, propagates to the right as an acoustic wave. Moreover, the same considerations as we have used in the discussion of steepening waves imply that the motion of the gas is a simple wave propagating to the right. The head of the wave, the initial disturbance created by the beginning of the piston motion, propagates along the $\overline{OA}$ characteristic with the speed of sound c_0. Even though all our qualitative conclusions will be general, we assume that the fluid is a perfect

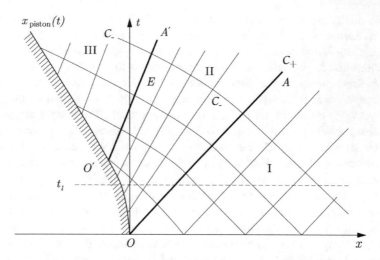

Fig. 6.8 A time-space diagram showing characteristics corresponding to a rarefaction wave. On the left there is a piston accelerating first to the left and from time t_1 continuing with constant velocity, $-V$. For details, see text

gas, having constant specific heats, doing this with the purpose of being able to write explicit formulae. Consider first the Riemann invariant J_-. As before, because this characteristic family originates on the x-axis, where this invariant is constant, it remains constant in all region I

$$J_- = u - \frac{2}{\gamma-1}c_s = -\frac{2}{\gamma-1}c_0, \qquad (6.213)$$

thus

$$u = -\frac{2}{\gamma-1}(c_0 - c_s), \quad c_s = c_0 + \frac{\gamma-1}{2}u. \qquad (6.214)$$

The $\mathscr{C}_+$ characteristics originate at the gas–piston interface so in the figure they are on the left of $\overline{OA}$. Therefore the relevant, for them, velocity of the gas is the same as the velocity of the piston $\dot{x}_{\text{piston}}(t) < 0$. Consequently, the sound speed as well as the pressure and density immediately adjacent to the piston are smaller than their initial values and decrease with increasing absolute value of the piston speed. The $\mathscr{C}_+$ characteristics originating at the piston are straight lines, remembering that J_- is constant everywhere, whose slopes vary. They are

$$\left(\frac{dx}{dt}\right)_+ = u + c_s = c_0 + \frac{\gamma+1}{2}u = c_0 - \frac{\gamma+1}{2}\left|\dot{x}_{\text{piston}}\right|. \qquad (6.215)$$

In the acceleration region (II), delineated on the left by the characteristic $\overline{O'A'}$, the characteristics fan out, i.e., are divergent. In region III, where the piston moves with constant velocity, the $\mathscr{C}_+$ characteristics become parallel again and their slope is

$$\left(\frac{dx}{dt}\right)_+ = c_0 - \frac{\gamma+1}{2}V. \qquad (6.216)$$

In that region the J_+ invariant is the same for all characteristics and is

$$J_+ = u + \frac{2}{\gamma-1}c_s = \frac{2}{\gamma-1}c_0 + 2u = \frac{2}{\gamma-1}c_0 - 2V. \qquad (6.217)$$

Also, in region III the flow variables are constant with $u = -V$, and $c_s = c_0(\gamma - 1)V/2 \equiv c_1$. It is possible to obtain an analytical solution for this problem for a perfect gas with constant specific heats if the acceleration of the piston is specified. In Problem 6.23, it is shown that assuming the velocity of the piston is $dx_{\text{piston}}/dt = -V(1 - e^{-t/\tau})$, $\tau > 0$, the velocity distribution is implicitly given by

$$x = \left(c_0 + \frac{\gamma+1}{2}u\right)t - u\tau + \tau\left(c_0 + \frac{\gamma+1}{2}u + V\right)\log\left(1 + \frac{u}{V}\right). \qquad (6.218)$$

Using this result we can consider now the concept of a *centered* rarefaction wave as a limit of the previous problem. Imagine that the acceleration period of the piston $t_1 < t < 0$ tends to zero. This can happen, of course, only if we imagine an infinitesimal period of infinite acceleration with the final velocity, V, which remains then constant. The segment $\overline{O'A'}$ thus tends to zero and the points O and O' approach each other. At this limit of an instantaneous jump to the final velocity of the piston $-V$ the characteristic $\overline{O'A'}$ originates at the same point as the $\overline{OA}$ one. However, region II continues to contain fanning-out characteristics, as point A' does not approach point A. All the three lines, the one designating the piston path, the head of the wave $\overline{OA}$, and its tail $\overline{OA'}$, originate at the same point. Such a wave is called a *centered* rarefaction wave. The simple prescription of such a centered wave has the form

$$x = [u + c_s(u)]t, \qquad (6.219)$$

since the considerations here are very similar to the ones employed in our previous description of the breaking water wave. The constant $K(u)$ of Eq. (6.210) is however equal to zero because it is proportional to τ, if one formally obtains Eq. (6.219) by taking the limit $\tau \to 0$. We may write the explicit solution of a centered rarefaction wave for an ideal gas with constant specific heats. The constancy of J_- gives

$$c_s = c_0 - \frac{\gamma-1}{2}|u|, \quad u < 0. \qquad (6.220)$$

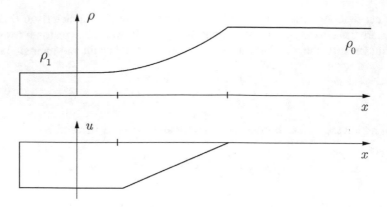

Fig. 6.9 The distribution of the velocity and density in a centered rarefaction wave moving to the right. For details, see text

Using

$$\rho = \rho_0 \left[1 - \frac{\gamma - 1}{2} \frac{|u|}{c_0} \right]^{2/(\gamma-1)},$$

$$P = \frac{P_0}{\rho_0} \rho^{\gamma},$$
(6.221)

we may eliminate c from Eqs. (6.219)–(6.220) to get an explicit solution of the variables as a function of the coordinates x and t. For the velocity we obtain

$$|u| = \frac{2}{\gamma + 1} \left(c_0 - \frac{x}{t} \right),$$
(6.222)

and the functions for ρ and P follow accordingly. It is interesting to note from the formula that as we have expected the head of the wave, where $u = 0$, should move along the line $x = c_0 t$, while its tail, where $u = -V$, should move along the line $x = (c_1 - V)t = [c_0 - (\gamma + 1)V/2]$. In Fig. 6.9, the density and velocity of a centered rarefaction wave are shown at some instant. Actually $|u|$ depends on the coordinates only through the ratio x/t, it is therefore self-similar (see also below in Sect. 6.4.6.2).

6.4.5 The Mathematical Analogy with Shallow Water Waves and Its Consequences

We have already discussed the fact that a remarkable analogy exists between the equations describing one-dimensional compressible gas flow and the equations describing the flow of incompressible fluid in a constant vertical gravitational field, with a free surface at some height from the bottom. We have considered the latter problem as a part of *nonlinear* surface water waves theory and discussed it in Chap. 4 in, e.g., Sect. 4.1.4 et seq. We recall here the definition of a key quantity, having the dimensions of velocity, in water waves theory: $c_g \equiv \sqrt{gh}$, where $h(x,t)$ is the water depth, and g is the constant acceleration, directed downward. The setting is identical to that of Chap. 4 and so is the notation. We shall, from now and on in this discussion of water waves, use $c_g \mapsto c$. Note that here c is *not* the speed of sound. In incompressible fluids sound speed is nominally infinite, while c is an important physical velocity scale in the problem, as we have seen in Chap. 4 and as we shall see below. Rewriting the shallow water equations (4.77), (4.78), and (4.76) for a flat bottom, $L_b = $ const., water height $h(x,t)$ and using $c^2(x,t) = gh(x,t)$ with g pointing in the $-\hat{z}$ direction,

$$\frac{\partial h}{\partial t} + \frac{\partial (hu)}{\partial x} = 0, \tag{6.223}$$

$$\frac{\partial u}{\partial t} + u\frac{\partial u}{\partial x} = -\frac{\partial (c^2)}{\partial x}, \tag{6.224}$$

$$w(x,z,t) = -z\frac{\partial u}{\partial x}, \tag{6.225}$$

we can see that the first two equations may easily be put into the more revealing form (verified in Problem 6.24)

$$\left[\frac{\partial}{\partial t} + (u+c)\frac{\partial}{\partial x}\right]\left[u+2c\right] = 0, \tag{6.226}$$

$$\left[\frac{\partial}{\partial t} + (u-c)\frac{\partial}{\partial x}\right]\left[u-2c\right] = 0, \tag{6.227}$$

while the third equation (6.225) is decoupled and provides information on the vertical velocity w.

Being aware of the formulations for a compressible gas flow in Sects. 6.4.1–6.4.4, including the mathematical and physical insights developed there, we see that there exist characteristic curves upon which combinations of u and c remain constant. In particular, we choose here to represent these two families of curves $\mathscr{C}_\pm$, denoting positive/negative characteristics, by a coordinate parametric representation, using the parameter s. In the plane (x,t) the characteristic curves are $x(s)$ and $t(s)$, such that

$$\frac{dx}{ds} = u(x,t) \pm c(x,t), \qquad \frac{dt}{ds} = 1,$$

respectively. Our presentation of characteristic curves here differs slightly from that done in one-dimensional gas dynamics above, but the formal difference in parametrization is not essential. A parametric representation of the present kind could have been done also in Sects. 6.4.1.2–6.4.4 and the reason of our choice here is to show the reader two possible equivalent representations.

In the discussion of the shallow water equations in Sect. 4.3, the functions corresponding to Riemann invariants, as defined in this chapter, are given by $u \pm 2c$, where $c = \sqrt{gh}$ in which h is the layer thickness. Very loosely speaking, similarity to the gas dynamics formulation can be achieved by setting formally $\gamma \mapsto 2$. Most importantly, along each of these characteristic curves, the quantities $u \pm 2c$ remain invariant. When we do not limit ourselves to a linear problem the dependence of the curve's coordinates with respect to the parameter is no longer simple and, in general, the characteristic curves cease to be straight lines. To make this clearer let us take the example of a positive characteristic $\mathscr{C}_+$. Although the quantity $u + 2c$ is constant along $\mathscr{C}_+$, the value $u + c$ might not be and so the way the location of the characteristic's spatial position changes with respect to the parameter s depends upon the value of $u(x,t) + c(x,t)$ evaluated at its given position $[t(s), x(s)]$—the challenge here being that neither u nor c are known a priori. It would seem that the elegant reformulation of the nonlinear equations of shallow water into the form found in Eqs. (6.226)–(6.227) does not seem to have made things easier. With hindsight, of course, we have presented this equivalent form of the shallow water equations because for certain problems the solutions can be determined with relative ease. For example, the well-known problem of the flow resulting from a sudden dam break is analogous to the problem of a rarefaction wave in gas dynamics. We shall end this foray into shallow water theory by fully working out one particular example. The choice of this example is motivated by the analogy to shock wave formation in gas dynamics. The problem setup is simple: a semi-infinite tank of water filled to a level equilibrium height h_0. At the position $x = 0$ is a moveable plate and at $t = 0$ the plate starts to be accelerated rightward so that its velocity is $U(t) = \alpha t$, with $\alpha > 0$ an arbitrary constant. The position of the plate X_p is, therefore, equal to $(1/2)\alpha t^2$. A schematic drawing of the situation is presented in Fig. 6.10. The equations of motion we aim to solve are (6.226)–(6.227). In terms of the time-space diagram properties for this situation (Fig. 6.12), there exists a natural causal boundary given by $x(t) = c_0 t$, where $c_0 = \sqrt{gh_0}$, indicating the point out beyond which the fluid remains undisturbed by the events around the moving boundary. In the undisturbed ($u = 0$) region of the fluid, the positive/negative characteristic lines are all of constant slope and equal to $\pm c_0$, respectively. Along each of these characteristics the Riemann invariant quantities $u \pm 2c = \pm 2c_0$ are preserved. All of the positive characteristics anchored to any point $x > 0$ at $t = t_0 = 0$ never influence the problem so there is no interest in them. All the negative characteristics, anchored to $x > 0$ at $t = 0$, begin to curve about one as they enter the disturbed region of the space-time diagram. The region is above the causal line and below the plate's space-time trajectory. On the negative characteristics the invariant $u - 2c = -2c_0$ is preserved. Thus the equation of motion satisfied in the disturbed interior region is the one for the positive characteristic with $2c$ replaced by $u + 2c_0$, that is,

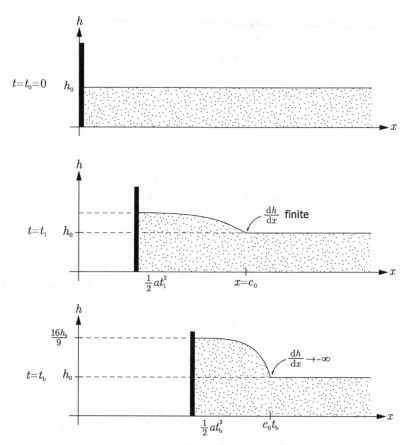

Fig. 6.10 Hydraulic jump schematic figure. For $t < 0$ the configuration is static with fluid at a level height h_0. At $t = 0$ the vertical plate begins to move with the velocity $U = \alpha t$ causing the fluid in front of it to pile up. The moment at which the hydraulic jump occurs is indicated as the first instant the negative gradient of the fluid height becomes infinite. This time occurs at $t = t_b = 2c_0/3\alpha$ as given in the text. The height of the layer at the plate at this instant is $16h_0/9$

$$\left[\frac{\partial}{\partial t} + \left(c_0 + \frac{3}{2}u \right) \frac{\partial}{\partial x} \right] u = 0. \tag{6.228}$$

This equation is none other than the Burgers equation in one dimension, which we discussed in Chap. 4. It has an *implicit* solution given by

$$u = F \left(x - c_0 t - \frac{3}{2}ut \right),$$

for some function F of a *single* argument. As it stands, all we know is that this function exists, but we do not yet know its form. However, we have information that will lead us to its form and, from it, determine an explicit solution for u and c, as a result. The velocity field and the plate's location play the role of a boundary conditions of sorts and may be exploited to infer the functional form of $F(\xi)$. Since at any time t the plate's position is given by $(1/2)\alpha t^2$ and its velocity is αt, the functional form for u should be consistent with this, thus

$$\alpha t = F\left(\frac{1}{2}\alpha t^2 - c_0 t - \frac{3}{2}\alpha t^2\right) = F\left(-\alpha t^2 - c_0 t\right). \tag{6.229}$$

In the purpose of finding the explicit form of F the following coordinate transformation $\xi \equiv -\alpha t^2 - c_0 t$ proves itself useful, since we can manipulate this identification to get our desired result. Specifically, we write t as a function of ξ to find

$$2\alpha t(\xi) = -c_0 + \sqrt{c_0^2 - 4\alpha\xi}.$$

Now the answer reveals itself: for $t > 0$, which is the same as $\xi < 0$, we find that the functional form of F as it is in Eq. (6.229) can be written in a more transparent form

$$F(\xi) = \frac{1}{2}\left[-c_0 + \sqrt{c_0^2 - 4\alpha\xi}\right]. \tag{6.230}$$

The velocity field can be expressed implicitly as

$$u = F\left(x - c_0 t - \frac{3}{2}ut\right) = \frac{1}{2}\left[-c_0 + \sqrt{c_0^2 - 4\alpha\left(x - c_0 t - \frac{3}{2}ut\right)}\right], \tag{6.231}$$

but it is a simple matter to solve for u *explicitly* as a function of x and t:

$$2u(x,t) = -\left(c_0 - \frac{3}{2}\alpha t\right) + \sqrt{\left(c_0 - \frac{3}{2}\alpha t\right)^2 - 4\alpha(x - c_0 t)}, \tag{6.232}$$

and for c

$$4c(x,t) = 4c_0 + 2u = 4c_0 - \left(c_0 - \frac{3}{2}\alpha t\right) + \sqrt{\left(c_0 - \frac{3}{2}\alpha t\right)^2 - 4\alpha(x - c_0 t)}. \tag{6.233}$$

As we know from our examination of the Burgers equation in Chap. 4, we expect that there is a possibility in which wave breaking occurs. Sea and ocean waves break obviously as well (Fig. 6.11), but those waves are mainly "pushed" by wind and therefore are more complicated than our simplistic description. In this problem this is the *hydraulic jump*. It occurs when the above solutions for u and/or c cease being single-valued and that is, of course, when the argument inside the square-root operation passes through zero. Another, perhaps more intuitive, way to understand the hydraulic jump is to say that it coincides with the place and time where the gradient of the function c

Fig. 6.11 A photograph of a breaking wave on the ocean. Even though these waves are mainly "pushed" by surface wind, this picture is very similar and suggestive to demonstrate wave breaking as in a hydraulic jump. (*Author: Shalom Jacobovitz. Image cropped. Licensed under the Creative Commons Attribution-Share Alike 2.0 Generic—http://creativecommons.org/licenses/by-sa/2.0/ deed.en*)

$$\frac{\partial c}{\partial x} = \frac{\alpha/2}{\sqrt{\left(c_0 - \frac{3}{2}\alpha t\right)^2 - 4\alpha(x - c_0 t)}}.$$

becomes infinite, which occurs when the expression inside the square-root sign is zero, i.e., at $t = t_b = 2c_0/3\alpha$ and $x = c_0 t_b$. The height of the fluid at this moment in time evaluated at the plate is given by $c^2(x,t)/g = c^2(\alpha t_b^2/2, t_b)/g = 16h_0/9$. Interestingly, all these conclusions could have been determined without appealing to the implicit solution of the nonlinear wave equation but, rather, by applying the geometrical reasoning, using the method of characteristics. As we had stated, the negative characteristics entering the disturbed domain begin to deviate from straight lines, but they all carry the same value of their respective invariant quantity, namely, $u - 2c = -2c_0$. On the other hand, the positive characteristics pinned to the plate are straight lines because u and c are constant on them. One may easily write the equation of this positive characteristic line. Choosing a reference point t_1, say, we see that on the space-time diagram the plate's coordinates are represented by $(x,t) = (\alpha t_1^2/2, t_1)$. The corresponding characteristic line must emerge from the plate's position with a constant slope given by $u + c = 3u/2 + c_0$. However, because the plate's speed at time t_1 is given by $u = \alpha t_1$, the slope of the emergent positive characteristic line has to be $3\alpha t_1/2 + c_0$. Therefore, the equation of the line containing that point is (see Fig. 6.12)

$$x(t) = \frac{1}{2}\alpha t_1^2 + \left(\frac{3}{2}\alpha t_1 + c_0\right)(t - t_1), \tag{6.234}$$

where u has the same value all along this line and is equal to αt_1. Furthermore, an explicit expression for $t_1(x,t,\alpha,c_0)$ is derivable by solving (6.234) directly for t_1. Therefore putting these together yields an expression for u, that is, $u = \alpha t_1(x,t,\alpha,c_0)$. The resulting expression for u is identical to the one found in Eq. (6.232).

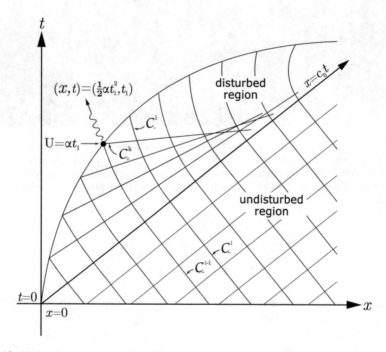

Fig. 6.12 Hydraulic jump space-time diagram. The undisturbed region is depicted for $x > c_0 t$. A plate with velocity $U = \alpha t$ has a position given by $x = (1/2)\alpha t^2$ and the disturbed region is found between the two lines shown. All positive characteristics in disturbed region are straight lines following the same arguments of the dam break problem uniqueness

6.4.6 Examples of Additional Specific Topics

6.4.6.1 Steady Isentropic Flow Through a Nozzle

In Chap. 2 we have summarized in Sect. 2.3.5 the various forms of Bernoulli's theorem, valid for different conditions. To demonstrate the power of these theorems we gave a number of examples and among them deviated briefly from the usual incompressible flows into the compressible flow regime. In Sects. 2.3.6.1 and 2.3.6.2, the nozzle and the Bondi problems, respectively, were described. Both involve compressible flows and may contain critical sonic points, but those Bernoulli's theorems that are valid also for compressible flows were sufficient to obtain important features of these examples, without actually solving them explicitly. Here we shall devote some space to discussing, in more detail, steady flow through a nozzle. Equation (2.32) may be transformed to an approximately one-dimensional form, whose variables depend only on x, say, due to the assumption that the nozzle is narrow, thus

$$\frac{1}{u}\frac{du}{dx} = \frac{1}{M^2 - 1}\frac{1}{A}\frac{dA}{dx},$$
(6.235)

where $A(x)$ is the cross-sectional area of the nozzle. We may summarize the cases of diverging ($dA > 0$) and converging ($dA < 0$) nozzle, or parts thereof thus

- Diverging nozzle:
  For $M < 1$ we have $du < 0 \to dP > 0$ as obtained from the equation of motion $\rho u du/dx + dP/dx = 0$. This case is referred to sometimes as a subsonic diffuser. For $M > 1$ we have $du > 0 \to dP < 0$.
- Converging nozzle:
  For $M < 1$ we have $du > 0 \to dP < 0$.
  For $M > 1$ we have $du < 0 \to dP > 0$, (supersonic diffuser).

A transonic flow, proceeding continuously from subsonic to supersonic speeds, must at some point pass through $M = 1$ and it is mathematically evident from Eq. (6.235) that it can only happen for an extremum of $A(x)$. Exploiting the relation (6.147) we get, by substituting into Eq. (6.235)

$$\frac{1}{M}\frac{dM}{dx} = \frac{1 + (\Gamma - 1)M^2}{M^2 - 1}\frac{1}{A}\frac{dA}{dx},$$
(6.236)

where Γ is the fundamental gasdynamic derivative, defined in relation (6.144). The case which interests us, $M = 1$, $dA/dx = 0$, cannot be directly treated by this equation because we have an indeterminate singularity of the sort $0/0$, but one may take recourse to l'Hopital's rule, yielding

$$\left(\frac{dM}{dx}\right)^2 = \frac{\Gamma}{2A}\frac{d^2A}{dx^2}.$$
(6.237)

This is a tractable equation and in fact guarantees that Γ must have the same sign as d^2A/dx^2. Nevertheless, two possibilities still exist, in principle, $dM/dx > 0$ and $dM/dx < 0$. The case $\Gamma > 0$ is the physically acceptable one as all normal gases in usual circumstances have a positive fundamental gasdynamic derivative. Cases of unusual $\Gamma < 1$ include the area variation of a transonic passage, the behavior of adiabatic flow with friction, some nonlinear wave propagation, special fluid composition, and only few other cases. In the case we adopt as being usual, A attains a minimum at the central throat and we have subsonic velocity, increasing in the converging part of the nozzle (upstream), up to the sonic point, where it crosses the sound speed and continues to grow. The case of a maximal cross-sectional area in the middle of the nozzle requires a negative second derivative of A there and thus negative Γ, which is usually unphysical. Thus we may summarize that a subsonic $\rightarrow$ supersonic transition of normal ($\Gamma > 0$) fluids can occur in a nozzle only at the area minimum of the throat. Such a converging–diverging nozzle is, as we may recall from Sect. 2.3.6, called *de Laval*, or supersonic nozzle. This is the normal operating mode of such nozzles, however the case of reversed flow, in which supersonic $\rightarrow$ subsonic transition occurs at the throat, is also allowed, as we saw, by the equations. It can be approximately realized in some practical situations, which we shall not discuss here. Finally, it is worth remarking that although a transonic transition occurs only at the throat, it does not mean that it must occur there always, in any flow. Assume that in Eq. (6.235) at the nozzle's throat point ($dA/dx = 0$) $M \neq 1$. This means that the throat point is not a critical point any more and either the flow will be subsonic through the whole nozzle or supersonic throughout.

The results discussed here are correct for any normal gas, not only for one obeying the perfect gas EOS. We would like to state, in particular, that the result displayed in Fig. 6.4, which depicted the normalized mass flux density as a function of M_*, remains qualitatively correct for any normal gas and circumstances. In particular it provides the existence of a maximal mass flux (discharge), i.e., mass per unit time, through a nozzle (tube of variable cross section, but narrow enough so that the flow can be assumed one-dimensional). As long as we assume a steady flow with no body forces, we get from Euler's equation $udu = -dP/\rho$. The isentropic flow assumption guarantees $dP = c_s^2 d\rho$ and this gives

$$\frac{d\rho}{du} = -\rho\frac{u}{c_s^2} \implies \frac{d(\rho u)}{du} = \rho\left(1 - \frac{u^2}{c_s^2}\right), \tag{6.238}$$

where the obvious relation $d(\rho u) = \rho du + u d\rho$ was used. Evidently, the mass flux density is maximal at the sonic point $u = c_s$, which is the throat of the nozzle and there we have the flux density $\rho_* u_*$, because if it would have been attained elsewhere it would be larger that the maximum. Therefore the mass discharge rate is determined there and is

$$Q_m = \rho_* u_* A_{\min}. \tag{6.239}$$

6.4.6.2   One-dimensional Similarity Flow

Similarity flows occur often when a problem is devoid of a typical physical dimensional scale, for example, a nonsteady one-dimensional flow in which there are typical velocities but no typical lengths. If a quantity having the dimensions of length cannot be constructed from the typical physical quantities of the flow and a typical velocity parameter exists, it follows that the dependence of the physical variables on the coordinate and time must always be throughout the ratio x/t, which has the dimensions of velocity. This causes the flow to be *self-similar* because all lengths must be measured in a unit which increases proportionally with time and this means that the flow pattern does not change in time. Defining the only relevant independent variable as $\xi \equiv x/t$ we get

$$\frac{\partial}{\partial x} = \frac{1}{t}\frac{d}{d\xi}; \quad \frac{\partial}{\partial t} = -\frac{\xi}{t}\frac{d}{d\xi}. \tag{6.240}$$

The condition for an isentropic flow can be easily found for such flows to read $(u - \xi)s' = 0$, where the prime denotes there the derivative with respect to the similarity variable ξ. This guarantees constancy of entropy with respect to the similarity variable, since the other choice, $u = \xi$, contradicts other equations of motion. Regarding those, it is easy to find that the other components of the velocity, besides the x one, are constants and, without limiting generality, may be chosen as zero. In addition, the equation of continuity and the x component of the Euler equation yield, respectively,

$$\frac{\partial \rho}{\partial t} + u\frac{\partial \rho}{\partial x} + u\frac{\partial \rho}{\partial x} = 0 \implies (u - \xi)\rho' + \rho u' = 0, \tag{6.241}$$

$$\frac{\partial u}{\partial t} + u\frac{\partial u}{\partial x} = -\frac{1}{\rho}\frac{\partial P}{\partial x} = 0 \implies (u - \xi)u' = -\upsilon P' = -\upsilon c^2\rho'. \tag{6.242}$$

The first equation here contradicts the previously mentioned choice $u = \xi$, because then u must be constant and we do not have any significant flow in the current frame of reference and zero flow in the fluid frame.

We discard the trivial solutions: $u =$ const., $\rho =$ const., and eliminate u' and ρ' from these equations. This results in $(u - \xi)^2 = c^2$ and thus $\xi = u \pm c$. A choice of $+$ out of the two means that we select the positive x direction in a manner explained below. Substituting thus $u - \xi = -c$, this choice in Eq. (6.241) yields $\rho du = cd\rho$. We now decide that for this homentropic gas, ρ will be the second independent thermodynamic variable and so the velocity is formally

$$u = \int \frac{c}{\rho}d\rho = \int \frac{1}{c\rho}d\rho = \int \sqrt{-dP d\upsilon}, \tag{6.243}$$

and the notation in the third integral means that one of the differentials is first expressed by the variable of the other differential, leaving open the choice. Our equation is $x/t = u + c$ and one of the first two integrals above give the implicit dependence of ρ on x/t, provided $c(\rho)$ is given. For a perfect gas with constant specific heats, this is soluble explicitly (Problem 6.25). In that problem we also discuss the implication of the choice of the x-axis direction, as made above, when we took $\xi = u + c$.

6.5   Shock Waves

6.5.1   General Shock Conditions

A shock is a very thin region, across which there is fluid flow and there are rapid variations of state variables. Across such a shock, one state of local thermodynamic equilibrium transits to another one. The transition is rapid and the width of the shock is actually the small distance needed for thermodynamic relaxation. A shock is usually approximated as a surface of discontinuity in space, this idealization being the result of the inviscid approximation of gas dynamics and the assumption that there is one kind of fluid on both sides. Real physical shocks have a non-zero thickness δ_{sh}, depending on the nature of the gas and the physical condition of the flow through the shock, allowing for an internal structure of the shock. We start our discussion, however, by using the discontinuous idealization. It should also be clear that a shock front is not necessarily stationary in the fluid and may propagate in it. A supersonically moving blunt object creates a bow shock in front of it, which usually smoothly joins the Mach cone. The idea behind shock formation is connected with the fact that the sound speed is the maximal speed of information propagation in a gas. Thus in a supersonic flow onto a stationary wall, for example, it is impossible to communicate the existence of the wall upstream. Thus the flow hits the wall and as a result a piling up of the fluid, a density jump (a shock) is created and it may propagate upstream. Another typical example in which shocks occur are explosions idealized as sudden injection of energy in a point of a gas. These, be it a supernova or a nuclear explosion, create, under idealized conditions, a spherical outgoing shock wave, which propagates into the undisturbed medium. Such shocks are usually called *blast* waves. To define the shock conditions we need a great deal of nomenclature and we face this at the outset. Let the flow take place in a plane and the velocity vector function, in an inertial frame, be denoted by $\mathbf{w}$. We consider now a shock in the form of a surface perpendicular to the flow plane and moving in the inertial frame with velocity $\mathbf{U}_{sh}$, say. The gas flows through the shock, and we denote the upstream side by the index 1, while the quantities on other side have the index 2, as in Fig. 6.13a.

The definition of the regions on the two sides of the discontinuity is best done similarly to the procedure usually performed in electromagnetic theory when there

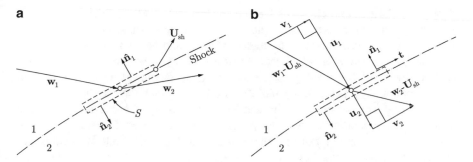

Fig. 6.13 Definition of the relevant directions and velocities in the inertial (**a**) and shock (**b**) frame. For details, see text

is a discontinuity in the dielectric constant and/or the magnetic permeability. One imagines a control volume, composed of two control surfaces S_1 and S_2 each parallel to the shock, but on the opposite sides and very close to it, say at a distance εL, where the flow's typical scale is L and $\varepsilon \ll 1$. The surfaces are equal, $S_1 = S_2 = S$, and also considered small, i.e, $\sqrt{S}\tilde{\varepsilon}L$. Thus formally with $\varepsilon \to 0$ we achieve the shock jump conditions *at a point*. Let $\hat{\mathbf{n}}_i$ be the unit normals to the control surfaces pointing *into* the side labelled i. Moving now into the *shock frame*, we have the positive, i.e., inflow into the control volume from side 1, perpendicular component of the fluid velocity, called u_1, and the positive, i.e., outflow out of the control volume into side 2, perpendicular component of the velocity, called by u_2 as in Fig. 6.13b. Denoting, as usual, a vector by a bold typeface and its size by a regular typeface we obtain the following relations

$$u_1 \equiv -(\mathbf{w}_1 - \mathbf{U}_{\text{sh}}) \cdot \hat{\mathbf{n}}_1$$
$$u_2 \equiv (\mathbf{w}_2 - \mathbf{U}_{\text{sh}}) \cdot \hat{\mathbf{n}}_2, \tag{6.244}$$

which can be verified by examining Fig. 6.13. Applying now physical conservation laws of mass, momentum, and energy plus the second law of thermodynamics to the control volume, that is, integrating the Eulerian inviscid equations (see Chap. 1), over the control volume, and remembering that a limit of its shrinking to zero, as explained above, is actually taken, we get

$$\rho_2 u_2 - \rho_1 u_1 = 0, \tag{6.245}$$

$$\rho_2 \mathbf{w}_2 u_2 - \rho_1 \mathbf{w}_1 u_1 = -P_2 \hat{\mathbf{n}}_2 - P_1 \hat{\mathbf{n}}_1, \tag{6.246}$$

$$\rho_2 \left(e_2 + \frac{1}{2} w_2^2\right) u_2 - \rho_1 \left(e_1 + \frac{1}{2} w_1^2\right) u_1 = -P_2 \hat{\mathbf{n}}_2 \cdot \mathbf{w}_2 - P_1 \hat{\mathbf{n}}_1 \cdot \mathbf{w}_1, \tag{6.247}$$

$$\rho_2 s_2 u_2 - \rho_1 s_1 u_1 \geq 0. \tag{6.248}$$

A physical clarification is in order. Despite the fact that the shock thickness has been effectively equated to zero, that is, the gradients in the shock are formally infinite, we have still allowed ourselves to neglect viscous and heat transport effects. This is actually an *assumption*, which posits that it is always possible to select a control surface, such that the velocity gradients are small enough to justify the inviscid approximation despite the very small thickness over which the changes occur. It is an interesting fact that the viscosity and heat conductivity of gases are typically so small that these conditions are usually satisfied in practice. Still, there are cases in which the inviscid assumption cannot be used inside the shock front. We will discuss an example of such a case below.

We shall now give a number of hints for Problem 6.27, in which Eqs. (6.245)–(6.248) are shown to allow for a very significant simplification.

- Mass: The first equation is already as simple as it can be.
- Momentum: Adding $\mathbf{U}_{sh}(\rho_1 u_1 - \rho_2 u_2) = 0$ to Eq. (6.246) and using $\hat{\mathbf{n}}_2 = -\hat{\mathbf{n}}_1$ yields a vector equation

$$\rho_2 u_2(\mathbf{w}_2 - \mathbf{U}_{sh}) - \rho_1 u_1(\mathbf{w}_1 - \mathbf{U}_{sh}) = \hat{\mathbf{n}}_1(P_2 - P_1). \qquad (6.249)$$

If one chooses the perpendicular unit vectors $\hat{\mathbf{n}}_1$, $\mathbf{t}$, in the plane of $\hat{\mathbf{n}}_1$ and $\mathbf{w}_1 - \mathbf{U}_{sh}$, and a third unit vector, $\hat{\mathbf{b}}$, perpendicular to that plane, completing the unit vector triad, the components of Eq. (6.249) yield simplified equations.
- Energy: Adding $P_2 u_2 - P_1 u_1$ to both sides of Eq. (6.247), with $\rho u \equiv \rho_2 u_2 = \rho_1 u_1$ gives

$$\rho u \left[h_2 + \frac{1}{2}u_2^2 - \left(h_1 + \frac{1}{2}u_1^2 \right) \right] = (P_2 - P_1)\mathbf{U}_{sh} \cdot \hat{\mathbf{n}}_1, \qquad (6.250)$$

which can be transformed to

$$h_2 + \frac{1}{2}u_2^2 - \left(h_1 + \frac{1}{2}u_1^2 \right) = (\mathbf{w}_2 - \mathbf{w}_1) \cdot \mathbf{U}_{sh}. \qquad (6.251)$$

A notational convention is that the jump in any quantity across a shock is represented by writing it in square brackets, e.g., $[Y] \equiv Y_2 - Y_1$. Using this notation and the fact that $\mathbf{v} = |\mathbf{v}_1|, = |\mathbf{v}_2|$ is the component of the fluid velocity parallel to the shock (can you show it?), we may summarize, using Problem 6.27, the shock jump conditions in the following simple equations:

$$[\rho u] = 0, \qquad (6.252)$$

$$[P + \rho u^2] = 0, \qquad (6.253)$$

$$[v] = 0, \qquad (6.254)$$

$$[h + \frac{1}{2}u^2] = 0, \tag{6.255}$$

$$[s] \geq 0. \tag{6.256}$$

The above shock conditions involve only velocities *relative* to the shock $w_i - U_{sh}$. Moreover, the only motion of the shock front that has physical or geometrical significance is the motion normal to the shock. The tangential to the shock front component of U_{sh}, which we called v is just an invariant across the shock. Thus we can always choose U_{sh} in such a way as to make $v_1 = v_2 = 0$. Thus the shock is a *normal* shock. This discussion can be summarized by the following important statements:

1. The shock jump conditions (6.252)–(6.256) are correct for *any* observer.
2. The above conditions hold irrespective of the constancy of the shock velocity or the fluid velocity ahead of the shock.
3. Every shock can be reduced to an equivalent *normal* shock.

Formally, the last statement is accomplished in the following way—we start by adding a tangential motion to the shock front, which as we have seen has no physical effect, defining effectively a new shock velocity

$$U'_{sh} = U_{sh} + vt, \tag{6.257}$$

where **t** is the unit tangent vector to the shock.

Then we can write (see Fig. 6.13b)

$$\mathbf{w}_1 - \mathbf{U}_{sh} = -u_1\hat{\mathbf{n}}_1 + v\mathbf{t} \implies \mathbf{w}_1 - \mathbf{U}'_{sh} = -u_1\hat{\mathbf{n}}_1, \tag{6.258}$$

$$\mathbf{w}_2 - \mathbf{U}_{sh} = -u_2\hat{\mathbf{n}}_1 + v\mathbf{t} \implies \mathbf{w}_2 - \mathbf{U}'_{sh} = -u_2\hat{\mathbf{n}}_1. \tag{6.259}$$

6.5.1.1 Example: Equivalent Normal Shock for a Bow Shock

In Fig. 6.14 we see a bow shock, created by a relatively compact object, a high speed massive star, moving supersonically in the interstellar medium. The speed of sound in the interstellar medium is low due to its low temperature, so the supersonic motion is not because of the extremely high velocity of the object, it is rather because of the relatively low sound speed. Assume, for simplicity, that the compact object is traveling with constant velocity $-W_\infty$ into a uniform medium. The bow shock wave, as we understand it, has a stationary nature, relatively to the object as it precedes it, moving with the same velocity. For an observer on the compact object, which produces the bow shock, $U_{sh} = 0$ and $w_1 = W_\infty$. For an observer on the ground $U_{sh} = -W_\infty$ and $w_1 = 0$. The procedure of reduction to an equivalent normal shock at each point proceeds naturally in the following way. Pick a point on the bow shock, P, say, and let the bow tangent at that point make an angle α with the apparent axis of symmetry of the bow pointing ahead. Then, a little

Fig. 6.14 The high velocity star (24 km/s) ζ Oph, 20 times more massive than the sun, and its wind compress the interstellar matter in front of it, creating a bow shock (seen in IR). The distance to the object is $\sim$450 light years. *Public Domain. Author: NASA/JPL-Caltech/Spitzer ST. Courtesy of NASA via Wikimedia commons—http://eol.jsc.nasa.gov/Info/use.htm*)

reflection will help us to understand that in the frame of the moving star and the bow shock we will have a perpendicular inflow into the bow shock, at a point P, with a velocity $u_1 = W_\infty \sin \alpha$. In the frame of an observer on earth we will see the shock moving into the interstellar medium with a velocity having the following velocity components at point P: $-W_\infty \sin \alpha$ perpendicular to the shock and $W_\infty \cos \alpha$ in parallel. The directions of the normals (to both sides) are obviously related to α. Thus all shocks can be treated *at a point* as a normal shock as there is no local physical distinction between any shock and its normal equivalent. This perhaps is the reason for singling out the shock conditions, given by Eqs. (6.252)–(6.256), with the trivial exception of Eq. (6.254), i.e., the *normal* shock conditions.

6.5.2   Rankine–Hugoniot Adiabat and Jump Conditions

The principal shock conditions, omitting the trivial $[v] = 0$, are given explicitly according to Eqs. (6.252)–(6.256) as

$$\rho_2 u_2 = \rho_1 u_1,$$

$$P_2 + \rho_2 u_2^2 = P_1 + \rho_1 u_1^2,$$

$$h_2 + \frac{1}{2} u_2^2 = h_1 + \frac{1}{2} u_1^2,$$

$$s_2 \geq s1, \tag{6.260}$$

and may be perceived as a problem, in which the shock velocity, U_{sh}, and the velocities expressed in the shock frame, ahead, i.e., on side 1, of the shock are known and the question is whether these are enough for the unique determination of the conditions behind the shock, i.e., on side 2. For example, a shock front advancing at a known velocity into a known stationary fluid would be a problem of this type. An already mentioned assumption has to be applied if we want to answer this question. It is that the fluid ahead and behind the shock is essentially the same, that is, having the same equations of state and composition, precluding the consideration of any chemical reaction at the shock. We shall assume this here, as the simplest case, and proceed to investigate this question.

Exploiting the constancy of the mass flux ($\rho_1 u_1 = \rho_2 u_2$) we may call this constant J. Combining the first (mass) and second (momentum) equation in (6.260) gives

$$u_1 u_2 [\rho] = [P]. \tag{6.261}$$

Now, multiplication by $\rho_1 \rho_2$ gives an equation which can be written using specific volumes, $v = 1/\rho$, instead of densities, thus

$$J^2 = -\frac{[P]}{[v]}. \tag{6.262}$$

This formula, together with the relations $u_i = J v_i$, (i=1,2), relates the perpendicular entry and exit speeds into and out of the shock wave, in the shock frame, to the pressures and densities on the two sides of the shock surface. As $J^2 > 0$ either $P_2 > P_1$ and $v_1 > v_2$ or vice versa. As we shall see below only the former case can occur. Finally, substituting $[P + J^2 v] = 0$, which is another way of writing the momentum conservation in the obvious relation $[u] = J[v]$ we obtain the important relation

$$[u]^2 = -[P][v]. \tag{6.263}$$

Moving on now to the energy equation (third equation in 6.260), we write it in the form $[h + \frac{1}{2} J^2 v^2] = 0$ and substituting J^2 from Eq. (6.262) we get

$$h_2 - h_1 = \frac{1}{2} (P_2 - P_1)(v_2 + v_1), \text{ or} \tag{6.264}$$

$$e_2 - e_1 = -\frac{1}{2}(P_2 + P_1)(v_2 - v_1), \tag{6.265}$$

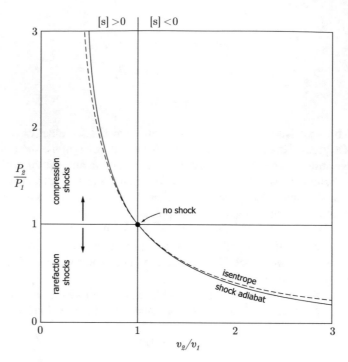

Fig. 6.15 The shock adiabat. The isentrope $P \propto \upsilon^{-\gamma}$ is also displayed (*dashed line*)

where the second equation follows from the definition of enthalpy in terms of energy $h = e + P\upsilon$. Equations (6.264)–(6.265) have considerable importance for the understanding of shocks and are called interchangeably *Rankine–Hugoniot* equations or *Hugoniot* adiabat or the *shock* adiabat. We shall discuss them now, and for the sake of clarity will do it for a perfect gas, for which it will also be possible to derive the celebrated Rankine–Hugoniot *jump ratio conditions*. First and foremost we define some useful quantities. The first of these will be the shock Mach number $M_1 \equiv u_1/c_1$, while the nondimensional *pressure jump* is defined as $\Pi \equiv [P]/(\rho_1 c_1^2)$. The numerical value of Π is, by definition, the measure of the strength of the shock. Shocks are also classified according to the sign of $[P]$, where $[P] > 0$ (< 0) corresponds to *compression (rarefaction)* shocks. Limiting the discussion now to a perfect gas with constant specific heats, we write $h = c_P + \text{const.} = \{\gamma/(\gamma-1)\} P\upsilon + \text{const.}$ Thus Eq. (6.264) reduces to the pressure jump ratio condition

$$\frac{P_2}{P_1} = \left(\frac{\gamma+1}{\gamma-1} - \frac{\upsilon_2}{\upsilon_1}\right)\left(\frac{\gamma+1}{\gamma-1}\frac{\upsilon_2}{\upsilon_1} - 1\right)^{-1}. \tag{6.266}$$

The isentrope adiabat is $P_2/P_1 = (\upsilon_2/\upsilon_1)^\gamma$ and one may consider a collection of such functions $P(\upsilon)$, taken as the values of these quantities with index 2, while those

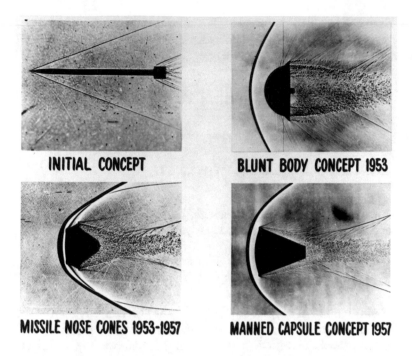

Fig. 6.16 Four shadowgraph images represent early re-entry vehicle concepts. A shadowgraph is a process that makes visible the disturbances that occur in a fluid flow at high velocity, in which light passing through a flowing fluid is refracted by the density gradients in the fluid resulting in bright and dark areas on a screen placed behind the fluid. A blunt body produces a shockwave in front of the vehicle that actually shields the vehicle from excessive heating. It was used in the Mercury, Gemini, and Apollo projects. (*Public Domain. Author: NASA/Ames research center. Courtesy of NASA via Wikimedia commons—http://eol.jsc.nasa.gov/Info/use.htm*)

with index 1 are regarded as constant parameters. Points lying above $P_2/P_1 = 1$ in Fig. 6.15 are compression shocks. The Gibbs equation dictates that for a perfect gas $(\partial s/\partial P)_\rho = c_V/P$ thus points above the isentrope all have $s > s_1$, if s_1 is the isentrope value. We see in Fig. 6.15, which reflects the true equations, that compression shocks have $s_2 > s_1$, which is in agreement with the shock condition $[s] \geq 0$. It follows that only compression shocks are possible, at least in this case of a perfect gas, but actually also in all non-exotic fluids. We show here, in passing, some bow shocks experimented by NASA for re-entry vehicles (Fig. 6.16). Details can be found in the figure caption.

In order to reach some general statements on shocks we start by rewriting the first of the Rankine–Hugoniot equations in the following form

$$[h] = v_1[P] + \frac{1}{2}[v][P].$$ (6.267)

We wish to obtain a relation between $[P]$ and $[s]$ and expand $h(s,P)$ and $\upsilon(s,P)$ in Taylor series. Using the thermodynamic identities $T = (\partial h/\partial s)_P$ and $\upsilon = (\partial h/\partial P)_s$ and retaining terms up to third order we get

$$[s] = \frac{1}{12T_1}\left(\frac{\partial^2\upsilon}{\partial P^2}\right)_s [P]^3 + \mathcal{O}\left([P]^4\right), \qquad (6.268)$$

which, when rewritten in nondimensional form,

$$\frac{[s]}{c_1^2/T_1} = \frac{1}{6}\Gamma_1\Pi^3 + \mathcal{O}\left(\Pi^4\right), \qquad (6.269)$$

stresses that it is valid for Π small enough, i.e., only for weak shocks. We repeat here, for convenience, the fundamental gasdynamic derivative

$$\Gamma = \frac{c^4}{2\upsilon^3}\left(\frac{\partial^2\upsilon}{\partial P^2}\right)_s \qquad (6.270)$$

as was defined in Eq. (6.144). $\Pi = [P]/(\rho_1 c_1^2)$ was also defined before. Now since $[s] \geq 0$, $[P]$ necessarily has the same sign as the fundamental derivative. Thus in this limit of weak shocks and for $\Gamma > 0$ (all normal fluids), we may safely conclude compression $[P] > 0$ as a result that is valid not only for perfect gases. From $J^2 = -[P]/[\upsilon]$, we obtain that $[\upsilon] < 0$ and thus $[\rho] > 0$ and $[u] < 0$. The shock adiabat in Fig. 6.15 is for a weak shock. In such a weak shock, because $[s] \propto [P]^3$, the entropy jump becomes negligibly small, and practically such shocks are adiabatic and their adiabat becomes almost indistinguishable from the isentrope adiabat. Consider the shock adiabat A in Fig. 6.17, which was drawn for a strong shock. Let the upstream gas, that is preceding the shock, be represented by point 1 and typical downstream gas, after passing a strong shock, be represented by the state 2 on the shock adiabat. Also, on the adiabat, the point $2w$ represents the conditions after a weak shock. For typical fluids the shock adiabat A will lie above the isentrope ($[s] \geq 0$). The chord connecting points 1 and 2 has the slope $-J^2(= [P]/[\upsilon])$. For the weak shock between 1 and $2w$ the slope is smaller and very close to that of the isentrope adiabat, so from the geometry of the figure we get

$$\left(\frac{\partial P}{\partial \upsilon}\right)_{s(2)} \leq -J^2 \leq \left(\frac{\partial P}{\partial \upsilon}\right)_{s(1)} \qquad (6.271)$$

Now, for the isentrope

$$\left(\frac{\partial P}{\partial \upsilon}\right)_s = -\rho^2 c^2. \qquad (6.272)$$

Fig. 6.17 A shock adiabat
(A) allowing for strong
shocks. The isentrope adiabat
is the *dashed* curve (I). For
details, see text

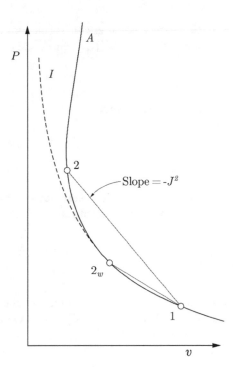

Thus because $J = \rho_2 u_2 = \rho_1 u_1$

$$\rho_2^2 c_2^2 \geq (\rho_2^2 u_2^2 = \rho_1^2 u_1^2) \geq \rho_1^2 c_1^2. \tag{6.273}$$

Hence $u_1 \geq c_1$, and $u_2 \leq c_2$, which is a very important result—the relative to the shock normal upstream inflow is supersonic, while the corresponding outflow on the other side is subsonic. The slight inaccuracy we have allowed ourselves here stems from the use of point 2 which represents post strong shock condition, while actually we have used relation (6.269), formally approximately correct for only weak enough shocks. It turns out, however, that these inequalities hold for strong shocks as well. This will be the subject of Problem 6.29.

Imagine that a strong shock has its strength progressively diminished by $[P]$ getting smaller. We thus have $u_i \to c_i$ from both sides and because $[u^2] = -[P][v]$, $[u] \to 0$. So at the limit of vanishing strength we have that the two expressions:

$$u_1 u_2 = \frac{[P]}{[\rho]} \quad \text{and} \quad c^2 = \left(\frac{\partial P}{\partial \rho}\right)_s \tag{6.274}$$

become identical. Thus what is left from the shock is an acoustic disturbance propagating with speed c relative to the fluid. We may thus summarize the important points in the following way:

1. Since only compression shocks are allowed ($[P] > 0$) for fluids which have positive $(\partial^2 v / \partial P^2)_s$, as almost all usual fluids do, then

$$P_2 > P_1 \tag{6.275}$$

and correspondingly

$$\rho_2 < \rho_1, \tag{6.276}$$

$$u_1 > u_2. \tag{6.277}$$

2. The upstream flow is supersonic with respect to the shock and the post-shock downstream flow is subsonic with respect to the shock, i.e.,

$$M_1 > 1 \quad \Rightarrow \quad (u_1 > c_1)$$
$$M_2 < 1 \quad \Rightarrow \quad (u_2 < c_2). \tag{6.278}$$

Thus disturbances downstream cannot have any effect on the upstream flow.
3. For weak shocks $[s] \propto [P]^3$, shocks thus become quickly close to isentropic with diminishing strength.

6.5.2.1   Example: Rankine–Hugoniot Jump Conditions for a Perfect Gas

This example summarizes the ratios of variables on the two sides of a shock, in a perfect gas with constant adiabatic exponent, by just γ and the upstream Mach number $u_1 / c_1 \equiv M$. There are several ways of deriving these formulae, each involving considerable algebra. We encourage the reader to obtain the form below from the original jump conditions and perfect gas relations (Problem 6.30):

$$\frac{\rho_2}{\rho_1} = \frac{(\gamma+1)M^2}{(\gamma+1)+(\gamma-1)(M^2-1)} = \frac{u_1}{u_2}, \tag{6.279}$$

$$\frac{P_2}{P_1} = \frac{(\gamma+1)+2\gamma(M^2-1)}{(\gamma+1)}, \tag{6.280}$$

$$\frac{T_2}{T_1} = \frac{[(\gamma+1)+2\gamma(M^2-1)]\,[(\gamma+1)+(\gamma-1)(M^2-1)]}{(\gamma+1)^2 M^2}. \tag{6.281}$$

Note that from these relations it follows that if $M > 1$, that the pressure, temperature, and density increase in a shock and the velocity decreases. It is important to see that the density jump has an upper limit even for a very strong (formally infinitely strong: $M \to \infty$) shock, which is $(\gamma+1)/(\gamma-1)$, while the pressure and temperature jumps have no bound.

6.5.3 Radiative Shocks

Shocks, which occur in the atmosphere, are usually dominated by short timescale processes, for example, the post-shock heated gas does not have enough time to cool by radiation, before it cools by expansion to the ambient medium.

The situation may be different in astrophysical applications, for example, we devote this section, in conjunction with Problem 6.34, to a brief discussion of radiating shocks. To deal with this physical situation in as simple a way as possible (but not simpler!) we assume a one-dimensional problem, in which cool gas is streaming and for some reason is shocked at a certain point. The pre-shock gas is assumed to be at thermal equilibrium, that is, its cooling is balanced by heating and therefore it stays at a constant temperature, T_1, say. The crucial assumption that the shock wave has a very small extent, indeed it can be approximated by a discontinuity, needs no justification at this point. It is also obvious that very far downstream, after the gas rids itself, in this case radiatively, of the extra thermal energy it will reasonably return to some conditions of thermal equilibrium in the sense $\mathscr{L} = 0$, where $\mathscr{L}(\rho, T)$ is the total cooling function per unit mass. This far region is not necessarily identical in its thermodynamic conditions to the pre-shock conditions and we shall denote the variables there by the index 3, e.g., temperature T_3. We have thus three regions, the intermediate one, denoted by index 2, is of hot gas not in thermal equilibrium, that is, it has $\mathscr{L} > 0$, which indicates cooling. We shall concentrate on the transition from the values marked with subscript 2 to those marked by 3. Using steady, one-dimensional flow equations:

$$\rho u = \text{const.},$$

$$\rho u^2 + P = \text{const.},$$

$$\rho u \frac{de}{dx} = -P \frac{du}{dx} - \rho \mathscr{L}, \qquad (6.282)$$

and the relation

$$e = \frac{1}{\gamma - 1} \frac{P}{\rho}, \qquad (6.283)$$

correct for a perfect gas with constant specific heats, as assumed, one can derive, and it is the essence of Problem 6.34, a single equation

$$\frac{1}{(\gamma - 1)} (c_s^2 - u^2) \frac{du}{dx} = -\mathscr{L}(\rho, T), \qquad (6.284)$$

where the speed of sound for a perfect gas is as usual $c_s^2 = \gamma P / \rho$. Collecting all the equations and adding the perfect gas equation of state,

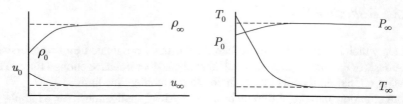

Fig. 6.18 The structure of the radiatively cooling region behind a one-dimensional steady shock. For details, see text

$$P = \frac{\mathscr{R}}{\mu} \rho T, \tag{6.285}$$

where the symbols have their aforementioned usual meaning, we can isolate four equations for u, ρ, T, and P. Starting from arbitrary *initial* values, u_0, ρ_0, T_0, and P_0, for this problem, one can solve these equations numerically, getting results that resemble those plotted in Fig. 6.18, where the constant value of the functions, marked by the subscript ∞ is achieved relatively fast, after only a few cooling times (what has to be known to assess the cooling time and how would you do it?). It is also possible to reason heuristically that the solution should qualitatively look like in this figure. We shall not discuss in any detail the topics of oblique shocks with the inclusion of specific radiative cooling behind the shock, to be expanded upon in Problems 6.31–6.34.

6.6   Explosions, Blast, and Detonation Waves

We move on to discuss a topic that is often important in the observation of astrophysical or other shocks, notably stellar or man-made nuclear explosions, which have been tested and used in the atmosphere in the times before a ban treaty was signed between the superpowers, who were the only participants in the "nuclear club."

6.6.1   *Strong Explosion at a Point: Blast Wave*

The search for a reliable fluid dynamical solution to the problem, in which a large amount of energy, say $\mathscr{E}$, is released instantaneously at a point (actually, a very small volume) of an extended body of gas, naturally arose with the development of first nuclear weapons in the 1940s. G.I. Taylor and L. Sedov worked on and solved this problem in the U.S. and U.S.S.R., respectively. It is obvious that an explosion of this sort will induce a strong shock wave to propagate outward. Assuming that we are dealing with a gas of constant specific heats we concentrate first on the, so-called, *blast wave* phase, in which energy is conserved. Thus the assumption is that the wave is not too far from the source, so as to preserve the strong shock conditions. Consequently, we may safely assume that the pressure discontinuity is very large so that we may neglect P_1 with respect to P_2 and that the post-shock density reaches its maximal limiting jump values

$$\rho_2 = \rho_1 \frac{\gamma+1}{\gamma-1}. \tag{6.286}$$

Dimensional arguments can readily convince one that the gas flow pattern, i.e., the solution, must be determined by only two-dimensional physical parameters, $\mathscr{E}$ and ρ_1, in addition to the radial coordinate r and time t. From these, only one nondimensional combination can be formed and it is written here as $\xi = r(\rho_1/\mathscr{E}t^2)^{0.2}$. When the governing equations are nondimensionalized they thus must depend on r and t only through the similarity variable ξ. A dimensionless constant can obviously be incorporated as multiplying ξ. We shall call it ξ_0, say, and it will determine the position of the shock R_{sh} thus

$$R_{sh}(t) = \xi_0 \left(\frac{\mathscr{E}t^2}{\rho_1} \right)^{1/5}, \tag{6.287}$$

and the velocity of the blast wave, which is also equal to u_1, is

$$U_{sh} = \frac{dR_{sh}(t)}{dt} = \frac{2R_{sh}(t)}{5t} = \frac{2}{5} \xi_0 \left(\frac{\mathscr{E}}{\rho_1 t^3} \right)^{1/5}. \tag{6.288}$$

With time the blast wave weakens, as a shock, because this is the phase of conservation energy, i.e., we have the constancy of $\rho_1 R_{sh}^3 U_{sh}^2$. In addition, ξ_0 is expected to be of order 1.

The challenge now is to solve, in detail, for the interior of the blast wave. Starting with the Rankine–Hugoniot conditions in the limit of a very strong shock wave, expressed in the frame in which the explosion point is at rest, we get

$$\rho_2 = \frac{\gamma+1}{\gamma-1} \rho_1, \tag{6.289}$$

$$u_2 = \frac{2}{\gamma+1} U_{sh}, \tag{6.290}$$

$$P_2 = \frac{2}{\gamma+1} \rho_1 U_{sh}^2. \tag{6.291}$$

Define now the dimensionless physical dependent variables β, v, p as functions of ξ in the following way, expressing the fluid dynamical variables $\rho(r,t), u(r,t), P(r,t)$ for which we have to solve:

$$\rho(r,t) = \left(\frac{\gamma+1}{\gamma-1}\right)\rho_1\beta(\xi) = \rho_2\beta(\xi), \tag{6.292}$$

$$u(r,t) = \frac{4}{5(\gamma+1)}\frac{r}{t}rv(\xi) = \frac{2}{\gamma+1}\frac{U_{sh}}{R_{sh}}rv(\xi) = u_2\frac{r}{R_{sh}}v(\xi), \tag{6.293}$$

$$P(r,t) = \frac{8}{25(\gamma+1)}\rho_1\left(\frac{r}{t}\right)^2 p(\xi) = P_2\left(\frac{r}{R_{sh}}\right)^2 p(\xi). \tag{6.294}$$

We have chosen the coefficients in front of the nondimensional dependent variables to be such that they are normalized for ξ_0 behind the shock, i.e., outside the spherical blast wave $\beta(\xi_0) = v(\xi_0) = p(\xi_0) = 1$.

To get (implicitly) ξ_0, which is the dimensionless position of the front, we write the energy conservation equation for the flow behind the shock, in spherical coordinates,

$$0 = \frac{\partial}{\partial t}\left[\rho\left(e+\frac{1}{2}u^2\right)\right] + \frac{1}{r^2}\frac{\partial}{\partial r}\left[r^2\rho u\left(e+\frac{P}{\rho}+\frac{1}{2}u^2\right)\right], \tag{6.295}$$

where the specific internal energy is $e = P/[\rho(\gamma-1)]$. Integrating now Eq. (6.295) over all space, from $r = 0$ to $r = \infty$, and noticing that at both ends $ru = 0$ we split the space integration into two

$$0 = 4\pi\frac{d}{dt}\int_0^{R_{sh}(t)}\left[\frac{P}{\gamma-1}+\frac{1}{2}\rho u^2\right]r^2 dr + 4\pi\frac{d}{dt}\int_{R_{sh}(t)}^{\infty}\frac{P_1}{\gamma-1}r^2 dr. \tag{6.296}$$

It is easy to see (show it!) that the second integral is negligible with respect to the first one. Integrating now over time, the time derivative of the dominant integral gives the total explosion energy

$$4\pi\int_0^{R_{sh}(t)}\left[\frac{P}{\gamma-1}+\frac{1}{2}\rho u^2\right]r^2 dr = \mathcal{E}, \tag{6.297}$$

which in nondimensional form reads

$$\frac{32\pi}{25(\gamma^2-1)}\int_0^{\xi_0}[p(\xi)+g(\xi)v^2(\xi)]\xi^4 d\xi = 1. \tag{6.298}$$

It is simple to understand that the first term in the integral is the fraction of $\mathcal{E}$ that is in the form of thermal energy and the second term is the kinetic energy fraction. This equation implicitly supplies ξ_0.

Now the change of variables from x and t in the fluid dynamical equations to ξ is effected by the replacements

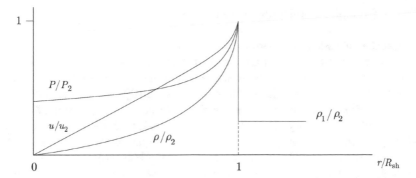

Fig. 6.19 The Taylor–Sedov solution of the constant energy phase of a strong point explosion, showing the inside region of the spherical blast wave for the case $\gamma = 5/3$

$$\frac{\partial}{\partial t} \mapsto -\frac{2}{5}\frac{\xi}{t}\frac{d}{d\xi}, \quad \frac{\partial}{\partial r} \mapsto \frac{\xi}{r}\frac{d}{d\xi}. \tag{6.299}$$

In Problem 6.35 it is shown that the fluid dynamical equation in the nondimensional variables are

$$-\xi\frac{d\beta}{d\xi} + \frac{2}{\gamma+1}\left[3\beta v + \xi\frac{d(\beta v)}{d\xi}\right] = 0, \tag{6.300}$$

$$-v - \frac{2}{5}\xi\frac{dv}{d\xi} + \frac{4}{5(\gamma+1)}\left(v^2 + v\xi\frac{dv}{d\xi}\right) = -\frac{2}{5}\left(\frac{\gamma-1}{\gamma+1}\right)\frac{1}{\beta}\left(2p + \xi\frac{dp}{d\xi}\right), \tag{6.301}$$

$$2(p+gv^2) + \frac{2}{5}\xi\frac{d}{d\xi}(p+gv^2) = \frac{4}{5(\gamma+1)}\left\{5v(\gamma p + \beta v^2) + \xi\frac{d}{d\xi}[(v(\gamma p + \beta v^2)]\right\}. \tag{6.302}$$

This system has to be solved from ξ_0 to 0 with the normalization condition from which, as explained before, ξ_0 is given implicitly by the integral relation (6.299). Using a computer of his vintage, Taylor obtained a complete solution to the problem. The value of $\xi_0 = 1.17$ was obtained for $\gamma = 5/3$. This allowed the yield of the first fission explosion in New Mexico to be computed by observing the motion of the blast wave. The number thus obtained was $\sim 10^{21}$ ergs. In the Soviet Union the problem was lack of electronic computers, but this did not prevent Sedov from obtaining a solution to these equations analytically and correctly, in a closed form. The interior temperature reaches its maximum at the center, where $P/\rho \to \infty$, and is minimal at the edge, in this phase. This is obvious as the shock velocity U_{sh} starts formally from extremely high values, but as the blast wave decelerates $U_{sh} \propto t^{-3/5}$ the temperature drops, because $T \propto U_{sh}^2$. In Fig. 6.19 the profiles inside the blast wave, as given by the Taylor–Sedov solution, are displayed as a function of r/R_{sh}.

6.6.1.1   Example: Application to a Supernova Explosion

An explosion of a star as a supernova (of any kind, for that matter) is for all practical purposes very similar to the problem treated above. As the star is practically a point source, relative to the vast expanses of the interstellar medium, its possible explosion as a supernova injects, nearly instantaneously, an enormous amount of energy ($\sim 10^{51}$ ergs) in what can be considered a point. The interior of the blast wave is well described by the Taylor–Sedov solution, with the physical quantities scaled appropriately so as to be given in physical units, as long as the assumption of constant energy holds. With the expansion of the blast wave it causes a denser shell to form around it and radiative cooling becomes important ($\mathscr{L} > 0$). The shell becomes cooler and denser and the expanding supernova remnant pushes the shell because of momentum conservation. This is referred to as a *snow plow* phase. This is an idealized picture, assuming tacitly that the interstellar medium is uniform. The truth is, however, very different. The interstellar medium is typically composed of clouds, which are significantly denser than the rare intercloud medium. It is thus logical that the blast wave shock propagates at different speeds, according to the outer density and achieves the cooling stage at different times. Thus the blast waves from different supernovae may merge and this is feasible, in particular when the positions of the supernovae are correlated, as is the case in supernovae type II, which are correlated with molecular clouds from which massive stars form. On the other hand, it has been realized that pre-supernova stars blow strong winds, which tend to homogenize the medium surrounding them. In any case, the picture is more complicated than just a relatively simple Taylor–Sedov blast wave solution.

We shall conclude this short discussion by quoting some data on a typical supernova. As we have already mentioned the typical energy input is of the order of 10^{51} ergs. We expect the blast wave approximation to hold only for $t \lesssim 100$ years, because the wave cannot move faster than the speed of the original ejecta $\sim 10^4$ km/s. At $t \sim 100$ years after the explosion, the radius of the blast wave is approximately 2 pc. Estimates of the total amount of radiation emitted by the supernova in time $t \sim 10^5$ years after the explosion, when the remnant size is of the order of 30 pc, should reach the total available energy. At the time of writing of this book, controversy still exists whether a fluid treatment of supernova remnant evolution has validity. This stems from the fact that a typical supernova emits 2 MeV protons, which are stopped by the ambient medium only after 10^3 pc! Thus magnetic fields (as small as a few μG should suffice) have to be invoked, in order to tie the protons to the ambient fluid and make their effective mean free path much smaller than the typical system size. The alternative is the formation of collisionless shocks, in which chaotic magnetic fields play a decisive role in creating discontinuities.

6.6.2 Detonation Waves

So far in this book, except perhaps for very few cases, we have dealt with flows
of fluids which are homogeneous, that is, their *composition* in terms of chemical,
nuclear, or other species is uniform. In this section we shall abandon this rule
and take into account the possibility of exothermic reactions, e.g., chemical or
thermonuclear, that may take place during gas flows and may influence considerably
the flow, by releasing extra energy. We do not wish to discuss the complex subject
of combustion under a variety of circumstances, but rather mention just the basic,
but important in our view to students of FD topic of a *detonation* wave. To observe
such a wave we must have an exothermic reaction between reactants that are not
in stable, chemical, or other reactive, equilibrium. The wave is physically described
as being headed by an ordinary shock wave, which raises the reactants to a high
enough temperature and pressure, so that the ingredients which are prone to react
acquire enough activation energy to start the reaction, which may proceed, more or
less, to completion, in a narrow region behind the shock. The overall thickness of
the detonation front, i.e., the reaction zone is typically very narrow, e.g., ~ 1 cm in
the case of hydrogen–oxygen combustion. The idea is that in the gas with potential
reactants in it, it may happen that the reaction starts and the front of combustion
propagates because heat from the reaction is conducted forward into the gas which
has not yet reacted. If this is the mechanism of propagation of the flame, its speed
is usually slow and subsonic and we call it *deflagration*. Our aim is to describe
qualitatively, and as much quantitatively as we can here, *detonation* waves which
differ from the above-mentioned slow deflagration. Assume that a shock wave
passes through the gas and ignites the reaction at a point it passes, the reaction
continuing until its completion during a time τ_{sr}. This value clearly depends not
only on the reaction kinetics but also on the shock's strength, through the post-shock
conditions imposed on the gas. The reaction layer will thus follow the shock and its
width will be $\Delta_d = u_1 \tau_{sr}$. We have already seen that Δ_d is small as compared to
typical problem scales and therefore it is possible, when our system is large enough,
to consider the shock *plus* the combustion layer following it as single surface of
discontinuity. Since the detonation front may be taken as a discontinuity between
the post-shock and the post-detonation regions, the conservation formulae as given
for shock waves before hold as long as we pay attention to endow the physical
quantities with the right indices. Thus the detonation adiabat is given implicitly by

$$h_1 - h_2 + \frac{1}{2}(v_1 + v_2)(P_2 - P_1) = 0. \tag{6.303}$$

Note that this formula is identical to the shock adiabat (6.264), but we have to
remember that the suffix 1 refers now to the post-shock conditions. The detonation
adiabat is, in fact, P_2 expressed as function of v_2 in the above equation with
P_1, v_1, and $h_2 - h_1$ considered as constant parameters. This will give the full
line in Fig. 6.20, where P_2 and v_2 are marked P and v, respectively. Note that
the detonation adiabat does not pass through the point (v_1, P_1) and this is due to

Fig. 6.20 The detonation adiabat (*full line*) and the shock adiabat (*dashed*)

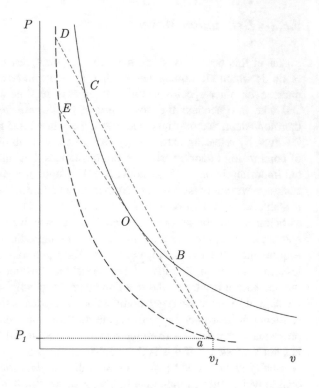

the simple fact that h_1 and h_2 are here not the same functions of v_i and P_i, because there are the enthalpies of the gases before (subscript 1) and after (subscript 2) the *combustion*, causing a change in the composition of the gas. In contrast, if Eq. (6.303) is perceived as a shock adiabat, the index 1 refers to the conditions ahead of the shock and 2 to those behind the shock, just ahead of the detonation. We plot in Fig. 6.20 also the shock adiabat which does pass through the point (v_1, P_1). The detonation adiabat always lies above the shock adiabat, because in the gas after the reaction the temperature and thus the pressure, for a given specific volume, is higher than that in the gas before the reaction.

The following additional relations, which we have used in the shock adiabat, hold also in the detonation adiabat

$$h_2 - h_1 + \frac{1}{2}(u_2^2 - u_1^2) = 0, \tag{6.304}$$

and

$$J^2 = \frac{P_2 - P_1}{v_1 - v_2}. \tag{6.305}$$

A possible detonation process can be represented by the route from the point (v_1, P_1) along the shock adiabat, all the way to point D. Then the state is changed along the cord $\overline{DC}$ because of the reaction. The first process is an ordinary shock of the mixed but not reacting reactants and is followed by the beginning of the reaction and concomitant expansion to point C. In principle, the reaction may perhaps proceed along this cord downwards, however only those points may be reached which lie above the point O. This is so because the point E is the lowest point on the shock adiabat from which a detonation can follow, through the cord $\overline{EO}$. The point O is called the *Chapman–Jouguet* point and no shock $\rightarrow$ reaction sequence is possible to a lower point than O. It turns out that Chapman–Jouguet detonations are the detonations that most often occur in nature.

In order to discuss in more depth the properties of a detonation, let us assume that h_1, P_1, v_1, the upstream thermodynamic variables, are considered as fixed quantities. The possible downstream states are on the detonation adiabat, *above* point O. As the slope of the cord between the initial and final states varies with the final point, J^2 must vary too, but the mass flux is $\rho_1 u_1$ so with ρ_1 fixed u_1 must be allowed to vary. The tangent to the shock adiabat at point (v_1, P_1) has slope $(\partial P / \partial v)_s = -\rho_1^2 c_1^2$. On the other hand, the slope of any cord passing through that point is equal to $-J^2$ and is *smaller* than the tangent slope as we are dealing with negative numbers. Thus

$$-\rho_1^2 c_1^2 > -J^2 = -\rho_1^2 u_1^2 \Longrightarrow u_1 > c_1, \tag{6.306}$$

so that the detonation velocity is always supersonic. Rewriting Eq. (6.305) as

$$\frac{P_2 - P_1}{v_1 - v_2} = J^2, \tag{6.307}$$

and holding P_1 and v_1 fixed, we differentiate to get

$$dP_2 + J^2 dv_2 = (v_1 - v_2)dJ^2. \tag{6.308}$$

Putting $u_i = v_i J$ in Eq. (6.304) gives

$$h_1 - \frac{1}{2}J^2 v_2^2 = h_2 - \frac{1}{2}J^2 v_1^2. \tag{6.309}$$

Differentiating it, holding the thermodynamic state variable of point 1 fixed gives

$$dh_2 + J^2 v_2 dv_2 = \frac{1}{2}(v_1^2 - v_2^2)dJ^2. \tag{6.310}$$

Remembering the Gibbs identity

$$dh_2 = T_2 ds_2 + v_2 dP_2, \tag{6.311}$$

and substituting it into the above equation, while using Eq. (6.308), yields

$$T_2 ds_2 = \frac{1}{2}(v_1 - v_2)^2 dJ^2. \tag{6.312}$$

Now since both T_2 and the square of specific volumes are positive we get

$$\frac{ds_2}{dJ^2} > 0. \tag{6.313}$$

Looking at the detonation adiabat we see that J^2 increases in either direction from the Chapman–Jouguet point O. From this it follows that so does the entropy, i.e., the Chapman–Jouguet point is the *minimum* of the entropy for the detonation products. In addition, the entropy is stationary in the vicinity of O. Regarding the velocity of the products in a Chapman–Jouguet detonation we use for the tangent at O

$$\left(\frac{\partial P_2}{\partial v_2}\right)_s = -\rho_2^2 c_2^2 = -J^2. \tag{6.314}$$

Thus

$$J^2 = \rho^2 v_2^2 = \rho c_2^2 \Longrightarrow u_2 = c_2, \tag{6.315}$$

implying that in the Chapman–Jouguet detonation the velocity of the products with respect to the wave is exactly sonic.

To sum up, we stress that every point on the detonation adiabat represents a possible end state of a reaction (combustion). One of the conventions is that if the end state is above O it is called a *strong* detonation. In such a case we have $u_1 > c_1$ and $u_2 < c_2$. At O we have the Chapman–Jouguet detonation with $u_1 > c_1$ and $u_2 = c_2$. The region in the vicinity of B produces what is called a *weak* detonation while still lower on the adiabat we have what is called *slow burning* or deflagration. Heat conduction may also occur in some of the cases (see below) complicating the picture. The subject of combustion in different limits is very complex and constitutes an important topic of its own. We have only touched it in the limit of detonation waves and only mentioned the name *deflagration* for cases in which the combustion is not propagating so violently. We shall conclude this chapter with an astrophysical example, which gives rise to supernovae type Ia.

6.6.2.1   Example: Deflagration to Detonation Transition in Thermonuclear Burning of White Dwarfs

Compact objects, that is, stars having incredibly high density, are usually the late results of stellar evolution. Medium mass stars with masses of the order of magnitude several Suns evolve after exhausting the light elements in nuclear fusion burning stages, which take billions of years. If the stellar mass is high enough to fuse elements up to carbon and oxygen, but not high enough to achieve temperatures needed to fuse these elements, the star will evolve into a red giant with an ever shrinking C/O core. Ultimately the envelope will be ejected due to various instabilities and the dead core remnant, slowly cooling, will have no choice but to shrink to a sphere whose radius is of the order of the Earth's and which is supported by electron degeneracy pressure. Assuming that such a stellar core has a mass of the order of the Sun, its central density will reach enormous values $\sim 10^9 \, \text{g cm}^{-3}$! A C/O core, half-half in abundance, say, supported against self-gravity by the electron degeneracy pressure is then a very interesting object, a *white dwarf* close in size to the Earth and close in mass to the Sun, with its supporting force originating in quantum mechanics. The astrophysical theory of structure and evolution of white dwarfs is certainly outside the scope of this book. Instead of dealing with the details of a white dwarf structure, we shall use a simplistic order of magnitude calculation to obtain its approximate structure and conditions inside such a star. Our ultimate goal will be to describe under what conditions in a star of this kind a thermonuclear detonation occurs, creating what is known as supernova Ia. For the sake of simplicity we assume that the star is composed of N protons and N electrons and is totally ionized because of the enormous density and hence pressure. Its mass is M and radius R. The approximate gravitational energy of the object (per proton) is

$$\mathscr{E}_{\text{grav}} \approx -\frac{GNm_p}{R}, \tag{6.316}$$

where m_p is the proton mass. Regarding the electron degeneracy, which gives rise to a Fermi energy per particle, we shall estimate the extreme case of ultra-relativistic electrons. This will also help us to obtain approximately the upper limit, the Chandrasekhar mass, for an electron degeneracy supported star. Using the Fermi momentum

$$p_F \approx \hbar n^{1/3} \tag{6.317}$$

where $n \equiv N/R^3$ is the approximate number density, in a formula for the extreme relativistic limit of the electron Fermi energy, we get

$$\mathscr{E}_F \approx p_F c \approx \hbar n^{1/3} c \frac{\hbar c N^{1/3}}{R}. \tag{6.318}$$

Now the Fermi energy of the electrons plays the role of kinetic energy and we get for the total energy

$$\mathscr{E}_{\text{tot}} = \frac{1}{R} \left(N^{1/3} \hbar c - GNm_p^2 \right). \tag{6.319}$$

This expression becomes more and more negative with growing N. Thus the limit of a stable state corresponds to $\mathscr{E}_{\text{tot}} = 0$ and gives the maximum number of baryons, if one wants the star not to collapse. The maximal number is

$$N_{\text{max}} \left(\frac{\hbar c}{Gm_p^2} \right)^{3/2} \sim 2 \cdot 10^{57} \tag{6.320}$$

and the maximal mass $M_{\text{max}} \sim N_{\text{max}} m_p = 2 \cdot 10^{57} \times 1.67 \cdot 10^{-24} \approx 3.3 \cdot 10^{33} \sim 1.5 M_{\odot}$.

The equilibrium radius associated with this maximal mass star is determined by the onset of relativistic degeneracy of the electrons $\mathscr{E}_F \geq m_e c^2$, which gives, using formulae (6.318) and (6.320)

$$R_{WD} \leq \frac{\hbar}{m_e c} \left(\frac{\hbar c}{G m_p^2} \right)^{1/2} \sim 5000 \, \text{km}. \tag{6.321}$$

The life of a white dwarf of this sort, if it is isolated from mass donors, proceeds through a very long cooling stage when the white dwarf cools indefinitely and approaches, in principle $T = 0^\circ K$ remaining forever as a stellar corpse, a *black* dwarf. However a nonnegligible fraction of such white dwarfs actually are a remnant of only one star in a close binary, and conditions are often such that matter, coming from the other star, can be accreted onto the white dwarf, driving its mass perhaps over the Chandrasekhar limit. The hydrodynamic and other physical processes involved in the binary evolution and in accretion are interesting and complicated. We have seen, in this book, only one aspect of accretion in binaries—the formation and mass transfer through an accretion disk. In any case, with the star's mass being over its Chandrasekhar limit, the degenerate material can no longer fully support the weight of the stellar material. The result is further contraction at the center whose density and temperature increase. This may give rise to the ignition of thermonuclear fusion reactions. Of course, the energy liberated at a point heating the material may also be conducted by degenerate electron conduction, and convected, so as to cool the ignition point. It turns out that the typical cooling time *exceeds* the nuclear energy production time when the density is $\rho_i \approx 3 \cdot 10^9 \, \text{g cm}^{-3}$ and the temperature is $T \approx 6.6 \cdot 10^8$ K. Skipping many details about what we understand about the location of the ignition and how it proceeds, we may safely state that observationally such a thermonuclear runaway appears as a supernova type Ia. These objects play a central role in our understanding of the expansion of the Universe, because they are surprisingly uniform in their peak luminosity. Thus they can serve as good standard "candles" to measure distances to the galaxies they are located in. Based on this, the accepted theory at the time of writing this book seems to firmly indicate that the Universe's expansion accelerates. This acceleration does not have currently an agreed upon source of energy and it is attributed to vacuum itself and called *dark* energy. Observations of supernovae of type Ia indicate that most of the white dwarf material is converted during the thermonuclear runaway into the most stable (iron peak = Ni, Co, Fe) nuclei. Some of the material, which is discovered in the outer layers of the ejecta, also contain intermediate mass elements. The abundance of these elements imposes limits on the nature of the thermonuclear runaway. As we have seen in this section, the thermonuclear runaway giving rise to supernovae Ia may spread as a deflagration or detonation wave. There is a significant difference between those two regimes—a spherical deflagration causes an expansion of the material before it as the propagation speed is subsonic. In contrast, a detonation proceeds supersonically and the material in front of the wave undergoes the reaction at the original high densities.

The decision which thermonuclear burning regime is chosen has import as to the composition of the nuclear ashes and on the total energy liberated. In order to fit theory to observations one must take recourse to numerical simulations. However the whole exploding star's scale is thousands of kilometers, while the widths of the deflagration or detonation front are ~ 0.1 and ~ 0.001 cm, respectively. Thus, it should be clear that it is impossible to resolve the front and determine the regime of nuclear burning. Therefore, the decision on deflagration or detonation is made ad hoc and the calculation is done accordingly. The problem is that neither model has reproduced the observations very well. The detonation is too violent to leave any non-iron peak elements and the deflagration is too weak to account for the observed mass ejection. Good agreement is obtained if the detonation is initiated on a pre-expanded object, say by partial deflagration. At this time the phenomenological *delayed* detonation model is accepted, whereby the nuclear burning in supernova Ia white dwarf starts as a deflagration that later strengthens to a detonation. There remains, however, a major unresolved problem—the detailed physics of the deflagration-detonation transition.

Problems

6.1.
Show, if ϕ_i decay to zero at infinity, that $\int (\phi_1 \nabla^2 \phi_2 - \phi_2 \nabla^2 \phi_1) d^3 x = 0$, if the integration is performed over all space.

6.2.
Find the formula [based on Eq. (6.26)] for a sound wave, whose velocity potential is dependent on only one spatial coordinate, y, say.

6.3.
Show that $F(\hat{\mathbf{n}} \cdot \mathbf{x} \pm c_s t)$, with $\hat{\mathbf{n}}$ being any unit vector, satisfies the wave equation (6.12) for any F and describes a wave propagating in the unit vector direction (for the $-$ sign) and in the opposite direction (for the $+$ sign). Show also that the magnitude of the wave is constant on planes perpendicular to $\hat{\mathbf{n}}$.

6.4.
Consider a plane sinusoidal wave, whose velocity potential is given by $\phi(\mathbf{x}, t) = \phi_0 \sin \left[\frac{2\pi}{\lambda}(x - c_0 t) \right]$, where ϕ_0, λ, and c_0 are constant. Assume also that $c_0 t > x$.

1. What is the physical meaning of these three constants and the assumed inequality?
2. Consider paths in the (x, t) plane of fluid particles and waves paths (the latter are *characteristics*). Give the differential equations with the help of which these two kind of paths can be found in the (x, t) plane.
3. Calculate the particle path $\xi(t)$, which at $t = 0$ was at $x_0 > 0$, and several positive characteristics, and plot it all schematically in the relevant part of the upper half of the $x - t$ plane. Plot schematically a low amplitude ϕ wave for $x < 0$.

6.5.

Let a linear plane wave of compression be given by

$$\rho(x,t) = \frac{\rho_0}{c_s}\left[f_1(x-c_st)+f_2(x+c_st)\right], \tag{6.322}$$

where f_i are certain unknown functions. Note that we have dropped the subscript 0 from unperturbed values of all quantities.

a) Show that the corresponding velocity wave has the form

$$u(x,t) = f_1(x-c_st)-f_2(x+c_st).$$

b) Let the initial conditions be given, the functions $\alpha(x)$ and $\beta(x)$ are specified:

$$\delta\rho(x,0)\equiv\rho-\rho_0 = \alpha(x) \quad\text{and}\quad u(x,0)=\beta(x).$$

Express $f_1(x,0)$ and $f_2(x,0)$ in terms of the initial conditions only.

6.6.

Show that the speed of sound in an ideal gas is $c_s = \sqrt{\gamma(P/\rho)}$ and thus it is a function of temperature only. γ is the constant adiabatic exponent.

6.7.

Perform a linearization of the fluid equations around a state of an isothermal, plane parallel atmosphere of an ideal compressible gas, as described in the text (6.36) and obtain the PDE (6.40).

6.8.

Consider the dispersion relation (6.43) as in the text. Find, for an isothermal atmosphere composed of an *ideal gas*, the numerical value of the nondimensional expression $4N^2H_z^2/c_s^2$. Use this result in order to show that the branch of internal gravity waves is well separated from the acoustic branch, if *either* one consists of short enough, with respect to the scale-height, waves ($k_zH_z \gg 1$ or $k_xH_z \gg 1$) and that the dispersion relation for acoustic waves is then well approximated by $\omega_a^2 \approx c_s^2[k_x^2 + k_z^2 + 1/(4H_z^2)]$.

6.9.

Consider the case $w = 0$ and show that such waves are also admissible in the situation described in the text, see Eq. (6.44), and in the last two problems. They are the *Lamb* waves. Indicate in the frequency–wavenumber (both dimensionless) plane, the location of a few members of the acoustic branch and of the Lamb wave.

6.10.

Consider a spherically symmetric initial value problem, in which the pressure inside a radius a, say, is $P = P_0+P'$, i.e., slightly higher than the pressure for $r > a (= P_0)$. To effect such a situation one may imagine a balloon, whose radius is a. The gas is initially at rest everywhere and at $t = 0$ the gas is released (the balloon bursts, say). Determine and describe the shape of the outgoing spherical wave.

6.11.
Prove (6.62) and (6.63). Are these relations exact always? If not, when do they hold or are good approximations? What is the physical meaning of $c_{s,0}^2 \langle \rho^2 \rangle / \rho_0$—discuss its role in force transfer by acoustic radiation.

6.12.
Prove formula (6.75), which is valid for a simple acoustic wave.

6.13.
Consider a sound wave which at any given time is localized only in a finite region of unbounded space, in other words, a wave-packet. Show that the total momentum of the wave-packet is $(1/c_s^2) \int_{\mathscr{V}'} P' \mathbf{u} \, d^3x$, where the integral is taken over the volume in which the wave-packet resides at a given moment, $\mathscr{V}'$. Wave-packets travel at the group velocity V_g.

6.14.
Give an implicit closed formula for the characteristic frequencies of spherically symmetric sound waves in a circular vessel of radius R. Hint: Use a form for the velocity potential of the standing wave given by Eq. (6.58).

6.15.
A circular pipe of length ℓ and radius $r = \ell/10$ is fixed to an aperture of a cubical cavity (a resonator), whose linear dimension is $L = 100\ell$. Assume that when air moves into and out of the resonator, its velocity can be assumed negligible inside the resonator, but it is non-zero in the pipe and the conditions in the pipe may be assumed uniform. Find the characteristic frequency created by the air motion in the tube (in and out of the resonator) and compare it with the lowest characteristic frequency of the cavity, which serves as the resonator.

6.16.
Assume that two sound point sources, located at the Cartesian coordinates $(0,0,d/2)$ and $(0,0,-d/2)$, emit acoustic waves of amplitude B and are relatively $180°$ out of phase. Find the energy flux at a distance $r \gg d$ and the total average power. Compare the last expression with the one derived in the text (6.110) and thus determine $\mathbf{A}$ for this case.

6.17.
Show that the time average of $T\nabla \cdot (\kappa \nabla T / T)$ can be ignored relative to the other terms in Eq. (6.128), when we consider a harmonic traveling wave.

6.18.
Express the temperature, pressure, and density along a streamline of an ideal gas in terms of their critical, i.e., sonic point, values, the Mach number, and the constant adiabatic exponent.

6.19.
Show that the fundamental gasdynamic derivative Γ is given for a perfect gas by $\Gamma = \gamma + 1/2$, where γ is the adiabatic exponent.

6.20.

Consider steady isentropic flow of a perfect gas and show that the mass flux density can be written as

$$\rho u = \frac{\rho_* u_* M}{[2/(\gamma+1) + M^2(\gamma-1)/(\gamma+1)]^{(\gamma+1)/[2(\gamma-1)]}}, \tag{6.323}$$

where M is the Mach number, u_* is the velocity at the *sonic point*, and all other symbols have their usual meaning.

6.21.

Write the gas dynamics equations in spherically symmetric flow by expressing their form along the two families of characteristics $dr/dt = u \pm c$.

6.22.

Consider *small* isentropic perturbations on a homentropic gas in a uniform unperturbed state: moving with velocity u_0 and whose pressure and density are P_0 and ρ_0. Find the linear approximate equations of the characteristics and the Riemann invariants, using the notation u_1, P_1, and ρ_1 for the perturbations. Prove that if a wave propagates to the right (left) only for the above linear approximation of the Riemann invariant, then $J_-(J_+)$ is constant in space and time. Hint: use Problem 6.5.

6.23.

Evaluate formula (6.218), under the assumptions given in the text.

6.24.

Starting with Eqs. (4.77)–(4.78) and assuming no bottom topography derive the equation set (6.226)–(6.227).

6.25.

Show that the centered rarefaction wave, discussed as it is depicted in Fig. 6.9, is an example of a similarity solution developed in the paragraph ending with formula (6.243). Use a perfect gas with $\gamma = 5/3$ and discuss also the issue of choice of the direction of the x-axis. See text.

6.26.

Show that a centered compression wave is impossible by attempting to construct it in the same way that was done for the centered rarefaction wave.

6.27.

Using the hints in the text simplify Eqs. (6.245)–(6.248) to the five simple shock conditions Eqs. (6.252)–(6.256). Show also that the velocity change across a shock is normal to the shock front and that for stationary shocks ($U_{sh} = 0$) the stagnation enthalpy $h_0 = h + w^2/2$ is invariant across the shock.

6.28.

Acoustic streaming is a steady current in a fluid driven by the absorption of high amplitude acoustic oscillations. This effect is nonlinear and therefore does not lend itself to elegant analytical treatment. Writing $u = u_{sound} + \bar{u}$, where the first term is the one contributed by the sound wave and the second is the steady, non-oscillatory part, derive the form of the Navier Stokes equation with these two contributions explicitly noted. Identify the steady body force that drives the streaming and show that if $u_{sound} = \varepsilon \cos(\omega t)$, then the quadratic nonlinearity generates a steady force proportional to the time average $\langle \varepsilon^2 \cos^2(\omega t) \rangle = \frac{1}{2}\varepsilon^2$. Attempt to derive the estimate (it is quite a challenge) that if viscosity is responsible for acoustic streaming, the value of viscosity disappears from the resulting streaming velocities, and the order of magnitude of the streaming velocities near a boundary (outside of the boundary layer) is: $U \sim -3/(4\omega) = u_{sound}^0 \times d(u_{sound}^0)/dx$, where one approximates the boundary layer by a straight wall in the direction of x and the superscript 0 denotes that this is the value of the velocity of sound *at* the wall.

6.29.
Show that the inequalities (6.275)–(6.278) hold for shocks of any intensity, if it is assumed that $(\partial^2 v/\partial P^2)_s > 0$, as it was for weak shocks.

6.30.
Show that the Rankine–Hugoniot conditions can be written for a perfect gas (with constant specific heats) in the form of Eqs. (6.279)–(6.281).

6.31.
Let the shock in Fig. 6.13a be planar and stationary $U_{sh} = 0$. Assume that the acute angle that the upstream velocity $\mathbf{w}_1$ makes with the shock (in the plane of the page) is β and the angle that the downstream velocity $\mathbf{w}_2$, leaving the shock, makes with the direction of $\mathbf{w}_1$ is θ. We are then discussing an *oblique shock* and want to find relations between the oblique velocities and the angles (θ is called the *turning angle*). Show that for a perfect gas with constant specific heats, an oblique shock relation between the angles gives

$$\tan \theta = \frac{(2\cot\beta)(M_1^2 \sin^2\beta - 1)}{(\gamma+1)M_1^2 - 2(M_1^2 \sin^2\beta - 1)}. \tag{6.324}$$

6.32.
The previous problem gives implicitly $\beta(\theta)$ for an oblique shock whose upstream Mach number is M_1. In general, this function is doubly valued. With a stationary shock it is possible to find, after considerable algebra, that the perfect gas which passes the oblique shock satisfies the important *Prandtl relation*, which with $u_1 = w_1 \sin\beta$, $u_2 = w_2 \sin(\beta - \theta)$, and $v = w_1 \cos\theta$ is

$$u_1 u_2 = c_*^2 - \frac{\gamma-1}{\gamma+1}v^2. \tag{6.325}$$

Show that the Prandtl relation indeed holds. Note that the sonic point sound speed is known by means of the upstream conditions $c_*^2 = (2c_1^2)/(\gamma+1) + u_1^2(\gamma-1)/(\gamma+1)$.

6.33.

Prove that in an oblique shock in a perfect gas the, so-called, equation of the *shock polar* is

$$U_{2y}^2 = \frac{(U_1 - U_{2x})^2 (U_1 U_{2x}^2 - 1)^2}{\frac{2}{\gamma+1} U_1^2 - U_1 U_{2x} + 1}, \tag{6.326}$$

where U stands for the Cartesian components of the normalized velocities *perpendicular* to the shock, viz.

$$U_{i\alpha} \equiv \frac{u_{i\alpha}}{c_*}, \quad \text{with } i = 1, 2; \ \alpha = x, y.$$

6.34.

Under the condition specified in Sect. 6.5.3 show that the differential equation for u is Eq. (6.284). Using the assumptions specified there, derive a set of self-contained equations for the desired variables. Plot *qualitatively*, using heuristic considerations, the run of ρ, u, P, and T along the cooling layer, until $\mathcal{L}$ it reaches zero again, starting with arbitrary post-shock values. Actually they should be found from the Rankine–Hugoniot conditions.

6.35.

Show that the fluid dynamical equations of conservation of mass momentum and energy, when transformed to the nondimensional form according to the dependent variable definitions (6.292)–(6.294), result in the system (6.300)–(6.302).

6.36.

Consider a Chapman–Jouguet detonation in a perfect gas with constant specific heats. Show that the jump conditions in such a detonation are

$$\frac{P_2}{P_1} = \frac{u_1^2 + (\gamma_1 - 1)c_{V,1}}{(\gamma_2 + 1)(\gamma_1 - 1)c_{V,1}T_1}, \tag{6.327}$$

$$\frac{v_2}{v_1} = \frac{\gamma_2 \left[u_1^2 + (\gamma_1 - 1)c_{V,1}T_1 \right]}{(\gamma_2 + 1)u_1^2}, \tag{6.328}$$

where T_1 is a (given) temperature of the original gas mixture and u_1 is the velocity of the detonation wave, which can be found (do it!) to read

$$u_1 = \left\{ \frac{\gamma_2 - 1}{2} \left[(\gamma_2 + 1)\mathcal{Q}_0 + (\gamma_1 + \gamma_2)c_{V,1}T_1 \right] \right\}^{1/2}$$

$$+ \left\{ \frac{\gamma_2 + 1}{2} \left[(\gamma_2 - 1)\mathcal{Q}_0 + (\gamma_2 - \gamma_1)c_{V,1}T_1 \right] \right\}^{1/2}. \tag{6.329}$$

Here $\mathcal{Q}_0 \equiv (h_1 - h_2)_0$, the index zero meaning that the heat is computed as the difference of enthalpies for temperature T_0. Obviously, the reactants mixture is denoted by index 1, while the gas containing the reaction products is denoted by index 2.

Bibliographical Notes

General

The most comprehensive and the best, in our opinion, book on compressible fluid dynamics is Thompson's important work, the only problem being that it is difficult and certainly not inexpensive to purchase. Another good, but challenging to understand, source is the classic book from the Landau and Lifshitz series.

Section 6.1

1. P.A. Thompson, *Compressible Fluid Dynamics*. Advanced Engineering Series (Rensselaer, Troy, 1988)
2. L.D. Landau, E.M. Lifshitz, *Fluid Mechanics*, chaps. VIII-XV, 2nd edn. (Elsevier, Amsterdam, 2004)

Section 6.2

Lord Rayleigh's classic two-volume work on sound has been reprinted by Dover and is recommended for the student who wishes to investigate the theory of sound in depth. We mention only this book here as the topic of sound waves is treated almost in any book that includes compressible fluids.

3. J.W.S. Rayleigh, *The Theory of Sound*, vols. I and II (Dover, New York, 1945)

Section 6.3

The material covered in this section is described in a good and comprehensive way in references [1] and [2].

Section 6.4

Out of the many books dealing with the use of characteristic curves to understand the steepening and formation of shock waves and other topics, we find the work of Zel'dovich and Raizer as the clearest and most physically transparent.

4. Ya.B. Zel'dovich, Yu.P. Raizer, *Physics of Shock Waves and High-Temperature Hydrodynamic Phenomena*, vol. I. (Academic Press, London, 1966)
   For those who are more interested in the mathematical, albeit not strictly rigorous, side of the method of characteristics we recommend:
5. P.M. Morse, H. Feshbach, *Methods of Theoretical Physics*, chap. 6 (Feshbach Publishing, Minneapolis, 1981)
   A very good treatment of the problems arising in shallow water theory, which are treatable by methods analogous to those of 1-D gas dynamics, that is, the method of characteristics, can be found in reference [2], and also in the concise and clear presentation in
6. D.J. Acheson, *Elementary Fluid Dynamics* (Oxford University Press, Oxford, 1999)

Section 6.5

In addition to the reasonable treatments of shock waves in our main references (numbers [1] and [2]), we recommend the following two excellent books, which treat the subject in depth.

7. R. Courant, K.O. Friedrichs, *Supersonic Flow and Shock Waves* (Springer, Berlin, 1976)
8. G.B. Whitham, *Linear and Nonlinear Waves* (Wiley, London, 1999)

Section 6.6

The point explosion problem is treated well in reference number [2], however it is explicitly linked to a supernova explosion and the notation used is more transparent than in reference number [2], in the book

9. F.H. Shu, *The Physics of Astrophysics II*, chap. 17 (University Science Books, Mill Valley, 1992)
   A recent book by Needham contains a wealth of information for the reader interested in blast waves
10. C.E. Needham, *Blast Waves* (Springer, Berlin, 2010)
   Finally, we recommend an entire book devoted to the theory and experiment of detonation, which can naturally broaden the perspective of those interested particularly in this topic
11. W. Fickett, W.C. Davis, *Detonation* (University of California Press, Berkeley, 1979)

Chapter 7
Hydrodynamic Stability

There is nothing stable in the world;
uproar's your only music.

John Keats (1795–1821); 'Letters of J.K'

7.1 Introduction: The Experiments of Reynolds and Taylor

Stability is an important consideration for fluid flow just as it is for any mechanical system. It is often said that in order to be observable, a flow must be stable since an unstable flow is merely a state of transition to some other flow, or possibly to turbulence. But if this were the whole story, hydrodynamic stability would be a dull pursuit in which we restrict our attention to only stable results. Instead, we turn our attention to the instabilities themselves, hoping that their form will help us understand the flows that support them.

In nature, fluids are subject to various forms of forcing and are put into configurations, which correspond to mathematically viable solutions. It is natural and logical that we usually begin our study by seeking *steady* solutions to these problems. Two examples that may come to mind are a steady ocean circulation, forced by wind and confined to a basin, or the overall hydrostatic balance of the Sun, thermally forced by its central energy source and confined by gravity. It is when solutions such as these are tested for stability that we discover a steady flow's true character. Ocean flows may develop jets, which develop meanders which develop eddies, and so on. Cells of convection form in the Sun which give rise to its pervasive turbulence. By following the growth of instability we discover the true origin of the complex time-dependent flows that we find in nature. But where to begin? Early researchers, including O. Reynolds, H. Bénard, Lord Rayleigh, Lord Kelvin, and H. von Helmholtz, set the stage for the way that stability is studied yet today. In short,

© Springer Science+Business Media, LLC 2016
O. Regev et al., *Modern Fluid Dynamics for Physics and Astrophysics*,
Graduate Texts in Physics, DOI 10.1007/978-1-4939-3164-4_7

we first seek steady solutions that depend on a characteristic, or control, parameter and then examine whether small perturbations to these steady solutions are predicted to grow or decay. In practice, small disturbances are present in any real system and forcing due to buoyancy or moving boundaries is almost unavoidable. *Transition*, in this context, is the change to another flow, when a control parameter attains a critical value.

In this opening section, we have decided to look at two renowned experiments and examine them in view of *linear stability theory*, the subject of this chapter. In the experiments of Reynolds, instability is found despite a mathematical result to the contrary, and at the experiments of Taylor and Couette, the onset of instability matches the theoretical prediction. Turbulence, whether we desire its presence or seek to delay it, is a ubiquitous state of fluid motion and instability figures prominently in the route by which a flow becomes turbulent. The phenomenon of turbulence remains as one of the last of the great unsolved problems of classical physics (see Chap. 9) and among our only means of insight into its nature is an understanding of the fluid processes taking place in the flow, often involving more than one concurrent instability mechanism. O. Reynolds's publication of 1895, describing his careful experiments of a flow through a pipe now known as the pipe Poiseuille, or Poiseuille–Hagen flow (see Sect. 3.2.2), is the first modern account of instability and the turbulence phenomenon, even though Reynolds did not use this word explicitly. He called laminar flow *direct* and a turbulent one *sinuous*, however it is clear that he understood some main physical differences between the two. The experiment consisted of forcing (by pressure) water into a pipe of constant radius and observing the flow (Fig. 7.1). Reynolds concluded that the outcome

Fig. 7.1 A drawing from the original paper by O. Reynolds, describing the experimental setup

depends on the value of a nondimensional *control parameter*, bearing his name, the Reynolds number Re $= Ud/v$. We have defined it already in Eq. (1.80) and saw, later on, its importance. Here the pipe diameter d is taken as the typical length scale of the flow, U as the typical velocity along the pipe (either the mean or the maximum), and the constant v is the fluid kinematical viscosity. Reynolds observed the emergence of turbulent flow as Re is increased and noticed that it is marked by *eddies* (closed loops of velocity) some of them very small, but still many orders of magnitude larger than molecular motions between collisions. He understood that such turbulent flow provides the fluid with the ability to transport heat, momentum, chemical species, etc., many orders of magnitude more effectively than molecular diffusion. Sure enough, there were some researchers, notably G.H.L. Hagen, who observed similar phenomena a few years before Reynolds, but the Reynolds paper of 1895 was remarkable in its completeness and its physical insights. Equally significant is the nature of the transitions, the route to turbulence, in this case: small turbulent patches appear along the length of the pipe, increasing in extent as Re grows, until by merging they *fill* the pipe with turbulence. One of the most important observations is the dependence of the transition Reynolds number on the size of the initial perturbations at the entrance to the pipe, which today is a subject of theoretical and experimental research. Instability leading to turbulence is now known to be dependent on the *size* of the initial perturbation. As we shall see, in *linear* stability analysis the perturbation is formally *infinitesimal*, a mathematical idealization that cannot be realized in a laboratory fluid. Reynolds succeeded to minimize the perturbations at the inlet to such a small amplitude that the flow remained laminar until Re $\approx$ 13,000, while if no special effort was made the flow became turbulent already above Re $\approx$ 2,000. Some recent experiments with very special inlet conditions achieved laminar flow persisting up to Re $\approx$ 90,000 prompting some researchers to declare that this flow is stable to infinitesimal perturbations. Here the caveat mentioned in the beginning of this book is in order. The description of a fluid as a continuum is an approximation, therefore the meaning to the expression "infinitesimal" perturbation is problematic. The perturbation has a lower finite limit, simply because of the fact that the use of differential calculus is only an approximation. The "differential" is only a small, but finite, distance, which must, at the same time, be very large as compared to the intermolecular mean free path. Obviously, to ensure laminar flow for every Reynolds number the initial perturbations must be smaller than some value. Thus the question arises what is this value for very large systems, having enormous—indeed astronomical!—Reynolds numbers, like astrophysical ones, which are obviously not too similar to Reynolds's pipes? Is it practical to expect incredibly small perturbations in nature? The phenomenon of transient growth (see Sect. 7.6) may also be instrumental in transition to turbulence, a possible example of a *subcritical* transition. This means that the transition may occur at a *lower* than critical control parameter, which is the Reynolds number. The accepted picture for the original Reynolds experiment is that when care is taken to minimize the initial perturbations of flow fed through a straight inlet, turbulence may be prevented until Re $\sim$ 2,000. If the inlet flow is sufficiently noisy, then turbulent flow can become manifest starting with chaotic

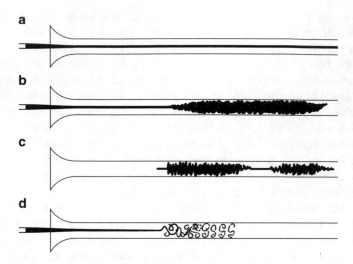

Fig. 7.2 The various types of flows, presented here schematically by the dye trace, in the Reynolds experiment. As Re is increased the flow changes from laminar (**a**) through transitional (**b-c**) to turbulent (**d**). Only one *turbulent slug* is visible in (**b**), while for successively higher Re values a sequence of such slugs appears, separated by laminar flow and merging finally into a fully turbulent flow along the entire pipe. The exact values of Re at the transition depend strongly on the disturbances at the entrance into the pipe (see text)

flow patches both near the pipe boundary and close to the inlet and, eventually, the flow developing into turbulent "slugs" filling the pipe (see Fig. 7.2). For Re > 2,000 the turbulent slugs are at first separated by laminar flow, but they tend to grow, merging finally into fully turbulent flow. The pace of this process depends on Re, wherein higher values drive more prompt transition to fully developed turbulence.

In 1923, G.I. Taylor published the description of a simple experiment giving rise to what was referred to in Chap. 5 as the Taylor–Couette (TC) flow. As we have described in that chapter, see in particular Sect. 5.2.2, the experiment consists of fluid flow between two concentric cylinders, whose radii are R and $R+d$. Note that previously, in Chap. 5, the radii were called simply R_1 and R_2, but now we examine a case of narrow gap d between the cylinders. Taylor found that depending on the value of suitably chosen control parameters, this flow is prone to instabilities once the control parameter reaches a critical value. It was found that once instability sets in, a new flow arises, which itself becomes unstable for a definite value of the control parameter, higher than the critical. As we shall discuss below, this is an instance of *supercritical* transition of an unstable flow to another flow. In this example we wish to discuss only the particular case in which the outer cylinder does not rotate, while the inner cylinder does, with prescribed angular velocity Ω. In addition, we limit ourselves to the case of a narrow gap, viz. $d \ll R$. One of the possibilities to define a dimensionless number that would characterize the flow described here is to define what is called the *Taylor number*. Note that the definition of the Taylor number is not unique, and in other books a slightly different definition is preferred. We define it here as

$$\text{Ta} \equiv \frac{d^3 \, \Omega^2 R}{\nu^2}, \tag{7.1}$$

where ν is the kinematic viscosity. We should remember also that the results may depend on the additional nondimensional ratios d/R and L/R, where L is the cylinder's height. When L is much larger than R, and the gap very narrow, the sole control parameter is Ta. The experiment is actually conducted by gradually increasing the cylinder's angular velocity from zero since R, d, L and the fluid viscosity ν are fixed. For low rotation rates the flow is laminar, the fluid being dragged around by the rotating inner cylinder. Once a critical rotation rate is reached, toroidal vortices appear, superimposed on the rotational motion, as shown in Fig. 7.3a. The critical value of the Taylor number for narrow annuli is $\text{Ta}_c = 1708$. The wavenumber of the Taylor cells at the onset of instability is $k_c = 3.12/d$. Increasing Ta by $\sim 25\%$ above the value of Ta_c changes the nature of the flow: the vortices lose their axi-symmetry and develop a *wavy* structure in the angular direction. As can be seen in Fig. 7.3b the flow is now more complex, but it is still ordered, and therefore laminar. Unsteady structures start to appear if Ta is increased yet further until eventually the flow becomes turbulent as in Fig. 7.3d. The time averaged flow pattern resembles the steady Taylor vortices, albeit the cells are somewhat larger, and superimposed on this mean flow we find an irregular fluctuating velocity component of the fluid particles. They are no longer confined to toroidal surfaces and as they are swept by the mean flow while they jostle seemingly randomly.

This is not the first time that we have encountered a flow that exhibits transitions that will be called *supercritical*. A sequence of these will lead to a particular route to turbulence. The case of a flow past a cylindrical obstacle (see Fig. 2.2 and the discussion there) was an earlier example. There we chose to treat the flow as potential but understood that in real flows the boundary layers, forming on the body, break the assumption of irrotational flow and create a wake. It turns out that the pictures we had are appropriate for relatively low velocities, or rather, Reynolds numbers, which involve also the diameter of the cylinder and the viscosity. Thus in Fig. 2.2, case (b) occurs for $\text{Re} \lesssim 1$, case (c) for $5 \lesssim \text{Re} \lesssim 40$, and in case (d) $100 \lesssim \text{Re} \lesssim 200$. Experiments show that increasing Re to the vicinity of 10^4 causes the vortices behind the body to look like patches of turbulence which get detached and travel downstream. If Re is increased even further, to 10^6 or so, the entire wake behind the body becomes turbulent. Later in this chapter, we will discuss the Bénard problem of the instability of a fluid layer heated from below and its control parameter, the Rayleigh number Ra. As we shall see, the transition there will also be supercritical.

Next, the basic mathematical framework of linear stability analysis will be presented. In the subsequent sections, the stability of sheared and buoyancy-driven flows will be discussed, including centrifugal instabilities. Following this, we shall

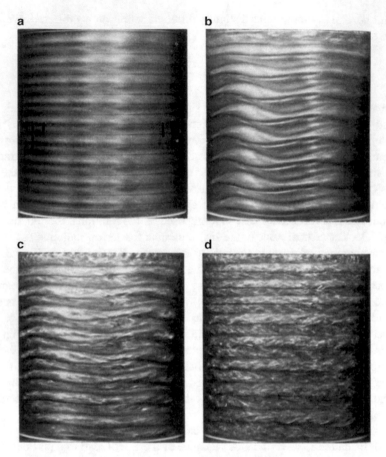

Fig. 7.3 Taylor–Couette cell; the inner cylinder is rotating, while the outer is stationary. As the rotation rate of the inner cylinder increases, the Taylor number Ta, see text, increases until vortices (**a**) appear, followed by wavy vortices (**b**)–(**c**), and eventually turbulence (**d**). *Reproduced from Fenstermacher, Swinney and Gollub, J. Fluid Mechanics 94, 103, 1979 (used with permission)*

present additional types instabilities, some of them connected to motion resulting from *self-gravity*. These should serve a preview for the reader of the vast domain of this topic.

7.2 Fundamentals of Linear Stability Theory

In simple terms, a flow is stable if any initially small perturbation remains small for all time. If even one perturbation to a flow grows to be large after some finite time, that flow is unstable. To make the above statements more quantitative we will need precise definitions of the terms *small, large*, and *perturbation*. From these, we will define the important concepts, in the stability context, of the *base state* and *bifurcation*. Note that a *base* state is sometimes referred to, equivalently, as *basic*. We have already treated some types of *waves* (e.g., in Chaps. 4, 5, 6) as linear disturbances, therefore much of the mathematical machinery developed and utilized in those chapters will be used here as well. The examination of linear stability or instability of a flow is a problem of linear analysis. A solution to the governing fluid equations, representing the flow under investigation, is perturbed by adding disturbances that are assumed to be small in the sense that products of disturbance quantities can be neglected. To this extent, the linear stability of a flow is studied in the same way as linear stability of a fixed point of a dynamical system, as we will soon illustrate in detail. The flow that is perturbed is typically a steady solution of the fluid equations and, as remarked above, is referred to as the basic state or the base flow. Steadiness is by no means a prerequisite as we will see in the cases of boundary layer and mixing layer flows, which are base states that evolve in time, albeit slowly enough to allow stability results to be meaningful. Rarely, a periodic base state can be found and its stability studied by linear perturbations, in effect performing a Floquet analysis. Should the periodic state itself represent the unstable development of a simpler, steady flow, then the Floquet analysis amounts to a nonlinear stability calculation, of the *secondary instability* type.

As a first example, consider the incompressible, $\nabla \cdot \mathbf{u} = 0$, Navier–Stokes equation, written here in a nondimensional form:

$$\frac{\partial \mathbf{u}}{\partial t} + \mathbf{u} \cdot \nabla \mathbf{u} = -\nabla P + \mathrm{Re}^{-1} \nabla^2 \mathbf{u} + \mathbf{b}, \qquad (7.2)$$

where $\mathbf{b}$ is some body force per unit mass, Re is the Reynolds number, and $\rho = 1$ has been assumed. The velocity and pressure fields, $\mathbf{u} = \mathbf{u}(\mathbf{x}, t)$ and $P = P(\mathbf{x}, t)$ will, in general, depend on the values of parameters of the system, here simply just Re. For convenience, we introduce the complete state vector $\mathbb{U}(\mathbf{x}, t; \mathrm{Re}) = [\mathbf{u}(\mathbf{x}, t; \mathrm{Re}), P(\mathbf{x}, t; \mathrm{Re})]^{\mathrm{T}}$, where T denotes transpose, making clear the dependence on the control parameter Re. In a dynamical system, the control parameter describes the evolution of a solution through a point of *bifurcation*, usually given by a specific value of the parameter. Although this word has an intuitive meaning, we shall have to wait until the first section of the next chapter to elucidate it, using a more formal language. The spatial domain of the flow is denoted by $\mathcal{V}(t)$ and its boundary by a surface $\mathcal{S}(t) = \partial \mathcal{V}$. Consider solutions $\mathbb{U}_0$ to (7.2) that are steady, such that $\partial \mathbb{U}_0 / \partial t = 0$, and which satisfy boundary conditions $\mathbf{F}(\mathbb{U}) = 0$, for some function $\mathbf{F}$, on $\partial \mathcal{V}$. Typical boundary conditions are that the fluid velocity $\mathbf{u}$ equals the velocity of the boundary $\mathbf{u}(\partial \mathcal{V})$ although commonly the boundary is fixed, having zero velocity. On other occasions we have reason to treat the flow as unbounded and require that $\mathbf{u}, P$ remain bounded as $|\mathbf{x}| \to \infty$. We refer to the steady state $\mathbb{U}_0 = (\mathbf{u}_0, P_0)$ as the basic state, and to $\mathbf{u}_0$ as the base flow. While many steady state solutions may exist for a given problem, there is no guarantee that our analytical methods will find any of them, except for the simplest possible flow configurations. Such simple configurations form the framework of the stability analyses that make up the remainder of this chapter.

The linear stability problem governing small perturbations, denoted by a prime, $\mathbb{U}' = (\mathbf{u}', P'; \text{Re})^{\text{T}}$ is formulated by first defining the following ansatz:

$$\mathbb{U} \mapsto \mathbb{U}_0 + \mathbb{U}' \quad \Longleftrightarrow \quad \mathbf{u}_0 \mapsto \mathbf{u}_0(\mathbf{x}) + \mathbf{u}'(\mathbf{x}, t), \quad P \mapsto P_0(\mathbf{x}) + P'(\mathbf{x}), \quad (7.3)$$

and then substituting it into the governing equations (7.2) and boundary conditions. Dropping products of perturbation terms leads to equations describing linearized perturbations:

$$\nabla \cdot \mathbf{u}' = 0, \tag{7.4}$$

$$\frac{\partial \mathbf{u}'}{\partial t} + \left(\mathbf{u}_0 \cdot \nabla\right)\mathbf{u}' + \left(\mathbf{u}' \cdot \nabla\right)\mathbf{u}_0 = -\nabla P' + \mathbf{b}\left(\mathbb{U}_0, \mathbb{U}'\right) + \text{Re}^{-1}\nabla^2\mathbf{u}'. \tag{7.5}$$

Our definition of stability can be made more formal as follows. A base state is referred to as *Liapunov stable* if, for all $\varepsilon > 0$, there exists a $\delta(\varepsilon)$ such that if

$$\max \|\mathbb{U}'(\mathbf{x}, 0)\| < \delta$$

then

$$\max \|\mathbb{U}'(\mathbf{x}, t)\| < \varepsilon, \qquad \text{for all } t > 0,$$

where the norm of $\|\mathbb{U}\|$ is a weighted combination of the norms of the components $\|\mathbf{u}'\|$ and $\|P'\|$. The actual norm $\|\cdots\|$ can be chosen from, for example, the 1-norm, 2-norm, or max-norm. We shall use norms based on the so-called Euclidean norm, also known as the 2-norm or L^2 norm, for example:

$$\|\mathbf{u}'\| \equiv \left[\frac{1}{\mathscr{V}} \int_{\mathscr{V}} \left(\mathbf{u}'\right)^2 d^3\mathbf{x}\right]^{1/2}.$$

When a state is stable based on this norm we sometimes refer to it as *stable in the mean*. Alternatively, we say that a state is *asymptotically stable* when it is Liapunov stable and in addition

$$\lim_{t \to \infty} \|\mathbb{U}'\| = 0.$$

Asymptotic stability is, perhaps, what is most commonly meant by *stability* in FD and also in dynamical systems theory, where a stable state is known as an attractor. Linear stability theory, as we have presented it so far, is formulated for steady, i.e., time-independent base state. We treat the perturbations' spatial structure and temporal behavior as separable, giving the familiar form

$$\mathbb{U}'(\mathbf{x}, t) = \hat{\mathbb{U}}(\mathbf{x})e^{pt} + \text{c.c.}, \tag{7.6}$$

with p a complex constant. It follows that we are able to express the solution to the initial value problem for the perturbations as a linear superposition of these normal modes, where p is the eigenvalue and the complex structure function $\hat{U}$ the eigenfunction. The task is to solve the linear system posed by (7.5) and determine the set of all p values, the *spectrum*, and the associated solutions for $\hat{U}$. Now we can refer to a base state as stable if $\Re(p) < 0$ for all p or as unstable if even a single p has $\Re(p) > 0$, since this implies exponential growth. The instability is also characterized by that mode of the system which has the largest value of the real part of the exponent p, indicating fastest-growing mode. If $\max[\Re(p)] = 0$, then the system is stable, but not asymptotically stable and we refer to this as *neutrally stable*. In our example problem, the value of the Reynolds number for which the fastest-growing mode has $\Re(p) = 0$ is the definition of the critical Reynolds number, Re_c, and we refer to this as the marginally stable state.

A significant number of other definitions of stability are possible and sometimes warranted for particular problems. It is not practical to present too many alternatives beyond those we defined above. Nevertheless, a few comments are in order. Sometimes disturbances that are generated at a fixed position $\mathbf{x}$ do not grow in time, but instead grow in space, usually as they travel downstream. This calls for a *spatial stability theory* for which we seek complex p representing growth in perturbations of the form e^{px}, for example. Often it is sufficient to apply Gaster's transformation (see below, at the conclusion of Sect. 7.3.3.3), obtaining a temporal stability analysis of a spatial instability. Related to this concept, we point out that linear stability theory as presented here is based on the Eulerian framework of the governing fluid equations. An Eulerian theory is not capable of determining the Lagrangian stability of fluid parcels, a question of their finite deformation properties along their trajectories. Finally, asymptotic stability is believed to be an overly stringent measure because many perturbed fluid systems exhibit a large but transient growth of energy, a case that we will explore in more detail below and in Sect. 7.6.

We have alluded several times to similarities between the linear analysis of fluid dynamical stability and the analysis of the stability of a fixed point of a dynamical system. The following example illustrates these similarities more precisely. Consider the 2×2 system of first order autonomous ODEs

$$\frac{d\mathbf{u}}{dt} = \vec{\mathcal{N}}(\mathbf{u}) \text{ where } \mathbf{u} = (u, \mathsf{v}), \tag{7.7}$$

having a fixed point $\mathbf{u}_0$, satisfying $d\mathbf{u}_0/dt = 0$. $\mathcal{N}$ is a nonlinear vector operator, but for small enough perturbations $\mathbf{u}'$, the system can be approximated near $\mathbf{u}_0$ as follows:

$$\frac{d\mathbf{u}'}{dt} = \mathbb{J}(\mathbf{u}_0)\mathbf{u}' \text{ where } J_{ij} = \frac{\partial \mathcal{N}_i}{\partial u_j}. \tag{7.8}$$

Evaluated at $\mathbf{u} = \mathbf{u}_0$, $\mathbb{J}$ takes the form of a constant coefficient matrix, $\mathbb{J}_0$, allowing perturbations having e^{pt} form and leading to the eigenvalue problem

Table 7.1 Summary of stability behavior. The first two cases are *hyperbolic* and amenable to linear stability analysis. The last case also allows the possibility of $\text{Tr}(\mathbb{J}) = 0$, for which zero is a double eigenvalue and a Jordan block form is required, leading to algebraic, rather than exponential growth. Refer also to Fig. 7.4

Det($\mathbb{J}$)	Condition	Stability Type (Name)
Det($\mathbb{J}$) > 0	4Det($\mathbb{J}$) $\leq$ Tr^2($\mathbb{J}$)	p_1, p_2 real, same sign (node, stable or unstable)
Det($\mathbb{J}$) < 0	—	p_1, p_2 real, opposite sign (saddle)
Det($\mathbb{J}$) > 0	4Det($\mathbb{J}$) $\leq$ Tr^2($\mathbb{J}$)	p_1, p_2 complex, $\Im(p) \neq 0$ (spiral, stable or unstable)
Det($\mathbb{J}$) > 0	Tr($\mathbb{J}$) = 0	p_1, p_2 imaginary (center, neutrally stable)
Det($\mathbb{J}$) = 0	Tr($\mathbb{J}$) $\neq$ 0	p_1, p_2 zero, simple (no name)

$$(\mathbb{J} - p\mathbb{I})\mathbf{u}' = \mathbf{0} \tag{7.9}$$

whose eigenvalues, $p = p_1, p_2$ are solutions to the characteristic polynomial

$$\text{Det}(\mathbb{J} - p\mathbb{I}) = p^2 - \text{Tr}(\mathbb{J})p + \text{Det}(\mathbb{J}). \tag{7.10}$$

The types of stability behavior depending on the eigenvalues are summarized in Table 7.1 and in Fig. 7.4. Note that only for hyperbolic behavior, the first three cases on the table, can we learn about stability properties using linear theory; the other cases require a nonlinear approach.

As we have mentioned, the restriction to asymptotic stability and to small perturbations, in the linearized sense, is not appropriate for every stability problem. Real world perturbations may be large and they may originate from noise that is not easily controlled nor measured. One example of the possible relevance of noise is found in the transient amplification of initial perturbation energy in systems whose linearized operator, i.e., $\mathbb{J}$, is non-normal.[1] Following Charru ([4] in the *Bibliographical notes*), we present the canonical example

$$\mathbb{J} = \begin{pmatrix} -\varepsilon & 1 \\ 0 & -2\varepsilon \end{pmatrix} \text{ where } 0 < \varepsilon \ll 1, \tag{7.11}$$

having eigenvalues $p_1 = -\varepsilon$, $p_2 = -2\varepsilon$ and eigenvectors $\hat{u}_1 = (1, 0)$ and $\hat{u}_2 = (1, -\varepsilon)$, which become nearly parallel as $\varepsilon \to 0$. The general solution $\mathbf{u}(t)$ for arbitrary initial conditions will lead to the growth of $|\mathbf{u}(t)|^2$, and thus energy, that may reach a relatively large value before it eventually decays asymptotically. Such behavior is not well captured by examining the modes of the problem individually because transient growth originates from the time-evolution of two or more *non-orthogonal modes*, resulting from non-normal operators, in superposition. We expand on these concepts and this example in Sect. 7.6, while further details on transient growth can be found in the book by Schmid and Henningson (reference [8] in the *Bibliographical Notes*).

---

[1] An operator, or matrix, is *non-normal* when it does not commute with its adjoint.

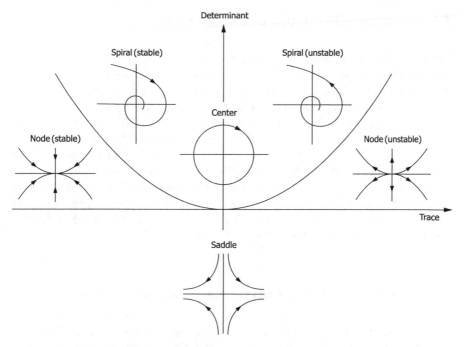

Fig. 7.4 Schematic diagram of the possible stability types of fixed points in linear dynamical systems, as described in the text and in Table 7.1

Once the linear stability or instability of a fluid state is determined, then what? How might we follow the flow as it begins to traverse into the nonlinear regime, where the assumption of small perturbations breaks down? At what point a perturbation becomes too large depends specifically on the problem examined. In most cases we are left to cope with the general strongly nonlinear behavior, using either numerical simulations, laboratory experiments, or observations of nature. Aside from a few advanced methods, which we mention here, but will not discuss them in any detail, such as the *Energy method*, *Serrin's theorem*, and *Liapunov's direct method*, which typically give only bounds on stability properties, few rigorous mathematical approaches to strongly nonlinear stability exist. In some limited cases in which a linear instability may be isolated, certain analytical techniques are available to assess the nonlinear fate. These techniques are grouped into a broad heading called *weakly nonlinear theory*, which we shall discuss and give examples of in the following chapter.

7.3   Stability of Plane-Parallel Shear Flows

Before turning our attention to instabilities driven by buoyancy or other forces, we examine the intrinsic stability properties of an inertial shear flow. The study of shear flows and their transition is one of the most difficult problems of fluid dynamics that has remained largely unsolved. Many readers will have heard that even Heisenberg struggled, in his doctoral thesis, with the stability of a basic shear flow. Since the time of Lord Rayleigh, in fact, countless researchers have considered the linear theory of plane-parallel shear flow, sketched generally in Fig. 7.5. Flows of this sort play a central dynamical role in almost every branch of science, with important examples found in astrophysical disks, meteorology, oceanography, geological flows, petroleum flows, turbo-machinery and blood flow . . . and the list goes on. While other physical processes may be involved in such examples, the role of the shear flow is undisputed.

7.3.1   Three-Dimensional Disturbances in Shear Flows

To appreciate the nature of shear flow instability, we turn our attention first to the fundamental case of an incompressible fluid exhibiting steady plane parallel shear flow of the form

$$\mathbf{u} = u_0(y)\hat{\mathbf{x}}. \tag{7.12}$$

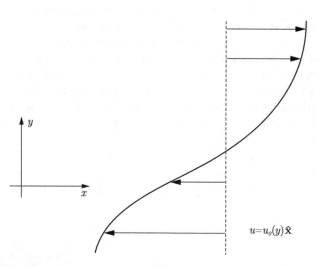

Fig. 7.5 Plane parallel shear layer velocity profile. The vertical centerline is arbitrary due to Galilean invariance. In the absence of parallel walls the flow should remain bounded at large $|y|$

Following convention, we refer to the x, y, z directions, respectively, as the *stream-wise, normal,* or *shear-wise* and *span-wise* directions. The general linearized equations of motion are given by

$$\left(\frac{\partial}{\partial t} + u_0 \frac{\partial}{\partial x}\right) u' + v' u_{0y} = -\frac{\partial \Pi'}{\partial x} + \frac{1}{\text{Re}} \nabla^2 u', \qquad (7.13)$$

$$\left(\frac{\partial}{\partial t} + u_0 \frac{\partial}{\partial x}\right) v' = -\frac{\partial \Pi'}{\partial y} + \frac{1}{\text{Re}} \nabla^2 v', \qquad (7.14)$$

$$\left(\frac{\partial}{\partial t} + u_0 \frac{\partial}{\partial x}\right) w' = -\frac{\partial \Pi'}{\partial z} + \frac{1}{\text{Re}} \nabla^2 w', \qquad (7.15)$$

$$\frac{\partial u'}{\partial x} + \frac{\partial v'}{\partial y} + \frac{\partial w'}{\partial z} = 0, \qquad (7.16)$$

where $u', v',$ and w' represent the stream-wise, normal, and span-wise velocity perturbations, respectively. $\Pi' \equiv P'/\rho_0$, where ρ_0 is the constant density. In addition $u_{0y} \equiv du_0/dy$, thereby using in the following a subscript notation for derivatives. The Reynolds number is, as usual, $\text{Re} = UL/v$, where U and L are characteristic velocity and length scales that are defined based on the flow $u_0(y)$ properties, or based on the boundaries, if any, and v is, as usual, the kinematic viscosity.

Now, taking the x, y, and z derivatives of equations (7.13–7.15), respectively, and making use of the incompressibility equation (7.16) gives a Poisson equation relating the pressure and velocity perturbations,

$$\nabla^2 \Pi' + 2u_{0y} \frac{\partial v'}{\partial x} = 0, \qquad (7.17)$$

that can be used to the eliminate the pressure perturbation Π', yielding the single equation

$$\left(\frac{\partial}{\partial t} + u_0 \frac{\partial}{\partial x}\right) \nabla^2 v' - u_{0yy} \frac{\partial v'}{\partial x} - \frac{1}{\text{Re}} \nabla^4 v' = 0. \qquad (7.18)$$

To complete the system we obtain the perturbation of normal vorticity, defined as

$$\zeta'_Y = \partial u'/\partial z - \partial w'/\partial x. \qquad (7.19)$$

Note the use of the unusual notation for the vorticity component ζ with index Y. This is so because the index-less ζ had previously denoted the simple vertical (z) vorticity component in a general (x, y, z) Cartesian frame. Operation with $\nabla \times$ on Eqs. (7.13–7.15) gives

$$\left(\frac{\partial}{\partial t} + u_0 \frac{\partial}{\partial x}\right) \zeta'_Y - \frac{1}{\text{Re}} \nabla^2 \zeta'_Y + u_{0y} \frac{\partial v'}{\partial z} = 0. \qquad (7.20)$$

The pair of equations (7.18) and (7.20) may be written in matrix notation as

$$\frac{\partial}{\partial t}\begin{pmatrix} v' \\ \zeta'_Y \end{pmatrix} = \begin{pmatrix} \mathbb{L}_{vv} & 0 \\ \mathbb{L}_{v\zeta} & \mathbb{L}_{\zeta\zeta} \end{pmatrix}\begin{pmatrix} v' \\ \zeta'_Y \end{pmatrix},\tag{7.21}$$

where $\mathbb{L}_{vv}, \mathbb{L}_{v\zeta}, \mathbb{L}_{\zeta\zeta}$ are the corresponding linear operators. The stream-wise and spanwise fields, u' and w', may be recovered from the solutions of perturbations v' and ζ'_Y.

Equation (7.21) together with physically appropriate boundary conditions completely describes the linear stability of the system. For a channel infinite in the stream-wise and span-wise directions, the viscous or no-slip boundary conditions imply that the perturbed field velocity quantities are zero, i.e., $u' = v' = w' = 0$. A shear flow also unbounded in the y direction would require these conditions in the so-called far field, as $y \to \pm\infty$. It is straightforward to show that no-slip conditions also lead to

$$\frac{\partial v'}{\partial y} = \zeta'_Y = 0\tag{7.22}$$

for the perturbation derivative and normal vorticity at the boundaries. In the inviscid case, $Re = \infty$, the above set of boundary conditions is replaced by the condition of impenetrability at the boundaries, i.e., $\hat{v} = 0$ only. It is this case that we examine first. We begin by examining the linear stability of equations, which summarize from the works of W.M. Orr and the acclaimed A. Sommerfeld, done in the beginning of the twentieth century, and the important insight of H.B. Squire from the 1930s. Equations (7.18) and (7.20) are assumed to describe normal mode form perturbations

$$\begin{pmatrix} v' \\ \zeta'_Y \end{pmatrix} = \begin{pmatrix} \hat{v} \\ \hat{\zeta}_Y \end{pmatrix} e^{ik_x(x-V_pt)+ik_z z} + \text{c.c.}\tag{7.23}$$

For the investigation of temporal stability, V_p is a *complex* quantity, as yet to be determined, and the eigenfunctions are $\hat{v} = \hat{v}(y)$, $\hat{\zeta}_Y = \hat{\zeta}_Y(y)$. Insertion of this normal mode ansatz into (7.18)–(7.20) results in the important and well-known *Orr–Sommerfeld equation*

$$[u_0(y) - V_p]\left(\frac{d^2}{dy^2} - k^2\right)\hat{v} - \frac{d^2u_0}{d^2y}\hat{v} + \frac{1}{ik_x Re}\left(\frac{d^2}{dy^2} - k^2\right)^2\hat{v} = 0,\tag{7.24}$$

for $\hat{v}(y)$, and in the *Squire equation*

$$[u_0(y) - V_p]\hat{\zeta}_Y - \frac{1}{ik_x Re}\left(\frac{d^2}{dy^2} - k^2\right)\hat{\zeta}_Y = -\frac{k_z}{k_x}\frac{du_0}{dy}\hat{v},\tag{7.25}$$

for ζ_Y, where $k^2 \equiv k_x^2 + k_z^2$. It is convenient to recognize two types of solutions: the first type consists of the solutions $\hat{v} = \hat{v}^{(OS)}$ that solve the Orr–Sommerfeld equation (7.24) along with its corresponding value of a complex V_p. These Orr–Sommerfeld modes are the main object of study in linear stability of shear flows. The second type of solutions, the so-called Squire modes, are $\hat{\zeta}_Y$, satisfying (7.25) with $\hat{v} = 0$. An analysis of Squire's equation shows that Squire modes are always damped. The proof of this assertion is left as an exercise (see Problem 7.5). The Squire equation is solved by inserting the solution $\hat{v}^{(OS)}$ and V_p into Eq. (7.25) from which the solution $\hat{\zeta}_Y = \hat{\zeta}_Y^{(OS)}$ is determined by solving the resulting linear differential equation. In the literature these two solutions are often denoted as $(\hat{v}, \hat{\zeta}_Y)^T = \left(0, \hat{\zeta}_Y^{(SQ)}\right)^T$ and $(\hat{v}, \hat{\zeta}_Y)^T = \left(\hat{v}^{(OS)}, \hat{\zeta}_Y^{(OS)}\right)^T$, i.e., Squire modes and Orr–Sommerfeld modes, respectively, where, as before, T denotes transpose.

Linear stability of shear flows is commonly studied within a two-dimensional framework based on the so-called Squire transformation. To paraphrase the result, a parallel shear flow becomes linearly unstable to two-dimensional normal mode disturbances at a smaller Reynolds number than for any three-dimensional disturbances. Consider two solutions of the Orr–Sommerfeld equation. Let the first one be the solution of the strictly two-dimensional problem in which the input wave vectors are given and labelled as $k_x = \alpha_{(2)}, k_z = 0$. Let the corresponding solution of the two-dimensional problem be written as $\hat{v}_{(2)}$, at Reynolds number $\text{Re}_{(2)}$, the subscript "(2)" indicating two-dimensional. The corresponding expression of the Orr–Sommerfeld equation is

$$\left(u_0 - V_p\right)\left(\frac{d^2}{dy^2} - \alpha_{(2)}^2\right)\hat{v}_{(2)} - u_{0yy}\hat{v} + \frac{1}{i\alpha_{(2)}\text{Re}_{(2)}}\left(\frac{d^2}{dy^2} - \alpha_{(2)}^2\right)^2\hat{v}_{(2)} = 0. \quad (7.26)$$

Compare this to a three-dimensional solution of the Orr–Sommerfeld equation where the wavenumbers are $k_x = \alpha_{(3)}, k_z = \beta_{(3)}$ together with Reynolds number $\text{Re}_{(3)}$. The corresponding solution $\hat{v}_{(3)}$ satisfies

$$\left(u_0 - V_p\right)\left(\frac{d^2}{dy^2} - \alpha_{(3)}^2 - \beta_{(3)}^2\right)\hat{v}_{(3)} - u_{0yy}\hat{v}_{(3)} + \frac{1}{i\alpha_{(3)}\text{Re}_{(3)}}\left(\frac{d^2}{dy^2} - \alpha_{(3)}^2 - \beta_{(3)}^2\right)^2\hat{v}_{(3)} = 0. \quad (7.27)$$

An inspection of both governing equations (7.26) and (7.27) shows that the two solutions $\hat{v}_{(2)}$ and $\hat{v}_{(3)}$, and their respective eigenvalues V_p, are identical when the following equivalences are admitted,

$$\alpha_{(2)}^2 \iff \alpha_{(3)}^2 + \beta_{(3)}^2, \qquad \alpha_{(2)}\text{Re}_{(2)} \iff \alpha_{(3)}\text{Re}_{(3)}. \quad (7.28)$$

An instability in the strictly two-dimensional case at the stream-wise wavenumber $\alpha_{(2)} > 0$ and $\text{Re}_{(2)}$ thus corresponds to an instability for three-dimensional disturbances with a corresponding stream-wise wavenumber $\alpha_{(3)} < \alpha_{(2)}$ and a correspondingly larger Reynolds number

$$\mathrm{Re}_{(3)} = \mathrm{Re}_{(2)} \frac{\alpha_{(2)}}{\alpha_{(3)}} < \mathrm{Re}_{(2)}.$$

In other words, *for any three-dimensional linear instability of the Orr–Sommerfeld equation at a given value of* Re, *an analogous two-dimensional linear instability also exists, but always at a lower value of* Re.

The consequence of this important statement is clear but sometimes misinterpreted: parallel shear flows first become unstable to two-dimensional disturbances. In shear flows with additional physical effects, such as rotation or stratification, Squire's theorem cannot be applied without demonstrating that the above properties are preserved. In other words, the existence of a Squire theorem for each class of flow must be separately established. We strongly emphasize that Squire's theorem holds at marginal stability, but not generally. Thus at Re $>$ Re$_c$ it is certainly possible that three-dimensional disturbances are more unstable than two-dimensional ones. In practice, we may therefore consider strictly two-dimensional disturbances in our study of the Orr–Sommerfeld equation. This means then we may restate the Orr–Sommerfeld equation in terms of the more familiar perturbation of $\hat{\mathbf{z}}$ directed vorticity ζ' and related stream function ψ':

$$\zeta' \equiv \frac{\partial v'}{\partial x} - \frac{\partial u'}{\partial y} = \left(\frac{\partial^2}{\partial x^2} + \frac{\partial^2}{\partial y^2} \right) \psi', \qquad \text{because} \qquad u' = -\frac{\partial \psi'}{\partial y}, \quad v' = \frac{\partial \psi'}{\partial x}. \tag{7.29}$$

By assuming again normal mode solutions, now for the stream function perturbation, of the form $\psi' = \hat{\psi}(y)e^{ik\left(x - V_{\mathrm{p}}t\right)} +$c.c., we find that the Orr–Sommerfeld equation is reexpressed as an ODE of the eigen function $\hat{\psi}(y)$:

$$\left[u_0(y) - V_{\mathrm{p}} \right] \left(\frac{d^2}{dy^2} - k^2 \right) \hat{\psi} - \frac{d^2 u_0}{dy^2} \hat{\psi} - \frac{1}{\mathrm{Re}\, ik} \left(\frac{d^2}{dy^2} - k^2 \right)^2 \hat{\psi} = 0. \tag{7.30}$$

7.3.2 Inviscid Shear

By setting Re $= \infty$, which formally corresponds to inviscid flow, in the above Orr–Sommerfeld equation (7.30), we obtain the important *Rayleigh equation* for inviscid disturbances to parallel shear flow:

$$\left(\frac{d^2}{dy^2} - k^2 \right) \hat{\psi} - \frac{u_{0yy}}{u_0 - V_{\mathrm{p}}} \hat{\psi} = 0. \tag{7.31}$$

In the absence of viscosity, we require that the stream function, which is the relevant variable instead of the perturbation normal velocity, $\hat{v} = ik\hat{\psi}$, is zero at $y = \pm 1$, or that it tends to zero as $y \to \pm\infty$. In temporal stability theory, our task is to determine the complex eigenvalue V_{p} (an extension of the real phase velocity, as known from

Chap. 4, to the complex number domain) as a function of k, for a given input parallel flow $u_0(y)$. The mathematical relative simplicity of equation (7.31) has led to a number of generalized criteria regarding the necessary conditions for the existence of unstable modes and the scale and magnitude of any instability. Three of these statements are:

- *Rayleigh's inflection point criterion*:
  If a solution of (7.31) in an open domain $\mathscr{V}$ supports an unstable mode, i.e., where $\Im(V_\mathrm{p}) > 0$, then $u_{0yy}(y)$ must vanish at some point $y = y_\mathrm{I}$ in $\mathscr{V}$, i.e., for $y_\mathrm{I} \in \mathscr{V}$, and not on its boundary. In other words, the shear profile must contain a point of inflection for there to be instability. This is a necessary but not sufficient condition for instability.
- *Fjortoft's criterion*:
  When considering a monotonic velocity profile $u_0(y)$ on a finite domain $\mathscr{V}$ of a shear-wise scale $\sim 2L$, i.e., $y \in [-L, L]$, if there exists an unstable solution of (7.31), then it must be true that

$$u_{0yy}(y)\left[u_0(y) - U_{0\mathrm{I}}\right] < 0$$

for some point y in the domain $\mathscr{V}$, where $U_{0\mathrm{I}} \equiv u_0(y_\mathrm{I})$ with y_I defined as the location of the inflection point in the basic shear flow, that is, where $u_{0yy}(y_\mathrm{I}) = 0$.
- *Howard's semicircle theorem*:
  Suppose a velocity profile $u_0(y)$ attains maximum value $U_{\max}$ and minimum value $U_{\min}$ on the y-axis within the domain of $\mathscr{V}$. Suppose also that (7.31) admits *instability*, such that $\Im(V_\mathrm{p}) > 0$ is true. The bound of real and imaginary parts of the mode's phase velocity must satisfy

$$\left[\Re(V_\mathrm{p}) - \frac{1}{2}\left(U_{\max} + U_{\min}\right)\right]^2 + \Im(V_\mathrm{p})^2 \leq \left[\frac{U_{\max} - U_{\min}}{2}\right]^2. \tag{7.32}$$

The theorem of Howard is important because it puts bounds on the growth rates expected in unstable flows, making it a valuable guide in assessing the reasonableness of predicted eigenvalues when these are obtained by numerical methods, for example.

We now sketch the proof of Rayleigh's inflection point theorem; proofs of the other criteria follow a similar approach and can be found in the sources given in *Bibliographical Notes*. Given that the boundaries of the flow are at $y = \pm L = \pm 1$, we know that if the flow is unstable $\Im(V_\mathrm{p}) > 0$. Writing $V_\mathrm{p} = V_{pr} + iV_{pi}$, where $V_{pr} \equiv \Re(V_\mathrm{p})$ and $V_{pi} \equiv \Im(V_\mathrm{p})$, Eq. (7.31) may be reexpressed as

$$\left(\frac{d^2}{dy^2} - k^2\right)\hat{\psi} - u_{0yy}\left[\frac{u_0 - V_{pr} + iV_{pi}}{(u_0 - V_{pr})^2 + V_{pi}^2}\right]\hat{\psi} = 0.$$

Multiplying the above expression by $\hat{\psi}^*$, the complex conjugate of $\hat{\psi}$, and integrating the resulting expression from $y = -1$ to $y = 1$ results in

$$\int_{-1}^{1} \left[\left| \frac{d\hat{\psi}}{dy} \right|^2 + k^2 |\hat{\psi}|^2 + \frac{u_{0yy}(u_0 - V_{pr})}{(u_0 - V_{pr})^2 + V_{pi}^2} |\hat{\psi}|^2 \right] dy + iV_{pi} \int_{-1}^{1} \frac{u_{0yy}|\hat{\psi}|^2}{(u_0 - V_{pr})^2 + V_{pi}^2} dy = 0,$$

(7.33)

after applying the boundary conditions. Each of the two integrals in the above complex expression are real quantities. Therefore, the real and imaginary components of (7.33) must separately be equal to zero. Rayleigh's criterion is the content of the second term, namely,

$$V_{pi} \int_{-1}^{1} \frac{u_{0yy}|\hat{\psi}|^2}{(u_0 - V_{pr})^2 + V_{pi}^2} dy = 0,$$

(7.34)

and in order for this statement to be true in general for $V_{pi} \neq 0$ it must be that u_{0yy} changes sign somewhere in the domain of integration. This establishes the Rayleigh criterion, as it was named in this discussion: stating the necessity of an inflection point in the velocity profile for instability to occur. Such *inflectional instabilities* have been explained by C.C. Lin in terms of vorticity, based on the observation that an inflection point implies an extremum in the vorticity distribution. A fluid element displaced from a vorticity extremum is not forced back to its point of origin by the background vorticity, implying instability (work it out!).

7.3.2.1 Kelvin–Helmholtz Instability

Perhaps the most elementary shear flow problem that is capable of providing some insight into the instability is the piecewise uniform profile

$$u_0 = \begin{cases} -u_{00}; \, y < 0, \\ u_{00}; \, y > 0, \end{cases}$$

(7.35)

where u_{00} is a constant. To examine this problem, we shall apply the methods employed in the study of surface gravity waves in Chap. 4. That strategy involves solving (7.31) in separate regions followed by applying physically meaningful formulae connecting the solutions found in the separate regions. In the above example, solutions of (7.31) shall be constructed separately for $y < 0$ and $y > 0$. With these in hand, the solutions across the boundary $y = 0$ will be determined by enforcing the conditions of velocity and pressure continuity across the layer. The inviscid Orr–Sommerfeld equation, for $y \neq 0$, is given by

$$\left(u_0 - V_p \right) \left[\frac{d^2 \hat{\psi}}{dy^2} - k^2 \hat{\psi} \right] = 0.$$

(7.36)

Since we consider this problem on an infinite domain with solutions decaying to zero as $y \to \pm\infty$, we find

$$\hat{\psi} = \begin{cases} A_- e^{ky}; & y < 0, \\ A_+ e^{-ky}; & y > 0. \end{cases} \tag{7.37}$$

Now we relate the two solutions across the boundary. The linearized vertical displacement in the normal mode formulation is simply $(u_0 - V_\text{p})\hat{\eta} = \hat{\psi}$, from the Lagrangian definition $d\eta/dt = v$. Writing $\hat{\eta}_\pm$ and interpreting it as the layer's position as evaluated above and below $y = 0$, respectively, we have that

$$\hat{\eta}_- = \left[\frac{\hat{\psi}}{u_{00} - V_\text{p}} \right]_{y \to 0^-} = -\frac{A_-}{u_{00} + V_\text{p}}, \quad \text{and,} \quad \hat{\eta}_+ = \left[\frac{\hat{\psi}}{u_{00} - V_\text{p}} \right]_{y \to 0^+} = \frac{A_+}{u_{00} - V_\text{p}}. \tag{7.38}$$

Clearly, there can be no lack of continuity of the fluid itself; this is a purely kinematic condition which leads to

$$A_- \left(V_\text{p} - u_{00} \right) = A_+ \left(V_\text{p} + u_{00} \right), \tag{7.39}$$

establishing a relationship between $A_\pm$ and V_p. Because mean pressure is uniform, the condition that the total pressure across the surface is continuous reduces to the continuity of the perturbation pressure, thus

$$\left[\left(u_0 - V_\text{p} \right) \frac{d\hat{\psi}}{dy} \right]_{y \to 0^-}^{y \to 0^+} = 0. \tag{7.40}$$

The above conditions follow directly from the perturbation pressure expression found in the normal mode reexpression of (7.13) for Re $\to \infty$, viz.:

$$-ik \left(u_0 - V_\text{p} \right) \frac{d\hat{\psi}}{dy} = -ik\hat{\Pi}.$$

Applying the solutions developed for $\hat{\psi}$ we see that

$$-k \left(u_{00} - V_\text{p} \right) A_+ - k \left(u_{00} + V_\text{p} \right) A_- = 0, \quad \Longrightarrow \quad kA_+ \left(V_\text{p}^2 + u_{00}^2 \right) = 0. \tag{7.41}$$

In obtaining the right-hand side of the above expression we have made explicit use of (7.39). For nontrivial solutions, $A_\pm \neq 0$, it follows that the complex phase velocity is $V_\text{p} = \pm iu_{00}$ implying the presence of a stable/unstable pair of modes. What is more, the result shows that instability is expected *for every horizontal wavenumber k*. In fact, the growth rate $\omega = kV_\text{p} = \pm iku_{00}$ is infinite as $k \to \infty$ indicating that the mathematical problem is ill-posed. For the moment, we shall say only that the ill-posed nature of the solutions results from the poor definition of the problem. With no length scale set either by a finite channel width or by the thickness of the shear

a   **b**

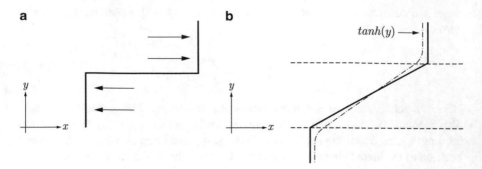

Fig. 7.6 Profiles for: (**a**) piecewise constant Kelvin–Helmholtz flow, and (**b**) a piecewise mixing layer—Rayleigh profile, with a continuous mixing layer—a *tanh* layer profile, superimposed upon it

layer, and no physical length scale set by viscosity or other effects, the timescale and thus the instability growth rate are ill-defined. We will discuss this in more detail in Sect. 7.4 when examining the Rayleigh–Taylor instability.

In the meantime, we point out that the well-known Rayleigh shear, or *piecewise mixing layer*, profile:

$$u_0 = \begin{cases} u_{00} & , y > H; \\ \Lambda y & , -H \le y \le H; \\ -u_{00} & , y < -H; \end{cases} \tag{7.42}$$

is, possibly, the simplest resolution of the ill-posedness of the Kelvin–Helmholtz piecewise constant profile, see Fig. 7.6a. In (7.42), u_{00} and H are constants, the latter always being greater than zero. Because the base flow u_0 is assumed to be continuous for all y, it follows that $\Lambda = u_{00}/H$. The procedure in calculating the solution to (7.31) for the piecewise mixing layer uses the same matching techniques as those used for the Kelvin–Helmholtz profile, but it is more involved because u_{0yy} is characterized by delta functions, viz.:

$$u_{0yy} = \Lambda \delta(y+H) - \Lambda \delta(y-H). \tag{7.43}$$

Nevertheless, after performing that calculation we find that the relation for $V_p^2 \equiv \omega^2/k^2$ is given by

$$V_p^2 = \frac{u_{00}^2}{4(kH)^2} \left[(1 - 2kH)^2 - e^{-4kH} \right], \tag{7.44}$$

indicating that only for values of kH less than the critical value, i.e., for $kH \le 0.6392$, will the disturbances be exponentially growing. The existence of a length scale, the cutoff scale for the instability, has rectified the ill-posedness of the Kelvin–Helmholtz profile. The Rayleigh profile is a canonical example of inviscid shear instability, the mechanism often referred to as the Rayleigh mechanism. This

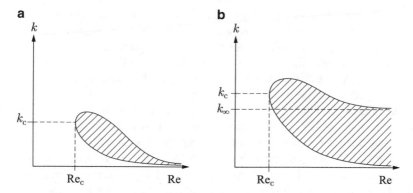

Fig. 7.7 Neutral stability curves and unstable parameter regions for shear flows indicating values of Re_c and k_c. (**a**) The Blasius boundary layer, unstable due to viscous effects, the unstable region shrinks to zero width as $Re \to \infty$. (**b**) A typical shear flow with an inflection point, unstable by the inviscid instability mechanism, as evident by the persistence of the unstable region as $Re \to \infty$

inviscid mechanism is prominent in real shear flows with viscosity. The region of inviscid instability, as described by the neutral stability curve, then identifies both the critical Reynolds number Re_c and the critical wavenumber k_c or its cutoff k_∞, as shown in Fig. 7.7b.

7.3.2.2 Smooth Shear and Mixing Layers

In this section we consider flows u_0 which are smooth, also having continuous first and second derivatives u_{0y} and u_{0yy}. We begin by examining the well-known mixing layer velocity profile as described using the hyperbolic tangent function, or $u_0 = u_{00} \tanh(y/H)$, valid for all y. Analytical solutions may be found relatively simply for exactly marginal modes for which $V_p = 0$. The velocity field is anti-symmetric with respect to $y = 0$ and contains an inflection point at $y = 0$ since $u_{0yy} = -(2/H^2)\text{sech}^2[y/H]\tanh[y/H]$. The transition to instability occurs for $V_p = 0$ and one may attempt to solve Eq. (7.31) subject to the conditions $\hat{\psi} \to 0$ as $y \to \pm\infty$, resulting in an eigenvalue problem for the critical value of k. A useful approach is to seek solutions of equation (7.31), with $V_p = 0$, on a new, stretched, coordinate defined by $\tau \equiv \tanh y$, noting that $\text{sech}^2(y/H) = 1 - \tanh^2(y/H) = 1 - \tau^2$. In such a coordinate system, it is straightforward to show that all derivative operators transform according to

$$\frac{d}{dy} \mapsto \frac{1}{H}(1 - \tau^2)\frac{d}{d\tau}, \qquad \frac{d^2}{dy^2} \mapsto \frac{1}{H^2}(1 - \tau^2)\frac{d}{d\tau}(1 - \tau^2)\frac{d}{d\tau}.$$

For a base shear given by $u_0 = u_{00}\tanh y$, Eq. (7.31) becomes

$$(1 - \tau^2)\frac{d^2\hat{\psi}}{d\tau^2} - 2\tau\frac{d\hat{\psi}}{d\tau} + \left[2 - \frac{k^2 H^2}{1 - \tau^2}\right]\hat{\psi} = 0, \qquad (7.45)$$

where we have taken $V_p = 0$. Equation (7.45) has boundary conditions $\hat{\psi}(\tau = \pm 1) = 0$, reflecting that we have effectively mapped the infinite domain onto $\tau = \pm 1$. Equation (7.45) is in the form of a general Legendre equation with bounded solutions $P_1^{kH}(\tau)$. In order to avoid singular solutions of the associated Legendre polynomials at $\tau = \pm 1$, it is required that $|kH| \leq 1$. The boundary conditions $\hat{\psi} = 0$ for $\tau \rightarrow \pm 1$ tell us that the only possible solution is $\hat{\psi} = P_1^1(\tau) \propto (1 - \tau^2)^{1/2}$. Solving in terms of y, we find that the disturbance eigenfunction is $\hat{\psi}(y) = \text{sech}(y/H)$ and, consequently, $kH = 1$. The result indicates that the only possible marginal solution corresponds to a mode with horizontal wavenumber $k = 2\pi/H$. What is the stability for values of kH not equal to one? Under these more general circumstances, one must appeal to numerical methods to assess the stability, which is outside the scope of this book. In that case, it is an unsurprising result that the *tanh* layer is unstable for all values of $kH \leq 1$.

7.3.3   Viscous Shear

The normal mode stability problem for viscous shear flows returns us to the Orr–Sommerfeld equation (7.24). Intuition says that viscosity ought to stabilize disturbances. While this is generally true, we shall also find that viscosity can destabilize flows that are otherwise stable to inviscid disturbances. This counterintuitive behavior comes about because viscosity plays two distinct roles in the stability of shear flows. The first role is that it places stringent constraints on the form of the base flow profile. A steady parallel profile, $u_0(y)$, must satisfy the steady Navier–Stokes equations and the no-slip boundary conditions. For this reason, among others, it makes little sense to examine unbounded flows, as we did in the previous section. Indeed, the stability properties of viscous flows are, to a great extent, determined by the shape of the base flow profile that is demanded by a particular forcing and set of boundaries. The demands of the viscous equations and boundary conditions are so restrictive that we generally resort to some approximation, such as a base flow that is not precisely steady, or use an approximate form of the Navier–Stokes equations. The exact steady solutions of viscous shear are so few that the well-known named examples constitute the majority of cases. Examples include the pressure driven Poiseuille flows: plane Poiseuille, between parallel walls, and pipe Poiseuille (a.k.a., Poiseuille–Hagen) in the interior of a cylinder; we will discuss pipe flow later in this chapter. Other examples include boundary layer flows, such as the Blasius flow, mixing layers and the Taylor–Couette (TC) flow between rotating concentric cylinders, which we shall discuss in detail in the following section in relation to the role of centrifugal "buoyancy" in that case.

The second role that viscosity plays is more subtle: the presence of viscosity allows some perturbations to become unstable by extracting energy from the mean flow. In flows which are stable to inviscid, or Rayleigh, instability, there may still be a viscous instability. The region of instability in the wavenumber–control parameter plane, identified by the neutral curve for a viscous instability, diminishes as Re grows, leading to the characteristic tapering of the instability region, as depicted in Fig. 7.7a. Plane Poiseuille flow and the Blasius boundary layer both lack an inflection point and their stability diagrams have the general form shown in Fig. 7.7a. The destabilization by viscosity is often described in terms of a *Reynolds stress* contribution to the disturbance energy equation. We examine this in more detail through the Reynolds–Orr equation below.

7.3.3.1   Reynolds–Orr Equation

The emergence of shear instability involves the transfer of energy from the basic shear into perturbations. Irrespective of the mechanism responsible, the instability can be seen to arise as a consequence of stresses induced by the perturbations, which possess appropriate correlated properties. As an introduction to this idea, consider the linearized equations of motion written as in Eqs. (7.13–7.16). Multiplying each of the Eqs. (7.13–7.15) by u', v', w', respectively, and integrating over the volume, $\mathcal{V}$, gives

$$\int_{\mathcal{V}} \left(\frac{\partial}{\partial t} + u_{0y} \frac{\partial}{\partial x} \right) \frac{\mathbf{u}'^2}{2} d^3x = -\int_{\mathcal{V}} \mathbf{u}' \cdot \nabla \Pi' d^3x - \int_{\mathcal{V}} u_{0y} u'v' d^3x + \frac{1}{\mathrm{Re}} \int_{\mathcal{V}} \mathbf{u}' \cdot \nabla^2 \mathbf{u}' d^3x$$

$$= -\int_{\partial \mathcal{V}} \Pi' \mathbf{u}' \cdot \hat{\mathbf{n}} dS - \int_{\mathcal{V}} u_{0y} u'v' d^3x + \frac{1}{\mathrm{Re}} \int_{\mathcal{V}} \mathbf{u}' \cdot \nabla^2 \mathbf{u}' d^3x$$

$$\tag{7.46}$$

where $\hat{\mathbf{n}}$ is the unit normal of the bounding surface $\partial \mathcal{V}$, using incompressibility to simplify the integral of the mechanical work. Provided that perturbation quantities are zero on the domain boundaries, or are periodic, the surface integral becomes zero. Supposing that the x direction is periodic, the term involving the integral of $u_0 \partial / \partial x$ identically integrates to zero (why?). Further integration by parts of the viscous term and imposition of the boundary conditions gives that

$$\frac{d}{dt} \int_{\mathcal{V}} E'_k d^3x = -\int_{\mathcal{V}} u_{0y} u'v' d^3x - \frac{1}{2\mathrm{Re}} \int_{\mathcal{V}} \left[(\nabla u')^2 + (\nabla v')^2 + (\nabla w')^2 \right] d^3x,$$

$$\tag{7.47}$$

in which $E'_k \equiv (u'^2 + v'^2 + w'^2)/2$ is the kinetic energy per unit volume. Equation (7.47) is the *Reynolds–Orr equation* for the perturbation energy, which shows two general properties. First, that the direct action of viscosity acts to damp modes, as given by the last integral on the right-hand side of (7.47), which is always less than zero. If there is growth in disturbance energy, then it must be true that the first term on the right-hand side of (7.47) has to be sufficiently greater than zero, i.e.,

$$-\int_{\mathcal{V}} u_{0y} u'v' d^3x > 0.$$

$$\tag{7.48}$$

The positivity of this stress, the so-called Reynolds stress, in (7.48) is one way to see that the perturbation vortex structure is tilted against the prevailing shear in an unstable flow.

7.3.3.2 Blasius and Other Boundary Layer Flows

Prior to the Blasius boundary layer, researchers studied a number of viscous boundary layer type problems, one of which was the, so-called, *first Stokes problem*, described briefly below. None of these flows is steady, as we discussed briefly in Chap. 3, but these flows generally have a self-similar form. Strictly steady plane-parallel flow must satisfy $\mathrm{Re}^{-1}(d^2 u_0/dy^2) = 0$, whose only solution is plane Couette flow. This disparity in approach to the viscous as compared to the inviscid stability problem calls into question the relevance of each to the other. This difficulty was acknowledged early on by many physicists working on this problem, including W. Heisenberg.

We start by examining the initially static fluid above a solid wall, taken to be the $(x-z)$ plane, which suddenly begins to move at constant speed U_0 in its own plane, in the $+x$ direction. Sometimes referred to as the *Rayleigh profile*, this problem was first solved by Stokes which is why it is also called the *first Stokes problem*. The solution has constant pressure and it is a solution of the parabolic PDE

$$u_t = \nu u_{yy} \tag{7.49}$$

with boundary conditions on $u(y)$ for $t > 0$ given by $u(0) = U_0$ and $u(\infty) = 0$. This is simply a diffusion, or heat, equation in one dimension. By introducing the similarity variable $\eta = y/2\sqrt{\nu t}$ we seek nondimensional solutions for $f(\eta) = u/U_0$ governed by

$$f'' + 2\eta f' = 0 \tag{7.50}$$

with boundary conditions $f(0) = 1$ and $f(\infty) = 0$. The solution is simply $u(y) = U_0(1 - \mathrm{erf}\,\eta)$ where

$$\mathrm{erf}\,\eta = \frac{2}{\sqrt{\pi}} \int_0^\eta e^{-\eta^2} d\eta \tag{7.51}$$

is the well-known error function. Flow solutions $u(y)$ at different times are similar and the boundary layer thickness $\delta = \sqrt{\nu t}$ develops in time. While seldom done, it is possible to perform a stability analysis of this flow. The results so closely resemble those of the Blasius layer that we do not discuss them further, except to point out that the relevance of instability is upheld when a mode's growth rate is not outstripped by the development of the boundary layer itself. Study of the stability of boundary layers demands some approximation, either by adopting unsteady base flow, as above, or in the case of flow over a flat plate, the use of a simplified form of the Navier–Stokes equations. In general, as we have already seen in a number of places in this book, a *boundary layer* in FD is a flow region, in many instances adjacent to a solid boundary, in which the flow-dependent variables, usually the velocity parallel to the boundary, vary rapidly with respect to their usual variation scale in the fluid. *Boundary layer theory* is an approximation technique in applied mathematics, applicable to problems that exhibit rapid variation in some small region and thus is appropriate for FD boundary layers. We have discussed a basic example of such an approximation problem, from its mathematical point of view, in Sect. 3.5.1. The theory was also applied to several flows, in which matched asymptotic expansions were employed. The so-called Prandtl boundary layer equations (mentioned in Chap. 3) are also a highly simplified model, albeit physical, in which the Navier–Stokes equations are reformulated in terms of a stretched, or boundary layer coordinate, $\tilde{y} = y/\delta$. Again, δ is the boundary layer thickness, a small number here given by $\delta = L/\sqrt{\text{Re}}$.

Prandtl's student, P. R. Blasius, solved this problem for the case of a uniformly flowing fluid encountering a plate edge-on. Without loss of generality, we can take the plate to be the half $(x - z)$ plane for $x \geq 0$. Because the flow for $y < 0$ and $y > 0$ is symmetric, we consider only the half space above the plate. The formulation of this boundary layer flow was presented in detail in Sect. 3.5. The Blasius equation must be solved numerically, but is amenable to standard[2] methods. Like the plane Poiseuille flow profile, the Blasius boundary layer has no inviscid type instability, its neutral stability curve resembling Fig. 7.7a. The sharp jump in the velocity profile of the Kelvin–Helmholtz flow is not a solution to the viscous equations of motion. In a viscous fluid, a flow profile initially resembling a Kelvin–Helmholtz jump would rapidly develop into a pair of matched boundary layers, in other words a mixing layer flow of the *error function* form. The Prandtl boundary layer equations can also be solved for flows with a weak imposed transverse flow, v, sometimes called a suction flow, for v < 0 or blowing one, when v > 0. Driven by industrial, primarily aerodynamic, applications, the stability of viscous boundary layer flows with suction has received much attention. While boundary layer flow over a wing, for example, may be stabilized by suction, it is almost never an economically practical method to control flow stability. In astrophysical problems, on the other hand, such as the flow of an accretion disk, suction is a reasonable approximation of the accretion component of the disk flow. The stability properties of these flows remains the subject of ongoing research.

7.3.3.3 Convective and Absolute Instability

Boundary layer instability, like the base flow itself, develops downstream. In this sense, a spatial, rather than temporal, linear stability analysis is warranted. For

---

[2]The Blasius functions, as the solutions are called, and their generalization to flow over wedges, known as Falkner–Skan solutions, are built-in functions in MATLAB, for example.

convenience, we temporarily adopt a standard wave notation where disturbances take the form $\propto e^{i(kx-\omega t)}$, where k is the wavenumber and ω the frequency. In spatial stability analysis, one normally fixes a purely real value of the perturbation frequency ω, and solves for a generally complex disturbance wavenumber, k. Spatially developing instabilities must be carefully examined for their growth properties. Two distinct forms of growth are possible: *convective* instability and *absolute* instability. A convective spatial instability experiences growth as it is carried, or convected, downstream by the base flow. Its growth may be catastrophic, as in the breakup of a wind driven liquid layer, exemplified fabulously by sea spray or in industrial atomization. But the influence of a convective instability does not extend back upstream to its point of first development. Absolute instability, on the other hand, leads to eventual growth of disturbances at all spatial values. To conclude, we clarify that in many cases, spatial instability can be treated using temporal theory together with the previously mentioned *Gaster's transformation*. Valid primarily near points of neutral stability, Gaster's transformation permits temporal stability results to be transformed to spatial stability results. Gaster showed that the imaginary component of k, for a spatial instability, is related to the imaginary component of ω, from a temporal stability result, by the disturbance group velocity, or:

$$\frac{\Im(\omega)}{\Im(k)} = -\frac{\partial \Re(\omega)}{\partial \Re(k)}. \tag{7.52}$$

The real components of the wavenumber, k, and frequency, ω, on the other hand, are approximately equal for spatial and temporal unstable modes, which makes the above right-hand side unambiguous. A diagnosis of the convective versus absolute character and of spatial instability techniques in general is beyond the scope of this book, but can be found in other sources, e.g., exercise 8.34 in reference [3] in the *Bibliographical Notes*.

7.4   Buoyant Instabilities

A wide range of instabilities are driven by forces of buoyancy, broadly defined. A commonly imagined setting is of a body of volume $\mathcal{V}$ completely immersed in stationary fluid; a uniform gravitational field of magnitude g points downward. As translated, the time immemorial principle of Archimedes says ... *any object, wholly or partially immersed in a fluid, is buoyed up by a force equal to the weight of the fluid displaced by the object*.[3] Suppose that the body has a uniform density ρ_b and the density of the fluid in which it is immersed is given by ρ_f. Thus the weight of the fluid displaced by the body is $\mathcal{V}\rho_f g$ and so the "net" weight of the body becomes

---

[3] Found in the work *On floating bodies* written by Archimedes of Syracuse circa 250 B.C.

$\mathcal{V}(\rho_b - \rho_f)g$. Thus if a lesser density body enters an ambient medium, that is one of higher density, the body continues to rise.

7.4.1   Rayleigh–Taylor Instability

Archimedean instability is also manifest in superposed layers of fluid, where it is known as the *Rayleigh–Taylor instability*. Consider two fluids, each with uniform densities $\rho_\pm$, arranged on top of one another in a plane-parallel configuration, in which a constant gravitational acceleration points downward, in the $-z$ direction. The two fluids are separated by an interface at $z = 0$; the case of a sharp interface is not required, and we will discuss other cases later. To start, consider that each layer has infinite vertical extension in the direction away from the interface and the domain is horizontally unbounded. Repeating the calculation detailed in Sect. 4.2 for this arrangement, assuming that the shape of the perturbed interface is given by $\eta = \hat{\eta} e^{ikx - i\omega t} + \text{c.c.}$, leads to the dispersion relation, expressed here for frequency squared $\omega^2(k)$:

$$\omega^2 = -gk\frac{\rho_+ - \rho_-}{\rho_+ + \rho_-}. \tag{7.53}$$

The familiar result for surface gravity waves examined in Sect. 4.2 is recovered when the overlying fluid is removed ($\rho_+ = 0$). When the overlying fluid is denser ("heavier," so to speak) than the fluid below ($\rho_+ > \rho_-$), ω^2 gives a pair of exponential modes, one of which grows with time and represents the instability. The appearance of a damped companion mode is a consequence of this problem's invariance to time reversal, $t \mapsto -t$. When dissipative effects such as viscosity are included, this invariance is lost.

At moderate growth rates, the Rayleigh–Taylor instability leads to a pattern of finger-like plumes inter-penetrating and eventually mixing the two layers. At large growth rates, the process becomes catastrophic and disordered, in other words: the two fluids become well mixed. It is therefore of interest to know the wavenumber of the most unstable, or fastest growing, mode. But (7.53) does not provide this. We now twice witnessed evidence that the problem defined above is not well posed. In the absence of surface tension, viscosity, or any explicit finite length scale, whether the thickness of the density jump or the size of a box, this instability also lacks a characteristic length scale and, consequently, lacks a timescale as well. Another consequence of the ill-posedness of the problem is the unbounded growth rate in (7.53) as $k \to \infty$. In terrestrial applications, surface tension is normally included, in which case the dispersion relation becomes

$$\omega^2(k) = \frac{k^3 \gamma_{\text{st}} - gk(\rho_+ - \rho_-)}{\rho_+ + \rho_-}, \tag{7.54}$$

where γ_{st} is the constant coefficient of surface tension. In Chap. 4, gravo-capillary waves were discussed in considerable depth as expressed in Eq. (4.62) and Fig. 4.15. However, there we were interested in surface gravity waves on a horizontally unbounded, finite depth layer of constant density liquid and the effect of surface tension on them. Here the setting is very different, and surface tension removes the pathologically unbounded growth rate for large k by defining a cutoff $k_c = 1/L_c$ where

$$L_c = \left(\frac{\gamma_{st}}{g|\rho_+ - \rho_-|} \right), \tag{7.55}$$

is the *capillary length*. For air–water $L_c \approx 2.5$ mm and on scales smaller than this the interface is stable, explaining how water can remain suspended at the top of a drinking straw but not as a layer adhered to the underside of a ceiling. In cases where fluids are miscible, including geophysical or astrophysical gases, a cutoff length must originate from a different physical source, such as the scale-height of density variation, finite layer thickness, or dissipation. A similar problem arises in the Kelvin–Helmholtz instability (Sect. 7.3.2.1); here, too, the introduction of a physical length scale eliminates the ill-posedness.

Any acceleration is equivalent to the gravitational one in the formulation of the Rayleigh–Taylor instability, as demonstrated by Taylor, making this instability relevant to supernovae, where dense material is rapidly accelerated outward in the direction of less dense gas. The structure of the instability becomes imprinted onto the hot gas of the young supernova remnant, forming visually spectacular nebulae such as that of the Crab (Fig. 7.8). *Buoyancy oscillations* are another manifestation of the Archimedes principle. Here we consider the response of the ambient fluid on vertically displaced fluid elements within a vertically stratified atmosphere. Consider an atmosphere in a constant gravitational field g per unit mass, directed in the $-z$ direction, or downward. The atmosphere has an equilibrium pressure and density profile given as $P = P(z)$ and $\rho = \rho_a(z)$, where the subscript a denotes "ambient." Equilibrium then dictates that

$$\frac{dP}{dz} = -\rho_a g. \tag{7.56}$$

Consider a small fluid element of volume $\delta\mathcal{V}$ at equilibrium height z. Assuming that this fluid element adjusts to its environment only by isentropic processes means that if the fluid element changes its density or pressure, its entropy must not change. Physically, the isentropic response occurs because the dynamic equilibration time with the surroundings, via sound waves, is much shorter than the time for any heat exchange. For this illustration let us say that the internal equation of state in the element is adiabatic, $P = K\rho^\gamma$, where γ is the ratio of specific heats. If this fluid element is displaced vertically to a new position $z + \xi$, where ξ is small in

Fig. 7.8 The Crab nebula, a young supernova remnant for which the finger-like structures of Rayleigh–Taylor instability are visible in the glowing gas. *Public Domain, Author: J. Hester and L. Loll, ASU. Courtesy of NASA and ESA via STScI*

the linearization sense, it will encounter a pressure and density different than the pressure and density at z. At its new position the element's density will be ρ, say, meaning that a net vertical force may act on it. Using the Archimedes principle and Newton's second law gives for its acceleration

$$\ddot{\xi} = g(\rho_a - \rho). \tag{7.57}$$

The ambient density ρ_a and the element density ρ at the new position can be approximated by Taylor series around their values at z. Thus

$$\rho_a = \rho_a(z) + \left(\frac{\partial \rho}{\partial P}\right)_s [P - P(z)] + \left(\frac{\partial \rho}{\partial s}\right)_P [s - s(z)] + \text{HOT}$$

$$\rho = \rho(z) + \left(\frac{\partial \rho}{\partial s}\right)_P [P - P(z)] + \text{HOT}. \tag{7.58}$$

Noticing that $P - P(z)$ appears in both of the above expressions, their subtraction gives

$$\rho_a - \rho = \left(\frac{\partial \rho}{\partial s}\right)_P \frac{ds}{dz} \xi, \tag{7.59}$$

where we have put $\rho_a(z) = \rho(z)$ since the initial density of the element is equivalent to the ambient value, and approximated $s - s(z) \approx ds/dz\,\xi$. Now we write the equation of motion (7.57) of the element, using Eq. (7.59) as

$$\ddot{\xi} + N^2 \xi = 0,$$

where the quantity N, defined by the following equation

$$N^2 \equiv -\frac{g}{\rho^2} \left(\frac{\partial \rho}{\partial s}\right)_P \frac{ds}{dz} \tag{7.60}$$

is again, as introduced in Chap. 5, e.g., in Eq. (5.30) in geophysical waves context, the *Brunt–Väisälä* or *buoyancy* frequency. It is elementary to see that if $N^2 < 0$, the fluid element will accelerate, exhibiting instability, while the converse is true if $N^2 > 0$, in which case the fluid element executes vertical oscillations at frequency N.

It is possible to put N^2 in terms of more conveniently observable quantities and connect the above finding to the well-known, from the study of stellar structure (see, e.g., reference [5] in the *Bibliographical Notes* to this chapter), *Schwarzschild criterion*. First, it is evident that

$$\left(\frac{\partial \rho}{\partial s}\right)_P = \left(\frac{\partial \rho}{\partial T}\right)_P \left(\frac{\partial T}{\partial h}\right)_P \left(\frac{\partial h}{\partial s}\right)_P, \tag{7.61}$$

where h is the specific enthalpy. Well-known thermodynamic relations give $c_P = (\partial h/\partial T)_P$, where c_P is the specific heat per unit mass at constant pressure, and $T = (\partial h/\partial s)_P$, thus

$$\left(\frac{\partial \rho}{\partial s}\right)_P = \frac{T}{c_P} \left(\frac{\partial \rho}{\partial T}\right)_P.$$

The vertical gradient of the entropy, $s(T, P)$, can be written as

$$\frac{ds}{dz} = \left(\frac{\partial s}{\partial T}\right)_P \frac{dT}{dz} + \left(\frac{\partial s}{\partial P}\right)_T \frac{dP}{dz}, \tag{7.62}$$

where

$$\left(\frac{\partial s}{\partial T}\right)_P = \left(\frac{\partial s}{\partial h}\right)_P \left(\frac{\partial h}{\partial T}\right)_P = \frac{c_P}{T},$$

as before. Now $dP/dz = -\rho_a g$ from (7.56) and we also have the thermodynamic Maxwell relation $(\partial s/\partial P) = (\partial \rho/\partial T)_P/\rho^2$, thus

$$\frac{ds}{dz} = \frac{c_P}{T}\frac{dT}{dz} - \frac{g}{\rho}\left(\frac{\partial \rho}{\partial T}\right)_P$$

and from Eq. (7.60) we conclude that

$$N^2 = -\frac{g}{\rho}\left(\frac{\partial \rho}{\partial T}\right)_P\left[\frac{dT}{dz} - \frac{gT}{\rho c_P}\left(\frac{\partial \rho}{\partial T}\right)_P\right] > 0 \qquad (7.63)$$

is sufficient for stability. The thermal expansion coefficient

$$\frac{1}{\upsilon}\left(\frac{\partial \upsilon}{\partial T}\right)_P = -\frac{1}{\rho}\left(\frac{\partial \rho}{\partial T}\right)_P,$$

where $\upsilon \equiv 1/\rho$, the specific volume, is positive for a great majority of fluids and so $N^2 > 0$ if the expression in square brackets is positive. For example, in *ideal gas* $\upsilon = \mathscr{R}T/(P\mu)$, with μ the mean molecular weight, and hence the condition for stability becomes:

$$N^2 = \frac{g}{T}\left(\frac{dT}{dz} + \frac{g}{c_P}\right) > 0 \quad \text{or} \quad \frac{dT}{dz} > -\frac{g}{c_P}.$$

This stability condition is sometimes called a *sub-adiatic temperature gradient*, in particular in the Schwarzschild criterion for convective stability in stellar structure theory. In any case, the entropy gradient is the crucial quantity in determining stability against buoyant motions and so it forms the most lucid statement of the Schwarzschild criterion. The Brunt–Väisälä frequency N has been defined most naturally in (7.63) in terms of the entropy gradient, thus the above criterion can be re-stated in the following way: An atmosphere in which entropy increases everywhere with height $(ds/dz > 0)$ is linearly *stable* to buoyancy motions, while a necessary condition for *instability*, not proved here, is that

$$\frac{ds}{dz} < 0 \qquad (7.64)$$

somewhere in the atmosphere. Put another way, internal gravity waves (Sect. 4.4.2) are the flip-side of unstable buoyancy oscillations. The study of buoyant instability is simplified by filtering out oscillations of different physical character, such as acoustic and Lamb modes, generally achieved by assuming a priori an incompressible equation of state together with some prescription relating density fluctuations to those of entropy (or temperature) like the Boussinesq approximation. This is the basis of study of Rayleigh–Bénard convection detailed next.

7.4.2 Rayleigh–Bénard Convection

In school, every young student learns about convection as one of nature's most effective methods to transport heat. Without convection, many natural phenomena would not exist, not least of which the stable equilibrium of the Sun, and many industrial processes would be impossible. In an atmosphere, Rayleigh–Bénard convection is the manifestation of what would otherwise be unstable gravity waves, but accounting for the effects of both viscosity and thermal conductivity. The actual problem examined experimentally by H. Bénard (1900) and theoretically by Lord Rayleigh (1916) was motivated by the behavior of slabs of wax heated from below. As a result, their formulation is very specific in construction and is considered to be of limited applicability to circumstances found in nature. In particular, to make the analysis amenable, the Rayleigh–Bénard problem is constructed so as to filter out sound waves, by assuming an incompressibility, yet still retaining the influence of buoyancy, by assuming the Boussinesq approximation.

The basic physical setting of Rayleigh–Bénard (RB) convection is the following. A plane parallel layer of incompressible Boussinesq fluid in constant gravitational field is held between horizontal, parallel, rigid, and impenetrable boundaries. The temperature values on the two boundaries are held fixed by some external means, with higher temperature on the lower surface. The gravitation force per unit mass g points downward. The fluid layer would thermally equilibrate, exhibiting a linear increase of temperature with depth and this might be perceived as the base state. However, a vertically displaced fluid element finds itself warmer than the ambient temperature, so that in the absence of heat diffusion and owing to the Boussinesq approximation, the fluid element will be less dense than its immediate ambient environment, making it buoyantly unstable. In the absence of viscosity the displaced element should continue to rise. However if the element is allowed to diffuse its heat, then it will lose it to its cooler environment, causing its density to increase. If the heat diffusion is efficient, then the rate of increase in the density of the fluid element may be faster than the increase of ambient density, in which case buoyant instability can be suppressed. Buoyant instability may also be suppressed by the viscosity of the fluid since a large enough viscosity will impede, by frictional work, the rate at which a fluid element rises. By taking longer to rise, the fluid element has more time to diffuse its heat and, correspondingly, adjust its density to become stable against buoyant instability.

The study of Rayleigh–Bénard convection is perhaps the most widely examined fluid phenomenon next to the Taylor–Couette problem (Sect. 7.4.3). The success of

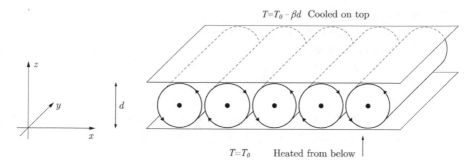

Fig. 7.9 Diagram of simple Rayleigh–Bénard convection problem. The temperatures are set to T_0 and $T_0 - \beta d$ where d is the separation of the layers and β is the vertical gradient of basic state temperature

the Rayleigh–Bénard (RB) model is in part due to the elegant way it represents the essential physics, but, without a doubt, also to the tractability of the mathematical statement of the problem. RB convection has thus led to advances in applied mathematics such as the universality of iterated quadratic maps and to the enormous interest in deterministic chaos with its various applications. More generally, RB convection is an excellent analog system with which to study the transition of many different physical systems from stability to instability and, possibly, to turbulence.

7.4.2.1 Formulation and Boundary Conditions

Rayleigh–Bénard convection refers to the problem of buoyantly driven flow between two horizontal plates maintained at two different temperatures, T_0 and $T_1 = T_0 - \beta d$, and separated by a distance d (see Fig. 7.9). The equation of state for the fluid is assumed to be Boussinesq, such that density fluctuations are directly related to thermal (entropy) fluctuations via the relationship

$$\rho = \rho_0 \left[1 - \beta_T (T_0 - T) \right], \tag{7.65}$$

where β_T is the coefficient of thermal expansion at constant pressure (see Chap. 4). The equations of motion of this problem are

$$\frac{D\mathbf{u}}{Dt} = -\nabla \left(\Pi + gz \right) + \hat{\mathbf{z}} g \beta_T (T - T_0) + \nu \nabla^2 \mathbf{u} \tag{7.66}$$

$$0 = \frac{\partial u}{\partial x} + \frac{\partial v}{\partial y} + \frac{\partial w}{\partial z}, \tag{7.67}$$

where we have defined $\Pi \equiv P/\rho_0$ as an effective enthalpy and where v is the kinematic coefficient of viscosity. These equations are supplemented by an appropriate evolution equation for the temperature variable including diffusion:

$$\frac{DT}{dt} = \kappa \nabla^2 T,$$

where κ is the coefficient of thermal conductivity. In the static steady state, denoted by over-lines, we reasonably assume thermal equilibrium, such that the temperature follows a linear distribution

$$\overline{T} = T_0 - \beta z,$$

where $\beta \equiv (T_0 - T_1)/d$ is the vertical gradient of the basic state temperature and $0 \leq z \leq d$. For a static fluid it follows from the vertical hydrostatic equilibrium condition that the steady enthalpy must be given by

$$\overline{\Pi} = \Pi_0 - gz - \tfrac{1}{2} g \beta_T \beta z^2.$$

Next we will introduce perturbations, linearize the resulting equations of motion, and examine normal mode solutions, subject to specified boundary conditions. The relevant variables are thus expressed as

$$\mathbf{u} = \mathbf{u}'(x,y,z,t), \qquad T = T_0 - \beta z + \theta'(x,y,z,t), \qquad \Pi = \overline{\Pi} + \Pi'(x,y,z,t),$$

where the primed quantities are small perturbations (note that θ' denotes the temperature perturbation). The linearized equations for the perturbations become

$$\frac{\partial \mathbf{u}'}{\partial t} = -\nabla \Pi' + \beta_T \theta' \hat{\mathbf{z}} + v \nabla^2 \mathbf{u}', \tag{7.68}$$

$$0 = \frac{\partial u'}{\partial x} + \frac{\partial v'}{\partial y} + \frac{\partial w'}{\partial z}, \tag{7.69}$$

and

$$\frac{\partial \theta'}{\partial t} = \beta w' + \kappa \nabla^2 \theta'. \tag{7.70}$$

Inspection of the perturbation equations suggests elimination of Π' from the momentum perturbation equation. This is achieved by operating with $\nabla \times$ on Eq. (7.68) to form the equation for the perturbed vorticity $\omega' \equiv \nabla \times \mathbf{u}'$

$$\frac{\partial \omega'}{\partial t} = g \beta_T \nabla \times (\theta' \hat{\mathbf{z}}) + v \nabla^2 \omega' \tag{7.71}$$

Operating again with $\nabla\times$, now on Eq. (7.71), making use of Eq. (7.69) and writing only the vertical component of the resulting equation gives

$$\frac{\partial \nabla^2 w'}{\partial t} = g\alpha\beta \left(\frac{\partial^2 \theta'}{\partial x^2} + \frac{\partial^2 \theta'}{\partial y^2} \right) + \nabla^4 w'. \tag{7.72}$$

Remember w' is the vertical *velocity* perturbation, while the vertical component of the *vorticity* perturbation is given by $\zeta' \equiv \hat{\mathbf{z}} \cdot \boldsymbol{\omega}'$. One may extract the vertical component of the vorticity perturbation equation from Eq. (7.71), finding:

$$\frac{\partial \zeta'}{\partial t} = \nabla^2 \zeta'. \tag{7.73}$$

Before proceeding it is advantageous to nondimensionalize the perturbation equations (7.70), (7.72–7.73). We begin by scaling lengths by d and time by the thermal conduction timescale d^2/κ. The relevant velocities are scaled by κ/d. Thermal fluctuations can be scaled by the steady state temperature difference between the upper and lower plates, βd. In order to remain consistent with these scalings, the enthalpy is measured in units of κ^2/d^2. In summary, we perform:

$$\mathbf{x} \mapsto d\,\mathbf{x}, \qquad t \mapsto (d^2/\kappa)t, \qquad \mathbf{u}' \mapsto (\kappa/d)\mathbf{u}', \qquad \theta' \mapsto (\beta d)\theta'$$

so that the quantities $\mathbf{x}, t, \mathbf{u}', \theta'$ are to be *nondimensional* representations of the corresponding spatial coordinates, temporal coordinate, velocity, and temperature perturbations, respectively.

From now on we shall use the same symbols to represent both the dimensional and nondimensional variable primarily to avoid clutter. The perturbation equations thus become

$$\frac{1}{\text{Pr}} \frac{\partial \zeta'}{\partial t} = \nabla^2 \zeta' \tag{7.74}$$

$$\frac{1}{\text{Pr}} \frac{\partial \nabla^2 w'}{\partial t} = \text{Ra} \left(\frac{\partial^2 \theta'}{\partial x^2} + \frac{\partial^2 \theta'}{\partial y^2} \right) + \nabla^4 w' \tag{7.75}$$

$$\frac{\partial \theta'}{\partial t} = w' + \nabla^2 \theta' \tag{7.76}$$

where we identify the Prandtl number Pr, defined in Sect. 1.6,

$$\text{Pr} \equiv \frac{\nu}{\kappa}$$

and add one more important nondimensional number, bearing Rayleigh's name Ra,

$$\text{Ra} \equiv \frac{g\alpha\beta d^4}{\kappa\nu}. \tag{7.77}$$

The Rayleigh number Ra is the primary parameter determining the onset of linear instability in this problem. It also happens to be the universal number by which buoyant convection is characterized in almost all applications. As we saw, the Prandtl number Pr measures the relative timescales between the action of viscosity and conduction. In most terrestrial applications Pr tends to be of $\mathcal{O}(1)$, but in many astrophysically relevant settings, e.g., in the Sun or in accretion disks, Pr can be extremely small, 10^{-5} or less. On the other end of the spectrum, Pr can be very large in some geophysical applications, such as in ice convection, where the viscosity can vary by several orders of magnitudes due to its dependence on temperature.

To simplify the mathematical manipulations, we apply the Helmholtz decomposition theorem[4] to the perturbation velocity, leading to a stream function, ψ', and a velocity potential, ϕ', so that the *horizontal* components of the velocity may be written as the sum of the appropriate derivatives:

$$u' = \frac{\partial \phi'}{\partial x} - \frac{\partial \psi'}{\partial y}, \qquad v' = \frac{\partial \phi'}{\partial y} + \frac{\partial \psi'}{\partial x}. \tag{7.78}$$

Note that both ϕ' and ψ' are functions of the vertical coordinate z. By taking the horizontal divergence of the above forms and making use of the incompressibility condition of Eq. (7.69), we find:

$$\frac{\partial w'}{\partial z} = -\left(\frac{\partial^2 \phi'}{\partial x^2} + \frac{\partial^2 \phi'}{\partial y^2}\right), \qquad \zeta' = \left(\frac{\partial^2 \psi'}{\partial x^2} + \frac{\partial^2 \psi'}{\partial y^2}\right). \tag{7.79}$$

The divergence of the velocity components in Eq. (7.78) and the relations in Eq. (7.79) give:

$$\left(\frac{\partial^2 u'}{\partial x^2} + \frac{\partial^2 u'}{\partial y^2}\right) = -\frac{\partial^2 w'}{\partial x \partial z} + \frac{\partial \zeta'}{\partial y}, \qquad \left(\frac{\partial^2 v'}{\partial x^2} + \frac{\partial^2 v'}{\partial y^2}\right) = -\frac{\partial^2 w'}{\partial y \partial z} - \frac{\partial \zeta'}{\partial x}. \tag{7.80}$$

The utility of this equation form should be clear: because the system of equations (7.74–7.73) is expressed in terms of the perturbation vertical vorticity and perturbation vertical velocity, one can reconstruct the perturbation horizontal velocity fields from knowledge of the former quantities. This formulation is especially suited to problems involving horizontally periodic or horizontally open boundaries.

In contrast, in treating the confining plates as rigid walls which are no-slip for non-zero values of the viscosity, the simplest boundary conditions are that the velocities should match that of the plate at the location of the plate. This amounts to imposing zero perturbation velocity at the boundaries, for non-moving plates, viz.

$$u' = v' = w' = 0, \qquad \text{at} \quad z = 0, 1. \tag{7.81}$$

---

[4]Allowing to decompose any sufficiently smooth vector field to the sum of a curl ($\nabla \times$) free component plus a divergence ($\nabla \cdot$) free component

These are the *rigid boundaries* originally introduced in Sect. 3.3.1. Another set of boundary conditions is referred to as *stress-free*; these conditions were briefly mentioned in the discussion of TC flow in Chap. 5. As the name implies, stress-free conditions would require the three components of stress on any boundary to be zero, as described in Sect. 1.4.2. In the context of classical Rayleigh–Bénard convection calculations, we follow the simplifying rationale employed by Rayleigh by assuming that the bounding plates are fixed but that the two *tangential stresses* are zero on $z = 0, 1$. Referring to Sect. 1.4.2, the two tangential stresses on the plates are symbolically given by σ_{31}, σ_{32}, or equivalently by σ_{zx}, σ_{zy}. Setting these to zero means

$$\frac{\partial u'}{\partial z} + \frac{\partial w'}{\partial x} = 0, \qquad \frac{\partial v'}{\partial z} + \frac{\partial w'}{\partial y} = 0, \qquad \text{at} \quad z = 0, 1, \tag{7.82}$$

with the requirement of no normal flow $w' = 0$ there, as well. The payoff from these boundary conditions is mathematical tractability: it is with these boundary conditions that analytical results may be found. Separately, we have to supply boundary conditions for the temperature. The physical assumption that boundaries are perfect thermal conductors translates to the temperature fluctuations being zero on the surfaces, or:

$$\theta' = 0, \qquad \text{at} \quad z = 0, 1, \tag{7.83}$$

referred to as either *fixed-temperature* or *conducting* boundary conditions.

One may instead impose the requirement that vertical heat flux, dimensionally given by $-\hat{z} \cdot \kappa \nabla T$, be held fixed on both boundaries of the system by an external agent working on a timescale that is much faster than any other thermal timescale characterizing the system. These conditions are often times referred to, depending upon the context, either as *fixed-flux* or *insulating* boundary conditions. Such conditions represent the behavior of a fluid layer with an internal heat source whose total thermal content is in some statistically steady state. In practice, fixed-flux conditions are a theoretical construction that allows for some mathematical expediency. Another result is that the normal gradient of the thermal fluctuations should be zero on the surface, or

$$\frac{\partial \theta'}{\partial z} = 0, \qquad \text{at} \quad z = 0, 1. \tag{7.84}$$

7.4.2.2 Normal Mode Analysis of Rayleigh–Bénard Convection

We now analyze Eqs. (7.74–7.76) assuming normal mode disturbances of the form

$$w' = W(z)e^{ik_x x + ik_y y}e^{pt} + \text{c.c.}$$
$$\theta' = \Theta(z)e^{ik_x x + ik_y y}e^{pt} + \text{c.c.}$$
$$\zeta' = Z(z)e^{ik_x x + ik_y y}e^{pt} + \text{c.c.} \tag{7.85}$$

Inserting these into (7.74–7.76) we find

$$\frac{1}{\text{Pr}}p\left(\frac{d^2}{dz^2}-k^2\right)W = -\text{Ra}k^2\Theta + \left(\frac{d^2}{dz^2}-k^2\right)^2 W, \qquad (7.86)$$

$$p\Theta = W + \left(\frac{d^2}{dz^2}-k^2\right)\Theta, \qquad (7.87)$$

$$\frac{1}{\text{Pr}}pZ = \left(\frac{d^2}{dz^2}-k^2\right)Z, \qquad (7.88)$$

where $k^2 \equiv k_x^2 + k_y^2$. Inspection of the above equations immediately indicates that the perturbation vertical vorticity decouples from the perturbation vertical velocity and temperature. It should be kept in mind that the normal-mode form of the horizontal velocities

$$u' = U(z)e^{ik_x x + ik_y y}e^{pt} + \text{c.c.}, \qquad v' = V(z)e^{ik_x x + ik_y y}e^{pt} + \text{c.c.}, \qquad (7.89)$$

taken together with the incompressibility condition leads to

$$\frac{dW}{dz} + ik_x U + ik_y V = 0. \qquad (7.90)$$

Given that u' and v' are related to w' and ζ', as found in Eq. (7.80), it follows that the vertical structure functions U and V relate to W and Z according to

$$-k^2 U = -ik_x\frac{dW}{dz} + ik_y Z, \qquad -k^2 V = -ik_y\frac{dW}{dz} - ik_x Z. \qquad (7.91)$$

Combining Eqs. (7.86–7.87) produces an equation for the quantity W, or equivalently for Θ, that is sixth order in z. After some rearrangement, the decoupled equation reads:

$$\left(\frac{d^2}{dz^2}-k^2\right)\left(\frac{d^2}{dz^2}-k^2-p\right)\left(\frac{d^2}{dz^2}-k^2-\frac{p}{\text{Pr}}\right)W = -\text{Ra}k^2 W. \qquad (7.92)$$

The nature of the solutions to the set (7.86–7.87), or equivalently of Eq. (7.92), will depend on six boundary conditions. It is possible, using Eq. (7.91), to show that the boundary conditions appropriate for stress-free conditions at both bounding surfaces are

$$\frac{dZ}{dz} = \frac{d^2W}{dz^2} = 0, \qquad \text{at} \quad z = 0, 1. \qquad (7.93)$$

Other boundary conditions for the normal modes are left as an exercise for the reader.

As we alluded to earlier, only the case of stress-free boundaries on both surfaces admits analytically tractable solutions, often identified as "free-free" conditions in the literature. In this case, the combination of the boundary conditions implies that

$$W = \frac{d^2W}{dz^2} = \frac{d^4W}{dz^4} = 0, \qquad z = 0, 1. \tag{7.94}$$

Examination of Eq. (7.92) shows that the operators appearing in it are all even powered applications of d/dz. Thus the parity of any solution substituted into that equation is preserved in each of the terms appearing therein. With the boundary conditions of Eq. (7.94) in mind, we see that $W = W_n(z) = \sin n\pi z$, where n is any natural number and would be a solution of (7.92) which naturally satisfies the stress-free boundary conditions as long as

$$\left(n^2\pi^2 + k^2\right)\left(n^2\pi^2 + k^2 + p\right)\left(n^2\pi^2 + k^2 + \frac{p}{\mathrm{Pr}}\right) - \mathrm{Ra}\,k^2 = 0. \tag{7.95}$$

The index n measures the number of interior nodes minus 1 in the vertical direction of the velocity structure function W. Borrowing from the language of vibrations, we shall refer to the $n = 1$ mode as the *fundamental* and all of the subsequent modes as *overtones*. The above displayed equation is in a quadratic form and is one of the rare instances in which an analytic expression for the temporal behavior can be written down without appeal to numerical calculations:

$$p = p(k, n, \mathrm{Pr}, \mathrm{Ra}) = -\tfrac{1}{2}(1 + \mathrm{Pr})\left(n^2\pi^2 + k^2\right) \pm \sqrt{\frac{1}{4}(1 + \mathrm{Pr})^2\left(n^2\pi^2 + k^2\right)^2 + \frac{\mathrm{Ra}\,\mathrm{Pr}\,k^2}{\left(n^2\pi^2 + k^2\right)}}$$

$$\tag{7.96}$$

where we have indicated the eigenvalue p as a function of all of the parameters characterizing the perturbation and the problem. A short examination shows that transition to exponentially growing modes, i.e., $\Re(p) > 0$, occurs for $\Im(p) = 0$. This is an instance of *exchange of stabilities*. If one examines (7.96) it is possible to conclude that the transition into instability happens as Ra is increased. There is a critical value of Ra, referred to as $\mathrm{Ra_c}$, in which there is exactly one mode at a specific wavenumber k, referred to as k_c, that neither grows nor decays (it is neutral), while all other modes are decaying (are asymptotically stable). Once Ra exceeds this *critical* Rayleigh number there appears a band of wavenumbers centered on the critical value k_c which are all unstable, while the rest, residing outside of this range, are stable.

Returning to the characteristic polynomial equation (7.95), we set $p = 0$ there and ask what values of Ra correspond to the marginal condition of zero growth rate. This is an expression for the *marginal* Rayleigh number Ra_m

$$\mathrm{Ra}_m(n, k) = \frac{\left(n^2\pi^2 + k^2\right)^3}{k^2}.$$

Table 7.2 Critical Rayleigh number for various models. Note that in all cases Ra_c occur for the fundamental mode ($n = 1$). "Free-free" indicates stress-free boundary conditions on both boundaries while "rigid-rigid" is similarly understood. "Rigid-free" indicates rigid conditions at $z = 0$ and stress-free at $z = 1$. "Stellar" conditions indicate fixed-flux at the base of the layer and fixed-temperature conditions at the top

Boundary conditions		Critical parameters	
Thermal	Kinematic	Ra_c	k_c
Fixed-temperature	Rigid-rigid	1708	3.117
	Free-free	657.5	2.221
	Rigid-free	1101	2.682
Fixed-flux	Rigid-rigid	720.0	0.0
	Free-free	120.0	0.0
	Rigid-free	320.0	0.0
"Stellar"	Rigid-rigid	1565	2.535
	Free-free	384.7	1.755
	Rigid-free	816.7	2.212

Given k and n, it is the value of Ra for which at least one mode is neither growing nor decaying. We note immediately two things. First, that the marginal Ra_m does not depend upon Pr and second, that for a given value of n the marginal curve $Ra_m(n, k)$ has a minimum value for some critical value of $k = k_c$ corresponding to a minimum (critical) value of the Rayleigh number Ra_c (see Fig. 7.10 below). Since increasing n increases the value of Ra_m, it follows that the absolute minimum of Ra should be for the fundamental mode, $n = 1$. It therefore follows that

$$\left(\frac{\partial Ra_m}{\partial k^2}\right)_{n=1} = 0 \quad \Longrightarrow \quad k_c = \frac{\pi}{\sqrt{2}} \quad \& \quad Ra_c = \frac{27\pi^4}{4}. \qquad (7.97)$$

It is important to emphasize the interpretation of this calculation. We have essentially identified the value of Ra below which there is no possibility of instability for any value of the parameters of the system, including the overtones $n > 1$. At $Ra = Ra_c$ a single mode, $n = 1$, exhibits at $k = k_c$ zero growth rate $\Re(p) = 0$. Holding $k = k_c$ fixed and increasing Ra leads to instability. A selection of the results are presented in Table 7.2 for a variety of applied boundary conditions. In addition to the fixed-temperature boundaries we have included the Ra_c and k_c for models exhibiting fixed-flux conditions and so-called stellar conditions, referring to a model of a stellar surface characterized by fixed-flux boundary conditions on the lower boundary and fixed-temperature conditions on the top.

We note that for the fixed-flux boundary models, the critical Rayleigh numbers occurs, for the fundamental mode, in the limit of zero horizontal wavenumber. However Ra_c occurs at non-zero k values for all subsequent overtones, $n \geq 2$. This

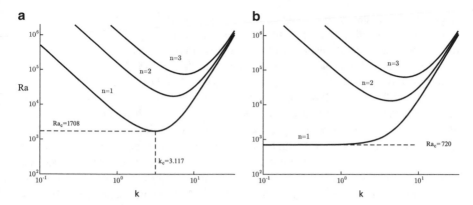

Fig. 7.10 Marginal Rayleigh numbers as a function of horizontal wavenumber k for two different thermal boundary conditions with rigid-rigid boundary conditions and for the fundamental and first two overtones: (**a**) fixed-temperature, (**b**) fixed-flux. Notice how in the fixed-flux case the minimum marginal Ra of the fundamental mode occurs asymptotically at $k = 0$. This is not the case for the overtones

is indicated in Fig. 7.10b. We shall re-examine some of these features in Chap. 8, performing asymptotic analysis of the fixed-flux and fixed-temperature problems, as shown in Fig. 7.10.

7.4.3 Taylor–Couette Centrifugal Instability

Centrifugal force acts analogously to buoyancy, so we examine it here in the wake of our analysis of Rayleigh–Bénard convection. Centrifugal effects are generic whenever flow is constrained to follow curved streamlines. In the laboratory, curved flow is normally constrained by walls, while in astrophysical problems gravity can achieve a similar result. Along a curved wall, for example, centrifugal instability leads to the so-called Görtler vortices while vortices resulting from an instability in a curved pipe are usually named after Dean. The scientists H. Görtler and W. R. Dean studied such flows in 1950 and the late 1930s correspondingly. In both cases they are essentially variations of the Taylor vortex flow. The flow between two coaxial cylinders, also known as Taylor–Couette (TC) flow, is, however, the best known and most widely studied example of centrifugal instability. We have already constructed the TC steady solution in Sect. 5.2.2 and discussed its instability in the introduction of this chapter. Unstable when the rotation speed of the inner cylinder is fast enough, a supercritical bifurcation leads to the appearance of a regular stack of axi-symmetric toroidal *Taylor vortices*. After stability is lost in this first bifurcation of the base flow, further increase of the inner cylinder speed ushers in a carnival of subsequent bifurcations, leading to increasingly complex flow patterns and eventually to turbulence (see Fig. 7.3). The stability of TC flow was the first case where experimental results matched the predictions of linear stability analysis.[5] It thus serves as a controlled experimental setting in which to perform detailed studies of instabilities, bifurcations, and of the transition to turbulence.

Consider a pair of coaxial infinitely long cylinders with radii r_1 and r_2 ($r_1 < r_2$) centered on the z-axis. The TC problem neglects gravity, but we assume that the flow is viscous with constant kinematic coefficient v. The traditional TC base flow is of constant density and steady. The geometry is such that the flow is purely azimuthal and takes place in the annular region between the two cylinders. Equation (5.51) of Chap. 5 is this base flow, but with a slightly different notation $r_i \mapsto R_i$. The flow is driven by rotation of the inner and outer cylinders at angular velocities Ω_1 and Ω_2, respectively. Corresponding to the steady flow is a steady pressure solution, whose radial gradient is given by Eq. (5.49). The equations of momentum balance and incompressibility in terms of the cylindrical-polar coordinates (r, φ, z) are

$$\frac{\partial u}{\partial t} + \mathbf{u} \cdot \nabla u - \frac{v^2}{r} = -\frac{\partial \Pi}{\partial r} + v \left(\nabla^2 u - \frac{2}{r} \frac{\partial v}{\partial \varphi} - \frac{u}{r^2} \right) \tag{7.98}$$

$$\frac{\partial v}{\partial t} + \mathbf{u} \cdot \nabla v + \frac{uv}{r} = -\frac{\partial \Pi}{\partial \varphi} + v \left(\nabla^2 v + \frac{2}{r} \frac{\partial u}{\partial \varphi} - \frac{v}{r^2} \right) \tag{7.99}$$

$$\frac{\partial w}{\partial t} + \mathbf{u} \cdot \nabla w = -\frac{\partial \Pi}{\partial z} + v \nabla^2 w \tag{7.100}$$

$$\frac{1}{r} \frac{\partial (ru)}{\partial r} + \frac{1}{r} \frac{\partial v}{\partial \varphi} + \frac{\partial w}{\partial z} = 0, \tag{7.101}$$

where u, v, and w are the radial, azimuthal, and vertical component of velocity, respectively, and $\Pi \equiv P/\rho_0$ is the pressure divided by the constant density. The steady solution for the azimuthal flow found in Eq. (5.51), now identified as v_0, will be recast in this section into a form more amenable to the analysis of TC flow centrifugal instability, viz.:

$$v_0(r) = \Omega_1 r_1 \left[\frac{\mu - \eta^2}{1 - \eta^2} \left(\frac{r}{r_1} \right) + \frac{1 - \mu}{1 - \eta^2} \left(\frac{r_1}{r} \right) \right], \qquad \text{with} \qquad \mu \equiv \frac{\Omega_2}{\Omega_1}, \qquad \eta \equiv \frac{r_1}{r_2}. \tag{7.102}$$

The two new nondimensional parameters μ and η in (7.102) are those most commonly used to identify points of instability in TC flows. The reader is warned not to confuse these with previously used same symbols. The condition for *stability* to axi-symmetric disturbances is

---

[5]The reader is referred to the stability texts by Drazin, Drazin and Reid, or Charru listed in the *Bibliographical Notes* to examine in detail G.I. Taylor's graphical comparison of theoretical and experimental results.

$$\mu > \eta^2, \qquad \Longleftrightarrow \qquad \frac{\Omega_2}{\Omega_1} > \frac{r_1^2}{r_2^2}. \tag{7.103}$$

This condition, often referred to as the *Rayleigh criterion* for rotating flows, will be examined in more detail in the following sections.

7.4.3.1 Stability of Inviscid TC Flows to Axi-symmetric Disturbances

We repeat here a heuristic calculation due to Rayleigh, and later von Kármán, who looked for some insight into the instability mechanism. For simplicity of the argument, viscosity is neglected and the flow is assumed axi-symmetric, i.e., $\partial/\partial\varphi = 0$. From (7.99) it follows that

$$\frac{\partial v}{\partial t} + (\mathbf{u} \cdot \nabla)v + \frac{uv}{r} = 0, \qquad \Longrightarrow \qquad \frac{dl_z}{dt} = 0, \quad \text{where } l_z \equiv rv. \tag{7.104}$$

l_z is recognized as the specific (i.e., per unit mass) angular momentum. This quantity is expected to be conserved, in the Lagrangian sense, due to absence of external torques. Kelvin's circulation theorem expresses this as well, because the circulation is just $2\pi l_z$. The following energy argument illustrates the dynamical consequences of this conservation. Consider two rings of very small rectangular cross section, labelled 1 and 2 and located at the coordinates (R_1, Z_1) and $(R_2 = R_1 + \Delta R, Z_2 = Z_1 + \Delta Z)$, respectively. Let these rings have side area $\delta_r \times \delta_z$, that is, they effectively are hoops of radial and vertical dimensions δ_r and δ_z, respectively, both very much smaller than any other scale of the system, in particular ΔR and ΔZ.

The above rings have kinetic energies per unit volume given by $\varepsilon_1 = (\rho_0/2)v(R_1)^2$ and $\varepsilon_2 = (\rho_0/2)v(R_2)^2$, respectively. In terms of their specific angular momenta, i.e.,

$$l_1 \equiv l_z(R_1) = R_1 v(R_1) \quad \text{and} \quad l_2 \equiv l_z(R_2) = R_2 v(R_2),$$

the corresponding rings' kinetic energies per unit volume

$$\varepsilon_1 = \frac{\rho_0}{2} \left(\frac{l_1}{R_1} \right)^2, \qquad \varepsilon_2 = \frac{\rho_0}{2} \left(\frac{l_2}{R_2} \right)^2,$$

so that the total energy of the rings is (the reason for the superscript "ini" will become clear shortly)

$$\varepsilon^{\text{ini}} \equiv \varepsilon_1 + \varepsilon_2 = \frac{\rho_0}{2} \left[\left(\frac{l_1}{R_1} \right)^2 + \left(\frac{l_2}{R_2} \right)^2 \right]. \tag{7.105}$$

Suppose now that the positions of the rings are switched, so that ring 1 is moved to (R_2, Z_2), while ring 2 to (R_1, Z_1). During this exchange, the angular momentum of each ring remains invariant, so that the ring originally at (R_1, Z_1) preserves its angular momentum, l_1, while occupying the new position (R_2, Z_2), and vice versa for the other ring. The kinetic energies of the two rings in their new positions, $\tilde{\varepsilon}_1$ and $\tilde{\varepsilon}_2$, are used to define the *final* total energy, i.e., after the exchange

$$\varepsilon^{\text{fin}} = \tilde{\varepsilon}_1 + \tilde{\varepsilon}_2 = \frac{\rho_0}{2}\left[\left(\frac{l_1}{R_2}\right)^2 + \left(\frac{l_2}{R_1}\right)^2\right]. \tag{7.106}$$

From an energetic point of view, if the system is in a state of larger energy in the final configuration, we may say that the system is stable in the initial configuration. The condition for stability is $\varepsilon^{\text{fin}} - \varepsilon^{\text{ini}} > 0$. When the difference in the positions of the two rings are small compared to their average position, that is, if

$$\max\left\{\frac{\Delta R}{2}, \frac{\Delta Z}{2}\right\} \ll \min\{R_1, Z_1\},$$

one can Taylor expand the energies around (R_1, Z_1) to show that stability should be guaranteed for

$$l_r^2 \frac{\rho_0}{r^3}\frac{\partial(r^2 v^2)}{\partial r} > 0, \quad \text{or} \quad l_z\frac{\partial l_z}{\partial r} = rv\frac{\partial(vr)}{\partial r} > 0. \tag{7.107}$$

If we adopt the Taylor–Couette profile (7.102) as the base flow, then the condition for stability takes the form $\mu > \eta^2$, provided $\Omega_1 > 0$. This is the same expression as found in Eq. (7.103).

7.4.3.2 Stability of Viscous TC Flows to Axi-symmetric Disturbances

We now introduce small disturbances into the equations of motion (7.98–7.101) and linearize around the steady Taylor–Couette profile (7.102). Applying the familiar substitutions

$$u \mapsto u'(r,z,t), \quad v \mapsto v_0(r) + v'(r,z,t), \quad w \mapsto w'(r,z,t), \quad \Pi \mapsto \Pi_0(r) + \Pi'(r,z,t),$$

where the steady pressure field $\Pi_0 = P_0/\rho_0$ is given in Eq. (5.49) with $\rho = \rho_0$. After linearizing, we find

$$\frac{\partial u'}{\partial t} - 2\frac{v_0}{r}v' = -\frac{\partial \Pi'}{\partial r} + v\left(\nabla^2 u' - \frac{u'}{r^2}\right), \tag{7.108}$$

$$\frac{\partial v'}{\partial t} + \frac{1}{r}\left[\frac{d(rv_0)}{dr}\right]u' = v\left(\nabla^2 v' - \frac{v'}{r^2}\right), \tag{7.109}$$

$$\frac{\partial w'}{\partial t} = -\frac{\partial \Pi'}{\partial z} + \nu\nabla^2 w', \tag{7.110}$$

$$\frac{\partial u'}{\partial r} + \frac{u'}{r} + \frac{\partial w'}{\partial z} = 0. \tag{7.111}$$

Assume now that the boundary conditions supplementing these equations are periodic in the vertical direction and rigid at the two radial boundaries, i.e., obeying $u' = v' = w' = 0$ at $r = r_1$ and $r = r_2 = \eta/r_1$. The ansatz for normal mode solutions has the form

$$\left(u', v', w', \Pi'\right)^{\mathrm{T}} = \left[U(r), V(r), iW(r), \tilde{\Pi}(r)\right]^{\mathrm{T}} e^{ikz+pt} + \text{c.c.},$$

where k is real but p can be complex. The perturbation equations simplify to a pair of equations for U and V:

$$\nu\left[\frac{d}{dr}\left(\frac{d}{dr}+\frac{1}{r}\right)-k^2\right]^2 U - p\left[\frac{d}{dr}\left(\frac{d}{dr}+\frac{1}{r}\right)-k^2\right]U = 2k^2\frac{v_0}{r}V, \tag{7.112}$$

$$\nu\left[\frac{d}{dr}\left(\frac{d}{dr}+\frac{1}{r}\right)-k^2\right]V - pV = \frac{1}{r}\left[\frac{d(rv_0)}{dr}\right]U, \tag{7.113}$$

where the boundary conditions here are $U = V = dU/dr = 0$ at $r = r_1, r = r_2$.[6] The combined set of equations cannot be solved analytically and in most cases one must take recourse to numerical computations.

Analytical treatment is possible only in some limiting cases. When the equations are considered in the *inviscid* limit, we find that they combine into a single ODE for U:

$$p^2\left[\frac{1}{r}\frac{d}{dr}\left(r\frac{d}{dr}\right)-k^2-\frac{1}{r^2}\right]U = \varpi^2 k^2 U, \qquad \varpi^2 \equiv \frac{2v_0}{r^2}\left[\frac{d(rv_0)}{dr}\right], \tag{7.114}$$

with the boundary conditions $U = 0$ at $r = r_1, r_2$. Multiplication of the above expression by rU^*, where U^* is the complex conjugate of U, and integration from $r = r_1$ to $r = r_2$ results in

$$p^2 \int_{r_1}^{r_2}\left\{\left|\frac{dU}{dr}\right|^2 + \left(rk^2+\frac{1}{r}\right)|U|^2\right\}dr = -k^2\int_{r_1}^{r_2}\varpi^2 r|U|^2 dr. \tag{7.115}$$

The integrand on the left-hand side is positive definite, while the sign of the right-hand side integrand can be either positive or negative, depending on ϖ^2. If $\varpi^2 > 0$

---

[6]Note that the condition that the vertical velocity is zero at the boundaries is equivalent to $dU/dr = 0$ at $r = r_1, r_2$ since $U = 0$ at those points too. This is a consequence of the continuity equation.

everywhere in the domain of integration, the integral will be positive and so $p^2 < 0$, indicating stability. This stability criterion can be expressed as

$$\frac{2v_0}{r^2}\frac{d(rv_0)}{dr} > 0, \qquad \text{for} \qquad r_1 < r < r_2,$$

which is simply the Rayleigh criterion for rotating flows that we obtained in the previous analysis (of an inviscid flow). It can be shown that this condition is necessary, but not sufficient, for stability of incompressible flow to axi-symmetric perturbations, although the flow may be unstable to non-axisymmetric perturbations. On the other hand, instability (i.e., $p^2 > 0$) is possible, but not guaranteed, if $\varpi^2 < 0$ somewhere in the domain of integration. Rayleigh showed that this instability is the analog of the Rayleigh–Taylor instability, where the radial pressure gradient plays the role of buoyancy. In other words, an unstable stratification of centrifugal force acts like an unstable stratification of density.

Rayleigh's criterion allows for instability for any non-zero rotation of the inner cylinder, Ω_1, in the case of a fixed outer cylinder, $\Omega_2 = 0$. In actual experiments, instability occurs at a finite value Ω_1 because viscous damping must be overcome. Taylor, motivated by careful observations of his experiments, obtained a marginal stability curve in terms of Ω_1 by assuming axi-symmetric perturbations and a narrow gap configuration $(r_2 - r_1) \ll r_1$. His calculations also assumed exchange of stabilities at marginal conditions, as in Rayleigh–Bénard convection, which allows the simplification $p = 0$ in the stability equations. We will not repeat Taylor's calculations here, except to get another expression for what is also referred to as the *Taylor number*:

$$\text{Ta} \equiv \frac{4\Omega_1^2 r_1^4}{v^2} \cdot \frac{(1-\mu)\left(1-\frac{\mu}{\eta^2}\right)}{(1-\eta^2)^2}, \tag{7.116}$$

which emerges as the single governing parameter of the system. Note that it differs from the Taylor number as defined in the problem of Rayleigh–Bénard convection in Sect. 7.4.2.1. Essentially the square of a Reynolds number based on the azimuthal velocity, Ta, as written here, plays the role of the Rayleigh number of Rayleigh–Bénard convection. In the small gap limit and for nearly equal rotational speeds $\Omega_1 \approx \Omega_2$, the Taylor–Couette problem is the exact analog of Rayleigh–Bénard convection so that the critical Taylor number for instability is also found to be $\text{Ta}_c = 1708$.

7.5   Additional Instabilities

In this section we present two examples from the many additional flow instabilities, that in our opinion bear enough importance to be explicitly considered in an FD graduate text. We begin by giving a brief summary of the breakup, due to surface tension, of a cylindrical liquid jet, studied by Lord Rayleigh, F. Savart, and J. Plateau, who, losing his eyesight, used a cello to excite disturbances in laboratory jets. We then expand our scope to the largest of scales and present the collapse, due to self-gravity, of an astrophysical gas as studied by Sir J. Jeans and others, thought to have a crucial role in star formation and probably of larger object as well.

7.5.1   *Rayleigh–Plateau–Savart Breakup of a Liquid Jet*

Everyone is familiar with the pattern of droplets that forms during the breakup of a stream of liquid falling from a slightly open tap. Already Leonardo da Vinci considered it in his notebook (1506). Leading mathematical physicists like T. Young and P.S. Laplace came close to a theoretical explanation of the phenomenon in the early nineteenth century but it was first studied experimentally by F. Savart (1833). It was not until later work of J. Plateau (1850) that the mechanism driving the breakup was determined to be surface tension. Plateau recognized that symmetric deformations would lead to reduced surface area, releasing surface energy, while Lord Rayleigh performed the first linear stability analysis in 1892, successfully predicting the wavenumber of the most unstable, or fastest growing, mode. It is this mode which sets the size of the droplets that form. In the case of a liquid–water, say, forming a laminar jet (a cylindrical thread) in air, we may justifiably neglect the presence of the surrounding air in first approximation, on the basis of the negligible air–water density ratio. We may also ignore any effect of flow non-uniformity in the interior of the liquid jet, modeling it as a cylinder of radius R, along which the fluid is moving at uniform velocity. In uniform flow, viscosity is inactive, so it too can be ignored. After a Galilean transformation to the jet frame, we can analyze a cylinder of liquid with base state $v = 0$, $P = P_a + \gamma_{st}/R$ where γ_{st} is the surface tension coefficient and P_a is the ambient air pressure. Here we have applied the well-known Young–Laplace condition, which is

$$\Delta P = \gamma_{st} \left(\frac{1}{R_1} + \frac{1}{R_2} \right), \qquad (7.117)$$

giving the pressure difference across an interface at a point, where the interface's radii of curvature are $R_i, i = 1, 2$.

To begin a linear stability analysis, we start with the incompressible Euler equations, with constant density ρ:

$$\rho \left(\frac{\partial \mathbf{u}}{\partial t} + \mathbf{u} \cdot \nabla \mathbf{u} \right) = -\nabla P$$

$$\nabla \cdot \mathbf{u} = 0. \qquad (7.118)$$

Let the general jet surface be located at $r = Y(\varphi, z)$, naturally adopting cylindrical coordinates. The outward normal to the interface, $\mathbf{n}$, satisfies $\nabla \cdot \mathbf{n} = 1/R_1 + 1/R_2$. In general the Young–Laplace condition is $P = P_a + \gamma_{st} \nabla \cdot \mathbf{n}$ on $r = Y$ and is actually the normal stress balance, here acting as a boundary condition. The second boundary condition is the kinematic condition of the moving interface, or $u = DY/Dt$ at $r = Y$. It is evident that perturbations to the velocity, pressure, and radial jet surface, respectively, can be written as $\mathbf{u}', P'$, and $Y' = Y - R$. Linearizing equation (7.118) in the perturbations on the base state, we obtain

$$\rho \frac{\partial \mathbf{u}'}{\partial t} = -\nabla P', \tag{7.119}$$

$$\nabla \cdot \mathbf{u}' = 0, \tag{7.120}$$

while the linearization of the Young–Laplace condition and the kinematic boundary constraint give

$$P' = -\gamma_{st} \left(\frac{Y'}{R^2} + \frac{\partial^2 Y'}{\partial z^2} + \frac{1}{R^2} \frac{\partial^2 Y'}{\partial \varphi^2} \right) \quad \text{on } r = R, \tag{7.121}$$

$$u' = \frac{\partial Y'}{\partial t} \quad \text{on } r = R, \tag{7.122}$$

where γ_{st} is the surface tension coefficient on the interface, whose units are [force/length]=[energy/area].

We can now introduce normal modes ansatz for $\mathbf{u}' = \hat{\mathbf{u}}(r) e^{pt + i(kz + n\varphi)}$, and similarly for P' and Y'. After some algebra, we obtain a single perturbation equation

$$\frac{d^2 \hat{P}}{dr^2} + \frac{1}{r} \frac{d\hat{P}}{dr} - (k^2 + n^2/r^2)\hat{P} = 0, \tag{7.123}$$

a modified Bessel equation of order n. Assuming bounded pressure at the interface, Eq. (7.123) has solution $\hat{P} = A I_n(kr)$ where A is an arbitrary constant and $I_n(x)$ is a Bessel function (note that the complementary Bessel function $K_n(x)$ is not a solution because it is unbounded as $r \to 0$). Using Eqs. (7.120)–(7.122) leads to an equation for the growth rate,

$$p^2 = \frac{\gamma_{st}}{\rho R^2} \frac{I_n'(kR)}{I_n kR} (1 - k^2 R^2 - n^2), \tag{7.124}$$

from which it follows that the jet is unstable only to axi-symmetric disturbances, $n = 0$; in this case, unstable disturbances must have wavenumber $kR < 1$. In other words, for instability, the wavelength $\lambda = 2\pi/k$ of axi-symmetric disturbances must be larger than the circumference of the jet $2\pi R$. For a water jet in air, the mode of fastest growth has wavelength $\lambda_{max} = 2\pi/k_{max} \approx 9R$, or about 4.5 times the jet diameter, a result which may be roughly but easily verified with the help of the kitchen sink faucet. The breakup of a liquid jet into droplets can be understood physically as a transition to a lower energy state for a system where surface energy is dominant.

7.5.2 Jeans Instability

Here we consider the role that self-gravity plays in causing a linear instability, which leads to the beginning of a gravitational collapse. This problem was originally discovered by Sir James H. Jeans and is now commonly known as the *Jeans instability*. J. H. Jeans (1877–1946) was a prominent physicist, astronomer, and mathematician, who contributed to several disciplines, but is perhaps most famous for writing in his popular book "…the Great Architect of the Universe now begins to appear as a pure mathematician." Jeans considered an inviscid compressible fluid medium (gas) having uniform density ρ and entropy s, at rest. The medium was examined under the ideal assumption that it is of *infinite* extent, in three dimensions. The governing equations are

$$\frac{\partial \mathbf{u}}{\partial t} + \mathbf{u} \cdot \nabla \mathbf{u} = -\frac{1}{\rho} \nabla P + \mathbf{b}, \tag{7.125}$$

in which the body force, per unit mass, is provided by the self gravity of the gas

$$\mathbf{b} = -\nabla \Phi,$$

where Φ is the Newtonian gravitational potential, satisfying the Poisson equation,

$$\nabla^2 \Phi = 4\pi G \rho, \tag{7.126}$$

and the continuity equation

$$\frac{\partial \rho}{\partial t} + \nabla \cdot (\rho \mathbf{u}) = 0. \tag{7.127}$$

We assume that the gas is perfect and processes that we shall be interested in are fast enough that they may be considered *adiabatic* which, when there are no sources or sinks of heat such as viscous dissipation or local gains or losses, is equivalent to *constant s*. This is conveniently expressed by the adiabatic relation $P = K(s)\rho^\gamma$, where γ is the ratio of specific heats and $K(s)$ is a constant, depending only on the value of the uniform entropy of the system.

When a static uniform base state is considered, on which a perturbation is performed, a physical inconsistency immediately arises. Expressing the variables of the problem as composed of their value in the base state plus a small perturbation, e.g., $\rho = \rho_0 + \rho'$, $P = P_0 + P'$, and $\Phi = \Phi_0 + \Phi'$, we see that the base state is simply impossible. It has to be a uniform medium extending to infinity, satisfying the unperturbed Poisson equation

$$\nabla^2 \Phi_0 = 4\pi G \rho_0,$$

resulting in a gradient of the mean gravitational potential $\nabla \Phi_0$ which cannot be identically zero. This, in turn, means that the unperturbed momentum equations (see (7.125)) cannot be satisfied. Hence for this stability analysis, the suggested base state is impossible!

Physics is a science of approximation and not mathematical rigor. In this problem, the "infinite" extension of the base state should be understood as a size *very much larger* than the particular system whose stability we wish to examine. It is certain that a scientist of Jeans's stature was well aware of that. So it seems a reasonable approximation to allow the size of the base state to be infinite in three dimensions and to ignore the Poisson equation in it and thus the gravitational potential, for the mean steady state, including it only when perturbations are introduced. In practice, this amounts to having a constant mean gravitational potential Φ_0, where the value of the constant can be chosen to be zero if desired. We have already introduced perturbations in the usual way above and to that we add the velocity $\mathbf{u} = \mathbf{u}'$, leading to the perturbation equations

$$\frac{\partial \rho'}{\partial t} + \rho_0 \nabla \cdot \mathbf{u}' = 0, \qquad \frac{\partial \mathbf{u}'}{\partial t} = -c_s^2 \frac{1}{\rho_0} \nabla \rho' - \nabla \Phi', \qquad c_s^2 = \left(\frac{\partial p}{\partial \rho} \right)_{s, \rho = \rho_0}.$$

$$(7.128)$$

Adiabaticity of the pressure perturbations has been used in writing the perturbation momentum equations, i.e., $P' \to c_s^2 \rho'$ where c_s^2 is the adiabatic sound speed as introduced in Sect. 6.2. As anticipated, there is no reference to a mean gravitational potential. These equations are supplemented by the linearization of equation (7.126)

$$\nabla^2 \Phi' = 4\pi G \rho'.$$

$$(7.129)$$

Taking the divergence of perturbation momentum equation in (7.128) and combining it with the perturbation continuity relation, we find the single equation for the perturbation ρ':

$$\frac{\partial^2 \rho'}{\partial t^2} - c_s^2 \nabla^2 \rho' - \left(4\pi G \rho_0 \right) \rho' = 0.$$

$$(7.130)$$

Normal mode solutions imply the ansatz of the form $\rho' = e^{pt + i \mathbf{k} \cdot \mathbf{x}}$, leading immediately to the following relation for the growth rate $p(k)$

$$p^2 = 4\pi G \rho - \mathbf{k}^2 c_s^2 = c_s^2 \left(k_J^2 - k^2 \right), \qquad k_J^2 = \frac{4\pi G \rho_0}{c_s^2}.$$

$$(7.131)$$

The *Jeans wavenumber* k_J, related to the *Jeans wavelength* $\lambda_J = 2\pi / k_J$, gives the condition for marginal stability. On the one hand, if the disturbance wavenumber is larger than k_J then, not only are such disturbances stable, but also for $k \gg k_J$ the disturbances can be identified with propagating adiabatic sound waves. On the other hand, if the disturbances have wavelengths longer than the Jeans length scale, $k < k_J$, then instability sets in. We note here that instability occurs by passing through zero frequency ($p^2 = 0$) indicating an exchange of stabilities, a concept we have encountered before in Rayleigh–Bénard and Taylor–Couette instabilities. While it is unsatisfying that maximal growth occurs at zero wavenumber, we should expect that for the approximations used to define this problem, any results will become increasingly inaccurate at large enough scales.

Jeans undoubtedly understood this, and also that it should be possible to obtain an equilibrium solution for a self-gravitating mass of *finite* size, but still significantly larger than the Jeans length, to serve as the base state. Perturbations would then yield a different size for the Jeans length than that obtained for an infinite medium. L. Spitzer did just that in his classic text in 1978, whose newer edition is listed as reference [6] in the *Bibliographical Notes*. He first found an equilibrium state of a two-dimensional slab (still infinite in the third direction) and found the Jeans length by performing a linear stability analysis. The result is similar to the Jeans result within a factor of $\mathcal{O}(1)$. The Jeans instability is a basic consideration in trying to develop a theory of star formation. Such a theory is bound to be very complex because of the many[7] physical processes involved and the influence of the neighboring objects. However the simplest mechanism involved is the Jeans instability. In Fig. 7.11, an active region of star formation in the Orion nebula is depicted. In the picture several relatively dense objects are singled out. These are thought to be flat gravitationally collapsed objects, whose further collapse is halted by rotation that is important in the formation of proto-planetary, relatively thick, disks (proplyds). Such proplyds are thought to be ultimately responsible for the accumulation of mass onto a central star, surrounded by a leftover disk, out of which planets may possibly form.

In the absence of a basic state, the physical interpretation of the Jeans instability can come from the perspective of a simple timescale argument. Consider, for simplicity, a spherical disturbance of a particular initial size R say and of uniform initial density ρ_0. Let $r(m,t)$ be the radius of a particular mass shell (a Lagrangian variable). Assume that entire sphere gravitationally collapses, as its pressure support vanishes. The Lagrangian equation of motion for $r(m,t)$ is

$$\ddot{r} = -\frac{Gm(R)}{r^2} = -\frac{4\pi G\rho_0 R^3}{3r^2}, \qquad (7.132)$$

where $m(R)$ is the mass interior to R, that is, the mass of the sphere and an over-dot denotes, as usual in this Lagrangian formulation, the time derivative. Multiplying Eq. (7.132) by $\dot{r}$ yields the energy integral familiar from mechanics, from which the infall velocity follows:

$$\dot{r} = -R\sqrt{\frac{8\pi G\rho_0}{3}\left(\frac{R}{r}-1\right)}.$$

---

[7]One of the simplest effects to consider is rotation. In Chandrasekhar's book (reference 1) the effect of uniform rotation on Jeans instability is given.

Fig. 7.11 The celebrated Orion Nebula, which appears to be a site of star formation and, in particular, hosts what is thought to be proto-planetary disks, thought to be the formation sites of planets around a central star. It is reasonable that the Jeans instability (see text) is instrumental in the formation of these collapsing, while rotating, objects (*Public domain. Author: M. Robberto (Space Telescope Science Institute/ESA), the Hubble Space Telescope Orion Treasury Project Team and L. Ricci (ESO), Courtesy: NASA/ESA/ESO/StSci*)

Now we would like to estimate the timescale for a total sphere collapse, the so-called *free-fall time*, τ_{ff}. This timescale is important in the theory of stellar structure, among other considerations. To this end we make the substitution

$$\frac{r}{R} \equiv \cos^2 \vartheta.$$

This yields after a nontrivial algebraic manipulation (show it) that

$$\vartheta + \frac{1}{2} \sin 2\vartheta = t \sqrt{\frac{8\pi G \rho_0}{3}},$$

where at $t = 0$, $\dot{r}$ has been set to zero (a reasonable assumption: the collapse begins from rest). We see that ϑ is the same for all mass shells at a given time. When $\vartheta = \pi/2$ all mass shells reach the center. This happens after what we call here the free-fall time:

$$\tau_{ff} = \sqrt{\frac{3\pi}{32 G \rho_0}}. \tag{7.133}$$

If the pressure gradient does not vanish totally, but it supports even a significant portion of the self-gravity, say $1/q$ with $q > 1$, then we may effectively multiply the right-hand side in Eq. (7.132) by a factor of $f = 1 - 1/q = (q-1)/q < 1$, which would then give the effective acceleration of the collapse. This is equivalent to "absorbing" f in G, so to speak, and τ_{ff} would be in such a case a factor of $\sqrt{1/f}$ longer.

In order to halt the collapse the gas sphere has to build up a pressure gradient and this can happen only on a sound crossing timescale. This can be estimated by dividing the initial radius by the sound speed, viz.

$$\tau_{sound} = \frac{R}{c_s}. \tag{7.134}$$

If $\tau_{ff} < \tau_{sound}$, the sphere will be unable to communicate its internal motion so as to halt the gravitational collapse, which will then proceed. In the opposite case the collapse would be impossible. This is the physical essence of the Jeans instability and it agrees, within an order of magnitude with the Jeans criterion (check it, by a short calculation taking as R the Jeans length). Incidentally, for q as small as 1.2, say (only ≈ 0.17 of the self-gravity not supported by pressure gradient) one finds $f \approx 0.17$ and the free-fall time would be longer by a factor of ≈ 2.4.

7.6 Instability Due to Transient Growth

7.6.1 Pipe Poiseuille Flow and the Failure of Normal Mode Analysis

Normal mode analysis is a powerful method to discover an instability, but it has limitations. In some cases, normal modes may not exist, as in the inviscid plane Couette problem, where only a continuous spectrum of modes is found. Taken at face value, the exponential in time character of a normal mode amounts to a statement about its asymptotic stability only. More general responses to perturbations, including transient growth may, nevertheless, result from a superposition of exponentially decaying *non-orthogonal* modes (see below). The interest in re-examining shear instability from the perspective of an initial value problem (IVP) analysis was motivated in part by pipe Poiseuille (a.k.a. Poiseuille–Hagen) flow, predicted to be always stable by linear normal mode theory, but which can be unstable in practice.

The axi-symmetric parallel flow inside a cylindrical pipe is both of great practical interest and attractive in its simplicity. Consider a cylinder whose radius is given by r_1 with its centerline along the z-axis. Incompressible viscous fluid of uniform density flows through the cylinder driven by a gradient of pressure (the pressure defines also Π) $P_0 = \rho_0 \Pi_0$, dependent only on z and having a constant gradient. The equations governing this system are (7.98–7.101). If we identify the steady base flow as having velocity along the axis, w_0, then the base pressure satisfies $dP_0/dz = P_{0z} =$ constant. Assuming that the steady axial velocity depends on r only, we find it by solving

$$0 = -\frac{P_{0z}}{\rho_0} + v\left(\frac{d^2w_0}{dr^2} + \frac{1}{r}\frac{dw_0}{dr}\right) \qquad \Longrightarrow \qquad w_0(r) = w_{00}(r_1^2 - r^2), \qquad (7.135)$$

subject to no-slip boundary conditions for w_0 on the cylinder wall, $r = r_1$. We assume perturbations that are, in general, not axi-symmetric. Substituting the dependent variables plus the perturbations in the relevant equation is equivalent, as usual, to the following replacements:

$$u \mapsto u', \qquad v \mapsto v', \qquad w \mapsto w_0(r) + w', \qquad \Pi \mapsto \Pi_0(z) + \Pi', \quad (7.136)$$

for which

$$(u', v', w', \Pi') = \left[\hat{u}(r), \hat{v}(r), \hat{w}(r), \hat{\Pi}(r)\right]e^{im\varphi + ikz - ikV_\mathrm{p}t} + \text{c.c..} \qquad (7.137)$$

The azimuthal wavenumber m is an integer, axial wavenumber k is real, and we write kV_p for ω. The perturbations are nondimensionalized using for the length and velocity scales r_1, and w_{00}, respectively, thus defining a Reynolds number $\mathrm{Re} = w_{00}r_1/v$. These nondimensional perturbations are substituted into the nondimensional form of the governing equations (7.98–7.101) and upon linearizing we find:

$$ik\left(w_0 - V_\mathrm{p}\right)\hat{u} = -\frac{d\hat{\Pi}}{dr} + \frac{1}{\mathrm{Re}}\left[\frac{d^2\hat{u}}{dr^2} + \frac{d\hat{u}}{dr} - \left(\frac{1+m^2}{r^2} + k^2\right)\hat{u} - \frac{2im}{r^2}\hat{v}\right]$$

$$(7.138)$$

$$ik\left(w_0 - V_\mathrm{p}\right)\hat{v} = -\frac{im\Pi}{r} + \frac{1}{\mathrm{Re}}\left[\frac{d^2\hat{v}}{dr^2} + \frac{d\hat{v}}{dr} - \left(\frac{1+m^2}{r^2} + k^2\right)\hat{v} + \frac{2im}{r^2}\hat{u}\right]$$

$$(7.139)$$

$$ik\left(w_0 - V_\mathrm{p}\right)\hat{v} + w_{0r}\hat{u} = -ik\Pi + \frac{1}{\mathrm{Re}}\left[\frac{d^2\hat{w}}{dr^2} + \frac{d\hat{w}}{dr} - \left(\frac{m^2}{r^2} + k^2\right)\hat{w}\right], \qquad (7.140)$$

$$0 = \frac{d\hat{u}}{dr} + \frac{\hat{u}}{r} + \frac{im}{r}\hat{v} + ik\hat{w}. \qquad (7.141)$$

In nondimensional form, the pipe boundary is located at $r = 1$ and the mean axial velocity profile is simply $w_0 = 1 - r^2$ with $w_{0r} \equiv dw_0/dr$. Perturbations are subject to no-slip boundary conditions at $r = 1$ together with a regularity and symmetry condition at $r = 0$.

There are no known analytical solutions for normal mode perturbations of pipe Poiseuille flow when the viscosity is non-zero. Numerically computed solutions are, however, easily found. These show that the flow is asymptotically stable for all values of the Re, irrespective of k and m. The prediction of stability is puzzling because this system is well documented as becoming unstable and even strongly turbulent in the lab. What is the origin of the instability if it is not due to normal mode dynamics, i.e., starts from a linear instability? One suggestion that is attractive, but remains unproven, is that pipe Poiseuille flow is unstable as a result of the, so-called, non-normal growth, which we discuss next.

7.6.2 Energy Growth Due to Non-normality

Open flows, such as boundary and mixing layers or pipe Poiseuille flow, are known to act like FD noise amplifiers. Might it be possible that small amplitude noise in a fluid be amplified enough to cause a global instability and ultimately perhaps a transition to turbulence? In the 1990s, it was identified that the non-normal character of the linearized differential operator, generating the equations of motion for the perturbations equations of motion, could cause temporarily large growth in the perturbation energy, referred to as *transient growth* or as non-normal growth due to the fact the eigenvectors need not be orthogonal to each other. It has been suggested that such growth may be large enough to promote nonlinear interaction among modes or to move the system to a new base state, perhaps less stable than that which was originally perturbed. As an explanation for certain turbulent flows, this is the idea known as *bypass transition* route to turbulence (see [8] in the *Bibliographical Notes*).

The non-normality of systems arising from perturbing shear flows can be inferred from a cursory study of an example set of equations (7.18–7.20). For the rest of this discussion, we shall consider the IVP version of these equations, starting from the assumption that all the disturbances are periodic in both the $\hat{x}$ and $\hat{z}$ directions. Specifically,

$$v' = \hat{v}(y,t)e^{ik_x x + ik_z z} + \text{c.c.}, \qquad \zeta'_Y = \hat{\zeta}_Y(y,t)e^{ik_x x + ik_z z} + \text{c.c.}, \qquad (7.142)$$

remembering that ζ_Y is the vorticity component in the Y direction. Similar expressions are found for the u' and w' velocity perturbations. Equations (7.18–7.20) can now be restated as

$$\frac{\partial}{\partial t}\left(\frac{\partial^2}{\partial y^2} - k^2\right)\hat{v} = -u_0 ik_x\left(\frac{\partial^2}{\partial y^2} - k^2\right)\hat{v} + u_{0yy} ik_x \hat{v} + \frac{1}{\text{Re}}\left(\frac{\partial^2}{\partial y^2} - k^2\right)^2 \hat{v},$$

$$(7.143)$$

$$\frac{\partial \hat{\zeta}_Y}{\partial t} = -U_{0y} ik_z \hat{v} - u_0 ik_x \hat{\zeta}_Y + \frac{1}{\text{Re}}\left(\frac{\partial^2}{\partial y^2} - k^2\right)\hat{\zeta}_Y. \qquad (7.144)$$

Written using matrices, the above pair of equations is

$$\begin{pmatrix} k^2 - \partial^2/\partial y^2 & 0 \\ 0 & 1 \end{pmatrix}\frac{\partial \mathbf{q}}{\partial t} = \begin{pmatrix} \hat{\mathbb{L}}_{vv} & 0 \\ -i\beta u_{0y} & \hat{\mathbb{L}}_{\zeta\zeta} \end{pmatrix}\mathbf{q}, \qquad (7.145)$$

where $\mathbf{q} = (\hat{v}, \hat{\zeta}_Y)^T$ and the matrix on the right hand, $\mathbb{M}$, is easily shown to be non-normal, or $\mathbb{M}\mathbb{M}^{\maltese} - \mathbb{M}^{\maltese}\mathbb{M} \neq 0$. Due to the non-uniformity of symbols in the literature, an *adjoint* matrix is usually marked by two symbols (often different in different textbook). As a result we have chosen to introduce and original notation: the superscript $\maltese$ indicates in this book *adjoint* which is *transpose* (superscript T) and *complex conjugate* (∗) at the same time. An adjoint *operator* is usually defined with the help of inner products of a particular Hilbert space, as should be known to the student of this book; we do not wish to devote space to the related mathematical details in this book.

Define E' to be the volume integrated kinetic energy of the perturbations. We assume for the rest of this discussion that the domain is finite and is restricted to $-1 < y < 1$. No-slip boundary conditions imply $\hat{v} = \partial \hat{v}/\partial y = \hat{\zeta}_Y = 0$ at $y = \pm 1$. For perturbations of the periodic form, it is a short exercise to show that nondimensional kinetic energy of the perturbation is

$$E'(t) = \frac{1}{2} \int_0^{L_x} \int_0^{L_z} \int_{-1}^{1} (u'^2 + v'^2 + w'^2) dx dy dz = \frac{1}{2k^2} \int_{-1}^{1} \left(\left| \frac{\partial \hat{v}}{\partial y} \right|^2 + k^2 |\hat{v}|^2 + |\hat{\zeta}_Y|^2 \right) dy$$

(7.146)

in which L_x and L_z are the wave lengths associated with the wavenumbers k_x and k_z, respectively, and where we have assumed that the density is 1.

An evolution equation for E' may be constructed directly from Eqs. (7.143) and (7.144) upon multiplication by $\hat{v}^*$ (the complex-conjugate of $\hat{v}$) and integrating in y

$$-\frac{d}{dt} \int_{-1}^{1} \left(\left| \frac{\partial \hat{v}}{\partial y} \right|^2 + k^2 |\hat{v}|^2 \right) dy = 2 \int_{-1}^{1} u_{0y} k_x \Im \left(\hat{v} \frac{\partial \hat{v}^*}{\partial y} \right) dy$$

$$+ \frac{1}{\mathrm{Re}} \int_{-1}^{1} \left(\left| \frac{\partial^2 \hat{v}}{\partial y^2} \right|^2 + 2k^2 \left| \frac{\partial \hat{v}}{\partial y} \right|^2 + k^4 |\hat{v}|^2 \right) dy.$$

(7.147)

The same procedure may be executed on Eq. (7.144) and, upon combining it with Eq. (7.147), we arrive at:

$$\frac{dE'}{dt} = -\frac{k_x}{k^2} \int_{-1}^{1} u_{0y} \Im \left(\hat{v} \frac{\partial \hat{v}^*}{\partial y} \right) dy - \frac{k_z}{k^2} \int_{-1}^{1} u_{0y} \Im \left(\hat{v} \hat{\zeta}_Y^* \right) dy$$

$$- \frac{1}{k^2 \mathrm{Re}} \int_{-1}^{1} \left(\left| \frac{\partial^2 \hat{v}}{\partial y^2} \right|^2 + 2k^2 \left| \frac{\partial \hat{v}}{\partial y} \right|^2 + k^4 |\hat{v}|^2 \right) dy - \frac{1}{\mathrm{Re}} \int_{-1}^{1} \left(\left| \frac{\partial \hat{\zeta}_Y}{\partial y} \right|^2 + k^2 |\hat{\zeta}_Y|^2 \right) dy.$$

(7.148)

This is similar in content to the Reynolds–Orr equation developed for the Orr–Sommerfeld equation (7.47). The last two integrands on the right-hand side of Eq. (7.148) include only positive definite quantities and represent the decaying action of viscosity. Perturbation energy growth may arise from the first two terms on the right-hand side: the first of these is the integrated Reynolds stress, cf. Eq. (7.48), while the second represents a correlation between the shear of the basic flow and the *helicity*, a concept to be defined in a general way in Chap. 9, in one of Eqs. (9.137). It is possible to find, accordingly, the expression $\hat{v}\hat{\zeta}_\gamma$ to be the helicity *of the perturbations*. In the literature these two effects are commonly identified as:

$$\text{Orr tilting mechanism}: \quad -\frac{k_x}{k^2}\int_{-1}^{1}u_{0y}\,\Im\left(\hat{v}\frac{\partial\hat{v}^*}{\partial y}\right)dy,$$

$$\text{Lift up effect}: \quad -\frac{k_z}{k^2}\int_{-1}^{1}u_{0y}\,\Im\left(\hat{v}\hat{\zeta}_\gamma^*\right)dy. \tag{7.149}$$

Orr tilting is associated with ideally closed streamline perturbations, or vortices, initially tilted against the background shear, leading to algebraic growth in time as these vortices are tilted by the shear. Lift-up, on the other hand, leads to the growth of horizontal velocity perturbations as a result of a normal velocity disturbance, v'. In the presence of a mean boundary layer type shear, normal velocity perturbations with span-wise wavenumber k_z will lift low speed fluid upward and push high speed fluid downward, leading to alternating high and low speed streamwise streaks that are also associated with streamwise oriented vortices. Qualitatively, Orr tilting is present in disturbances that are two-dimensional owing to the presence of k_x in its definition. On the other hand, lift-up is primarily a three-dimensional effect since it requires $k_z\neq0$ for it to operate. In most shear flows, transient growth of disturbances is primarily driven by lift-up.

Most importantly, further analysis of the perturbation energy, whether in integral form, using matrix exponentials, or by numerical computations, reveals that transient growth in shear flows exhibits scaling relations in the Reynolds number. In particular, the maximum value of the energy at any time, $E'_{\max}$, scales as

$$E'_{\max} = C\,\mathrm{Re}^2 \tag{7.150}$$

for some constant, C, while there is a similar scaling relation that predicts the three-dimensional wavenumber of maximum energy growth. This result has been validated in experiments and in numerical simulations and supports the role of transient growth as a noise amplifier. In astrophysical problems, where Re values are very large, transient growth may be of even greater importance. These topics form the subject of *optimal growth* a topic that we do not examine here (but may be learned from references, e.g., [8] in the *Bibliographical Notes*).

7.6.2.1 A Transient Growth Model

To more concretely demonstrate non-modal growth we return to the system (7.11) and develop this model problem in more detail. In this example, one can think of the parameter ε as representing Re^{-1}. The structure of the matrix associated with this model is the analog of the equations as similarly expressed in matrix form in

Eq. (7.145). Recall that normal mode analysis found the two eigenvalues $p = -1, -2$ and nearly parallel eigenvectors. According to our definitions, the system is both stable and asymptotically stable. For initial conditions given by $u(0) = u_0$ and $v(0) = v_0$, the general solution is

$$u(t) = (u_0 + \varepsilon^{-1}v_0)e^{-t} - \varepsilon^{-1}v_0 e^{-2t}, \qquad v(t) = v_0 e^{-2t}.$$

Defining the symbols U and V corresponding to $|u|$ and $|v|$ scaled by their initial values, we reexpress the solution quoted above in terms of the variable $\chi(t) \equiv e^{-t}$, such χ has the value 1 at $t = 0$ and approaches value 0 as $t \to \infty$, finding

$$U = \frac{1}{|Q-1|}|\chi Q - \chi^2|, \qquad V = \chi^2, \text{ where } Q \equiv \frac{\varepsilon u_0}{v_0} + 1.$$

Looking at V, we see that its maximum value of 1 occurs at $\chi = 1$, or $t = 0$; V then monotonically decreases to zero as $t \to \infty$. This is expected of a quantity that is asymptotically stable. Similarly, U is also asymptotically stable, but closer inspection of its time evolution shows that although it has initial value $U = 1$, it can achieve a value much greater than one for times $t > 0$. A local extremum of U is easily found to occur at $\chi_{max} = Q/2$ for which the maximum value is given by:

$$U_{max} = \frac{1}{4}\frac{Q^2}{Q-1}.$$

The significance of this can be appreciated almost immediately: non-normality may lead to arbitrarily large growth in the amplitude of u before it asymptotically decays to zero. If transient growth is large enough, the amplitude achieved from a tiny perturbation could be sufficient to push the system a nonlinear regime.

In this example, growth in the amplitude is controlled by the value ε, which plays the role here of Re^{-1}. This type of mechanism leads, in shear flows, to the $\mathcal{O}(Re^2)$ energy scaling. The above is an example of *transient growth* resulting from the *non-normality* of the linear operators of the system. The normality of matrix $\mathbb{J}$ is assessed, as was mentioned above by evaluating whether or not $\mathbb{J}$ commutes with its adjoint. In the above example it is a short exercise to show that $\mathbb{J}$ is non-normal.

Problems

7.1.

Consider the linear system

$$\frac{du}{dt} = -\alpha u + \gamma v$$

$$\frac{dv}{dt} = -\beta u \qquad (7.151)$$

in which $\alpha, \beta > 0$ and with initial conditions $v(0) = v_0, u(0) = u_0$. Examine the transient growth properties of this system. In particular, define $\mathscr{U} = |u/u_0|$ and determine (i) for what conditions transient growth is possible and when it is possible, at what $t = t_{tg}$ does this occur? (ii) What is the maximum value attained by $\mathscr{U}$ at $t = t_{tg}$? Lastly, examine the behavior of the solution when $\alpha \to \beta > 0$.

7.2.
Complete the steps leading to Eq. (7.107). Make sure to expand out to second order in powers of ℓ_r.

7.3.
The inviscid Goldreich–Schubert–Fricke instability is the analog of the axi-symmetric Taylor–Couette instability. It had originally been invoked to study the possibility of instability in rotating stars whose profiles show gradients along both the rotating star's (cylindrical) radial and vertical (aligned with the star's rotation axis) coordinates. In such cases the mean swirling state is a function of these coordinates, i.e., $v_0 = v_0(r, z)$. Following a similar energy argument like that found in Sect. 7.4.3.1 show that linear stability is expected if

$$\left| \frac{\partial(r v_0)}{\partial r} \right| > \left| \frac{\ell_r}{\ell_r} \frac{\partial(r v_0)}{\partial z} \right|, \qquad (7.152)$$

where ℓ_r and ℓ_z are the corresponding radial and vertical length scales of the system. Conclude that if the vertical gradient of the swirling flow is weak that instability is expected only for very short radial wavelength disturbances. The criterion found in Eq. (7.152) is the condition for the *Goldreich–Schubert–Fricke* instability in the limit of no viscosity and no stratification.

7.4.
Prove the *Toomre stability criterion* for a self-gravitating gaseous thin disk $Q_T \equiv \frac{c_s \kappa}{\pi G \Sigma} > 1$, where Σ is the surface density. Consult the book of Binney and Tremaine, *Galactic Dynamics*, Princeton 2008, especially Sect. 6.2.3. κ is the *epicyclic frequency* defined earlier in this book.

7.5.
Assuming normal mode type solutions in and $\hat{v} = 0$ show that the solution $\hat{\zeta}_Y$ of (7.25) is always damped in time, i.e., $\Im(V_p) < 0$. Construct the solution for the case of impenetrable and no-horizontal slip boundaries at $y = \pm 1$. Construct a similar argument for the case where the boundaries extend to $y \to \pm\infty$. What additional caveats are there upon the mathematical form of the structure function $\hat{\zeta}_Y(y)$ as this limit is approached.

7.6.

Consider a plane-parallel sheared fluid lying between walls located at $y = \pm L$. Suppose furthermore that the shear profile is similar to the Rayleigh problem in which

$$u_0 = \begin{cases} u_{00} & , H < y < L \\ \Lambda y & , |y| < H \\ -u_{00} & , -L < y < -H, \end{cases} \tag{7.153}$$

where $0 < H < L$ and with $\Lambda = u_{00}/H$. Using the techniques developed, show that this profile supports normal modes only if $L > 3H/2$.

7.7.

The "density confined" Rayleigh problem: Consider a shear layer a profile like the Rayleigh one (7.42). However suppose that the density is constant *except* for sudden "jumps" across $y = \pm H$, i.e.,

$$\rho_0 = \begin{cases} \rho_{00} & |y| \leq H \\ \rho_{01} & |y| \geq H \end{cases} \tag{7.154}$$

and where ρ_{00} and ρ_{01} are constants greater than zero. Assume that in the mean state the global pressure is constant. Find the normal modes of the system by imposing the continuity of the total pressure of the system across the moving boundaries. Determine the growth rates and comment on how results compare to the classical Rayleigh shear problem as some function of $\rho_{00} - \rho_{01}$. Are there conditions in which normal modes cease to exist in this case, similar to the plane-Couette problem?

7.8.

Consider the problem on the finite domain $|y| \leq H$ in which the mean velocity field is zero. Explain how Orr–Sommerfeld equation (7.24) simplifies into

$$\left(\frac{d^2}{dy^2} - k^2 \right) \left(\frac{d^2}{dy^2} - k^2 + ikV_p\mathrm{Re} \right) \hat{\psi} = 0,$$

together with boundary conditions that $\hat{\psi} = d\hat{\psi}/dy = 0$ at $y = \pm H$. Argue that due to the symmetries along y the solutions of this system may be analyzed in terms of even and odd symmetric modes. Show that

$$\text{(sinuous modes)} \quad V_p = -\frac{i(k^2 + \kappa^2)}{k\mathrm{Re}}, \quad k\tanh k = -\kappa \tan \kappa$$

$$\text{(varicose modes)} \quad V_p = -\frac{i(k^2 + \kappa^2)}{k\mathrm{Re}}, \quad k\coth k = \kappa \cot \kappa$$

with $\kappa^2 = ik\mathrm{Re}V_p - k^2$. (Hint: examine the space of possible solutions for each κ relationship shown.)

7.9.
Starting from the inviscid equations of motion for a plane parallel shear flow, show that the total kinetic energy of the flow (composed of basic background velocity shear plus its perturbations) is, both linearly and nonlinearly, conserved. Assume that the perturbations are zero on the boundaries $y = \pm 1$ and that they are otherwise periodic in the streamwise and spanwise directions. Given that the perturbation energy shows fluctuations due to the lift-up effect and Orr tilting, identify the *other* energy quantity that is the source of the energy fluctuations driving the perturbations. Convince yourself that this is the energy reservoir discussed in the text. Note that it involves the streamwise perturbation velocities.

7.10. Consider gas at $T = 10^3 K$ and particle density of $n = 10^2$ cm^{-3}. What is its Jeans wavelength λ_J? If a sphere of such gas has the radius $R = \lambda_J$, what is its mass, M_J, referred to as the Jeans mass. Give a general expression of the Jeans mass as a function of temperature and density. Find its value for average approximate conditions ($T \sim 10^6$ K, $\rho = 1$ g cm^{-3}) as well as those in a molecular cloud ($T \sim 10$ K, $n \sim 10^6$ cm^{-3}).

7.11.
Starting from the inviscid equations of motion for a plane parallel shear flow, show that the total kinetic energy of the flow (composed of basic background velocity shear plus its perturbations) is, both linearly and nonlinearly, conserved. Assume that the perturbations are zero on the boundaries $y = \pm 1$ and that they are otherwise periodic in the streamwise and spanwise directions. Given that the perturbation energy shows fluctuations due to lift-up and Orr tilting, identify the *other* energy quantity that is the source of the energy fluctuations driving the perturbations. Convince yourself that this is the energy reservoir discussed in the text. Note that it involves the streamwise perturbation velocities.

7.12.
Infinite plane Couette flow in which $u_0 = \Lambda y$ can be analyzed in the so-called *shearing sheet* setting often invoked in astrophysical problems. It means analyzing Eqs. (7.18–7.20) in a new coordinate frame defined by

$$X = x - \Lambda y t, \qquad Y = y, \qquad Z = z, \qquad T = t.$$

With this in hand, and assuming all perturbations are periodic in the new (X, Y, Z) coordinate system with scales $L_x = 2\pi/k_x$, $L_y = 2\pi/k_y$, $L_z = 2\pi/k_z$, show that the total energy of the system E_{vol} integrated over the volume is given by

$$E_{vol} = \frac{1}{2} \int_{\text{stretched volume}} (u'^2 + v'^2 + w'^2) d^3x = \frac{L_x L_y L_z}{2k^2} \left\{ \left| i(k_y - k_x \Lambda t) \tilde{v} \right|^2 + k^2 |\tilde{v}|^2 + |\tilde{\zeta}_Y|^2 \right\}.$$

$$(7.155)$$

Start this calculation by proving to yourself that the Jacobian of the coordinate transformation is a constant, i.e., $dxdy = dXdY$.

Bibliographical Notes

Sections 7.1 and 7.2

The general principles of stability can be found in wide variety of books on dynamical systems and applied mathematics, too numerous to list here. We shall mention here at the outset just the classic book

1. S. Chandrasekhar, *Hydrodynamic and Hydromagnetic Stability* (Dover, New York, 1961)

Section 7.3

The study of the stability of sheared hydrodynamic flows covers a hundred (plus) year span of time. The following three books are the most relevant in our opinion,

2. P.G. Drazin, W.H. Reid, *Hydrodynamic Stability* (Cambridge University Press, Cambridge, 1982)
3. P.G. Drazin, *Introduction to Hydrodynamic Stability* (Cambridge University Press, Cambridge, 2002)
   The narrative in this section largely follows the organization found in these two studies. Finally, a comprehensive and modern monograph on stability is
4. F. Charru, *Hydrodynamic Instabilities* (Cambridge University Press, Cambridge, 2007)

Section 7.4

The discussion regarding Rayleigh–Bénard convection and the Taylor–Couette systems analysis is a modernized retelling of this story based Chandrasekhar's classic book [1].

Some of the astrophysical discussion is taken from another important book, by Kippenhahn and Weigert

5. R. Kippenhahn, A. Weigert, *Stellar Structure and Evolution* ("study edition") (Springer, New York, 1990)

Section 7.5

Lyman Spitzer's indispensable *Diffuse matter in space* has fortunately been re-edited by Wiley's classics. It is a short but very clever treatise:

6. L. Spitzer Jr, *Physical Processes in the Interstellar Medium* (Wiley Classics Library, New York, 1998)
   In addition to the above and S. Chandrasekhar's work, we recommend the book by F. Shu:
7. F.H. Shu, *Gas Dynamics* (University Science Books, Mill Valley, 1992)
   The physical discussions found therein are especially revealing.

Section 7.6

The text listed below (and references therein) possesses a detailed examination of initial value problems and non-normal growth. Readers are encouraged to use these resources for further study. Additionally, a slightly more pedestrian account (but no less mathematically rigorous or enlightening) may be found in both [3] and [2] listed above. A wide class of exercises on eigenvalue analysis as well as the IVP may be found in both those books.

8. P.J. Schmid, D.S. Henningson, *Stability and Transition in Shear Flows* (Springer, New York, 2001)

Chapter 8
Weakly Nonlinear Instability

And speak not of anything with certainty without adding,
Allah be willing!

The Quran 18:23–24 (Surah al-Kahf)

8.1 Introduction

In the preceding chapter, we examined the stability of a wide range of flows, most of them steady solutions to their governing fluid equations. In every case, our approach was to perturb the steady flow in order to define a new problem for the evolution of the perturbations themselves, or in other words the robustness of the steady base flow. We then drew conclusions about the stability properties of the original flow based on the evolution of the perturbations. To make these problems tractable, we assumed that the perturbations were small, neglecting all nonlinear terms, that is, terms involving products of the small perturbation quantities. As a linear instability grows, it will eventually become large enough in magnitude that nonlinear terms can no longer be neglected. In this case, nonlinearity plays several roles, but their main role, and of greatest interest to us, is the ability to saturate the exponential growth.

It is the saturation of an instability that leads to the pattern of Taylor–Couette flow depicted in Fig. 7.3a, itself a steady, spatially periodic flow that can be considered a large amplitude manifestation of the unstable linear mode. Recall that the instability in Taylor–Couette flow arises as the Taylor number exceeds a critical value, a *bifurcation* point at which the eigenvalue of the least stable mode passes through zero, leading to an exchange of stabilities. Thus, the word "bifurcation" illustrates the appearance of a "fork" in the abstract route delineated by the change of a parameter. The Rayleigh–Bénard convection undergoes a similar stationary-type instability as its control parameter, in this case the Rayleigh number, passes through

© Springer Science+Business Media, LLC 2016
O. Regev et al., *Modern Fluid Dynamics for Physics and Astrophysics*,
Graduate Texts in Physics, DOI 10.1007/978-1-4939-3164-4_8

a critical value, the value of which depends on the boundary conditions. When a control parameter, λ, say, is varied to pass from a stable to an unstable state, we refer to this situation as being near the instability threshold. From here on, it will be useful to introduce a dimensionless control parameter

$$\varepsilon \equiv \frac{\lambda - \lambda_c}{\lambda_c}, \tag{8.1}$$

where λ_c is the critical value and $\varepsilon = 0$ defines the instability threshold above which, for $\lambda > \lambda_c$, we may examine the onset of instability. Since the onset occurs at the critical value of control parameter(s), a "near onset" situation is synonymous to "near criticality."

Aside from laboratory or numerical experiments, there are no general mathematical techniques that are broadly applicable to nonlinear flow stability, by which we mean when deviations from steady flows are of large amplitude and/or the parameters are far from threshold values. The nonlinear stability that we examine in this chapter is confined to the neighborhoods near onset and to small amplitudes, a case known as *weakly nonlinear theory*. In spite of its restrictions, weakly nonlinear theory is a powerful tool that may provide an accurate description of the finite amplitude stage of linear instability, thus offering insight into the physical mechanisms for the quenching of exponential growth. Furthermore, weakly nonlinear theory provides a framework for the description of spatiotemporal patterns and for the study of the stability of saturated nonlinear states, which may become unstable to the so-called secondary instabilities. As we have seen in the example of Taylor–Couette flow, further increase of the control parameter may lead to the eventual destabilization of all available stationary and periodic states. In a dynamical system, this behavior leads to chaos as first proposed by D. Ruelle and F. Takens. In a fluid flow, this scenario leads to spatiotemporal disorder characterized by a broad spectrum of fluctuations and may therefore explain one route to turbulence. This will be discussed in considerable detail in Chap. 9, whose title is simply Turbulence.

At this point it is important to preview some of the important questions and goals that motivate a weakly nonlinear stability analysis. First is the issue of wavenumber selection. As we have seen from linear stability theory, the onset of an instability is characterized by both a critical control parameter value and a critical wavenumber, k_c. What determines this scale in the nonlinear regime and is it unique? In many cases, these questions lead to the issue of pattern formation, a topic too vast to include in this chapter, but covered in detail in some of the references of the *Bibliographical Notes*. We may also examine a saturated nonlinear state in turn for its stability to small perturbations of generic form. Finally, there are important practical issues, such as the effect of a nonlinear instability on transport or the impact of a nonlinear mode on the control properties of a flow. We will breach some of these topics here, but the main aim of this chapter is to provide a solid foundation for performing calculations and understanding their limitations.

This chapter is organized as follows: in Sect. 8.2 we introduce weakly nonlinear stability by presenting three examples of amplitude equations, followed by a description of the multiple scale method of weakly nonlinear analysis. In Sect. 8.3 we perform a weakly nonlinear analysis on Rayleigh–Bénard convection. This is followed by an analysis of amplitude equations and a discussion of nonlinear instability. In the final section of the chapter, we present several other examples of systems amenable to weakly nonlinear analysis and present a brief survey of well-known model nonlinear equations.

8.2 Amplitude Equations and Multiple Scale Analysis

We begin our study of weakly nonlinear stability with three examples from ODE theory. While elementary, these examples illustrate several important concepts and serve to introduce terminology.

Supercritical Pitchfork Consider the first-order ordinary differential equation for the quantity u:

$$\frac{du}{dt} = \varepsilon u - u^3, \tag{8.2}$$

in which ε is a control parameter and u typically represents the amplitude of an unstable mode. This equation is commonly known as the Landau equation, bearing the name of the acclaimed physicist L.D. Landau (1908–1968), who contributed to virtually all branches of physics pursued at his time, and is one of the simplest and most common examples of an amplitude equation.[1] The general solution depends on the parameter, ε, so we write it as $u(t; \varepsilon)$ and solve for steady solutions, denoted by $\bar{u}$, by solving $\varepsilon \bar{u} - \bar{u}^3 = 0$. Clearly $\bar{u} = 0$ is the only real solution when $\varepsilon < 0$, while for $\varepsilon > 0$ there are three solutions, $\bar{u} = \pm\sqrt{\varepsilon}$ and $\bar{u} = 0$. In this way the point $\varepsilon = 0$ corresponds to a bifurcation of the solutions; this type of bifurcation is commonly referred to as a *supercritical pitchfork* bifurcation.

The linear stability of each steady solution may be evaluated by introducing perturbations $u = \bar{u} + u'(t)$ and assuming the normal mode solution form $u' = \hat{u}e^{pt} + \text{c.c.}$, for which we find

$$p - \varepsilon + 3\bar{u}^2 = 0. \tag{8.3}$$

We see that for the state $\bar{u} = 0$ the growth rate is simply $p = \varepsilon$, indicating that this steady state is stable for $\varepsilon < 0$ and unstable for $\varepsilon > 0$. On the other hand, the states $\bar{u} = \pm\sqrt{\varepsilon}$ are stable, since for real $\bar{u}$ we must have $\varepsilon > 0$, leading to $p = -2\varepsilon^2$. We conclude that for $\varepsilon < 0$, perturbations away from $\bar{u} = 0$ return to the state $u = 0$, while for $\varepsilon > 0$, perturbations drive the solutions towards the steady states $\bar{u} = \pm\sqrt{\varepsilon}$

---

[1]The Landau equation first appeared as a model for first-order phase transitions for which u represents an order parameter.

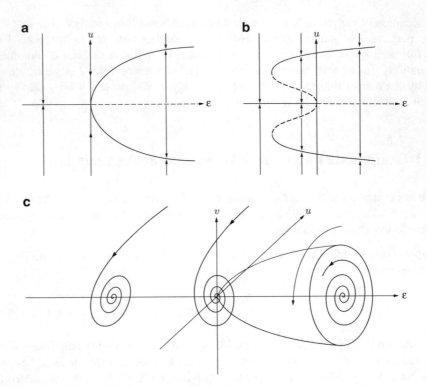

Fig. 8.1 Illustrations of the solutions of various types of bifurcations of one-dimensional amplitude equations: (**a**) supercritical pitchfork, $\dot{u} = \varepsilon u - u^3$, (**b**) subcritical pitchfork $\dot{u} = \varepsilon u + u^3 - \gamma u^5$, and (**c**) Hopf bifurcation, $\dot{u} = -v + (\varepsilon - u^2 - v^2)u$, $\dot{v} = u + (\varepsilon - u^2 - v^2)v$. In (**a**) and (**b**) linearly stable steady states are denoted by a *continuous line*, while linearly unstable ones by *dashed lines*. (**c**) depicts a stable limit cycle

depending upon the sign of the perturbation at the initial time. This behavior is sketched in Fig. 8.1a. The qualitative behavior sketched in this example is supported by an exact solution (Problem 8.1).

Subcritical Pitchfork   The behavior of previous example is referred to as supercritical because the instability of the steady solution $\bar{u} = 0$ occurs for $\varepsilon > 0$, in other words when the control parameter exceeds the critical value. The next example demonstrates a richer behavior thanks to the presence of a higher order nonlinearity. Consider the model system

$$\frac{du}{dt} = \varepsilon u + u^3 - \gamma u^5, \quad \text{where} \quad \gamma > 0. \tag{8.4}$$

The steady states of this system are a bit more complicated and are given by

$$
\bar{u} = \begin{cases}
0, & \varepsilon < -\dfrac{1}{4\gamma}; \\[2ex]
0, \pm\left[\dfrac{1}{2\gamma}\left(1 \pm \sqrt{1+4\varepsilon\gamma}\right)\right]^{1/2}, & -\dfrac{1}{4\gamma} < \varepsilon < 0; \\[2ex]
0, \pm\left[\dfrac{1}{2\gamma}\left(1 + \sqrt{1+4\varepsilon\gamma}\right)\right]^{1/2}, & \varepsilon > 0.
\end{cases}
\tag{8.5}
$$

Again, the $\bar{u} = 0$ state is linearly stable for $\varepsilon \leq 0$ and unstable otherwise. The complication occurs when $-1/4\gamma < \varepsilon < 0$, for which there are five possible steady states, as illustrated in Fig. 8.1b (see also Problem 8.3). Of these, the upper branches, which are selected by choosing $+\sqrt{1+4\varepsilon\gamma}$ in Eq. (8.5), are stable, while the middle branches, associated with the choice $-\sqrt{1+4\varepsilon\gamma}$, are unstable. As a result, a large enough perturbation of the $\bar{u} = 0$ state may drive the system towards the upper branch. This is an example of a *subcritical pitchfork bifurcation* of which there are many examples in fluid dynamics, including wave trains in water flowing down an inclined surface. The possibility of a subcritical bifurcation is limited to the narrow range of ε values given above, since for $\varepsilon < -1/4\gamma$ the system is stable no matter how large the initial perturbation. Because a finite amplitude perturbation is needed to move from one stable solution branch to another, the model described in Eq. (8.4) may also exhibit hysteresis as ε is moved across the interval $-1/4\gamma < \varepsilon < 0$, the solution not switching between simultaneously stable branches until the varying parameter forces a switch to occur (can you think how this might happen?).

Hopf Bifurcation In two dimensions different forms of bifurcation are possible. Consider the system

$$
\frac{du}{dt} = -v + (\varepsilon - u^2 - v^2)u,
\tag{8.6}
$$

$$
\frac{dv}{dt} = u + (\varepsilon - u^2 - v^2)v.
\tag{8.7}
$$

The only steady state possible in this system is $\bar{u} = \bar{v} = 0$. Introduction of perturbations around the zero state yields a system whose linearized matrix $\mathbb{M}$ is

$$
\mathbb{M} = \begin{pmatrix} \varepsilon & -1 \\ 1 & \varepsilon \end{pmatrix},
$$

which has eigenvalues given by $p = \varepsilon \pm i$. In the event that $\varepsilon > 0$, the system is unstable with growing oscillations. Thus, as ε passes through zero from below, the system undergoes a bifurcation: a quiescent state at $\varepsilon < 0$ executes growing oscillations when $\varepsilon > 0$. The amplitude at which this oscillatory instability saturates can be found by means of a transformation into polar coordinates:

$$
u = A(t)\cos\Theta(t), \qquad v = A(t)\sin\Theta(t),
$$

which puts Eqs. (8.6)–(8.7) into the form

$$\frac{dA}{dt} = A\left(\varepsilon - A^2\right), \qquad \frac{d\Theta}{dt} = 1,$$

which are decoupled. With $\Theta = t$, it is evident that u and v oscillate with frequency 1. The amplitude $A \geq 0$ is governed by a Landau equation, as examined in the first example above. We can thus form a picture of the fate of small amplitude linear perturbations when $\varepsilon > 0$: the amplitude rises from nearly zero and approaches the steady amplitude $\overline{A} = \sqrt{\varepsilon}$ as $t \to \infty$. As the amplitude rises, the functions u and v cycle as they asymptote to their limiting amplitudes, illustrated in Fig. 8.1c. Similarly, if the basic state were given by $\overline{A} = \sqrt{\varepsilon}$, a linear stability analysis about this state predicts exponentially decaying oscillations of u and v. The stable oscillatory solution $\overline{A} = \sqrt{\varepsilon}$ is an example of a *limit cycle*.

8.2.1   Multiple Scale Analysis Near Onset

Once a control parameter exceeds its critical value and a linear instability sets in, the resulting exponential growth will usually violate the assumption of linearization. Nonlinear terms then become important and, in the cases of greatest interest here, determine a new state according to their growth limiting mechanism. In Rayleigh–Bénard convection, for example, the static conduction state transitions to a state of steady rolls. In general, such a transition may lead from one steady state to another, perhaps of different form, as in the pitchfork examples, or it may send the system into a state with time variability but whose spatial structure is nearly uniform and regular, as for the Hopf bifurcation. The general picture is more varied than these few cases and may include the interaction of multiple modes at onset, controlled by more than one parameter. In these more complex cases, we may see transition to an unsteady pattern, weak chaos, or even turbulence.

For chaotic or turbulent states, there are few analytic tools and generally little hope of developing useful approximate descriptions. However, when the linear growth rate is modest, as is true close to the instability threshold, a natural small parameter is available in which to perform an expansion of the nonlinear amplitude. In this approximation, the lowest order nonlinear term is small, but not negligible with respect to the linear one. We refer to this as the *weakly nonlinear* case. In this chapter we restrict our attention to transitions to new steady or spatially regular states when the nonlinearity is weak, meaning than the nonlinear terms are significantly smaller than the linear ones. We focus on the question: what amplitude the instability grows to as a new state is achieved? In addition, we seek to understand the mechanism of amplitude saturation when the latter occurs, a result that is inaccessible from linear theory, where the amplitude is indeterminate.

In the next section, we use the example of Rayleigh–Bénard convection to demonstrate the implementation of weakly nonlinear analysis and its application to saturation mechanisms. As we shall see, the technique allows one to unambiguously identify the physical effect that leads to saturation. In the case of Rayleigh–Bénard convection, the vertical advection of the background temperature gradient adjusts the temperature profile so as to shut off the linear instability.

But first we must outline the method of weakly nonlinear multiple scale analysis. Consider the nonlinear evolution of a state vector, $\mathbf{V}(\mathbf{x},t)$, dependent on space and time. The evolution of this vector is prescribed by the set of coupled PDEs valid on some domain $\mathscr{D}$,

$$\frac{\partial \mathbf{V}}{\partial t} = \mathbb{P}\left(\mathbf{V};\lambda\right), \tag{8.8}$$

with boundary conditions on $\mathbf{V}$ enforced on $\mathscr{S} = \partial \mathscr{D}$, the boundary of $\mathscr{D}$. The operator $\mathbb{P}$ may involve any combination of spatial partial derivatives and is nonlinear. The solution $\mathbf{V}$ also includes the tunable control parameter, λ, which will become the focus of much of our attention. Suppose that there exists a state $\mathbf{V}_0$ which is a steady solution of the system, viz., $\mathbb{P}(\mathbf{V}_0;\lambda) = 0$. In the remainder of this discussion, we will examine the behavior of such a system using perturbative expansion about this steady state. The method is sufficiently general that it may be equally applied to states which involve time periodic behavior.

Let the solution of (8.8) be comprised of the steady state plus a perturbation, which may be time dependent, $\mathbf{V}(\mathbf{x},t) = \mathbf{V}_0(\mathbf{x}) + \mathbf{V}'(\mathbf{x},t)$. In this case the evolution equation reads

$$\frac{\partial \mathbf{V}'}{\partial t} = \mathbb{L}\left(\mathbf{V}_0,\mathbf{V}';\lambda\right) + \mathbb{N}\left(\mathbf{V}_0,\mathbf{V}';\lambda\right), \tag{8.9}$$

where we have deliberately written $\mathbb{P}$ as a sum of two operators, $\mathbb{L}$ and $\mathbb{N}$, where the former is a strictly linear operator on the state vector $\mathbf{V}'$ and the latter is a nonlinear operator acting on the same. Note that in this formulation, the sum $\mathbf{V}_0 + \mathbf{V}'$ satisfies the requisite boundary conditions on $\mathscr{S}$. In Chap. 7 we have captured the linear stability of a given system in terms of normal modes, solutions whose spatial and temporal dependence is separable. In these terms, the linear system is an eigenvalue problem described by the sequence of eigenvalues $p = \{p_j\}$, which depend upon the value of the parameter λ, viz.,

$$\frac{\partial \mathbf{V}'}{\partial t} = \mathbb{L}\left(\mathbf{V}_0,\mathbf{V}';\lambda\right) \quad \Longrightarrow \quad p\hat{\mathbf{V}} = \mathbb{L}\left(\mathbf{V}_0,\hat{\mathbf{V}};\lambda\right) \quad \text{where} \quad \mathbf{V}' = \hat{\mathbf{V}}e^{pt}. \tag{8.10}$$

Suppose that the sequence $\{p_j\}$ is ordered according to a decreasing magnitude of the real part of p, such that $\Re(p_1) > \Re(p_2) > \Re(p_3) > \cdots$ and that associated with each eigenvalue is a structure function, its eigenvector, with the same label, $\hat{\mathbf{V}}_j$. We will refer to the leading eigenvalue and associated eigenvector, both labelled by the index $j = 1$, as the least stable, or *most dangerous* mode. In writing a solution, the mode labelled by j would be written as $A_j\hat{\mathbf{V}}_j e^{p_j t}$, where the amplitudes, A_j, are

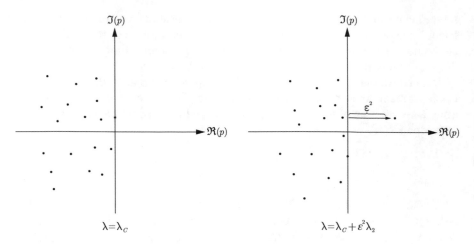

Fig. 8.2 A schematic plot of the spectrum of eigenvalues p_j. The *left panel* shows the situation where the parameter of the system has its critical value, $\lambda = \lambda_c$, and the least stable eigenvalue resides on the imaginary p-axis. On the *right panel*, the parameter is slightly increased to $\lambda = \lambda_c + \varepsilon^2 \lambda_2$, so that the least stable mode now has growth rate of order ε^2

arbitrary.[2] The presence of this arbitrary amplitude in the solution of the normal mode problem is very important for what follows.

We now define the critical value of the parameter λ, which we denote by λ_c, such that $\Re(p_1) = 0$ and $\Re\ (p_j) \leq 0$ for $j \geq 2$. We suppose that a small deviation of λ from λ_c destabilizes the least stable eigenvalue while leaving all remaining eigenvalues stable. That is, for $\lambda = \lambda_c + \varepsilon^2 \lambda_2$ with $\lambda_2 = \mathscr{O}(1)$, we have $\Re(p_1) > 0$, while $\Re(p_j) \leq 0$ for $j \geq 2$. We assume that the growth rate associated with this deviation from the marginal value of the parameter λ scales as some power of ε and we shall, in this discussion, assume that it is ε^2. The choice of ε^2 follows from a trial and error that ultimately gives rise to correctly balanced equations. Figure 8.2 graphically depicts the eigenvalue behavior described above. The least stable eigenvalue has a positive real part (is unstable) and is separated by a *finite* distance difference (in its real part) from all the other eigenvalues, which are situated to the left of the imaginary axis. To describe the saturation of the linear instability, we seek an approximate solution $\mathbf{V}'(x,y,z,t;\lambda)$ for λ deviating from λ_c, as just described. Under these circumstances, we write more precisely $\mathbf{V}' = \mathbf{V}'(x,y,z,t;\lambda_c,\varepsilon)$, indicating explicitly that the solution is a function of ε. The weakly nonlinear multiple time and space scale approach demands that the solution for $\mathbf{V}'$ be written as an asymptotic series of the form

$$\mathbf{V}'(x,y,z,t;\lambda_c,\varepsilon) = \mathbf{V}'(x,y,z,\varepsilon x,\varepsilon y,\varepsilon z,\varepsilon t,\varepsilon^2 t,\cdots;\lambda_c)$$
$$\approx \alpha(\varepsilon)A(X,Y,Z,\tau,T)\hat{\mathbf{V}}(x,y,z,t) + \text{HOT in } \varepsilon. \quad (8.11)$$

---

[2]In an initial value problem, the amplitudes A_j would be determined through projection of the initial condition $\mathbf{V}'(\mathbf{x},0)$ onto the set of eigenvectors $\hat{\mathbf{V}}_j$.

The key feature of this ansatz is that the amplitude, $A(X, Y, Z, \tau, T)$, now depends (implicitly) on the parameter ε. For generality, A is also scaled by $\alpha(\varepsilon)$, a function of ε. Usually, the functional form of $\alpha(\varepsilon)$ will be some power of ε, for instance, ε or ε^2, although in some cases the leading order dependence may be $\varepsilon \ln \varepsilon$. These forms are problem specific and are usually found through trial and error.

The amplitude A now depends also upon the newly defined space and timescales $X \equiv \varepsilon x, Y \equiv \varepsilon y, Z \equiv \varepsilon z, \tau \equiv \varepsilon t, T \equiv \varepsilon^2 t$, which are slowly varying by definition. Reminiscent of the modulational amplitude familiar as "beats" of nearby acoustic frequencies, the nonlinear amplitude restricts the growth of a linear instability to an envelope function. Based on this framework, we must now derive an evolution equation for the amplitude A in terms of these new variables X, Y, Z, τ, T and use the result to estimate the amplitudes achieved by this unstable mode. The steps that are required to define a PDE for A almost always require that we ensure the existence of higher order solutions (HOT) at each order of the perturbation series.

8.3   Stationary Rolls in Rayleigh–Bénard Convection I: Fixed Temperature

Employing the nondimensionalization used to examine classical linear RB convection, detailed in Sect. 7.4.2, leads to the following equations of motion:

$$\frac{\partial \mathbf{u}}{\partial t} + \left(u \frac{\partial}{\partial x} + v \frac{\partial}{\partial y} + w \frac{\partial}{\partial z} \right) \mathbf{u} = -\nabla \Pi + \mathrm{Pr} \nabla^2 \mathbf{u} + \mathrm{Ra} \mathrm{Pr}\, \Theta \hat{\mathbf{z}} \qquad (8.12)$$

$$\frac{\partial \Theta}{\partial t} + \left(u \frac{\partial}{\partial x} + v \frac{\partial}{\partial y} + w \frac{\partial}{\partial z} \right) \Theta = w + \nabla^2 \Theta \qquad (8.13)$$

$$\nabla \cdot \mathbf{u} = 0. \qquad (8.14)$$

Recall from Sect. 7.4.2 that the vector $\mathbf{u} = (u, v, w)$ represents deviations from a mean steady state of zero velocity, while Θ represents thermal deviations from a steady thermal state $T = 1 - (\beta d/T_0)z$. In the convection problems examined in this section and in Sect. 8.4, we require that the fluid remain confined between $z = 0$ and $z = 1$, that is, there is no flow across the bottom and top boundaries. The characteristic dimensionless parameters are the Rayleigh and Prandtl numbers, Ra and Pr, respectively. But only the Rayleigh number serves as a control parameter in determining the linear instability. The state vector for this system is $\mathbf{V} \equiv (u, v, w, \Theta)^{\mathrm{T}}$ and the steady state, of no motion, is $\mathbf{V}_0 = \mathbf{0}$.

As a first example, we shall present in detail the calculation of the weakly nonlinear description of unstable rolls, that is, cylindrically shaped convection cells having axis of symmetry along one of the horizontal directions, $\hat{\mathbf{y}}$, say, (see Fig. 7.9). Thus, the spatial dependence of the system is on x and z only. In this case we may also set $v = 0$ in Eqs. (8.12)–(8.14). The equations of motion may be simplified by introducing a stream function. In this case we define the stream function $\psi = \psi(x, z, t)$ via

$$w = -\frac{\partial \psi}{\partial x}, \qquad u = \frac{\partial \psi}{\partial z}, \tag{8.15}$$

with the continuity equation automatically satisfied. If we define the $\hat{\mathbf{y}}$ component of the vorticity by

$$\zeta_Y' \equiv \frac{\partial u}{\partial z} - \frac{\partial w}{\partial x} = \nabla^2 \psi,$$

where the Laplacian, ∇^2, is understood to be of two-dimensional form, in the coordinates x, z, for the remainder of this analysis. Now operating with $\nabla \times$ on Eq. (8.12), we find

$$\frac{\partial \nabla^2 \psi}{\partial t} + \{\nabla^2 \psi, \psi\} = \Pr \nabla^4 \psi - \Pr \mathrm{Ra} \frac{\partial \Theta}{\partial x}, \tag{8.16}$$

where $\{\cdot\}$ is here, and in what follows, the classical Poisson bracket operation for canonical coordinates x and z:

$$\{f, g\} \equiv \frac{\partial f}{\partial x} \frac{\partial g}{\partial z} - \frac{\partial f}{\partial z} \frac{\partial g}{\partial x}. \tag{8.17}$$

Equation (8.16) is supplemented by Eq. (8.13) rewritten in terms of the stream function:

$$\frac{\partial \Theta}{\partial t} + \{\Theta, \psi\} = -\frac{\partial \psi}{\partial x} + \nabla^2 \Theta. \tag{8.18}$$

To keep the calculations relatively simple, we will consider here the fate of roll solutions under stress-free, fixed temperature boundary conditions, i.e., $w = \partial u/\partial z = \Theta = 0$ on $z = 0, 1$. In the previous chapter, this case was referred to as {"free-free," fixed temperature}. In the present stream function formulation, this amounts to enforcing that

$$\frac{\partial \psi}{\partial x} = 0, \qquad \frac{\partial^2 \psi}{\partial z^2} = 0, \qquad \Theta = 0, \qquad \text{on} \quad z = 0, 1. \tag{8.19}$$

Thus, the equations which will concern us for the remainder of this subsection are (8.16) and (8.18) together with the boundary conditions (8.19). Recall that in Sect. 7.4.2, the minimum value of Ra required for the onset of convection using these boundary conditions was found to be $\mathrm{Ra}_c = 27\pi^4/4$, occurring at a critical horizontal wavenumber $k = k_c = \pi/\sqrt{2}$. The reader is invited to review the calculations presented there for details. We are now interested in the behavior of the least stable mode in the vicinity of these critical values. A Taylor series expansion around the critical values results in the following expression for the least stable growth rate:

$$p_1 \approx \left(\pi^2 + k_c{}^2\right)\left(\frac{\text{Pr}}{1+\text{Pr}}\right)\frac{\text{Ra} - \text{Ra}_c}{\text{Ra}_c} + \left(\frac{\partial^2 p_1}{\partial k^2}\right)_{k=k_c}(k-k_c)^2 \qquad (8.20)$$

where

$$\left(\frac{\partial^2 p_1}{\partial k^2}\right)_{k=k_c} = -\frac{3\pi^2}{2}\left(\frac{\text{Pr}}{1+\text{Pr}}\right)\frac{4\sqrt{3}}{2\text{Ra}_c}.$$

We may now identify precisely our slow time and space variables. As discussed in Sect. 7.4.2 and depicted in Fig. 7.10, the growth rate, near onset, scales by the greater of $(k-k_c)^2/k_c^2$ and $(\text{Ra} - \text{Ra}_c)/\text{Ra}_c$. Accordingly, we define a new control parameter $\varepsilon \ll 1$, assuming that the departure from the critical wavenumber scales approximately as ε and that the departure from the critical value of Ra scales as ε^2. In this way, the two terms comprising the right-hand side of (8.20) are comparable in their influence on the magnitude of the growth rate p (i.e., appropriate balance, sometimes referred to as "distinguished limit," is achieved). This insight is the crux of the weakly nonlinear approximation that we shall apply. Thus, we may formally define ε via

$$\text{Ra}_2 = \frac{1}{\varepsilon^2}\left(\frac{\text{Ra}}{\text{Ra}_c} - 1\right) \implies \text{Ra} = \text{Ra}_c\left(1 + \varepsilon^2 \text{Ra}_2\right),$$

additionally defining the new $\mathcal{O}(1)$ control parameter for the instability, Ra_2. The assumption of small departures from the critical wavenumber k_c leads to the two distinct, but coexisting, length scales x and X, where $x \sim 2\pi/k_c$ is the primary scale associated with the critical wavenumber and $X = \varepsilon x$ is the long horizontal length scale. The long length scale can be interpreted as an amplitude variation of the primary waveform of wavenumber k_c, just as in our discussion in Chap. 4 on the derivation of the KdV equation. Functionally we write this dependence upon the secondary length scale as $f(x) \mapsto f(x, \varepsilon x = X)$. It is implicitly assumed, of course, that the long length scale is not larger than the horizontal size of the physical domain, L_x.

In a similar fashion, we observe that the only timescale available to the system, near instability onset, scales as T/ε^2 where T is also $\mathcal{O}(1)$. These observations lead us to the ansatz in which the evolution of the system depends on these multiple space and timescales. Or, in terms of the stream function, ψ:

$$\psi(x, z, t; \text{Ra}, \text{Pr}) \mapsto \psi\left(x, \varepsilon x, z, \varepsilon^2 t; \text{Ra}_c, \text{Ra} - \text{Ra}_c, \text{Pr}\right) = \psi\left(x, X, z, T; \text{Ra}_c, \text{Ra}_2, \text{Pr}, \varepsilon\right).$$

There is now explicit dependence on εx and $\varepsilon^2 t$, indicating that these new dependencies are felt once $x \sim 1/\varepsilon$ and $t \sim 1/\varepsilon^2$. Generally, any error in these assumed dependencies will be revealed during the development of the perturbation solutions as an inconsistency.

We are compelled to now express the partial derivative operators, appearing in Eqs. (8.16), (8.18), and (8.19) as follows:

$$\frac{\partial}{\partial t} \mapsto \varepsilon^2 \frac{\partial}{\partial T}, \qquad \frac{\partial}{\partial x} \mapsto \frac{\partial}{\partial x} + \varepsilon \frac{\partial}{\partial X}, \qquad \nabla^2 \mapsto \nabla^2 + 2\varepsilon \frac{\partial}{\partial x}\frac{\partial}{\partial X} + \varepsilon^2 \frac{\partial^2}{\partial X^2}.$$

$$(8.21)$$

and apply the newly defined operators in the governing equations (8.16), (8.18), and boundary conditions (8.19), which become:

$$\varepsilon^2 \frac{\partial}{\partial T} \left(\nabla^2 \psi + 2\varepsilon \frac{\partial}{\partial x}\frac{\partial \psi}{\partial X} + \varepsilon^2 \frac{\partial^2 \psi}{\partial X^2} \right) + \left\{ \nabla^2 \psi + 2\varepsilon \frac{\partial}{\partial x}\frac{\partial \psi}{\partial X} + \varepsilon^2 \frac{\partial^2 \psi}{\partial X^2}, \psi \right\}$$

$$+ \varepsilon \left\{ \nabla^2 \psi + 2\varepsilon \frac{\partial}{\partial x}\frac{\partial \psi}{\partial X} + \varepsilon^2 \frac{\partial^2 \psi}{\partial X^2}, \psi \right\}_X + \mathrm{PrRa}\frac{\partial \Theta}{\partial x} + \varepsilon\mathrm{PrRa}\frac{\partial \Theta}{\partial X}$$

$$= \mathrm{Pr} \left(\nabla^4 \psi + 4\varepsilon \nabla^2 \frac{\partial}{\partial x}\frac{\partial \psi}{\partial X} + 2\varepsilon^2 \nabla^2 \frac{\partial^2 \psi}{\partial X^2} \right.$$

$$\left. + 4\varepsilon^2 \frac{\partial^2}{\partial x^2}\frac{\partial^2 \psi}{\partial X^2} + 4\varepsilon^3 \frac{\partial}{\partial x}\frac{\partial^3 \psi}{\partial X^3} + \varepsilon^4 \frac{\partial^4 \psi}{\partial X^4} \right),$$

$$(8.22)$$

$$\varepsilon^2 \frac{\partial \Theta}{\partial T} + \{\Theta, \psi\} + \varepsilon\{\Theta, \psi\}_X$$

$$= -\frac{\partial \psi}{\partial x} - \varepsilon \frac{\partial \psi}{\partial x} + \nabla^2 \Theta + 2\varepsilon \frac{\partial}{\partial x}\frac{\partial \Theta}{\partial X} + \varepsilon^2 \frac{\partial^2 \Theta}{\partial X^2},$$

$$(8.23)$$

where the subscript X on the Poisson bracket indicates the use of X, z rather than x, z as canonical coordinates. The equations are supplemented by the boundary conditions

$$\frac{\partial \psi}{\partial x} + \varepsilon \frac{\partial \psi}{\partial X} = \frac{\partial^2 \psi}{\partial z^2} = \Theta = 0, \qquad \text{on} \qquad z = 0, 1. \qquad (8.24)$$

8.3.1   The Amplitude Equation

The strategy now is to write the desired solution as an asymptotic expansion. The expansion must be constructed so that nonlinearities are introduced at the correct order of the expansion scheme. There is no general prescription for choosing the correct leading order scalings, as this will vary from problem to problem. As we saw at the beginning of this chapter, trial and error will usually yield the correct relative scalings. The guiding principle is to arrange the relative scalings so that the simplest nonlinearity gets introduced at some order of the perturbation expansion beyond the lowest one. For this classic Rayleigh–Bénard convection problem for rolls, we find that the asymptotic series solution ansatz that works is

$$\psi(x,X,z,T) = \varepsilon\psi_1(x,X,z,T) + \varepsilon^2\psi_2(x,X,z,T)$$

$$+\varepsilon^3 \psi_3(x,X,z,T) + \text{HOT} \qquad (8.25)$$

$$\Theta(x,X,z,T) = \varepsilon\Theta_1(x,X,z,T) + \varepsilon^2\Theta_2(x,X,z,T)$$

$$+\varepsilon^3\Theta_3(x,X,z,T) + \text{HOT} \qquad (8.26)$$

$$\text{Ra} = \text{Ra}_c(1 + \varepsilon^2\text{Ra}_2). \qquad (8.27)$$

It is based on our scaling of the departures from the critical values of Ra. The first nontrivial nonlinearity, the vertical advection of the perturbed temperature gradient, will appear at order ε^2. Following the spirit of the calculation that led to the KdV equation at the end of Chap. 4, we substitute expansions (8.25)–(8.27) directly into the governing equations (8.22)–(8.24) and solve the resulting equations order by order in powers of ε. Next, we summarize this procedure for the first three orders of the expansion.

At order ε the equations to be solved are:

$$\text{PrRa}_c \frac{\partial\Theta_1}{\partial x} = \text{Pr}\nabla^4\psi_1 \qquad (8.28)$$

$$\frac{\partial\psi_1}{\partial x} = \nabla^2\Theta_1, \qquad (8.29)$$

which are, by construction, the same system that we encountered in Sect. 7.4.2, albeit in vorticity-temperature form, as opposed to the vertical velocity-temperature form. Combining these into a single equation for ψ_1, we find

$$\mathbb{L}_1\psi_1 \equiv \left[\left(\frac{\partial^2}{\partial z^2} + \frac{\partial^2}{\partial x^2}\right)^3 + \text{Ra}_c\frac{\partial^2}{\partial x^2}\right]\psi_1 = 0,$$

where we have defined $\mathbb{L}_1$ as the lowest order linear partial differential operator. It is no surprise that $\mathbb{L}_1$ is independent of the slow variables X and T. The underlying mode of the solution has horizontal wavenumber k_c and the system is separable, allowing us to write the solution to ψ_1 as

$$\psi_1 = A(X,T)\psi_{11}(z)e^{ik_cx} + \text{c.c.}, \qquad (8.30)$$

where the *amplitude* $A(X,T)$ is, as yet, an unknown function of the slow variables (X,T). It will be determined at order ε^3. We now have the ODE

$$\mathbb{L}_{11}\psi_{11} \equiv \left[\left(\frac{d^2}{dz^2} - k_c^2\right)^3 - \text{Ra}_ck_c^2\right]\psi_{11} = 0, \qquad (8.31)$$

and the boundary conditions

$$\psi_{11} = 0, \qquad \frac{d^2 \psi_{11}}{dz^2} = 0, \qquad \frac{d^4 \psi_{11}}{dz^4} = 0, \qquad \text{on} \quad z = 0, 1.$$

As we expect, the solution to this system, with $\mathrm{Ra}_c = 27\pi^4/4$ and $k_c = \pi/\sqrt{2}$, is simply $\psi_{11} = \sin \pi z$, as we have already found. The corresponding temperature solution is

$$\Theta_1 = A(X, T)\Theta_{11}(z)e^{ik_c x} + \text{c.c.} \qquad \text{with} \qquad \Theta_{11}(z) \equiv -\frac{ik_c}{\pi^2 + k_c^2} \sin \pi z. \qquad (8.32)$$

At the next order in ε, the governing equations are

$$\mathrm{PrRa}_c \frac{\partial \Theta_1}{\partial X} + \mathrm{PrRa}_c \frac{\partial \Theta_2}{\partial x} + \{\nabla^2 \psi_1, \psi_1\} = \mathrm{Pr}\nabla^4 \psi_2 + 4\mathrm{Pr}\nabla^2 \frac{\partial^2 \psi_1}{\partial x \partial X}, \qquad (8.33)$$

$$\frac{\partial \psi_2}{\partial x} + \frac{\partial \psi_1}{\partial X} + \{\Theta_1, \psi_1\} = \nabla^2 \Theta_2 + 2\frac{\partial^2 \Theta_1}{\partial x \partial X}, \qquad (8.34)$$

and the corresponding boundary conditions at this order are

$$\frac{\partial \psi_2}{\partial x} + \varepsilon \frac{\partial \psi_{11}}{\partial X} = 0, \qquad \frac{\partial^2 \psi_2}{\partial z^2} = 0, \quad \Theta_2 = 0, \qquad \text{on} \quad z = 0, 1. \qquad (8.35)$$

The first of these simplifies to $\partial \psi_2/\partial x = 0$ since $\psi_{11} = 0$ on $z = 0, 1$. Following the same procedure used at the previous order, we may combine the two equations into a single one for ψ_2. We begin by noting that the solution for ψ_1, found in Eq. (8.30), implies that $\{\nabla^2 \psi_1, \psi_1\} = 0$. We then combine Eqs. (8.33) and (8.34) into a single equation for ψ_2, by operating on Eq. (8.33) with ∇^2 followed by replacing $\nabla^2 \Theta_2$ using Eq. (8.34). The result may be expressed in the form

$$\mathbb{L}_1 \psi_2 = \mathrm{Ra}_c \frac{\partial \nabla^2 \Theta_1}{\partial X} + \mathrm{Ra}_c \frac{\partial}{\partial x}\left[\{\Theta_1, \psi_1\} - 2\frac{\partial^2 \Theta_1}{\partial x \partial X} + \frac{\partial \psi_1}{\partial X}\right] - 4\nabla^4 \frac{\partial^2 \psi_1}{\partial x \partial X},$$

$$= \frac{\partial}{\partial X}\left(\mathrm{Ra}_c \nabla^2 \Theta_1 - 2\mathrm{Ra}_c \frac{\partial^2 \Theta_1}{\partial x^2} + \mathrm{Ra}_c \frac{\partial \psi_1}{\partial x} - 4\nabla^4 \frac{\partial \psi_1}{\partial x}\right) + \mathrm{Ra}_c \frac{\partial \{\Theta_1, \psi_1\}}{\partial x}.$$

$$(8.36)$$

One can show that the right-hand side of the above expression is equal to zero. Using the solution found for ψ_1 and Θ_1 in $\{\Theta_1, \psi_1\}$ indicates, after a short calculation, that

$$\{\Theta_1, \psi_1\} = -\frac{2k_c^2 \pi}{k_c^2 + \pi^2}|A|^2 \sin(2\pi z) \qquad (8.37)$$

which is clearly independent of the coordinate x. We shall see below that $\{\Theta_1, \psi_1\}$ is the sole source of saturation in the system near instability onset and that this term indicates how the convective roll flattens out the mean temperature gradient. But first, we note that insertion of the solutions for ψ_1 and Θ_1 into the expression

$$\mathrm{Ra}_c \nabla^2 \Theta_1 - 2\mathrm{Ra}_c \frac{\partial^2 \Theta_1}{\partial x^2} + \mathrm{Ra}_c \frac{\partial \psi_1}{\partial x} - 4\nabla^4 \frac{\partial \psi_1}{\partial x},$$

yields 0, since $\mathrm{Ra}_c = 24\pi^4/4$, and $k_c = \pi/\sqrt{2}$. The vanishing of the right-hand side of (8.36) is no accident: it is indicative of the fact that we are considering the dynamics of the Rayleigh–Bénard system in the vicinity of onset, that is, at values of Ra and k where the system is extremal, so that the rate of change of the linear growth rate is necessarily zero. Thus, we find that ψ_2 vanishes at this order[3] and it is easy to check that the solutions are

$$\psi_2 = 0, \qquad \Theta_2 = \Theta_{20}(z, X, T) + \left[\Theta_{21}(z, X, T) e^{ik_c x} + \text{c.c.} \right], \tag{8.38}$$

where

$$\Theta_{20}(z, X, T) = -\frac{k_c^2}{2\pi(\pi^2 + k_c^2)} |A(X, T)|^2, \qquad \Theta_{21}(z, X, T) = \frac{k_c^2 - \pi^2}{(\pi^2 + k_c^2)^2} \frac{\partial A}{\partial X}. \tag{8.39}$$

The first nonlinear corrections appear in the expression for Θ_2. Suppose we are interested in calculating the horizontal average of the total temperature profile $\overline{T}$, that is, the horizontal average of the combination of the base temperature profile plus the disturbances. We have seen that this horizontal mean will be modified by nonlinearity only at order ε^2, that is,

$$\overline{T} \sim 1 - (\beta d/T_0)z - \left(\frac{\mathrm{Ra} - \mathrm{Ra}_c}{\mathrm{Ra}_c} \right) \frac{k_c^2 |A(X, T)|^2}{2\pi(\pi^2 + k_c^2)} \sin(2\pi z).$$

Inspection of the above equation's form shows clearly that the base temperature gradient $1 - (\beta d/T_0)z$ will be reduced in the region around $z = 1/2$ due to the $-\sin(2\pi z)$ form of the correction, no matter what the form of A. Once we know A, then we may estimate by how much the temperature profile is flattened as the system reaches saturation.

---

[3]It is true that $\mathbb{L}_1 \psi_2 = 0$ implies that there is a solution of arbitrary amplitude, say, $B(X, T)$, which is of the same form as expressed in (8.30). We are not concerned with these solutions, because they do not influence the evolution equation for A that is determined later. It is common practice to absorb this solution implicitly into the one appearing at the previous order.

Next, we turn to the task of finding the amplitude A, which requires analysis of the order ε^3 system. At order ε^3 the equations are the following:

$$
\frac{\partial \nabla^2 \psi_1}{\partial T} + \left\{ 2\frac{\partial^2 \psi_1}{\partial x \partial X}, \psi_1 \right\} + \{\nabla^2 \psi_1, \psi_1\}_X + \text{PrRa}_c \frac{\partial \Theta_3}{\partial x} + \text{PrRa}_2 \frac{\partial \Theta_1}{\partial x}
$$

$$
+ \text{PrRa}_c \frac{\partial \Theta_2}{\partial X} = \text{Pr}\left(\nabla^4 \psi_3 + 2\varepsilon^2 \nabla^2 \frac{\partial^2 \psi_1}{\partial X^2} + 4\varepsilon^2 \frac{\partial^2}{\partial x^2}\frac{\partial^2 \psi_1}{\partial X^2}, \right) \tag{8.40}
$$

$$
\frac{\partial \Theta_1}{\partial T} + \{\Theta_2, \psi_1\} + \{\Theta_1, \psi_1\}_X = -\frac{\partial \psi_3}{\partial x} + \nabla^2 \Theta_3 + 2\varepsilon \frac{\partial}{\partial x}\frac{\partial \Theta_2}{\partial X} + \varepsilon^2 \frac{\partial^2 \Theta_1}{\partial X^2}, \tag{8.41}
$$

with the boundary conditions

$$
\frac{\partial \psi_3}{\partial x} = 0, \qquad \frac{\partial^2 \psi_3}{\partial z^2} = 0, \qquad \Theta_3 = 0, \qquad \text{on} \quad z = 0, 1,
$$

in which we have applied that $\psi_2 = 0$. Once again we repeat the same procedure as for the previous orders, combining the two partial differential equations into a single one for the quantity ψ_3, inserting solutions for ψ_1 and Θ_2 and simplifying to find

$$
\mathbb{L}_1 \psi_3 = N_{30}(X,T)\sin(2\pi z) + \left[N_{31}(X,T)e^{ik_c x} + \text{c.c.} \right]\sin(\pi z)
$$

$$
+ \left[N_{33}(X,T)e^{3ik_c x} + \text{c.c.} \right]\sin(3\pi z), \tag{8.42}
$$

where the quantities N_{3j} are explicit expressions involving operators acting on A. We are interested here only in $N_{31}(X,T)$, which is given by

$$
N_{31}(X,T) = (\pi^2 + k_c^2)^2\left[\frac{\partial A}{\partial T} - \mu A - \alpha\frac{\partial^2 A}{\partial X^2} + \beta|A|^2 A \right], \tag{8.43}
$$

where

$$
\mu \equiv (\pi^2 + k_c^2)\frac{\text{Pr}}{1+\text{Pr}}\text{Ra}_2, \qquad \alpha \equiv 3\pi^2\left(\frac{\text{Pr}}{1+\text{Pr}}\right)\frac{\sqrt{3}}{\text{Ra}_c}, \qquad \beta \equiv \frac{k_c^2}{2}\frac{\text{Pr}}{1+\text{Pr}}.
$$

The equation for ψ_3 indicates that we should seek solutions of the form

$$
\psi_3 = A_{30}(X,T)\psi_{30}(z) + \left[A_{31}(X,T)\psi_{31}(z)e^{ik_c x} + \text{c.c.} \right] + \left[A_{33}(X,T)\psi_{33}(z)e^{3ik_c x} + \text{c.c.} \right], \tag{8.44}
$$

where $A_{3j}(X,T)$ are unknown functions to be determined. Substituting the above solution into Eq. (8.42) leads, after equating like powers of $e^{ik_c x}$ to zero, to a series of ordinary differential equations for each of the functions $\psi_{30}, \psi_{31}, \psi_{33}$. Our main concern is with a solution for ψ_{31}. The associated ODE for it is

$$A_{31}(X,T)\mathbb{L}_{11}\psi_{31} = N_{31}(X,T)\sin(\pi z). \tag{8.45}$$

The boundary conditions on ψ_{31} must satisfy

$$\psi_{31} = 0, \qquad \frac{d^2\psi_{31}}{dz^2} = 0, \qquad \frac{d^4\psi_{31}}{dz^4} = 0, \qquad \text{on} \quad z = 0,1.$$

We note that these boundary conditions and the ordinary differential operator acting on ψ_{31} are identical to those for the function ψ_{11}. Because we are considering the growth of a mode at onset, the linear operator clearly has a zero eigenvalue, the growth rate, and thus the operator has a non-empty null space. In solving an inhomogeneous equation like (8.45), it is a general result that finite solutions exist only when the inhomogeneous term has no components in that null space. Usually, this is expressed by the so-called solvability condition statement that the right-hand side must be orthogonal to the null space of $\mathbb{L}^\dagger$, the operator's adjoint.[4] The more general statement of the solvability condition is a direct result of the so-called Fredholm alternative for linear operators, $\mathbb{L}$, acting on functions, f,g, in a Hilbert space with inner product $\langle f,g\rangle$.[5] The Fredholm alternative says that **either** $\mathbb{L}f = 0$ has no nontrivial solutions **or** the adjoint homogeneous equation, $\mathbb{L}^\dagger g = 0$, has a nontrivial solution $g \neq 0$. If the latter condition is met, it follows that $\mathbb{L}f = b$ can have a solution if and only if $\langle b,f\rangle = 0$ for all nonzero g satisfying $\mathbb{L}^\dagger g = 0$. A more in-depth exposition of the Fredholm alternative can be found in many graduate-level PDE texts or in B. Friedman's excellent book, reference [6] in the *Bibliographical Notes*. Returning to our problem, we multiply Eq. (8.45) by $\psi_{11} = \sin(\pi z)$. Integrating by parts and applying the boundary conditions leads to the result $N_{31}(X,T) = 0$ which leads directly to the *Ginzburg–Landau* equation:

$$\frac{\partial A}{\partial T} = \mu A + \alpha \frac{\partial^2 A}{\partial X^2} - \beta |A|^2 A. \tag{8.46}$$

Because the coefficients μ α and β are real, this is the real Ginzburg–Landau equation, or RGL for short. Steady solutions of the RGL for $\mu > 0$ may be found once horizontal boundary conditions are specified. When the X domain is infinite, but A remains bounded at infinity, we find the steady solutions:

$$A_0(X) = \pm \sqrt{\frac{\mu}{\beta}} \tanh\left[\sqrt{\frac{\mu}{2\beta\alpha}}(X - X_0)\right], \tag{8.47}$$

---

[4]Because we are dealing here with differential operators, the definition of $\mathbb{L}^\dagger$ has to include a choice of a Hilbert space with a well-defined inner product. In the current calculation, the relevant operators are self-adjoint; the reader is referred to references in the *Bibliographical Notes* for more detailed discussion of adjoint operators.

[5]A typical inner product for square integrable functions f,g defined on the interval $[x_L, x_R]$ is $\langle f,g\rangle \equiv \int_{x_L}^{x_R} f^* g\, dx$.

where the positive root may usually be selected without loss of generality. The presence of the constant X_0 in the solution is a consequence of the translational invariance of the RGL. It is this invariance that also accounts for the lack of quadratic nonlinearities in RGL (use complex exponentials to show why!). The transition region of the tanh has a length scale given by $\Delta X \sim \sqrt{2\alpha\beta/\mu}$. There are also X-independent solutions given by $A_0 = \pm\sqrt{\mu/\beta} \equiv \pm A_{00}$, which are also valid on a periodic domain with periodicity L_X, say. We may test the linear stability of these solutions by examining the RGL subject to linear perturbations of the form $A(T) = \pm A_{00} + A'(T)$ where A' is small. A simple analysis shows that for solutions of the form $A' \propto e^{rT+ikL}$, the growth rate

$$r = -2\mu - k^2\alpha,$$

indicating linear stability. This suggests that the final state for the pattern of constant relative phase rolls will have maximum amplitude given by

$$\max(|A|) \le A_{00} = \sqrt{\frac{\mu}{\beta}} = \sqrt{\frac{\pi^2 + k_c^2}{k_c^2/2}\mathrm{Ra}_2} = \frac{1}{\varepsilon}\sqrt{6\left(\frac{\mathrm{Ra}}{\mathrm{Ra}_c} - 1\right)},$$

where we have explicitly replaced $k_c^2 = \pi^2/2$ and the expression of Ra_2 has been restored in terms of deviations from the critical value of Ra.

We may now complete the full solution for the rolls and write the leading order solution for ψ by restoring variables in terms of their original definitions. Doing this for the tanh solution, we find modulated rolls of the form

$$\psi(x,z,t) = \varepsilon A_0(X)e^{ik_c x}\sin(\pi z) + \text{c.c.} + \mathscr{O}(\varepsilon^2)$$

$$\approx \underbrace{2\sqrt{6\left(\frac{\mathrm{Ra}}{\mathrm{Ra}_c} - 1\right)}\tanh\left(\frac{x-x_0}{\Delta x}\right)}_{\text{slow envelope variation}} \times \underbrace{\cos\left(\frac{\pi}{\sqrt{2}}x\right)\sin(\pi z)}_{\text{fundamental roll profile}}, \qquad (8.48)$$

where the transition scales are related by $\Delta X = \varepsilon\Delta x = \varepsilon\sqrt{\mu/2\alpha\beta}$, defining the scale in physical units.

8.3.2 Patterns in Three Dimensions

The laboratory study of three-dimensional convection shows that roll solutions are selected when the horizontal dimensions are disparate in size. However, when this restriction is relaxed, that is, when the horizontal scales L_x and L_y are comparable, patterned states may emerge, such as the squares and hexagons seen in laboratory and numerical experiments (Fig. 8.3). Under these conditions a weakly nonlinear

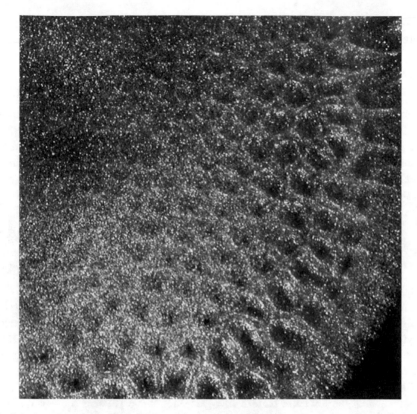

Fig. 8.3 Experimental depiction of convection cells in a Rayleigh–Bénard experiment. This image shows the pattern of convection cells in the NOAA exhibit. The fluid is silicone oil with a density similar to water. Pearl essence, a shimmery powder, is mixed in to the oil so you can see the movement of the fluid and the patterns formed by convection *Public Domain, courtesy NOAA, Earth System Research Laboratory; http://www.esrl.noaa.gov/psd/outreach/education/science/convection/Table.html*

analysis becomes more involved especially if one is interested in multiple space and timescale dynamics. By considering horizontal scales similar in magnitude, however, each of order λ_c in size, the weakly nonlinear analysis may be simplified to introducing only multiple timescales. This is what is shown in Fig. 8.3.

A linear analysis of the problem shows that the onset of stationary convection occurs for the same values of Ra_c and k_c as in the two-dimensional case. We assume small departures from Ra_c, as before, given by $\mathrm{Ra} - \mathrm{Ra}_c = \mathrm{Ra}_c \varepsilon^2$, and assume, for the sake of this example, that the two horizontal scales of the domain L_X, L_Y are of the same order of magnitude as the critical wavelength $2\pi/k_c$. This latter condition means that we do not need to seek solutions involving long spatial variables, simplifying the calculation. As before, however, the system responds temporally on a scale $\sim 1/\varepsilon^2$. We expect, therefore, that a set of coupled Landau-type equations will describe the resulting dynamics. Based on the weakly nonlinear solution for steady rolls, the leading order solution for, say, the temperature perturbation representing squares may be given by

$$\Theta \sim \varepsilon \left[\{ A(T)e^{ik_c x} + \text{c.c.} \} + \{ B(T)e^{ik_c y} + \text{c.c.} \} \right] + \varepsilon^2 \left[\cdots \right] + \varepsilon^3 \left[\cdots \right] + \text{HOT} \tag{8.49}$$

where A, B are amplitudes depending on the slow time variable. Similar expansions would apply to the velocity fields u, v, and w. The leading order terms above appear as squares, and it is our task to calculate their amplitudes. We anticipate that this evolution will involve cross-amplitude interactions but otherwise follow a similar strategy as in our examination of rolls. The detailed calculation is not shown here, but the result leads to the coupled system:

$$\frac{dA}{dT} = \mu A - \beta |A|^2 A - \gamma |B|^2 A, \tag{8.50}$$

$$\frac{dB}{dT} = \mu B - \gamma |A|^2 B - \beta |B|^2 B, \tag{8.51}$$

where μ, γ, and β are real coefficients and μ is the control parameter. It is a straightforward exercise to show that stationary squares are stable and have equal amplitude, $|A| = |B|$, when the cross-mode coupling constant is less than β, or $|\gamma| < \beta$. For values of $|\gamma|$ larger than this, the system becomes unstable and returns to a roll state of one orientation or the other.

Similar considerations involve the analysis of hexagons. It is important to remember that the hexagon pattern involves three wave patterns $\mathbf{k}_i$, $i = 1, 2, 3$, in which $|\mathbf{k}_i| = k_c$ and the three wave vectors add to zero: $\mathbf{k}_1 + \mathbf{k}_2 + \mathbf{k}_3 = 0$. Assuming that at leading order the perturbations have a form like

$$[A(T)e^{i\mathbf{k}_1 \cdot \mathbf{x}} + \text{c.c.}] + [B(T)e^{i\mathbf{k}_2 \cdot \mathbf{x}} + \text{c.c.}] + [C(T)e^{i\mathbf{k}_3 \cdot \mathbf{x}} + \text{c.c.}] \tag{8.52}$$

a similar procedure leads to the following coupled amplitude equations:

$$\frac{dA}{dT} = \mu A - \beta |A|^2 A - \gamma \left(|B|^2 + |C|^2 \right) A, \tag{8.53}$$

$$\frac{dB}{dT} = \mu B - \beta |B|^2 B - \gamma \left(|C|^2 + |A|^2 \right) B, \tag{8.54}$$

$$\frac{dC}{dT} = \mu C - \beta |C|^2 C - \gamma \left(|A|^2 + |B|^2 \right) C. \tag{8.55}$$

Further analysis can be done on these equations as well, and this is left as an exercise for the reader (Problem 8.6).

8.4 Stationary Rolls in Rayleigh–Bénard Convection II: Fixed Flux

We now examine Rayleigh–Bénard convection under fixed thermal flux boundary conditions. The equations of motion that we shall investigate are the same, Eqs. (8.16)–(8.18); however, the boundary conditions are now

$$\frac{\partial \psi}{\partial x} = 0, \qquad \frac{\partial^2 \psi}{\partial z^2} = 0, \qquad \frac{\partial \Theta}{\partial z} = 0, \qquad \text{on} \quad z = 0, 1. \tag{8.56}$$

We shall consider the same departure from Ra, viz., $\mathrm{Ra} = \mathrm{Ra}_c (1 + \varepsilon^2 \mathrm{Ra}_2)$. This corresponds to the case {"rigid-rigid," fixed flux} in the previous chapter.

The expansion procedure is similar to that of the previous section. Recall from the linear stability study in Sect. 7.4.2 that the onset of instability in this case occurs at nearly infinite horizontal wavelength. The weakly nonlinear study of fixed flux convection must therefore proceed in a slightly different manner. We make the following ansatz regarding the solutions: $\Theta = \Theta(z, \varepsilon X, \varepsilon^4 t) = \Theta(z, X, T)$ and similarly for ψ. This reflects the fact that we do not consider the dynamics of $\mathcal{O}(1)$ values of a horizontal wavenumber, instead; we restrict our attention to the largest horizontal scales. Notice how in this case the timescales are much longer than those assumed in the previous example ($\sim 1/\varepsilon^4$). It follows that the operators now take the forms

$$\frac{\partial}{\partial t} \to \varepsilon^4 \frac{\partial}{\partial T}, \qquad \frac{\partial}{\partial x} \to \varepsilon \frac{\partial}{\partial X}, \qquad \nabla^2 \to \frac{\partial^2}{\partial z^2} + \varepsilon^2 \frac{\partial^2}{\partial X^2}.$$

Applying these to the governing equations reveals the following scaled equations governing the evolution of rolls under conditions of fixed flux:

$$\varepsilon^4 \frac{\partial}{\partial T} \frac{\partial^2 \psi}{\partial z^2} + \varepsilon^4 \frac{\partial}{\partial T} \frac{\partial^2 \psi}{\partial X^2} + \varepsilon \left\{ \frac{\partial^2 \psi}{\partial z^2} + \varepsilon^2 \frac{\partial^2 \psi}{\partial X^2}, \psi \right\}_X$$
$$= \mathrm{Pr} \left(\frac{\partial^4 \psi}{\partial z^4} + 2\varepsilon^2 \frac{\partial^4 \psi}{\partial X^2 \partial z^2} + \varepsilon^4 \frac{\partial^4 \psi}{\partial z^4} \right) - \varepsilon \mathrm{Pr} \mathrm{Ra}_c (1 + \varepsilon^2 \mathrm{Ra}_2) \frac{\partial \Theta}{\partial X}, \tag{8.57}$$

$$\varepsilon^4 \frac{\partial \Theta}{\partial T} + \varepsilon \{\Theta, \psi\}_X = -\varepsilon \frac{\partial \psi}{\partial X} + \frac{\partial^2 \Theta}{\partial z^2} + \varepsilon^2 \frac{\partial^2 \Theta}{\partial X^2}. \tag{8.58}$$

In setting the relative amplitude scalings of the dependent variables, it turns out that the expansion which produces a consistently balanced (of distinguished limit) approximation for small ε^2 is the following:

$$\Theta = \Theta_0 + \varepsilon^2 \Theta_2 + \varepsilon^4 \Theta_4 + \mathrm{HOT} \tag{8.59}$$

$$\psi = \varepsilon \psi_1 + \varepsilon^3 \psi_3 + \mathrm{HOT}. \tag{8.60}$$

As we have done before, these expansions are substituted into the equations of motion (8.57)–(8.58) and separated into a sequence of equations in powers of ε. At $\mathscr{O}\left(1\right)$ of (8.58) and order $\mathscr{O}\left(\varepsilon\right)$ of (8.57), we have

$$\frac{\partial^2 \Theta_0}{\partial z^2} = 0, \quad \frac{\partial^4 \psi_1}{\partial z^4} = \text{Ra}_c \frac{\partial \Theta_0}{\partial X}; \quad \text{with} \quad \frac{\partial \Theta_0}{\partial z} = \frac{\partial \psi_1}{\partial X} = \frac{\partial^2 \psi_1}{\partial z^2} = 0 \text{ on } z = 0, 1.$$
(8.61)

The first of the above equations indicates that the lowest order temperature perturbation must be a constant in the vertical direction, viz., $\Theta_0 = A(X, T)$, a consequence of the zero flux perturbation condition at the bottom and top boundaries. Physically, we must infer that a vertically uniform temperature perturbation results from the thin layers where the perturbations are insulated. We may now find that the solution for ψ_1 is a fourth-order polynomial in z, given by

$$\psi_1(z, X, T) = z(z^3 - 2z + 1)\frac{\text{Ra}_c}{24}\frac{\partial A}{\partial X}.$$
(8.62)

Moving now to $\mathscr{O}\left(\varepsilon^2\right)$ terms of the temperature evolution equation (8.58), we obtain

$$\frac{\partial^2 \Theta_2}{\partial z^2} = -\frac{\partial^2 \Theta_0}{\partial X^2} + \frac{\partial \psi_1}{\partial X} + \frac{\partial \Theta_0}{\partial X}\frac{\partial \psi_1}{\partial z}$$
(8.63)

with boundary conditions $\partial \Theta_2 / \partial z = 0$ on $z = 0, 1$. This expression may be integrated and simplified to find the solvability condition:

$$\frac{\partial^2 A}{\partial X^2}\left(\frac{\text{Ra}_c}{120} - 1\right) = 0.$$
(8.64)

Thus, the asymptotic procedure has elegantly determined the critical value of the Rayleigh number for these conditions, $\text{Ra}_c = 5! = 120$. As an exercise, one may repeat the same calculation for a problem with rigid kinematic boundary conditions and fixed flux boundaries to find that the corresponding critical value becomes, instead, $\text{Ra}_c = 6! = 720$. The equation for ψ at order ε^3 becomes

$$\frac{\partial^4 \psi_3}{\partial z^4} = \frac{1}{\text{Pr}}\left\{\frac{\partial^2 \psi_1}{\partial z^2}, \psi_1\right\}_X - 2\frac{\partial^4 \psi_1}{\partial X^2 \partial z^2} + \text{Ra}_c\frac{\partial \Theta_2}{\partial X} + \text{Ra}_c \text{Ra}_2\frac{\partial \Theta_0}{\partial X},$$
(8.65)

with boundary conditions $\psi_3 = \partial^2 \psi_3 / \partial z^2 = 0$ on $z = 0, 1$. It is a straightforward, but tedious, calculation to show that the solution for ψ_3, which satisfies the requisite boundary conditions, is

$$\psi_3(z, X, T) = \text{Ra}_2 \psi_{30}(z)\frac{\partial A}{\partial X} + \psi_{31}(z)\frac{\partial^3 A}{\partial X^3} + + \left[\frac{1}{\text{Pr}}\psi_{33}^{(o)}(z) + \psi_{33}^{(e)}(z)\right]\frac{\partial}{\partial X}\left(\frac{\partial A}{\partial X}\right)^2,$$
(8.66)

where the functions $\psi_{3j}(z)$ for $(j = 0, 1, 2, 3)$ are polynomials in z of varying degree (see Problem 8.5) and the superscripts (o) and (e) indicate an *odd* and *even* function, respectively.

To reach the point where we may determine a weakly nonlinear evolution equation for the amplitude $A(X, T)$ requires examination of the order ε^4 terms resulting from the expansion of (8.58). At this order we find

$$\frac{\partial^2 \Theta_4}{\partial z^2} = \frac{\partial \Theta_0}{\partial T} + \{\Theta_2, \psi_1\}_X + \{\Theta_1, \psi_1\}_X + \varepsilon \frac{\partial \psi_3}{\partial X} - \frac{\partial^2 \Theta_2}{\partial X^2}, \tag{8.67}$$

together with the boundary conditions $\partial \Theta_4 / \partial z = 0$ on $z = 0, 1$. As before, our interest lies mainly in the solvability condition for the existence of solutions, more than in finding a solution for Θ_4 itself. Integrating over $z = [0, 1]$, we find this solvability condition governing the evolution of the amplitude A, namely,

$$\frac{\partial A}{\partial T} + \text{Ra}_2 \frac{\partial^2 A}{\partial X^2} + \frac{37}{168} \frac{\partial^4 A}{\partial X^4} - \frac{155}{189} \frac{\partial}{\partial X} \left(\frac{\partial A}{\partial X} \right)^3 = 0. \tag{8.68}$$

This amplitude equation is one form of the *Cahn–Hilliard* equation, an equation used to model phase separation.[6] A number of insights into the behavior of fixed flux convection may be uncovered from the properties and solutions of this equation. One feature that is immediately notable is the linear behavior of the amplitude around $A = 0$. Assuming solutions of the form $\propto e^{pT + ikX}$ leads to the following leading behavior of the eigenvalue p

$$p \sim k^2 \left(\text{Ra}_2 - \frac{37}{168} k^2 \right)$$

as is depicted in Fig. 8.4. The maximal growth rate occurs at a horizontal wavenumber $k_m = \sqrt{42 \text{Ra}_2 / 37}$. Steady solutions of this equation on a domain of scale L_X are expressed in terms of elliptic functions of the first kind, which can support any number of roll solutions. However, the ultimate nonlinear fate of these rolls is a steady sequence of transitions in which neighboring roll states consume each other, ultimately resulting in a single roll state filling out the whole of the domain with horizontal length scale L_X. This behavior is precisely the coarsening dynamics commonly seen in phase separation and which motivated J. Cahn and J. Hilliard.

---

[6] A similar equation would have been found had one adopted, instead, rigid boundary conditions throughout.

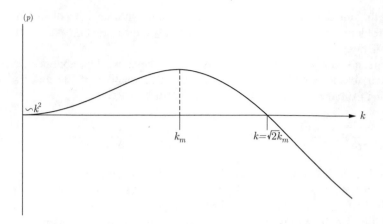

Fig. 8.4 Graphical representation of the growth rate associated with the linear perturbations around $A = 0$ of the amplitude equation (8.68). Growth occurs in a band out to a maximum wavenumber $= \sqrt{2}k_m$, with maximum growth rate at $k_m = \sqrt{42\mathrm{Ra}_{2_2}/37}$. Despite the fact that the maximal growth rate occurs at k_m, the ultimate configuration is for a roll of horizontal scale equal to the horizontal domain size L_x

8.5 Additional Systems

In this section, we quote, without performing the full necessary calculations, some qualitative nonlinear behavior of additional example systems and circumstances. The discussion below is by no means exhaustive and is meant to give merely a flavor of the mentioned systems.

8.5.1 Subcritical Transitions

Much of our discussion in Chap. 7 addressed cases where the linear instability leads to a supercritical transition. There are, however, many examples in fluid dynamics in which the linear instability indicates the presence of a subcritical bifurcation. A classic example is the transition to instability of Tollmien–Schlichting waves in plane Poiseuille flow. Plane Poiseuille flow is the plane-parallel analog of pipe Poiseuille flow, discussed in Sect. 7.6.1, having a base velocity profile, directed along the x-axis and given by

$$u_0(y) = 1 - y^2, \tag{8.69}$$

for a flow confined between the values of $y = \pm 1$, in dimensionless form. The stability boundaries for plane Poiseuille flow show the same qualitative shape as that for Blasius boundary layer flow (see Fig. 7.7a), but for plane Poiseuille flow, the critical Reynolds number is given by $\mathrm{Re}_c = 5774$. Experiments, on the other hand, show that there exists a transition into an unstable state for values of Re around 2000.

J. T. Stuart and J. Watson, independently in 1960, undertook a weakly nonlinear analysis to examine the amplitudes and nonlinear coefficient in a corresponding Landau equation, without investigating the effects of long trains like we have done for RB rolls by invoking multiple spatial scales. The marginal curve plots for plane Poiseuille flow feature a "nose," such that for values of $\mathrm{Re} > \mathrm{Re}_c$ there exist two marginal wavenumbers k_m. One can compute amplitude equations in terms of powers of small departures from this marginal state, $\varepsilon = \mathrm{Re} - \mathrm{Re}_m$, at either k_m value. Specifically, if A represents the amplitude of the marginal mode, then an amplitude equation is found of the form

$$\frac{\partial A}{\partial t} = \sigma\left(\mathrm{Re} - \mathrm{Re}_m, k_m\right)A - \beta\left(\mathrm{Re}_m, k_m\right)|A|^2 A. \tag{8.70}$$

For values of Re_m in the vicinity of $\mathrm{Re}_c \approx 5770$, the sign of the real part of the function β is negative, indicating that a supercritical transition is not saturated by the cubic nonlinearity. The instability is therefore subcritical and one expects that the series expansion must be expanded to at least quintic terms, giving, instead:

$$\frac{\partial A}{\partial t} = \sigma\left(\mathrm{Re} - \mathrm{Re}_m, k_m\right)A - \beta_2\left(\mathrm{Re}_m, k_m\right)|A|^2 A - \beta_4\left(\mathrm{Re}_m, k_m\right)|A|^4 A + \cdots \tag{8.71}$$

under the assumption that the next Landau coefficient, i.e., β_4, is greater than zero. In Sect. 8.2 we saw how the quintic term plays a stabilizing role in the Landau equation. Extensive examinations of this problem (see relevant references in the *Bibliographic Notes*) for situations where the basic state is not even linearly unstable, as when $\mathrm{Re} < 5774$, showed that the value of β_2 does not become positive again until about values of $\mathrm{Re} \approx 2000$.

Given the difficulty of obtaining a stable steady amplitude profile, investigators cleverly asked the question of whether or not there might exist stable propagating wave solutions to describe subcritical transition for this problem. After much effort a class of stable and unstable traveling wave solutions were discovered, while further investigations showed that even the stable propagating wave solution may undergo a secondary three-dimensional instability known as an elliptic instability. We do not discuss elliptical instability further, but references to it are found in the *Bibliographic Notes* of this chapter while secondary instabilities of steady Rayleigh–Bénard rolls are briefly discussed in Sect. 8.6.

8.5.2 Faraday Instabilities

M. Faraday first observed crispations on the surface of a vertically vibrated layer of fluid, recognizing these as standing wave deformations of the fluid–air interface. These parametrically excited surface waves exhibit a rich variety of wave crest patterns, including striped, square, hexagonal, and circular patterns, depending on the properties of the vibrational forcing and of the experimental design, including the liquid itself. Faraday experimented with both ordinary liquids, such as water and alcohol, and more exotic liquids, such as egg white and mercury, in which he reported patterns of "extreme beauty."

In an actual experiment, when a shallow fluid layer in an open container is vibrated vertically with amplitude δ and frequency ω_p, such that the effective gravitational acceleration felt by the layer is $g(t) = g_0(1 + \delta \cos \omega_p t)$, where g_0 is the acceleration due to gravity. For forcing of small amplitude, the surface of the liquid remains flat. Crispations begin to form once δ increases beyond a threshold value, δ_c, determined by the point at which dissipation, mainly due to viscosity, is overcome. The response is subharmonic, meaning that the surface waves, which oscillate at frequency ω_0, do so at half the driving frequency, or $\omega_p = 2\omega_0$. A full calculation of the linear and nonlinear Faraday instability is complicated, although we may employ the methods demonstrated in this chapter. Instead, we consider a linear time-dependent model equation for a surface displacement function η, describing the propagation of linear surface gravity waves (Chap. 4):

$$\frac{\partial^2 \eta}{\partial t^2} + \tau_d^{-1} \frac{\partial \eta}{\partial t} + \omega^2(k)\eta = 0, \qquad \omega^2(k) = g(t)k \tanh kL, \qquad (8.72)$$

where $\omega^2(k)$ is the frequency behavior in a uniform layer of fluid with depth L and horizontal wavenumber k. We have introduced a crude model for damping using the parameter τ_d^{-1}, and we further define a wavenumber k_0 and frequency ω_0 such that $\omega_0 = \omega(k_0) = \sqrt{g_0 k_0 \tanh k_0 L}$. The above model equation has the form of a *Mathieu* equation:

$$\frac{\partial^2 \eta}{\partial t^2} + \tau_d^{-1} \frac{\partial \eta}{\partial t} + \omega^2(k)\left[1 + 2\delta \cos(2\omega_0 t)\right]\eta = 0. \qquad (8.73)$$

When both τ_d^{-1} and δ are zero, then the above system exhibits oscillations with frequency ω. Otherwise, we may take $k = k_0 + k_1$ and assume that k_1, τ_d^{-1}, δ, and η are all equally small quantities scaled by the parameter $\varepsilon \ll 1$ and use a two-term perturbation series for $\eta = \eta_0 + \varepsilon \eta_1$. If we further assume that

$$\eta_0 = A(\varepsilon t)e^{-i\omega_0 t} + \text{c.c.} \qquad (8.74)$$

and carry the procedure out to second order, we are led to a solvability condition which, in turn, generates the following equation for the amplitude A:

$$\frac{\partial A}{\partial t} = (-\tau_d^{-1} + i\omega_1)A - i\delta A^*, \qquad \omega_1 \equiv \left(\frac{\partial \omega}{\partial k}\right)_{k=k_0} k_1. \qquad (8.75)$$

Solving for A's real and imaginary components, separately, and assuming $\propto e^{p_1 t}$ behavior of the solutions gives, for the growth rate,

$$p_1 = -\tau_d^{-1} \pm \sqrt{\delta^2 - \omega_1^2}, \tag{8.76}$$

clearly indicating that there must be a minimum vertical driving amplitude δ to trigger an instability. Generalizations of this equation can be found to study two-dimensional patterns, much in the same way that we discussed earlier for Rayleigh–Bénard convection. If the system is sufficiently horizontally extended, further increase in the driving amplitude δ may induce transverse secondary instabilities and, eventually, chaos.

8.5.3   Model Equations

Model equations refer to dimensionally reduced or otherwise highly simplified mathematical models of typically much more complicated systems. The Navier–Stokes equations themselves are a model for the collective behavior of a large number of molecules, such as those which make up a liquid or gas on macroscopic scale. In this book we have largely pursued approximate or highly simplified forms of solutions, themselves models of Navier–Stokes.

In Chap. 4 we used asymptotic methods to derive the KdV equation, a description which is asymptotic to the original Euler equations describing surface water waves. Despite the success of the KdV model, its validity breaks down for high-frequency wave disturbances, for which short length scales violate the asymptotic ordering assumed from the outset. In the case of the KdV equation, the term that causes problems is the triple spatial derivative dispersion inducing term $\partial_{xxx}\eta$. We can modify this term by noting the symmetry observed for solitary wave structures propagating at some phase velocity, viz., $\partial_t \eta + c\partial_x \eta \approx 0$. From this, one of the three derivatives in the problematic dispersive term may be replaced by a time derivative generating the *Bona–Benjamin–Mahoney* (BBM) equation describing the amplitude propagation of water waves:

$$\frac{\partial \eta}{\partial t} + \sqrt{gL}\,\frac{\partial \eta}{\partial x} + \frac{3}{2}\frac{\sqrt{gL}}{2L}\eta\frac{\partial \eta}{\partial x} - \frac{1}{6}L^2\frac{\partial}{\partial t}\left(\frac{\partial^2 \eta}{\partial x^2}\right) = 0, \tag{8.77}$$

a variation of the original KdV equation, cf. (4.107). The advantage of this equation is that it has clear bounds on the group and phase velocity of short wavelength waves. The one-dimensional version of BBM has single solitary wave solutions (Problem 8.10).

Another type of model equations may be found by a truncated approximation. Perhaps the most well-known example of this are the *Lorenz equations*, a truncated model of Rayleigh–Bénard convection. A complete Fourier/Galerkin expansion for the stream function and temperature fluctuations may be represented by

$$\psi(x,z,t) = \sum_{j,k} \psi_{jk}(t) e^{ikx} \sin j\pi z + \text{c.c.}, \qquad \theta(x,z,t) = \sum_{j,k} \theta_{jk}(t) e^{ikx} \sin j\pi z + \text{c.c.}$$

$$(8.78)$$

Substituting these expansion into the governing Boussinesq equations of motion generates an infinite set of coupled nonlinear ordinary differential equations for the time-dependent amplitudes ψ_{jk} and θ_{jk}. This procedure is referred to as a Galerkin projection. By retaining a large enough number of these modes, we end up with an acceptably accurate approximation of the complete system. Historically, E. Lorenz, mathematician and meteorologist, chose in the early 1960s to focus on three amplitudes: the $k = 1$ and $j = 1$ components of the temperature and stream function fluctuations and the $k = 0$ and $j = 2$ temperature fluctuation. The resulting truncated set, appropriately rescaled, is

$$\frac{d\psi_{11}}{dt} = \sigma(\theta_{11} - \psi_{11}), \qquad \frac{d\theta_{11}}{dt} = \rho\psi_{11} - \theta_{11} - \psi_{11}\theta_{20}, \qquad \frac{d\theta_{20}}{dt} = -\beta\theta_{20} + \theta_{11}\psi_{11}.$$

$$(8.79)$$

where σ, ρ, and β are positive constants.[7] The Lorenz equations exhibit deterministic chaotic solutions and are historically one of the first nonlinear dynamical systems, along with the Moore–Spiegel oscillator, used to study the properties of chaos in natural systems. We refer readers to examine some of the reference texts at the end of the chapter, e.g., Regev's book [10], for a more detailed examination of Lorenz and similar systems.

Several other model systems have been used to study pattern dynamics in Rayleigh–Bénard convection. One of these is the well-known *Swift–Hohenberg* equation for the real variable $u(x,y,t)$:

$$\frac{\partial u}{\partial t} = \varepsilon u - (\nabla^2 + 1)^2 u - u^3.$$

$$(8.80)$$

Similar in form to the amplitude equations derived for describing square and hexagon patterns, Swift–Hohenberg equation has the advantage of including spatially extended dynamics via the ∇^2 operator and is much simpler to deal with than a three-dimensional system. The disadvantage, of course, is that much physics has been lost in the construction of this model. Another equation frequently encountered model is the *Kuramoto–Sivashinsky* (KS) equation:

$$\frac{\partial \phi}{\partial t} = -\nabla^2 \phi - \nabla^4 \phi - \frac{1}{2}(\nabla\phi)^2,$$

$$(8.81)$$

one of the simplest models exhibiting spatiotemporal chaos. The Kuramoto–Sivashinsky equation is often used to study the phase dynamics of nearly steady but obviously chaotic patterns in convection. For example, $\phi(x,y,t)$ would be phase

---

[7]In the literature these equations often appear in the following form: $\dot{x} = \sigma(x - y)$, $\dot{y} = \rho x - y - xz$, $\dot{z} = -\beta z + xy$.

field of long train of rolls, as used by Kuramoto to study Eckhaus instability (see below). Independently, Sivashinsky used (8.81) in a slightly different form to describe patterns of a combustion front. The above equation has also been found in studying surface waves on a thin liquid film flowing down a sloped surface. Notice the similarity of this equation with the Cahn–Hilliard equation.

8.6 Beyond Weak Nonlinearity

In this chapter we have studied aspects of the nonlinear development of instabilities in fluid systems, primarily involving buoyant convection. We have chosen the perspective of weakly nonlinear theory, focusing on growth near the threshold of a linear instability, as controlled by a parameter, for instance, Ra in Rayleigh–Bénard convection. In this way, we have found a solution that is asymptotically correct as the distance above marginality is reduced, or $Ra - Ra_c \rightarrow 0$. Many interesting instabilities in fluid dynamics do not conform to this restriction and these require alternate approaches, while other instabilities demand numerical simulation. The main purpose of this chapter was to provide, through examples, a computational framework for weakly nonlinear theory of instability near threshold and the derivation of amplitude equations. It is the nature of weakly nonlinear theory to build a low dimensional model that, while easy to manipulate, is physically highly approximate. The hope is that, by following a formal approximation procedure, the relevant physics is not only preserved but lends itself to being studied.

We have only touched on the issues of pattern theory and secondary instabilities, both rich topics which demand far more attention than is space for in this book. In the case of Rayleigh–Bénard convection, for example, it is clear that to determine the physical relevance of any steady nonlinear state, we must also test its stability with respect to all forms of perturbation. One of the most important secondary instabilities involves resonant, or side-band, perturbations of a trio of unstable waves. The resulting process, known as Eckhaus instability, leads to a changed roll wavelength. Other examples of secondary instabilities include the zigzag instability, to perturbations acting only along the rolls' axes, and the cross-roll instability of two distinct roll patterns aligned at an oblique angle to one another. Taken together, the secondary instabilities help define a region of nonlinear stability in (Ra, Pr, k)-parameter space known as the Busse balloon, named following the extensive work on this topic by F. Busse and collaborators. For a more detailed description, see, e.g., reference [10].

As we first saw in Chap. 7, instability is the first step on the route to turbulence. Far beyond the power of weakly nonlinear analysis, turbulent flows force us to rely, in part at least, on statistical tools even for their description. Turbulence is our next destination.

Problems

8.1.
Show that the solution to the Landau equation (8.2), with $\varepsilon > 0$ and with initial condition $u(0) = u_0$, is given by

$$u(t) = \text{sgn}(u_0) \, \frac{\sqrt{\varepsilon} e^{\varepsilon t}}{\sqrt{A + e^{2\varepsilon t}}},$$

and find A.

8.2.
Consider the model equation for *transcritical* bifurcations:

$$\frac{du}{dt} = \varepsilon u - u^2.$$

Analyze the steady states and stability properties, including a sketch of the form in Fig. 8.1. Show that solutions of this system can be unstable depending upon the size and sign of the initial perturbation.

8.3.
Show that the steady states of (8.4) are correctly given by (8.5). Test the linear stability of each of these possible steady states.

8.4.
Consider the system

$$\frac{du}{dt} = \varepsilon u - u^2 + \gamma v, \qquad (8.82)$$

$$\frac{dv}{dt} = -v, \qquad (8.83)$$

with $u(0) = u_0$ and $v(0) = v_0$. Are there conditions in the parameters of the system, including the initial conditions, in which transient growth might cause a transition from one steady state to another?

8.5.
Complete the solution process going from (8.65) to (8.66). Show that the structure functions $\psi_{3j}(z)$ are given by

$$\psi_{32}(z) = -2\psi_{31}(z) = -z/2 + 5z^3/6 - z^5/2 + z^6/6, \quad \psi_{30}(z) = 5z - 10z^3 + 5z^4,$$

$$\psi_{33}^{(o)}(z) = -5z/504 + 5z^3/84 - 5z^6/12 + 5z^7/7 - 25z^8/56 + 25z^9/252,$$

$$\psi_{33}^{(e)}(z) = -29z/252 + 5z^3/28 - z^6/6 + 5z^7/42 - 5z^9/252 + z^{10}/252. \qquad (8.84)$$

Verify that $\psi^{(o)}$ is an odd function with respect to the point $z = 1/2$ and that $\psi^{(e)}$ is similarly even. Use this observation to explain why there is no explicit dependence of the amplitude A upon Pr in the case of fixed flux convection, yet the appearance of the aggregate solution will depend upon Pr.

8.6.
Analyze Eqs. (8.53)–(8.55) assuming the coefficients appearing are real. What are the stability boundaries designating the transition from hexagons to rolls?

8.7.
Assuming a small aspect ratio of a rectangular convection container and a weak Couette shear $u = \varepsilon \Lambda_1 z$, with $\varepsilon \ll 1$ develop the amplitude equation for fixed flux convection in the case of rigid boundary conditions. Do this by following the derivation procedure of Sect. 8.4. Derive the following modified Cahn–Hilliard equation:

$$\frac{\partial A}{\partial T} = -\left(\text{Ra}_2 - \frac{\Lambda_1^2}{120}\right)\frac{\partial^2 A}{\partial X^2} - \Lambda_1 \frac{9}{28}\frac{\partial}{\partial X}\left(\frac{\partial A}{\partial X}\right)^2 - \frac{17}{462}\frac{\partial^4 A}{\partial X^4} + \frac{10}{7}\frac{\partial}{\partial X}\left(\frac{\partial A}{\partial X}\right)^3.$$
(8.85)

Recast the above equation into a friendlier form by defining $\tau \equiv 17\,T/462$, $X = (17/462)^{1/2}\,\xi$, $F = (17/660)^{-1/2}\,A$, and $\lambda_1 = (9/56)\sqrt{7/10}\,\Lambda_1$ together with the new redefined control parameter $\mu^2 \equiv \text{Ra}_2 - \Lambda_1^2/120$. What does this equation look like in these variables?

8.8.
Show that the fixed point $\psi_{11} = \theta_{11} = \theta_{20} = 0$ of the Lorenz equations (8.79) is stable to infinitesimal perturbations if $\rho < 1$. Develop an approximate solution in the near vicinity of $\rho = 1$ and demonstrate that the system collapses into a Landau equation, i.e.,

$$\frac{d\theta_{11}}{dt} \approx (\rho - 1)\theta_{11} - \beta^{-1}\theta_{11}^3.$$
(8.86)

Convince yourself that the Lorenz system is mathematically consistent with weakly nonlinear theory of classic Rayleigh–Bénard convection only in the limit where $|\rho - 1| \ll 1$.

8.9.
Identify all fixed points of the Lorenz system (8.79) assuming the constant parameters ρ, β, σ are all positive. Examine the stability of each fixed point and categorize them according to the types discussed at the beginning of the chapter.

8.10.
Show that the following is a solution of the BBM equation (8.77):

$$\frac{3A^2}{1-A^2}\text{sech}^2\left(Ax - \frac{At}{1-A^2} + \delta\right)$$

where δ is an arbitrary constant and A is the (squared) amplitude.

8.11.

Consider the following nonlinear model of a parametrically excited system like surface water wave. The surface position variable η evolves according to the following model PDE:

$$\frac{\partial^2 \eta}{\partial t^2} + 2\tau_d^{-1}\frac{\partial \eta}{\partial t}(1 + \delta_0 \sin 2\omega_0 t)\sin \eta = \frac{\partial^2 \eta}{\partial x^2}. \qquad (8.87)$$

Analyze this system by considering linear perturbations around the $\eta = 0$ state. Show that the dispersion relation for $\propto \exp(i\omega t - ikx)$ types of modes is $\omega^2 = 1 + k^2$ for negligible values of the linear damping τ_d^{-1} and driving amplitude δ_0. Suppose there exists a value of $k = k_0$ such that $\omega(k_0) = \omega_0$, i.e., half the driving frequency of the time-dependent driving found above. Assume that the domain is periodic on a length scale L_x (much larger than $2\pi/k_0$) and perform a weakly nonlinear analysis on the system. Assume that the parameters δ_0 and τ_d^{-1} are scaled by the small parameter ε^2 and assume a series solution for u $u = \varepsilon u_1 + \varepsilon^2 u_2 + \varepsilon^3 u_3 + \text{HOT}$, where

$$u_1 = A(\varepsilon x, \varepsilon t, \varepsilon^2 t)e^{i\omega_0 - ikx} + B(\varepsilon x, \varepsilon t, \varepsilon^2 t)e^{i\omega_0 + ikx} + \text{c.c.}$$

Recalling our terminology from Chap. 4, A and B are the amplitudes the corresponding carrier wave moving with velocity $c_0 = \pm\omega_0/k_0$ (respectively). Show that by taking the calculation out to order ε^3, one gets the coupled amplitude equations

$$\frac{\partial A}{\partial T} = -\tau_{d2}^{-1}A + \delta_{02}\left\langle B^* \right\rangle + i\alpha\frac{\partial^2 A}{\partial X_+^2} - \left(\beta|A|^2 + \gamma\left\langle |B|^2 \right\rangle\right)A, \qquad (8.88)$$

$$\frac{\partial B}{\partial T} = -\tau_{d2}^{-1}B + \delta_{02}\left\langle A^* \right\rangle - i\alpha\frac{\partial^2 B}{\partial X_-^2} - \left(\beta^*|B|^2 + \gamma^*\left\langle |A|^2 \right\rangle\right)B, \qquad (8.89)$$

with

$$\tilde{c}_g \equiv \left(\frac{\partial \omega}{\partial k}\right)_{k=k_0}, \qquad \alpha \equiv \left(\frac{\partial^2 \omega}{\partial k^2}\right)_{k=k_0}, \qquad \text{and} \qquad \left\langle \bullet \right\rangle \equiv \frac{1}{L_X}\int_0^{L_X} \bullet \, dX,$$

and where $T \equiv \varepsilon^2 t$ and $X_\pm = \varepsilon(x \pm \tilde{c}_g)$, together with $L_x = L_X/\varepsilon$, $\delta_0 = \varepsilon^2\delta_{02}$ and $\tau_d^{-1} = \varepsilon^2\tau_{d2}^{-1}$. Note that the derivation implies that $A(\varepsilon x, \varepsilon t, \varepsilon^2 t) \mapsto A(T, X_+)$ and $B(\varepsilon x, \varepsilon t, \varepsilon^2 t) \mapsto B(T, X_-)$, which indicates that each amplitude function drifts at the group velocity in the local frame of the basic carrier wave with which it is associated. Analyze the spatially uniform state and show that the supercritical transition is a Hopf bifurcation of the oscillatory waves.

This is only a model not to be confused with proper analysis of the original equations of motion—for more references on this, see the review article by Cross and Hohenberg listed in the *Bibliographical Notes*.

Bibliographical Notes

Section 8.1

The main focus of this chapter has been on the techniques of weakly nonlinear theory. This is covered in several texts, but the two main ones that provide deeper development of the ideas and extensions to several problems include the important work of Maneville and the monograph of Cross and Hohenberg:

1. P. Maneville, *Dissipative Structures and Weak Turbulence* (Academic Press, Boston, 1990)
2. M.C. Cross, P.C. Hohenberg, Pattern formation outside of equilibrium. Part II, American Physical Society. Rev. Mod. Phys. **65**(3), 851 (1993)

These above texts are highly recommended for the reader interested in much deeper analysis of pattern theory and in the development of the techniques used in this chapter.

Section 8.2

In addition to Drazin's and Reid's book on hydrodynamic stability (see the previous chapter's *Bibliographical Notes*), which has a very good introduction to nonlinearity and weakly nonlinear analysis, there is also Drazin's stand-alone book on nonlinear systems:

3. P.G. Drazin, *Nonlinear Systems* (Cambridge University Press, Cambridge, 1992)
   From the perspective of ordinary differential equations, the study of nonlinear systems and its analysis and various classifications is discussed with considerable mathematical rigor in the text:
4. J. Guckenheimer, P. Holmes, *Nonlinear Oscillations, Dynamical Systems and Bifurcations of Vector Fields* (Springer, New York, 1983)

Sections 8.3 and 8.4

The presentation of the weakly nonlinear techniques is inspired by the chapter entitled "Pattern Forming Instabilities" authored by S. Fauve in

5. C. Godreche, P.C. Hohenberg (eds.), *Hydrodynamics and Nonlinear Instabilities* (Cambridge University Press, Cambridge, 1998)
   The theory of linear operators and in particular the Fredholm alternative property of these operators, which is the basis of solvability conditions, can be found in the very valuable concise book:

6. B. Friedman, *Principles and Techniques of Applied Mathematics* (Dover, New York, 1990) reprint of the original Wiley book from 1956 (in particular, Chap. 1)

Section 8.5

Nonlinear instability in shear flows is reviewed in

7. B.J. Bayly, S.A. Orszag, T. Herbert, Instability mechanisms in shear-flow transition. Ann. Rev. Fluid Mech. **20**, 487 (1988)
   This reference also has a very thoughtful discussion of subcriticality, including references to the original discussion of the plane Poiseuille flow problem, separately in the papers by Stuart and Watson:
8. J.T. Stuart, On the non-linear mechanics of wave disturbances in stable and unstable parallel flows. Part 1. The basic behavior in plane Poiseuille flow. J. Fluid Mech. **9**, 353–370 (1960)
9. J. Watson, On the non-linear mechanics of wave disturbances in stable and unstable parallel flows. Part 1.The development of a solution for plane Poiseuille flow and for plane Couette flow. J. Fluid Mech. **9**, 371 (1960)

Section 8.6

A discussion of chaos and model equations can be found in the first part of

10. O. Regev, *Chaos and Complexity in Astrophysics* (Cambridge University Press, Cambridge, 2012)
    The solitary wave solution of the BBM solution is found in the original work:
11. P.J. Olver, Euler operators and conservation laws of the BBM equation. Math. Proc. Camb. Philos. Soc. **85**, 143 (1979)

Chapter 9
Turbulence

Don't despair, not even over the fact
that you don't despair

Franz Kafka (1883–1924), 'Diary'

9.1 Introduction

The Oxford English Dictionary defines *turbulent* flow of a fluid as one in which
the velocity at any point fluctuates irregularly and there is continual mixing, rather
than a steady flow pattern. As the first example of the use of this adjective, the
dictionary quotes a sentence from the classical book by H. Lamb, *Hydrodynamics*
(1895): "The resistance, in the case of turbulent flow, is found to be sensibly
independent... of the viscosity of the fluid."

Nowadays, a substantial number of scholarly books exist, devoted entirely to
fluid turbulence, the noun derived from the adjective *turbulent* and thus indicating
the phenomenon itself. We shall mention several of them in the *Bibliographical
Notes*. The phenomenon, as it appears in nature, has been known for many
centuries—probably already since ancient times. The Latin word *turbulentia* originally refers to the disorderly motion of a crowd (*turba*), but Lucretius, a Roman
poet and philosopher, who lived in the first century BC, mentioned it not by this
name in his masterpiece *De rerum natura*. Over five centuries ago, Leonardo da
Vinci was probably the first to use the Italian word *turbolenza*, when observing
and describing the slow decay of eddies formed behind the pillars of a bridge.
Turbulent flow appeared depicted masterfully in Leonardo's celebrated drawings as
well as in the paintings of Japanese masters (nineteenth century). Despite the fact
that significant experimental, observational, and theoretical efforts have been made
to gain understanding of this classical physics phenomenon, much has remained

© Springer Science+Business Media, LLC 2016
O. Regev et al., *Modern Fluid Dynamics for Physics and Astrophysics*,
Graduate Texts in Physics, DOI 10.1007/978-1-4939-3164-4_9

unknown. According to an apocryphal story, Sir Horace Lamb once said: "I am an old man now, and when I die and go to heaven there are two matters on which I hope for enlightenment. One is quantum electrodynamics, and the other is the turbulent motion of fluids. And about the former I am rather optimistic." Lamb considered turbulence a subject of "outstanding difficulty" (1895), and Richard Feynman, a Nobel laureate for his work on quantum electrodynamics, stated in the 1960s that "turbulence is the most important unsolved problem of classical physics." In spite of over a century of ever-growing insights from theory, experiment, and computation, Feynman's assessment is unfortunately still largely with us. In addition to the quest for a theoretical understanding of turbulence's fundamental aspects and an appropriate mathematical formalism to describe these aspects, the phenomenon is associated with some important practical problems in engineering: aerodynamics, hydraulics, combustion, acoustics, as well as astrophysics and geophysics—in particular, the "hot" topic of climate change. Thus, there seems to exist an urgent need to develop models that can allow the understanding of experiments and observations, as well as provide tools for the prediction and control of turbulent flows.

Fluid turbulence usually arises following the onset of a hydrodynamical or magnetohydrodynamical instability in a laminar flow, which itself occurs as a *control parameter* changes so as to pass through a critical value. Chapter 7 discussed, in detail, some such linear instabilities, and we have seen at its outset (Sect. 7.1) experimental examples of two different transitions. The development of turbulent flow from an instability is not a trivial matter. Some instabilities saturate and give rise to a new laminar flow, while others develop secondary instabilities as they transition to a turbulent flow. According to the once well-regarded L. Landau's conjecture, these transitions may give rise to an infinite sequence of unstable modes, which is the essence of turbulence. D. Ruelle and F. Takens showed in 1971, in contrast to the above, that only a limited number of unstable modes may be needed for turbulence to arise. They found that even as few as three unstable modes may be sufficient to be involved in the emergence of turbulence, as long as they have incommensurate frequencies and lead to quasiperiodic behavior. Additional "roads" (or "routes") to turbulence have been proposed, and we have already mentioned most of them in this previous chapter. They are still referred to a "scenaria," a word whose meaning hints that the theoretical understanding of the mechanisms is not complete, to say the least. Another way of producing a turbulent flow and learning about its experimental properties is to allow a laminar stream of fluid, usually gas, to flow past a grid of solid bars. What is obtained is usually regarded as *freely evolving* turbulence, but closer scrutiny shows that FD instabilities are induced close to the obstacles of the grid. Similarly, appropriate stirring, shaking, or sloshing of a quiescent fluid may produce turbulence, but once the forcing is stopped (or in the case of the grid, far enough from the disturbance), the turbulence freely decays. Its energy is being dissipated with the smallest irregularities (eddies) decaying first. We would like to stress at the outset that we shall be dealing with *incompressible* turbulence (except if stated explicitly otherwise). This assumption simplifies the problems posed by the topic, and we shall adopt it because of the considerable difficulty of the subject matter.

9.1.1 The Origin of Turbulence, Its Ubiquity, and Some Basic Properties

As we have seen, turbulent motion may appear first in patches and only later, with a further change of the control parameter, do these patches merge and fully developed turbulence engulfs the flow, as in Reynolds's experiments. Another type of transition is when spatiotemporal chaotic behavior, i.e., turbulence, occurs when a certain threshold is exceeded. It may start as an instability of the mean flow, leading to development of a secondary flow, which subsequently loses its stability with appropriate change of the control parameter until reaching a state of fully turbulent motions, as in the Taylor–Couette experiments.

In any case, without formally defining what a turbulent flow is, we may safely state some of its properties. Turbulent flow is invariably spatiotemporally complex. Focusing, for a moment, on the velocity field in a turbulent flow, we may say that the velocity fluctuates seemingly randomly, in time at any fixed position, and, in addition, it is highly disordered in space at a given instant, say, with a multitude of length scales. Another typical property of the velocity is that it is chaotic, in the dynamical systems sense of this word, that is, a very small change in the initial conditions will produce a very large change to the subsequent motion. An alternative definition, involving *vorticity*, is probably better—turbulence may be "defined" as a flow in which a spatiotemporal complex vorticity field is advected in a chaotic manner. An important point worth stressing when trying to characterize turbulence is that in many flows laminar regions exist alongside turbulent ones and there is a clearly apparent separation between the two. Another point worth making at the outset is that the early studies of turbulence focused on its emergence in *shear flows*, which, as we already know, are essentially flows in one direction with the magnitude of the velocity changing in the direction perpendicular to the flow.

Turbulent flows are not restricted to the laboratory or just human scale and appear in nature on a wide variety of scales, from the truly vast scales encountered in extragalactic astrophysics through the turbulent flows on the surface of stars and our Sun and, in particular, in the atmosphere of our Earth and its oceans. Turbulence is all around us and is instrumental in shaping the climate. It appears also on smaller scales, from the turbulent "pockets" encountered while flying on a jetliner down to turbulent winds created by obstacles to air or water flow, such as buildings or bridge piers. Engineers must deal with turbulence and their concern is less theoretical and more practical, from the design of automobiles and airplanes, through manufacturing internal combustion engines and to the challenge of minimizing the dispersion of pollutants.

It is tempting to speculate that turbulence originates due to some nature-imposed upper limit on the content of disordered kinetic energy that may be contained in a flow with given microscopic fluid viscosity. If such a limit does indeed exist, the flow will develop so as to dissipate and increase the entropy as effectively as possible for any kinetic energy beyond this limit. Assuming that the microscopic viscosity of a fluid is constant, the only way for the fluid to achieve dissipation is to lower its effective Reynolds number, so to speak. Flows without solid boundaries have no ability to create boundary layers, so the only option for emergence of very small scales is spatial (and thus also temporal) fluctuations in the flow, so that an *effective* "turbulent" viscosity, which may be written $v_{turb} \sim \alpha c_s l_{turb}$, will arise. Here l_{turb} is some effective scale for turbulent transport, presumably linked to the large turbulent eddies, those responsible for transport, and $\alpha = \mathcal{O}(1)$ is a constant. Thus, v_{turb} will be the approximately correct viscosity coefficient for turbulent momentum transport, instead of the microscopic one. This viscosity coefficient is orders of magnitude larger than the molecular one and therefore also allows for highly increased dissipation rate. All this provided that the flow *somehow* creates fluctuations on a small scale. This "somehow" is the heuristic essence of turbulence. Obviously, to maintain turbulence, there must be a source of kinetic energy in the flow. This may be, for example, the mechanical source driving the flow (shear flow, say) without which the instabilities would be unable to feed the field of steady turbulent fluctuations. Although the idea behind this conjecture is crude and imprecise, we feel that it may one day be refined by correct statistical considerations for systems out of equilibrium to a general statement, perhaps similar in spirit to the celebrated *fluctuation–dissipation* relation. The latter is a powerful tool in statistical physics for predicting the behavior of nonequilibrium thermodynamical systems. Such systems involve the dissipation of energy into heat from the reversible thermal fluctuations at thermodynamic equilibrium in irreversible microscopic statistical physics. Of course, the above relation relies on the system of microscopic particles being in thermodynamic equilibrium and involves the Boltzmann constant and thus is applicable to systems that seem very different from turbulent fluids. However, we are aware of recent efforts to obtain instead a *fluctuation–response* relation, which could be applicable to turbulence theory. Details of the progress towards such a relation are outside the scope of this book.

What is our expectation when we look for a *theory* of turbulence? The obvious hope is, probably, to find something resembling the tremendously successful statistical mechanics theory. There we dealt with a system in which the small (microscopic) constituents behaved apparently randomly. From this grew the theoretical edifices of kinetic theory and statistical mechanics which ultimately offered a macroscopic, well-defined, physical theory. The result of such efforts for turbulence, conducted for over a century, was largely disappointing. To be sure, there exist a multitude of pieces of theory, applicable to particular cases, e.g., stratified flows, boundary layers, etc. The two exceptions, which, despite being applicable to a particular case, do offer some *universal*, or at least nearly such, laws, are shear flows near a smooth wall and the Kolmogorov's analysis of isotropic, fully developed turbulence. The eminent mathematician, A.N. Kolmogorov (1903–1987) who made significant contributions to the mathematics of probability theory, topology, classical mechanics, algorithmic information theory, and computational complexity, also found time to tackle turbulence. This state of affairs pushes the applied researcher, say, an astrophysicist or engineer, to either parameterize the turbulent contributions or use numerical studies, which are often unable to resolve the full range of scales and still require some turbulence modeling of the smallest scales. In this chapter of our book, we would like to review for the interested reader some of the new and not-so-new ideas on turbulence, so that he or she will be able to see at least what is available. Our descriptions will often be short and uncomplicated, by necessity, but the *Bibliographical Notes*, at the end of the chapter, list some very good books, in our view, for further reading. Among the many topics discussed in this chapter, three topics will be merely summarized in the final section of the chapter together with references to the best, in our opinion, sources for reading on them.

We move now to some of the basic properties of turbulent flows, stressing at the outset that the discussion here and indeed a large majority of the chapter concerns only *freely evolving* turbulence, e.g., as a result of instability—that is, not forced by mechanical energy injection (e.g., grid, stirring). Turbulence, as we saw, is perhaps best defined as a spatially complex distribution of *vorticity* which is being advected chaotically by the flow. The vortices interact with each other until they create what is called a *fully developed* turbulence, containing a variety of length scales (vortex sizes). The vortices go under the name of turbulent *eddies*, which can be perceived as the basic entities of a turbulent flow, an image evoked beautifully by L. F. Richardson (1881–1953) in his famous poem: *Big whorls have little whorls/That feed on their velocity,/And little whorls have lesser whorls/And so on to viscosity.* The "whorls" were later called eddies, and Richardson's poem of 1920 elegantly captures the view of turbulence as a *cascade*. The biggest eddies, which are considered to be the ones from which the cascade starts, have a linear dimension of the order of what is called the *integral scale* and denoted by ℓ_0. According to Richardson's vision, the cascade terminates at the smallest size eddies, measuring what is called the *Kolmogorov microscale*, this consisting of eddies so small that the microscopic viscosity is important. This microscale's typical length will be denoted by η or η_K, on account of its being sometimes called Kolmogorov's microscale. We shall use these two symbols interchangeably.

Decompose now the flow variables, notably the velocity field, into an *average* part and a fluctuating part. The word *mean* is sometimes used instead of "average," and in what follows, we shall use these two expressions interchangeably. Thus, the velocity is written as

$$\mathbf{u}(\mathbf{x},t) = \bar{\mathbf{u}}(\mathbf{x}) + \mathbf{u}'(\mathbf{x},t). \tag{9.1}$$

This definition of an average is done essentially over time, thus the mean is steady. As mentioned above, we ask the reader to be patient with this definition of the mean, as will be made more precise later. We remind the reader that we denote averages by angle brackets, $\langle \cdot \rangle$, or by an over-bar $\bar{\cdot}$. In this chapter the latter notation will be usually used. Later on, in Sect. 9.3 we shall give a more general discussion and definition of the average (9.58); in particular, an *ensemble* average will be introduced. Assume now a flow whose Reynolds number, characterized by typical large scales, $l \lesssim \ell_0$, is $\text{Re} = ul/v$, with u being of the order of magnitude of the fluctuating part of the velocity, and let this $\text{Re} \sim 10^4$, say. The idea behind the turbulent cascade is the following. The large eddies, most likely created by an instability of the flow, are themselves subject to instabilities and break up to eddies of smaller scale. The lifetime of a typical eddy of this sort is of the order of its turnover time. For the large eddies, this time is $\tau_l \sim l/u$, after which it breaks up to smaller eddies and the latter break up to even smaller ones and so on. Microscopic viscosity plays no part in this process since Re remains large. The cascade is driven by inertial forces. The process comes to its completion, the end of the cascade, when $Re_K = u_\eta \eta/\nu \sim 1$, where u_η is the velocity scale of the fluctuating velocity of the smallest eddies and the index K denotes the Kolmogorov microscale. This end of the cascade arises simply because we reach scales of microscopic viscous dissipation. Assuming now steady state, based on dimensional arguments only, we can easily estimate the rate at which energy is being transferred into smaller and smaller eddies, until ultimate physical dissipation by molecular viscosity; this rate is given by

$$\varepsilon \sim \frac{u^2}{l/u} = \frac{u^3}{l}. \tag{9.2}$$

The units of this quantity are energy per unit mass per unit time, and it is based on the eddies just between the integral scales. The assumption of a steady-state cascade makes it valid for each scale. Denote the general typical scale value by ℓ say, replacing u by u_ℓ and l by ℓ in Eq. (9.2), yielding what is called the *Kolmogorov–Obukhov* law:

$$u_\ell = (\varepsilon \ell)^{1/3}. \tag{9.3}$$

Now, the viscous dissipation on the smallest scales follows from $Re_K = u_\eta \eta/\nu \sim 1$, giving

$$\varepsilon_K \sim \nu \frac{u_\eta}{\eta^2}. \tag{9.4}$$

Using also this statement as above, and not forgetting that the energy cascade is steady, from which it follows that $\varepsilon_K = \varepsilon$, we get

$$\frac{\eta}{l} \sim Re^{-3/4} \quad \Rightarrow \eta \sim \left(\frac{\nu^3}{\varepsilon}\right)^{1/4}, \tag{9.5}$$

$$\frac{u_\eta}{u} \sim Re^{-1/4} \quad \Rightarrow u_\eta \sim (\nu\varepsilon)^{1/4}. \tag{9.6}$$

9.1.2 Topics Discussed in This Chapter

The formidability of our subject can be perhaps appreciated by the large number of existing books on turbulence, each giving the same basic dimensional arguments due to Kolmogorov (see above) but advocating its own approach to more advanced issues. Here we shall try to summarize what are, in our opinion, the essentials in just *one* chapter. We think that we owe our readers a presentation of the different approaches to the problem, and consequently we shall first discuss the classical approaches, which developed from the Reynolds equations. Osborne Reynolds was active both as an experimenter and theoretician close to the end of the nineteenth century. The Reynolds equations and their associated closure problem will be our first topic. In the same Sect. 9.2, we also describe important results on shear flows, notably the logarithmic velocity profile in turbulent boundary layers. We shall then move to Sect. 9.3 and discuss the statistical and probabilistic approach to turbulence, leading to the ergodic hypothesis and the statistical formulation of the fundamental problem of turbulence. Correlation and structure functions will also be introduced in that section. The subsequent section will be devoted to the work due mainly to Kolmogorov and his followers, containing several basic topics like conservation laws, definition of Fourier spectra, and the energy contained in various scales. These topics will then be used in a review of Kolmogorov's 1941 theory and some important phenomenological results, including the important spectral power laws and the dependence of the number of grid points needed to resolve the turbulence, which goes as $\propto (\text{Re})^{(3d)/4}$, with d being the dimension. Section 9.5 is devoted to two-dimensional turbulence and its oddities, after which we shall make, in a short section, some comments on turbulence in astrophysics. The discussions in the final Sect. 9.7 are so brief that it would be fair to say that we merely list the somewhat more advanced topics of intermittency, dynamical system approach, and numerical methods and recommend bibliographical sources for each topic.

9.2 Reynolds Equations and Shear Flows

It is natural to approach a seemingly randomly fluctuating fields, like the velocity in a turbulent flow, in statistical way. Some assert that our ignorance in turbulence and the phenomenon's complexity are so large that one does not even know what questions to ask. While this may be true in the context of a quest for a complete theoretical formalism, there are still many, perhaps somewhat naive, questions like "what is the spatial distribution of the mean velocity?" or "what are the transport properties of turbulent flows?" or "how does the intensity of turbulence vary from place to place in different flows?" or "what are the spatial spectra of turbulent fluctuations?". Actually, the systematic statistical approach, discussed in the next section, will give rise to the (statistical) formulation of the so-called fundamental problem of turbulence. It is evident that to achieve progress one must start with some restrictive assumptions. One of the central assumptions is that we are dealing with statistically steady flows. This allows us to identify *ensemble* averages with temporal and spatial averages, in agreement with the ergodic hypothesis (see Sect. 9.3).

9.2.1 Reynolds Stress and the Closure Problem

Consider the incompressible Navier–Stokes equations (1.53), with $\nabla \cdot \mathbf{u} = 0$. In addition, we consider a case without any body forces, so that the main physical content of the equation is the fluid internal stresses: pressure and viscosity. In any case, writing the vector equation of motion (1.53) in component form brings us back to a form which may be easily derived from Eq. (1.50), with the condition of incompressibility, which in component language reads $\partial u_m / \partial x_m = 0$, where the Einstein convention of summation over a double index is applied here as elsewhere in this book. Actually it will be the simplest to posit here that the density is a constant (in time and space):

$$\rho \left(\frac{\partial u_i}{\partial t} + u_j \frac{\partial u_i}{\partial x_j} \right) = -\frac{\partial P}{\partial x_i} + \frac{\partial \tau_{ij}}{\partial x_j}. \tag{9.7}$$

Here τ_{ij} is the viscous (or deviatoric) stress, and as shown in Eq. (1.42) it can be expressed, for incompressible flow, in the following way:

$$\tau_{ij} = \nu \rho \left(\frac{\partial u_i}{\partial x_j} + \frac{\partial u_j}{\partial x_i} \right). \tag{9.8}$$

We note, in passing, that the expression in parentheses is directly related to what we have called the deformation rate tensor (see Eq. (1.15))

$$\mathscr{D}_{ij} = \frac{1}{2} \left(\frac{\partial u_i}{\partial x_j} + \frac{\partial u_j}{\partial x_i} \right), \tag{9.9}$$

and thus we have $\tau_{ij} = 2\rho \nu \mathscr{D}_{ij}$.

Assuming turbulent flow, we separate the variables into an average and fluctuating part, as in Eq. (9.1). Here we simply understand *average*, denoted by overline, as *time* average, that is, for the velocity, say,

$$\bar{\mathbf{u}}(\mathbf{x}) \equiv \frac{1}{T} \int_{t_0}^{t_0+T} \mathbf{u}(\mathbf{x}, t)dt, \tag{9.10}$$

where our interest in the system is between times t_0, an arbitrary value, and $t_0 + T$, with T being long enough interval in which we follow the system. The fluctuating part is time dependent, and we assume that its time average is zero, because the variable is perceived as changing randomly and on a much shorter timescale than T. After separating each variable into its average and fluctuating parts, we average the complete equations. Performing this average, as we have just explained it, on the equation of motion (9.7), the following equation for the averaged functions and the fluctuations emerges:

$$\rho \bar{u}_j \frac{\partial \bar{u}_i}{\partial x_j} = -\frac{\partial \bar{P}}{\partial x_i} + \frac{\partial}{\partial x_j} \left(\bar{\tau}_{ij} - \rho \overline{u'_i u'_j} \right). \tag{9.11}$$

This result follows from the fact that $\overline{gq} = \overline{(\bar{g}+g')(\bar{q}+q')} = \bar{g}\bar{q} + \overline{g'q'}$ for any two functions g and q whose fluctuations average to zero.

Focusing now on the term $\rho \overline{u'_i u'_j}$, we see that this is the contribution from the postulated turbulent flow. The mean quantities behave as could be expected; however, there is an additional quadratic contribution involving turbulent fluctuations. Formally, this contribution, which by its structure forms a flux of momentum density by the velocity fluctuations of the momentum density in the fluctuations, appears as a turbulent correction to the mean viscous stress $\bar{\tau}_{ij}$. If we write $\tau_{ij}^R \equiv -\rho \overline{u'_i u'_j}$, we obtain the definition of the celebrated *Reynolds stress tensor*, which, if substituted into the averaged equation of motion, gives

$$\rho \bar{u}_j \frac{\partial \bar{u}_i}{\partial x_j} = -\frac{\partial \bar{P}}{\partial x_i} + \frac{\partial}{\partial x_j} \left(\bar{\tau}_{ij} + \tau_{ij}^R \right). \tag{9.12}$$

This equation, supplemented by the incompressibility condition for the fluctuations (a result of averaging the incompressibility condition for the total velocity), that is,

$$\frac{\partial u'_j}{\partial x_j} = 0, \tag{9.13}$$

is usually referred to as the *Reynolds equations*.

Even though this form may naively be interpreted as true progress towards formulation of the correct equation replacing the Navier–Stokes equation for turbulent flow (for constant density fluids and with no action of body forces), it is certainly not so because the Reynolds stress is not only unknown and requires an additional equation, but also it is not a stress in its usual meaning (see Problem 9.3). The nomenclature is perhaps not too important, but deriving an equation for the evolution τ_{ij}^R is a necessity, if we want to make any progress using these ideas. In Problem 9.4 we derive:

$$\bar{u}_k \frac{\partial}{\partial x_k} \left(\rho \overline{u'_i u'_j} \right) = \tau_{ik}^R \frac{\partial \bar{u}_j}{\partial x_k} + \tau_{jk}^R \frac{\partial \bar{u}_i}{\partial x_k} + \frac{\partial}{\partial x_k} \left(-\rho \overline{u'_i u'_j u'_k} \right) - \frac{\partial}{\partial x_i} \left(\overline{P' u'_j} \right) - \frac{\partial}{\partial x_j} \left(\overline{P' u'_i} \right)$$

$$+ \overline{2P' \mathscr{D}'_{ij}} + \nu \frac{\partial^2}{\partial x_k \partial x_k} \left(\rho \overline{u'_i u'_j} \right) - 2\nu\rho \left(\frac{\partial u'_i}{\partial x_k} \frac{\partial u'_j}{\partial x_k} \right). \tag{9.14}$$

It is worthwhile to note that

$$\frac{\overline{D}}{Dt} \left(\rho \overline{u'_i u'_j} \right) = \bar{u}_k \frac{\partial}{\partial x_k} \left(\rho \overline{u'_i u'_j} \right), \tag{9.15}$$

because the partial time derivative of an averaged quantity is zero.

As is apparent, the equation governing the Reynolds stress evolution is not only very complicated but also its derivation has introduced a new unknown: $\overline{u'_i u'_j u'_k}$. Thus, in seeking an equation for the Reynolds stress, which is an average of a product of velocity fluctuations, we find that the average of the *triple* product of velocity fluctuations is needed. If we try in a straightforward, but algebraically complex, way to derive an equation for this triple product, in the lengthy equation that follows, we shall have a term with averages of quadruple products of fluctuations:

$$\frac{\overline{D}}{Dt}\left(\rho\overline{u'_i u'_j u'_k}\right) = \bar{u}_n \frac{\partial}{\partial_n}\left(\rho\overline{u'_i u'_j u'_k}\right) = \cdots + \frac{\partial}{\partial x_m}\left(-\overline{u'_i u'_j u'_k u'_m}\right) + \cdots, \qquad (9.16)$$

where the ... indicate other terms, whose form is not essential to our discussion. What is essential is that the general tendency of having more unknowns than equations will continue without end. This is known as the *closure problem of turbulence* and is similar in principle to the problem encountered in kinetic theory, where the so-called BBGKY hierarchy of equations is obtained and is infinite, if not somehow truncated. We mention, in passing, that the above averages of different fluctuation products are a special case of *correlation functions*, a concept we shall deal with more fully in Sect. 9.3. The Navier–Stokes equation, from which we have started, is a deterministic equation, but it produces seemingly chaotically fluctuating solutions for some conditions. Utilizing a statistical approach, we turn the averages of different fluctuations products—the abovementioned correlation functions—into well-defined nonrandom variables; however, we pay the price of inability to find a closed set of equations for them. It is impossible to close the system by algebraic manipulation and *additional* information is needed. Several types of closures exist, and it seems that they are not more than ad hoc dimensionally correct statements that are reasonable for the problem, usually a physical or engineering application. This difficulty of a statistical approach appearing in the Reynolds equations had been realized quite early, and the first attempts were to introduce closure already at the first stage, that is, posit an expression for the Reynolds stress. Boussinesq proposed around 1870 a closure, written here in Cartesian coordinates as follows:

$$\bar{\tau}_{xy} + \tau^R_{xy} = \rho(\nu + \nu_{\text{turb}})\frac{\partial\bar{u}_x}{\partial y}. \qquad (9.17)$$

This total shear-stress relationship seems to be adequate for a shear averaged flow, containing a turbulent contribution to the viscosity with the so-called turbulent (or eddy) viscosity ν_{turb}, being unspecified at this point, but from the general understanding that turbulent momentum mixing is dominant over the molecular one, the reasonable assumption would be $\nu_{\text{turb}} \gg \nu$. Now following formally the usual expression for the viscous stress in a fluid as in Eq. (1.42), but replacing the molecular stress tensor with the Reynolds one and the kinematic viscosity with the turbulent one, we get by analogy a three-dimensional expression

$$\tau_{ik}^R = \rho \, v_{\text{turb}} \left(\frac{\partial \bar{u}_i}{\partial x_k} + \frac{\partial \bar{u}_k}{\partial x_i} \right) + \quad ? \tag{9.18}$$

because the divergence of the average velocity is zero and therefore the terms, which in (1.42) contain $(\partial u_m / \partial x_m)$, vanish. The question mark in Eq. (9.18) indicates that there must be some defect in this equation, which hopefully can be corrected by an additive term. Remembering that the trace of the Reynolds stress tensor must be

$$\tau_{ii}^R = - \sum_i \rho \overline{(u_i')^2}, \tag{9.19}$$

we see that, in contrast to this, Eq. (9.18) gives for this quantity $0 + \ ?$. This is so because the average velocity derivatives add up to twice the divergence of the average velocity, which is zero for incompressible flows. Thus, for consistency Boussinesq decided to write instead of the "?" in Eq. (9.18) the expression as below:

$$\tau_{ik}^R = \rho \, v_{\text{turb}} \left(\frac{\partial \bar{u}_i}{\partial x_k} + \frac{\partial \bar{u}_k}{\partial x_i} \right) + \frac{\rho}{3} \overline{u_m' u_m'} \delta_{ik}. \tag{9.20}$$

This is called the Boussinesq equation, and we leave it for the reader to verify that the extra term satisfies the trace requirement.

So far the analogy between microscopic transfer and that arising from random motion of turbulent eddies was exploited in order to express the Reynolds stress. The quantity v_{turb} remained, however, unknown. In the early twentieth century, L. Prandtl proposed to extend the abovementioned analogy in order to write an expression for the turbulent viscosity as well. The kinetic theory of gases suggests the approximate size of the microscopic viscosity as

$$\nu \sim \Lambda_{\text{mfp}} \, U_{\text{rms}}, \tag{9.21}$$

where Λ_{mfp} is the mean free path of the microscopic constituents (atoms or molecules) of the gas and U_{rms} is their root mean square velocity. Exploiting this physical intuition, Prandtl reasoned that the turbulent viscosity should be expressible as a typical mixing length scale of the turbulence times its typical fluctuating velocity; he thus proposed

$$v_{\text{turb}} \sim l_{\text{m}} \, U_{\text{turb}}, \tag{9.22}$$

where U_{turb} is some measure of a suitably defined average of the velocity fluctuation, e.g., $\sim \sqrt{\overline{(u')^2}}$, and l_{m}, called the *mixing length*, has to be defined in some way. In what follows we shall explain Prandtl's mixing length model for the case of transport of linear momentum in one-dimensional average shear flow, that is, for a case in which $u_x(y)$, say, is known. Assume that we are interested in vertical turbulent transport of horizontal average momentum per unit mass. We choose a distance l smaller than the typical vertical scale height of the velocity, which is

$|\bar{u}_x||\partial \bar{u}_x/\partial y|^{-1}$, and then we may formally write the spread of $\bar{u}_x(y)$ over l to be $\pm l|\partial \bar{u}_x/\partial y|$. The approximate identification of the velocity fluctuations and their approximate equality $u'_x \sim u'_y$, because there is no reason to assume that they be significantly different, are undoubtedly the "heaviest" assumptions of the procedure. We thus put

$$|u'_y(y)| \sim |u'_x(y)| \sim l \left| \frac{\partial \bar{u}_x(y)}{\partial y} \right|, \qquad (9.23)$$

where we would like to eventually identify l with the mixing length. The problem with this is that a Taylor expansion was assumed over a distance which is similar to the mixing length, that is, of the order of the largest eddies. There is also a difficulty to understand how blobs of fluid retain their momentum over a mixing length l_m, and then they give up their momentum suddenly to their new surroundings. Indeed, Prandtl himself had doubted whether his theory was adequate, but because of the lack of alternative ideas, the mixing length theory began to be used for calculating turbulent transport in practical applications, and the results were not too bad. If one accepts Eq. (9.23), one next notes that if the derivative $\partial_y \bar{u}_x(y) > 0$, there should be a negative correlation between u'_x and u'_y, and conversely if this derivative is negative, it is very likely that this correlation be positive (can you see why?). As mentioned above, there is no reason to expect that these fluctuations should be different in their *absolute* value; thus, we can estimate

$$-\overline{u'_x u'_y} = l_m^2 \left| \frac{\partial \bar{u}_x(y)}{\partial y} \right| \frac{\partial \bar{u}_x(y)}{\partial y}, \qquad (9.24)$$

where any constants have been absorbed in the definition of l_m. Remembering that on the left-hand side we have the Reynolds stress over the density, we find for this simple one-dimensional flow, using the appropriate one-dimensional Boussinesq equation

$$\nu_{turb} = l_m^2 \left| \frac{\partial \bar{u}_x(y)}{\partial y} \right|, \qquad (9.25)$$

where any constants have been absorbed into l_m.

This model is called the *mixing length* theory, and with suitably defined mixing length and turbulent viscosity, it remains a useful tool in approximating the effects of turbulence in real systems, among them astrophysical objects. For example, the turbulent energy transport in the convective regions of the interior of a star (discussed in Chap. 7) and the angular momentum transport in accretion disks (discussed in Chap. 5) are usually modeled based on the mixing length idea and a turbulent viscosity. Thus, this crude theory, based essentially on dimensional arguments, works reasonably well, at least better than it may be expected, in particular in engineering applications, where l_m may be determined experimentally. Prandtl's mixing length theory is a *one-equation* closure. *Two-equation* closure models exist as well and are somewhat more complicated. They are frequently used in engineering applications, in particular as a tool for numerical computations of the LES type (see below). Among them the most popular and influential ones are the $k - \varepsilon$ and $k - \omega$ models (see the book of Wilcox, referenced in the last section of this chapter). In engineering applications, the parameters needed for the models are usually inferred from experiments.

9.2.2   Energy Transfer from the Mean Flow to the Turbulence

The Reynolds stress is also responsible for the transfer of energy from the mean flow to turbulence. In laminar flow we have found that the rate of working of the viscous stress on a fluid element, whose surface is $S(t)$, is according to Eq. (1.44)

$$\oint_{S(t)} \mathbf{u} \cdot \mathbf{T} dS \tag{9.26}$$

with $\mathbf{T}$, the traction, being the force resulting from the stress, which can be expressed as $T_k = \sigma_{ki} n_i$. Since $\sigma_{ki} = -P\delta_{ki} + \tau_{ki}$, it is not difficult to express the rate of working of the *viscous stress only* on a unit volume fluid element as

$$\frac{\partial}{\partial x_j} (u_i \tau_{ij}) . \tag{9.27}$$

Now, by analogy to the laminar case, we may write for the rate of working of the Reynolds stress on the average flow:

$$\frac{\partial}{\partial x_j} \left(\bar{u}_i \tau_{ij}^R \right) = \bar{u}_{ij} \frac{\partial \tau_{ij}^R}{\partial x_j} + \tau_{ij}^R \bar{\mathscr{D}}_{ij} . \tag{9.28}$$

We note here that the second term is not a contribution to the *internal* energy, but rather to the kinetic energy contained in the turbulent fluctuations of the velocity.

Integrate Eq. (9.28) over a volume $\mathscr{V}$ which contains the turbulent patch of the flow. On the left the integrand is a divergence and thus becomes a surface integral, which vanishes since the Reynolds stress on this surface vanishes. We are thus left with the two terms on the right side of that equation, which have to balance, viz.,

$$\int_{\mathscr{V}} \left(-\bar{u}_i \frac{\partial \tau_{ij}^R}{\partial x_j} \right) d^3 x = \int_{\mathscr{V}} \tau_{ij}^R \bar{\mathscr{D}}_{ij} d^3 x . \tag{9.29}$$

We have already seen that $\partial \tau_{ij}^R / \partial x_j$ is the net force per unit volume acting on the *mean* flow as a result of the Reynolds stress, that is, turbulent fluctuations. Thus, this term multiplied by $-\bar{u}_i$ is just minus the rate of working on the mean flow by the turbulence. So this expression is simply the loss of the kinetic energy of the *mean* flow. On the other side, we get the gain of the kinetic energy of the turbulence. In laminar flow this would be the gain of the thermal energy, but here it is the gain of the *kinetic* energy of the turbulence. To gain deeper understanding, consider a simplistic situation in which we have a flow, say, through a pipe, that is steady on average and is governed by the averaged Navier–Stokes equation, which we write

$$\rho(\bar{\mathbf{u}} \cdot \nabla)\bar{u}_i = \frac{\partial \bar{P}}{\partial x_i} + \frac{\partial \tau_{ij}^R}{\partial x_i} + \bar{F}_i^{\text{visc}}, \tag{9.30}$$

where the last term is the average viscous "force." We shall also use the abbreviation $f_i \equiv \partial(\tau_{ij}^R)/\partial x_j$. Using Eq. (9.28), we can express $f_i \bar{u}_i$ as

$$-f_i \bar{u}_i = \tau_{ij}^R \bar{\mathscr{D}}_{ij} - \frac{\partial}{\partial x_j}\left(\tau_{ij}^R \bar{u}_i\right). \tag{9.31}$$

The second term above is a divergence and integrates to a surface integral, often zero, depending on the surface boundary conditions, as we have seen. The first term is the local rate of the increase of turbulent kinetic energy, as can be readily seen. It is also called the *deformation work* and expresses the fact that the mean shear tends to stretch and increase turbulent vorticity. From Eq. (9.31) it follows, as we have seen, that $-f_i \bar{u}_i$ and $\tau_{ij}^R \bar{\mathscr{D}}_{ij}$ always balance globally, that is, kinetic energy removed from the main flow feeds the turbulent kinetic energy. Locally, however, the divergence term need not be zero. This means that the energy removed from the mean motion by the Reynolds stress need not appear as turbulent kinetic energy at the same location. There is here a subtle point—energy fluxes are involved. We conclude this section by obtaining the equation for the kinetic energy of the turbulence. For that we shall have to exploit the lengthy Eq. (9.14). Setting $i = j$ yields after some algebra:

$$\bar{\mathbf{u}} \cdot \nabla \left[\frac{1}{2}\rho \overline{(\mathbf{u}')^2}\right] = \frac{\partial}{\partial x_k}\left[-\overline{P'u_k'} - \overline{u_i'\tau_{ik}'} - \frac{1}{2}\rho \overline{u_i'u_i'u_k'}\right] + \tau_{ik}^R \bar{\mathscr{D}}_{ik} - 2\rho \nu \overline{\mathscr{D}_{ij}\mathscr{D}_{ij}}. \tag{9.32}$$

In Problem 9.7 the following equation is derived for the *average* kinetic energy:

$$\bar{\mathbf{u}} \cdot \nabla \left(\frac{1}{2}\rho \bar{u}^2\right) = \frac{\partial}{\partial x_k}\left[-\bar{u}_k \bar{P} + \bar{u}_i\left(\bar{\tau}_{ik} + \tau_{ik}^R\right)\right] - \tau_{ik}^R \bar{\mathscr{D}}_{ik} - 2\rho \nu \bar{\mathscr{D}}_{ik}\bar{\mathscr{D}}_{ik}. \tag{9.33}$$

Equations (9.32)–(9.33) allow a simple physical interpretation to the various terms. Noting that all the generation rates as well as sinks are everywhere per unit volume, we can write for the average mean flow kinetic energy the following interpretation:

$$\left\{\begin{array}{c} \text{rate of change} \\ \text{of} \\ \text{mean flow kinetic energy} \end{array}\right\} = \left\{\begin{array}{c} \text{influx} \\ \text{of the} \\ \text{mean flow kinetic energy} \end{array}\right\}$$

$$-\left\{\begin{array}{c} \text{loss of} \\ \text{mean flow kinetic energy} \\ \text{to turbulence} \end{array}\right\} - \left\{\begin{array}{c} \text{viscous} \\ \text{dissipation} \end{array}\right\},$$

while the terms in the turbulent kinetic energy equation (9.32) lend themselves to the following physical interpretation:

$$\left\{\begin{array}{c} \text{rate of change} \\ \text{of} \\ \text{turbulent kinetic energy} \end{array}\right\} = \left\{\begin{array}{c} \text{influx (by transport)} \\ \text{of} \\ \text{turbulent kinetic energy} \end{array}\right\}$$

$$+\left\{\begin{array}{c} \text{generation} \\ \text{of} \\ \text{turbulent kinetic energy} \end{array}\right\} - \left\{\begin{array}{c} \text{viscous} \\ \text{dissipation} \end{array}\right\}.$$

Note that the term $\tau_{ik}^R \bar{\mathcal{D}}_{ik}$ appears in both equations with opposite signs. This is natural since this is the term responsible for transferring the mean flow's kinetic energy to the turbulent one. This term enters the turbulence and is ultimately dissipated by molecular viscosity, progressing downward through the energy cascade. It is possible to mark the various terms in our equation of turbulent kinetic energy (9.32) by single symbols:

$$\Gamma = -\overline{u_i' u_j'}\mathcal{D}_{ij} \quad \text{Generation,}$$

$$\varepsilon = 2\nu\overline{\mathcal{D}_{ij}\mathcal{D}_{ij}} \quad \text{Dissipation.} \tag{9.34}$$

Notice that the flux of the kinetic energy that appears, with a minus sign, in the square-bracketed term can be defined to be ρT_k, thus giving the vector $\mathbf{T}$ (do not confuse with traction!) the meaning of kinetic energy flux, allowing us to summarize the turbulent kinetic energy equation as

$$\bar{\mathbf{u}} \cdot \nabla\left[\frac{1}{2}\rho\overline{(\mathbf{u}')^2}\right] = -\nabla \cdot \mathbf{T} + \Gamma - \varepsilon. \tag{9.35}$$

In statistically homogeneous turbulence, the expressions after the ∇ sign vanish (can you see why?) and we have then the simple, physically comforting, equation

$$\Gamma = \varepsilon, \tag{9.36}$$

that is, the local rate of generation of turbulent energy is equal to the rate of viscous dissipation. We know that viscous dissipation occurs on the microscopic scale, while the generation of turbulent energy occurs usually on much larger scales. Thus, a *cascade* of turbulent energy from large scales (eddies) to smaller and smaller ones must take place. In a homogeneous and steady state, the rate that the energy "flows" through the cascade is Π, and in steady homogeneous turbulence, necessarily $\Gamma = \Pi = \varepsilon$ and as we have already seen $\varepsilon \sim u^3/l$. Turbulence is often inhomogeneous and sometimes unsteady. This quality often also depends on scale. However, even though one might expect that Π is not equal to Γ and to ε, the following remains a remarkably good approximation $\Gamma \sim \Pi \sim \varepsilon \sim u^3/l$, which obviously is wrong when $\Gamma = 0$ (freely decaying turbulence). Still, even in this case, the other approximate relations seem to hold.

As mentioned above there exist more sophisticated closures than the one-equation mixing length models, but they are often idiosyncratic and work well only in particular flows, for which there exist experimental data, that are then used in the model. Such is the aforementioned $k - \varepsilon$ model, the details of which can be found in specific literature, which can be found at the *Bibliographical Notes*. The dependence of these models on experimental data and their successful applicability to particular flows are often referred to as *engineering* models of turbulence. The underlying physical idea remains the turbulent viscosity.

9.2.3   Wall-Bounded Flows and the Log-Law

If a "wall," that is, a flat motionless surface, say, for simplicity, bounds a flow, it forces specific boundary conditions on the solution, viz., zero velocity for a viscous flow. For inviscid flow, only the vertical velocity component has to be zero. As is well known, boundary conditions are often not less important than the differential equations themselves. Thus, the presence of walls has profound consequences, also on turbulent flows. Near such boundaries, *boundary layers* form. We have already discussed boundary layers in this book (in Chap. 2), stressing mainly the structure of a laminar boundary layer in a direction perpendicular to the wall, as well as explaining in considerable detail the mathematical approximate asymptotic technique of *matching*, which results in the formation of boundary layers. Here we would like to discuss a turbulent flow near a flat surface, which gives rise to the noted logarithmic profile, which, in turn, is applicable to *turbulent boundary layers*. But before doing so, we distinguish between *internal* flows, which are closed by a boundary, e.g., in a channel or pipe, and *external* ones, limited by just one boundary.

9.2.3.1   Turbulent Flow in a Channel

Let there be a one-dimensional, fully developed, turbulent flow, whose averaged velocity lies in the (x, y) surface, say. We assume that this time-averaged plane parallel flow is in the x direction positing that the average y and z components of the velocity are zero. Let the flow $\bar{\mathbf{u}} = (\bar{u}_x(y), 0, 0)$ be confined to a channel $2w > y > 0$. Assuming steady state and all the other conditions that led to the Reynolds equation (9.12), we may write the two relevant components (x, y) of that equation, substituting also the relevant fluctuations in the expression detailing the Reynolds stress. We thus have

$$\rho \frac{\partial}{\partial y} \left[v \frac{\partial \bar{u}_x}{\partial y} - \overline{u'_x u'_y} \right] = \frac{\partial \bar{P}}{\partial x}, \tag{9.37}$$

$$\rho \frac{\partial}{\partial y} \left[-\overline{u'_y u'_y} \right] = \frac{\partial \bar{P}}{\partial x}. \tag{9.38}$$

The assumption that the turbulence is fully developed implies that all fluctuations are x independent. Now we see that the value of $\bar{P} + \overline{u'_y u'_y}$ at $y = 0$ is $\bar{P}(0) \equiv \bar{P}_w$, i.e., the wall pressure. Thus, Eq. (9.37) can be rewritten with $\bar{P}_w$ replacing $\bar{P}$, all other terms being identical. The right-hand side of this equation is independent of x

$$\rho \frac{d}{dy} \left[v \frac{d\bar{u}_x}{dy} - \overline{u'_x u'_y} \right] = f_1(y), \tag{9.39}$$

while the left-hand side is y independent

$$\frac{d\bar{P}_w}{dx} = f_2(x).$$

This way of writing separates the variables of the original equation and guarantees that both previous equations are equal to a negative constant $-C = \rho^{-1} d\bar{P}/dx$. Integrating Eq. (9.39) with the appropriate constant on the right-hand side, one may easily show (see Problem 9.8) that the average viscous plus Reynolds, or total stress, is

$$\tau = \bar{\tau}_{xy} + \tau_{xy}^R = \tau_w \left(1 - \frac{y}{w} \right), \tag{9.40}$$

where τ_w is the stress at the wall $(y = 0)$. Define now what is called the *friction velocity*, u_*, by the relation defining its square: $u_*^2 \equiv \tau_w/\rho = Cw$. The physical meaning of the ρu_*^2 is the x-momentum flux given to the wall by the fluid, and because of steady state, it is this flux that also flows in the fluid in the *negative x* direction. Straightforward algebra then gives

$$u_*^2 = \frac{\tau_w}{\rho} = Cw \implies \tau = \rho \left(v \frac{d\bar{u}_x}{dy} - \overline{u'_x u'_y} \right) = \rho \left(u_*^2 - Cy \right). \tag{9.41}$$

9.2.3.2 Turbulent Boundary Layer

We shall now concentrate on the lower boundary and try to analyze the structure of the turbulent flow in its vicinity. Equation (9.41) as written above cannot be solved for $\bar{u}_x(y)$, because the y-dependence of the Reynolds's stress is unknown. This is a good example of the utility of a dimensional argument, which dictates that the gradient $d\bar{u}_x/dy$ at any y must be dependent only on the abovementioned momentum flux, the density, and the coordinate y. To economize the notations, we define this flux ρu_*^2 to be written as the indexless σ. Bearing in mind that our result is intended to hold for very large Re and that the momentum transport away from the wall is turbulent and not molecular, we refrain from using the fluid viscosity v as one of our dimensional building blocks of $d\bar{u}_x/dy$. We have to remember, however, that v will become important very close to the wall.

Since the only combination of the above building blocks which has the dimensions of velocity gradient is

$$\left[\frac{d\bar{u}_x}{dy}\right] = \left[\left(\frac{\sigma}{\rho y^2}\right)^{1/2}\right], \tag{9.42}$$

where the square brackets denote here "the dimensions of...," we get

$$\frac{d\bar{u}_x}{dy} = \frac{u_*}{\kappa} \cdot \frac{1}{y}, \tag{9.43}$$

where κ is the von Kármán constant and must be found experimentally. It turns out that $\kappa = 0.4$ and the integration of Eq. (9.43) immediately gives the velocity profile

$$\bar{u}_x(y) = \frac{u_*}{\kappa} \log(y + \mathrm{d}), \tag{9.44}$$

where d is a constant of integration. This formula reveals the well-known logarithmic profile of turbulent boundary layers near planar surfaces, but the determination of the integration constant d requires some nontrivial physical insight. The value of d cannot be found from the boundary condition at $y = 0$ because the formula (9.43) is mathematically singular there. Physically, that formula was derived for far enough distances from the wall's surface. At sufficiently small distances from that surface, the microscopic viscosity begins to be important. Denoting this small distance by y_0, we attempt to determine its value as follows. The length scale of the turbulence at such close distance to the wall is y_0, while the velocity is of order u_*. This gives for the Reynolds number near the wall $\mathrm{Re_w} \sim u_* y_0 / v$. The microscopic viscosity is important when this Reynolds number is close to unity, and this gives $y_0 \sim v/u_*$. At distances smaller than y_0, we have a thin layer in which the mean velocity varies linearly with y, similarly to the boundary layer found in Chap. 2. We are not interested in this viscous sublayer, whose transition to the turbulent regime is gradual. We just have to ensure that for $y \sim y_0$ we get $u \sim u_*$. This is achieved if we take $\mathrm{d} = -\log y_0$, and this yields finally the logarithmic velocity profile

$$\bar{u}_x(y) = u_* \frac{1}{\kappa} \log\left(\frac{u_*}{\nu} y\right),\tag{9.45}$$

which is correct for y substantially larger than y_0, as defined above.

Having understood that this approximate derivation, due to L. Landau, can have only logarithmic accuracy, we rewrite the formula by adding a constant term inside the logarithm, which in practice has the effect of adding a constant of the order unity times u_*. Indeed the logarithmic mean velocity law, which holds for $y \gg y_0$ and y much smaller than some external length scale L, is usually written as

$$\bar{u}_x(y) = u_* \left[A \log\left(\frac{u_*}{\nu} y\right) + B\right],\tag{9.46}$$

where $A = \kappa^{-1} = 2.5$ is the reciprocal of the von Kármán's constant and B can be determined experimentally as $B = 5.1$. The results of a proper matched asymptotic expansion, given in Problem 9.9, are summarized in Fig. 9.1. We describe now in words this result: the outer region is for $y^+ \gg 1$—the rightmost portion of the range. The viscous sublayer is for $y^+ \lesssim 5$: the leftmost part of the range. The other two intermediate ranges constitute the inner region, $\eta = y/w \ll 1$, and the overlap regions, $y^+ \gg 1$ and $\eta = y/w \ll 1$, are not marked in the figure but are easily identified around y^+. The log-law of the wall is valid where indicated on the figure and obeys:

$$\bar{u}_x(y) = \frac{1}{\kappa} u_* \ln\left(\frac{u_* y}{\nu} + \text{const}\right).\tag{9.47}$$

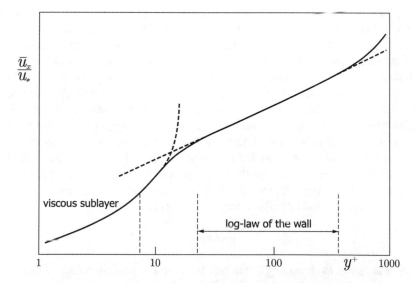

Fig. 9.1   A plot of the averaged horizontal velocity in the channel $\bar{u}_x$ in units of u_* as a function of the nondimensional coordinate $y^+ (\equiv y u_*/\nu)$. The log-law of the wall and the region of its validity is clearly apparent

We should perhaps also mention that various alternative formulations of this problem exist, such as the *velocity defect* and the *law of the wall*, but we feel that these are beyond the scope of the present discussion and can be looked up in the references, mentioned in the *Bibliographical Notes*.

9.2.4 Free Shear Flows

Turbulent and laminar regions are sometimes found side by side in the same flow. Jets, wakes, and mixing layers are three well-known examples of free shear flows that lead to such heterogeneous states. Turbulent jets are found in many forms in the atmosphere, in industrial applications, and in astrophysics, but it turns out that astrophysical jets usually involve magnetic forces on ionized fluids and so their discussion is deferred to Chap. 10. Turbulent hydrodynamic jets obviously exist also, e.g., in the atmosphere or produced by machinery, including jet engines. We have also already seen one example of a wake, which may become turbulent, when discussing the flow past a cylinder (see Fig. 2.2 and the description of transition to turbulence, in that case). This was a classical phenomenon known as *separation*, where one sees turbulent regions break away from a boundary to become interspersed with laminar flow regions.

9.2.4.1 The Turbulent Jet

This problem was first tackled by L. Prandtl, who started in 1925 to investigate its properties, using simple similarity arguments. We shall not follow Prandtl's work, rather we shall consider a single description and treatment of a turbulent jet. Consider a turbulent jet of fluid emanating from a thin cylindrical pipe into an infinite space filled with the same fluid. This problem was solved in its laminar regime in Problem 3.13. We concentrate on distances larger than the diameter of the mouth of the tube, and hence we allow ourselves to assume that the jet is axially symmetric. We assume that for whatever reason (e.g., large enough Re) the jet is turbulent and proceed to determine the form of the turbulent region, using dimensional considerations. Let the x-axis be the axis of the jet, with $x = 0$ at the mouth of the tube, and denote by $R(x)$ the radius of the jet cone. We wish to obtain the function $R(x)$ but do not have at our disposal any significant parameter with the dimension of length and thus $R \propto x$ and geometry dictates

$$R = x \tan \theta \tag{9.48}$$

where θ is half of the cone angle, the one between x and the jet's edge. Let r be the distance from the jet axis and a point. Clearly, for the same dimensional argument as before, the mean velocity distribution in the jet must be

$$\bar{u}_x(r,x) = \bar{u}_0(x)f\left[\frac{r}{R(x)}\right],\qquad(9.49)$$

where f is an unknown function and $\bar{u}_0(x)$ the velocity at the axis, and note that we are talking about *time-averaged* values of the velocity, hence the notation including the bar, since, as we know, in a turbulent flow the velocity fluctuates. Thus, we may say that the turbulent jet structure is self-similar, depending only on the similarity variable $\xi \equiv r/R(x)$. It turns out (experimentally) that $f(\xi) = 1$ for $\xi = 0^+$, and it diminishes rapidly as the argument increases. We have $f(0.4) = 0.5$ but the value of f reaches 0.01 as the boundary of the turbulent region is reached. There is also transverse velocity which is responsible for the inflow into the turbulent region; at its inner edge, it has the value $-0.025\bar{u}_0$. The flow outside the turbulent region can be calculated analytically (see Problem 9.10). The dependence of the velocity behavior inside the jet as x changes can, however, be determined using simple physical considerations. The conservation of momentum implies, in steady state, the conservation of its total flux along the jet. Using spatial (on r) averages of the functions and averages with respect to time, here we have

$$\Phi_p^{\text{tot}} \sim \rho u^2 R^2 \hat{\mathbf{x}},\qquad(9.50)$$

where Φ_p^{tot} is the properly averaged total momentum flux in the jet, as explained above. We have seen an expression for such a flux in a sound wave in Sect. 6.2.4, but the expressions are obviously not identical. We leave for the reader to convince herself or himself that this is the correct averaged momentum flux in the jet. Now using Eq. (9.48), the velocity estimate becomes

$$u(x) \sim \frac{1}{x}\sqrt{\frac{P}{\rho}}.\qquad(9.51)$$

That is, the surface averaged velocity diminishes like x^{-1} from the point of emergence. Obviously we have to avoid a singularity at $x = 0$ and this is physically guaranteed by the fact that we are looking at distances $x \gg d$. Let Q be the mass transported per unit time, through a cross section of this turbulent jet. Clearly Q must be of the order $\sim \rho u R^2$. Substituting from Eqs. (9.51) and (9.48) for u and R, respectively, and understanding that the leading dependence of Q must be $\propto x$, we finally get

$$Q = \delta Q_0 \frac{x}{d}\qquad(9.52)$$

where Q_0 is the mass of the fluid that emanates from the aperture at unit time and δ is a numerical coefficient depending only on the form of the aperture (for a circular aperture, the value is $\delta = 1.5$). Note that the amount of fluid in the turbulent jet increases with x; thus, it *entrains* fluid which flows in a laminar way into the turbulent zone (see Fig. 9.2).

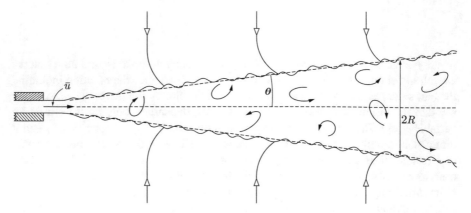

Fig. 9.2 A submerged jet emanating from a very thin (of size $\sim d \ll R$, for R in any region of interest) opening; for details, see text

9.2.4.2   Homogeneous Shear Flow

Perhaps the simplest example of free shear flows is the one in which the shear is homogeneous on large scales. Such flows can be approximations to certain flows in nature, e.g., regions in an interstellar cloud, or in the lab. We take this opportunity to expound on the discussion of a turbulent homogeneous shear flow, in contrast to our description of jets, which was mainly based on dimensional arguments.

Assume that the *mean* flow velocity field can be simply written as $\bar{\mathbf{u}} = (\bar{u}_x(y), 0, 0)$ with $\bar{u}_x(y) = \mathscr{S}y$. $\mathscr{S}$, here, a newly introduced constant, is obviously the partial derivative of $\bar{u}_x(y)$ with respect to y and is equal to $2\mathscr{D}_{xy}$, where $\mathscr{D}_{\alpha\beta}$ is the deformation rate tensor, as defined in general, for Cartesian indices in formula (1.15). Here, with this velocity profile, the constant $\mathscr{S}$ may be called the *shearing rate*. The *turbulent fluctuations* in $\mathscr{S}$, denoted by $\mathscr{S}'$, are however not constant! Assuming that the spatial gradients in the turbulent fluctuations are negligible and that the rate of change of these turbulent fluctuation is small enough, we can suppose that time averages will capture faithfully enough what we may call "mean time evolution."

The quantities that we are interested to evaluate here are

$$\tau_{xy}^R \quad \text{and} \quad \overline{(u_\alpha')^2} \quad \text{for} \quad \alpha = (x, y, z), \tag{9.53}$$

because our simplistic model, if useful for anything, is for expressing the energetic interaction between the mean shear and the turbulence and perhaps also for assessing the role that pressure forces play in this redistribution. Symmetry arguments allow us to neglect the Reynolds stresses whose components include z (do you see why?). For *homogeneous* turbulence, we can derive time-dependent generalizations of Eq. (9.14) as follows:

$$\partial_t \left[\rho \overline{(u'_x)^2} \right] = 2\overline{P'\mathcal{S}'_{xx}} - 2\rho \nu \overline{\frac{\partial u'_x}{\partial x_k} \frac{\partial u'_x}{\partial x_k}} + 2\tau^R_{xy}\mathcal{S}$$

$$\partial_t \left[\rho \overline{(u'_y)^2} \right] = 2\overline{P'\mathcal{S}'_{yy}} - 2\rho \nu \overline{\frac{\partial u'_y}{\partial x_k} \frac{\partial u'_y}{\partial x_k}}$$

$$\partial_t \left[\rho \overline{(u'_z)^2} \right] = 2\overline{P'\mathcal{S}'_{zz}} - 2\rho \nu \overline{\frac{\partial u'_z}{\partial x_k} \frac{\partial u'_z}{\partial x_k}}$$

$$\partial_t \left[\tau^R_{xy} \right] = 2\overline{P'\mathcal{S}'_{xy}} + 2\rho \nu \overline{\frac{\partial u'_x}{\partial x_k} \frac{\partial u'_y}{\partial x_k}} + \rho \overline{(u'_y)^2}\mathcal{S}, \tag{9.54}$$

where $k = 1, 2, 3$ correspond to (x, y, z) and, as usual, summation over a double index is tacitly assumed. Now, assuming that the small-scale eddies, which contribute to the viscous terms above, are approximately isotropic, i.e., independent of direction, allows the viscous terms to be replaced by the isotropic values:

$$2\nu \overline{\frac{\partial u'_i}{\partial x_j} \frac{\partial u'_k}{\partial x_j}} = \frac{2}{3}\varepsilon \delta_{ik}. \tag{9.55}$$

Remembering that incompressibility of the turbulence dictates $\mathcal{S}'_{\beta\beta} = 0$ for $\beta = x, y, z$, Eq. (9.54) reduces to

$$\partial_t \left[\frac{1}{2}\rho \overline{(u'_x)^2} \right] = -\frac{1}{3}\rho \varepsilon + \tau^R_{xy}\mathcal{S},$$

$$\partial_t \left[\frac{1}{2}\rho \overline{(u'_y)^2} \right] = -\frac{1}{3}\rho \varepsilon,$$

$$\partial_t \left[\frac{1}{2}\rho \overline{(u'_z)^2} \right] = -\frac{1}{3}\rho \varepsilon,$$

$$\partial_t \left[\tau^R_{xy} \right] = \rho \overline{(u'_y)^2}\mathcal{S}. \tag{9.56}$$

This immediately gives that the mean kinetic energy per unit mass of the turbulence, E, develops as

$$\frac{d\mathrm{E}}{dt} = \frac{\tau_{xy}}{\rho}\mathcal{S} - \varepsilon = \Gamma - \varepsilon. \tag{9.57}$$

This means that the turbulence is driven by the rate of working of the Reynolds stress, Γ, and destroyed, as usual, by viscous dissipation. When $\Gamma = \varepsilon$ we have steady-state turbulence, as it was expressed in Eq. (9.36) and the related discussion. Here we allow for the growth or decay of the turbulent kinetic energy, as the driving and dissipating terms are not necessarily equal.

9.3   Rudiments of a Statistical Description of Turbulence

As we already know, a defining characteristic of turbulence is the presence of disordered fluctuations in the dynamic flow variables. The instantaneous *spatial* as well as the local *temporal* dependence of the fluid dynamical functions has a complex and seemingly random nature. Moreover, two turbulent systems prepared as identically as possible in the laboratory will develop differently. This property of sensitivity to initial conditions (known as SIC) is accepted as characterizing *deterministic chaos*. But the hope that the (re-)discovery of deterministic chaos in the 1980s will ultimately solve the turbulence enigma was largely premature. True, dynamical system theory and chaos have provided new tools and improved understanding, but it did not bring about a conceptual breakthrough (see Sect. 9.7.2), at least until some new ideas were proposed in 2004 by a number of authors (see reference [9] in the *Bibliographical Notes*). In this section we shall attempt to summarize the statistical approach to turbulence. As said before the obvious motivation for this approach is the analogy: randomly moving microscopic particles ↦ seemingly randomly moving turbulent eddies, and the unusual success to deal with the former by methods of statistical mechanics.

9.3.1   Averaging

The *average*, or *mean*, is a fundamental tool of statistical analysis. Averaging should produce smoother *mean* values of the dynamic fields, which can then be investigated by usual analytical methods. Although averaging does not have a unique definition, there is little question that the different methods should be, at the end, equivalent.

In a homogeneous turbulent field, it is useful to consider a cube with periodic boundary conditions (PBC), as we shall discuss in Sect. 9.4. Then averaging spatially over this cube may be a good prescription for spatial averaging in this case. On the other hand, following Eq. (9.1) a time average over a sufficiently long time may be an equivalently good definition of an average. As mentioned already several times, it appears that we are lucky that under a wide range of mathematical conditions, these averages are the same. This is the essence of the ergodic hypothesis. In what follows we shall touch upon a considerably general approach to the definition of the mean or average.

9.3.1.1   Weighted Space-Time Average

For the sake of generality, we may define a *weighted* space-time average of a function of space and time thus

$$\overline{f(\mathbf{x},t)} = \int \int f(\mathbf{x} - \mathbf{x}', t - \tau) W(\mathbf{x}', \tau) d^3 x' d\tau, \qquad (9.58)$$

where the integrals are from $-\infty$ to $+\infty$ in all $x'_j, (j = 1, 2, 3)$ and in τ, W is a weighting function, generally nonnegative, with which one may change this definition in a desired way. It satisfies a normalization condition

$$\int \int W(\mathbf{x}', \tau) d^3 x' d\tau = 1. \tag{9.59}$$

It is unavoidable that this general definition (9.58) may give rise to many possible mean values, because it will depend on W and, through it, on the particular region in time and space over which the average is done. In choosing the averaging rule, we will opt for the one that is most suitable to the problem at hand, and in this chapter, we shall always remark what kind of averaging we are using. The need for idiosyncratic averaging leads to a lot of analytical difficulties that cannot be avoided. Moreover, for every problem a suitable weight function has to be found. Whatever the averaging procedure, we postulate after Reynolds that for any two functions f and g of space and time, the following five relationships, known as Reynolds conditions, are satisfied. Average is denoted here by overline:

$$\overline{f + g} = \bar{f} + \bar{g}, \tag{9.60}$$

$$\overline{af} = a\bar{f}, \quad \text{if } a = \text{const}, \tag{9.61}$$

$$\bar{a} = a, \quad \text{if } a = \text{const}, \tag{9.62}$$

$$\overline{\frac{\partial f}{\partial s}} = \frac{\partial \bar{f}}{\partial s}, \quad \text{for } s = x_j, t. \tag{9.63}$$

$$\overline{\bar{f}g} = \bar{f}\bar{g}. \tag{9.64}$$

It would be highly desirable to find a method yielding a well-defined unique mean, instead of worrying about its dependence on the method of averaging. As we already know, the *ergodic hypothesis* is the central tool in this direction. So let us proceed and clarify what we can about it, once and for all. A *probabilistic* approach is required, and we shall now sketch its basic concepts.

9.3.1.2 Probability Average, Random Variables

The central concept in a *probability-theoretical* approach to turbulence theory is that of an *ensemble*. Instead of considering a single flow, we imagine a collection of all similar flows, created by some fixed set of external conditions. Such a set of flows, a statistical ensemble, serves as the *probability sample space*, a necessary tool in probability theory. As an example, which undoubtedly may clarify this, consider a short and thin cylindrical probe, which we shall later put in a wind tunnel, containing a flow past a spherical obstacle. Now immerse our probe in a particular position $\mathbf{x}$, say, of a laminar region of such a flow. It will record, after a time t, say, a velocity $u_x(\mathbf{x}, t)$. A second experiment in the same setting, or an identical one, will also give a reading $u_x(\mathbf{x},t)$, by the identically placed probe after the same time which is identical to the first measurement. In principle, if appropriate initial and boundary conditions are applied, the velocity will simply be the unique solution of the Navier–Stokes equations, although such solutions are often not easily found. In contrast, in *turbulent* regions of the flow, small uncontrollable perturbations in the initial conditions lead to a situation that each identical experiment results in different values of the velocity $u_x(\mathbf{x},t)$ at the specified place and time. One may talk of an *ensemble* of values of $u_x(\mathbf{x},t)$, and also other flow variables, as will be seen below, at this point and time, which in laminar regions gives a unique value. Each result in the ensemble is referred to as a *realization* chosen, at random, from this ensemble. In principle, such identical experiments may be performed many times, giving the same values for the flow variables at a time and in a place when and where the flow is laminar, but in the turbulent regions and times, the results will be somewhat different, for each experiment, even though they are measured at the same $(\mathbf{x},t)$. Performing now an *arithmetic* mean of the large number of the results in the ensemble, we shall get a value $\overline{u_x(\mathbf{x},t)}$, which we may call the *ensemble average* or *probability mean*. There is one important proviso, however—the number of ensemble members is large enough so that the ensemble average does not change too much upon adding some more members. In practice, as the number of ensemble members increases, the ensemble average may oscillate somewhat around the true mean and finally converge to it. This definition of average or mean will from now on be used by us and denoted by an overline or $\langle \cdot \rangle$. The averages of other dynamical variables can also be defined this way, if we can be confident that their means over large ensembles remain stable and deviate only slightly from some constant value.

Consider, in the context of the previous discussion, what is called the *indicator function* $\chi_{u_x(\mathbf{x},t)}(u',u'')$ with $u'' > u'$. The above indicator function is defined by

$$\chi_{u_x}(u',u'') = \begin{cases} 1 \text{ if } u' \leq u_x \leq u'' \\ 0 \text{ otherwise.} \end{cases}$$

Now imagine evaluating the indicator function for all values of u_x in the ensemble. The number $P(u',u'')$ about which the arithmetic mean of the indicator function value oscillates is just equal to the frequency of the occurrence of those experiments in the ensemble, in which the value of $u_x(\mathbf{x},t)$ is between u' and u'', that is, $u' \leq u_x(\mathbf{x},t) \leq u''$. This number $P(u',u'')$ so defined may be called the *probability* that u_x will take the value between u' and u'' in this particular ensemble. Under normal conditions, this positive number can be represented as an integral

$$P(u',u'') \equiv \int_{u'}^{u''} p(u)du, \tag{9.65}$$

of a nonnegative function $p(u)$—the *probability density* function of $u_x(\mathbf{x},t)$. It is assumed that this definition is normalized, i.e., the probability density is chosen so as to guarantee $P(-\infty,+\infty) = 1$. Definition (9.65) expressed above can also be written in a differential form, assuming that the variable u is continuous,

$$P\{u < u_x(\mathbf{x},t) < u + du\} = p(u)du, \tag{9.66}$$

where $P\{\wp\}$ indicates the probability of the condition marked by $\wp$ is being satisfied. It is thus only natural to define the *probability* mean, or average, and denote it by an overline, say, as

$$\overline{u_x(\mathbf{x},t)} = \int_{-\infty}^{\infty} up(u)du. \tag{9.67}$$

The knowledge of the probability density $p(u)$ allows the determination of the probability mean of any function of the velocity $F[u_x(\mathbf{x},t)]$, as in

$$\overline{F[u_x(\mathbf{x},t)]} = \int_{-\infty}^{\infty} F(u)p(u)du. \tag{9.68}$$

In general, variables w, say, having a definite probability density, are called in probability theory *random variables*, and the set of all possible probabilities $p(w',w'') \equiv P\{w' < w < w''\}$ corresponding to the random variable w is called its *probability distribution*.

9.3.2 Correlation Functions, Velocity Increments, and Structure Functions

Our wish is to physically characterize the complicated, seemingly chaotic or random field of variable fluctuations. One of the most important notions in this characterization is the introduction of the concepts of *correlation* and *structure* functions, similarly to what is done in disordered macroscopic media in condensed matter theory and statistical physics. Consider some general turbulent field $\mathbf{u}(\mathbf{x},t)$. Sticking to our notation, we separate this field variable into a mean part and a fluctuation, where we do not specify the method of averaging, which nevertheless must obey the averaging rules (9.60)–(9.64). We shall divide our discussion into a general case, which will mainly be devoted to definitions, and the homogeneous and isotropic cases, in which simplifications and progress towards a theory can be made.

9.3.2.1 General Case, Mainly Definitions

We start by defining the *longitudinal velocity increment*

$$\delta u_{\parallel}(\mathbf{x},\mathbf{r}) \equiv [\mathbf{u}(\mathbf{x}+\mathbf{r}) - \mathbf{u}(\mathbf{x})] \cdot \frac{\mathbf{r}}{r}, \tag{9.69}$$

where $\mathbf{r}$ indicates a vector whose magnitude is r. This allows us to immediately define an important characteristic of the velocity field, the so-called longitudinal structure function of order p. It is

$$S_p^{\|}(\mathbf{x},\mathbf{r}) \equiv \overline{[\delta u_{\|}(\mathbf{x},\mathbf{r})]^p}. \qquad (9.70)$$

Similarly, a *transverse velocity increment* is a component of the vector increment $\delta \mathbf{u} = [\mathbf{u}(\mathbf{x}+\mathbf{r}) - \mathbf{u}(\mathbf{x})]$ that is perpendicular to the vector $\mathbf{r}$. Obviously such a component is not unique, call a particular one of them $\delta u_{\perp}$, but at least one of this kind exists, e.g., the following can be written

$$\delta u_{\perp} = [\mathbf{u}(\mathbf{x}+\mathbf{r}) - \mathbf{u}(\mathbf{x})] \times \frac{\mathbf{r}}{r}, \qquad (9.71)$$

with the *transverse structure function (of order p)* defined, similarly to formula (9.70):

$$S_p^{\perp}(\mathbf{x},\mathbf{r}) \equiv \overline{[\delta u_{\perp}(\mathbf{x},\mathbf{r})]^p}. \qquad (9.72)$$

Any progress, more than just the definitions, is difficult in this topic of structure functions, when the turbulent flow is of general nature. So we turn now to the discussion of correlation functions and related topics. Following Taylor's insight, we consider the averaged quantity

$$\mathscr{C}_{\mathbf{u}'\mathbf{u}'}[(\mathbf{x},t),(\mathbf{x}+\mathbf{r},t)] = \overline{\mathbf{u}'(\mathbf{x},t) \cdot \mathbf{u}'(\mathbf{x}+\mathbf{r},t)} \qquad (9.73)$$

and understand that it has to yield a number expressing some average measure of the mutual spatial interdependence between the fluctuations in the functions $\mathbf{u}(M)$ and $\mathbf{u}(M_+)$, where, in this case, the functions' arguments (the space-time points) are $M \mapsto (\mathbf{x},t)$ and $M_+ \mapsto (\mathbf{x}+\mathbf{r},t)$. It is interesting to see that in our example if $\mathbf{r} = 0$, that is, the sought correlation function is between the same functions at the same point, the result is simply $\mathscr{C} = \overline{|\mathbf{u}'(\mathbf{x},t)|^2}$, i.e., twice the average kinetic energy per unit mass in the fluctuations, that is, in the turbulence. Now we define a correlation function in a general form and examine some specific examples:

$$\mathscr{C}_{\{u_i\}}[\{M_j\}] = \overline{\prod_i u_i(M_j)}, \qquad (9.74)$$

where in this general definition we have a finite sequence of functions, which can be vector functions, and then the products between them are scalar $\{u_i\}_{i=1,N}$, and a finite sequence of space-time points, the functions' arguments, $\{M_j\}_{j=1,N}$. A simple example of a correlation function has already been discussed above—the case of Eq. (9.73), when we arrive at it from the general formula (9.74) considering the correlation of the full velocity vector perturbation between just two points with distance $\mathbf{r}$ between them where the "origin" point $\mathbf{x}$ is of general nature. We also consider the velocity perturbations at the same time t and will do so from now on in all cases discussed, unless specifically indicated otherwise. Choosing to use the notation $Q(\mathbf{x}, \mathbf{r}, t)$ for this correlation function, we have

$$Q(\mathbf{x}, \mathbf{r}, t) \equiv \overline{\mathbf{u}'(\mathbf{x}, t) \cdot \mathbf{u}'(\mathbf{x} + \mathbf{r}, t)}. \tag{9.75}$$

As it was already noted, for $\mathbf{r} = \mathbf{0}$ we simply get twice the value of the average turbulent kinetic energy at the point $\mathbf{x}$, i.e., the *strength* of the turbulence:

$$Q(\mathbf{x}, \mathbf{0}, t) = \overline{|\mathbf{u}'(\mathbf{x}, t)|^2}. \tag{9.76}$$

The opposite case, namely, if $|\mathbf{r}| \gg |\mathbf{x}|$, is also of interest because then velocity fluctuations at distant points $\mathbf{x}$ and $\mathbf{x} + \mathbf{r}$ should be very weakly correlated, if at all, that is,

$$Q(\mathbf{x}, \mathbf{r}, t) = \overline{\mathbf{u}'(\mathbf{x}, t) \cdot \mathbf{u}'(\mathbf{x} + \mathbf{r}, t)} \to \overline{\mathbf{u}'(\mathbf{x}, t)} \; \overline{\mathbf{u}'(\mathbf{x} + \mathbf{r}, t)} = 0, \tag{9.77}$$

since the mean of a fluctuation is zero. Thus, the correlation function has values that are appreciably different from zero if $r = |\mathbf{r}|$ lies within some range. An approximate length, whose order of magnitude is within this range, is called the *correlation length* of the turbulence. It turns thus out that the correlation function, as introduced above, contains information about the strength of the velocity fluctuations as well as about the distance over which these velocity fluctuations are correlated.

We now drop altogether the time dependence from all the velocity fluctuations and will do so in all cases discussed, unless specifically indicated otherwise. This is exactly justified if we assume stationarity, as we will see also in the homogeneous isotropic case below. So far, the velocity correlation function examples we considered were scalar, but originating from scalar products of vector functions. Another important example is obtained when we consider velocity fluctuation components at different locations: $u_i'(\mathbf{x})$ and $u_j'(\mathbf{x} + \mathbf{r})$, *say*. The velocity correlation will in this case become a *tensor*:

$$Q_{ij}(\mathbf{x}, \mathbf{r}) = \overline{u_i'(\mathbf{x}) u_j'(\mathbf{x} + \mathbf{r})}. \tag{9.78}$$

The number of different components of the tensor obviously depends on symmetries of the system. Finally, we shall mention that there are more general possibilities, like a three-point correlation function, a tensor with three indices

$$Q_{ijk}(\mathbf{x}_1, \mathbf{x}_2, \mathbf{x}_3) = \overline{u_i'(\mathbf{x}_1) u_j'(\mathbf{x}_2) u_k'(\mathbf{x}_3)},$$

and even more general cases regarding the correlation functions and structure functions exist. We have, however, limited ourselves to what can be studied in the next section, when the turbulence is homogeneous and isotropic.

9.3.2.2 Homogeneous and Isotropic Turbulence

When turbulence is homogeneous, no physical property depends on position.
i.e., there is translational invariance. Isotropy guarantees that no such property
depends on direction (rotational invariance). These nontrivial symmetries allow for
significant simplifications in the understanding and calculation of various properties
of the system. In our context, the first thing to note is that it is possible to show, due
to homogeneity and isotropy, that the second-order longitudinal structure function
does not depend on $\mathbf{x}$ and depends only on the absolute value of the vector $\mathbf{r}$. Thus,
we have

$$S_2^\|(r) = \overline{[\delta u_\|(r)]^2}. \tag{9.79}$$

Similarly to the above, the independence on $\mathbf{x}$ and on $\mathbf{r}$'s *direction* is correct for the
perpendicular structure function. Moreover, it is also possible to show that it does
not matter which $\delta u_\perp$ is chosen and in the transverse structure function of order two
the argument $\delta u_\perp$ is a function of $r = |\mathbf{r}|$ only:

$$S_2^\perp(r) \equiv \overline{[\delta u_\perp(r)]^2}. \tag{9.80}$$

As not much can be easily deduced from the structure functions at this point,
we move to investigate correlation functions, which, as we shall see, can lead
to some progress in understanding when the turbulence is homogeneous and
isotropic. We shall deal with Q_{ij}, because its form in the homogeneous and isotropic
turbulence will lead to important insight. Thus, consider the defining Eq. (9.78):

$$Q_{ij}(\mathbf{x}, \mathbf{r}) = \overline{u_i'(\mathbf{x})u_j'(\mathbf{x}+\mathbf{r})}.$$

First, it should be noted that the existence of a mean flow, $\bar{\mathbf{u}}(\mathbf{x}, t) \neq 0$, which is a
vector, is inconsistent with isotropy. We must have $\bar{\mathbf{u}} = 0$ and consequently the field
consists of fluctuations only. Therefore, we may drop the primes from the velocity
functions without risk of ambiguity. Next, since the correlation function should be
only a function of r, as said before, homogeneity forbids its dependence on $\mathbf{x}$ and
isotropy forbids its dependence on $\hat{\mathbf{r}}$, we may write

$$Q_{ij}(r) = \overline{u_i(\mathbf{x})u_j(\mathbf{x}+\mathbf{r})}, \tag{9.81}$$

which is formally valid for any $\mathbf{x}$. Note that

$$\frac{\partial Q_{ij}}{\partial r_j} = \overline{u_i(\mathbf{x})\frac{\partial u_j(\mathbf{x}+\mathbf{r})}{\partial r_j}} = 0, \tag{9.82}$$

because of incompressibility, as the second term is actually $\nabla_r \cdot \mathbf{u}$. The variable of
the ∇ operator is $\mathbf{r}$ as this is the variable of the function it acts on, and $\mathbf{x}$ may
be considered as a constant shift. Also, because of the symmetries of isotropy and
homogeneity, we have

$$\frac{\partial Q_{ij}}{\partial r_i} = \frac{\partial Q_{ij}}{\partial r_j} = 0. \tag{9.83}$$

This result will be needed in the derivation of an important equation below (9.167).

It can be shown that the most general form of a tensor function with the properties of symmetry, as given above, is

$$Q_{ij} = A(r)r_ir_j + B(r)\delta_{ij}, \tag{9.84}$$

where δ_{ij} is the Kronecker delta and $A(r)$, $B(r)$ are appropriate scalar functions. Trying now to express $A(r)$ and $B(r)$ in terms of two other scalar functions, because they are more appropriate to express as clearly as possible the correlation function in homogeneous and isotropic turbulent field, we consider two points at $\mathbf{x}$ and at $\mathbf{x}+\mathbf{r}$. Introducing a Cartesian system of coordinates, so that the index n denotes a vector component *along* $\mathbf{r}$ and the index m one of the components *perpendicular* to $\mathbf{r}$, one sees immediately that $r_n = r$ and $r_m = 0$. So we can write

$$Q_{nn}(r) = A(r)r^2 + B(r) \equiv \frac{1}{3}\overline{u^2}f(r),$$

$$Q_{mm}(r) = B(r) \equiv \frac{1}{3}\overline{u^2}g(r), \tag{9.85}$$

where there is no summation on n and m, as these are components. As noted in the above expressions, this is the *definition* of $f(r)$ and $g(r)$, consistently with Eq. (9.84) and with $Q_{ii}(0) = \overline{u^2}$, where there is a summation on the double index. The latter is forcing the constraint $f(0) = g(0) = 1$, and where $f(r)$ and $g(r)$ are actually the longitudinal and transversal correlations, so to speak, and $Q_{ij}(r)$ can be expressed in terms of the new functions, using Eqs. (9.85) and (9.84), thus

$$Q_{ij}(r) = \frac{1}{3}\overline{u^2}\left[\frac{f(r) - g(r)}{r^2}r_ir_j + g(r)\delta_{ij}\right]. \tag{9.86}$$

Problem 9.19 shows that using incompressibility and symmetry, a relation between $f(r)$ and $g(r)$ can be found from this equation. Its form is

$$g(r) = f(r) + \frac{1}{2}r\frac{df}{dr}. \tag{9.87}$$

Already in the 1930s, von Kármán and Howarth made this point and it became clear that only one function is needed. It is reasonable to assume that the longitudinal part, $f(r)$, must be a function starting from 1 at $r = 0$, decaying slowly at first but ultimately developing an exponential tail, as correlation functions usually do.

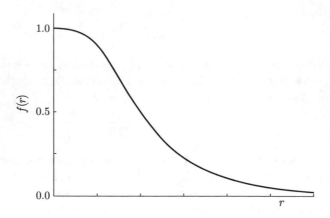

Fig. 9.3 A physically motivated qualitative shape of the longitudinal correlation $f(r)$

In Fig. 9.3 a function of this sort is shown. We now consider the spatial Fourier transform of the correlation tensor $Q_{ij}(r)$, which reads

$$\Phi_{ij}(\mathbf{k}) = \frac{1}{(2\pi)^3} \int Q_{ij}(r) e^{-i\mathbf{k}\cdot\mathbf{r}} d^3 r. \qquad (9.88)$$

Now, spherical symmetry of a function of $\mathbf{r}$, that is, its dependence on r only, implies similar symmetry of its transform in the $\mathbf{k}$-space (can you show it?). Thus, the inverse Fourier transform gives back $Q_{ij}(r)$, which has the form

$$Q_{ij}(r) = \int \Phi_{ij}(k) e^{i\mathbf{k}\cdot\mathbf{r}} d^3 k. \qquad (9.89)$$

From incompressibility and symmetry (see Eq. (9.83)), it follows that

$$k_i \Phi_{ij} = k_j \Phi_{ij}, \qquad (9.90)$$

and similarly to the form (9.84) it is now possible to invoke the most general form of $\Phi_{ij}(k)$ as follows:

$$\Phi_{ij}(k) = C(k) k_i k_j + D(k) \delta_{ij}. \qquad (9.91)$$

Here one can immediately see from Eq. (9.90) that the scalar functions $C(k)$ and $D(k)$ are related in the following way:

$$D(k) = -k^2 C(k).$$

This implies that Φ_{ij} can be written as dependent on just one scalar function

$$E(k) = -4\pi k^4 C(k), \qquad (9.92)$$

as follows:

$$\Phi_{ij}(k) = \frac{E(k)}{4\pi k^4}(k^2\delta_{ij} - k_i k_j). \tag{9.93}$$

$E(k)$ can be shown to be the energy spectrum, that is, $E(k)dk$ is the turbulence energy between wavenumbers k and $k + dk$, if we calculate $Q_{ii}(0)$, which, as we know, is equal to $\overline{u^2}$. So

$$\frac{1}{2}\overline{u^2} = \frac{1}{2}Q_{ii}(0) = 2\pi\int\Phi_{ii}(k)k^2 dk, \tag{9.94}$$

where the Fourier transform (9.89) has been used as has spherical symmetry in k. It remains to substitute $\Phi_{ii}(k)$ from expression (9.93) into this integral to get the important result

$$\frac{1}{2}\overline{u^2} = \int_0^\infty E(k)dk, \tag{9.95}$$

relating the constituent of the correlation function in k-space (9.92) to the energy spectrum.

We have already mentioned the contribution of von Kármán and Howarth to the investigations of correlation functions. However, we have to remark that the material on the correlation functions, which has been reviewed above for the homogeneous and isotropic case, was based almost solely on geometrical arguments. The only physical consideration was incompressibility. Later on we shall derive the Kármán and Howarth equation (9.167), as it is called, for the simplest case of homogeneity and isotropy. The equation is usually the *equation of motion* of scalar functions related to correlation functions and as such adds the element of dynamics to the theory.

9.3.3   Random Fields

From the probabilistic point of view, we may consider the value of a velocity component at a fixed point M, whose coordinates are $(\mathbf{x}, t)$, say, as a random variable, described by some definite probability distribution. There is no a priori reason not to apply the same approach to the whole *field* of a velocity component. Thus, we may look at $(\mathbf{x}, t)$ as a continuously variable coordinate. The statistical ensemble used is then the one of *all possible fields* $u_x(\mathbf{x}, t)$. Now, for the entire field to be random, it is required that it has to yield a variable at any point $M_i - (\mathbf{x}_i, t_i)$, say. Each has to have its own probability density, because we cannot expect that for even 2 points, $u_x(M_1)$ and $u_x(M_2)$, their arithmetic mean is statistically stable. The correct approach when dealing with only two points is to posit that there exists a *two-dimensional* probability density, $p_{M_1 M_2}(u_1, u_2)$, defined by the probability relationship:

$$P\{u_1 < u_x(M_1) < u_1 + du, \quad u_2 < u_x(M_2) < u_2 + du\} = p_{M_1 M_2}(u_1, u_2) du_1 du_2.$$
$$(9.96)$$

The generalization to any number of points should be straightforward. The existence of a positive N-dimensional probability density justifies the statement that corresponding velocities are N random numbers, and the existence of *all* possible such probability densities, for all positive integers and all positive choices of N points in space-time, guarantees that the field $u_x(\mathbf{x}, t)$ is random. We cannot dwell here on a more rigorous definition of a random field, especially since it can be found in almost any probability theory book. The one suited best for us is, in our opinion, the two-volume expansive book of Monin and Yaglom (reference [7]), as it deals explicitly with the statistical theory as applied to fluid dynamics and hence to turbulence.

9.3.3.1 Moments of Random Variables and Fields

Even though a rigorous probabilistic definition of random fields is nontrivial, it turns out that in practice it is sufficient to restrict oneself to simpler statistical parameters which describe, in a way that is sufficient for our purposes, the statistical properties of turbulent flows. We shall discuss here the most important of those parameters—the *moments* of the random fields. Consider a system having N random variables, u_i, where $i = 1, N$, say, associated with an N-dimensional probability density $p(u_1, u_2, \ldots, u_N)$. Then the moments of the variables are defined as

$$\mathscr{M}_{k_1, k_2, \ldots, k_N} \equiv \overline{u_1^{k_1} u_2^{k_2} \ldots u_N^{k_N}}$$

$$= \int \int_{-\infty}^{\infty} \ldots \int u_1^{k_1} u_2^{k_2} \ldots u_N^{k_N} p(u_1, u_2, \ldots, u_N) du_1 du_2 \ldots du_N \qquad (9.97)$$

where k_i are nonnegative integers whose sum is the *order* of the moment. Particularly useful are the *central moments* defined as

$$\mathscr{M}_{k_1, k_2, \ldots, k_N}^{\text{cen}} \equiv \overline{(u_1 - \bar{u}_1)^{k_1} (u_2 - \bar{u}_2)^{k_2} \ldots (u_N - \bar{u}_N)^{k_N}}. \qquad (9.98)$$

A little reflection is sufficient to understand the meaning of the central moment. Moreover, Problem 9.17 uncovers some familiar statistical quantities: variance, standard deviation, covariance, etc., related to what we prefer to call here moments.

Moving now to the discussion of moments of random fields, we consider functions of points M_i, in space time, and define first the K-order moment of such a field by

$$\mathscr{M}_{uu \ldots u}(M_1, M_2, \ldots M_K) = \overline{u(M_1), u(M_2) \ldots u(M_K)}, \qquad (9.99)$$

where it is understood that some of the points M_j may be identical. The number of different points M_i defines whether the moment is a one-point type, two-point type, etc. Now, as before, there are higher order moments: the order is the sum of the powers to which each field is raised. The averages of the product of values of several *different* albeit statistically related, random fields constitute ever more complicated entities: *joint moments*, which are, in general, tensors. A particularly simple but useful combination may arise if we consider two points, for example, but of one random field only. Then we have a symmetric tensor with respect to two indices $i = j$. It is easy to identify that this is a particular case of the general construction that we have called before the correlation function:

$$\mathscr{C}_{u,u}(M_1, M_2) = \overline{u(M_1)u(M_2)}, \tag{9.100}$$

with $u = u_i = u_j$. Note that this function is a *two-point, second-order* moment, using the nomenclature of the discussion here above. The two-point correlation function is evidently symmetric in its arguments and also it possesses the property, proven in Problem 9.18:

$$\sum_{i=1}^{N}\sum_{j=1}^{N}\mathscr{C}_{u,u}(M_i, M_j) \geq 0. \tag{9.101}$$

It turns out, but we shall not show it here, that any function $\mathscr{C}_{u,u}(M_1, M_2)$, satisfying Eqs. (9.100)–(9.101), may be the correlation function of some random field. Finally, we should mention here the joint two-point moment $\mathscr{C}_{u,v}(M_1, M_2) = \overline{u(M_1)v(M_2)}$, called the *cross correlation* function of the two different fields u and v. The properties of the cross correlation function are very similar to those of the correlation function (cf. 9.100)–(9.101). We shall stop here, although higher order correlation and cross correlation functions may obviously be defined. Due to the limitations of space, we shall try to be brief yet comprehensive.

9.3.3.2   Gaussian Fields: Random Fields with a Gaussian Probability

Even though this section may seem like a diversion on our way to the ergodic theorem (or rather, ergodic hypothesis, in turbulence), the topic of Gaussian fields is very important not only in turbulence theory but also in any statistical study, e.g., of matter distribution in the universe. We wish, thus, to define this special and important case of *Gaussian* fields, which are actually random fields with normal probability distribution, defined for the random variables u_i with $i = 1, 2 \dots, N$ as

$$p(u_1, u_2, \dots, u_N) = C_N \exp\left\{ -\frac{1}{2}\sum_{j,k=1}^{N} g_{jk}(u_j - \alpha_j)(u_k - \alpha_k) \right\}, \tag{9.102}$$

where α_i and g_{jk} are real constants. The latter are such that the matrix $\mathbb{G}$, whose elements are g_{jk}, is a positive definite one and C_N is a constant determined from the normalization of p, that is,

$$\int_{-\infty}^{\infty} \dots \int_{-\infty}^{\infty} p(u_1, u_2, \dots u_n)du_1u_2 \dots.du_N = 1.$$

It is not difficult to understand that the constants α_j and g_{jk} are simply connected to the first two moments of the normal distribution (see Problem 9.21), and so we see that for a normal probability distribution, the first and second moments completely define the probability density. This property is a very strong one, as essentially all the statistical characteristics of the corresponding random variables may be determined by just those two moments, which are a single number each.

We conclude this discussion of the normal distribution by considering Gaussian random *fields*, that is, in general $\mathbf{u}(M)$, where the probability distribution of any finite number of its values is normal. Here too, the complete statistical determination of the fields reduces to the knowledge of their means and correlations, for example, a one-dimensional field $u(M)$, where we remind the reader that M denotes a point of space-time. It may be shown that if $\mathcal{M}_{uu}^{\text{cen}}(M_1, M_2)$ denotes the correlation function of the field fluctuations, then for any choice of points $M_1, M_2, \ldots M_N$, an N-dimensional normal distribution may be found with an appropriate probability density, giving the mean values $\overline{u(M_i)}$ and the second moments $\mathcal{M}_{uu}(M_i)(M_j) = \mathcal{M}_{uu}^{\text{cen}}(M_i, M_j) + \overline{u(M_i)u(M_j)}$. This will also apply in multidimensional random fields. Finally, we would like to note that often the random fields of the FD variables are actually close to Gaussian in many respects. In general, the Gaussian random fields with their convenient properties often serve as a paradigm on which various hypotheses are checked.

9.3.4   Sketch of the Proof of the Ergodic Theorem in Turbulence Theory

We now return to the question of the conditions under which the appropriate time and space mean values of a random field $u(\mathbf{x}, t)$ will converge to the probability average as the averaging intervals tend to infinity. In short, this is the question of the applicability of the ergodic theorem in turbulence. The rigorous proof of the ergodic theorem in general turbulent flows is difficult and lengthy. We shall limit ourselves here to a survey of the considerations on the way to the proof. Moreover, we shall consider only a simplistic version of the theorem, valid for *stationary* random functions and *homogeneous* random fields. In the general case, we shall posit that the ergodic theorem is just a hypothesis and accept it.

For simplicity we deal with time averaging only. Considering a field $u(\mathbf{x}, t)$, we limit the discussion to a function of *time* only, as according to our choice we deal with the case in which the dependence on $\mathbf{x}$ has no significance. We would like to investigate time averaging of $u(t)$ over an infinite interval. Defining a new function of time, we shall call it $u_T(t)$, as it is, in principle, dependent on T, because it is the mean of the function over the interval T around 0, viz.,

$$u_T(t) \equiv \frac{1}{T} \int_{-T/2}^{T/2} u(t + \tau) d\tau. \tag{9.103}$$

Our interest is to investigate if $u_T(t)$ will converge to the probability average $\overline{u(t)}$ as $T \to \infty$. One relatively easy conclusion stems from the fact that $u(t)$ is a *bounded* function, and therefore, for an arbitrary interval $[t, t_1]$, we have

$$u_T(t) - u_T(t_1) \equiv \frac{1}{T} \left[\int_{-T/2}^{T/2} u(t + \tau) d\tau - \int_{-T/2}^{T/2} u(t_1 + \tau) d\tau \right]. \tag{9.104}$$

Shifting now the limits of the integrations by adding to the upper limit of the left integral $t_1 - t - T$ and subtracting the same amount from the *lower* limit of the right integral should not change the result, but we may write the integrals in other appropriate variables

$$u_T(t) - u_T(t_1) = \frac{1}{T} \left[\int_{t-T/2}^{t_1-T/2} u(s) ds - \int_{t+T/2}^{t_1+T/2} u(s) ds \right], \tag{9.105}$$

where t and t_1 are fixed numbers. The expression (9.105) will thus become infinitely small as $T \to \infty$. Thus, in the limit as $T \to \infty$, $u_T(t)$ and $u_T(t_1)$ must become equal. It follows that the time mean (9.103), which is, generally speaking, a function of t, must actually be independent of t. Therefore, for the time mean to be equal to the probability mean, $\overline{u(t)}$, it is sufficient that

$$\overline{u(t)} = U = \text{const.} \tag{9.106}$$

This line of reasoning is good for a particular case, namely, for *stationary* random fields, and proves the ergodic theorem for turbulence for this limited case.

A similar idea may be used for homogeneous random fields. In this case we are looking at random fields following the same line of reasoning as for stationary fields, but with space variable instead of time. We wish that the space average

$$u_{A,B,C}(\mathbf{x}) = \frac{1}{ABC} \int_{-A/2}^{A/2} \int_{-B/2}^{B/2} \int_{-C/2}^{C/2} u(x_1 + \xi_1, x_2 + \xi_2, x_3 + \xi_3) d\xi_1 d\xi_2 d\xi_3, \tag{9.107}$$

in a three-dimensional box whose sides are of sizes A, B, and C, with $A, B, C \to \infty$, will tend to the probability mean $\overline{u(\mathbf{x})}$. As in the temporal case, it turns out that for this to be true, it is sufficient that the fields are *homogeneous*, that is, satisfy

$$\overline{u(\mathbf{x})} = U = \text{const.} \tag{9.108}$$

It is evident that not all random fields in turbulent flows are stationary and/or homogeneous, but we shall be satisfied with these cases and their discussion, in order to justify the use of the ergodic hypothesis in the statistical approach to turbulence. The hypothesis is summarized as follows: *usual time and space averages approach, for large averaging domains, the ensemble average.*

We conclude this section by citing, without proof, some important results of the ergodic theorem for random fields, in the stationary and homogeneous fields' case.

The derivation is technical and lengthy and includes integral Fourier transforms, which will be discussed in some detail later, and also some probabilistic material not covered in this book. The results can be formulated simply and they refer to the case when the energy spectrum of a random, stationary or homogeneous, field is a power law in the Fourier space. The two results are:

1. *Stationary random field*
   The temporal Fourier spectrum of the energy having the form of a power law

$$E(\omega) \propto |\omega|^{-n}, \quad 1 < n < 3 \quad \text{yields} \quad \overline{[u(t') - u(t)]^2} \propto |t' - t|^{n-1}. \quad (9.109)$$

2. *Homogeneous random field*
   The spatial Fourier spectrum of the energy having the form of a power law

$$E(k) \propto k^{-n}, \quad 1 < n < 3 \quad \text{results in} \quad \overline{|\mathbf{u}(\mathbf{x}') - \mathbf{u}(\mathbf{x})|^2} \propto |\mathbf{x}' - \mathbf{x}|^{n-1}. \quad (9.110)$$

9.3.5 The Statistical Formulation of the Fundamental Problem of Turbulence

Treating turbulent fluid mechanical variables as random fields, whose meaning is based on probability theory and statistics, which was initiated by Kolmogorov and his school, is now generally accepted. There is, however, one important problem related to this issue. In short, the theory has to be scrutinized experimentally, and in experiments, usually only one flow is observed and the spatiotemporal values of various flow fields measured. Thus, averaging over spatial and/or temporal segments is the only possibility in experiments. It is impossible to realize in practice the basic concept of an *ensemble* and define the necessary averages in the probability theoretical sense, as discussed in the previous section.

This situation is analogous to the study of statistical mechanics, in which averages are defined with the help of different realizations of an ensemble of the same system. There it is usually asserted that for a system with a finite number of degrees of freedom, one may replace the *ensemble average* by a directly observable average over long time. This is based on the assumption that as the averaging time interval becomes infinitely great, the time averages converge to the corresponding ensemble averages. This assertion can be strictly proven under some well-defined mathematical conditions, yielding the ergodic *theorem*, proved by G.D. Birkhoff. If not all (but most) conditions necessary for the mathematical rigor of the proof are satisfied, the result is adopted as well as the highly likely ergodic *hypothesis*.

Now we shall try to formulate, in principle, what we shall call the fundamental, or central, problem of turbulence. The necessary proviso is that the problem will be spelled out in statistical language. Let us accept the existence and uniqueness conjecture of laminar incompressible flows of the Navier–Stokes equations. This means that the same flow should be expected for the same initial and boundary conditions, if all the parameters are the same. We can assume this to be correct only in some range of the parameters, including initial and boundary conditions. It is well known that some simple laminar flows lose the possibility of mathematical existence if the Reynolds number (Re) is above some critical value. In place of the laminar flow, we will see a turbulent one, in which, to the best of our understanding, the values of the variables, such as velocity, will depend considerably on extremely small, uncontrollable, disturbances in the initial and/or boundary conditions. Turbulent flow exhibits sensitivity to initial conditions (SIC) as we have mentioned above, which, as far we understand, is analogous to SIC of simple chaotic dynamical systems. We adopt the usual assertion that the fluid dynamical equations remain applicable in turbulent flow, as well. Still, because of SIC, the changes in the velocity and other variables in a turbulent flow are so complex that they are practically unpredictable. Under these circumstances, the flow will be best described by the probability distributions for the corresponding fluid dynamical fields, while their exact value, at a given point and time, and its connection to the equations are of no interest. Thus, the fluid dynamical equations will be used, in turbulent flows, only for investigating the relevant probability distributions or particular values thereof.

Since fluid dynamics is described by PDEs in space and time, there are obvious mathematical requirements on the random fields, which must be applied to the PDEs. Also, since the random fields are defined by probability distributions, actually the mathematical requirements are that the probability distributions should be regular enough, so that the random fields they define are continuous and sufficiently smooth. Now we ask the reader for a little patience, we are about to try to define, in principle at least, *what is the question of turbulence*. How many times have we heard that in turbulence we don't even know what to ask? Well, let us say that at some initial time t_0, the unspecified, abovementioned, regularity conditions are obeyed by all the probability distributions. Thus, our necessary actual realizations of all relevant random fields will vary smoothly in time until any $\tau > t_0$. Thus, the whole set of the random fields will be a well-defined set of functions of space, at time $t = t_0 + \tau$. This means that the probability density for any fluid dynamical function at any $t > t_0$ may be determined, in principle at least, from the initial probability densities. In other words, a choice of initial probability distributions, satisfying certain regularity conditions, will determine through the equations of motions applied to the suitable random fields the probability distributions at all times. We may thus summarize and formulate statistically *the central problem of turbulence* for incompressible flows in a semi-heuristic way, as follows:

Given the probability distribution of the three velocity vector components at time t_0 at various space points, so that the velocity function subspace is divergenceless (solenoidal), it is required to determine the probability distribution of the velocity and pressure functions at all subsequent times. In the case of compressible flow, we have to take into account the probability distributions of all the independent fluid dynamical functions. It is a very difficult problem—also in the incompressible case!—and there is no known way to find the solution. However, we thought that formulating the problem already constitutes some progress.

9.4 Some Theory, Mostly Kolmogorov

9.4.1 Introduction

We have already discussed briefly in Sect. 9.1.1 in a qualitative way a possible general approach to fully developed turbulence. In this section we shall discuss this subject in more detail. Taylor, Richardson, and, of course, Kolmogorov were the main contributors to the subject. Here we shall try to explain some of their work. Along with the discussion of phenomenological and theoretical spectral laws, which constitute the greatest achievements of this topic, we shall spend some time on technical aspects and Fourier spectra in turbulence.

9.4.1.1 Fourier Representation of a Flow

When a turbulent flow is statistically homogeneous, it is useful to work in Fourier space. The simplest way to then introduce a Fourier representation is to imagine a fictitious box in real space, a cube of side size L, say, chosen in such a way that it contains all the physical spatial structures we want to study. Obviously this limits one to the ability to study structures not larger than L. We focus thus on the flow in the box and assume periodic boundary conditions (PBC) on its faces. One can then fill the whole space of a homogeneous turbulent flow with an infinite number of such identical boxes. Let $\mathbf{u}(\mathbf{x},t)$ be a periodic velocity field of the same period, L, say, in all three spatial Cartesian directions: $x_i,(i=1,2,3)$. A theorem proved by J. Fourier guarantees then the validity of the expansion:

$$\mathbf{u}(\mathbf{x},t) = \left(\frac{2\pi}{L}\right)^3 \sum_{n_1,n_2,n_3=-\infty}^{+\infty} \hat{\mathbf{u}}_{\mathrm{box}}(\mathbf{n},t)\, e^{\frac{2\pi}{L}i\mathbf{n}\cdot\mathbf{x}}. \tag{9.111}$$

The constant in front of the expression takes care of the normalization, as we shall see, and the vector $\mathbf{n}$ is defined by its Cartesian components $\mathbf{n} = (n_1,n_2,n_3)$ which allows to define the discrete normalized *wave vector* as follows:

$$\mathbf{k} \equiv \frac{2\pi}{L}\mathbf{n}. \tag{9.112}$$

The above discrete Fourier series (9.111) can then be written as

$$\mathbf{u}(\mathbf{x},t) = (\Delta k)^3 \sum_{n_j=-\infty}^{+\infty} \hat{\mathbf{u}}_{\mathrm{box}}(\mathbf{k},t)\, e^{i\mathbf{k}\cdot\mathbf{x}}, \tag{9.113}$$

where $\Delta k = (2\pi)/L$. Note that here the wave vectors attain only discrete values, integer multiples of $2\pi/L$, and so the Fourier transform $\hat{\mathbf{u}}_{\mathrm{box}}(\mathbf{k},t)$ of the periodic velocity $\mathbf{u}(\mathbf{x},t)$ is only defined for wave vectors whose components are integral multiplies of Δk. To answer the question how can the discrete transform $\hat{\mathbf{u}}_{\mathrm{box}}(\mathbf{k},t)$ be calculated from the periodic function $\mathbf{u}(\mathbf{x},t)$, we consider a more general case.

Moving now to such a case, we examine a flow $\mathbf{u}(\mathbf{x},t)$, not necessarily periodic and defined in whole space. The *integral Fourier transform* $\hat{\mathbf{u}}(\mathbf{k},t)$ of the above velocity field is defined, provided some mathematical restriction on the original functions are valid, in particular regarding their behavior at infinity, viz.,

$$\hat{\mathbf{u}}(\mathbf{k},t) = \left(\frac{1}{2\pi}\right)^3 \int \mathbf{u}(\mathbf{x},t)e^{-i\mathbf{k}\cdot\mathbf{x}}d^3x. \tag{9.114}$$

The relation between the function and its integral transform is given, as usual, by the *inverse transform*, so assuming that some sufficient conditions of mathematical rigor can be satisfied, we obtain the inverse transform as

$$\mathbf{u}(\mathbf{x},t) = \int e^{i\mathbf{k}\cdot\mathbf{x}} \hat{\mathbf{u}}(\mathbf{k},t)\,d^3k, \tag{9.115}$$

even though a homogeneous turbulent field may not decrease to zero sufficiently fast as $|\mathbf{x}| \to \infty$. These integral expressions for the Fourier and inverse Fourier transform hold also for scalar functions. In fact, one of the possible expressions of the (triple) Dirac delta function involves the Fourier transform (see Problem 9.22). We shall now deduce some conclusions relating transforms of discrete periodic functions to general functions. For example, if one uses Eq. (9.113) in (9.114), the following expression is obtained:

$$\hat{\mathbf{u}}(\mathbf{k},t) = \left(\frac{1}{2\pi}\right)^3 \int e^{-i\mathbf{k}\cdot\mathbf{x}} \left(\frac{2\pi}{L}\right)^3 \sum_{\mathbf{k}'} e^{i\mathbf{k}'\cdot\mathbf{x}}\hat{\mathbf{u}}_{\mathrm{box}}(\mathbf{k}',t)\,d^3x. \tag{9.116}$$

In Eq. (9.116) the components of the vector $\mathbf{k}$ are *not* necessarily integer multiples of $\Delta k = 2\pi/L$; however, the components of $\mathbf{k}'$ *must* satisfy this condition. Thus, it is easy to see that Eq. (9.116) may be written, exploiting Eq. (9.115), in the concise notation

$$\hat{\mathbf{u}}(\mathbf{k},t) = \left(\frac{1}{2\pi}\right)^3 \sum_{\mathbf{k}'} \delta(\mathbf{k}-\mathbf{k}')\hat{\mathbf{u}}_{\mathrm{box}}(\mathbf{k}',t), \tag{9.117}$$

and can be visualized like a three-dimensional "comb" of nonzero intensities. Averaging this expression in Fourier space over the box proves (see Problem 9.23) that the discrete Fourier transform is the average of the integral Fourier transform over the box. This can be formulated mathematically

$$\hat{\mathbf{u}}_{\mathbf{b}m}(\mathbf{k}_m,t) = \left(\frac{1}{V_{\mathbf{b}m}}\right)\int_{\mathbf{b}m}\left(\frac{1}{2\pi}\right)^3 d^3k \int \mathbf{u}(\mathbf{x},t)e^{-\mathbf{k}\cdot\mathbf{x}}d^3x, \tag{9.118}$$

where $\mathbf{k}_m$ is the location, in Fourier space, of the mth cube. The latter is the box designated as $\mathbf{b}m$, so that the integral over $\mathbf{b}m$ is over its volume in Fourier space and $V_{\mathbf{b}m}$ is the normalized box volume in real space: $(2\pi/L)^3$.

Equation (9.118) connects the integral, continuous Fourier representations, which are the natural choice for analytical work, to discrete transforms. The latter must be used in numerical calculations and hence the importance of this equation. This book is not expansive enough to include all useful mathematical or physical formulae that may have application in theoretical efforts to understand turbulence. Still, we have decided to group here and in the problems several well-known results, relating mainly to the Fourier space, that may sometime and somewhere be useful. We shall denote a *Fourier pair*, that is, a function $u(\mathbf{x},t)$ say and its integral Fourier transform in space only, $\hat{u}(\mathbf{k},t)$, where the Fourier operator $\mathbb{F}$ transforms between them. Thus, the pair can be transformed back using the inverse operator. Obviously, there are mathematical restrictions on the functions, especially their behavior at infinity, as well as demands on the existence of inverse transforms, but we shall not be bothered by such considerations. We shall use (9.115) and (9.114) for the case of a scalar function and equal normalization in both. Thus, our pair $u(x,t)$ and $\hat{u}(k,t)$ satisfies

$$\hat{u}(\mathbf{k},t) = \mathbb{F}\left[u(\mathbf{x},t)\right], \tag{9.119}$$

$$u(\mathbf{x},t) = \mathbb{F}^{-1}\left[\hat{u}(\mathbf{k},t)\right], \tag{9.120}$$

and the operation of the transform and its inverse is as explained above.

Among the basic statements which we state now without proof is the Fourier integral theorem in one space and k dimension:

$$u(x) = \left(\frac{1}{2\pi}\right) \int_{-\infty}^{+\infty} \left[\int_{-\infty}^{+\infty} u(\xi)e^{-ik\xi}d\xi\right] e^{ikx}dk, \tag{9.121}$$

valid if the functions satisfy appropriate mathematical conditions. We continue by listing several statements that can be checked by straightforward inspection: using the definition (9.119), we readily get

$$\mathbb{F}\left[\frac{\partial u}{\partial x_j}\right] = ik_j\,\hat{u}(\mathbf{k},t),$$

$$\mathbb{F}\left[\nabla u\right] = i\mathbf{k}\,\hat{u}(\mathbf{k},t),$$

$$\mathbb{F}\left[\nabla^2 u\right] = -|\mathbf{k}|^2\,\hat{u},$$

$$\mathbb{F}\left[\nabla \cdot \mathbf{u}(\mathbf{x},t)\right] = i\mathbf{k}\cdot\hat{\mathbf{u}}(\mathbf{k},t),$$

$$\mathbb{F}\left[\nabla \times \mathbf{u}(\mathbf{x},t)\right] = i\mathbf{k}\times\hat{\mathbf{u}}(\mathbf{k},t),$$

$$\mathbb{F}\left[u(\mathbf{x},t)\mathrm{v}(\mathbf{x},t)\right] = \{\hat{u}*\hat{\mathrm{v}}\}(\mathbf{k},t), \tag{9.122}$$

where $*$ denotes the *convolution* product $\int \hat{u}(\mathbf{p},t)\hat{\mathrm{v}}(\mathbf{k}-\mathbf{p},t)d^3p$.

Examining now the incompressible Navier–Stokes equations, we see that $\nabla \cdot \mathbf{u} = 0$ together with the above fourth statement in the list implies that the Fourier transform of the velocity is in a plane *perpendicular* to $\mathbf{k}$. We shall call this plane Π and, in words, the incompressibility plane. Noticing that $\partial \hat{\mathbf{u}}(\mathbf{k}, t)$ and $\nu k^2 \hat{\mathbf{u}}$ also lie in this plane, while the pressure gradient transform $\hat{P}\mathbf{k}$ is perpendicular to it, helps to realize that the Fourier transform of

$$(\mathbf{u} \cdot \nabla)\mathbf{u} + \rho_0^{-1} \nabla P, \tag{9.123}$$

is the projection of $(\mathbf{u} \cdot \nabla)\mathbf{u}$ onto the incompressibility plane. Introducing the tensor

$$\chi_{ij} = \delta_{ij} - \frac{k_i k_j}{k^2}$$

allows the projection of any vector $\mathbf{a}$ on Π to be written as $a_j \chi_{ij}$. As the index j is contracted (summed upon according to Einstein's convention), this is the *vector's* ith component. This discussion should suffice for the reader to be able to solve Problem 9.24.

9.4.2 A Theoretical Simplification for Periodic Boundary Conditions

It is *probably* safe to say that the Navier–Stokes equations (1.52)–(1.53) contain everything that is essential to obtain turbulent behavior. The reservation embodied in *probably* stems from the fact that the long-standing failure to understand and solve the problem of turbulence may have something to do with the continuum hypothesis itself, as given in Sect. 1.1, which leads to the Navier–Stokes equation. Singularities of various kind are perhaps the ignored culprit.

Here we shall take a simple case of incompressible flow and no body forces, in which turbulence is observed, and try to achieve some theoretical progress, in this case an elimination of one variable, here the pressure. First, we write down the FD equations in indical form

$$\frac{\partial u_i}{\partial t} + u_j \frac{\partial u_i}{\partial x_j} = -\frac{\partial P}{\partial x_i} + \frac{1}{\mathrm{Re}} \frac{\partial^2 u_i}{\partial x_j \partial x_j},$$

$$\frac{\partial u_i}{\partial x_i} = 0. \tag{9.124}$$

All the symbols have their usual meaning, introduced before in this book, and we stress that the equation is written in its nondimensional form. Note that the relative magnitude of the nonlinear term, with respect to the dissipative one, grows with the Reynolds number, Re. This nonlinear partial differential equation will be well posed only if suitable boundary and initial conditions are specified. Unfortunately, we cannot quote here any exact mathematical conditions, which would guarantee existence and uniqueness of a solution. The existing mathematical results on the subject, which is being continuously pursued, are clearly beyond the scope of this book.

The simplest approach is to consider a maximally symmetrical system with no boundaries at all. Assuming it is homogeneous, that is, any two different points in space are indistinguishable physically, and isotropic, that is, any two directions in space are indistinguishable physically, we may avoid the difficulties that infinite systems pose by looking at the fluid in a cube of size L and imposing periodic boundary conditions on the cube surfaces. This allows the following simplification of Eq. (9.124). Taking the divergence of the first equation in indical form, and using the second one, we get

$$\frac{\partial}{\partial x_i}\left(u_j\frac{\partial u_i}{\partial x_j}\right) = \frac{\partial^2}{\partial x_i \partial x_j}(u_i u_j) = -\frac{\partial^2 P}{\partial x_i \partial x_i}. \tag{9.125}$$

It is not difficult to see that this means that the pressure satisfies the Poisson equation

$$\nabla^2 P = \varsigma, \tag{9.126}$$

where ς *is defined* as the second term of the above equation. It turns out that the above Poisson equation can be solved within the class of divergence-free L-periodic functions, provided that

$$\overline{\varsigma} \equiv \frac{1}{L^3}\int_{\mathcal{V}_L}\varsigma(\mathbf{x})d^3x = 0, \tag{9.127}$$

where we shall henceforth denote the volume of the $L \times L \times L$ box, over which integrals are performed, by $\mathcal{V}_L$. Equation (9.127) shows that the spatial average of the integrand vanishes. Since ς satisfies this condition (see Problem 9.1), we may try to solve the Poisson equation by expanding both ς and P, which are both L-periodic in space, in Fourier series as

$$\varsigma(\mathbf{x}) = \sum_{\mathbf{k}}\hat{\varsigma}(\mathbf{x})e^{i\mathbf{k}\cdot\mathbf{r}}, \tag{9.128}$$

$$P(\mathbf{x}) = \sum_{\mathbf{k}}\hat{P}_{\mathbf{k}}(\mathbf{x})e^{i\mathbf{k}\cdot\mathbf{r}}, \tag{9.129}$$

where $|\mathbf{k}| = 2\pi/L$. The Fourier coefficients are

$$\hat{\varsigma}_{\mathbf{k}} = \overline{\varsigma(\mathbf{x})_{\mathbf{k}}e^{-i\mathbf{k}\cdot\mathbf{r}}}, \tag{9.130}$$

$$\hat{P}_{\mathbf{k}} = \overline{P(\mathbf{x})e^{-i\mathbf{k}\cdot\mathbf{r}}}. \tag{9.131}$$

It may be concluded from Eq. (9.127) that $\hat{\varsigma}_0 = 0$ (can you prove it?), and from the Poisson equation (9.126), it follows that

$$\hat{P}_\mathbf{k} = -\frac{\hat{\varsigma}_\mathbf{k}}{k^2}, \quad k \neq 0. \tag{9.132}$$

$\hat{P}_0$ is arbitrary and indeed its inclusion does not change the Navier–Stokes equations. This solution (is it unique?) will be denoted with the help of a nonlocal operator. Thus, $\nabla^{-2}\varsigma$, the explicit form of which in physical space involves a convolution. In any case, formally, the Navier–Stokes equation can be written without the pressure function; thus:

$$\frac{\partial u_i}{\partial t} + \left(\delta_{ij} - \frac{\partial^2}{\partial x_i \partial x_j}\nabla^{-2}\right)\frac{\partial}{\partial x_k}(u_k u_j) = \frac{1}{\text{Re}}\frac{\partial^2 u_i}{\partial x_j \partial x_j}. \tag{9.133}$$

Since pressure is absent, it is sufficient to use an initial condition—an L-periodic, divergence-free, velocity function. Now, because Eq. (9.133) is an equation for this function only, if solved it will furnish the chosen solution.

9.4.2.1 Some Definitions, Identities, Symmetries, and Conservation Laws

We know from Noether's theorem that symmetries induce conservation laws and this is, strictly speaking, correct only for Hamiltonian systems. The Navier–Stokes equations are dissipative, not Hamiltonian. Still the discussion of conservation laws is best done when symmetries are discussed, even though Noether's theorem is not strictly valid.

A few definitions are important and we give them here, before the identities and conservation laws:

$$E \equiv \overline{\frac{1}{2}|\mathbf{u}|^2} \quad \text{energy}, \tag{9.134}$$

$$Z \equiv \overline{\frac{1}{2}|\omega|^2} \quad \text{enstrophy}, \tag{9.135}$$

$$H \equiv \overline{\frac{1}{2}\mathbf{u}\cdot\omega} \quad \text{helicity}, \tag{9.136}$$

$$H_\omega \equiv \overline{\frac{1}{2}\omega\cdot(\nabla\times\omega)} \quad \text{vortical helicity}. \tag{9.137}$$

In all these definitions, we should explicitly indicate that it is the *mean* energy, enstrophy, etc., but we drop this adjective for the sake of economy. Also, in the case of energy, it is actually the kinetic energy *per unit mass*, which is the same as *per unit volume* if we talk of fluid of constant density, set by choosing appropriate units to be equal to 1. We move now to a list of identities and conservation laws.

Periodic box identities: We shall focus first on the translational symmetry and list identities for periodic functions. Using the imaginary periodic box that we have already introduced in Sect. 9.4.2, one may evaluate the average of a periodic function u over that fundamental periodic box to be

$$\bar{u} = \frac{1}{L^3} \int_{\mathscr{V}_L} u(\mathbf{x},t)d^3x, \tag{9.138}$$

where, owing to ergodicity, it is also the ensemble average. The following identities can also be shown to hold for any sufficiently smooth scalar u, v or vector $\mathbf{u}, \mathbf{v}$ arbitrary function pairs, as long as they are periodic. All that is needed to prove the relations is integration by parts:

$$\overline{\frac{\partial u}{\partial x_i}} = 0,$$

$$\overline{\left(\frac{\partial u}{\partial x_i}\right) v} = \overline{u\left(\frac{\partial v}{\partial x_i}\right)},$$

$$\overline{(\nabla^2 u)v} = -\overline{\left(\frac{\partial u}{\partial x_i}\right)\left(\frac{\partial v}{\partial x_i}\right)},$$

$$\overline{\mathbf{u}\cdot(\nabla\times\mathbf{v})} = \overline{(\nabla\times\mathbf{u})\cdot\mathbf{v}},$$

$$\overline{\mathbf{u}\cdot(\nabla^2\mathbf{v})} = -\overline{(\nabla\times\mathbf{u})\cdot(\nabla\times\mathbf{v})}, \tag{9.139}$$

where the last identity is correct only for solenoidal $\mathbf{v}$, i.e., $\nabla\cdot\mathbf{v} = 0$.

We can now write the three principal conservation laws of Newtonian physics in the form of

Conservation of momentum

$$\frac{d}{dt}\bar{\mathbf{u}} = 0, \tag{9.140}$$

Conservation of energy

$$\frac{d}{dt}E = -2vZ, \tag{9.141}$$

Conservation of helicity

$$\frac{d}{dt}H = -2vH_\omega. \tag{9.142}$$

The additional important quantity—dissipation rate, which was already mentioned—can now be written as

$$\varepsilon \equiv -\frac{dE}{dt}. \tag{9.143}$$

The proof of all these identities and conservation laws is a technicality, and we leave it for Problem 9.25. Note that the time derivative here has its usual meaning: indicating the rate of change of a time-dependent function.

We conclude with two remarks:

1. As already mentioned in the text, the application of calculus techniques requires conditions of sufficient smoothness of the functions. We assume that these conditions are met.
2. In *two dimensions* there is an additional balance equation for the enstrophy

$$\frac{d}{dt}Z = -2\nu P,\tag{9.144}$$

where $P = \frac{1}{2}\overline{|\nabla \times \omega|^2}$, not to be confused with pressure, is called the mean *palinstrophy*.

9.4.2.2 Energy Contained in Different Scales

The energy conservation statement above does not distinguish between length scales. Note that statement does not include any contributions from the nonlinear terms of the Navier–Stokes equation, as can be seen during the solution of Problem 9.25. To include the effect of nonlinearity, we have to understand that the effect of the nonlinear term is to *redistribute* energy among the different scales of motion, without having any effect on the *global* energy budget. One of the ways to bring out the energy budget scale by scale is to define scale *filtering* of a variable function. Consider an L-periodic function of space $u(\mathbf{x})$. Its discrete Fourier series is

$$u(\mathbf{x}) = \sum_{\mathbf{k}} \hat{u}_{\mathbf{k}} e^{i\mathbf{k}\cdot\mathbf{x}}.\tag{9.145}$$

Define now *low-* and *high*-pass filtered functions, in the Fourier space, where the cutoff defining the filter is q, in the following way:

$$u_{<q} \equiv \sum_{k\leq q} \hat{u}_{\mathbf{k}}\, e^{i\mathbf{k}\cdot\mathbf{x}},$$

$$u_{>q} \equiv \sum_{k>q} \hat{u}_{\mathbf{k}}\, e^{i\mathbf{k}\cdot\mathbf{x}}.\tag{9.146}$$

The wavenumber q defines a length scale $\ell \equiv q^{-1}$ which is the scale of the filtering. Obviously

$$u(\mathbf{x}) = u_{<q} + u_{>q},\tag{9.147}$$

but one should note that $u_{<q}$ or $u_{>q}$ are not Fourier transforms of u. If the function u is actually the three-dimensional turbulent velocity field, $\mathbf{u}$, the functions $\mathbf{u}_{<q}(\mathbf{u}_{>q})$ are quantitatively regarded as eddies larger (smaller) than the scale ℓ, as defined above.

We shall now define some concepts and quote a few results without proof, leaving it to Problem 9.26. Let $\mathbb{P}_q : u(\mathbf{x}) \mapsto u_{\leq q}$ be an operator which annihilates all Fourier components of wavenumber greater than q. Its properties, to be proven in the above mentioned problem, are listed below:

1.

$$\mathbb{P}_q\left[\mathbb{P}_q\right] = \mathbb{P}_q \tag{9.148}$$

2.

$$\mathbb{P}_q\left[\nabla\right] = \nabla\left[\mathbb{P}_q\right] \quad \text{and} \quad \mathbb{P}_q\left[\nabla^2\right] = \nabla^2\left[\mathbb{P}_q\right] \tag{9.149}$$

3.

$$\overline{(f_{>q}g_{<q})} = 0 \quad \text{orthogonality, for any } f, g. \tag{9.150}$$

Turning now to the Navier–Stokes equation, our purpose is to find the *cumulative equation budget*, in other words its conservation, up to a certain scale. After rewriting here the equation, including a *space periodic* body force term $\mathbf{b}$, and for constant unit density, for convenience,

$$\frac{\partial \mathbf{u}}{\partial t} + \mathbf{u} \cdot \nabla \mathbf{u} = -\nabla P + \nu \nabla^2 + \mathbf{b},$$

$$\nabla \cdot \mathbf{u} = 0, \tag{9.151}$$

we operate with $\mathbb{P}_q$ on the equation and use (9.147) and the above item 1. The result is

$$\frac{\partial \mathbf{u}_{<q}}{\partial t} + \mathbb{P}_q\left[(\mathbf{u}_{<q} + \mathbf{u}_{>q})\right] \cdot \nabla\left(\mathbf{u}_{<q} + \mathbf{u}_{>q}\right) = -\nabla P_{<q} + \nu \nabla^2 \mathbf{u}_{<q} + \mathbf{b}_{<q},$$

$$\nabla \cdot \mathbf{u}_{<q} = 0. \tag{9.152}$$

This equation can be simplified by some nontrivial algebraic manipulations (see Problem 9.27 and hints therein), to obtain the *cumulative*, up to a scale $\ell = 1/q$, energy equation:

$$\frac{\partial E_q}{\partial t} + \Pi_q = -2\nu Z_q + F_q. \tag{9.153}$$

In this equation there appears the cumulative energy, between wavenumber 0 and q, that is *down* to scale ℓ:

$$E_q \equiv \frac{1}{2}\overline{|\mathbf{u}_q|^2} = \frac{1}{2}\overline{\sum_{k \leq q}|\hat{\mathbf{u}}_\mathbf{k}|}, \tag{9.154}$$

the cumulative enstrophy:

$$Z_q \equiv \frac{1}{2}\overline{|\mathbf{u}_q|^2} = \frac{1}{2}\overline{\sum_{k \leq q}k^2|\hat{\mathbf{u}}_\mathbf{k}|}, \tag{9.155}$$

the cumulative energy input into the spectral range $0 \leq k \leq q$, by the work of periodic force $\mathbf{b}$:

$$F_q \equiv \overline{\mathbf{b}_{<q}\mathbf{u}_{<q}} = \overline{\sum_{k \leq q}\hat{\mathbf{b}}_k \cdot \hat{\mathbf{u}}_{-\mathbf{k}}}, \tag{9.156}$$

and, finally, the average energy flux:

$$\Pi_q \equiv \overline{\mathbf{u}_{<q}(\mathbf{u}_{<q} \cdot \nabla \mathbf{u}_{>q})} + \overline{\mathbf{u}_{<q}(\mathbf{u}_{>q} \cdot \nabla \mathbf{u}_{>q})}. \tag{9.157}$$

This expression looks cumbersome, but we shall give, later on, another form of it that will be easier to perform calculations with.

  To sum up—we have reached the very important conclusion that even though the total energy balance does not depend on the nonlinear terms and thus it remains the same for each Fourier component (length scale), the nonlinearity, as expected, transfers energy from scale to scale. The definition of cumulative quantities allows one to find energy in different spectral ranges. In particular, the physical meaning of Eq. (9.153) has become evident: the rate of change of energy in all length scales *down to* $\ell = q^{-1}$ is equal to the energy *injected* at these scales by a force, if it exists, F_q, which can drive the activity, *minus* the energy dissipated at these scales ($2\nu Z_q$), which is dominated by dissipation at the smallest scales $\ell_d \sim \eta_K$ and minus the flux of energy towards smaller than the current scales Π_q. The flux results from nonlinearities. We finally confirm the traditional picture of a turbulent cascade, whereupon energy is injected at large scales, ℓ_0 say, as is reasonable for a large Re number, followed by its transfer to smaller and smaller scales, until scales $\ell_d \sim \eta_K \ll \ell_0$, where viscous dissipation degrades it to heat.

9.4.3   Two Experimental Findings

We shall now give two fundamental experimental laws of fully developed turbulence, They are:

1. In a very high Reynolds number turbulent flow the mean square velocity increment, i.e., second-order longitudinal structure function: $\overline{[\delta u_\parallel(r)]^2}$ depends on distance to the two-third power, or $r^{2/3}$.
2. In an experiment on turbulent flow, in which all the parameters are kept fixed, save the viscosity, which is lowered as much as possible, the energy dissipation per unit mass has a finite positive limit.

9.4.3.1   The Two-Thirds Law

Consider the second-order longitudinal structure function, as we have defined it in Eq. (9.79) for the homogeneous and isotropic case. In this case we can drop the dependence on the point $\mathbf{x}$ and get a function of the distance r only. Additionally, $\delta u_\parallel(r)$ is actually the velocity difference between two points at a distance $\mathbf{r}$ projected on the line of separation. Thus, the velocity increment is scalar, and because of isotropy, the second-order structure function depends on the scalar r:

$$S_2^\parallel(r) = \overline{[\delta u_\parallel(r)]^2}. \tag{9.158}$$

As it was found experimentally, $S_2(r)$ behaves at least over ~ 2 decades of r $(0.5 \lesssim \log r \lesssim 2.5)$ as a power law $S_2(r) \propto r^{2/3}$. These experimental results refer to the work of Y. Gagne and E. Hopfinger in the late 1980s, using the S1 wind channel of ONERA with a very high Reynolds number, Re $\sim 10^7$, as can be seen in reference [10] in the *Bibliographical Notes*. It is possible to see in experiments that the second-order transverse function shows a substantial $r^{2/3}$ dependence as well. We now move to the way in which the $2/3$ law is expressed in the energy spectra (frequency and wavenumber). Consistently with (9.109)–(9.110) when $n - 1 = 2/3$ (our case), the energy and wavenumber Fourier spectra behave over a substantial range as $\propto \omega^{5/3}$ and $\propto k^{-5/3}$, respectively. Indeed experimental results display a region of a least two decades in which $\log E(k)$ and $\log E(f)$ both go like $-5/3$, where $f = \omega/2\pi$ is the frequency (see reference [10]).

9.4.3.2   Energy Dissipation in a Flow Around an Obstacle

In Chap. 2 we have discussed potential flow around symmetrical obstacles in two and three dimensions, paying attention to the fact that the fluid may exert force on the obstacle and, as Newton's third law dictates, the obstacle will then exert an equal and opposite force on the fluid. We have explained the d'Alembert paradox by asserting that a totally inviscid fluid does not exist in nature, so that exactly potential flow around a body is an idealization. A boundary layer, however small, must form, and it is clear that energy is dissipated near the body and therefore the body must suffer a drag—i.e., work has to be done in order to move it at constant speed. When the Reynolds number is increased, the boundary layer will eventually separate, as we have seen in discussing the two-dimensional flow behind a cylindrical obstacle in Chap. 2 and a wake is created.

The description of such a flow can be assumed to be one of an ideal fluid flow everywhere else. The width of the wake depends on the position of the line of separation on the body. While it is true that this position depends on the properties of the boundary layer, it is possible to show by solving for the flow near the line of separation (see reference [1], Chap. 4) that it is independent of the Reynolds number. This is correct for large enough Reynolds numbers ($\text{Re} \gtrsim 1000$). Thus, we can say that the whole flow pattern for large enough but not too large value Re (as long as the wake is laminar; see below) is almost independent of the viscosity. Having at our disposal only three different dimensional numbers—the fluid velocity of the main stream (unperturbed by the body, because of the distance), u_∞, the fluid density ρ, and the body surface cross section S—we may define the dimensionless *drag coefficient*, C_D, by the force exerted by the fluid on the body:

$$F = \frac{1}{2} C_D \rho S u_\infty^2. \tag{9.159}$$

One can easily calculate the work done per unit time by the fluid, in moving the object with velocity u_∞ against this force: it is simply $F u_\infty$. Thus, the dissipation per unit time, per unit mass of the kinetic energy must, to maintain steady state, be

$$\varepsilon \approx \frac{1}{2} C_D \frac{u_\infty^3}{d}, \tag{9.160}$$

where $d = \sqrt{S}$. The important thing to observe here is that the dissipation depends on viscosity *only* through the dependence of C_D on Re.

In Fig. 9.4, which is a qualitative drawing based on experimental data, we see that there is indeed approximately a two-order magnitude (in Re) region, in which C_D is almost constant. For low enough Reynolds number, the independence of the flow, and hence the drag, on Re and thus on viscosity breaks down, and we enter the Stokes flow regime, which, as we have seen in Problem 3.7, gives the following result for the drag force in the present notation: $F \propto \rho \nu U_0 R$, with R being the radius

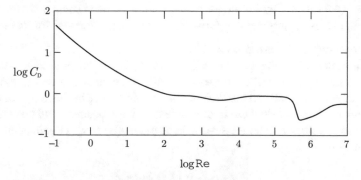

Fig. 9.4 Experimental results of the log of the drag coefficient C_D as a function of $\log \text{Re}$. These results are for a circular cylinder of radius d and thus a Reynolds number of $\text{Re} = u_\infty d / \nu$

of the sphere and U_0 its velocity. Comparing with Eq. (9.159), we readily get that in this regime $C_D \propto \text{Re}^{-1}$, which is consistent with the experimental results for small enough Re. For $\text{Re} \gtrsim 10^5$, the drag coefficient appears approximately constant, *except* for the drag crisis region.

Thus, we can say that for $3 \lesssim \log \text{Re} \lesssim 5.5$, C_D is approximately independent of Re and thus the dissipation does not depend on viscosity. The sudden considerable drop in C_D for $5.5 \lesssim \log(\text{Re}) \lesssim 6.7$ is sometimes called the drag *crisis*. The reason for this drop is the transition to turbulence and a concomitant reduction of the wake width, hence the drop. After the drop, C_D rises again with Re and flattens out justifying (a posteriori) formula (9.4) for very large Reynolds numbers. Recent experimental results show that indeed the curve is approximately flat for $\text{Re} \gtrsim 10^7$, and this means that these results are supported by numerical simulation which are consistent with *finite* dissipation rate for $\nu \to 0$. This result is nontrivial and is not fully understood, since as we have seen in Eq. (9.141), the mean dissipation rate of kinetic energy is equal to $2\nu Z$. Thus, for very small ν, the enstrophy Z must become large, i.e., vortex stretching should be considerable and rapid.

9.4.4 Rudiments of Kolmogorov's 1941 Theory

As is well known, a full theory of turbulence is still missing. Nevertheless, it is possible to use hypotheses, which are compatible with the abovementioned experimental findings and even predict some phenomena. This is largely due to the Kolmogorov 1941 theory, which we shall try to summarize in this section, in an elementary way, by listing Kolmogorov's hypotheses and then discussing briefly their consequences, mainly various power laws. Before going into details, we have to understand that the emergence of turbulence breaks all the symmetries of the macroscopic Navier–Stokes equations, but it is remarkable that when the turbulence becomes *fully developed*, these symmetries are restored *in the statistical sense* at small scales and far from the boundaries. This interesting fact is probably behind the hypotheses put forward by Kolmogorov. We shall try now to summarize the essence of these hypotheses, stressing that *formally* they were put forward for $\text{Re} \to \infty$:

1. *Homogeneity and isotropy on small scales*
   Small scales are defined as $\ell \ll \ell_0$, where ℓ_0 is the integral scale—the scale characteristic of the turbulence excitation. As we shall see this is the largest scale of what we shall call the *Richardson cascade*. Let ℓ be the length of a vector increment which we shall use in the context that *homogeneity* will be understood in terms of velocity increments $\delta \mathbf{u}(\mathbf{x}, \vec{\ell}) \equiv \mathbf{u}(\mathbf{x} + \vec{\ell}) - \mathbf{u}(\mathbf{x})$. Homogeneity is thus assumed if

$$\delta \mathbf{u}(\mathbf{x} + \mathbf{r}, \vec{\ell}) = \delta \mathbf{u}(\mathbf{x}, \vec{\ell}), \tag{9.161}$$

for all increments $\vec{\ell}$ and displacements **r**, whose absolute value is also much smaller than the integral scale. This is therefore sometimes called microscopic homogeneity. By the same token, we can talk of microscopic isotropy which, in this context, means that the velocity increments are invariant under simultaneous rotations of $\delta\mathbf{u}$ and $\vec{\ell}$. Do you see why both vectors have to be rotated simultaneously?

2. *Self-similarity*

   This assumption seems plausible when the flow satisfies hypothesis 1 above. The fully developed turbulent flow is self-similar at small scales, that is, there exists a scaling exponent, call it s, such that

   $$\delta\mathbf{u}(\mathbf{x},\lambda\vec{\ell}) = \lambda^s\delta\mathbf{u}(\mathbf{x},\vec{\ell}). \tag{9.162}$$

   Structure functions would then scale, for example, as $S_p[\delta\mathbf{u}(\lambda\vec{\ell})] \mapsto \lambda^{sp}S_p[\delta\mathbf{u}(\vec{\ell})]$, for all real λ.

3. *Finite constant dissipation rate*

   In a turbulent flow, having the properties as described in hypothesis 1, there exists on the average a finite and constant rate of dissipation per unit mass, ε. For the integral scale, it is $\varepsilon = u_0^3/\ell_0$ and similarly for smaller scales. This hypothesis is correct, in particular, for vanishing viscosity, but technically, ε has to be nondimensionalized first by its integral scale value

   $$\varepsilon_{\text{n.d.}} \equiv \frac{\varepsilon}{(u_0^3/\ell_0)},$$

   and it is this quantity that remains finite and fixed even when $v \to 0$.

One additional hypothesis, dealing with *universality*, has also been put forward by Kolmogorov, but it was not easily accepted by the community of turbulence researchers (notably by L. Landau) and raised a controversy. In this book universality is introduced mainly for historical reasons and matters of principle, as we shall not use this hypothesis for any practical problems. One way to state this hypothesis is:

Universality hypothesis:

For $\infty > \text{Re} \gg 1$, the small-scale statistical properties are uniquely and universally determined by the energy dissipation rate ε, scale ℓ, and the dissipation or Kolmogorov microscale η_K. The quantities mentioned above were all defined in Sect. 9.1.1.

9.4.4.1 The Kármán–Howarth Equation and the 4/5 Law

Kolmogorov used homogeneity, isotropy, and the hypothesis of the finiteness of the dissipation in order to exactly *derive* the so-called 4/5 law. We shall now sketch the Kárman–Howárth (from now on KH) equation, which is essentially a relation for time evolution of various functions related to correlation functions for $\ell \neq 0$, which Kolmogorov exploited. For the sake of simplicity, we shall consider only the *isotropic* case. The new ingredient in the KH equation is the inclusion of *dynamics*, using the Navier–Stokes equations.

Before embarking on the derivation, we remind the reader of some notation that we have already used in our discussion of correlations in Sect. 9.3.2.1 and add some more notation. We dealt in considerable detail with the two-point correlation function or tensor Q_{ij} and only mentioned the triple correlation tensor Q_{ijk}. We now rename the latter, for the convenience of compatibility with the nomenclature of many books, in the following way $S_{ijk} \equiv Q_{ijk}$. The full definition is

$$S_{ijk} \equiv \overline{u_i(\mathbf{x})\, u_j(\mathbf{x})\, u_k(\mathbf{x}+\mathbf{r})}, \tag{9.163}$$

and we add now two more definitions, for the new quantities R and K,

$$R(r) \equiv \frac{1}{2} Q_{ii} \quad \text{and} \quad u^3 K(r) \equiv \overline{u_x^2(\mathbf{x})\, u_x(\mathbf{x}+r\hat{\mathbf{x}})}.$$

We are now ready for the KH equation, stressing again that we shall derive it here only for the *isotropic* case. The Navier–Stokes equation, assuming constant density $\rho = 1$ and no body forces, can be written twice, in indical form, thus

$$\frac{\partial u_i}{\partial t} = -\frac{\partial (u_i u_k)}{\partial x_k} - \frac{\partial P}{\partial x_i} + v\nabla_x^2 u_i,$$

$$\frac{\partial u_j'}{\partial t} = -\frac{\partial (u_j' u_k')}{\partial x_k'} - \frac{\partial P'}{\partial x_j'} + v\nabla_x'^2 u_j', \tag{9.164}$$

where we have abbreviated $\mathbf{x}' \equiv \mathbf{x}+\mathbf{r}$, $\mathbf{u}' \equiv \mathbf{u}(\mathbf{x}')$ and likewise for P. Multiplying the first equation in (9.164) by u_j' and the second by u_i, adding the two and averaging gives the complicated looking equation:

$$\frac{\partial \overline{u_i u_j'}}{\partial t} = -\overline{\left(u_i \frac{\partial (u_j' u_k')}{\partial x_k'} + u_j' \frac{\partial (u_i u_k)}{\partial x_k} \right)} - \overline{\left(u_i \frac{\partial P'}{\partial x_j'} + u_j' \frac{\partial P}{\partial x_i} \right)} + v\overline{\left(u_i \nabla_x'^2 u_j' + u_j' \nabla_x^2 u_i \right)}.$$
$$\tag{9.165}$$

Significant simplification can be achieved immediately, noticing the simple facts: (a) the derivatives over x_i and x_j', when operating on averages, may be replaced by the derivatives over $-r_i$ and r_j, respectively, and (b) $\nabla_{x'} u_i = 0$ and $\nabla_x u_j' = 0$; also, $\overline{u_i P'} = 0$, because of incompressibility and isotropy (do you understand why?). Thus, Eq. (9.165) acquires a more compact form

$$\frac{\partial}{\partial t} Q_{ij} = \frac{\partial}{\partial r_k} (S_{ijk} + S_{jki}) + 2v\nabla^2 Q_{ij}. \tag{9.166}$$

It is relegated to Problem 9.32 to fill in some nontrivial algebra on the way to the KH equation for isotropic turbulence, which is the following:

$$\frac{\partial R}{\partial t} = \frac{1}{2r^2}\frac{\partial}{\partial r}\left[\frac{1}{r}\frac{\partial}{\partial r}\left(r^4 u^3 K\right)\right] + 2\nu\nabla_r^2 R. \tag{9.167}$$

Other cases and formulations can be found in the literature, for example, in references [10] and [8] in the *Bibliographical Notes*. Kolmogorov used the isotropic version of the KH equation, and his derivation of the 4/5 law was very concise (see the second part of Problem 9.32). It reads

$$\overline{(\delta u_\parallel(\mathbf{x},\vec{\ell}))^3} = -\frac{4}{5}\varepsilon\ell, \tag{9.168}$$

where ε is, as usual, the dissipation rate and the result is formally correct for $\mathrm{Re} \to \infty$, $\eta_K \ll \ell \ll \ell_0$, and if the turbulence is homogeneous and isotropic. This is an exact and nontrivial result and may be used as an "acceptance test" for turbulence theories. They must satisfy this law in order to be correct, or violate the assumptions, and thus it constitutes a *necessary* condition for a turbulence theory.

9.4.4.2 Some More on Kolmogorov's 4/5 Law

Equation (9.168), which may be written in the more compact form,

$$S_3(\ell) = -\frac{4}{5}\varepsilon\ell, \tag{9.169}$$

is invariant under random Galilean transformations, yet isotropy cannot be preserved because the Galilean boost introduces a preferred direction. To correct this, R. H. Kraichnan proposed in the 1960s to apply Galilean boosts to be random and isotropically distributed. R. H. Kraichnan was perhaps the most prominent contributor to modern turbulence theory, from the second half of the twentieth century on. We also observe that in deriving the 4/5 law, several limits were taken: $t \to \infty$, to give a statistical steady state; $\nu \to 0$, to eliminate any dissipation at any scales; and $\ell \to 0$, to eliminate large-scale forcing. Thus, the correct formulation of the 4/5 law should actually be

$$\lim_{\ell\to 0}\lim_{\nu\to 0}\lim_{t\to\infty}\frac{S_3(\ell)}{\ell} = -\frac{4}{5}\varepsilon. \tag{9.170}$$

In your opinion, is the order of the limits important? Explain.

We shall now show the consequences that the 4/5 law implies on the moments of the longitudinal velocity increments at the inertial range separation. The *inertial range* is the segment of scales between the integral scale and the dissipation scale, that is, ℓ satisfying $\ell_0 > \ell > \eta$ is in the *inertial range*. Assuming homogeneity and isotropy and that the following moments are finite, we look at the structure function of order p, written in a special way, which can be easily shown to be equivalent to our usual definition:

$$S_p(d) = \overline{[(\mathbf{u}(\mathbf{x}) + d\hat{\mathbf{r}} - \mathbf{u}(\mathbf{x})) \cdot \hat{\mathbf{r}}]^p}, \qquad (9.171)$$

where d is a distance variable and $\hat{\mathbf{r}}$ a unit vector in an arbitrary direction. On the other hand, we have the scaling assumption (9.162), asserting that under rescaling in Eq. (9.171) if the distance $d \mapsto \lambda d$, the structure function changes by a factor λ^{ps}. Using now Problem 9.28, we put $s = 1/3$, which as shown in the problem is the result of the 4/5 law. We readily obtain, replacing d by ℓ,

$$S_p \propto \ell^{p/3}, \qquad (9.172)$$

and since $(\varepsilon\ell)^{p/3}$ has exactly the same dimension as S_p, we have

$$S_p(\ell) = C_p \varepsilon^{p/3} \ell^{p/3}, \qquad (9.173)$$

with the constants C_p being dimensionless for various p. These constants cannot depend on Re as the limit of infinite Reynolds number was assumed before. The 4/5 law is obtained for $p = 3$ and then $C_3 = -\frac{4}{5}$. As said before, we shall omit here the discussion, mainly between Landau and Kolmogorov, on the question of universality of these and other results. A whole Sect. 6.4 of reference [10] is devoted to this interesting, but nontrivial, topic. Two observations are worth making, however. First, extensive experimental studies by K. R. Sreenivasen showed in 1995 that $C_2 = 2 \pm 0.4$ for a large range of the Reynolds number $50 < \text{Re} < 10^4$. Second, the expression for the structure function does not depend on the integral scale, nor on ν. Thus, formally taking the limit of $\ell_0 \to \infty$ and $\nu \to 0$, while holding $\varepsilon > 0$ fixed, all the structure functions have finite limits.

Returning now to the consequences of the structure function dependence given in Eq. (9.173) for $p = 2$, which is proportional to energy, we get that the energy obeys an $\varepsilon^{2/3} \ell^{2/3}$ law. It gives $n - 1 = 2/3$ and thus the energy spatial spectrum behaves as

$$E(k) \propto \varepsilon^{2/3} k^{-5/3}, \qquad (9.174)$$

according to Eq. (9.110) and the above. At the beginning of this section, it was shown that a small viscosity gives rise to the emergence of an inertial range of length scales on which there is neither energy injection nor dissipation. It is obvious, e.g., using a dimensional argument, that the length scale on which viscous dissipation occurs is the Kolmogorov dissipation scale or microscale. For the sake of stressing the importance of this concept we repeat the expression defining it:

$$\eta_K = \left(\frac{\nu^3}{\varepsilon}\right)^{1/4}. \qquad (9.175)$$

The range of scales comparable and smaller than η_K is called the dissipation *range*. In this range the energy input and viscous dissipation are in exact balance. Energy is input by external or other nonlinear macroscopic sources on the integral scale. It is transferred to smaller and smaller scales, without any source or sink in the inertial range as heuristically demonstrated in the *Richardson cascade* discussed in Sect. 9.4.5.1. Finally, when the dissipation range is reached, all energy arriving at the microscale through the cascade, as well as energy created by nonlinearities *at* this scale, is dissipated by viscosity. We stress here that assuming the existence of nonlinear sources of energy in the dissipation range is *wrong*.

A historical remark is now in order. This section has in its name "mostly Kolmogorov," but it would be wrong to attribute all findings described here to this influential mathematician. The $k^{-5/3}$ dependence of the energy spectrum on wavenumber was actually found first by Obukhov in 1941, and Kolmogorov himself asserted that Obukhov's derivation, using closure and dimensional arguments, is independent of his. Among the scientists, some very prominent who also independently found the Kolmogorov–Obukhov $-5/3$ power law, are Heisenberg and von Weissäcker, who were, after the Second World War, detained by the British. They remained in Germany until the end of the war. The revered L. Onsager also found independently, in 1945, a $k^{-11/3}$ law, but he worked with a three-dimensional spectrum $E(\mathbf{k})$. For one dimension the power reduces to $-5/3$. Onsager's discovery was only made in 1949, but independently of Kolmogorov's and Obukhov's findings, as well as those of the German detainees.

9.4.5 Phenomenology Revisited, in View of Kolmogorov's Work

Phenomenological approaches to turbulence often yield the same results as more systematic approaches, e.g., the Kolmogorov $4/5$ law discussed above, but the phenomenological approach is based more on conjectures, mental pictures, and plausibility arguments. The price for that is the lack of reasonably rigorous, physical and mathematical foundation. The main points in the phenomenological approach to turbulence can be listed as follows:

- A particular length scale ℓ in the inertial range, which is below the integral scale ℓ_0 can be chosen in forced turbulence—that is, created by outside intervention, like stirring or a grid. This is an *injection scale*, which has to be above the microscopic viscous dissipation scale η or η_K.
- A typical velocity scale u_ℓ is associated with the length scale ℓ. A plausible definition of u_ℓ is

$$u_\ell = \left[\overline{\delta u_\parallel^2(\ell)}\right]^{1/2}, \tag{9.176}$$

an object familiar from earlier discussions.

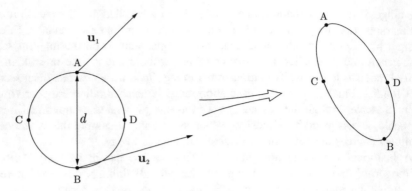

Fig. 9.5 A circle of diameter d being distorted to roughly an ellipsoid by fluid motion. While some points (e.g., A and B) separate, other (e.g., C and D) approach

- The typical timescale emerges as $\tau_\ell \sim \ell/u_\ell$, typical for a structure of size ℓ to undergo a significant distortion because of relative motion, and is referred to as eddy *turnover time*. In Fig. 9.5 we show this kind of distortion in two dimensions, suffered by a circular loop of diameter d which becomes of roughly elliptical shape after time:

$$\tau_d \sim \frac{d}{|\mathbf{u}_1 - \mathbf{u}_2|}.$$

- The r.m.s. velocity fluctuation is plausibly taken as $\sim u_{\ell_0}$ which is the flow velocity fluctuation on the integral scale.

Note the use of the symbols $\sim$ and $\approx$ in the definitions and equalities in the phenomenological approach. While $\sim$ means "of the order of," the best explanation of the meaning of $\approx$ is perhaps "equal within a constant factor of order unity," and it is only natural that we cannot aim at a better accuracy within a phenomenological theory. In the inertial range, as there is no energy input or loss, the energy flux is independent of ℓ and equal to the constant dissipation rate; thus,

$$\Pi \sim \frac{u_\ell^3}{\ell} \sim \varepsilon, \quad \Rightarrow \quad u_\ell \sim (\varepsilon\ell)^{1/3}. \tag{9.177}$$

So we have found in a phenomenological way the value of the scaling exponent $s = 1/3$. From the definition of the eddy turnover time above and the second relation here, we get

$$\tau_\ell \sim \varepsilon^{-1/3} \ell^{2/3}. \tag{9.178}$$

In the integral scale, near the top of the inertial range, when $\ell \sim \ell_0$ the velocity becomes $u_0 \sim (\varepsilon\ell_0)^3$, which gives $\varepsilon \sim u_0^{1/3}/\ell_0$. This equation is actually used frequently in the empirical modeling of turbulence. The smallest scales, where dissipation releases the energy, can be phenomenologically estimated as follows. The typical time to viscously attenuate activity at scale ℓ must be $\tau_{\text{visc}} = \ell^2/\nu$. This times goes to zero with ℓ faster than the turnover time (9.178). Equating these two times shows that however small the viscosity, a length scale emerges

$$\ell_{\text{visc}} \sim \left(\frac{\nu^3}{\varepsilon}\right)^{1/4}, \tag{9.179}$$

which is nothing other than the Kolmogorov dissipation scale η_K, previously introduced. In fact viscous effects become relevant experimentally at scales larger than this by a factor of ten or more. It is to be expected that the phenomenological theory cannot be exact, and, in addition, this finding indicated that nonlinear effects are probably weaker than those predicted by a phenomenological turnover time argument.

9.4.5.1 Richardson Cascade

The phenomenology described so far may be neatly illustrated by a schematic drawing of the Richardson cascade. This is shown in Fig. 9.6. The uppermost eddies are of scale ℓ_0 and the injection of energy is on this scale. Let us denote by $n = 0, 1, 2, 3 \ldots$ the generations of eddies, starting at the top ($n = 0$). The nth generation of eddies is of size $\ell_n = \ell_0 \xi^n$, where $0 < \xi < 1$. Obviously, the smallest, lowermost eddies' scale is η_K. Even though this figure shows the essence, it should be remembered that it is only schematic and qualitative. The cascade picture brings out two important features of Kolmogorov's 1941 theory:

1. *Scale invariance*—this is correct for the inertial range and comes out from the space-filling property of the eddies.
2. *Locality*—the meaning of this is locality in scale, not position. In other words, the energy flux in the inertial range at scales $\sim \ell$ involves mainly comparable scales, say, between $\xi\ell$ and ℓ/ξ. This property follows from the fact that an inertial scale eddy, of size $\sim \ell$, will be *swept* by the energy contained eddies of scale $\ell_0 \gg \ell$. This induces no distortion because of the Galilean invariance of the Navier–Stokes equations. See however the work of R. H. Kraichnan (reference [11]).

Consider now for a moment the eddies' *distortion*. The typical shearing rate on scale ℓ is

$$S_\ell \sim \frac{u_\ell}{\ell} \sim \varepsilon^{1/3}\ell^{-2/3}, \tag{9.180}$$

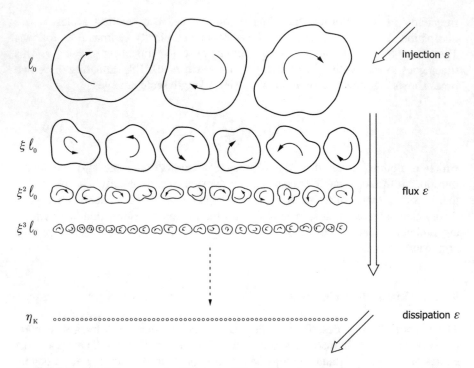

Fig. 9.6 The Richardson cascade. The space-filling eddies decrease in size downwards, in the direction of the constant energy flux

where Eq. (9.176) has been used. It follows that the smallest shear is at the top and the largest at the bottom of the cascade. The shearing of an eddy by eddies very much larger or smaller is ineffective (can you see why?). So the main shearing action on an eddy of scale ℓ is done by eddies of scale $\ell' \sim \ell$.

9.4.5.2 Reynolds Numbers and Numerical Consequences

It is trivial to express, using the phenomenological relations, the velocity and turnover time at scale ℓ, that is, u_ℓ and τ_ℓ, in terms of these variables at the integral scale, u_0 and τ_0, which themselves are related through the integral length scale $\ell_0 = u_0 \tau_0$; thus,

$$\frac{u_\ell}{u_0} \sim \left(\frac{\ell}{\ell_0}\right)^{1/3}, \text{ and } \frac{\tau_\ell}{\tau_0} \sim \left(\frac{\ell}{\ell_0}\right)^{2/3}. \tag{9.181}$$

Usually when one speaks of the Reynolds number of a turbulent flow, it is this quantity for the integral scale, that is,

$$\mathrm{Re} \sim \frac{\ell_0 u_0}{\nu}. \tag{9.182}$$

On the other hand, we easily derive the relation for the viscous cutoff:

$$\frac{\ell_0}{\eta_K} \approx \left(\frac{\nu^3}{\ell_0^3 u_0^3}\right)^{-1/4} \approx \mathrm{Re}^{3/4}. \tag{9.183}$$

This is a very important result, asserting that in the Kolmogorov 1941 theory, the *dynamical range*, between the integral and dissipation scales (so including the inertial range), requires $N_1 \sim \mathrm{Re}^{3/4}$ grid points in one spatial dimension in order to have a faithful numerical resolution. In three dimensions, the numerical mesh has to have at least $N_3 \sim \mathrm{Re}^{9/4}$. Here lies the bugbear of numerical simulation of huge Re turbulent flows. The storage grows like $\mathrm{Re}^{9/4}$, and since the timestep has usually to be taken proportional to the mesh size, the number of timesteps needed to integrate the equation in a fully resolved way for a fixed number of turnover times grows like Re^3, so for $\mathrm{Re} \sim 10^4$ we face an obstacle that is enormous $N_3 \sim 10^9$ and $\mathrm{Re}^3 \sim 10^{12}$ to capture the dynamic range. Some problems of numerical fluid dynamics will be discussed briefly in Appendix B.

An important remark here is in order. The habit to take the figure $\mathrm{Re}^{9/4}$ as the *number degrees of freedom* of the flow is justified within the theory that demands the flow in the inertial range to be totally disorganized. However, experiments with large Re often display patterns quite distinct from spatial disorganization. This perhaps points out that there are prospects to numerically simulate such flows more efficiently than the required $\mathrm{Re}^{9/4}$ seems to indicate, but not necessarily on uniform meshes. In any case, this issue brings us to the topic of the application of dynamical systems theory to turbulence. We shall briefly discuss this later in this book.

9.4.5.3 Energy Decay in Free Turbulence

Turbulent flows demand persistent energy injection; otherwise, they are bound to freely decay. This is an experimental fact and is consistent with the steady Richardson cascade picture, demanding a steady transfer of energy from large to small scales. In his later work, still in 1941, Kolmogorov found, by conjecture and by using qualitative and quantitative arguments, that:

If a turbulent flow is freely decaying, and initially possesses a self-similarity $u_\ell \sim C\ell^s$, as $\ell \to \infty$ (this is a formal statement) with the scaling exponent $-2.5 < s < 0$ and C constant, then this law persists for all later times, with the same s and C.

This a deep and nontrivial result. For example, if we assume that there is a time-dependent integral scale, we can now try to find it. The time-dependent $\mathrm{Re} \sim \ell_0 u_0 / \nu$ is assumed to be, as usual, very large. Using the above statement, we have $u_0 \sim C\ell_0^s$ and

$$\frac{d(u_0^2)}{dt} \sim -\varepsilon \sim -\frac{u_0^3}{\ell_0}. \tag{9.184}$$

The scaling statements quoted above help to obtain an ODE for the integral scale

$$\frac{d(C^2 \ell_0^{2s})}{dt} \sim -C\ell_0^{3s-1}. \tag{9.185}$$

This has the trivial solution, for constant C,

$$\ell_0 \propto (t+\theta)^{1/(1-s)} \quad \Rightarrow \quad u_0 \propto (t+\theta)^{s/(1-s)}, \tag{9.186}$$

and gives for the energy $E \propto (t+\theta)^{2s/(1-s)}$ and for the Reynolds number $\mathrm{Re} \propto (t+\theta)^{(1+s)/(1-s)}$. In all the above, θ is a constant and s is negative, as in the above scaling statement. Thus, the energy always *decreases* and the integral scale *increases*. Examining two example scalings, we find that for $0 > s > -1$ we have a Reynolds number increasing in time. Note that for $s < -1$, this solution breaks down when Re falls to ~ 1. For $s > 0 > -2.5$, we have, as we saw above, $u_\ell^2 \sim C^2 \ell^{2s}$, for $\ell \to \infty$ which translates to the following relation in Fourier space. If initially

$$E(k) \sim C'k^h, \quad h = -2s \quad \text{for } k \to 0, \tag{9.187}$$

this scaling persists. However, it is difficult to guarantee rigorously the constancy of C' and there are indeed values (like $h = 4$) for which it is not so.

9.5 Two-Dimensional Turbulence

Nature contains no example of an exact two-dimensional flow, but we are free to imagine such a case and to attack it with theory, just as we are free to create virtual experiments, in the form of computer simulations. So why study two-dimensional turbulence? If turbulence is, as Feynman says, "the most important unsolved problems in classical physics," it seems clear that the study of it in two dimensions may offer some insights into the general problem. The Earth's atmosphere and oceans are approximately two-dimensional because of the constraining effects of rotation, stratification, and thin geometry, and theories of two-dimensional turbulence have proved to be useful predictive tools in geophysical problems. We expect that two-dimensional turbulence will also be useful when applied to similarly constrained laboratory and astrophysical flows, like the one illustrated in Fig. 9.7. We may even hope that with a reduction in dimension comes a reduction in the difficulty of the problem. Instead, in two dimensions turbulence takes on an entirely new character—one that is, in many ways, more challenging for the practitioner. Turbulence in two dimensions is more visually striking, displaying richly convoluted vorticity fields and supporting long-lived coherent structures—features that are absent in three dimensions, but are often found in approximately two-dimensional flows in the laboratory or in nature. Whether freely decaying or forced, the dynamics of two-dimensional turbulence is typically described in terms of an *inverse cascade* of energy, as it is transported from the scales on which it is input up to the largest

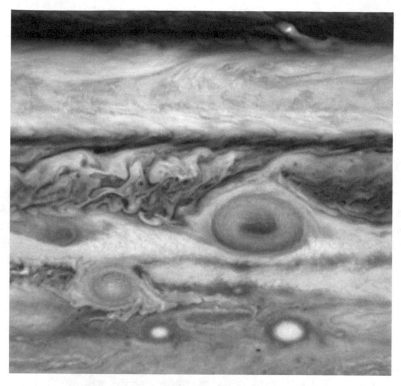

Fig. 9.7 A composite Hubble space telescope image of Jupiter's Red Spot and its approximately two-dimensional turbulent atmosphere, including several other large spots (vortices). *Public domain from NASA, Hubble Space Telescope. Courtesy NASA, HST*

scales that are available to system. Yet, like in three-dimensional turbulence, the energy is a conserved quantity in the absence of dissipation. How does this "backward" cascade happen? In the following, we will look in more detail at the mathematical descriptions of energy and enstrophy, and their inverse and direct cascades, respectively. In examining this so-called double cascade, we clarify the assumptions that underly the predictions of the theories. Unless noted otherwise, only cases of homogeneous and isotropic turbulence will be considered consistent with the majority of discussions throughout this chapter. We will briefly discuss the relevance of coherent structures and the importance of a *condensate* state and conclude by mentioning some relevant results from experiments and numerical simulations.

9.5.1   Governing Equations and Integral Constraints

A look at the mathematical description of the fluid dynamics governing two-dimensional turbulence readily shows that it is the presence of an additional invariant, mean square vorticity, that allows a simultaneous cascade of energy to large scales and of enstrophy to small scales.

We begin with the forced Navier–Stokes equations for a two-dimensional flow:

$$\frac{D\mathbf{u}}{Dt} = \frac{\partial \mathbf{u}}{\partial t} + \mathbf{u} \cdot \nabla \mathbf{u} = -\nabla P + \nu \nabla^2 \mathbf{u} + \mathbf{b}_f, \tag{9.188}$$

where $\mathbf{b}_f$ is an unspecified forcing (per unit mass) and we have taken $\rho = 1$ since we assume constant density, satisfying automatically the incompressibility condition $\nabla \cdot \mathbf{u} = 0$. Clearly, in this section we use $\mathbf{u}, \mathbf{x}$, and the ∇ operator in their two-dimensional forms, $\mathbf{u}(\mathbf{x}, t) = [u(\mathbf{x}, t), v(\mathbf{x}, t)]$. Now we can write $\mathbf{u} = (\partial_y \psi, -\partial_x \psi)$, using the stream function, ψ, and assuming the flow in the $x - y$ plane. By operating with $\nabla \times$ on (9.188), we obtain an equation for the scalar vorticity field, $\zeta = \nabla \times \mathbf{u} = -\nabla^2 \psi$:

$$\frac{D\zeta}{Dt} = \frac{\partial \zeta}{\partial t} + \mathbf{u} \cdot \zeta = \nu \nabla^2 \zeta + f, \tag{9.189}$$

where $f = \nabla \times \mathbf{b}_f$. No vortex stretching appears in (9.189), as distinct from the three-dimensional case. The lack of vortex stretching is the main physical difference between two- and three-dimensional turbulence, and the consequences of this property explain two-dimensional turbulence's unique characteristics, including the double cascade. The above equations of motion must be complemented by appropriate boundary conditions, which we assume to be periodic on a square domain of size L unless otherwise noted and except when discussing laboratory results, which involve particular boundary conditions for each experiment. By forming the appropriate products in (9.188)–(9.189), we are led to the energy and enstrophy equations:

$$\frac{D}{Dt}\left(\frac{1}{2}\mathbf{u}^2\right) = -\nabla \cdot (P\mathbf{u}) - \nu \left[\zeta^2 + \nabla \cdot (\zeta \times \mathbf{u})\right] + \mathbf{b}_f \cdot \mathbf{u}, \tag{9.190}$$

and

$$\frac{D}{Dt}\left(\frac{1}{2}\zeta^2\right) = -\nu \left[(\nabla \zeta)^2 - \nabla \cdot (\zeta \nabla \zeta)\right] + f\zeta, \tag{9.191}$$

where we have used a vector identity and incompressibility in rewriting the dissipation terms. Next, we perform an average of (9.190)–(9.191), defining the energy $\mathsf{E} = \frac{1}{2}\langle \mathbf{u}^2 \rangle$ and mean square vorticity, or enstrophy $\mathsf{Z} = \frac{1}{2}\langle \zeta^2 \rangle$, where the ensemble average $\langle \cdot \rangle$ is equivalent to a spatial average because of the ergodic theorem discussed in Sect. 9.3. All the divergences vanish due to homogeneity, giving, for the unforced, $\mathbf{b}_f = 0$ and $f = 0$, case:

$$\frac{dE}{dt} = -v\langle \zeta^2 \rangle = -vZ \equiv -\varepsilon, \tag{9.192}$$

and

$$\frac{dZ}{dt} = -v\langle (\nabla \zeta)^2 \rangle = -vP \equiv -\beta, \tag{9.193}$$

where the palinstrophy, P, already defined as the mean square curl of the vorticity, is in the two-dimensional case $P = \frac{1}{2}\langle (\nabla \times \zeta)^2 \rangle$ or, equivalently, as the mean square vorticity gradient $P = \frac{1}{2}\langle (\nabla \zeta)^2 \rangle$. The two-dimensional energy and enstrophy dissipation rates, ε and β, are defined by the above equations.

In the inviscid limit, (9.192)–(9.193) conserve both energy, E, and enstrophy, Z. Taking the limit, $\lim v \to 0$, the above equations provide a preview of the unusual character of two-dimensional turbulence. First, we see from (9.192) that the rate of change of kinetic energy vanishes in the inviscid limit of two-dimensional turbulence since both $v \to 0$ and the enstrophy, Z, are bounded from above according to (9.193), because of the lack of a vortex stretching contribution. While it is important to appreciate that energy conservation in the inviscid limit differs from the behavior of three-dimensional turbulence, it is more important to understand that this result implies that two-dimensional turbulence cannot dissipate energy at small scales, as happens in three-dimensional turbulence. In this case, as we will soon see, energy instead moves to larger scales. Finally, we see from (9.193) that the rate of change of the enstrophy, Z, does not necessarily vanish as $v \to 0$; rather, we need to look to the dynamics of the palinstrophy, P, which determines dZ/dt. As for the energy and enstrophy, we can derive an equation for the rate of change of P, finding

$$\frac{dP}{dt} = -v\langle (\nabla^2 \zeta) \rangle - \langle \mathscr{D}_{ij}(\nabla \zeta)_i (\nabla \zeta)_j \rangle, \tag{9.194}$$

where index notation has been adopted for convenience and $\mathscr{D}_{ij}$ has been defined previously in (9.9). Note that the final strain (or deformation) rate-dependent term in (9.194) is closely related to the third-order structure function $S_3^{\parallel}$. Thus, the rate of enstrophy dissipation does not necessarily vanish as $v \to 0$, and, based on the similarity to the energy in three-dimensional turbulence, we anticipate that enstrophy is able to directly cascade to smaller scales where it will be ultimately dissipated. The vorticity gradients (palinstrophy) that must come into play to support this direct cascade are therefore expected to grow in order to establish a stationary turbulent state.

Note that in the inviscid unforced case, one may define an infinity of invariants $Z_n = \frac{1}{2}\langle \zeta^n \rangle$, apparently overly constraining the system and presenting a conceptual challenge to understanding two-dimensional turbulence. The resolution of this challenge, of course, is that dissipation removes nearly all of these constraints. The reader interested in idealized truly inviscid systems is encouraged to consult the literature on point vortices or, for a physical manifestation, the literature on super-fluid helium; the review paper of G. Boffetta and R. E. Ecke, in the *Bibliographical Notes*, provides references to all these topics. The above conclusions, taken together, amount to an intuitive summary of a double cascade of two-dimensional turbulence first detailed by Kraichnan (1967) (see below and reference [14] and supplemented by the works of C.E. Leith in 1968 and G.K. Batchelor in 1969). To proceed in more depth, we must return to spectral notation and work with the energy and enstrophy spectra, $E(k)$ and $Z(k)$, respectively, for two-dimensional wavenumber $k = |\mathbf{k}|$. Note that $E = \int_0^\infty E(k)dk$ and $Z = \int_0^\infty k^2 E(k)dk$, while more generally

$$E(k) = \left(\frac{L}{2\pi} \right)^2 \pi k \langle |\hat{\mathbf{u}}(\mathbf{k})|^2 \rangle, \qquad (9.195)$$

just as we previously defined in Sect. 9.4, but here, for compactness, we have not included time, t, in the notation, even though the functions are generally time dependent. In the spectral formulation, it is convenient to discuss forcing at a particular input scale, ℓ_0, that we describe in terms of its wavenumber $k_0 = 1/\ell_0$, although in general forcing will be spread over some range of scales, with wavenumbers perhaps centered on k_0. Analogous to the self-similar inertial range in three-dimensional turbulence, we envision two inertial ranges in forced two-dimensional turbulence: a direct cascade of enstrophy for $k \gg k_0$ and an inverse cascade of energy for $k \ll k_0$. In each of these inertial ranges, self-similarity leads to an energy spectrum with a different wavenumber scaling exponent.

The abstract of Kraichnan's 1967 paper gives what is, perhaps, the most succinct summary of the double cascade of two-dimensional turbulence, viz.,

two-dimensional turbulence has both kinetic energy and mean square vorticity as inviscid constants of motion. Consequently it admits two formal inertial ranges, $E(k) \propto \varepsilon^{2/3} k^{-5/3}$ and $E(k) \propto \beta^{2/3} k^{-3}$, where ε is the rate of cascade of kinetic energy per unit mass, β is the rate of cascade of mean square vorticity, and the kinetic energy per unit mass is $\int_0^\infty E(k)dk$. The $-5/3$ range is found to entail backward energy cascade, from higher to lower wavenumbers k, together with zero vorticity flow. The -3 range gives an upward vorticity flow and zero energy flow. The paradox in these results is resolved by the irreducibly triangular nature of the elementary wavenumber interactions. The formal -3 range gives a nonlocal cascade and consequently must be modified by logarithmic factors. If energy is fed in at a constant rate to a band of wavenumbers $\sim k_i$ and the Reynolds number is large it is conjectured that a quasi steady state results with a $-5/3$ range for $k \ll k_i$ and a -3 range for $k \gg k_i$, up to the viscous cutoff. The total kinetic energy increases steadily with time as the $-5/3$ range pushes to ever lower k, until scales the size of the entire fluid are strongly excited. The rate of energy dissipation by viscosity decreases to zero if kinematic viscosity is decreased to zero with other parameters unchanged.

(Note that in the above quotation, k_i is equivalent to what we denote by k_0.) Next, we look in more detail at each part of the double cascade of two-dimensional turbulence.

9.5.2 *Inverse Energy Cascade*

The idea that energy must move to larger scales as a result of the high Reynolds number, near-inviscid conservation of both energy and entropy was already realized in 1951 by T.D. Lee, before the 1967 work of Kraichnan. At about this same time, Fjørtoft used the idea of a discrete wave triad to demonstrate the inverse cascade in 1953, a concept that Von Neumann generalized to the continuous case in 1967. As demonstrated in Sect. 9.5.1, the possibility of an inverse cascade can be seen intuitively by the fact that dE/dt vanishes in the inviscid limit. A more concrete illustration of the transport of energy to large scales can be made by examining the evolution of a wavenumber triad. To illustrate, we select k_1, $k_2 = 2k_1$, and $k_3 = 3k_1$; although other choices are also possible, not every triad $k_1 < k_2 < k_3$ gives a useful result. Associate with each wavenumber its energy $E_1 = E(k_1)$ and total energy $E = E_1 + E_2 + E_3$ and enstrophy $Z = k^2 E = k_1^2 E_1 + k_2^2 E_2 + k_3^2 E_3$. Conservation of energy and enstrophy requires that after some time, δt say, $\delta E = \delta E_1 + \delta E_2 + \delta E_3 = 0$ and $\delta Z = k_1^2 \delta E_1 + k_2^2 \delta E_2 + k_3^2 \delta E_3 = 0$, where the δ indicates change over time δt. It is a straightforward algebraic exercise to show that if the central mode, k_2, loses energy, $\delta E_2 < 0$, then more energy flows to k_1, the smaller scale, than to k_3. Likewise, more enstrophy will flow to the smallest scale, k_3 (show this). This result can be made more persuasive if it is repeated for successive iterations of δt, but it remains a discrete and simplistic model. If instead we define characteristic energy and enstrophy wavenumbers of a continuous spectrum:

$$k_E^2 = \frac{\int k^2 E(k) dk}{\int E(k) dk} = \frac{Z}{E}, \tag{9.196}$$

and

$$k_Z^2 = \frac{\int k^4 E(k) dk}{\int k^2 E(k) dk} = \frac{P}{Z}, \tag{9.197}$$

then it is straightforward to show that (9.192)–(9.193) imply that

$$\frac{d}{dt} k_E^2 = 2\nu k_E^2 (k_E^2 - k_Z^2) \leq 0, \tag{9.198}$$

invoking the Schwarz inequality. Therefore, the characteristic energy scale increases.

Additional insight about the inverse energy cascade comes from the energy transport equation in spectral form:

$$\frac{\partial E(k)}{\partial t} = T(k) + F(k) - \nu k^2 E(k), \tag{9.199}$$

where $T(k)$ represents energy transfer by nonlinear interactions, $F(k)$ is input due to forcing, and the final term is loss due to viscous dissipation. The nonlinear transfer term can be used to define the net energy flux across the scale k^{-1}:

$$\Pi_E(k) = \int_k^\infty T(k')dk', \qquad (9.200)$$

where we have $\Pi_E(0) = 0$ because energy is conserved in the unforced and inviscid case. Kolmogorov's cascade idea is based on local interactions in spectral space, i.e., interactions occurring on the same scale, and would imply that the flux is proportional to the energy in wavenumbers $\sim k$ and to a characteristic distortion frequency ω_k, or

$$\Pi_E(k) \propto \omega_k k E(k), \qquad (9.201)$$

where

$$\omega_k^2 = \int_0^k (k')^2 E(k')dk' \qquad (9.202)$$

defines ω_k as the effective shear rate of the distortion of flow structures (eddies) at scale k^{-1}. In the above definition, it has been assumed that the effect of wavenumbers larger than k is expected to average to zero over timescales ω_k^{-1}. With the form (9.200) of energy flux, a stationary scale-free state is described by $\Pi_E(k) = \varepsilon$, and the only solution of the form $E(k) \propto k^{-n}$ is found, by dimensional analysis, to be

$$E(k) = C\varepsilon^{2/3}k^{-5/3}, \qquad (9.203)$$

where C is a dimensionless Kolmogorov constant. We remind the reader that this solution does not predict the inverse sense of the energy cascade in two-dimensional turbulence, although it is consistent with it. To demonstrate the inverse nature of the cascade that is observed in nearly two-dimensional experiments and in simulations, we must resort to triad or integral arguments, as above. The inverse energy cascade solution is sketched in Fig. 9.8.

Other direct proofs are available showing the direction of energy transport is to larger scales; as expected, these are generally more complex to formulate. In the limited space of this discussion, we describe only one such approach, based on the calculation of third-order structure functions. In the case of two-dimensional turbulence, the longitudinal third-order structure function can be calculated under the assumption of constant fluxes of energy and enstrophy in their respective inertial ranges. One finds in that case and referring to the definition (9.70) that

$$S_3^{\|}(r) = \langle (\delta u_\|(r))^3 \rangle = \frac{3}{2}\varepsilon r, \qquad (9.204)$$

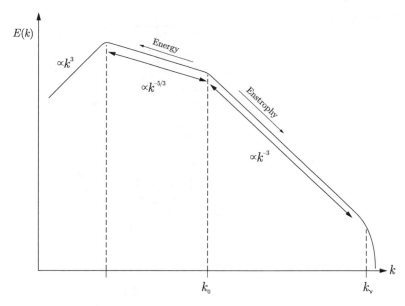

Fig. 9.8 Sketch of the energy spectra in the two inertial regimes of the Kraichnan–Batchelor double cascade model. Injection at the forcing wavenumber k_0 separates energy moving towards low wavenumbers ($-5/3$ scaling), while enstrophy moving towards high wavenumber leads to a -3 scaling and is ultimately dissipated by viscosity at the dissipation wavenumber k_v

a result that is the two-dimensional analog of the Kolmogorov $4/5$ law, except that $3/2$ replaces $-4/5$, the sign change implying a reversal of the direction of spectral transport: to large scales.

What is the physical picture of the above, mainly mathematical, description of a process occurring in spectral space? Observations of approximately two-dimensional turbulence provide a consistent picture, in which a patch of vorticity is distorted—extended in one direction, to a scale D. As a result of material conservation of vorticity, the patch is squeezed to a small scale d in the transverse direction, with d decreasing as D grows. Eventually, d's scale will shrink to the point that diffusion will begin to act. In the schematic Fig. 9.9, we see both the inverse cascade, to large D, and the direct cascade, to a dissipation scale d. One point of view associates the growth of D with mergers between like-signed vortices. However, while regions of like-signed vorticity do aggregate, there is mounting evidence that the merger process contributes little to the inverse cascade in forced two-dimensional turbulence.

9.5.2.1 Stationary State and Condensates

If the inverse cascade carries energy to large scales, effectively hiding it from viscous diffusion, then there is no mechanism to dissipate the accumulating large-

Fig. 9.9 A sketch of
distorted vorticity regions and
tendrils in two-dimensional
turbulence

scale energy. Forced two-dimensional turbulence therefore cannot reach a stationary
state. The resolution of this problem is to introduce an additional dissipation, a sink
of energy at large scales. Such dissipation is sometimes referred to as Ekman or
Rayleigh drag, or Hartman drag for magnetically two-dimensionalized flows, and
appears as a linear friction term in (9.188), viz.,

$$\frac{D\mathbf{u}}{Dt} = \frac{\partial \mathbf{u}}{\partial t} + \mathbf{u} \cdot \nabla \mathbf{u} = -\nabla P + \nu \nabla^2 \mathbf{u} + \mathbf{b}_f - \alpha \mathbf{u}, \tag{9.205}$$

where α is an appropriate constant. The introduction of such a term is similar
to what is done in dynamo theory, as we will see in the following chapter. This
friction can be used to define a (large) dissipation scale $l_\alpha = (\varepsilon/\alpha^3)^{1/2}$ and can
be considered to originate from the three-dimensional world in which the two-
dimensional flow is embedded. More practically, it is usually associated with friction
of the boundaries of the system.

As energy is pushed to larger and larger scales in the inverse cascade, a dominant
eddy forms, eventually filling the domain of the system. If the dissipation scale
is larger than the system size, the inverse cascade leads to a condensed state,
characterized by a scale comparable to the system size. Kraichnan already anticipated such a possibility and, based on its similarity to Bose–Einstein condensates,
concluded that as energy piles up onto the smallest wavenumbers, a condensation
process would occur, leading to a system-sized state consistent with its boundaries.
Condensates have been observed in both laboratory experiments and numerical
simulations of forced approximately two-dimensional turbulence. In freely decaying
two-dimensional turbulence, as we will discuss below, a condensate is not formed,
but at long times, a single vortex dipole is formed, compatible with the system's
initial conditions.

9.5.3 Direct Enstrophy Cascade

The direct cascade of enstrophy to smaller scales, as we saw in Sect. 9.5.1, is predicted based on the fact that the evolution of enstrophy, given by $dZ/dt = -\nu P = -\beta$, is controlled by the mean squared vorticity gradients, as measured by the palinstrophy, P. On scales in between that of the forcing and the dissipation, the evolution of enstrophy is thus analogous to the $5/3$ law of Kolmogorov 1941. As a result, the direct cascade is relatively well understood, both from statistical and physical points of view. Physically, the direct cascade of enstrophy is associated with the distortion of vorticity patches, in particular the thin filaments that are drawn out as fluid parcels are stretched. Figure 9.9 is a sketch illustrating the typical appearance of distorted vorticity patches and surrounding tendrils. These filaments are hallmarks of steepened vorticity gradients, $\nabla \zeta$, and the process proceeds until scales are reached for which viscosity is able to smooth these gradients. The evolution of the gradients is described mathematically by the Eq. (9.194) for palinstrophy, showing that the process is controlled by strain. The evolution of the enstrophy spectrum, $Z(k) = k^2 E(k)$, is described analogously to the energy spectrum in (9.199)–(9.200) but where the nonlinear transfer of enstrophy is $k^2 T(k)$ and the enstrophy flux, $\Pi_Z(k)$, is defined as

$$\Pi_Z(k) = \int_k^\infty (k')^2 T(k') dk'. \tag{9.206}$$

In the range of the direct cascade, $k \gg k_0$, the enstrophy flux is estimated to be $\Pi_Z(k) \propto \omega_k k^3 E(k)$, and the balance $\Pi_Z(k) = \beta$ allows us to again apply standard dimensional analysis to find a scale-free stationary solution of the form

$$E(k) = C' \beta^{2/3} k^{-3}, \tag{9.207}$$

where C' is another dimensionless Kolmogorov constant. Since viscosity is available to dissipate the vorticity where the local vorticity gradients are high enough (or, as we would say in spectral language, viscosity provides the large k cutoff of the cascade), a dissipation wavenumber can be found: $k_\nu \propto \beta^{1/6} \nu^{-1/3}$, depending only on the enstrophy dissipation rate, β, and the molecular viscosity, ν. The Reynolds number based on k_ν is

$$\mathrm{Re}_\nu \equiv \frac{u}{k_0 \nu} \propto \left(\frac{k_\nu}{k_0} \right)^2, \tag{9.208}$$

which shows that in two-dimensional turbulence, the number of degrees of freedom is $\mathscr{O}(\mathrm{Re})$, very different from the $\mathrm{Re}^{9/4}$ dependence of three-dimensional turbulence, as discussed in Sect. 9.4.5.2.

While the direct cascade of enstrophy follows the Kolmogorov theory closely, it is not difficult to find some inconsistencies in the resemblance. By substituting (9.207) into the definition (9.202), we see that the integral diverges and therefore the enstrophy transfer rate diverges, as well, as $k \to 0$. To remedy this problem, Kraichnan proposed a log-corrected spectrum and found, using the same arguments as for the -3 result, that a constant enstrophy flux now predicts:

$$E(k) = C'\beta^{2/3}k^{-3}[\log(k/k_0)]^{-1/3}, \tag{9.209}$$

which is illustrated alongside the indirect cascade solution in Fig. 9.8. In this case, we also have $\omega_k^2 \propto \beta^{2/3}[\log(k/k_0)]^{-2/3}$; if valid, this indicates that the enstrophy transfer is not so local in k-space, with most of the contribution to ω_k coming from wavenumbers $k' \ll k$. In other words, the dominant deformation motion is on a much larger spatial scale than the flow structures that are being deformed. For this reason, the enstrophy transfer is often likened to spectral transfer of a passive scalar, i.e., tracer, in the k^{-1} range. Batchelor looked primarily at the enstrophy inertial range in the case of freely decaying turbulence, which is not stationary. Nevertheless, the same direct cascade and spectral scaling relations were found, as we will discuss in more detail in the following section.

9.5.4 Freely Decaying Two-Dimensional Turbulence

For the most part, the study of two-dimensional turbulence is concentrated on cases in which there is forcing and dissipation, at least on small scales where viscosity acts. Nevertheless, freely decaying two-dimensional turbulence has received much attention and, more importantly, shares many features with the forced problem. While freely decaying two-dimensional turbulence does not support an inverse energy cascade, *per se*, two closely related processes are ubiquitous: the emergence of coherent structures and the growth of large vortices through mergers of smaller like-sign vortices. We will look at the freely decaying inverse cascade in more detail below, in the context of numerical simulations and experiments.

On the other hand, the direct enstrophy cascade in freely decaying two-dimensional turbulence is essentially equivalent to that of the forced cascade. Batchelor, working independently of Kraichnan, derived the same k^{-3} dependence for the energy spectrum of freely decaying enstrophy cascade. The evolving self-similar energy spectrum is assumed to take the form $E(k,t) = u^3tF(ukt)$, where the self-similarity requirement leads to the result

$$E(k,t) \propto t^{-2}k^{-3}, \tag{9.210}$$

in place of (9.207). Supporting this, experiments and simulations display the same physical picture in which vortex patches are distorted and drawn out into filaments, the filamentary structures representative of intensified vorticity gradients. Simulations performed by one of us show just this behavior; one example is shown in Fig. 9.10. Thanks to increasingly resolved numerical simulations, a more refined picture of freely decaying two-dimensional turbulence is continuously emerging.

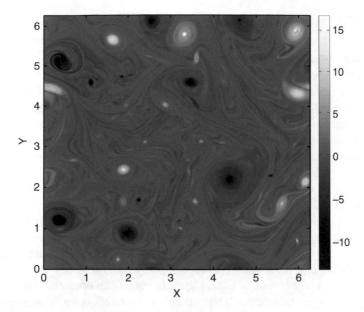

Fig. 9.10 Vorticity in decaying turbulence simulation, showing merger events and the early phases of condensate appearance. The simulation was performed by one of us, O. M. Umurhan, using a pseudospectral numerical method

To summarize, it is now believed that simulations initiate with a transient period characterized by shear-driven instabilities and axi-symmetrization around vorticity peaks. Details of the flow features in this initial stage are, however, strongly dependent on the resolution and initial conditions of the code. The following stage of evolution is an extended period during which the direct enstrophy cascade dominates. In this phase, strain leads to filamentation in the regions between vortices, while the number and mean radius of the vortices evolves self-similarly in time. Inside the coherent vortices, the direct cascade description fails, but it appears that peak vorticity is nearly conserved throughout the evolution. The final stage is characterized by a single large vortex dipole, which decays slowly due to dissipation. The terminal dipole is reminiscent of the condensate of the forced inverse cascade, but unlike in that case, the dipole is not determined by the boundaries of the system.

A number of challenges remain for the numerical simulation of decaying two-dimensional turbulence. For one, simulations tend to find energy spectra steeper than the predicted k^{-3} dependence. Also, simulations are not yet able to distinguish the logarithmic factors of (9.209), a challenge that also faces laboratory experiments. Beyond the issue of initial condition dependence mentioned above, numerical simulations also face the problem of distinguishing between the behavior of models based on ordinary viscosity (a Laplace operator) and those based on a hyper-viscosity (Laplacian squared).

9.5.5 Experiments and Simulations

The earliest laboratory experiments aimed at two-dimensional turbulence were performed in a thin layer of liquid mercury. A uniform magnetic field was used to establish a very thin Hartmann layer and the flow was then forced electro-magnetically. An inverse cascade was measured, albeit over a relatively narrow inertial range. Electromagnetic driving is more widely, and more recently, applied to electrolyte cells, typically based on salt water. These experiments are able to recover the inverse energy cascade over a wide inertial range, in addition to being able to examine freely decaying turbulence.

Another type of approximately two-dimensional experiment is performed on a soap film channel. The reader is urged to consult high-quality images available in the literature, cited and presented in the 2012 review paper, [12] in the *Bibliographical Notes*. In this paper there are depictions of soap film turbulence experiments, but in the references therein some are strikingly illustrated. The soap film configuration is especially notable for allowing a very large aspect ratio to be reached, $\mathcal{O}\left(10^4\right)$, and differs from other types of experiments in that, as an open flow, it is a natural example of decaying turbulence. Thus, while it is not as effective in investigating the inverse cascade, soap films are a useful way to study the direct enstrophy cascade. The very small scales that are supported in soap films as a result of their thinness allow a relatively large inertial range for which scaling exponents slightly steeper than the predicted -3 have been found. It remains to be known whether corrections to the theory of enstrophy transport can explain the steeper spectra seen in both experiments and simulations.

9.6   A Few Remarks on Turbulence Is Astrophysical Flows

Turbulence is ubiquitous in astrophysical flows. Suffice it to study observations of gaseous flows in the vicinity of stars and stellar nebulae, the regions of star formation and interstellar clouds, in order to discover the typical signatures of turbulence: the presence of great complexity and a large range of scales, in order to convince oneself that turbulence is the rule, rather than the exception in astrophysical flows.

9.6.1   Turbulent Convection

As we have seen in the previous chapter, there is a very simple criterion, named after Schwarzschild, for *convective instability*. It is mainly applied to stars and if it is satisfied ($ds/dr < 0$ or $|dT/dr| > |dT/dr|_{\text{ad}}$, where s is specific entropy and $|dT/dr|_{\text{ad}}$ is the *adiabatic temperature gradient*), hydrostatic equilibrium of a star is impossible and heat begins to be transported outward by fluid motions. This is similar to the well-known phenomenon of everyday life—hot air rises *up*, against the direction of gravity, replacing the colder air, which sinks.

In stellar structure and evolution theory, at least in its elementary form, the need to include the convective heat flux is usually answered by adopting a simplistic model. It is assumed that the instability saturates when $s(r)$ becomes constant, that is, the temperature gradient is adiabatic and that, conceivably, turbulent convection satisfies, on the average, a quasi-steady state. The idea is based on Prandtl's *mixing length theory*, which was already mentioned as the lowest means of closure. In contrast to Prandtl's purpose, which was to estimate turbulent viscosity, here we are interested in turbulent energy transport. The physical idea is to estimate a length scale l_m over which an element of hot fluid rises in turbulent convection before it dissolves and joins the ambient medium. In astrophysics, basing the heuristic argument on the schematic picture of Fig. 9.11, the *mixing length* (note the equality of names) is conjectured to be

$$l_m = \alpha H_P, \quad \text{where} \quad H_P \sim \frac{P}{dP/dr}, \tag{9.211}$$

with P being the pressure and H_P its scale height. α is a constant, usually between 1 and 2. This seems to be a reasonable choice and indeed gives a computational result for the Sun that is in agreement with that of helioseismology.

Now one has to obtain the convective energy flux through the star, denoted here by F_{con}, in order to add it to the fluxes of energy transport by other means, notably radiation. We explain the derivation of F_{conv} consulting the book by Kippenhahn and Weigert, reference [15] in the *Bibliographical Notes*, which in our opinion is by far the best source for understanding the theory of stellar structure and evolution. Before doing so, we have to define several symbols that have a special meaning in stellar structure theory. They are

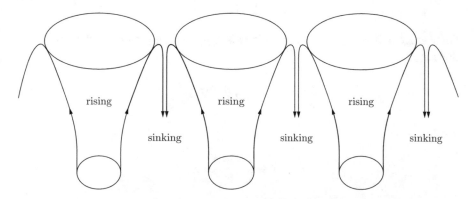

Fig. 9.11 The reasoning behind the conjecture for the expression of the convective mixing length. Note also the different shapes of the rising element region as opposed to one of the sinking elements

$$\nabla \equiv \left(\frac{d \ln T}{d \ln P} \right)_s ; \quad \nabla_e \equiv \left(\frac{d \ln T}{d \ln P} \right)_e , \tag{9.212}$$

where e is the specific thermal energy and we consider these relations at a stellar radius r, or the Lagrangian coordinate, m, the mass interior to the point r; thus, $dm = 4\pi \rho r^2 dr$ where $\rho(r)$ is the mass density.

To proceed we consider a blob, i.e., a convective mass element at a point, having an excess temperature δT compared to its surroundings at the same radius. Let it move in the spherically symmetric star with a velocity u which is much smaller than the speed of sound, and therefore it remains in pressure balance with its surroundings, i.e., $\delta P = 0$. This immediately gives the local flux of convective energy

$$F_{\text{conv}} = \rho u c_P \delta T, \tag{9.213}$$

where c_P is the specific heat at constant pressure. Now the central idea behind the *mixing length* concept is brought out: an element may start with a fluctuation, but because it is buoyantly unstable, the element rises (or sinks if $\delta T < 0$) a distance l_m, where it loses its identity. Performing now a simplistic statistical approximation, we consider a sphere of radius r in the star and assume that an element originates at a distance l_m from the sphere. Naturally, other elements each have a different starting point, between distance l_m down to zero, say. Thus,

$$\frac{\delta T}{T} = \frac{1}{T} \frac{\partial(\delta T)}{\partial r} \frac{l_{\text{m}}}{2} = (\nabla - \nabla_e) \frac{l_{\text{m}}}{2 H_P}. \tag{9.214}$$

The radial buoyancy force, per unit mass, is $b_r = -g \delta \rho / \rho$ and

$$\frac{\delta \rho}{\rho} = \left(\frac{\partial \ln \rho}{\partial \ln T} \right) \frac{\delta T}{T}. \tag{9.215}$$

On average half of this force acts on the element along a path $l_{\text{m}}/2$; thus, the work done is

$$\frac{1}{2} b_r l_{\text{m}}/2 = g\Delta(\nabla - \nabla_e) \frac{l_{\text{m}}^2}{8 H_P}, \tag{9.216}$$

where

$$\Delta \equiv \left(\frac{\partial \ln \rho}{\partial \ln T} \right). \tag{9.217}$$

Assume that half of this work goes to the kinetic energy of the element $u^2/2$, while the rest goes to the work needed to push other elements aside. Then

$$u^2 = g\Delta(\nabla - \nabla_e)\frac{l_m^2}{8H_P}. \tag{9.218}$$

Using this and Eq. (9.214) in (9.213), we finally get an approximate expression for the convective flux that can be used in calculations of stellar structure

$$F_{\text{conv}} = \frac{1}{4\sqrt{2}}c_P\rho Tl_m^2\sqrt{g\Delta}\left(\frac{(\nabla - \nabla_e)}{H_P}\right)^{3/2}. \tag{9.219}$$

9.6.2 More Turbulence in Astrophysics

As we have flows of astrophysical fluid objects are usually turbulent. Turbulence may play an important role in *rotating stars*, and it is thought to be the agent responsible for outward angular momentum transport in *accretion disks*, which as we have seen are very important objects, appearing whenever there is *accretion* of matter endowed with angular momentum with respect to the compact accreting object. We have discussed these topics in Chap. 5.

We find turbulence in astrophysics not only within stars and stellar objects, like accretion disks. The interstellar medium (ISM) seems also to be largely turbulent. The problem is further complicated by the fact that this is, usually, a magnetohydrodynamic turbulence (see Chap. 10). The dynamo, creating the magnetic field on a variety of scales in astrophysics, is thought to be intimately connected to turbulence. Remarkably, the kinetic energy density of the ISM turbulence, its magnetic energy density, and the energy density of *cosmic rays* satisfy an approximate *equipartition*. Those energy densities have the approximate value of $\sim 10^{-12}$ erg cm^{-3}.

It seems that we are still far from a viable theoretical model and understanding of turbulence in general and astrophysical turbulence, in particular. The main body of research relies on numerical simulations, which are also, as discussed above, far from trivial.

9.7 Very Brief Summaries on a Few Additional Topics

We may joke that the volume of research on turbulence seems to be inversely proportional to our understanding of the subject. In a book like the present one, we have only limited space and therefore we chose to devote the main portion of the space to topics that we consider as basic. What remains is to summarize only briefly other issues that, despite their possible importance, we cannot expound on.

9.7.1 Intermittency

Kolmogorov's theory enjoys the greatest success on small scales. L. Landau was not convinced from the outset that the statistical features of the small scales should be *universal*. The main problem, which has been known for a long time (it was anticipated by G.I. Taylor as early as 1917), is the fact that the vorticity, enstrophy, and dissipation are highly *intermittent*. In its simplest definition, intermittency means that a particular behavior at a point of a turbulent flow alternates between periods of statistical self-similarity and homogeneity and the lack thereof.

On the experimental side, velocity signals of certain setups, when subject to high-pass filtering, clearly show intermittent bursts (see Fig. 9.12 which is a typical example) in this case of the velocity fluctuations in a turbulent jet. It is therefore clear that the Kolmogorov 1941 theory cannot be complete. R.H. Kraichnan and others contributed a large number of papers on the fascinating topic of intermittency. We found Chap. 8 in the book by Frisch (reference [10]) and the references therein, the most complete and enlightening as an account of intermittency.

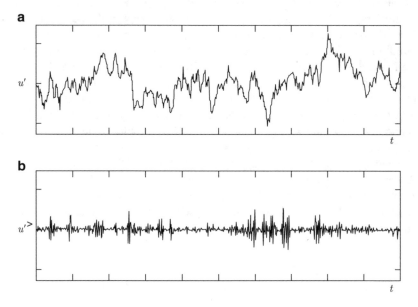

Fig. 9.12 An illustrative case of an intermittent signal in a turbulent flow, a jet in this case: (a) shows the fluctuating signal while (b) shows the high-pass filtered signal. After Y. Gagane's thesis (1980) cited in the *Bibliographical Notes*

9.7.2 Dynamical Systems Approach to Turbulence

We have indicated that the number of degrees of freedom in turbulent flow seems to grow according to Kolmogorov's theory, like $Re^{9/4}$. On the other hand, if one considers a turbulent flow as a dissipative dynamical system, the number of relevant degrees of freedom should be defined as the dimension of the *attractor* (see reference [17]). The problem is that methods to find such a dimension when it is relatively small exist, and they give reliable results. In contrast, no reliable methods have been found to measure attractor dimension of, say, $\gtrsim 10$.

Dynamical systems have been found that are much simpler than the full Navier–Stokes system, but still possess states which resemble turbulent behavior. It is hoped that the study of such systems and a broader definition of the notion of turbulence can pave the way for progress in fluid dynamical and MHD turbulence. A relatively recent book, reference [18], deals at length with this idea. Very recent new ideas, linking turbulence to dynamical system theory, were made in 2013 by P. Cvitanović and collaborators, and they seem to us promising (see reference [19] in the *Bibliographical Notes*).

9.7.3 Two Basic Approaches to Simulation

Numerical computations of the DNS (direct numerical simulations) type, or LES (large eddy simulations, including some turbulence model based on closure), have contributed much to visualization of turbulent flows. In the section on two-dimensional turbulence, we have shown some results of DNS, and in Wilcox's book (given the *Bibliographical Notes*), the reader can find several recipes for LES. In any case, we shall devote Appendix B to a short primer on methods in CFD (computation fluid dynamics), which includes extensive bibliography, as well as some important caveats related to this line of research, which actually is not among the subjects of this book.

Problems

9.1.
Show that the function of space $\varsigma(\mathbf{x})$, as defined after Eq. (9.127), satisfies the solvability condition (9.127).

9.2.
Work with the vorticity $\boldsymbol{\omega} \equiv \nabla \times \mathbf{v}$ and eliminate the velocity from the vorticity equation by solving the appropriate Poisson equation. Use the same nonlocal operator ∇^{-2}, as used in the velocity formalism (9.133), in order to get an equation solely in terms of vorticity.

9.3.
Explain the physical meaning of the Reynolds stresses, showing that they are the mean momentum fluxes induced by the turbulence. Hint: consider a small cube whose edges are along the Cartesian axes and identify the Reynolds stresses by examining the total momentum fluxes along and through the faces of the cube.

9.4.
Show that the evolution of the Reynolds stress is according to Eq. (9.14). Hint: subtract Eq. (9.11) from (9.7). After obtaining a second similar equation, multiply the first equation by $\bar{u}'_j$ and the second one by $\bar{u}'_i$ and after adding the two resulting equations, average them.

9.5.
Obtain an energy equation for the fluctuations by setting $i = j$ in Eq. (9.14). Show that this equation can also be rewritten as

$$\frac{\overline{D}}{Dt}\left[\frac{1}{2}\rho\overline{(\mathbf{u}')^2}\right] = u_k\frac{\partial}{\partial x_k}\left[\frac{1}{2}\rho\overline{(\mathbf{u}')^2}\right] = \tau_{ik}^R\mathscr{D}_{ik} - 2\rho\nu\overline{\mathscr{D}'_{ik}\mathscr{D}'_{ik}}$$

$$+ \frac{\partial}{\partial x_k}\left[-\overline{P'u'_k} + \overline{u'_i\tau'_{ik}} - \frac{1}{2}\rho\overline{(\mathbf{u}')^2 u'_k}\right]. \qquad (9.220)$$

Hint: manipulate the viscous term of Eq. (9.14) so that for $i = j$ they reduce to

$$2\left[\frac{\partial}{\partial x_k}\left(\overline{u'_i\tau'_{ik}}\right) - 2\rho\nu\overline{\mathscr{D}'_{ik}\mathscr{D}'_{ik}}\right]. \qquad (9.221)$$

9.6. Consider a steady, on average, flow in which the forces acting on a fluid element are $\mathbf{f}_{tot}$ = the total *pressure and viscous* forces and assume that no other force acts on the flow. Derive from the Navier–Stokes equation the kinetic energy equation. Next, split the velocity into a steady and fluctuating part $\mathbf{u} = \bar{\mathbf{u}} + \mathbf{u}'$ and time average the kinetic energy equation to get

$$\bar{\mathbf{u}}\cdot\nabla\left(\frac{1}{2}\rho\bar{\mathbf{u}}^2 + \frac{1}{2}\rho\overline{(\mathbf{u}')^2}\right) + \overline{\mathbf{u}'\cdot\nabla\left(\frac{1}{2}\rho(\mathbf{u}')^2\right)} = \frac{\partial}{\partial x_j}\left(\bar{u}_i\tau_{ij}^R\right) + \overline{\mathbf{f}_{tot}\cdot\mathbf{u}}. \qquad (9.222)$$

Now average the equation of motion and only then derive from it the following kinetic energy equation:

$$\bar{\mathbf{u}}\cdot\nabla\left(\frac{1}{2}\rho\bar{\mathbf{u}}^2\right) = \bar{u}_i\frac{\partial\tau_{ij}^R}{\partial x_j} + \overline{\mathbf{f}}_{tot}\cdot\bar{\mathbf{u}}. \qquad (9.223)$$

Subtract the two equations and interpret the result physically.

9.7.
Derive the kinetic energy equation for the mean flow

$$\bar{\mathbf{u}} \cdot \nabla \left(\frac{1}{2} \rho \bar{\mathbf{u}}^2 \right) = \frac{\partial}{\partial x_k} \left[-\bar{u}_k \bar{P} + \bar{u}_i \left(\bar{\tau}_{ik} + \tau_{ik}^R \right) \right] - \tau_{ik}^R \bar{\mathscr{D}}_{ik} - 2 \rho v \bar{\mathscr{D}}_{ik} \bar{\mathscr{D}}_{ik},$$

using Eq. (9.30). Note that $\bar{\tau}_{ik}$ is the average viscous stress.

9.8.
Show by integrating Eq. (9.39) with the appropriate constant on the right-hand side that the total stress is given by Eq. (9.40).

9.9.
Use the technique of *matched asymptotic expansions* (introduced in Chap. 2 for laminar boundary layers) to show that in a turbulent boundary layer above a flat plate

$$\bar{u}_x = u_* f(y^+) \quad y \ll \delta \quad \text{inner region,}$$
$$\Delta \bar{u}_x = \bar{u}_\infty - \bar{u}_x = u_* g(y/\delta) \quad y^+ \gg 1 \quad \text{outer region,}$$
$$\bar{u}_x = u_* \left[\frac{1}{\kappa} \ln y^+ + A \right], \quad y^+ \gg 1, y \ll \delta \quad \text{overlap region,} \qquad (9.224)$$

where u_* is the width of the mean velocity departure from zero, $y^+ = u_* y / v$, and δ is the local boundary layer thickness; f and g are appropriate functions and κ and A constants.

9.10.
Determine the mean flow in a circular jet outside the turbulent region. Use polar spherical coordinates and remember that outside the turbulent jet we have potential flow.

9.11.
Using the form of the average momentum flux in the turbulent axi-symmetric jet, find the average mass flux Q up to a constant coefficient; thus, $Q = \beta Q_0 x/a$, where Q_0 is the mass of the fluid which issues from the pipe aperture in unit time, a is a linear dimension of the aperture, and β depends on its actual geometrical shape. Q_0/a is the only external parameter which determines the flow in the jet. Write the Reynolds number in terms of this parameter.

9.12.
Consider a flow over a semi-infinite thin plate, placed on the $x - z$ plane, starting with at $x = 0$. The flow in which the plate is placed is $\mathbf{u} = U\hat{\mathbf{x}}$ for $y > 0$ and $\mathbf{0}$ otherwise. Assume that a turbulent boundary layer is formed over the plate. What is the solution $u_x(y)$? Does it differ from the logarithmic profile of formula (9.45) and if so, in what way?

9.13.
Show that Eq. (9.14) reduces to (9.54) for the model of homogeneous shear turbulence we have been using in the text.

9.14.
We see from the first equation in (9.56) that the Reynolds stress feeds only turbulent fluctuations of kinetic energy contained in the x component of the velocity. Yet in experiments it is found that after sufficient time this energy is redistributed to also be contained in u'_y and u'_z. Identify qualitatively the physical mechanism which may cause this redistribution. Explain.

9.15. Using the general definition of an average (9.58), propose a structure or weight function $W_s(\xi, \tau)$ so that the definition will yield a space average, over some parallelepiped, and also a $W_s(\xi, \tau)$, which will yield a time average over a time segment $[0, T]$.

9.16. Using the averaging operation properties (9.60)–(9.64), prove the equalities which follow:

1. $\bar{\bar{f}} = \bar{f}$
2. $\overline{f'} = \overline{f - \bar{f}} = 0$
3. $\overline{\bar{f}\bar{g}} = \bar{f}\bar{g}$
4. $\overline{\bar{f}g'} = \bar{f}\bar{g}'$

9.17. Show that for $N = 1$ in Eqs. (9.97)–(9.98) $\mathcal{M}_1^{\text{cen}} = 0$, $\mathcal{M}_2^{\text{cen}} = \mathcal{M}_2 - \mathcal{M}_1^2$, and $\mathcal{M}_3^{\text{cen}} = \mathcal{M}_3 - 3\mathcal{M}_1\mathcal{M}_2 + 2\mathcal{M}_1^3$. Conclude that $\mathcal{M}_2^{\text{cen}}$ is the *variance* (σ_2^2) of u. What is the standard deviation and what is the covariance, in terms of the moments and/or central moments?

9.18.
Prove the property (9.101) of the correlation function.

9.19.
Using Eq. (9.86) and the conditions (9.83), show that $g(r) = f(r) + \frac{1}{2}r \, df/dr$, indicating that only one function is needed to obtain the full correlation tensor $Q_{ij}(r)$.

9.20.
Let $Q(r) \equiv \frac{1}{2}Q_{ii}(r)$. Prove the following relations between $Q(r)$ and $E(k)$:

$$E(k) = \frac{2}{\pi} \int_0^\infty Q(r)kr\sin(kr)\,dr,$$

$$Q(r) = \int_0^\infty E(k)\frac{\sin(kr)}{kr}\,dk.$$

9.21.
Show that the constants in the normal distribution (9.102) are related to the moments of the distribution; thus, $\bar{u}_j$ (the first moment) is equal to the constant α_j, while the central second moment $\overline{(u_j - \bar{u}_j)(u_k - \bar{u}_k)}$ is equal to G/G_{jk}, where $G = \det(g_{jk})$ and

$G_{jk} = \partial G/\partial g_{jk}$ is the cofactor of the element g_{jk} in determinant G. Find also an expression for the ordinary (noncentral) second moments of the distribution $\mathcal{M}_{jk}$.

9.22.

Using Eqs. (9.114)–(9.115), show that the (triple) Dirac delta function $\delta\mathbf{k}$, satisfying $\delta(\mathbf{k}) = 0$ for $\mathbf{k} \neq 0$ and $\int \delta(\mathbf{k})d^3k = 1$, can be expressed as

$$\delta(\mathbf{k}) = \left(\frac{1}{2\pi}\right)^3 \int e^{-i\mathbf{k}\cdot\mathbf{x}} d^3x = \left(\frac{1}{2\pi}\right)^3 \int e^{+i\mathbf{k}\cdot\mathbf{x}} d^3x. \tag{9.225}$$

9.23.

Prove formula (9.118) and explain its physical meaning in words.

9.24.

Using the hints and discussion in the text (last part of Sect. 9.4), write the Fourier transform of the Navier–Stokes incompressible equations.

9.25.

Prove identities 9.139 (all that is needed is integration by parts) and the conservation laws (9.140)–(9.142).

9.26.

Prove the properties (9.148)–(9.150) of the low-pass filtering operator $\mathbb{P}_q \equiv u(\mathbf{x}) \mapsto u_{\leq q}$.

9.27.

Show that the cumulative (up to the wavenumber q) energy balance equation (9.152) holds, starting from expression (9.153). Hint 1: Some equalities here are identical to those used in the derivation of Eq. (9.141)—see Problem 9.25. Hint 2: The contributions from the nonlinear term do not vanish here, however, and when expanded they produce four terms. Two of these terms, listed below, vanish. Show that it is so. The vanishing terms are

$$\overline{\mathbf{u}_{<q} \cdot (\mathbf{u}_{<q} \cdot \nabla \mathbf{u}_{<q})} = 0, \quad \text{and} \quad \overline{\mathbf{u}_{<q} \cdot (\mathbf{u}_{<q} \cdot \nabla \mathbf{u}_{<q})} = 0.$$

9.28.

Show, using the 4/5 Kolmogorov law, that the scaling exponent in $\delta\mathbf{u}(\mathbf{x}, \lambda\bar{\ell}) = \lambda^s \delta\mathbf{u}(\mathbf{x}, \bar{\ell})$ has the value $s = 1/3$.

9.29.

Use the phenomenological expression for eddy turnover time (9.178) to find how the square of the separation distance between two fluid particles, released a distance ℓ apart in a turbulent flow, increases with ℓ. This result was found by Richardson as early as 1926.

9.30.

Consider the mean thermal speed of the microscopic particles of a fluid and, as is usual, assume that it is approximately equal to the sound speed c_s. Show that

$$\frac{\eta_K}{\Lambda_{\text{mfp}}} \sim \frac{\text{Re}^{1/4}}{M},\tag{9.226}$$

where Λ_{mfp} is the mean free path of the microscopic particles in the fluid and M is the Mach number. What strong inequality is necessary so that the fluid continuum approximation holds for the smallest (dissipative) scale and what does it mean physically?

9.31.
Show that the ratio of the velocity to length at the Kolmogorov dissipation scale, denoted here by η, can be estimated to be

$$\frac{u_\eta}{\eta} \sim \frac{v^{3/4}\ell_0^{1/4}}{u_0^{1/4}},\tag{9.227}$$

where the index 0 indicates the integral scale. This ratio is sometimes called the *velocity gradient*.

9.32.
- Complete the algebra in the text to bring Eq. (9.165) to the concise form Eq. (9.166). In addition, use the definitions we had to arrive at the explicit form of the first term on the right-hand side of the Kármán–Howarth equation for isotropic turbulence (9.167).
- Derive from the Kármán–Howarth equation (9.167) the 4/5 law (9.168) using the following hints: Show that $S_3^\parallel = 6u^3K(r)$, then write the Kármán–Howarth equation in terms of the second- and third-order structure functions, and neglect terms $\propto \ell$ with respect to similar terms $\propto \ell_0$.

9.33.
Two simple models of the inverse cascade of energy were presented in Sect. 9.5.2; referring to these: (a) Quantitatively complete the calculation of the example wavenumber triad, then make a different choice for the three initial k values, and repeat. (b) Show that relation (9.198) is true.

9.34.
Using the constant enstrophy flux solution (9.207), show that ω_k is, in fact, proportional to $\log(k)$ and not constant. Discuss the meaning of a nonconstant ω_k and how this leads to (9.209). Hint: deal with any improper integrals. Bonus: derive (9.209).

9.35.
Consider the linear evolution of vortex profile initially given by

$$\zeta_0 = \zeta_{00} \cos x \cos y$$

subject to the external velocity field $U = \Lambda y\hat{x}$ evaluated on an infinite domain in x and y. Show that the energy (per unit area) of the flow due to the initial vortex distribution asymptotically behaves like $\sim 1/t^2$ where t is time. Conclude that as the gradients of the vortex increase due to distortion, the energy contained within decreases.

Bibliographical Notes

General

A rather large number of accounts on turbulence exist. The Landau–Lifshitz classic book on fluid mechanics contains merely one short and idiosyncratic chapter, while Davidson's book holds 650 pages and this seems too much for our purposes. In any case, some parts of Davidson's book can be very useful. We estimate that the relatively recent proceedings of the Les Houches session LXXIV on turbulence contain topics that are close to the state of the art on the subject. We recommend, in particular, those lectures that deal with subjects we were unable to include here.

1. L.D. Landau, E.M. Lifshitz, *Fluid Mechanics*, 2nd edn. (Elsevier, New York, 2004)
2. P.A. Davidson, *Turbulence* (Oxford University Press, Oxford, 2004)
3. M. Lesieur, A. Yaglom, F. David (eds.), *Les Houches Session LXXIV. New Trends in Turbulence*. EDP Sciences (Springer, Berlin, 2000)
   A conference commemorating the Nice conference of 1961 on the mechanics of turbulence was held in the same venue 50 years later. A concise review article by H. K. Moffat opens the proceedings:
4. H.K. Moffatt, Homogeneous turbulence, an introductory review. J. Turbul. **13**(1), N39 (2012)

Section 9.1

The best introductory text which can serve us here is, in our opinion, that of Tennekes and Lumley, while Lesieur's book is more advanced mathematically and more comprehensive.

5. H. Tennekes, J.L. Lumley, *A First Course of Turbulence* (MIT Press, Cambridge, 1972)
6. M. Lesieur, *Turbulence in Fluids* (Springer, Berlin, 2008)

Section 9.2

In our opinion the clearest presentation of Reynolds equations, as well as results on shear flows, can be found in Davidson's book [2]. Landau and Lifshitz [1] also contain some useful discussion on the "log-law of the wall," as well as on turbulent jets.

Section 9.3

Here the best source remains, in our opinion, the classic work of Monin and Yaglom, recently re-edited by Dover. Lesieur's book [6] is also useful. Chapter 6 of Davidson's book [2] includes comprehensive material on correlation functions in an isotropic medium as well as a clear derivation of the Kármán–Howarth equation. This equation together with an interesting, quite advanced, look on the theory of turbulence can be found in E. A Spiegel's notes of S. Chandrasekhar's lectures.

7. A.S. Monin, A.M. Yaglom, *Statistical Mechanics of Turbulence*, vol. I, II (Dover, New York, 2007)
8. E.A. Spiegel (ed.), *The Theory of Turbulence*. S. Chandrasekhar's 1954 Lectures (Springer, New York, 2011)
   The following article in *Science* points out some new important observations on turbulent flow, which prompted the recent beginnings of linking dynamical systems theory to turbulent flow (see below, reference [19]).
9. B. Hof, C.W.H. van Doorne, J. Westerweel, F.T.M. Nieuwstadt, H. Faisst, B. Eckhardt, H. Wedin, R.R. Kerswell, F. Waleffe, Science, **305**, 1594 (2004)

Section 9.4

Kolmogorov and his school were early chief contributors to the theory of turbulence. There is probably no book which does not attempt to describe their work. We found the parts of the voluminous reference number [2] useful. Frisch's work, cited below, fits well the subject of this section, and in particular it contains the derivation of the Kármán–Howarth equation in the non-isotropic case and from it the $4/5$ Kolmogorov law. Some of Kraichnan's many contributions to turbulence theory can be found in his paper and references therein.

10. U. Frish, *Turbulence* (Cambridge University Press, Cambridge, 1995)
11. R.H. Kraichnan, The structure of isotropic turbulence at very high Reynolds number. J. Fluid Mech. **5**, 497 (1959)

Section 9.5

An excellent modern review of forced two-dimensional turbulence with many references to recent experiments can be found in

12. G. Boffetta, R.E. Ecke, Ann. Rev. Fluid Mech. **44**, 427 (2012)
    The following very good lecture series, which was presented at the Les Houches session LXXIV, is also highly recommended.
13. J. Sommeria, *Two-dimensional turbulence*, pp 387–442 of in number [3], is also an excellent source for the topic of two-dimensional turbulence.
    The original work of Kraichnan, whose reading is not easy, is recommended as the pioneering contribution.
14. R.H. Kraichnan, Phys. Fluids **10**, 1417 (1967)

Section 9.6

15. R. Kippenhahn, A. Weigert, *Stellar Structure and Evolution* (Springer, Berlin, 1990)

Section 9.7

16. Y. Gagne, *Contribution à l'étude expérimentale de l'intermittence de la turbulence à petitte échelle*. Thesis, Grenoble University (1980)
17. O. Regev, *Chaos and Complexity in Astrophysics* (Cambridge University Press, Cambridge, 2012)
18. T. Bohr, M.H. Jensen, G. Paladin, A. Vulpiani, *Dynamical Systems Approach to Turbulence* (Cambridge University Press, Cambridge, 1998)
    This following Internet website and paper are the most recent and significant development in the use of dynamical systems theory to turbulence.
19. J.F. Gibson, P. Cvitanović, www.ChaosBook.org/tutorials (2013) and P. Cvitanović, J. Fluid Mech. **726**, 1 (2013)
    Among the books describing some (mainly engineering) LES numerical approaches to turbulence is
20. C. Wilcox, *Turbulence Modeling for CFD*, II edn. (DCW Industries, Anaheim, 1998)
    Modern computational approaches are also discussed at considerable length in the Les Houches lectures (reference number [3]).

Chapter 10
Magnetohydrodynamics

The lodestone makes iron come or it attracts it.

Chinese Master of Demon (4th-century BC); 'The Master of Demon Valley'

10.1 Introduction

Magnetohydrodynamics (MHD), also called sometimes *magnetofluid dynamics* or *hydromagnetics*, is the study of the dynamics of electrically conducting fluids in the presence of a magnetic field. Electrical fields induced in such a fluid as the result of its motion are bound to drive electrical currents that modify the magnetic field which, in turn, imparts force on the fluid, affecting its motion. A correct description of such a complicated system can be achieved by combining FD equations, which we have been studying in this book, with the Maxwell equations of electromagnetism. J.C. Maxwell (1831–1879) unified, as expressed in his equations, electricity with magnetism giving rise to the understanding of electromagnetic waves. This respected mathematical physicist was also active in thermodynamics. In this chapter we shall embark on combining Maxwell's equations with those of FD, explaining in considerable detail the assumptions that are usually made in the purpose of formulating the MHD picture of conducting fluids. Liquid metals, like mercury at room temperature, and water containing a solution of chemicals dissociated into ions are also examples of conducting fluids. However, very hot *dilute* ionized gases may even not obey the continuum approximation as discussed in Sect. 1.1, and they go under the name of *plasma*. Some consider the state of plasma to be the fourth phase of matter, in addition to solid, liquid, and gas. This nomenclature seems today quite outdated, as in modern research the boundaries between these phases are no longer clear-cut. In any case, plasma physics is a vast subject, and as such, it is clearly outside the scope of a book on fluid dynamics.

© Springer Science+Business Media, LLC 2016

O. Regev et al., *Modern Fluid Dynamics for Physics and Astrophysics*,
Graduate Texts in Physics, DOI 10.1007/978-1-4939-3164-4_10

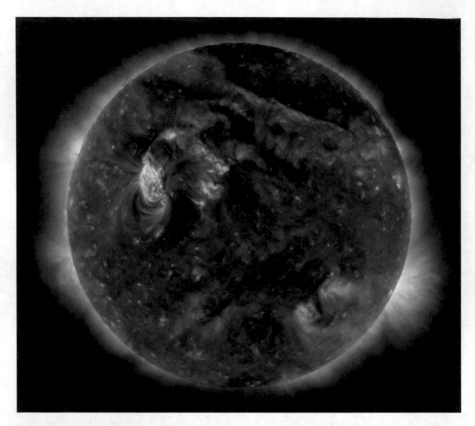

Fig. 10.1 A multi-wavelength extreme ultraviolet snapshot of the Sun from the Solar Dynamics Observatory, showing the Sun's northern hemisphere in mid-eruption. Different colors in the image represent different gas temperatures. *Public Domain, from the NASA picture of the day. Courtesy of NASA/SDO/AIA*

The physics of plasmas is however important to study, in the laboratory as well as in nature. This is due to the intensive laboratory efforts to achieve controlled fusion, which could be a vast source of clean energy, and to geophysical and astrophysical research. For example, our Sun appears to the naked eye as a perfect uniform disk. However, it turns out that the Sun's magnetic field, although not strong enough to cause significant disruptions of the star itself, is found to have energetically prominent and very variable behavior. In Fig. 10.1 the solar disk is shown in mid-2010, when almost the entire Earth facing side of the Sun erupted in a tumult of magnetic activity. We turn now to MHD, this chapter's subject. It is that part of plasma physics which lends itself to a fluid, that is, continuum, description as it was defined in Sect. 1.1. Moreover, the majority of our discussion will be based on MHD in its *narrower sense*, a model in which the following assumptions are adopted.

Assumption 1

The medium is considered to be a single neutral conducting fluid, which at each point **x** of space and at each time is endowed with well-defined hydrodynamic functions: density $\rho(\mathbf{x}, t)$, pressure $P(\mathbf{x}, t)$, and velocity $\mathbf{u}(\mathbf{x}, t)$. The fluid is considered to have a prescribed temperature at each point and time, because all the charged species separately are assumed to obey locally the Maxwell–Boltzmann distribution of their random (thermal) velocities, with the *same* temperature.

The conclusion of the last sentence above is that local thermodynamic equilibrium (LTE) prevails. MHD description of conducting fluids is a branch of plasma physics, and as such it is a relatively recent discipline in physics. H. Alfvén, who is considered a leading pioneer in plasma physics, received a Nobel Prize only in 1970, "for fundamental work and discoveries in magnetohydrodynamics with fruitful applications in different parts of plasma physics." It is true that already in 1928 I. Langmuir coined the term *plasma* for the ionized gas in a discharge tube, and even earlier cathode rays, whose nature was thought to be a special state of matter, were observed in Crookes tubes. In addition, already in 1919 J. Larmor proposed a theory for the generation of the solar magnetic field. However, systematic studies of plasma, created in the laboratory or existing in nature, became serious only in the second half of the past century.

A reasonably rigorous transition from plasma kinetic theory (collection of a large number of charged particles) to fluid description is formidable. Even in the case of neutral fluids, we have chosen not to follow the route to hydrodynamic equations from the kinetic ones, but have instead applied macroscopic conservation laws on a *fluid particle*. In MHD we are talking about electrically conducting fluids, acted upon electromagnetic force fields, and thus list explicitly our assumptions (along the text), wishing to follow a way similar to neutral fluids, towards a closed set of hydromagnetic equations. The above **Assumption 1** stems from the fluid being regarded as dominated by collisions between the microscopic particles, that is, any typical macroscopic frequency of interest in our system, marked by a tilde, $\tilde{\omega}$, must satisfy $\tilde{\omega} \ll \tilde{\nu}_c$, where $\tilde{\nu}_c$ is the average collision frequency between the microscopic particles: electrons, ions, and themselves, as well as electron–ion collisions. Introducing an average mean free path between collisions ℓ and a characteristic random thermal velocity $\tilde{w}$, we get $\tilde{\nu}_c \sim \tilde{w}/\ell$. Defining now a characteristic macroscopic length of our problem L, we note that typical macroscopic motions have a velocity of the order of a typical sound speed, $U \sim \tilde{c}_s$, which in turn is of the order $\tilde{c}_s \sim \tilde{w}$. Thus, the characteristic frequency for macroscopic phenomena is $\omega_L \sim U/L$. Clearly then $\omega_L \sim \tilde{\omega}$, and we shall use both notations interchangeably, so the relation $\tilde{\omega} L \sim U \sim \tilde{w}$ is self-evident. It is easy to see that the condition $\tilde{\omega} \ll \tilde{\nu}_c$ is equivalent to Kn $\ll 1$, where the Knudsen number is, as we have defined it in Chap. 1, the ratio between the mean free path ℓ and a typical macroscopic system length scale L. We shall look into the collisional processes in more detail below. It is clear that **Assumption 1**, and as we shall see below some of the other assumptions as well, forces a low-frequency limit for MHD. This is sometimes called, as mentioned at the outset, MHD in the narrow sense, but we shall prefer to call it the *one-fluid* MHD model. The assumptions, which we shall list as we proceed, are all necessary for the one-fluid model. Clearly, low frequency means long wavelength and thus this approximation is bound to work well for large bodies of conducting fluids— cosmic or geophysical systems, for example—while in the laboratory, e.g., in fusion experiments, the one-fluid equations usually fail to describe the system correctly and one must take recourse to MHD in some broader sense, e.g., Hall MHD, or two-fluid MHD, if not full-fledged plasma physics approach (see Sect. 10.9).

So far we have said nothing about the different charged species, which, for simplicity, we assume to be electrons and protons. It is however necessary that we have a *neutral* one-fluid, that is, in terms of particle densities $|n_e - n_p| \ll n$, where we are talking about particle densities. This is a good assumption, as we shall see below, because even a small charge imbalance creates a relatively huge, electric field. This field acts so as to restore neutrality. The equations governing MHD include the FD equations, including a magnetic source for the body force (see Chap. 1) and the Maxwell equations, which we shall give below, and couple them to the fluid equations, consistently with our previous and forthcoming assumptions. We have chosen the c.g.s. Gaussian e.s.u. system of units for electromagnetism, to fit the c.g.s. mechanical system, used so far. Before starting the algebraic manipulations necessary to derive the full set of MHD equations, we state the second assumption.

Assumption 2

Regarding the macroscopic electric and magnetic fields $E(\mathbf{x}, t)$ and $B(\mathbf{x}, t)$, which permeate the fluid, we stress that we deal with media in which the electric permittivity and magnetic permeability are, to excellent approximation, equal to their vacuum values. The usual transport coefficients, like HD dynamic shear viscosity η and bulk viscosity ζ, as well as heat (κ) and electric (σ_{el}) conductivity, are all assumed to be constant scalar positive quantities. That is, the fluid is assumed isotropic.

The isotropy of the coefficients is one of our simplifying assumptions, valid in most cases. We shall use an isotropic description, even though there are important systems in which the electric conductivity is not isotropic and Ohm's law (see below) is nonlinear. Also, the heat conductivity may depend on the direction, perpendicular or parallel to the magnetic field. We shall not deal with such complications. It is useful to conclude this introductory section by giving the definitions and estimates of a number of important quantities, known as the *kinetic coefficients* which are determined by taking into account also microscopic properties of the system. Some of these quantities are important mainly in dilute plasmas, where the MHD model is *not* applicable, but most others are helpful in understanding the right approximations and their underlying assumptions that may be used in one- or multi-fluid MHD.

10.1.1 Estimates of Some Kinetic Coefficients

For simplicity, we shall consider hydrogen plasma, in LTE, that is, the electron and proton temperatures are both equal T, say, at a specific location and time. We shall consider the kinetic coefficients at this position and time. This assumption will be sufficient for our purposes, while the one regarding composition can be trivially generalized for positive ions of other charges.

- *Microscopic thermal motion*
  The thermal random velocities can be estimated using the known fact that the kinetic energy of every degree of freedom in the random microscopic motion is $\frac{1}{2}k_B T$, where k_B is the Boltzmann constant ($k_B = 1.381 \times 10^{-16}$ ergs/deg). Thus, we have a thermal velocity component of a particle

$$w_\alpha \sim \sqrt{\frac{1.5 k_B T}{m_\alpha}}, \quad \text{with } \alpha = e,p \text{ for electrons and protons}$$

implying $\quad \dfrac{w_e}{w_p} = \sqrt{\dfrac{m_p}{m_e}} \approx 43.$ \hfill (10.1)

Numerically, the thermal velocities of the electrons are of the order $w_e \sim 5 \times 10^7 \sqrt{T_4}$ cm/s, where T_4 is the temperature in units of $10^4 K$, while those of the protons are smaller by a factor $\sqrt{m_p/m_e} \approx 40$.

Shifting now the attention to the mean free path between collisions ℓ, we use the general, self-evident, expression (for both ℓ_e and ℓ_p)

$$\ell \sim \frac{1}{n\sigma_{tot}}, \quad \text{with } \sigma_{tot} \sim r_{int}^2,$$ \hfill (10.2)

where we have approximately set the total collisional cross section between charged particles to be the square of the interaction range r_{int}. Formally, the Coulomb interaction has an infinite range, however, in a mixture of thermal electrons and protons, at distances larger than the *Debye length*:

$$\lambda_D = \sqrt{\frac{k_B T}{4\pi e^2 n}}.$$ \hfill (10.3)

The field of a point charge is totally screened by charges with the opposite sign. In Problem 10.1 this statement is proven and the Debye length is estimated for a number of conditions. To get an approximation for the interaction range, we may roughly equalize the kinetic and Coulomb potential energy of the particles; thus,

$$k_B T \approx \frac{e^2}{r_{int}}.$$ \hfill (10.4)

Remembering that in the derivation of the shielding distance we have assumed an exponential falloff $\propto \exp(-r/\lambda_D)$ multiplying the Coulomb potential, we understand that in the use of the above rough estimate for r_{int} we lose a factor of $\ln(\lambda_D/r_{int})$ which is called the *Coulomb logarithm*. The latter is a slowly varying function of temperature and density and for most conditions is of the order of $\sim$ *a few*. It follows from Eq. (10.2) that the following is a reasonable estimate:

$$\ell_e \sim \ell_p \sim \frac{(k_B T)^2}{ne^4} \equiv \ell. \tag{10.5}$$

Expressing the temperature in units of $10^4 K$ and the number density in 10^{22} cm^{-3}, we get the numerical estimate of $\ell \sim 5 \times 10^{-9} T_4^2/n_{22}$ cm. Turning now to the free flight times, between collisions, of the two particles, we get

$$\tau_\alpha \equiv \frac{1}{v_c^\alpha} = \frac{\ell}{w_\alpha}. \tag{10.6}$$

Using Eqs. (10.1–10.5), we obtain

$$v_c^\alpha \equiv \frac{1}{\tau_\alpha} \sim \frac{ne^4}{\sqrt{(k_B T)^3 m_\alpha}}. \tag{10.7}$$

The numerical value of electron collisional frequency can also be expressed using our fiducial value of the temperature and density (see Problem 10.2). Again, as in the case of the thermal velocity, the difference between the collision times and frequencies of electrons and protons results only from their mass differences

$$\frac{\tau_p}{\tau_e} \sim \sqrt{\frac{m_p}{m_e}} \approx 43, \tag{10.8}$$

with the ratio of the collisional frequencies trivially following.

If we wish to be more precise, we have to estimate a proton–electron collision frequency (so far we dealt only with v_c^α with $\alpha = e, p$, which are electron–electron and proton–proton collisional frequencies, respectively). It is a simple exercise to estimate the electron–proton collision frequency, using a nonrelativistic approximation of classical mechanics. Noting that the electron mass is much smaller than that of the proton and, obviously, of heavier ions as well as that the electron velocity $v_e \gg v_p$, the proton velocity, we conclude that the time needed to change the momentum of an ion because of a collision with an electron is appreciably longer than the one given by the reciprocal value of v_c^α in Eq. (10.7). Let us call that time τ_{e-p} and use the conservation of momentum and energy in the collision. The changes in the respective velocities in such a collision are approximately

$$\Delta v_e \sim v_e, \quad \Delta v_p \sim \frac{m_e}{m_p} v_e \sim \sqrt{\frac{m_p}{m_e}} v_p, \tag{10.9}$$

as the result of Problem 10.3 shows. Thus, $\sqrt{m_p/m_e}$ collisions of a proton with an electron are necessary to lose its energy and momentum, so that

$$\tau_{e-p} \sim \sqrt{\frac{m_p}{m_e}} \tau_p \quad \Longleftrightarrow \quad v_c^{e-p} \sim \sqrt{\frac{m_e}{m_p}} v_c^p. \tag{10.10}$$

So we may add to our estimates also

$$\tau_{e-p} \sim \frac{m_p}{m_e} \tau_e \quad \Longleftrightarrow \quad v_c^{e-p} \sim \frac{m_e}{m_p} v_c^e. \tag{10.11}$$

In what follows we shall call the frequency of electron–proton collision just v_c. The ordering $\tau_e : \tau_i : \tau_c \sim 1 : 43 : 1850$ evidently indicates that if hydrogen plasma is slightly disturbed out of thermodynamic equilibrium at a certain location, the electrons will settle into a local thermodynamic equilibrium, between themselves rather quickly—after a few τ_e, protons will do it only after a time a factor of ~ 40 larger, and not necessarily to the same equilibrium temperature, and both species will finally settle into local thermodynamic equilibrium with a *common* temperature only after $\sim 2000\tau_e$.

• *Plasma parameters*

We have already seen the Debye length λ_D and now wish to define what is sometimes called the *plasma parameter*. This is the reciprocal of the dimensionless combination $n_e\lambda_D^3$. Thus, the plasma parameter approximates the reciprocal of the number of electrons in the Debye sphere around a proton, say. We expect that the plasma parameter will be a small quantity, compared to 1, that is, a significant number of electrons are in the sphere. Since their mean distances are approximately $\sim n_e^{-1/3}$, the quantity $e^2 n_e^{1/3}$, which is the average pair electrostatic energy, should be very small, compared to $k_B T$, if the plasma parameter is a small number. Thus, the expansion of the exponentials in Problem 10.1 is a posteriori justified. If we take a characteristic dimension of the variation of physical properties in the conducting fluid system to be L and if $L \gg \lambda_D$, we may be confident that the electron density is to a large degree equal to the proton density or to the ion density. If the neutrality is violated, significant electric fields appear in the fluid frame, so as to neutralize it. It is instructive to understand that the violation of neutrality, that is, separation of electrons and protons, will give rise to very rapid oscillations in the effort to achieve charge neutrality, as said before. The neutralization timescale can be found easily. The associated frequency is called the *plasma frequency,* ω_P, and is different for the two charged species, due to their different mass. We have for the two species—electrons and protons, say. $\alpha = e, p$

$$\omega_{P\alpha} = \sqrt{\frac{4\pi n_e e^2}{m_\alpha}}. \tag{10.12}$$

Numerically, the associated timescale for electrons having number density of 10^{22} cm^{-3} is extremely short ($\sim 10^{-15}$ s)!

We are thus confident that for a collisionally dominated conducting fluid in LTE, if we are interested in length scales that greatly exceed ℓ and λ_D and low enough frequencies, the fluid may be perceived as neutral, albeit ionized, and the MHD description should work well, when the macroscopic velocity of the fluid **u** is approximately equal to the macroscopic velocity of the ions, because they are significantly more massive than the electrons. In contrast, the electrical *current*, if any, is carried by the electrons.

- *Gyromotion*

  Charged particles execute circular motions around the magnetic field lines, when such a field is present, due to the Lorentz force acting on them. It is of interest to examine whether these motions interfere with the MHD description. To calculate the cyclotron frequency of a single particle of mass m_α in a magnetic field B is a trivial exercise

  $$\omega_{C\alpha} = \frac{eB}{m_\alpha c},\tag{10.13}$$

  where c is the speed of light. The direction of this motion is in a circle lying in a plane perpendicular to a magnetic field line, with its center at the point in which the magnetic field line pierces the plane being the plane of the circle. To get an order of magnitude estimate here, we consider an electron moving in a magnetic field typical of the solar value (~ 1 kG) and get 3×10^9 rad/s. This value is significantly larger than the collision frequency at typical conditions (see Problem 10.2). In any case, we should not expect any macroscopic currents, as a result of gyro-motion (can you explain why?), and consequently we ignore the possible effects of such motions.

10.2   MHD One-Fluid Model Equations

10.2.1   Combining the Maxwell Equations with the FD Equations

Assumption 3

The equations are written for an inertial frame of reference, and it is assumed that the nonrelativistic approximation holds, that is, $U/c \ll 1$, where c is the velocity of light in vacuo and $U \sim L\omega_L$ is a typical macroscopic velocity in the problem. The latter can also be a typical sound speed $\tilde{c}_s$ or Alfvén speed $\tilde{c}_A$ (see below).

The above assumption allows us to slightly modify Maxwell's equations, in the way described below. In their original, unchanged form, remembering that we use vacuum values for magnetic permeability and electric permittivity, these equations are

$$\nabla \cdot \mathbf{E} = 4\pi \rho_q, \tag{10.14}$$

$$\nabla \times \mathbf{E} = = -\frac{1}{c}\frac{\partial \mathbf{B}}{\partial t}, \tag{10.15}$$

$$\nabla \cdot \mathbf{B} = 0, \tag{10.16}$$

$$\nabla \times \mathbf{E} = = \frac{4\pi}{c}\mathbf{j} + \frac{1}{c}\frac{\partial \mathbf{E}}{\partial t}, \tag{10.17}$$

where ρ_q is the electric charge density and $\mathbf{j}$ is the electric current density that in the one-fluid model of MHD results from a very slow (see below) drift of the electrons with respect to the positive ions. All the other symbols have their usual meaning. Equation (10.15) results from the Faraday induction law, and Eq. (10.17) is Ampére's law plus the *displacement current* found by Maxwell, sometimes referred to as post-Maxwellian Ampére's law. Exploiting **Assumption 3**, we may estimate the order of magnitude of the displacement current in Eq. (10.17); thus,

$$\frac{1}{c}\frac{\partial \mathbf{E}}{\partial t} \sim \frac{\tilde{E}}{c\tau}, \tag{10.18}$$

where $\tilde{E}$ is a typical value of the electric field and τ, as before, the relevant timescale. This has to be compared to the left-hand side expression

$$\nabla \times \mathbf{E} \sim \frac{\tilde{E}}{L}, \tag{10.19}$$

with L, as before, the significant length scale. Thus, the displacement current is of the order U/c times the left-hand side of this equation, because $U \sim L/\tau$. Consequently, we may neglect the displacement current, on account of assuming nonrelativistic speeds, e.g., $U/c \ll 1$, and obtain, instead of Eq. (10.17), its *pre-Maxwellian* form:

$$\nabla \times \mathbf{B} = \frac{4\pi}{c}\mathbf{j}. \tag{10.20}$$

All this means that the conduction current vastly exceeds the displacement current. Neglecting the displacement current is also consistent with the fact that in virtually all MHD models, as explained above, we are dealing with low-frequency phenomena (can you see why?). This is the only approximation we make in Maxwell's equation, and we have the original set, with (10.20) replacing (10.17), meaning that the magnetic fields are created by conduction currents only.

Regarding the velocity involved in a typical macroscopic drift of electrons, with respect to the ions, we can readily estimate it to have an almost absurdly low value. Using Ampère's law, one can get order of magnitude estimate for the conduction current $\tilde{j}$:

$$\tilde{j} \sim \frac{c\tilde{B}}{4\pi L}. \tag{10.21}$$

In Problem 10.4 it is found that for solar conditions such a current corresponds to electron drift velocity of v_e^{drift} 10^{-10} cm/s! This may be surprising, as this current is compatible with the magnetic field, which is certainly nonnegligible and in fact is very instrumental in the motion. The answer lies, of course, in the size L. In cosmic bodies L is huge and the magnetic fields are important, as we shall see, in the fluid motion. In smaller systems, as Eq. (10.21) shows, the conduction currents and drift velocities are proportionally larger. In any case, when writing the equation of motion, we shall be able to define nondimensional numbers measuring the importance of the magnetic (Lorentz) force on the conducting fluid. The smallness of the drift velocity is still another aspect of one-fluid MHD. It is certainly consistent with this model.

To derive the equations controlling the dynamics of matter, we begin with the mass conservation and Navier–Stokes equations (1.52–1.53) dividing the momentum equation by the mass density, ρ, for convenience:

$$\frac{\partial \rho}{\partial t} + \nabla \cdot (\rho \mathbf{u}) = 0, \tag{10.22}$$

$$\frac{\partial \mathbf{u}}{\partial t} + (\mathbf{u} \cdot \nabla)\mathbf{u} = \mathbf{b} - \frac{1}{\rho}\nabla P + \nu\nabla^2\mathbf{u} + \left(\nu_2 + \frac{1}{3}\nu\right)\nabla(\nabla \cdot \mathbf{u}), \tag{10.23}$$

with the tacit assumption that the viscosity coefficients are scalar constants. Splitting now the body force $\mathbf{b}$ into its magnetic contribution, $\mathbf{b}_L = \mathbf{j} \times \mathbf{B}/c$, the Lorentz force, with $\mathbf{j}$ taken from Ampère's law (10.20) plus the nonmagnetic part $\mathbf{b}_{\mathrm{nm}}$ (gravity say) and neglecting the second viscosity, as is customary for fluids in LTE, we get the following momentum equation, which is often called, the equation of motion:

$$\frac{D\mathbf{u}}{Dt} = \mathbf{b}_{\mathrm{nm}} + \frac{1}{4\pi\rho}(\nabla \times \mathbf{B}) \times \mathbf{B} - \frac{1}{\rho}\nabla P + \nu\left[\nabla^2\mathbf{u} + \frac{1}{3}\nabla(\nabla \cdot \mathbf{u})\right]. \tag{10.24}$$

Equations (10.22) and (10.24) are not a closed set for two reasons. First, which is already known from Chap. 1, there is a need to use an equation of state for P, which, in turn, calls for some form of heat or energy equation. Second, an equation for the development of the magnetic field is needed. We shall answer this need in the next subsection, but before that we make a few important remarks, regarding the Lorentz force (per unit mass), which appears in the equation of motion. This force is written in Eq. (10.24) in terms of the magnetic field and mass density only. Vector identities help one to write it in a somewhat different form

$$\mathbf{b}_L = \frac{1}{4\pi\rho}(\nabla \times \mathbf{B}) \times \mathbf{B} = -\frac{1}{\rho}\nabla\left(\frac{B^2}{8\pi}\right) + \frac{1}{4\pi\rho}(\mathbf{B}\cdot\nabla)\mathbf{B}, \qquad (10.25)$$

where $B \equiv |\mathbf{B}|$.

The above splitting of the Lorentz force allows us to understand the physical meaning of the two contributions. The first term is similar in its form to the pressure term in the usual FD equation of motion; thus, one may perceive the expression $B^2/(8\pi)$ as a *magnetic pressure*. Indeed, this is the magnetic energy density, similar to the thermal pressure, which is the internal energy density. Force is exerted on the fluid if the magnitude of the magnetic pressure varies with position, i.e., has a nonzero gradient. In contrast, the physical effect of the second term is one of a *magnetic tension* force, acting along the field lines. The magnetic pressure contribution vanishes when there is no gradient of the field strength, while the magnetic tension term vanishes if the field lines are straight. Problem 10.6 deals with some examples. Substituting Eq. (10.25) into the equation of motion (10.24) makes it more transparent physically

$$\frac{D\mathbf{u}}{Dt} = \mathbf{b}_{nm} - \frac{1}{\rho}\nabla(P + P_{mag}) + \frac{1}{4\pi\rho}(\mathbf{B}\cdot\nabla)\mathbf{B} + v\left[\nabla^2\mathbf{u} + \frac{1}{3}\nabla(\nabla\cdot\mathbf{u})\right], \qquad (10.26)$$

where $P_{mag} \equiv B^2/(8\pi)$.

We are now able to define some nondimensional numbers, whose values indicate the physical properties of an MHD flow. An important number of this sort is the so-called plasma beta, defined as

$$\beta_{pl} = \frac{\tilde{P}}{\tilde{P}_{mag}}, \qquad (10.27)$$

where tilde, as usual, indicates typical values in the flow. It should be evident that β_{pl} measures, in a reciprocal way, so to speak, the importance of the magnetic effects in an MHD flow. Just as the sound speed c_s is associated with the thermal pressure, there is also a speed associated with the magnetic pressure, called the *Alfvén speed*. We shall see later that this speed is also associated with a wave, whose MHD version we examine, but before that let us just define

$$c_A \equiv \sqrt{\frac{B^2}{4\pi\rho}}, \qquad (10.28)$$

thus

$$P_{mag} = \frac{1}{2}\rho c_A^2. \qquad (10.29)$$

Consequently, $\tilde{c}_A^2/\tilde{c}_s^2$, which is sometimes called the *Cowling number*, Co, indicates the importance of magnetic effects in an MHD flow and clearly $Co \sim \beta_{pl}^{-1}$. More specifically, it gives the ratio of magnetic to internal thermal energies. The *Alfvén number* $A = U/\tilde{c}_A$, where U is a typical fluid velocity, expresses the ratio of the of the inertial to magnetic stresses and its square the ratio of the kinetic energy to the magnetic energy. The last number, which we introduce in this subsection, is the *Alfvén Mach number*:

$$A_M \equiv \frac{\tilde{c}_A}{\tilde{c}_s}. \tag{10.30}$$

Usually, typical MHD speeds are $U \lesssim \tilde{c}_A$; thus, if $A_M \ll 1$, the incompressible approximation for our magnetofluid holds and the term which includes $\nabla \cdot \mathbf{u}$ may be neglected in the equation of motion (10.26), similarly to nonmagnetic incompressible flow.

10.2.2   The Induction Equation and the Fluid Heat Equations in MHD

In the effort to close the system of equations, we search for an additional relation which the magnetic field satisfies. It is clear that $\nabla \cdot \mathbf{B} = 0$ cannot be sufficient (think why); however, we remember that **Assumption 3** guarantees nonrelativistic motion, and thus we can transform the electromagnetic field to the fluid frame *locally*, using $\mathbf{u}$ as the velocity for the Galilean transformation. Denoting the values in the fluid frame by a prime, we get

$$\mathbf{B}' = \mathbf{B}, \tag{10.31}$$

$$\mathbf{j}' = \mathbf{j} \tag{10.32}$$

$$\mathbf{E}' = \mathbf{E} + \frac{1}{c}\mathbf{u} \times \mathbf{B}. \tag{10.33}$$

Now, we remember that we are dealing with a conducting fluid and the driver of the current is the electric field; consequently, we need an equation providing a connection between $\mathbf{j}$ and $\mathbf{E}$.

Assumption 4

We adopt the simplest form of the electric current–field relation, which is also consistent with the one-fluid MHD assumption. It is the linear Ohm's law which holds in the fluid frame

$$\mathbf{j}' = \sigma\mathbf{E}', \tag{10.34}$$

where the conductivity σ_{el} is assumed in this simplest case to be a constant scalar. Problem 10.7 uses simple arguments to obtain its value

$$\sigma_{el} = \frac{n_e c^2}{m_e v_c}, \tag{10.35}$$

where the symbols have their usual meaning, remembering that v_c is the mean electron–proton collisional frequency. More involved versions of Ohm's law complicate the situation, and the MHD model that arises is usually not of the one-fluid kind. Now, by effecting a Galilean transformation of $\mathbf{E}'$ into $\mathbf{E}$, by using (10.33) and substituting the simplest Ohm's law expression (10.34), we get

$$\mathbf{E} = \frac{\mathbf{j}}{\sigma_{el}} - \frac{\mathbf{u}}{c} \times \mathbf{B}. \tag{10.36}$$

Inserting now the expressions from Faraday's law (10.15) and pre-Maxwellian Ampére's law (10.20), replacing the triple vector product by its equivalent expression using a vector identity and remembering that $\nabla \cdot \mathbf{B} = 0$, we finally obtain the *induction equation* for $\mathbf{B}$

$$\frac{\partial \mathbf{B}}{\partial t} = \nabla \times (\mathbf{u} \times \mathbf{B}) + \eta_m \nabla^2 \mathbf{B}, \tag{10.37}$$

where the *resistivity* η_m is defined and expressed in the following way

$$\eta_m \equiv \frac{c^2}{4\pi \sigma_{el}} = \frac{c^2 m_e v_c}{4\pi n_e e^2} = \frac{c^2 v_c}{\omega_{Pe}}, \tag{10.38}$$

looking, by analogy, as if it is a "magnetic kinematic viscosity," but it should not be confused with the dynamic fluid viscosity η. In the spirit of previous assumptions, we take the coefficient of resistivity to be a constant. The numerical value of the resistivity in any physical units cannot hint at the strength of its physical effects. Instead, we try to achieve a physically meaningful description by making Eq. (10.37) nondimensional. This introduces an important MHD dimensionless number, the *magnetic Reynolds number*, Rm:

$$\text{Rm} \equiv \frac{LU}{\eta_m} = LU \frac{4\pi \sigma_{el}}{c^2}, \tag{10.39}$$

where L and U are typical length and fluid velocity scales of the problem at hand. Returning to the induction equation, one can immediately weigh the importance of its two terms on the right-hand side, by making the equation nondimensional. Using the typical values of L and U for length and velocity scales, respectively, this equation reads

$$\frac{\partial \mathbf{B}}{\partial t} = \nabla \times (\mathbf{u} \times \mathbf{B}) + \frac{1}{\text{Rm}} \nabla^2 \mathbf{B}. \tag{10.40}$$

It is not too difficult to understand the physical meaning of the following limits of the induction equation.

- The limit $\text{Rm} \gg 1$.
  This happens for fluids with very low resistivity, i.e., high conductivity for given L and U, remembering that the magnitude of these scales influences the size of Re as well. So if we go to the limit of Rm being so large, that the term containing it in the denominator can be neglected, we get

$$\frac{\partial \mathbf{B}}{\partial t} + \nabla \times (\mathbf{B} \times \mathbf{u}) = 0, \tag{10.41}$$

which gives an interesting physical phenomenon. This last equation is completely analogous to Eq. (2.24), with the vorticity ω in that equation, replaced here by $\mathbf{B}$, the magnetic field. Equation (2.24) was one of the expressions of Kelvin's theorem, which states that in the fluid motion the velocity circulation or the vorticity flux through any material surface $S(t)$ is conserved, i.e.,

$$\frac{d}{dt} \int_{S(t)} \omega \cdot \hat{\mathbf{n}} \, dS = 0. \tag{10.42}$$

As the magnetic field, in the case of infinite magnetic Reynolds number, satisfies in MHD the same equation as the vorticity, for an inviscid barotropic flow in FD, an identical flux conservation holds:

$$\frac{d}{dt} \int_{S(t)} \mathbf{B} \cdot \hat{\mathbf{n}} \, dS = 0. \tag{10.43}$$

The above phenomenon (10.43) is, in MHD, referred to as *field, or flux, freezing* and is a very important property of MHD flows with $\text{Rm} \gg 1$. For example, a heuristic argument can be made for very high Rm flows to understand that if there is magnetic field, say, uniform, in an initial state, then a strong compression, due to gravitational collapse, for example, creates an opposing magnetic pressure, which often halts the collapse. This is, by the way, one of the issues in understanding star formation. As the similarity of the equations for the vorticity and magnetic field holds in special conditions, it is recommended for the reader to review the properties of vorticity and vortex tubes in Sect. 2.5. As we shall see, they are similar in many ways to *magnetic* flux tubes.

- The opposite limit $\text{Rm} \ll 1$.
  Very small magnetic Reynolds number allows one to neglect the $\nabla \times (\mathbf{u} \times \mathbf{B})$ term with respect to the diffusive term in Eq. (10.40). Thus, we are left with a diffusion equation for the magnetic field. In dimensional units, it is

$$\frac{\partial \mathbf{B}}{\partial t} = \eta_m \nabla^2 \mathbf{B}, \tag{10.44}$$

where the resistivity plays the role of the diffusion coefficient. This is a well-studied equation, and we can write its general solution (see Problem 10.9) for an initial condition $\mathbf{B}_0(x)$, where for simplicity we take a one-dimensional case in an infinite domain, as

$$\mathbf{B}(x,t) = \sqrt{\frac{1}{4\pi\eta_m}} \int_{-\infty}^{\infty} dx' \mathbf{B}_0(x') \exp\left[-\frac{(x-x')^2}{4\eta_m t}\right]. \tag{10.45}$$

We shall not dwell here on the diffusion equation, but rather deduce from its solution the typical diffusion time for the magnetic field. This clearly is

$$\tau_{\text{Bdiff}} \sim \frac{L^2}{\eta_m}, \tag{10.46}$$

where L is a typical length scale. This expression allows one to estimate the timescale for the magnetic field to decay by diffusion as expressed here in a conducting fluid. The general conclusion from Problem 10.8 is that the magnetic field loss timescale depends crucially on the system. Generally large astrophysical systems have diffusion times of the magnetic field that are larger than their, and sometimes even the universe's, lifetime. Laboratory conditions, in which the size of the system is of the order of meters, say, usually give rise to quick dissipation of the magnetic field. For example, a cube of molten iron will lose its magnetic field in a matter of seconds. In the Earth core ($\tau_B \sim 10^4 - 10^5$ s) or a low mass magnetic star atmosphere ($\tau_B \sim 10^{14}$ s) are intermediate systems in this sense, as they should relatively rapidly lose their magnetic fields by diffusion, if some kind of *dynamo* action would not strengthen the field. We defer till later any discussion of the difficult topic of MHD dynamos, but remember that in any case in such systems, the first term in the induction equation cannot be neglected.

A rule of thumb can be formulated that the diffusive term rules in the laboratory, while field freezing is a good approximation in extended astrophysical bodies, at least in short enough timescales. Before turning to the discussion of MHD equilibria and simple flows, we wish to introduce yet another nondimensional number of significance, which measures the relative magnitude of fluid viscosity to "magnetic viscosity" so to speak, and actually resistivity coefficient. This number, termed the *magnetic Prandtl number* (Pm) and defined as

$$\text{Pm} \equiv \frac{\nu}{\eta_m} = \frac{\text{Rm}}{\text{Re}}, \tag{10.47}$$

can be expressed in c.g.s units using the values in reference [4] of this chapter's *Bibliographical Notes* as

$$\text{Pm} = 5.910 \times 10^{-29} \frac{T^4}{\rho}, \tag{10.48}$$

where T is in degrees K and ρ in c.g.s. It is apparent that this is a small number unless the temperature of the fluid is very high and/or the density very low. We shall give now a simple example of an MHD flow which demonstrates some of the principles set out above.

10.2.2.1   Example: Hartmann Flow

Consider a steady parallel flow of an incompressible viscous and resistive conducting fluid in the space between two parallel planes, actually two rigid plates. This is similar to plane Poiseuille flow in FD (see Problem 3.2), but here we add a uniform magnetic field applied in the direction perpendicular to the plates. There are no other body forces. Assume that the plates are parallel to the $x - y$ plane and note that in our hydrodynamical example in Chap. 2, we used $x - z$ as the plate plane. Let a uniform magnetic field be applied only in the z direction: $B_0\hat{\mathbf{z}}$. The problem is considered y-independent and so are all the functions relevant to it, that is, it possesses a y-symmetry which does not allow any vector to have a $\hat{\mathbf{y}}$ component, except the induced electric field and the current which, as it happens, have *only* this component, as is found in Problem 10.12.

The relevant equations, i.e., the MHD equation of motion (10.26) and the induction equation (10.37), are supplemented by the always correct $\nabla \cdot \mathbf{B} = 0$ and the incompressibility condition $\nabla \cdot \mathbf{u} = 0$. When looking for a solution, we use the same reasoning as was used in plane Poiseuille flow. The solution ansatz has a velocity distribution $\mathbf{u}(\mathbf{x},t) = u_x(z)\hat{\mathbf{x}}$, and the same functional form is assumed also for the longitudinal magnetic field $B_x(z)$, which results only from the fluid motion. $\nabla \cdot \mathbf{B} = 0$ implies then, together with what has already been assumed and symmetry arguments, that the vertical component of the magnetic field is constant, that is, $B_z(\mathbf{x},t) = B_0$. The pressure, however, must depend on x as it drives the flow. Consider now the z-component of the magnetic equation of motion (10.26), with the above assumption on the solution substituted:

$$-\frac{1}{\rho}\frac{\partial}{\partial z}\left(P + P_{\text{mag}}\right) = 0, \tag{10.49}$$

where $P_{\text{mag}} \equiv B^2/(8\pi)$.

The pressure P may be considered as x-dependent; we remember that there is no dependence on y of any of the functions, but because of the uniformity in the x direction, dP/dx is a constant. We shall call the pressure difference over a typical distance L along x as ΔP. We are thus left with the following pair of equations, since $\nabla \cdot \mathbf{u} = 0$ is satisfied identically

$$B_0\frac{du_x}{dz} + \eta_m\frac{d^2B_x}{dz^2} = 0, \tag{10.50}$$

$$v \frac{d^2 u_x}{dz^2} + \frac{B_0}{4\pi\rho_0} \frac{dB_x}{dz} = P_d, \qquad (10.51)$$

where $\eta_m = c^2/(4\pi\sigma)$ is the magnetic resistivity, v is the kinematic viscosity, and ρ_0 is the constant density of the fluid. The constant term

$$P_d \equiv -\Delta P/(L\rho_0)$$

is referred to as the pressure *drop*. These equations have to be solved subject to the boundary conditions $u_x = 0$, $B_x = 0$ at the walls (plates) $z = \pm w$ (2w is the width of the channel). The rigid walls are assumed to be insulating. Define now a length scale which will serve as a measure for the boundary layer in Hartman flow

$$\delta = \frac{c_A}{B_0} \sqrt{\frac{\rho\eta_m}{\sigma_{el}}}, \qquad (10.52)$$

where $c_A \equiv \sqrt{B_0^2/(4\pi\rho)}$ is the representative, appropriate Alfvén speed. Together with the channel half-width, we can define a nondimensional number for this flow, which will play an important role in the solution's self-similarity, namely, the *Hartman number* (Ha):

$$\mathrm{Ha} = \frac{w}{\delta} = \frac{wB_0}{c_A} \sqrt{\frac{\sigma_{el}}{\rho\eta_m}}. \qquad (10.53)$$

One can easily find from differentiating equation (10.50) and substituting into it from Eq. (10.51) a third-order linear differential equation for u_x, whose solution for $u_x(\pm w) = 0$ is

$$u_x(z) = U_0 \left[1 - \cosh\left(\mathrm{Ha}\frac{z}{w}\right) / \cosh(\mathrm{Ha})\right]. \qquad (10.54)$$

Similarly, a linear differential equation can be found for $B_x(z)$ and solved for the insulating boundary condition, and finally, the value of U_0, the velocity at the mid-plane, can be found in terms of the constant P_d; see Problem 10.10 (Fig. 10.2).

A moment of reflection leads to the realization that the Hartman number gives an estimate of how much the magnetic field affects the velocity profile, which for $B_0 = 0$ is determined by the regular fluid viscosity. Clearly, if z is measured in units of w, it is the Hartman number that determines a family of self-similar flows. It is also obvious that for $\mathrm{Ha} \to 0$ we obtain the nonmagnetic plane Poiseuille flow:

$$u_x(z) = U_0 \left(1 - \frac{z^2}{w^2}\right). \qquad (10.55)$$

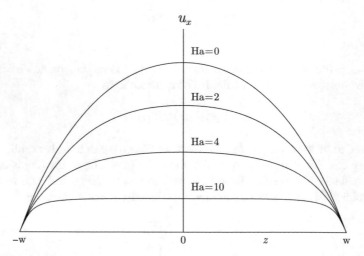

Fig. 10.2 Velocity $u_x(z)$ profiles in Hartman flow for different values of the Hartman number for a flow in a channel of half-width w under a fixed pressure drop P_d. The maximal value of the velocity U_0 for each case can be straightforwardly expressed in terms of the pressure drop

In the opposite case, i.e., Ha $\gg$ 1, the solution for the velocity is

$$u_x(z) = \left[1 - \exp\left(-\frac{\mathrm{w} - |z|}{\delta} \right) \right], \qquad (10.56)$$

that is, for stronger and stronger magnetic fields, the velocity loses its Poiseuille, parabolic shape profile, and it flattens for a growing size of the channel cross section, diminishing rapidly to zero only in the δ narrow boundary layers near the walls in accord with Eq. (10.56).

10.2.3   *Flux Tubes and Their Properties*

The concept of a flux tube arises naturally when one considers the magnetic field lines and imagines the volume encompassed by adjacent lines. There is a tacit good reason (does it have physical content?) for the similarity between magnetic flux tubes and vortex tubes, examined in Sect. 2.5. Equation (2.24) which reads

$$\frac{\partial \omega}{\partial t} + \nabla \times (\omega \times \mathbf{u}) = 0, \qquad (10.57)$$

and holds for an inviscid barotropic fluid, looks essentially similar to the field freezing statement, valid for zero resistivity

$$\frac{\partial \mathbf{B}}{\partial t} + \nabla \times (\mathbf{B} \times \mathbf{u}) = 0, \qquad (10.58)$$

Fig. 10.3 A schematic drawing of a flux tube, "cutting" surfaces S_1 and S_2 in two parallel planes

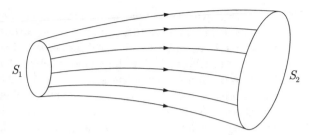

where ω and $\mathbf{B}$ are vorticity and the magnetic field, respectively. The identical equations that these two vector fields satisfy in non-dissipative flows, where in the case of the vorticity equation barotropic conditions are also needed, point to the formal similarity of flux tubes to vortex tubes. The vortex tubes satisfy the Helmholtz theorems on vortex motion, given in Sect. 2.5. We shall now see what can be said about the behavior of magnetic flux tubes. As should be obvious by now, a field line of $\mathbf{B}$ is obtained by the differential equations

$$\frac{dx}{B_x} = \frac{dy}{B_y} = \frac{dz}{B_z}, \tag{10.59}$$

where we have chosen for convenience to use the notation (x, y, z) for the Cartesian triad (x_1, x_2, x_3). A magnetic *flux tube* is the volume enclosed by the set of field lines, which intersect a simple closed curve (see Fig. 10.3). The *strength F* of a flux tube is the flux crossing the surface S, interior to the curve that defines the tube

$$F = \int_S \mathbf{B} \cdot \hat{\mathbf{n}} \, dS, \tag{10.60}$$

where F is always nonnegative due to the choice of the direction of the surface S, always in the direction of $\mathbf{B}$. Some of the basic properties of flux tubes are summarized here, and we apologize for mere repetition of some of the corresponding Helmholtz theorems on vortex tubes. A number of properties are almost self-evident.

1. *The strength of a flux tube remains constant along its length.*
   Taking an arbitrary flux tube, and closing it by two arbitrary different surfaces S_1 and S_2, we obtain a closed volume. Integrating $\nabla \cdot \mathbf{B} = 0$ over this volume gives

$$0 = \int_{\mathcal{V}} \nabla \cdot \mathbf{B} \, d^3x = \int_{\partial \mathcal{V}} \mathbf{B} \cdot \hat{\mathbf{n}} \, dS = \int_{S_2} \mathbf{B} \cdot \hat{\mathbf{n}} \, dS - \int_{S_1} \mathbf{B} \cdot \hat{\mathbf{n}} \, dS \tag{10.61}$$

   where the Gauss divergence theorem has been used. Thus, the strength of the tube is equal in two arbitrary points, i.e., is constant along the tube.
2. *The mean field strength of a flux tube is inversely proportional to its thickness.*
   Indeed, Eq. (10.60) may be used to define the mean field at a level where the tube thickness is S_0, say. $F = \int_{S_0} \mathbf{B} \cdot \hat{\mathbf{n}} \, dS \equiv \langle B \rangle S_0$ and the result follows.

3. *Compression of a flux tube while keeping its length constant increases the magnetic field and the density in the same proportion.*
   To prove this, consider a cylindrical flux tube and change its radius from R_0 to λR_0, $\lambda < 0$, without letting the cylinder height change, causing compression. The conservation of matter gives $\rho \propto \lambda^2 \rho_0$ and the conservation of magnetic flux, similarly $B \propto \lambda^2 B_0$, where the quantities with 0 subscript are the initial values. Thus, $B/\rho = $ const is correct in any given flux tube.
4. *Stretching of a flux tube, without compression, increases the field.*
   There is no compression in this case, so the volume and density of this flux tube part is unchanged. It becomes longer and thus thinner. The conservation of flux then gives an increase in the magnetic field.

10.3   Equilibrium Configurations

Equilibrium magnetohydrostatic configurations are, in principle, possible. The physical conditions required are that the mechanical forces are balanced by magnetic stresses. Examples of such equilibrium states exist in nature, e.g., sunspots, magnetic filaments in nebulae. In fusion research, the ability to hold a laboratory plasma in an equilibrium (or near equilibrium) state is a topic of intensive study. In the previous section, we defined the Alfvén speed as

$$c_A^2 \equiv \frac{2P_{\text{mag}}}{\rho} = \frac{B^2}{4\pi\rho}. \tag{10.62}$$

We shall see below that just as the sound speed plays a crucial role in sound waves so the Alfvén speed plays a similar role in MHD waves. We stress now that a condition for an equilibrium configuration to be possible is that the fastest speed that the magnetic field can drive the material to the static state, call it $\tau_B \sim L/U$, is actually $\tau_B \sim L/c_A$. This time must be shorter than the diffusion time of the magnetic field, which can be estimated from the induction equation (10.37), as found above $\tau_{B\text{diff}}$. This helps to define yet another nondimensional number, the *Lundquist number* Lu, which is

$$\text{Lu} \equiv \frac{\tilde{c}_A L}{\eta_m} = \frac{\text{Rm}}{\text{A}}. \tag{10.63}$$

Thus, the Lundquist number gives the ratio of magnetic dynamic phenomena to the magnetic field diffusion. Lu plays a role in a necessary condition for establishing an equilibrium configuration, but it is not a sufficient condition. This is the case because of a simple reason—in MHD the equilibrium configuration is more often unstable than stable.

Let us consider perhaps the simplest case of an equilibrium configuration, one in which the conductivity is extremely large, or more elegantly Lu $\to \infty$. We also assume that there are no body forces, save the Lorentz force. The appropriate equations for this problem, for a constant density case, which we set to unity are

$$\nabla \cdot \mathbf{B} = 0, \tag{10.64}$$

$$\nabla P = \frac{1}{c}\mathbf{j} \times \mathbf{B}, \tag{10.65}$$

$$\mathbf{j} = \frac{c}{4\pi}\nabla \times \mathbf{B}, \tag{10.66}$$

where we chose to write the equations so as to explicitly bring out the current density $\mathbf{j}$. Scalar multiplication of equation (10.65) by $\mathbf{B}$ and by $\mathbf{j}$ correspondingly yields

$$(\mathbf{B} \cdot \nabla)P = 0, \tag{10.67}$$

$$(\mathbf{j} \cdot \nabla)P = 0, \tag{10.68}$$

where P as before is the thermal pressure alone. These equations yield immediately a very important concept, that of *magnetic surfaces*. Constant pressure surfaces $P(\mathbf{x}) = $ const, assuming that some exist, are by virtue of Eqs. (10.67–10.68) magnetic surfaces. The magnetic field lines and the current lines lie on them. The magnetic surfaces are, in principle, possible boundaries of equilibrium configurations.

10.3.1 Example of Cylindrical Unbounded Configuration (Pinch)

Consider an unbounded cylindrical equilibrium MHD configuration, that is, one of unlimited length and which is uniform along it; we choose this direction as the z-axis. The configuration we shall present goes by the name of a *pinch*. Now express the dependence of all functions in cylindrical coordinates (r, z, φ). It is easy to see that the radial components B_r and j_r must be zero (show it!). The components of Eqs. (10.66) then give

$$j_\varphi = -\frac{c}{4\pi}\frac{dB_z}{dr} \tag{10.69}$$

$$j_z = \frac{c}{4\pi r}\frac{d}{dr}\left(rB_\varphi\right). \tag{10.70}$$

Integration of the second of these gives

$$B_\varphi(r) = \frac{2}{cr}J(r), \tag{10.71}$$

where we define the net or total z direction current $J(r)$ as

$$J(r) \equiv 2\pi \int_0^r j_z r dr. \tag{10.72}$$

Equation (10.65) then becomes

$$\frac{dP}{dr} = -\left\{ \frac{1}{2\pi c^2 r^2} \frac{d}{dr} \left[J^2(r) \right] + \frac{1}{8\pi} \frac{d}{dr} \left[B_z^2(r) \right] \right\}. \tag{10.73}$$

Examine two possible cases: the first having $B_z = 0$ and the other, in which $B_\varphi = 0$ and $j_z = 0$. In the first case, multiplying equation (10.73) by r^2 and integrating on r from zero to the pinch radius, a, say, we get the equilibrium condition

$$2\pi \int_0^a \frac{dP}{dr}(r) r^2 dr = \frac{J^2(a)}{2c^2}, \tag{10.74}$$

where $J(a)$ is the total current though the pinch. In the second case, we get from the same force equation

$$P(r) + \frac{B_z^2(r)}{8\pi} = \frac{B_{\text{out}}^2}{8\pi}, \tag{10.75}$$

where B_{out} is the $\hat{z}$ directed magnetic field, just outside the pinch and parallel to it, and we have assumed $P(0) = 0$.

10.3.2   The Grad–Shafranov Equation

Consider an axially symmetric MHD equilibrium for which, as shown in Problem 10.15, the following equations hold:

$$B_r = -\frac{1}{2\pi r} \frac{\partial \psi}{\partial z}, \quad B_z = \frac{1}{2\pi r} \frac{\partial \psi}{\partial r} \tag{10.76}$$

$$j_r = -\frac{1}{2\pi r} \frac{\partial J}{\partial z}, \quad j_z = \frac{1}{2\pi r} \frac{\partial J}{\partial r} \tag{10.77}$$

where $\psi(r,z)$ is the total magnetic flux and $J(r,z)$ the total current, both vertical, through a circle of radius r at height z, and perpendicular to the z-axis. Equations (10.76–10.77) imply (you may want to use Eqs. (10.68)-(10.69)) that ψ and J are constant on the magnetic surfaces ($P = $const) and thus J and P can be expressed as a function of ψ alone, that is,

$$P = P(\psi), \quad J = J(\psi). \tag{10.78}$$

Now, similarly to Problem 10.15, we may write also the φ component of Eqs. (10.64)–(10.66) in cylindrical coordinates. This gives

$$B_\varphi = \frac{2J}{cr}, \quad j_\varphi = -\frac{c}{8\pi^2 r}\left(\frac{\partial^2 \psi}{\partial r^2} - \frac{1}{r}\frac{\partial \psi}{\partial r} + \frac{\partial^2 \psi}{\partial z^2}\right). \tag{10.79}$$

Finally, substituting this and Eqs. (10.76–10.77) in the r-component of equation (10.65) and in the above equation, we obtain the celebrated *Grad–Shafranov* equation

$$\frac{\partial^2 \psi}{\partial r^2} - \frac{1}{r}\frac{\partial \psi}{\partial r} + \frac{\partial^2 \psi}{\partial z^2} = -16\pi^3 r^2 \frac{dP}{d\psi} - \frac{8\pi^2}{c^2}\frac{dJ^2}{d\psi}, \tag{10.80}$$

which gives some equilibrium configurations which are possible, in principle, if only particular functions $P(\psi)$ and $J(\psi)$ are specified. The magnetic field and the current follow from Eqs. (10.76–10.77) and (10.79).

10.3.2.1 Example

The Grad–Shafranov equation has many simple solutions, which can be studied analytically. It is a *linear* equation, if $dP/d\psi = a$ and $dJ^2/d\psi = b$, with a,b constants. By applying straightforward algebra and calculus, one can check that in this case the expression

$$\psi(r,z) = \alpha_z z^2 + \alpha_R(r^2 - R^2)^2, \tag{10.81}$$

with the α_i conveniently defined new constants, is a solution. R is a parameter, whose meaning will become clear soon. Explicitly, the constants are

$$\alpha_z \equiv -\frac{4\pi^2}{c^2}\frac{dP}{d\psi}, \qquad \alpha_R \equiv -2\pi^3 \frac{dP}{d\psi}. \tag{10.82}$$

It is possible to show that the surfaces $\psi =$const in this kind of a configuration are *toroids*, that is, topological distortions of tori, i.e., bodies having a doughnut-like form. The solutions (10.81) are a family of toroidal magnetic surfaces lying one inside the other. The innermost surface degenerates into a circle, called the *magnetic axis*, and is given by $z = 0, r = R$.

A detailed discussion of the solutions of the Grad–Shafranov equation and their properties is clearly outside the scope of an FD book. We shall be satisfied instead in a largely qualitative rudimentary discussion on few more features of the solutions family given above: magnetic surfaces as nested toroids. The pressure at the outermost toroid is minimal (in fact, zero), and it gradually increases to a maximum value at the magnetic axis ($z = 0, r = R$). When r is close to R, the solution (10.81) takes the approximate form

$$\psi(r,z) \approx \alpha_z z^2 + 4\alpha_R R^2 (r^2 - R^2). \tag{10.83}$$

In particular, if $a_z = 4R^2 a_R$, the toroids have a circular meridional cross section, that is, they are tori. Can you show that if a_z and a_R do not have the same sign, the magnetic surfaces close to the magnetic axis intersect the meridional plane along a hyperbola? This destroys the equilibrium configuration near the magnetic axis and there is no equilibrium possible.

10.3.3 MHD Equilibria in a Gravitational Field

So far we have treated cases without any body force, except for the magnetic one. In some cases, e.g., in astrophysical systems, the gravitational force cannot be ignored. Thus, we have to incorporate in the set of Eqs. (10.64)–(10.66), a term arising from a gravitational potential Φ. The appropriate equations for this problem are thus

$$\nabla \cdot \mathbf{B} = 0 \tag{10.84}$$

$$\nabla P + \rho \nabla \Phi = \frac{1}{c} \mathbf{j} \times \mathbf{B} \tag{10.85}$$

$$\mathbf{j} = \frac{c}{4\pi} \nabla \times \mathbf{B}. \tag{10.86}$$

In order to make some progress, it is convenient to assume that there is also a cooling–heating equilibrium and the gas is nearly isothermal. The assumption of perfect gas yields the following relation between the pressure and density $P = C\rho$, where $C = (\mathscr{R}/\mu)T_0$ is a positive constant and T_0 is the uniform, constant temperature. This allows us to write

$$\nabla P + \rho \nabla \Phi = \nabla P + \frac{P}{C} \nabla \Phi = \exp\left(-\frac{\Phi}{C}\right) \nabla Q, \tag{10.87}$$

where $Q \equiv P \exp(\Phi/C)$.

Following A. Dungey, who presented this way of reasoning in 1953, in an effort to make some analytical progress, we limit now ourselves to a two-dimensional case in the $y - z$ plane. As a result of equation $\nabla \cdot \mathbf{B} = 0$, we may introduce a vector potential $\mathbf{A}$, such that $\mathbf{B} = \nabla \times \mathbf{A}$. Since $\mathbf{B}$ lies in the $y - z$ plane, one may choose for $\mathbf{A}$ to have only an x component, $\mathbf{A} = A\hat{\mathbf{x}}$, and obtain (show it!) $\mathbf{B} = \hat{\mathbf{x}} \times \nabla A$. Now, Eq. (10.86) can be written as

$$\mathbf{j} = -\frac{c}{4\pi} \hat{\mathbf{x}} \nabla^2 A, \tag{10.88}$$

(see Problem 10.18). Using the second result of that problem finally transforms the force equation, for this case, into

$$-\frac{1}{4\pi}(\nabla^2 A)\nabla A = \exp\left(-\frac{\Phi}{C}\right)\nabla Q. \tag{10.89}$$

This means that the gradients of A and Q are parallel and the corollary is the functional relation $Q = Q(A)$. Thus, the above equation may be written as a scalar quasi-linear Poisson-like equation:

$$\nabla^2 A = -4\pi \exp\left(-\frac{\Phi}{C}\right)\frac{dQ}{dA}. \tag{10.90}$$

The idea now is that if one supplies arbitrary smooth functions $Q = Q(A)$, then each of them would provide a magnetostatic configuration, provided a pair of two-dimensional Poisson equations are solved, for given Φ, or calculated from ρ and the above equation for A (see however Problem 10.20).

10.3.4 Force-Free Fields

There are cases in which the magnetic configuration is such that a magnetic *force-free* configuration is a good approximation, that is, the condition $\mathbf{j} \times \mathbf{B} = 0$ holds in it. Problem 10.21 indicates that this indeed is a good approximation for low β_{pl}, that is, a situation in which the magnetic pressure is dominant over the thermal one. This is a typical case of the tongue in cheek rule that if a certain term in an equation is too big, then that it must be zero. Force-free condition demands that

$$\mathbf{j} \parallel \mathbf{B} \implies \nabla \times \mathbf{B} = \alpha(\mathbf{x})\mathbf{B}, \tag{10.91}$$

because neither $\mathbf{j}$ nor $\mathbf{B}$ can be zero, which leads to a trivial and uninteresting case. α is an arbitrary function of position. In the $\nabla \times$ free case, i.e., $\mathbf{j} = 0$, $\alpha = 0$, the configuration is also current free. Now, for $\alpha \neq 0$ in the static state $\nabla \cdot \mathbf{j} = 0$. But $\nabla \cdot \mathbf{B} = 0$ as well, so it follows that (see Problem 10.22)

$$\mathbf{B} \cdot \nabla \alpha = 0, \tag{10.92}$$

that is, α is constant along field lines.

10.3.4.1 Example: Twisted Field Lines

Consider a constant magnetic field, in a constant thermal pressure region, that is initially in the vertical direction ($B_0\hat{\mathbf{z}}$), and assume that it is twisted azimuthally by an angle ϑ per unit length, so that a field line becomes a helix with $B_\varphi = \vartheta r B_z$. Now, the total force equation (10.65) is satisfied for $P = $ const., provided that

$$B_z = \frac{B_0}{\sqrt{1 + \vartheta^2 r^2}} \tag{10.93}$$

$$B_\varphi = \frac{\vartheta r B_0}{\sqrt{1 + \vartheta^2 r^2}}. \tag{10.94}$$

Note that φ is the angular coordinate in a cylindrical coordinate system, while ϑ, as defined above, is the twist angle per unit length. This is a case in which $\mathbf{j} \parallel \mathbf{B}$ and so $\mathbf{j} \times \mathbf{B} = 0$ and is a simple example of a force-free field.

Returning now to the general discussion of force-free fields, we notice that the second equation in (10.91) implies that if one follows a field line, the neighboring field lines curl around this line. A force-free field is thus essentially a twisted field. There are perhaps two leading configurations of force-free fields. These are:

1. A torus confined by external pressure. The field lines spiral on the surface of the torus and are also present inside it, surrounding a central field line at the axis of the torus. For α not constant everywhere, we have nested tori, labelled by α =constant. This type of configuration has been usually used in laboratory efforts to obtain controlled fusion.
2. Essentially two types, albeit of the same sort, of force-free fields are related to the Sun's atmosphere, for example, a field that is generated in the inner, high β_{pl} regions, but extends well into the low β_{pl} outer region. It is assumed that outside the field is initially $\nabla \times$ free. The emerging fields can be twisted by vortices on the surface of the Sun or perhaps having their "feet" pushed together by subsurface motions.

10.4   MHD Waves

We have already devoted a whole Chap. 4 to FD waves in constant density and Boussinesq fluids and in an additional Chap. 6 to detailed discussions of waves in compressible fluids, where linear (acoustics) and nonlinear (shocks) waves were described. In this section we shall discuss waves in a fluid in which magnetic effect is important, within the MHD approximation. Naturally, we will uncover and discuss effects that go beyond the nonmagnetic case; however, we shall use the basic definition and results on waves in fluids that were already described in considerable detail in Chaps. 4 and 6.

Consider a nonrotating, inviscid, vertically stratified magnetofluid in equilibrium in a fixed gravitational field inducing acceleration directed in the $-\hat{\mathbf{z}}$ direction and subject to a uniform magnetic field $\mathbf{B} = B_0\hat{\mathbf{z}}$. By choosing this configuration, we abandon the simplest one, having no gravity, and endowed with a constant density and pressure, but remain in a simple enough setting. The additional, simplifying assumptions are thermal, the atmosphere is isothermal, and the fluid obeys the perfect gas EOS. This example is essentially similar to the one given in Chap. 6, where we found the way sound waves propagate in a stratified medium (see the example following Eq. (6.36)). The unperturbed equilibrium variables are marked by a zero subscript, which satisfy the hydrostatic equation

$$\frac{dP_0}{dz} + \rho_0 g = 0, \tag{10.95}$$

where the gravitational force per unit mass is $\mathbf{b}_{\mathrm{grav}} = -g\hat{\mathbf{z}}$. Evidently, the solutions for the density and pressure are

$$P_0(z) = P_s e^{-z/\Lambda_z}, \qquad \rho_0(z) = \rho_s e^{-z/\Lambda_c}, \tag{10.96}$$

where the constant P_s and ρ_s are the not independent boundary conditions at $z = 0$ and the additional constant Λ_c, the vertical scale height, is

$$\Lambda_c = \frac{P_0}{\rho_0 g} = \frac{\mathscr{R}}{\mu T_0 g}, \tag{10.97}$$

as the equation of state is $P = (\mathscr{R}/\mu)\rho T$. Note that in Chap. 6 a similar problem was solved, but there the scale height was called H_z (see Eq. (6.38)) and not Λ_c as here. We hope that this will not cause any unnecessary confusion. Impose now small departures from the equilibrium and mark them by prime, with the understanding that these departures are small in the sense that their squares are neglected, that is, we linearize the equations in these perturbations. Write

$$\mathbf{u} = \mathbf{0} + \mathbf{u}', \quad \rho = \rho_0 + \rho', \quad P = P_0 + P', \quad \mathbf{B} = \mathbf{B}_0 + \mathbf{B}' \tag{10.98}$$

and linearize the basic equations: continuity (1.52), motion (10.24), and all other relevant magnetohydrodynamical relationships given in this chapter when we are also assuming an adiabatic relation between the pressure and density *perturbations*. This is usual for fast enough perturbations as we have seen several times in this book. The following set emerges:

$$\frac{\partial \rho'}{\partial t} + (\mathbf{u} \cdot \nabla)\rho_0 + \rho_0(\nabla \cdot \mathbf{u}) = 0, \tag{10.99}$$

$$\rho_0 \frac{\partial \mathbf{u}}{\partial t} = -\nabla P' + \frac{1}{4\pi}(\nabla \times \mathbf{B}') \times \mathbf{B}_0 - \rho' g\hat{\mathbf{z}}, \tag{10.100}$$

$$\frac{\partial P'}{\partial t} + (\mathbf{u} \cdot \nabla)P_0 = c_s^2 \left[\frac{\partial \rho'}{\partial t} + (\mathbf{u} \cdot \nabla)\rho_0\right] = 0, \tag{10.101}$$

$$\frac{\partial \mathbf{B}'}{\partial t} = \nabla \times (\mathbf{u} \times \mathbf{B}_0), \tag{10.102}$$

$$\nabla \cdot \mathbf{B}' = 0. \tag{10.103}$$

We have dropped the prime from the velocity perturbation, for economy of notation, but without danger of confusion. The sound speed follows from the equation of state

$$c_s^2 = \gamma \frac{P_0}{\rho_0} = \gamma \frac{\mathscr{R}}{\mu} T_0, \tag{10.104}$$

where γ is the adiabatic exponent.

Problem 10.31 consists of the algebraic and differential manipulations at the end of which the above set (10.99–10.103) is reduced to the following single linear equation for $\mathbf{u}$:

$$\frac{\partial^2 \mathbf{u}}{\partial t^2} = c_s^2 \nabla(\nabla \cdot \mathbf{u}) - (\gamma - 1)g\hat{\mathbf{z}}(\nabla \cdot \mathbf{u}) - gu_z + [\nabla \times (\nabla \times (\mathbf{u} \times \mathbf{B}_0))] \times \frac{\mathbf{B}_0}{4\pi\rho_0}. \tag{10.105}$$

It should be by now a standard procedure how to proceed, as we have already performed it before, e.g., in Chaps. 4 and 6, and it is to substitute an ansatz of a plane wave, that is, a solution of the form

$$\mathbf{u} \propto \exp[i(\mathbf{k} \cdot \mathbf{x} - \omega t)], \tag{10.106}$$

where, as usual, $\mathbf{k}$ is the wave vector, while $\lambda = 2\pi/|\mathbf{k}|$ is the wavelength. We shall not be interested here in the limit $\mathbf{B}_0 = 0$. Suffice it to say that this case gives sound waves, which were discussed in detail in Chap. 6. When the magnetic field, even though it is z-independent, is present, a complication is bound to arise, as we shall see shortly. However, assuming that the wavelength of the waves in question is much shorter than the scale height Λ_z allows for an approximation, or at least qualitative reasoning, in which the vertically dependent unperturbed quantities are taken as constant. The idea behind the plane wave ansatz is to obtain the dispersion relation, i.e., the function $\omega(\mathbf{k})$, which allows to classify the different types of waves and their properties. In what follows, we shall focus only on waves in which the magnetic field plays a role.

10.4.1   Magnetic Waves

The Lorentz force, as we have explained, can be divided into two parts which express different physical effects. The term $(\mathbf{B} \cdot \nabla)\mathbf{B}/(4\pi)$ represents magnetic tension, giving in the unperturbed case a tension force of $T_B = B_0^2/(4\pi)$, while the other part gives magnetic pressure, i.e., force per unit area, $P_B = B_0^2/(8\pi)$. The tension force, acting along magnetic field lines, makes them analogous, at least in principle, to mechanical strings, which support transversal waves whose phase velocity is

$$V_p = \sqrt{\frac{T_B}{\rho_0}} = \frac{B_0}{\sqrt{4\pi\rho_0}}. \quad (10.107)$$

In Chap. 6 we have seen that isentropic gas, obeying $P_0/\rho_0^\gamma = $ const., produces longitudinal waves with phase velocity, which we have been marking in this book by V_p, is $c_s = \sqrt{\gamma P_0/\rho_0}$. In the case of magnetic frozen in field, we have the relation $B_0/\rho_0 = $ const, which translates to $P_M/\rho_0^2 = $ const. Taking this as analogous to the adiabatic law, we may obtain the phase velocity of longitudinal waves to be

$$V_p = \sqrt{\frac{2P_B}{\rho_0}} = \frac{B_0}{\sqrt{4\pi\rho_0}}. \quad (10.108)$$

The above heuristic analysis points to the existence of a purely magnetic wave and provides an equal phase velocity for two types of such waves, setting it in both cases to be equal to the Alfvén speed (see (10.28))

$$V_p = c_A \equiv \sqrt{\frac{B_0^2}{4\pi\rho_0}}, \quad (10.109)$$

whose order of magnitude in the Sun's photosphere is $c_A \approx 10^6$ cm/s, while in its dilute corona this speed may be larger by up to a factor of ~ 30.

This above heuristic expectation is supported by mathematical analytic calculation. If the magnetic force dominates the equilibrium, the following, purely magnetic wave dispersion relation, follows from Eq. (10.105)

$$\omega^2 \mathbf{u} = \left[\mathbf{k} \times (\mathbf{k} \times (\mathbf{u} \times \hat{\mathbf{B}}_0)) \right] \times \hat{\mathbf{B}}_0 c_A^2, \quad (10.110)$$

where terms with P_0, c_s^2, and g have been neglected, with respect to the magnetic terms. Also, $\hat{\mathbf{B}}_0$ is a unit vector in the direction of the undisturbed magnetic field. Straightforward algebraic manipulations, using formulae from vector analysis, give the following relation (show it!)

$$\omega^2 \frac{\mathbf{u}}{c_A^2} = \mathbf{u} k^2 \cos^2 \theta_B - (\mathbf{k} \cdot \mathbf{u}) \hat{\mathbf{B}}_0 \cos \theta_B + \left[(\mathbf{k} \cdot \mathbf{u}) - k(\hat{\mathbf{B}}_0 \cdot \mathbf{u}) \cos \theta_B \right] \mathbf{k}, \quad (10.111)$$

where θ_B is the angle between $\mathbf{k}$ and $\mathbf{B}_0$ and $k = |\mathbf{k}|$. Now, since $\nabla \cdot \mathbf{B}' = 0$ it follows $\mathbf{k} \cdot \mathbf{B}' = 0$, i.e., that the wavenumber is perpendicular to the perturbation in the magnetic field. Second, taking the scalar product of $\hat{\mathbf{B}}_0$ with Eq. (10.111) yields $\hat{\mathbf{B}}_0 \cdot \mathbf{u} = 0$, i.e., the velocity perturbation is normal to the unperturbed magnetic field. Third, taking the scalar product of the entire Eq. (10.111) with $\mathbf{k}$ gives

$$(\omega^2 - k^2 c_{\text{A}}^2)(\mathbf{k} \cdot \mathbf{u}) = 0. \tag{10.112}$$

This equation, as we shall shortly see, distinguishes between two sorts of magnetic waves, both having their phase velocity equal to the Alfvén speed.

- *Incompressible (shear) Alfvén waves*
  If the perturbation is incompressible, i.e.,

$$\nabla \cdot \mathbf{u} = \mathbf{k} \cdot \mathbf{u} = 0, \tag{10.113}$$

then using this in Eq. (10.111), together with $\mathbf{B}_0 \cdot \mathbf{u} = 0$, as shown above, gives

$$\omega = \pm k c_{\text{A}} \cos \theta_B. \tag{10.114}$$

The negative root indicates just the opposite direction of the wave vector, but the same physics. These incompressible Alfvén waves are sometimes called shear waves. The phase velocity of these waves is

$$V_{\text{p}} = \frac{\omega}{k} = \pm c_{\text{A}} \cos \theta_B. \tag{10.115}$$

The polar diagram in Fig. 10.4 shows clearly the phase velocity; it is the chord labelled $\frac{\omega}{k}$, which is the phase velocity, V_{p}, and its direction with respect to the direction of the unperturbed magnetic field, as well as the magnitude of the phase velocity, in comparison with the Alfvén speed. Note also that the cases of both signs, as explained above, are taken into account. The waves in this case are *transverse* and their group velocity can be found as well. To do that we may take the unperturbed field direction as the z-axis, say. Then Eq. (10.114) yields $\omega = k_z c_{\text{A}}$, and since the vector group velocity is

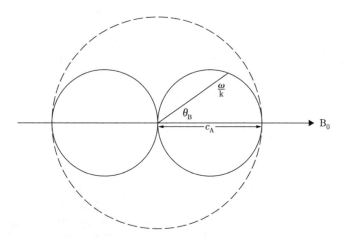

Fig. 10.4 Polar diagram of the phase velocity of shear Alfvén waves. For details see text

$$\mathbf{V}_g = \frac{d\omega}{d\mathbf{k}} = \nabla_k \omega(\mathbf{k}), \tag{10.116}$$

we get $\mathbf{V}_g = c_A \hat{\mathbf{B}}_0$, that is, energy is carried along the unperturbed magnetic field lines, in spite of any direction of the phase velocity. Returning to Eqs. (10.99) and (10.101), one can be convinced that there are no density or pressure changes induced by these waves. Finally, Eq. (10.102) becomes, for a plane wave solution,

$$\omega \mathbf{B} = -\mathbf{k} \times (\mathbf{u} \times \mathbf{B}_0) = (\mathbf{k} \cdot \mathbf{u})\mathbf{B}_0 - (\mathbf{k} \cdot \mathbf{B}_0)\mathbf{u}, \tag{10.117}$$

which with $\mathbf{k} \cdot \mathbf{u} = 0$ and using Eqs. (10.113) and (10.114) gives

$$\mathbf{u} = -\frac{\mathbf{B}'}{\sqrt{4\pi\rho_0}}. \tag{10.118}$$

Thus, $\mathbf{u}$ and $\mathbf{B}'$ are in the same direction and lie in the plane parallel to the wavefront. Since, as we have seen $\mathbf{B}_0 \cdot \mathbf{u} = 0$, it implies $\mathbf{B}_0 \times \mathbf{B}' = 0$, i.e., the field perturbation is perpendicular to the field.

- *Compressional Alfvén waves*
  The second solution to Eq. (10.112), arising when $\mathbf{k} \cdot \mathbf{u} \neq 0$ gives $\omega = kc_A$, and is obviously valid for *compressional* waves. Now, Eq. (10.111) and the relation quoted above, $\mathbf{k} \cdot \mathbf{B}' = 0$, imply that for these waves $\mathbf{u}$ lies in the $(\mathbf{k} - \mathbf{B}_0)$ plane and it is normal to $\mathbf{B}_0$. Thus, it has both parallel and perpendicular components to $\mathbf{k}$, giving rise to both density and pressure variations. The dispersion relation guarantees that the phase velocity is $V_p = c_A$, regardless of wave vector direction. The group velocity is $\mathbf{V}_g = c_A \mathbf{k}$ (can you explain the physical meaning of this finding with regard to energy propagation?). In addition, from Eq. (10.117) it follows that $\mathbf{B}'$ lies in the plane of $\mathbf{u}$ and $\mathbf{B}_0$ but is normal to $\mathbf{k}$. Using these observations in Eq. (10.111), we conclude that $\mathbf{u} \parallel \mathbf{k}$ so that these waves are *longitudinal*, as expected from our heuristic estimate.

10.4.2   Magneto-acoustic Waves

We consider now MHD waves in which most of the simplifying assumptions are dropped, keeping however the one stating that gravity is neglected $g = 0$, that is, in these waves the magnetic and the pressure forces play a rôle. In this case, an equation, which is a generalization of (10.111), has to be taken into account. The equation is

$$\omega^2 \frac{\mathbf{u}}{c_A^2} = \mathbf{u}k^2 \cos^2\theta_B - (\mathbf{k} \cdot \mathbf{u})\hat{\mathbf{B}}_0 \cos\theta_B +$$

$$+ \left[(1 + c_s^2/c_A^2)(\mathbf{k} \cdot \mathbf{u}) - k(\hat{\mathbf{B}}_0 \cdot \mathbf{u}) \cos\theta_B \right] \mathbf{k}. \tag{10.119}$$

Compare it to Eq. (10.111). Since $\mathbf{u}$ appears in this equation on the right-hand side, in $\mathbf{k} \cdot \mathbf{u}$ and $\hat{\mathbf{B}}_0 \cdot \mathbf{u}$ combinations, we proceed by taking scalar products of the equation with $\mathbf{k}$ and $\hat{\mathbf{B}}_0$, in turn. The following two equations result:

$$\left(-\omega^2 + k^2 c_s^2 + k^2 c_A^2\right)(\mathbf{k} \cdot \mathbf{u}) = k^3 c_A^2 \cos \theta_B (\hat{\mathbf{B}}_0 \cdot \mathbf{u}), \tag{10.120}$$

$$\omega^2 (\hat{\mathbf{B}}_0 \cdot \mathbf{u}) = k \cos \theta_B c_s^2 (\mathbf{k} \cdot \mathbf{u}). \tag{10.121}$$

If $\mathbf{k} \cdot \mathbf{u} = 0$, we find the Alfvén wave solution, as before. Otherwise, $(\mathbf{k} \cdot \mathbf{u})/(\hat{\mathbf{B}}_0 \cdot \mathbf{u})$ may be eliminated from Eq. (10.121) and substituted into Eq. (10.120), and the dispersion relation for the *magneto-acoustic*, sometimes called *magneto-sonic* waves, is obtained:

$$\omega^4 - \omega^2 k^2 (c_s^2 + c_A^2) + c_s^2 c_A^2 \cos^2 \theta_B = 0. \tag{10.122}$$

For $\omega/k > 0$ the above quartic dispersion relation has two solutions for the phase velocity of this magneto-acoustic wave:

$$V_p^{[\pm]} = \frac{\omega}{k} = \frac{1}{2} \left[(c_s^2 + c_A^2) \pm \sqrt{c_s^4 + c_A^4 - 2c_s^2 c_A^2 \cos 2\theta_B}\right]^{1/2}. \tag{10.123}$$

The solution with the $+$ sign has a higher frequency, for given k, and its phase velocity is larger in its absolute value. Consequently, it is called the *fast* magneto-acoustic wave. The other solution is, correspondingly, the *slow* magneto-acoustic wave. Note that $V_p^{[+]} > c_A > V_p^{[-]}$ and that is why the magnetic Alfvén wave is sometimes called the *intermediate mode*. The left panel of Fig. 10.5 is a polar sketch of the variation of the phase velocity, for both the fast and the slow modes, with the angle θ_B, determined by the direction of $\mathbf{k}$, relative to $\mathbf{B}_0$. A polar diagram for the group velocity can be seen in the right panel of Fig. 10.5. This is the result of performing the derivative $d\omega/dk$, using Eq. (10.123), and see Problem 10.33. The energy of the slow waves propagates within the narrow cones, while that of the fast waves propagates more isotropically. In the case of a weak magnetic field, such that $c_s \gg c_A$, the group velocities tend as follows:

$$V_g^{[+]} \to c_s, \text{ in the } \hat{\mathbf{k}} \text{ direction} \quad \text{and} \quad V_g^{[-]} \to c_A, \text{ in the } \hat{\mathbf{B}}_0 \text{ direction}.$$

10.5   Discontinuities and MHD Shock Waves

Just as in FD, also in MHD there is a possibility of the formation of discontinuities and shock waves. Consider an element of the surface of discontinuity $\hat{\mathbf{n}} dS$, which we shall assume to be at rest. We shall be using the nomenclature for jump conditions

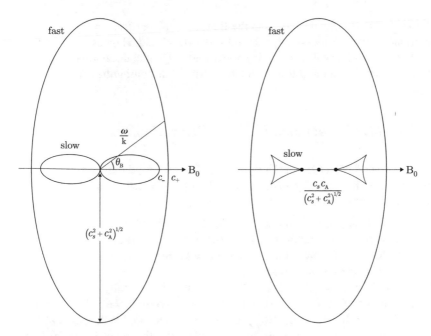

Fig. 10.5 Polar diagrams for fast and slow magneto-acoustic waves whose wave vector is at an angle θ_B to the unperturbed field. c_- and c_+ are the slower and faster, respectively, of c_s and c_A. The left panel depicts the phase velocity, while the right one the group velocity

as introduced in the fluid dynamical case (see Sect. 6.5), i.e., $[\Gamma] \equiv \Gamma_2 - \Gamma_1$, for any function Γ with the subscript denoting the side of the discontinuity. Subscript 2 is on the side into which the unit normal (here it is unique) points.

Let $\mathbf{u} \cdot \hat{\mathbf{n}} \equiv u_n$ and $j \equiv \rho u_n = u_n/\upsilon$, where similarly to the HD case j is here the momentum flux (do not confuse with electrical current density) and υ the specific volume. Using the results of Problem 10.35, remembering from the theory of electromagnetism that $[B_n] = 0$ and $[E_t] = 0$ and exploiting the fact that in *ideal* MHD, having zero resistivity, $\mathbf{E} = -\mathbf{u} \times \mathbf{B}/c$, one may obtain the fundamental system of equations for jump conditions in ideal MHD, in the frame of reference moving with the discontinuity:

$$j\left[h + \frac{1}{2}j^2\upsilon^2 + \frac{1}{2}v_t^2 + \frac{\upsilon\mathbf{B}_t^2}{4\pi}\right] = \frac{B_n[\mathbf{B}_t \cdot \mathbf{u}_t]}{4\pi}, \qquad (10.124)$$

$$[P] + j^2[\upsilon^2] + \frac{[\mathbf{B}_t^2]}{8\pi} = 0, \qquad (10.125)$$

$$j[\mathbf{u}_t] = \frac{B_n[\mathbf{B}_t]}{4\pi}, \qquad (10.126)$$

$$B_n[\mathbf{u}_t] = j[\upsilon\mathbf{B}_t]. \qquad (10.127)$$

The subscripts t and n denote, as in the fluid dynamical case, tangential and normal components and h is the enthalpy. MHD supports additional types of discontinuities, compared to ordinary FD. Not only shocks and tangential discontinuities, which are familiar from FD, but also the so-called rotational discontinuities are possible.

10.5.1   MHD Rotational (Alfvén) Discontinuities

Assume that an MHD discontinuity exists. If $j = 0$, i.e., the fluid moves parallel to the surface of discontinuity, then we have the first two cases enumerated below. The third case is for $[\rho] = 0$:

1. $B_n \neq 0$. It is possible to deduce from Eqs. (10.124–10.127) that $\mathbf{u}, \mathbf{B}$, and P must be continuous (show it!). However, ρ and therefore all the thermodynamic variables, save the pressure, may have any discontinuity. Such a discontinuity is referred to as a *contact* discontinuity.
2. $B_n = 0$. Then Eqs. (10.124–10.127) are satisfied identically. This is a *tangential* discontinuity (similar to FD) and may be summarized in the following way. First, $j = 0$, $B_n = 0$, $[\mathbf{u}_t] \neq 0$, $[\mathbf{B}_t] \neq 0$, $[v] \neq 0$. Second, the density discontinuity may take any value but the pressure discontinuity must be consistent with Eq. (10.125). Thus,

$$[v] \neq 0, \quad \left[P + \frac{\mathbf{B}_t^2}{8\pi} \right] = 0, \tag{10.128}$$

   and the discontinuities of other thermodynamic variables can be found with the help of those of v, P, and the equation of state[1].
3. The most interesting discontinuity is obtained, when the density is continuous; $[j] = 0$, as always, and since we posit that $[v] = 0$, we must have also $[u_n] = 0$. Since $j \neq 0$, we may use Eqs. (10.126–10.127) to get

$$j = \frac{B_n}{\sqrt{4\pi v}}, \tag{10.129}$$

$$[\mathbf{u}_t] = [\mathbf{B}_t] \sqrt{\frac{v}{4\pi}}. \tag{10.130}$$

Now we substitute $h = e + Pv$ in Eq. (10.124), using $[v] = 0$ and replacing B_n from Eq. (10.129). The result is

$$j[e] + jv \left[P + \frac{\mathbf{B}_t^2}{8\pi} \right] + \frac{1}{2} j \left[\left(\mathbf{u}_t - \sqrt{\frac{v}{4\pi}} \mathbf{B}_t \right)^2 \right] = 0. \tag{10.131}$$

---

[1] A square of a vector means the square of its absolute value.

Equation (10.130) shows that the third term in the above equation vanishes and the second term vanishes as a result of Eq. (10.128) and $[v] = 0$ as we have assumed above. Thus, the seemingly lengthy calculation finally gives the continuity of internal energy, $[e] = 0$. Since v is also continuous, every other thermodynamic variable, which can always be written as a function of the above two, is continuous. The pressure is among these variables and so $[P] = 0$, implying $[\mathbf{B}_t^2] = 0 \Rightarrow [B_t] = 0$. This, together with $[B_n] = 0$, following from Eq. (10.129), implies that the vector $\mathbf{B}$ itself and its angle to the surface are continuous—this means that the vector may be turned around through an angle about the normal, when crossing the surface, retaining its magnitude and its angle with the surface. Equations (10.129–10.131), along with our introductory comments above, summarize the properties of these rotational or Alfvén discontinuities. The *vector* $\mathbf{B}_t$ is *not* continuous, and thus by Eq. (10.130), neither is $\mathbf{u}_t$. However, $[u_n] = j[v] = 0$, i.e., the normal velocity is continuous and has the value

$$u_n = \frac{B_n}{(4\pi\rho)^{1/2}} = B_n \left(\frac{v}{4\pi}\right)^{1/2}. \tag{10.132}$$

Note that u_n is minus the velocity of propagation of the discontinuity, relative to the fluid, and that it is equal to the phase velocity of an Alfvén wave.

10.5.2 MHD Shock Waves

Assume a given value of j. In FD shock waves, the relations $[j] = 0$ and $[v] \neq 0$ hold (see Sect. 6.5); thus, $u_{n,1} \neq u_{n,2}$ and there is a jump in the normal velocity. Taking into account the MHD relations, we compare Eqs. (10.126) and (10.127), assuming for the time being $B_n \neq 0$. Clearly, the difference vectors $[\mathbf{B}_t]$ and $[v\mathbf{B}_t]$ are parallel to the same vector difference $[\mathbf{u}_t]$ and thus are parallel to each other. This is only possible if the vectors themselves $\mathbf{B}_{t,1}$ and $\mathbf{B}_{t,2}$ are *co-linear*; otherwise,

$$(\mathbf{B}_{t,1} - \mathbf{B}_{t,2}) \parallel (v_1\mathbf{B}_{t,1} - v_2\mathbf{B}_{t,2}) \tag{10.133}$$

is impossible for $v_1 \neq v_2$.

We are ready now to derive the MHD shock adiabatic relation (similar to the Hugoniot formula in hydrodynamic shocks). Using Eqs. (10.126–10.127), we get

$$j^2[vB_t] = \frac{1}{4\pi}B_n^2[B_t], \tag{10.134}$$

where we have written the vector magnitude of the tangential components of the field instead of the vectors because $\mathbf{B}_{t,i}$ for $i = 1$ and 2 are co-linear (10.133). Rewrite now Eq. (10.124) as

$$[h] + \frac{1}{2}j^2[\upsilon^2] + \frac{[\upsilon \mathbf{B}_t^2]}{4\pi} - \frac{B_n^2}{32\pi^2 j^2}[B_t^2] = \frac{1}{2}\left[\left(\frac{B_n}{4\pi j}\mathbf{B}_t - \mathbf{u}_t\right)^2\right]. \tag{10.135}$$

The right-hand side of this equation is equal to zero by virtue of Eq. (10.126). From Eq. (10.125), we can eliminate j^2 thus

$$j^2 = -\frac{[P + \mathbf{B}_t^2/(8\pi)]}{[\upsilon]}. \tag{10.136}$$

Replacing now the term j^2 in the second term of (10.135) by this expression and in its fourth term using Eq. (10.134), we obtain, after some algebra, the MHD shock adiabat:

$$e_1 - e_2 + \frac{1}{2}(\upsilon_2 - \upsilon_1)(P_1 + P_2) + \frac{1}{16\pi}(\upsilon_2 - \upsilon_1)(B_{t,1} - B_{t,2})^2 = 0. \tag{10.137}$$

Note that this differs from the hydrodynamic case only by the last term, which is the sole magnetic contribution.

Having shown this basic relation, we think that in a book on FD lengthy deliberations about MHD shock waves should be omitted, as this complex topic is far from our main subject matter. Instead, we give below what we consider the most important remarks, in the form of a short primer on MHD shocks, most without proofs.

- Repeating, for the sake of completeness, the equation for the discontinuity in $\mathbf{u}_t$ (10.126)

$$\mathbf{u}_{t,2} - \mathbf{u}_{t,1} = \frac{B_n}{4\pi}\left(\mathbf{B}_{t,2} - \mathbf{B}_{t,1}\right), \tag{10.138}$$

we may now add Eqs. (10.134), (10.136), (10.137), and (10.138) to form a complete set, which is able to fully describe MHD shock waves.
- Similarly to HD shocks, we shall choose the front to be stationary and the fluid passing it from side labelled by the number 1 to the other side (2). We remember from Sect. 6.5 that both the density and pressure *increase* in a shock, making it a compression wave. It is assumed that

$$\left(\frac{\partial^2 \upsilon}{\partial P^2}\right)_s > 0. \tag{10.139}$$

This is, strictly speaking, not a thermodynamic identity, but it is satisfied in all practical cases. Using this inequality, one may show that MHD shock waves are *compression* waves, provided they are *weak*. Then one may expand the shock adiabat (10.137) in powers of $(P_2 - P_1)$ and $(s_2 - s_1)$ (do it!), proving the fact that also in this case shock waves are compressive. If, in addition to the relation (10.139), the thermal expansion coefficient is positive, it is possible to show that rarefaction in a shock is impossible even if the changes are not small, the conclusion being that a *strong* shock must be a compression wave as well. In short, we may be confident that also MHD shocks are only compression waves, in all practical cases in which (10.139) holds.

- Consider the case in which the magnetic field lies on both sides in the tangential plane only ($B_n = 0$). In this case, using Eq. (10.138) we get that the tangential component of the velocity is continuous, i.e., $\mathbf{u}_{t,1} = \mathbf{u}_{t,2} \equiv \mathbf{u}_t$; this means that we can always choose coordinates in which $\mathbf{u}_t$ on the shock face is zero, and therefore the velocity is perpendicular to the shock front. Assuming this, Eq. (10.134) implies $v_2 B_2 = v_1 B_1$. Using this relation in Eqs. (10.136–10.137) provides, after a little reflection, the equations

$$j^2 = \frac{(P_2^{\mathrm{mag}} - P_1^{\mathrm{mag}})}{v_1 - v_2}, \tag{10.140}$$

$$e_2^{\mathrm{mag}} - e_1^{\mathrm{mag}} + \frac{1}{2}(P_2^{\mathrm{mag}} + P_1^{\mathrm{mag}})(v_2 - v_1) = 0. \tag{10.141}$$

Without the superscript "mag," for *magnetic*, these relations are identical to the HD case (see Sect. 6.5). With the magnetic field of the kind specified here, this differs from the hydrodynamic case only by changing the equation of state. The true equation of state $P = P(v,s)$ has to be replaced by the expression

$$P^{\mathrm{mag}}(v,s) = P(v,s) + \frac{m^2}{8\pi v^2}. \tag{10.142}$$

Here $m \equiv Bv$. The thermodynamic relation $(\partial e/\partial v)_s = -P$ must be satisfied by the quantities with the superscript "mag" as well, whence $e^{\mathrm{mag}} \equiv e + m^2/(8\pi v)$.

10.6   Some Common MHD Instabilities

The theory of hydrodynamic instability (see Chap. 7) includes a wealth of material, which is complicated already without the inclusion magnetic forces. It is only natural that conducting fluids, with magnetic fields permeating them, add additional complications. As we have seen, magnetic field lines are endowed with tension, which, among other things, opposes bending. They also create an additional pressure, the magnetic one, and thereby open the possibilities for many steady MHD flows to be actually linearly unstable. In this section we shall give a short review of the most significant MHD instabilities, remaining within the realm of linear instability and normal mode analysis, remembering that the problems lend themselves, in addition to full numerical computations, also to semi-analytic asymptotic and weakly nonlinear approximation techniques. It is worth mentioning that it is now widely accepted, also in MHD, that a flow may become turbulent without being linearly unstable, as discussed for ordinary fluids in Chap. 7.

To begin, we point out that instability of hydrodynamical flows is usually not quenched by magnetic forces, although the rigidity and the resistivity in MHD may reduce the tendency to instability. If anything, the rule of thumb is that the presence of magnetic fields in conducting fluids is usually a *destabilizing* agent. Thus, flows with large Re and which pass near rigid boundaries or obstacles, Kelvin–Helmholtz unstable hydrodynamical flows, convectively unstable flows, and other unstable HD flows have their unstable counterparts in MHD. In what follows, we shall discuss the qualitative aspects of the sausage and kink instabilities of the pinch, a more quantitative description of the MHD Kelvin–Helmholtz instability and the magneto-rotational instability in a cylindrical magnetic Taylor–Couette flow. A discussion of the astrophysically applicable Parker instability, which we shall bring at the end of the list, as an example, will follow. A large host of other MHD instabilities exist, many of them related to fusion research, but we cannot give any comprehensive review of them all. In addition, we shall not discuss at all the very important topic in solar physics—the influence of magnetic fields on convection. This subject is still poorly understood (and see below). We give here an (incomplete) list of MHD instabilities. Note that some of them do not have an HD counterpart.

- *Pinch instabilities*
  When discussing MHD equilibria, we have given in Sect. 10.3.1 an example of a pinch. Let us now examine, largely qualitatively, its stability. First, we recall the pinch structure. According to Eq. (10.70), the pinch current components in cylindrical coordinates are

$$j_\varphi = -\frac{c}{4\pi}\frac{dB_z}{dr} \tag{10.143}$$

$$j_z = \frac{c}{4\pi r}\frac{d}{dr}\left(rB_\varphi\right). \tag{10.144}$$

Substituting this explicitly to the equilibrium condition,

$$0 = -\nabla P + \frac{1}{c}\mathbf{j}\times\mathbf{B}, \tag{10.145}$$

yields

$$0 = \frac{d}{dr}\left[P + \frac{1}{8\pi}\left(B_z^2 + B_\varphi^2\right)\right] + \frac{B_r^2}{r}. \tag{10.146}$$

Examine now qualitatively the stability of a configuration created by a uniform current sheet $B_0 c/(4\pi)$ flowing in the z direction on the surface of a cylinder $r = r_0$. From the equilibrium equation given above, it is easy to solve for the magnetic field, if it is assumed that it has only an angular component:

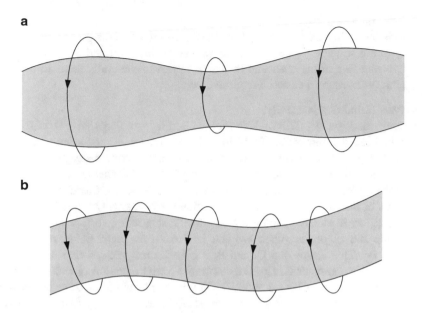

Fig. 10.6 Schematic drawing of the sausage and kink instabilities

$$B_\varphi(r) = B_0 \frac{r_0}{r} \quad \text{if } r \geq r_0,$$
$$= 0 \quad \text{otherwise.} \tag{10.147}$$

In this configuration, the pressure is uniform outside the cylinder but is greater by $B_0^2/(8\pi)$ from that inside the surface of the cylinder ($r = r_0 - \varepsilon$). The fluid inside the cylinder is *pinched* as a result of the magnetic tension along the field lines of the encircling field. It is not difficult to see that the pinch equilibrium is unstable to at least two simple perturbations:

– *The sausage instability*
   This instability, illustrated in Fig. 10.6a, can be readily understood when we imagine what happens if the pinch becomes at some point a little thinner than its adjacent region. As Eq. (10.147) indicates, the magnetic field and magnetic pressure increase in the thinner part, further constricting it. The sausage instability is a simple example of a general type of *exchange* instabilities, which plague trials of plasma confinement by concave (towards the plasma) magnetic field lines. It is conceivable that a bundle of field lines near the plasma surface straighten, while not changing the volume occupied by them. The gap left behind them will be occupied by plasma leaking to this low magnetic pressure domain. It is a nontrivial to show, but understandable heuristically, that the total energy of the new configuration is lower. Hence, the exchange is an instability.

– *The kink instability*

This instability, as depicted in Fig. 10.6b, occurs when the pinch is bent. Field lines are squeezed together on the inside of the bend and separated on the opposite side of the kink, and the excess pressure naturally tends to bend the pinch even more, a destabilizing feedback.

- *Kelvin–Helmholtz instability*

Among the simplest Kelvin–Helmholtz stability problems in FD is the one in which an incompressible inviscid fluid of density ρ_1 flows with horizontal velocity U over the horizontal surface of a second fluid of density ρ_2, at rest. The FD version of this problem was treated in Chap. 7. Solving a similar problem, with the effect of gravity replaced by magnetic effects, we consider two uniform streams of fluid on the two sides of the plane, $z = 0$, say. Let both velocities of the undisturbed steady state be parallel to the separating surface and be $\mathbf{U}_1$ on one side and $\mathbf{U}_2$ on the other. We add horizontal magnetic fields, respectively, $\mathbf{B}_1$ and $\mathbf{B}_2$. This means that $\mathbf{U}_i$ and $\mathbf{B}_i$ are *horizontal* vectors, that is, they have only x and y components. Consider now the MHD equation of motion and the induction equation—(10.24) and (10.37) with neglected dissipative terms—and add a small perturbation to the velocities $\mathbf{u}'_j$, to the magnetic fields $\mathbf{B}'_j$, and to the *total* pressure ($P_{\text{tot}} = P + B^2/(8\pi)$), P'_{tot}, where we have dropped the index j. Linearizing the equations for $j = 1, 2$, one obtains, with the indices j dropped for simplicity:

$$\rho \left[\frac{\partial \mathbf{u}'}{\partial t} + (\mathbf{U} \cdot \nabla) \mathbf{u}' \right] = -\nabla P'_{\text{tot}} + \frac{1}{4\pi} (\mathbf{B} \cdot \nabla) \mathbf{B}', \tag{10.148}$$

$$\frac{\partial \mathbf{B}'}{\partial t} + (\mathbf{U} \cdot \nabla) \mathbf{B}' = (\mathbf{B} \cdot \nabla) \mathbf{u}'. \tag{10.149}$$

These equations have to be solved for the perturbations subject to the solenoidal property of the velocity and magnetic field perturbations:

$$\nabla \cdot \mathbf{B}' = 0, \quad \nabla \cdot \mathbf{u}' = 0. \tag{10.150}$$

As it turns out in this case, we will be able to find a *sufficient* condition for *stability*. The appropriate way to achieve it is to try an ansatz

$$\propto \exp\left[i\omega t - i\mathbf{k}_h \cdot \mathbf{x}_h \pm kz\right], \tag{10.151}$$

where the vectors with the h index have only x and y components, so that $k_h \equiv \sqrt{k_x^2 + k_y^2}$ and the $+(-)$ signs are used for $z < 0, (z > 0)$ to prevent divergence for large $|z|$. Here we only see that this ansatz satisfies $\nabla^2 P'_{\text{tot}} = 0$, the latter following from taking the divergence of equation (10.149). In Problem 10.38 it is shown that a calculation, using this ansatz for all the perturbations, pressure continuity at the surface, $\zeta(\mathbf{x}_h, t)$ (using the nomenclature of Chap. 4 for surface HD waves), and the condition $\dot{\zeta} = u'_z$ at this surface yields the following quadratic equation for the value ω, after inserting back the indices $j = 1, 2$ and if $\mathbf{k}_h$ is assigned:

$$\rho_1 \left(\omega - \mathbf{k}_h \cdot \mathbf{U}_1\right)^2 + \rho_2 \left(\omega - \mathbf{k}_h \cdot \mathbf{U}_2\right)^2 = \frac{1}{4\pi} \left[(\mathbf{k}_h \cdot \mathbf{B}_1)^2 + (\mathbf{k}_h \cdot \mathbf{B}_2)^2\right].$$

$$(10.152)$$

It is a real equation for ω and therefore the solution must be real or contain only complex conjugate pairs. Real ω guarantees stability (constant amplitude oscillations in this ideal case). For real values of ω, the left-hand side of Eq. (10.152) has a minimum, corresponding to

$$(\rho_1 + \rho_2)\omega = \mathbf{k}_h \cdot (\rho_1 \mathbf{U}_1 + \rho_2 \mathbf{U}_2), \qquad (10.153)$$

for which the left-hand side attains the minimum (for ω real) value of

$$\frac{\rho_1 \rho_2 \left[\mathbf{k}_h \cdot (\mathbf{U}_1 - \mathbf{U}_2)\right]^2}{\rho_1 + \rho_2}. \qquad (10.154)$$

So the following inequality

$$\frac{\rho_1 \rho_2 \left[\mathbf{k}_h \cdot (\mathbf{U}_1 - \mathbf{U}_2)\right]^2}{\rho_1 + \rho_2} < \frac{1}{4\pi} \left[(\mathbf{k}_h \cdot \mathbf{B}_1)^2 + (\mathbf{k}_h \cdot \mathbf{B}_2)^2\right] \qquad (10.155)$$

guarantees that ω is real and hence stability. In Chap. 7 we discussed the ill-posedness of such a problem and how to remedy it.

- *Magnetorotational instability (MRI)*
  This *linear* instability, known from 1959 (E. Velikhov) for a cylindrical configuration, has become since the 1980s the subject of prolonged controversies in the astrophysical community. At the time of the writing of this book, there is no clear consensus, as far as the present authors know, regarding the results of perturbative analytical as well as extensive numerical studies of its *nonlinear* development. The reason for this instability's importance was the suggestion that it may explain the origin of a laminar $\rightarrow$ turbulent transition and hence the enhanced angular momentum transport outward in accretion disks (see Chap. 5). The issue is, for some reason, considered extremely important in theoretical astrophysics. The instability can be formulated simply: any magnetic Taylor–Couette flow, that is, a cylindrically symmetric of a conducting fluid permeated by magnetic field, whose rotation law, $\Omega(r)$, satisfies $d\Omega/dr < 0$, is linearly unstable. There is also a simple and intuitive explanation of the instability. Consider for a moment Fig. 10.7. Let the cylinder in question (which is permeated by a magnetic field in the vertical direction) be perturbed in such a way that along it we set up a small amplitude torsional wave. We concentrate on the upper half of the wave, above the shaded reflection symmetry plane. Now note that points marked by a and b are on the same field line. Also, at one of the wave's extrema (at height z_+), we spin up slightly the fluid annulus, by $\delta\Omega$, say. In contrast, half a wavelength of the torsional wave below (at height z_0) a similar small spin down (a perturbation

Fig. 10.7 Schematic drawing
illustrating the idea behind
the MRI (after F. Shu,
reference [8]); for details, see
text

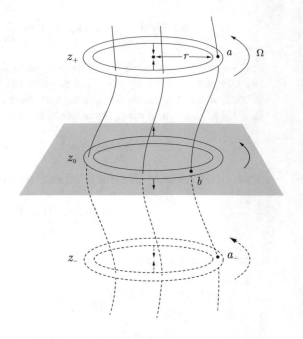

$-\delta\Omega$) is applied on the corresponding annulus. The two annuli are connected by
a field line and exert a torque (on each other), which removes angular momentum
from the advanced upper annulus and adds angular momentum to the lower,
retarded annulus. As a result, the upper annulus has to shrink, i.e., decrease its
radius and consequently speed up, in accord with the Keplerian rotation law.
Similarly, the lower annulus will expand in radius and slow down. So the initial
perturbation $\delta\Omega$ will simply grow *in the applied direction*, implying instability.
One should realize that this *linear* instability may be "choked" if the magnetic
field is too strong, so as not to allow any torsion to begin with, or the wavelength
of the torsional wave is too long so that the communication time between the
annuli cannot be faster than half a wavelength divided by the Alfvén speed,
allowing for dissipative and other processes to take over. The full linear analysis
of this instability has been performed, as mentioned before, a long time ago,
but we feel that it is beyond the scope of this book to bring it in its details.
The interested reader may find it in reference [10]. As mentioned before, we
have at present some approximate analytic asymptotic studies which seem to
indicate that the instability saturates for very small magnetic Prandtl numbers
Pm $\ll 1$. Regarding full numerical studies, calculations with sufficient resolution
have started to appear, but it is not clear if they support transition to turbulence
for very small Pm as thin disks have.

10.6.1 Example: Parker Buoyancy Instability, Interstellar Clouds, and Sunspots

In the second half of the 1960s, E. Parker studied in detail the nature of vertical equilibria, in which a conducting gas and *submerged magnetic field* are held by downward force of gravity. Among such situations, one may consider, for example, stellar atmospheres, as well as interstellar gas with horizontal magnetic fields, in the vertical gravitational field of the galaxy. The idea behind the instability Parker discovered is that the magnetic field may be thought of as a kind of "gas" as it does impart pressure, that is, has internal energy per volume. Parker devoted a whole chapter of his classic book, reference [1] in the *Bibliographical Notes*, to this instability and found that it is causing the breakup and escape of submerged magnetic fields. Imagine a horizontal magnetic field submerged below the surface of a star, say. Approximating the thin upper layer of a star by a plane parallel configuration perpendicular to the z direction (along which gravity acts), one can write the local condition for equilibrium in the layer as

$$\frac{d}{dz}\left(P + \frac{1}{8\pi}B^2\right) = -\rho g, \qquad (10.156)$$

where all the symbols have their usual meaning. If the field is fully submerged, the rapid decrease of the magnetic pressure (to practically zero), as the surface of the star is approached, must be compensated by an increase in gas pressure, if equilibrium is to be maintained. This, in turn, causes increase also in density in the upper direction (we may safely assume that the temperature does not change much in these layers). If the field decreases to zero over distances small compared to Λ, here the scale height, the density should increase upward by

$$\delta\rho \sim \left(\frac{1}{g\Lambda}\right)\frac{B^2}{8\pi}. \qquad (10.157)$$

Thus, we will have a situation of a gas of a higher density above one with lower density, an unstable situation (Rayleigh–Taylor instability; see Chap. 7), so that one may expect overturning with tongues of the dense, low-field fluid sinking inward while similar structures of the less dense, higher-field fluid rising (Fig. 10.8a, b).

There is also an additional effect valid not only in the very upper regions—the field is confined by the weight of the gas above it. It can be imagined that the magnetic field lines are "loaded" with the conducting fluid which, as it were, "hangs" on them. Now, if a horizontal field line is perturbed so as to create a small downward trough in the B-line, loaded matter will slide down and increase the weight of the growing undulations and at the same time allow for the unburdened portions to rise and expand upward further. It is thus clear that unless there is some robust process to quench this runaway, a situation of dense clouds with rarer intercloud regions will develop (see Fig. 10.8c). This explanation, however

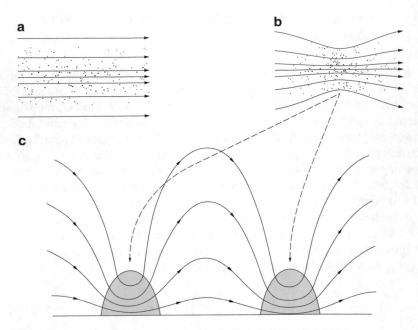

Fig. 10.8 Schematic drawing illustrating the idea behind the Parker instability

plausible, must be accompanied by an actual stability calculation and, as it turns out, the instability indeed works for long enough wavelengths. Also, its nonlinear development has been confirmed numerically. In addition, it can be shown that the Parker instability also belongs to the class of the *exchange* instabilities already mentioned in this chapter. This concept is best understood with the help of the *energy method* in stability theory (see reference [9]). The buoyancy of magnetic fields, as explained here, may give rise under certain conditions, e.g., in the Sun's upper layer, where there is magnetic convection, which we shall not discuss here, to a situation in which narrow flux tubes are formed below the surface of the object.

This process may be one of the ingredients in the formation of *sunspots*, which appear as small dark blemishes on its surface. Although the details of sunspot generation are still a matter of research, it appears that sunspots are the visible counterparts of magnetic flux tubes in the Sun's convective zone that get wound up by differential rotation. If the stress on the tubes reaches a certain limit, they curl up like a rubber band and puncture the Sun's surface. Convection is inhibited at the puncture points; the energy flux from the Sun's interior decreases, and with it surface temperature. This is the reason that the spots appear darker than the Sun. Observations, using the Zeeman effect to find the magnetic field value, indicate that often the sunspots appear in bipolar pairs. The spots themselves are transient phenomena; a typical spot appears and disappears on a timescale of days to weeks. However, the total number of sunspots, usually measured by the fractional area of the solar disk covered by spots, has a clear $\sim$ 11-year cycle, which still remains

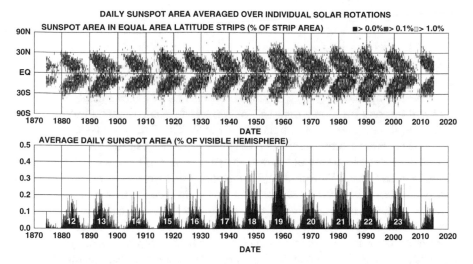

Fig. 10.9 Upper panel: the "butterflies" diagram—the fraction of the solar disk covered by sunspots as a function of time, the vertical axis is the latitude. Note that yellow color *inside* the "butterflies" indicates the highest area covered by sunspots, while outside the "butterflies" there is practically zero coverage by sunspots. The lower panel depicts the average portion of the visible solar hemisphere which is covered by sunspots, as a function of time. The enigmatic 11-year cycle is easily discerned *Public Domain, created by NASA Marshal Space Flight Center, Solar Physics, Courtesy of NASA*

a mystery. The point of highest sunspot activity during this cycle is known as solar maximum, and the point of lowest activity is solar minimum. Early in the cycle, sunspots appear in the higher latitudes and then move towards the equator as the cycle approaches maximum. This so-called Spörer's law created a typical "butterflies" diagram which can be seen in Fig. 10.9. The physical origin of this behavior is unfortunately unknown, despite intensive research. Magnetic convection simply belongs to chaotic turbulent phenomena and is still poorly understood.

10.7 Rudiments of Dynamo Theory

The need for a dynamo mechanism stems from the astrophysical finding that most stars and planets, as well as even galaxies, are endowed with magnetic fields. In the latter case, it may perhaps be claimed that the field is primordial, i.e., pushing the question into the unknown past; however, if one calculates the timescale for Ohmic dissipation of the magnetic field in stars and some planets, the conclusion is, as we have seen following the estimates made using Eq. (10.46), that dynamo action is necessary to explain the relatively high magnetic fields in at least some stars. We saw that the magnetic field diffusion time is

$$\tau_{\text{Bdiff}} \sim \frac{L^2}{\eta_m},$$ (10.158)

where L is a typical length scale. Problem 10.8 gives an estimate of the decay times of fields and points to a conclusion that a dynamo is needed in stars. Notwithstanding the decay time estimated, we also know that the magnetic field of the Earth reverses itself on a timescale of between tens and hundred million years. Moreover, the magnetic field of stars is variable. The Sun has 11-year cycles of magnetic spots, and other stars exhibit cyclical variability in their magnetic fields as well. In short, some way of creating magnetic fields in a conducting fluid is an absolute necessity, and the challenge to explain the mechanism of self-sustained field generation, the *dynamo problem*, as it is called, constitutes one of the most complex open questions remaining on the frontier of MHD and classical physics. We are unable to give here a full discussion of the problem, not only because of space limitations, but also since the understanding of MHD dynamos has not yet reached a satisfactory stage, which may be summarized in a textbook like this one. Instead, we choose to only review some ideas, sometimes disjoint from one another, and direct the interested reader to relevant references. Magnetic dynamo theory attempts to understand what patterns of MHD flows are needed to produce spatially coherent fields over long enough times. One thing that is clear is that even the most successful dynamo cannot grow a magnetic field from nothing, i.e., a configuration of $\mathbf{B} = \mathbf{0}$ everywhere. Thus, some kind of primordial "seed" field, however small, is always needed.

One very powerful statement can be stated at the outset, asserting that a viable dynamo is impossible in two-dimensional configurations. T.G. Cowling was able to show, as early as in 1934, that *it is impossible to construct a steady axi-symmetric dynamo*. Later on, a number of other *anti-dynamo theorems* have appeared, for example, the assertion that a two-dimensional planar dynamo is impossible. We shall give now a sketch of the proof for the axi-symmetric case and leave one simplified planar case for a problem (10.39). Assume that an axi-symmetric dynamo is possible even though the fluid is endowed with resistivity. For convenience we repeat the induction equation (10.37) here

$$\frac{\partial \mathbf{B}}{\partial t} = \nabla \times (\mathbf{u} \times \mathbf{B}) + \eta_m \nabla^2 \mathbf{B},$$ (10.159)

where, as before, we assume that the resistivity η_m is constant. If a dynamo is possible, the magnetic field would not decay totally, and its loss due to diffusion, in the second term on the right-hand side of this equation, has to be balanced by field growth induced somehow by the term before it. If this assumption holds, we must envisage an axi-symmetric configuration of a non-decaying magnetic field in a finite domain. This, in turn, requires also a system of electric currents of finite extent that create this axi-symmetric magnetic configuration of closed field lines. It is not impossible to become convinced that the generic configuration of the above-described sort has to be topologically equivalent to the torus depicted in

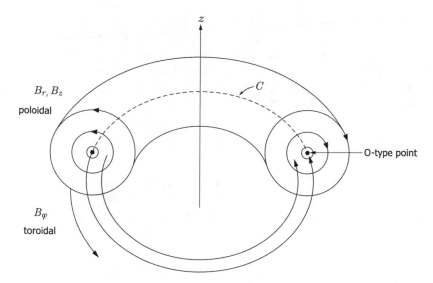

Fig. 10.10 An auxiliary drawing depicting a generic, nontrivial, axi-symmetric magnetic configuration. The circle C, referred to in the text, is the curve passing through the two O-type points
along the middle of the torus, depicted dashed when hidden (after F. Shu, reference [8])

Fig. 10.10. We use cylindrical coordinates, with the z-axis shown in the figure, the
angle variable is φ, and the radial coordinate is denoted by r. Thus, the magnetic
field has a toroidal B_φ and poloidal (B_r, B_z) contributions. The poloidal field has to
close on itself, and therefore there must be a neutral point, at least one, of the center
type inside the torus (see figure). At this O-point, the total poloidal field must be
zero. To support the poloidal field, there must be a current flowing along the axis of
the torus, and thus we have that the current density must satisfy

$$\oint_C \mathbf{j} \cdot \mathbf{dl} > 0 \equiv J_L \tag{10.160}$$

On the other hand, Ohm's law gives

$$\mathbf{j} = \sigma_{\text{el}} \left(\mathbf{E} + \frac{1}{c} \mathbf{u} \times \mathbf{B} \right), \tag{10.161}$$

and the line element in this case is $\mathbf{dl} = R\varphi\,\hat{\varphi}\,d\varphi$, where R is the radius of the
circle C. Now the poloidal field is zero along C and thus we have only $B_\varphi\,\hat{\varphi}$, i.e., a
contribution of a toroidal field component. This means that the $\mathbf{u} \times \mathbf{B}$ vanishes along
$\mathbf{dl}$. Thus, we are left with

$$0 < J_L = \oint_C \mathbf{j} \cdot \mathbf{dl} = \sigma_{\text{el}} \oint_C \mathbf{E} \cdot \mathbf{dl} = \sigma_{\text{el}} \int_0^R (\nabla \times \mathbf{E}) \cdot \hat{z}\, 2\pi r\, dr =$$

$$= -\frac{\sigma_{\text{el}}}{c} \frac{d}{dt} \int_0^R \mathbf{B} \cdot \hat{z}\, 2\pi r\, dr = -\frac{\sigma_{\text{el}}}{c} \frac{d\Phi_B}{dt}, \tag{10.162}$$

where Φ_B is the magnetic flux through the circle C. Thus, $d\Phi_B/dt < 0$. This decaying magnetic flux cannot support the poloidal field, as we have assumed, quite on the contrary. Thus, the conclusion is that a configuration of the sort we have envisaged is possible only if the field is purely toroidal, i.e., $\mathbf{B} = B_\varphi \hat{\varphi}$ to start with. We examine now this possibility. However, before substituting it into the induction equation (10.159), we use the vector identity

$$\nabla \times (\mathbf{u} \times \mathbf{B}) = (\mathbf{B} \cdot \nabla)\mathbf{u} - (\mathbf{u} \cdot \nabla)\mathbf{B} - \mathbf{B}\rho \frac{D}{Dt}\left(\frac{1}{\rho}\right), \tag{10.163}$$

where $\nabla \cdot \mathbf{B} = 0$ and $\nabla \cdot \mathbf{u} = \rho D/Dt\left(\rho^{-1}\right)$ (continuity) have been used. Now, remembering that the field is toroidal and the identity (10.163) in writing out the toroidal component of the induction equation (10.159), it is only a matter of some algebra to get

$$\frac{D}{Dt}\left(\frac{B_\varphi}{r\rho}\right) = \frac{\eta_m}{r\rho} \nabla_p^2 B_\varphi, \tag{10.164}$$

where ∇_p^2 is the poloidal part of the Laplacian operator, obtained by setting the φ derivatives to zero in the expression of the Laplacian in cylindrical coordinates. This last relation completes the proof of the anti-dynamo theorem in the axially symmetric case because the right-hand side is a diffusive term, acting so as to decrease the toroidal field, and the left-hand side has just passive advection, in addition to the partial time derivative of the purely toroidal field.

We move now to a review of potentially working dynamos, but first we define, in passing, the concept and will start our discussion of such viable dynamos with their simplest category—the so-called kinematic dynamos, in which the velocity field is *prescribed*. Examining the simplest case, where $\mathbf{u}$ is a constant flow independent of time, we immediately see that even then this problem is nontrivial. One has to solve the induction equation for $\mathbf{B}$, and since it is a linear problem, for given, time independent $\mathbf{u}$, we have to solve a nontrivial eigenvalue problem, which emerges upon using the ansatz $\mathbf{B} = \mathbf{B}_0(\mathbf{x})\exp st$. Using appropriate boundary conditions, e.g., $\mathbf{B}_0 \to 0$ as $|\mathbf{x}| \to \infty$, we have to analyze all the eigenmodes according to $s_k = \sigma_k + i\omega_k$ with k being here the index denoting the different eigenmodes. Usually, most of the eigenmodes are oscillatory and the diffusion term dominates them, that is, they decay fast as there is large negative contribution of σ_{el}. However, if even one mode exists with positive σ_{el}, however small, we have a dynamo, because this mode will become dominant at large times. This kind of analysis loses its meaning after a sufficiently long time, because it is a linear problem and therefore the magnetic field will grow without bound. In reality, we need to consider a *dynamic* dynamo, that is, one in which $\mathbf{B}$ is the solution of a nonlinear equation of motion. This will undoubtedly limit any exponential growth. Before mentioning some examples of kinematic dynamos, we would like to comment that even if the situation is such that none of the anti-dynamo theorems hold, there is still a lower bound for Rm, the magnetic Reynolds number, for a magnetic dynamo action to be possible. Basically, the bound arises from calculating the rate of change of the magnetic energy, using the induction equation, and then exploiting general vector inequalities. For example, solenoidal fields, like **B**, confined in a sphere of radius a, matched to a decaying potential outside satisfy

$$\int |\nabla \times \mathbf{B}|^2 d^3x \geq \frac{\pi^2}{a^2} \int |\mathbf{B}|^2 d^3x, \tag{10.165}$$

(try to prove it!).

There exists a lower bound based on the above mentioned considerations, called the *Childress bound*, named after its discoverer, which is $\text{Rm} \geq \pi$. Using another approach, G. Backus found a larger lower bound. The *Backus bound* is $\text{Rm} \geq \pi^2$. We turn now to two examples of viable kinematic dynamos.

1. *Roberts dynamo*

   The flow, proposed by G.O. Roberts in the early 1970s, belongs to the so-called *ABC* flows (after Arnold, Beltrami, and Childress). These time-independent, spatially periodic flows, which have the form

   $$\mathbf{u}(\mathbf{x}) = (C\sin z + B\cos y)\hat{\mathbf{x}} + (A\sin x + C\cos z)\hat{\mathbf{y}} + (B\sin y + A\cos x)\hat{\mathbf{z}}, \tag{10.166}$$

   are indeed Beltrami flows, i.e., have the property of vorticity being equal to velocity $\omega \equiv \nabla \times \mathbf{u} = \mathbf{u}$ (see Problem 10.40) and thus have nonzero *helicity* ($H \neq 0$). The latter concept has already been used in this book and was *formally* defined in one of the (9.137) equations. The Roberts flow is the case with $A = B = 1$ and $C = 0$ that is

   $$\mathbf{u}(\mathbf{x}) = \cos y\,\hat{\mathbf{x}} + \sin x\,\hat{\mathbf{y}} + (\sin y + \cos x)\,\hat{\mathbf{z}}. \tag{10.167}$$

   It is not too difficult to see that this flow can be written in terms of a stream function, $\psi(x,y) = \sin y + \cos x$; thus,

   $$\mathbf{u}(\mathbf{x}) = \frac{\partial \psi}{\partial y}\,\hat{\mathbf{x}} - \frac{\partial \psi}{\partial x}\,\hat{\mathbf{y}} + \psi\hat{\mathbf{z}}. \tag{10.168}$$

   Roberts wrote the magnetic field as $\mathbf{B}(\mathbf{x}) = \mathbf{B}_{xy}(x,y)\exp(st + ikz)$, inserted this form into the induction equation, and used an appropriately truncated double Fourier expansion for $\mathbf{B}_{xy}(x,y)$, assuming its double spatial periodicity (in x and y), following from the flow form. The coefficients of the Fourier series then form a linear matrix eigenvalue problem for s. Solving this numerically, Roberts found an optimal k (for each Rm), maximizing the growth rate. Later on, A. Soward used asymptotic theory for large Rm (small $\varepsilon = 1/\text{Rm}$), qualitatively confirming Roberts result.

2. *Ponomarenko dynamo*
   The flow driving this dynamo is given in cylindrical coordinates as

$$\mathbf{u} = r\Omega_0 \hat{\varphi} + U_0 \hat{\mathbf{z}}, \quad \text{for } r < R,$$

$$= 0, \qquad\qquad \text{otherwise.} \tag{10.169}$$

Ω_0 and U_0 are constant parameters. This flow, thus, has a discontinuity and clearly it also possesses helicity:

$$H = \mathbf{u} \cdot (\nabla \times \mathbf{u}) = U_0 \frac{1}{r} \frac{\partial}{\partial r} \left(r^2 \Omega_0 \right) = 2 U_0 \Omega_0. \tag{10.170}$$

We must have $U_0 \neq 0$, because otherwise one of the anti-dynamo theorems will be valid. The idea was to also seek a non-axi-symmetric form of the field, thus avoiding the Cowling anti-dynamo theorem as well:

$$\mathbf{B}(r, z, \varphi) = \mathbf{B}_{\mathrm{rad}}(r) \exp\left[st + im\varphi + ikz \right]. \tag{10.171}$$

Using this ansatz and the velocity field as in Eq. (10.169) reduces the induction equation to a tractable eigenvalue problem. It is also possible to get the critical mode that minimizes the integer m and $\mathrm{Rm} = a(U_0^2 + a^2\Omega_0^2)^{1/2}/\eta_m$. It yields a dynamo that is strongest near $r = a$, where the shear is greatest.

Other dynamos and various limits of the above two exist in the literature as well, but we would like to move on now to dynamos, which are not kinematic and arise from self-consistent MHD flow fields. One method in a search of such self-consistent flow consists of what is known as the *mean field dynamo theory*, and we shall give a short and rudimentary review thereof. The starting step reminds one of the usual statistical approaches to hydrodynamic turbulence, wherein the variables are split into an average laminar and a fluctuating part; thus, $\mathbf{V} = \overline{\mathbf{V}} + \mathbf{V}'$, denoting the average by an overline and the remaining fluctuation by a prime. Without specifying the exact definition of averaging, we, however, use the Reynolds relations between them (see Chap. 9). Averaging the induction equation and applying the Reynolds rules, we get

$$\frac{\partial \overline{\mathbf{B}}}{\partial t} = \nabla \times \overline{(\mathbf{u} \times \mathbf{B})} + \eta_m \nabla^2 \overline{\mathbf{B}}, \tag{10.172}$$

where it is clear that the important term is the nonlinear one $\overline{(\mathbf{u} \times \mathbf{B})}$. Using Reynolds relations, we can find that the following expression has to be evaluated when calculating the above nonlinear term

$$\overline{(\mathbf{u} \times \mathbf{B})} - \overline{(\mathbf{u}' \times \mathbf{B}')} \equiv \mathbf{X}, \tag{10.173}$$

where this serves as well as the definition of the auxiliary (so as to make the procedure clearer) vector $\mathbf{X}$. Subtracting the averaged induction equation (10.172) from the original one (10.37), we get

$$\frac{\partial \mathbf{B}'}{\partial t} + \nabla \times \left[(\overline{\mathbf{u}} \times \mathbf{B}') + (\mathbf{u}' \times \overline{\mathbf{B}}) \right] + \nabla \times \mathbf{D}, \tag{10.174}$$

with $\mathbf{D} \equiv \overline{\mathbf{u}' \times \mathbf{B}'} - \mathbf{u}' \times \mathbf{B}'$. The above is a linear equation for $\mathbf{B}'$ with a forcing term. If one thinks of it as an equation for the generation of the turbulent field $\mathbf{B}'$ by the fluctuating (turbulent) $\mathbf{u}'$ interacting with the mean magnetic field, it is plausible to write an expansion $\mathbf{X}$, which in indicial notation is

$$X_k = a_{kj}\overline{\mathbf{B}}_j + b_{kji}\frac{\partial \overline{\mathbf{B}}_j}{\partial x_i} + \text{HOT} \tag{10.175}$$

The tensors a_{kj} and b_{kji} are complicated. They are functions of the average and fluctuating velocity (which we have no way of knowing). We therefore posit a kind of closure by assuming that they are *constant and isotropic*, i.e.,

$$a_{kj} = \alpha \delta_{kj}, \quad b_{kji} = -\beta \varepsilon_{kji}. \tag{10.176}$$

In fact one can make α and β space dependent, but this does not enhance the model in principle and just makes the problem more involved, thus demanding more assumptions. Equation (10.176) with the assumption that α and β are (recycled) constants leads to the following induction equation for the mean field

$$\frac{\partial \overline{\mathbf{B}}}{\partial t} = \nabla \times \overline{(\mathbf{u} \times \mathbf{B})} + \alpha \nabla \times \overline{\mathbf{B}} + \eta_m^{\text{T}} \nabla^2 \overline{\mathbf{B}}, \tag{10.177}$$

where $\eta_m^{\text{T}} = \eta_m + \beta$ and is perceived as total resistivity, including the part arising from fluctuations. This is the reason for calling it turbulent resistivity.

Equation (10.177) contains the basic elements of the so-called turbulent α-dynamo. It contains a linear growth term, giving rise to the growth of mean magnetic field in the direction of $\nabla \times \overline{\mathbf{B}}$, that is, of the main current. Note that the Parker buoyancy instability (see Fig. 10.8) can be regarded as a special case of the α effect. Loops of magnetic field like those of Fig. 10.11 imply the presence of currents parallel to the azimuthal field. The motion introduces a necessary helicity by giving the rising material, with the loops in it, a systematic angular velocity. The α-dynamo is linear and only the mechanism of field growth is predicted by models based on this concept. Another model which combines the α-dynamo with rotation, usually differential, has been examined, but we shall not discuss it here at all, stating only that a dynamo relying on these processes is called an $\alpha - \Omega$ dynamo and remarking that the loops in Fig. 10.11 need to be rotated.

An additional meaningful distinction is made between a *slow* dynamo in which the field grows on resistive timescale and a *fast* dynamo, whose growth rate is

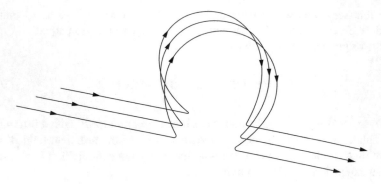

Fig. 10.11 Parker instability viewed as an α-dynamo

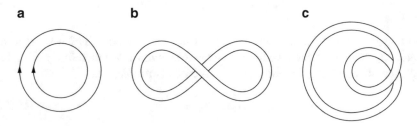

Fig. 10.12 An auxiliary drawing depicting an imagined *stretch, twist, fold* action

determined by the turnover time of the flow. To understand the meaning of this difference, let us now imagine qualitatively a situation in which we start from a planar concentric magnetic configuration, like the one depicted in Fig. 10.12a. Consider now an MHD flow that stretches the lines and then twists them to the form of a nearly planar "8-shape" (b). The third and crucial operation is effected by the flow folding the "8" in such a way that it becomes a three-dimensional configuration, reminiscent of the original one, but now doubled, with two loops lying one above the other, in a direction normal to the page, like in Fig. 10.12c. This discussion follows the qualitative idea of Vainshtain and Zel'dovich, and the development, as depicted in Fig. 10.12, has to include a very complicated flow. In contrast, as was shown in Problem 10.39, if we were to stay in the plane, that is, could not twist and fold, the dynamo would have been impossible.

In the book by S. Childress [12] the reader can find a wealth of high-level detailed material on the general principle of "Stretch, Twist & Fold," giving rise to a fast dynamo. We are unable to discuss this problem here but wish to mention that chaotic flows result from this kind of model. Several other ideas giving rise to fast dynamos contain flows whose characteristics give rise to chaotic Poincaré maps. We shall conclude this chapter without describing any of the analytical ideas to provide nonlinear limitation to the dynamo growth, like α or Ω quenching, and leave out the subject of the nonlinearity of dynamos. In the *Bibliographical Notes*, some references of MHD computer simulations are given, which naturally include the nonlinear turbulent dynamo action. Instead, we shall conclude by describing qualitatively one laboratory example and one astrophysical phenomenon and then describe briefly and qualitatively three advanced broad topics of interest, indicating good references for the reader who wishes to study them more deeply.

10.8 Short Overviews of Two Real Research Problems

10.8.1 Fusion Research

Main sequence stars produce their enormous energy by controlled fusion, converting hydrogen to helium. The extra mass Δm, coming from the nuclei binding energy, that is lost, so to speak, in these reactions is converted to energy according to the celebrated Einstein formula $E = \Delta mc^2$. The control or this process, so that the Sun, for example, does not explode as a gigantic hydrogen bomb, is effected by its *self-gravity*, which is physically important in massive enough cosmic bodies, both holding the high temperature plasma together and also keeping the control of the reaction, utilizing a "thermostat" that can be easily understood using the virial theorem. If the reaction proceeds a little too fast, the center of the Sun heats up and the star expands, lowering the temperature and the reaction, and vice versa. Controlled fusion in the lab has, for quite a long time, been the "holy grail" of plasma physicists. The hope had been that the energy problems of the world's population will soon be solved, but, unfortunately, the hope was premature. Today, we understand that the problems that a star "knows" how to solve so easily appear, at least today, as insurmountable in the lab. Starting from the 1960s, various machines, which included ingenious magnetic configuration, were built at a very large cost in various laboratories. These "tokamaks," as they are called have failed, so far, to hold Deuterium (a hydrogen isotope suitable for the reaction) at the enormous temperatures, which are necessary for fusion into helium to occur. It has been calculated by J.D. Lawson that for plasma containing hydrogen isotopes, whose number density is n, confined for a time equal or larger than τ, the energy output, due to the deuterium–tritium reaction, will be larger than the energy input on heating only if $n\tau \gtrsim 10^{-14} \mathrm{cm}^{-3}$ s. Typically, confinement times of only 0.1 s, at best, have been achieved for plasmas of $n \approx 10^{-15} \mathrm{cm}^{-3}$. In short, the best tokamaks still are two orders of magnitude below where the Lawson criterion would indicate positive energy yield. Recently, it has been reported, mainly by the media, that the various machines in the USA and France are, in the time of writing these words, closer to the breakeven that we explain here. While $\tau = 0.1$ s seems a short time, MHD instabilities are developing fast, ruining confinement. Even if most instabilities of this sort can be suppressed, there is a host of micro-instabilities, whose physics is on the microscopic level, which cause diffusion of the plasma from the inner part of the configuration, out. We hope for some good news, supported by peer-refereed scientific articles, in the reasonably near (sarcastic interpretation of this "near" is 10 years, which is a constant since the 1980s) future. It should be mentioned, however, that ITER (the France-based big international project) is currently running into severe budgetary and other difficulties.

Frustrated by the null results of the extreme investment in tokamaks, plasma scientists have proposed other ideas, i.e., achieving that Lawson breakeven by "zapping" a pellet of hydrogen by a powerful, high-energy laser. These efforts have not yet yielded published positive results, but research continues and the NIF (National Ignition Facility) in the USA. As of late 2015, this facility is understood to have achieved an important milestone towards commercialization of fusion, namely, for the first time a fuel *capsule* gave off more energy than was applied to it. This is a major step forward. A similar large-scale device in France is also being tested. The stakes are high and any success would be a true "jackpot."

10.8.2   Astrophysical Jets

Confined jets on various scales are observed in astrophysical objects, from close binary stars through nebulae containing forming stars and up to the grand scales of galaxies ejecting jets from their nuclei. The realization that jets are present almost always in conjunction with an accretion disk was important in the understanding that matter possessing angular momentum, which is being accreted on a central object, may, in part, be ejected along the axis of the disk. Energetic considerations and the unexpectedly high collimation of the jets point out that magnetic fields must be involved in jet ejection and collimation. Although there exist self-similar analytical or quasi-analytical solutions, most of the published results are numerical and depend strongly on initial and boundary conditions. Here we give a very short and largely qualitative discussion of one of the most likely mechanisms of ejection from disks—the *centrifugal wind*. Consider Fig. 10.13. On the right panel, we display a magnetic field line anchored at point P at a cylindrical radius R, say, of the disk. If the field line is fanning, at an angle θ, say, then a fluid particle at that point will experience (in its frame) two forces—the gravitational attraction towards the center $\mathbf{F}_g$ and the centrifugal force $\mathbf{F}_c$ outward. Magnetic stress does not influence motion along field lines. It is necessary (Ferraro law, see Problem 10.19) that the field line rotates at the same angular velocity as its base, so that

$$\Omega_P = \sqrt{\frac{GM}{r_P^3}}. \tag{10.178}$$

Instead of working with forces, we shall employ potentials. We have for the gravitational potential

$$\Phi_g = -\frac{GM}{(r^2 + z^2)^{1/2}} \tag{10.179}$$

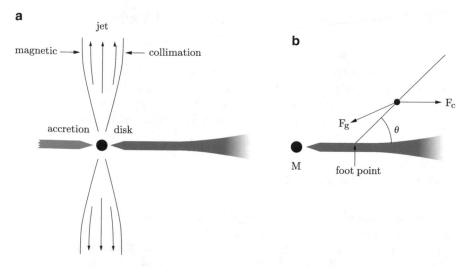

Fig. 10.13 The formation a centrifugal wind from and accretion disk and its collimation to form a jet

and for the centrifugal potential $\Phi_c = -\Omega_P^2 r^2/2$. Note that it depends on the point P. Substituting Ω_P from formula (10.178) and adding both potentials, we obtain the total potential for the particle at point R:

$$\Phi(r,z) = -\frac{GM}{r_P}\left[\left(\frac{r}{r_P\sqrt{2}}\right)^2 + \frac{r_P}{(r^2+z^2)^{1/2}}\right]. \tag{10.180}$$

Assume now that $r = r_P + \delta r$ and $z = 0 + \delta z$, where $\delta r, \delta z \ll r_P$. Expanding the above equation up to order $(\delta r/r_0)^2 \sim (\delta z/r_0)^2$ gives, where the index 0 indicates initial coordinates,

$$\Phi(r,z) = -\frac{GM}{r_P}\left[\frac{3}{2} + \frac{3}{2}\left(\frac{\delta r}{r_P}\right)^2 - \frac{1}{2}\left(\frac{\delta z}{r_P}\right)^2\right]. \tag{10.181}$$

Denoting now by ℓ the distance from the disk *along* the field line we have $\delta r = \ell\cos\theta$ and $\delta z = \ell\sin\theta$, we get that the force *out, along the field line* is

$$F_\ell = \frac{GM\ell}{r_P^3}(3\cos^2\theta - \sin^2\theta). \tag{10.182}$$

Outward force ejection is guaranteed if $3\cos^2\theta - \sin^2\theta > 0$, i.e., $\theta < \pi/3$. It is possible to show that once the ejection occurs, it removes angular momentum from the disk, and since the magnetic field lines spiral around the z-axis, the flow should be collimated.

10.9   Further Reading on Three Additional Topics

This is an FD book, with only one chapter on MHD; thus, we are naturally unable to cover many important and interesting advanced subjects of MHD. In this short section we make an attempt to list just three topics that certainly require further reading, if one wants to seriously study MHD. We shall say a few words on each topic and direct the interested reader to suitable, in our opinion, references.

10.9.1   MHD Turbulence

We have only mentioned the topic, in passing, when talking about the turbulent dynamo. It is relatively easy to obtain general relations (similar to Kolmogorov's hydrodynamical theory, based on dimensional arguments, of homogeneous turbulence) (see reference [5], pp 255–256). For a deeper study, the classical work by Parker (reference [1]) certainly comes to mind. We would like also to recommend the more modern self-contained book by Biskamp (reference [13]) and a volume of lecture notes of a conference, held in Nice in 1998 (edited by Passot and Sulem— reference [14]). Among other references we should single out the significant body of research by Goldreich and collaborators, whose theory is summarized in reference [15]. There exist also a large number of numerical works, from which we recommend the papers of Cattaneo and collaborators (e.g., [16] and references therein), as well as those of Schekochihin and collaborators (e.g., [17] and references therein). These authors have also been simulating numerically dynamo action, a topic which is clearly connected with MHD turbulence.

10.9.2   Reconnection

Magnetic reconnection, that is, changes, often explosive, that give rise to the annihilation of oppositely directed field lines, when they come to close contact is a phenomenon found in the laboratory, observed in the Sun and inferred in other astrophysical contexts. Most books on MHD devote some section to reconnection (e.g., references [2, 9]). However, the recent book of Priest and Forbes (reference [18]) is entirely devoted to this interesting phenomenon and its different aspects. We recommend the book as a good reference on the topic, including current sheet formation, steady and unsteady reconnection configurations, among them relevant instabilities, and astrophysical application of eruptions, flares, and particle acceleration.

10.9.3 Hall MHD

By using a more realistic than just the simplest linear Ohm's law, one may obtain a closed set of MHD equations, which is somewhat more general. It is named the *Hall MHD*. While it is relatively easy to find relevant equations in most comprehensive texts on plasma physics, which contain also various degrees of the MHD approximation, e.g., the book by Bittencourt (reference [19]), we find it useful to also see the use of the Hall MHD equations for explicit cases, in references [20] and [21], where actual analytical asymptotic equilibria and even instabilities were found for accretion disks.

Problems

10.1.
Solving the Poisson equation for the electrostatic potential (Φ), $\nabla^2 \Phi = -4\pi\rho_q$, where ρ_q is the charge density, find this potential assuming that at the center of coordinates there resides a positive ion (proton) find the potential, assuming that there are also charged particles around protons and electrons, in equal number densities ($n_e = n_p \equiv n$) in a thermodynamic equilibrium with temperature T. Exploit the assumption that $|e\Phi| \ll k_B T$, where k_B is the Boltzmann constant. Make sure that your solution goes like $\Phi \to 1/r$, when $r \to 0$ and $\Phi \to 0$ as $r \to \infty$.

10.2.
Express the proton collision frequency in Hz, writing T_4 for the temperature in units of $10^4 K$ and n_{22} for density in units of 10^{22} cm^{-3}. How can you get the electron collision frequency from that?

10.3.
Estimate the velocity change of electrons (whose initial velocity is v_e) and of protons (v_p) in an elastic collision (use classical mechanics) in one dimension.

10.4.
Find the approximate relation between the electron drift velocity v_e^{drift} and the conduction current. Now, using the approximate relation (10.21) and the conditions in the outer convective region of the Sun $\tilde{B} \sim 1$ kG, $n_e \sim 10^{21}$ cm^{-3}, $L \sim 10^{10}$ cm, estimate this velocity.

10.5. Using field estimates from Maxwell's equations, show that the condition for global charge neutrality ($|n_p - n_e| \ll n$) can be expressed in c.g.s. as $6 \times 10^5 U\tilde{B}/L \ll n$, where U and L are typical velocity and length scale, respectively, and $\tilde{B}$ is a typical magnetic field. Check it for values approximate for the solar outer layers—$U \sim 10^7$ cm/s, $\tilde{B} \sim 1$ kG, $L \sim 10^9$ cm.

10.6.

Consider the following magnetic fields and find the pressure and tension parts of the Lorentz force. Sketch roughly this force in each case:

1. $\mathbf{B} = B_0\hat{\mathbf{z}}$, with B_0 constant.
2. $\mathbf{B} = (B_0/y)\hat{\mathbf{z}}$, for $y > 0$.
3. $\mathbf{B} = B_0(-y\hat{\mathbf{x}} + \hat{\mathbf{x}})$. In this case find first the shape of the field lines for the drawing.

10.7.

Explain, term by term, why in an ionized hydrogen the following equation is a good approximation to the momentum conservation equation of the electrons without any gravitational and inertial terms:

$$
m_e\frac{d\mathbf{v}_e}{dt} = -e\left(\mathbf{E} + \frac{1}{c}\mathbf{v}_e\right) - m_e v_c \mathbf{v}_e.
\tag{10.183}
$$

v_c is the mean collision frequency with the ions, representing momentum transfer between the two species ("drag" on the electrons). Explain why it is reasonable to assume that the second term in the electromagnetic force can be dropped from the equation as well as the electron acceleration (when we are interested in the terminal velocity, in which the electric force is balanced by the "drag"). From this conclude that if the linear Ohm's law holds ($\mathbf{j} = \sigma\mathbf{E}$, remember that you are in the protons frame, which is approximately the fluid one), then $\sigma = n_e e^2/(m_e v_c)$.

10.8.

Show that in completely ionized hydrogen, the resistivity can be written as $\eta_m \sim 10^{13} T^{-3/2}$ cm^2 s^{-1}. It is given that molten iron has a resistivity of $\eta_m = 1.1 \times 10^4$ cm^2 s^{-1}. Estimate the magnetic Reynolds number Rm and the magnetic field diffusion time τ_{Bdiff} in

- The Earth outer core (assume it is composed of molten iron)—$L \sim 3000$ km, $U \sim 2\pi L/\tau_{\text{rot}}$, where $\tau_{\text{rot}} \sim 24$ h.
- An ionized nebula close to a massive star (an HII region)—$L \sim 10$ ly, $U \sim 10^6$ cm s^{-1}, $T \sim 10^4$K.
- Solar atmosphere—$L \sim 10^{10}$ cm, $U \sim 10^5$ cm s^{-1}, $T \sim 10^4$ K.
- Solar interior—$L \sim 10^9$ cm, $U \sim 10^7$ cm s^{-1}, $T \sim 10^7$ K.
- A rapidly rotating container of molten iron, in the lab—$L = 100$ cm, $U = 10^3$ cm s^{-1}.

Discuss your results, analyzing to what extent is field freezing a good assumption in each system and for how long.

10.9.

Consider a scalar one-dimensional diffusion equation

$$
\frac{\partial B}{\partial t} = \eta_m \nabla^2 B,
\tag{10.184}
$$

valid on an infinite spatial domain and with the initial condition $B(x,0) = f(x)$. Using the method of complex Fourier transform, show that the general solution of this equation is

$$B(x,t) = \int_{\infty}^{\infty} dx' f(x') G(x,t;t'),$$ (10.185)

where the Green function for this problem is

$$G(x,t;t') = \sqrt{\frac{1}{4\pi\eta_m}} \exp\left[-\frac{(x-x')^2}{4\eta_m t}\right].$$ (10.186)

Note that strictly speaking the initial condition is specified at $t = 0+$, so as to avoid singularities. Discuss the case of finite domain with $f(x)$ defined at the segment $-x_0 \le x \le x_0$, say, with $f(-x_0)$ and $f(x_0)$ playing now the role of initial conditions.

10.10.
Find $B_x(z)$ and the relation between U_0 and the constant P_d in the Hartman flow using Eqs. (10.50–10.51) and the boundary conditions in the text. Hartman flow is discussed in Sect. 10.2.2.1.

10.11.
Find the average velocity $u_{av} \equiv (1/2w) \int_{-w}^{w} u_x(z) dz$ for the plane Poiseuille flow of water at room temperature and for the Hartman flows with Ha $= 1$ and for Ha $= 50$.

10.12.
Find the *electric field* present in a Hartman flow.

10.13.
Show that Eqs. (10.65) and (10.66) can be written in tensorial form as $\frac{\partial \Pi_{jk}}{\partial x_k} = 0$, $\Pi_{jk} = P\delta_{jk} - (B_j B_k - \frac{1}{2}B^2 \delta_{jk}/4\pi)$. Using this, derive the following expression (due to Chandrasekhar and Fermi):

$$\int_{\mathcal{V}} \left(3P + \frac{B^2}{8\pi}\right) d^3x = \oint_{\partial\mathcal{V}} \left[\left(P + \frac{B^2}{8\pi}\right)\mathbf{x} - (\mathbf{B}\cdot\mathbf{x})\frac{\mathbf{B}}{4\pi}\right] \cdot \hat{n} dS,$$ (10.187)

where the second integral is on the surface containing the volume $\mathcal{V}$.

10.14.
Using the result of the previous problem (10.13), show that an equilibrium bounded in space is impossible, unless it is being influenced by *external* sources.

10.15.
Consider a magnetostatic toroidal configuration (φ independent, in cylindrical coordinates). Write Eqs. (10.64)–(10.66) in cylindrical coordinates taking into account azimuthal symmetry. Defining $\psi(r,z) \equiv 2\pi \int_0^r B_z r dr$ and $J(r,z) \equiv 2\pi \int_0^r j_z r dr$, i.e., the magnetic flux and the total current through a circle of radius r, perpendicular to the z-axis, show that

$$B_r = -\frac{1}{2\pi r}\frac{\partial \psi}{\partial z}, \quad B_z = \frac{1}{2\pi r}\frac{\partial \psi}{\partial r} \tag{10.188}$$

$$j_r = -\frac{1}{2\pi r}\frac{\partial J}{\partial z}, \quad j_z = \frac{1}{2\pi r}\frac{\partial J}{\partial r}. \tag{10.189}$$

10.16.
Show that any magnetic field can be separated to a poloidal part and toroidal one, which in cylindrical coordinates can be written as

$$\mathbf{B} = B_\varphi \hat{\varphi} + \frac{1}{2\pi r}\hat{\varphi} \times \nabla \psi, \tag{10.190}$$

where ψ is the poloidal flux defined in the previous problem.

10.17.
Describe the solution of the Grad–Shafranov equation in the example (10.82). Discuss the cases $b+1 < 0$, $a > 0$ only. Find the magnetic field and the current.

10.18.
Show that in the two-dimensional case discussed in the text, in Sect. 10.3.3, $\mathbf{A} = \hat{x}A(y,z)$ $\nabla \times \mathbf{B} = -\hat{x}\nabla^2 A$, where $\mathbf{A}$ is the vector potential of the magnetic field and $A = |\mathbf{A}|$. Show also that in this case $(\nabla \times \mathbf{B}) \times \mathbf{B} = -\nabla^2 A \nabla A$.

10.19.
Prove *Ferraro's iso-rotation law*: In an axially symmetric *ideal MHD* configuration, with no toroidal field, steady state is possible only if the angular velocity $\Omega(r,z)$ is constant along field lines.

10.20.
Consider two very near magnetic field lines in the plane (x,y) (a two-dimensional problem). Let the field lines be labelled a parameter λ. Thus, we have two lines λ and $\lambda + d\lambda$. Show that the mass per unit length in the x direction satisfies

$$dm = d\lambda \int_{-a}^{a} \rho \frac{\partial z}{\partial \lambda} dy, \tag{10.191}$$

where $\pm a$ are arbitrary limits of the y integration and we may apply periodic boundary conditions. From this show that

$$Q(\lambda) = C\frac{dm}{d\lambda}\left(\int_{-a}^{a} e^{-\Phi/C}\frac{\partial z}{\partial \lambda}dy\right)^{-1}, \tag{10.192}$$

where the nomenclature of Sect. 10.3.3 is used. Discuss the relation of this limitation to the free choice of $Q(\lambda)$ in the Dungey solution.

10.21.
Show that in the case of a low β_{pl} configuration, the force-free assumption is a good approximation.

10.22.
Using the equation of a force-free configuration, show that

$$\mathbf{B} \cdot \nabla \alpha = 0. \qquad (10.193)$$

10.23.
Show that a configuration in which the magnetic field is force-free *everywhere* is impossible.

10.24.
Assume that in a force-free-field α defined in Eq. (10.91) is constant everywhere. Show that the equation determining the magnetic field is $(\nabla^2 + \alpha^2)\mathbf{B} = 0$, an equation whose solutions are well known. Give an example.

10.25.
Show that in cylindrical flux tube (remember $B_r = 0$), in which all relevant variables depend only on the radial coordinate of the cylindrical coordinates set (r, z, φ), the equilibrium condition is

$$\frac{dP}{dr} + \frac{d}{dr}\left(\frac{B_\varphi^2 + B_z^2}{8\pi}\right) + \frac{B_\varphi^2}{4\pi r} = 0, \qquad (10.194)$$

where P is the fluid pressure.

10.26.
Assume that you are looking at a section of height H of an equilibrium flux tube, described in the previous problem. Show that the twist of the field line ϑ about the axis, on going from one end of the section of flux tube to the other, is

$$\vartheta(r) = \frac{H B_\varphi(r)}{r B_z(r)}. \qquad (10.195)$$

10.27.
Consider a cylindrical flux tube with free ends, in a force-free equilibrium. Let $B(r = a)$ for some a be held fixed. (a) Prove that the mean square value of B_i equals $B^2(a)$ and is invariant with respect to twisting. Show also that as the tube is twisted, the mean B_z across the tube decreases and is always less than $B(a)$, where the extra magnetic flux builds up B_φ. (b) Suppose that a twisted tube expands, but with its axial and azimuthal flux held fixed. Show that the tube becomes gradually more and more twisted and as B_φ increases and B_z decreases, the tube buckles, if it remains stable until it happens.

10.28.
Consider the $(y-z)$ plane and assume that the total pressure is continuous across the plane at $x = 0$. Assume also that the magnetic field can be written $\mathbf{B} = \hat{z}B$. Let for the two sides $x < 0$ and $x > 0$ the indices be 1 and 2, respectively. We require horizontal continuity of total pressure, giving equilibrium in the $\hat{x}$ direction:

$$P_1 + \frac{B_1^2}{8\pi} = P_2 + \frac{B_2^2}{8\pi}. \tag{10.196}$$

If the magnetic contributions are much larger than the pressure, we may neglect P_i. The B_i are then equal, but they may be opposite. In the latter case, we refer to the configuration as a *current sheet*. Assume that if the very small width of the current sheet is δ, the vertical fields are B_0 and $-B_0$ and the resistivity is $\eta_m = 1000\delta B_0/\sqrt{\rho_0}$. With $\rho_0 \sim 1$ g/cm^3, how long will it take for the current to decay to $1/100$ of its value, because of field diffusion across the sheet?

10.29.
Show that a finite current sheet converts magnetic energy into heat and the total (kinetic+internal) energy of the flow.

10.30.
There are MHD configurations, in which the vanishing of the current density everywhere, or at least in a simply connected region, $\mathscr{D}$, say, is a good approximation. Show that in such a case the magnetic field can be derived from a scalar magnetic potential, ϕ_B, say, which satisfies the Laplace equation $\nabla^2 \phi_B = 0$, in $\mathscr{D}$. Specify two examples of boundary conditions (on $\mathbf{B}$) at the bounding surface $\partial\mathscr{D}$ (or at infinity) for which this yields a unique solution.

10.31.
Show that the linearized set (10.99–10.103) can be reduced to a single equation for $\mathbf{u}_1$, Eq. (10.105) in the text.

10.32.
Using the formula for the two magneto-acoustic modes (10.123) and Fig. 10.5, show that

1. For propagation along field lines ($\theta_B = 0$), the phase velocity is either the sound speed or the Alfvén speed.
2. For propagation perpendicular to field lines ($\theta_B = \pi/2$), the phase velocity is either $\sqrt{c_s^2 + c_A^2}$ or zero.
3. As $\theta_B \to \pi/2$ for the slow wave

$$\frac{\omega}{k\cos\theta_B} \to v_{\text{cusp}} \equiv \frac{c_A c_s}{\sqrt{c_A^2 + c_s^2}} \tag{10.197}$$

Discuss the physical meaning of v_{cusp} in phase velocity.

10.33.

Use Eq. (10.123) to get the group velocity of the fast and slow magneto-acoustic modes. Find also the meaning of v_{cusp} (from the previous problem) in the group velocity plot.

10.34.

Using results from Chap. 7 and from this chapter, show that the dispersion relation for magneto-acoustic gravity waves, that is, where the acceleration due to gravity cannot be neglected, but the approximation that the wavelength is much shorter than the vertical density scale height can be used in an approximation leading to the buoyancy force, is

$$\omega^4 - \omega^2 k^2 (c_s^2 + c_A^2) + k^2 c_s^2 N^2 \sin^2 \theta_g + k^4 c_s^2 \cos^2 \theta_B = 0, \qquad (10.198)$$

where we use isothermal conditions and the Brunt–Väisälä frequency, N, where

$$N^2 = \frac{g^2}{c_s^2} \left(\gamma - \frac{c_s^2}{c_s^2 + c_A^2} \right), \qquad (10.199)$$

and the angles are defined by $\cos \theta_B = \hat{\mathbf{k}} \cdot \hat{\mathbf{B}}_0$, $\cos \theta_g = \hat{\mathbf{k}} \cdot \hat{\mathbf{z}}$.

10.35.

Using the equations of nondissipative (ideal) MHD, the continuity of the energy flux through a surface element, pointing in the $\hat{\mathbf{n}}$ direction, gives, with h being the enthalpy,

$$\left[\rho u_n \left(\frac{1}{2} u^2 + h \right) + u_n \frac{B^2}{4\pi} - B_n \frac{\mathbf{u} \cdot \mathbf{B}}{4\pi} \right] = 0, \qquad (10.200)$$

while the continuity of the momentum flux gives the two conditions

$$\left[P + \rho u_n + \frac{(B_t^2 - B_n^2)}{8\pi} \right] = 0, \qquad (10.201)$$

$$\left[\rho u_n \mathbf{u}_t - \frac{B_n \mathbf{B}_t}{4\pi} \right] = = 0. \qquad (10.202)$$

The suffix n denotes the $\hat{\mathbf{n}}$ component of a vector, $\mathbf{V}$, say, and the t suffix marks the tangential component of that vector, that is, $\mathbf{V}_t \equiv \mathbf{V} - V_n \hat{\mathbf{n}}$.

10.36.

Show, using the statement shown in the text (10.133), that the three vectors $\mathbf{B}_1$, $\mathbf{B}_2$, and the normal to the discontinuity surface are coplanar, irrespective of the size of B_n.

10.37.
Show also that for the previous problem setup, without the loss of generality, it is possible to assume that the vectors $\mathbf{v}_1$ and $\mathbf{v}_2$ lie in the same plane as $\mathbf{B}_1$ and $\mathbf{B}_2$. Using a coordinate system moving with a velocity $\mathbf{U}_f \equiv \mathbf{u}_t - (u_n/B_n)\mathbf{B}_t = \mathbf{u}_t - (j\upsilon/B_n)\mathbf{B}_t$ (for $B_n \neq 0$), show that $\mathbf{u}$ and $\mathbf{B}$ are collinear on each side of the discontinuity.

10.38.
In the discussion of the magnetic Kelvin–Helmholtz instability, it was asserted that Eq. (10.152) is derivable from the ansatz (10.151), substituted into the perturbation equations (10.148–10.149). Prove it.

10.39.
Prove that a planar, say, in the $x - y$ plane, dynamo is impossible. For simplicity, assume incompressibility and that all quantities depend on z only. Hint—use the diffusion equation for B_z to show that it must decay. Then concentrate on the surviving component of the vector potential $\mathbf{A}$ (defined by $\mathbf{B} = \nabla \times \mathbf{A}$) and obtain an induction equation for it.

10.40.
Show that an ABC flow, as given by Eq. (10.166), is indeed a Beltrami flow, that is, it satisfies $\omega = \mathbf{u}$.

10.41.
Find that $\overline{(\mathbf{V} \times \mathbf{W})} - \overline{(\mathbf{V}' \times \mathbf{W}')}$, where $\mathbf{V}$ and $\mathbf{W}$ are variable vector fields that are split (each) into an average part, denoted by an over-bar, and a fluctuating part, denoted by a prime. Use Reynolds relations.

10.42.
The *magnetic helicity* if there is field $\mathbf{B}$ in a volume $\mathcal{V}$ is defined as $H_{\mathrm{mag}} = \int \mathbf{A} \cdot \mathbf{B} d^3 x$, where $\mathbf{A}$ is the vector potential ($\mathbf{B} = \nabla \times \mathbf{A}$). Prove

1. H_{mag} is gauge invariant, that is, if $H'_{\mathrm{mag}} = \int \mathbf{A}' \cdot B d^3 x$ with $\mathbf{A}' = \mathbf{A} + \nabla\varphi$, for any φ, then $H'_{\mathrm{mag}} = H_{\mathrm{mag}}$.
2. The magnetic helicity for the configuration in Fig. 10.14 is $H_{\mathbf{mag}} = \Phi_1 \Phi_2$, where Φ_i are the fluxes of the interlaced tubes.
3. In a fixed volume cV,
   $dH_{\mathrm{mag}}/dt = 0$ in ideal MHD. This result is due to L. Woltjer.

Bibliographical Notes

General

The most comprehensive and the best, in our opinion, book on magnetohydrodynamics is Parker's work. The book is not easily acquirable as it is out of print (last edition was published in 1979). The relevant chapters of Priest's book on solar

Fig. 10.14 Two interlaced
magnetic flux tubes

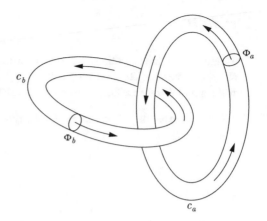

magnetohydrodynamics are also a very good source to MHD in general. Godbloed
and Poedts is a recommended modern textbook covering a variety of relevant topics.
The excellent treatise by Spitzer contains a wealth of information on the values of
various microscopic properties of conducting (fully ionized) gases.

Section 10.1

1. E.N. Parker, *Cosmical Magnetic Fields* (Clarendon Press, New York, 1979)
2. E.R. Priest, *Solar Magnetohydrodynamics* (D. Reidel, Dordrecht, 1982)
3. H. Goedbloed, S. Poedts, *Principles of Magnetohydrodynamics* (Cambridge University Press, Cambridge, 2004)
4. L. Spitzer Jr., *Physics of Fully Ionized Gases, 2nd edn.* (Wiley, New York, 1962)

Section 10.2

Landau and Lifshitz's classical book series contains a part, which gives a very good,
but (as usual) not too transparent account on the derivation of MHD equations. The
classical book by Cowling is also an excellent exposition of these equations. Its
conciseness is clearly a virtue. A flux tube approach to MHD is best understood
using reference number [1].

5. L.D. Landau, E.M. Lifshitz, L.P. Pitaevskii, *Electrodynamics of Continuous Media, 2nd edn.*, Chap. 8 (Elsevier, Amsterdam, 1993)
6. T.G. Cowling, *Magnetohydrodynamics* (Adam Hilger, Bristol, 1976)

Section 10.3

The material covered in this section is described in a very detailed way in Chap. 4 of reference number [2] and in the following book by Roberts. Good discussion of equilibria and force-free fields can be found also in Shu's book (see below).

7. P.H. Roberts, *An Introduction to Magnetohydrodynamics* (Longmans, London, 1967)

Section 10.4

Most MHD books treat hydromagnetic waves, with various degrees of detail. We find F. Shu's book especially useful for our purposes. A more concise description can be found in reference number [5].

8. F.H. Shu, *The Physics of Astrophysics II-Gas Dynamics* (University Science Books, Mill Valley, 1992)

Section 10.5

Discontinuities and MHD shock waves are not a trivial matter. Our aim was to present the main ideas without entering into too much details and algebraic complexity. We thus find references number [8] and [5] as suitable for our intentions.

Section 10.6

MHD instabilities are a vast topic. We are able, however, to discuss in this book only a few important ones. An interesting description of MHD instabilities, based on the energy principle, which we have not used in our book, can be found in

9. R.M. Kulsrud, *Plasma Physics for Astrophysicists* (Princeton University Press, Princeton, 2005)
   The influence of magnetic fields on hydrodynamic stability is also discussed in the following classic text
10. S. Chandrasekhar, *Hydrodynamic and Hydromagnetic Stability* (Oxford University Press, Oxford, 1961)

Section 10.7

It would be inappropriate not to recommend the unusually concise, but full of content, book by the pioneer of dynamo theory, T. G. Cowling, reference number [6]. The best modern source for dynamo theory is, in our view, the very comprehensive Les Houches lecture notes from 2008, containing a wealth of material. We also recommend the elegant book by Childress and Gilbert.

11. Ph. Carden, L.F. Cugliandolo, *Dynamos, Les Houches Session LXXXVII* (Elsevier, Amsterdam, 2008)
12. S. Childress, A. Gilbert, *Stretch, Twist, Fold: The Fast Dynamo* (Springer, New York, 1995)

Section 10.8

We find the following works useful for the topics mentioned in the text

13. D. Biskamp, *Magnetohydrodynamic Turbulence* (Cambridge University Press, Cambridge, 2003)
14. T. Passot, P.-L. Sulem (eds.), *Nonlinear MHD Waves and Turbulence* in *Lecture Notes in Physics*, vol. 536 (Springer, New York, 1999)
15. Y. Litwick, P. Goldreich, S. Sridhar, Imbalanced strong MHD turbulence. Astrophys. J. **665**, 269 (2007)
16. J. Mason, J. Perez, S. Boldyrev, F. Cattaneo, Numerical simulations of strong incompressible magnetohydrodynamic turbulence. Phys. Plasmas **19**, 055902 (2012)
17. A.A. Schekochihin, S.C. Cowley, in *Magnetohydrodynamics: Historical Evolution and Trends*, ed. by S. Molokov, R. Moreau, H.K. Moffatt (Springer, New York, 2007)
18. E. Priest, T. Forbes, *Magnetic Reconnection: MHD Theory and Applications* (Cambridge University Press, Cambridge, 2000)
19. J.A. Bittencourt, *Fundamentals of Plasma Physics* (Springer, New York, 2004)
20. Y.M. Shtemler, M. Mond, Asymptotic self-similar solutions for thermally isolated Z-pinches. J. Plasma Phys. **71**, 267 (2005)
21. E. Liverts, M. Mond, A.D. Chernin, The Hall instability of weakly ionized, radially stratified, rotating disks. Astrophys. J. **666**, 1226 (2007)

Appendix A
Vector Formulae

A.1 Identities

For any vectors $\mathbf{A}, \mathbf{B}, \mathbf{C}, \mathbf{D}$ and any scalar ψ:

$$\mathbf{A} \cdot (\mathbf{B} \times \mathbf{C}) = \mathbf{B} \cdot (\mathbf{C} \times \mathbf{A}) = \mathbf{C} \cdot (\mathbf{A} \times \mathbf{B}) \tag{A.1}$$

$$\mathbf{A} \times (\mathbf{B} \times \mathbf{C}) = (\mathbf{A} \cdot \mathbf{C})\mathbf{B} - (\mathbf{A} \cdot \mathbf{B})\mathbf{C} \tag{A.2}$$

$$(\mathbf{A} \times \mathbf{B}) \cdot (\mathbf{B} \times \mathbf{D}) = (\mathbf{A} \cdot \mathbf{C})(\mathbf{B} \cdot \mathbf{D}) - (\mathbf{A} \cdot \mathbf{D})(\mathbf{B} \cdot \mathbf{C}) \tag{A.3}$$

$$\nabla \times \nabla \psi = 0 \tag{A.4}$$

$$\nabla \cdot (\nabla \times \mathbf{A}) = 0 \tag{A.5}$$

$$\nabla \times (\nabla \times \mathbf{A}) = \nabla(\nabla \cdot \mathbf{A}) - \nabla^2 \mathbf{A} \tag{A.6}$$

$$\nabla \cdot (\psi \mathbf{A}) = \mathbf{A} \cdot \nabla \psi + \psi \nabla \cdot \mathbf{A} \tag{A.7}$$

$$\nabla \times (\psi \mathbf{A}) = \nabla \psi \times \mathbf{A} + \psi \nabla \times \mathbf{A} \tag{A.8}$$

$$\nabla(\mathbf{A} \cdot \mathbf{B}) = (\mathbf{A} \cdot \nabla)\mathbf{B} + (\mathbf{B} \cdot \nabla)\mathbf{A} + \mathbf{A} \times (\nabla \times \mathbf{B}) + \mathbf{B} \times (\nabla \times \mathbf{A}) \tag{A.9}$$

$$\nabla \cdot (\mathbf{A} \times \mathbf{B}) = \mathbf{B} \cdot (\nabla \times \mathbf{A}) - \mathbf{A} \cdot (\nabla \times \mathbf{B}) \tag{A.10}$$

$$\nabla \times (\mathbf{A} \times \mathbf{B}) = \mathbf{A}(\nabla \cdot \mathbf{B}) - \mathbf{B}(\nabla \cdot \mathbf{A}) + (\mathbf{B} \cdot \nabla)\mathbf{A} - (\mathbf{A} \cdot \nabla)\mathbf{B} \tag{A.11}$$

If $\mathbf{x}$ is the coordinate of an arbitrary point, P, say, with respect to some origin, O, say, and we denote the radial distance of this point from the origin by $r \equiv |\mathbf{x}|$, then $\mathbf{n} \equiv \hat{\mathbf{x}}$ is a unit radial vector pointing along the direction $\overline{OP}$. Assume that $f(r)$ is a sufficiently well-behaved function. The following identities hold:

© Springer Science+Business Media, LLC 2016
O. Regev et al., *Modern Fluid Dynamics for Physics and Astrophysics*,
Graduate Texts in Physics, DOI 10.1007/978-1-4939-3164-4

$$\nabla \cdot \mathbf{x} = 3 \tag{A.12}$$

$$\nabla \times \mathbf{x} = 0 \tag{A.13}$$

$$\nabla \cdot [\mathbf{n} f(r)] = \frac{2}{r} f + \frac{df}{dr} \tag{A.14}$$

$$\nabla \times [\mathbf{n} f(r)] = 0 \tag{A.15}$$

$$(\mathbf{A} \cdot \nabla) \, \mathbf{n} f(r) = \frac{f(r)}{r} [\mathbf{A} - \mathbf{n} (\mathbf{A} \cdot \mathbf{n})] + \mathbf{n} (\mathbf{A} \cdot \mathbf{n}) + \frac{df}{dr} \tag{A.16}$$

A.2   Integral Theorems from Calculus

In the following $\mathbf{A}(\mathbf{x})$ and $\phi(\mathbf{x})$, $\psi(\mathbf{x})$ are sufficiently well-behaved functions (vector or scalar) of the coordinates. We denote here by $\mathscr{V}$ a three-dimensional volume bounded by the surface $\partial \mathscr{V}$. The differential element of the volume is written as d^3x and that of the surface as dS; however, the latter has a direction (outward), and it is thus considered as a vector quantity, viz., $\mathbf{n} dS$, where n is the unit normal perpendicular to the element in the *outward* of the closed surface direction. Sometimes we also denote $\mathbf{n} dS = \mathbf{dS}$. The following theorems are important in vector calculus:

$$\int_{\mathscr{V}} \nabla \cdot \mathbf{A} \, d^3x = \oint_{\partial \mathscr{V}} \mathbf{A} \cdot \mathbf{n} \, dS \quad \text{(Divergence, or the Gauss theorem)} \tag{A.17}$$

$$\int_{\mathscr{V}} \nabla \psi \, d^3x = \oint_{\partial \mathscr{V}} \psi \mathbf{n} \, dS \tag{A.18}$$

$$\int_{\mathscr{V}} \nabla \times \mathbf{A} \, d^3x = \oint_{\partial \mathscr{V}} \mathbf{n} \times \mathbf{A} \, dS \tag{A.19}$$

$$\int_{\mathscr{V}} \left(\phi \nabla^2 \psi - \nabla \phi \cdot \nabla \psi \right) d^3x = \oint_{\partial \mathscr{V}} \phi \mathbf{n} \cdot \nabla \psi \, dS \quad \text{(Green's first identity)} \tag{A.20}$$

Let now S be an open surface and C the contour binding it having a line element $\mathbf{dl}$ along it. The normal $\mathbf{n}$ to S defines by the right-hand screw "law," the positive direction along the contour C. The following theorems hold:

$$\int_{S} (\nabla \times \mathbf{A}) \cdot \mathbf{n} \, dS = \oint_{C} \mathbf{A} \cdot \mathbf{dl} \quad \text{(Stokes' theorem)} \tag{A.21}$$

$$\int_{S} (\mathbf{n} \times \nabla) \, \psi \, dS = \oint_{C} \psi \mathbf{dl} \tag{A.22}$$

Appendix B
A Primer of Numerical Methods
for Computational Fluid Dynamics

B.1 Introduction

Computational fluid dynamics (CFD) has been advancing rapidly together with the development of increasingly powerful high-performance computing infrastructure and more sophisticated numerical methods. In the beginning, this field of study was mainly applied to high-tech engineering areas of aeronautics and astronautics. Gradually, however, CFD found its way to the study of *bona fide* theoretical and experimental fluid dynamical problems. The most general equations, governing the latter, are *nonlinear* PDE, and thus analytical approaches to the full problems were bound to fail.

The field of fluid dynamics is much older than computers. In this book we tried to describe and understand fluid dynamical problems, which usually included a significant number of simplifying assumptions, so as to lend themselves to be treatable without recourse to CFD calculations. We exhibited analytical or perturbative methods of solutions, with only few instances where numerical calculations were exploited, but it remained clear, we hope, that more general problems, e.g., certain complex flows, require numerical simulation. Numerical simulations serve as a proxy for difficult or dangerous experiments, as well as to prod our intuition. Indeed, recently numerical calculations acquired sometimes the name of *numerical experiments*, but obviously, the question of the CFD result's *reliability* has continued to loom. The fast development of computer hardware and numerical algorithms has also reduced the researcher's cost to conduct big computational flow simulations. On the other hand, the need to simulate more extreme physical conditions, higher Reynolds and/or Mach numbers, higher temperature, etc., has brought an increase in effort associated with experimental testing. Thus, it has become sometimes more economical to conduct big CFD calculation and treat them as "experiments."

© Springer Science+Business Media, LLC 2016
O. Regev et al., *Modern Fluid Dynamics for Physics and Astrophysics*,
Graduate Texts in Physics, DOI 10.1007/978-1-4939-3164-4

In this appendix, we shall try to give a short primer of available numerical methods, indicating their advantages and disadvantages together with some caveats. In an appendix to a book of this size, we are able to give only the basics. After all, this is not a book on CFD (on the contrary, if one may say so). There exists, today, an immense modern literature on computational methods in fluid dynamics. We shall mention in the text of this appendix some of the best (in our opinion) bibliographical sources. Clearly, a successful "attack" on a specific fluid dynamical problem should contain experiments, if possible, analytical and semi-analytical, methods (this book) to understand the basics of the problem, and numerical simulations, where it should be remembered that analytical, semi-analytical, and approximate methods may contribute to the *physical understanding* of the processes taking place in a given setting. Numerical simulation results in a solution to *one* case; however, if several such calculations are done so as to experiment with certain input parameters, it significantly increases the cost. Still the "sea" of numbers has to be visually analyzed, but the physics behind the results often remains obscure. A prominent professor of one of us (O.R.) insisted that *one should not learn physics from a computer*. Today, this may seem to some perhaps old-fashioned, but in the present state of affairs we find still much wisdom in this saying.

B.2   Short Summary and References for the Local Methods

We shall concentrate here on methods for the numerical solution of fluid dynamical equations. As we know, these equations are a nonlinear set of partial differential equations (PDEs) in space and time variables, valid in well-defined spatial domains, with appropriate *boundary* and *initial conditions*. Unfortunately, there are rigorous mathematical results on the existence and uniqueness of solutions in only special cases. This usually does not prevent the fluid numericists from applying their methods obtaining some result, which they call a solution. Again, how reliable is this result? Indeed a lot of analysis (mainly based on linear and quasilinear equations) is performed, in order to unravel what is being called a *well-posed problem*, one that hopefully will numerically yield a unique, well-defined solution. We shall end this short paragraph by saying that it is very often advantageous to write the equation in the form of conservation laws (cf. Randall J. LeVeque, *Numerical methods for conservation laws, 2nd ed.*, 1999, Bikhäuser). In this section we shall mention methods of expressing a PDE on a finite grid (or finite volumes) created in the domain of the calculation. Necessary derivatives (usually not of high order) are approximated using the (truncated) Taylor series expansion or by using other methods of calculus, e.g., integrating over the small, almost local volume and using the Gauss theorem.

B.2.1   *Finite Difference Methods*

The finite difference method is the oldest of the numerical methods for the solution of PDE. It is believed that already L. Euler had in 1768 "discretized" differential equation on a grid and sought their solution by hand calculations. The finite difference method, as it was already hinted above, consists of dividing the computational domain into a *grid*, that is, each coordinate domain is divided to a number of points. At each point of the grid, the Taylor series expansion is utilized to generate finite difference approximations. In this way at each grid point an *algebraic equation* arises. The method is most commonly applied to structured grids, even though the grid spacing need not be uniform (remembering that grid stretching and distortion may hurt accuracy). We shall give here some basic possibilities in two dimensions, where the function discretized is $\phi(x,y)$. We may write

$$\left(\frac{\partial \phi}{\partial x}\right)^{c}_{i,j} = \frac{\phi_{i+1,j} - \phi_{i-1,j}}{2\Delta x} + \mathcal{O}\left(\Delta x^2\right)$$

$$\left(\frac{\partial \phi}{\partial x}\right)^{f}_{i,j} = \frac{\phi_{i+1,j} - \phi_{i,j}}{\Delta x} + \mathcal{O}\left(\Delta x\right)$$

$$\left(\frac{\partial \phi}{\partial x}\right)^{b}_{i,j} = \frac{\phi_{i,j} - \phi_{i-1,j}}{\Delta x} + \mathcal{O}\left(\Delta x\right), \tag{B.1}$$

where for the sake of simplicity we have chosen a fixed grid spacing $\Delta x = x_j - x_{j-1} = x_{j+1} - x_j$. The remainder in each case estimates the accuracy of the scheme.

Different ways of differencing (in particular when also the time variable is discretized) exist, and we will not be able to bring even most of the accepted schemes (they are usually named) here. See, e.g., C.A.J. Fletcher, *Computational Techniques for Fluid Dynamics, vol 1, ed, 2, 1990*, Springer. Fletcher's book contains in Chap. 3 and especially in Sect. 3.5 most of the known finite difference schemes and their estimated accuracy. In addition, he discusses also the conditions for numerical stability of explicit methods, using the celebrated Courant-Friedrichs-Lewy (CFL) condition, which in its simplest formulation reads $C \equiv u\Delta t/\Delta x \leq C_m$. If the numerical calculation is stable (i.e., does not develop growing oscillations with a short, i.e., of very few grid points, "wavelength"), C the Courant number has to be less than C_m. In simplest explicit schemes $C_m = 1$. This effectively bounds from above the allowed timestep and makes calculations more expensive in computer time. Implicit methods also exist, which suffer from much less stringent conditions, but they require usually complex iterative calculations. Do you understand the *physical* basis for the CFL condition?

B.2.2 Finite Volume Methods

This method is not applied on PDE in their conservation form (as mentioned in the first paragraph of Sect. B.2) but rather on the integral form of the conservation equations in physical space. The computational domain is subdivided into a finite number of contiguous volumes, which go under the name of control volumes. The value of the relevant functions is formally evaluated at the centroid of the control volume. The surfaces bound the control volumes in the appropriate integral conservation relations. For example, we shall give here the appropriate approximation in two dimensions of the x and y derivative of the function $\phi(x, y)$. We shall apply the Gauss divergence theorem in the second equality of each formula below, with dS^x and dS^y denoting the *projected* areas, in the x and y directions, respectively, of the bounding surfaces of a volume element.

Thus,

$$\left(\frac{\partial \phi}{\partial x}\right) = \frac{1}{\Delta V} \int_{\Delta V} \frac{\partial \phi}{\partial x} d^3x = \frac{1}{\Delta V} \int_A \phi \, dS^x \approx \frac{1}{\Delta V} \sum_{j=1}^{N} \phi_j S_j^x, \tag{B.2}$$

where ϕ^j are the values of the function at the elemental surfaces, ΔV is the element volume, and N denotes the number of bounding surfaces of this volume. Similarly,

$$\left(\frac{\partial \phi}{\partial y}\right) = \frac{1}{\Delta V} \int_{\Delta V} \frac{\partial \phi}{\partial y} d^3x = \frac{1}{\Delta V} \int_A \phi \, dS^y \approx \frac{1}{\Delta V} \sum_{j=1}^{N} \phi_j S_j^y. \tag{B.3}$$

For a more detailed description of the finite volume method, including examples, see Sect. 5.2 of the book by C.A.J. Fletcher, *Computational Techniques for Fluid Dynamics, vol 1, ed, 2, 1990*, Springer.

B.3 Weighted Residual Methods

Weighted residual methods (WRM) are different in their basic concept from finite difference and volume methods. The approximations, necessary for numerical solutions to be extracted from, are defined by truncated series expansions, such that the *residual* (actually the error) is made as small as possible, at least in the mean. Consider the following approximation to the solution function $\phi(x)$ (all is done in one dimension, for simplicity, in some finite segment, (a, b), say):

$$\phi_N(x) = \sum_{j=0}^{N} \hat{\phi}_j \varphi_j(x), \tag{B.4}$$

where $\varphi_j(x)$ are a set of basis (a.k.a. "approximating") functions, assumed here orthogonal, for the sake of simplicity (however, see below).

The residual is thus $R_N(x) = \psi - \psi_N$ and we shall now show how it may be canceled, at least in the mean. One way of canceling the residual can be understood in the following sense: first form

$$(R_N, \Psi_j)_{w_*} = \int_a^b w_* R_N \overline{\Psi}_j \, dx, \qquad j = 0, 1, 2 \ldots, N \tag{B.5}$$

Note that the above is a kind of $w^*(x)$ weighted scalar product of the residual with a member of a set of *trial* functions: $\Psi_j(x)$. One of the possibilities ("traditional Galerkin approach") to proceed is to choose the trial functions to coincide with the basis functions, $\varphi_k(x) = \Psi_k(x)$. Orthogonality of the basis/trial functions now allows one to determine the coefficients $\hat{\phi}_j$; thus,

$$\hat{\phi}_j = \int_a^b \phi \overline{\Psi}_j \, w_* \, dx, \qquad j = 0, 1 \ldots, N. \tag{B.6}$$

Doesn't this remind one of the well-known "least squares method"? The above mentioned book by Fletcher devotes his Chap. 5 to WRM.

B.3.1 Spectral Methods

Spectral methods belong to the broad category of WRM. The first possibility of a *spectral* method uses a similar form as the Galerkin method, mentioned above. The approximating functions and the weight functions are nonzero throughout the computational domain. In this sense these are *global* methods. The important attribute of a *spectral* method that it uses *orthogonal*, under the constant weight w, functions for *both* the approximating functions and the weight functions, written here in one dimension, for simplicity. Note that we have chosen the basis function to be orthogonal for simplicity of presentation only, they need not be:

$$(\varphi_j, \varphi_k)_w = \int \varphi_j(x) \varphi_k(x) w \, dx = \delta_{jk} \tag{B.7}$$

Fourier series, Legendre polynomials, and Chebyshev polynomials are well-known examples of orthogonal functions.

Another possibility is to select the trial functions as

$$\overline{\Psi}_j = \delta(x - x_j), \qquad x \in (a, b), \tag{B.8}$$

where the points x_j are chosen in a nonarbitrary manner and the weight function is $w_* = w = 1$. This is called a *collocation* approach and the method is called *pseudo-spectral*. In general, a pseudo-spectral method is closely related to spectral methods, but it complements the basis by an additional pseudo-spectral basis, which allows one to represent functions on a quadrature grid. This simplifies the evaluation of certain operators and can considerably speed up the calculation when using fast algorithms such as the fast Fourier transform.

Spectral methods gained prominence in the 1970s. Substantial work started to appear in the professional literature, culminating by the seminal treatise of D. Gottlieb and S.A. Orszag, *Numerical Analysis of Spectral Methods*, 1977, SIAM-CMBS. Today spectral and pseudo-spectral methods are considered the most accurate and reliable and are routinely used for the most difficult direct numerical simulations (DNS) of complex and, in particular, turbulent flows. Today, there exist a substantial number of books on spectral methods. Suffice it to say that the huge work of C. Canuto, M.Y. Hussaini, A. Quateroni, and T.A. Zang, *Spectal Methods, 2 vols*, 2006, Springer, contains over a thousand pages in its two volumes. This excellent book, written by experts in the field, is recommended for any student who wishes to learn and perhaps use these powerful methods.

B.4 Summary and Some Caveats

We hope that it is clear to the reader that this appendix is *not* the extension of our book to include numerical methods in CFD. Rather than that, it is a short primer, mentioning the methods and their principles and citing the extensive bibliography that on the subject. Successful and careful application of the right CFD method (finding the most appropriate one is not an easy task) in physics, astrophysics, and indeed all the physical sciences has become an important part of modern research. However, it should not be forgotten that theoretical physics and astrophysics are not just computation. There is the danger that inaccurate or just wrong results of numerical calculations may mislead a generation of researchers. CFD *by itself* cannot lead, in our opinion, to proper physical understanding and progress. It can, if it is carefully and correctly executed, *simulate* nature, but as it is attributed to F. Dyson "We have already one nature (Universe), we want to understand it." This is one of the main reasons that this book was written.

Finally, some caveats: we think that it would be fair to say that the knowledge of the solution can go a long way in deciding on the proper numerical scheme. More seriously, *accuracy, scheme stability, and convergence* strengthen the computation reliability. By *convergence* we mean here sufficient resolution, which can be tested by examining how the result changes with the amount of grid points or spectral functions. Obviously, it is a must to perform tests on problems to which analytical solutions are known. Feasibility (as far as computer resources allow) is also an important factor in the choice of a CFD method for a particular problem. Here it seems that progress is unlimited, having perhaps the power to change our science, one day. Right now, it would be still fair to say that choosing the right problem plus a CFD method that is the best for it and conducting a successful simulation contain an element of "art," similarly to knowing how to employ analytical methods.

Index

© Springer Science+Business Media, LLC 2016
O. Regev et al., *Modern Fluid Dynamics for Physics and Astrophysics*,
Graduate Texts in Physics, DOI 10.1007/978-1-4939-3164-4

Printed in the United States
By Bookmasters